MW00851140

CMOS LOGIC CIRCUIT DESIGN

CMOS LOGIC CIRCUIT DESIGN

John P. Uyemura
Georgia Institute of Technology

KLUWER ACADEMIC PUBLISHERS
Boston / Dordrecht / London

Distributors for North, Central and South America:
Kluwer Academic Publishers
101 Philip Drive
Assinippi Park
Norwell, Massachusetts 02061 USA
Telephone (781) 871-6600
Fax (781) 871-6528
E-Mail <kluwer@wkap.com>

Distributors for all other countries:
Kluwer Academic Publishers Group
Distribution Centre
Post Office Box 322
3300 AH Dordrecht, THE NETHERLANDS
Telephone 31 78 6392 392
Fax 31 78 6546 474
E-Mail <orderdept@wkap.nl>

Electronic Services <http://www.wkap.nl>

Library of Congress Cataloging-in-Publication Data

Uyemura, John P.
CMOS logic circuit design/John P. Uyemura
p.cc.
Includes bibliographical references and index.
ISBN 0-7923-8452-0 (alk. paper)
1. Metal oxide semiconductor, complimentary--Design and construction. 2. Logic design. Digital integrated circuits--Design and construction. 3. Integrated circuits--Very large Scale integration--Design and construction. I. Title.

TK7871.99.M44 U92 1999
621.39/732 21 2001038650

Copyright © 2001 by Kluwer Academic Publishers. Second Printing 2002.

All rights reserved. No part of this publication may be reproduced, stored in a retrieval system or transmitted in any form or by any means, mechanical, photo-copying, recording, or otherwise, without the prior written permission of the publisher, Kluwer Academic Publishers, 101 Philip Drive, Assinippi Park, Norwell, Massachusetts 02061

Printed on acid-free paper.

Printed in the United States of America

Dedication

This book is dedicated to

Christine and Valerie

for all of the joy and happiness
that they bring into my life

植　村

Preface

This book is based on the earlier Kluwer title **Circuit Design for CMOS/VLSI** which was published in 1992. At that time, CMOS was just entering the main-stream as a technique for high-speed, high-density logic circuits. Although the technology had been invented in the 1960's, it was still necessary to include Section 1.1 entitled *Why CMOS?* to justify a book on the subject. Since that time, CMOS has matured and taken its place as the primary technology for VLSI and ULSI digital circuits. It therefore seemed appropriate to update the book and generate a second edition.

Background of the Book

After loading the old files and studying the content of the earlier book, it became clear to me that the field is much more stable and well-defined than it was in the early 1990's. True, technological advances continue to make CMOS better and better, but the general foundations of modern digital circuit design have not changed much in the past few years. New logic circuit techniques appearing in the literature are based on well-established ideas, indicating that CMOS has matured.

As a result of this observation, the great majority of the old files were abandoned and replaced with expanded discussions and new topics, and the book was reorganized to the form described below. There are sections that didn't change much. For example, Chapter 1 (which introduces MOSFETs) includes more derivations and pedagogical material, but the theme is about the same. But, many items are significantly different. For example, the earlier book contained about 60 pages on dynamic logic circuits. The present volume has almost three times the number of pages dedicated to this important area. In addition, the book has been written with more of a textbook flavor and includes problem sets.

Contents

Chapter 1 introduces the MOS system and uses the gradual-channel approximation to derive the square-law equations and basic FET models. This sets the notation for the rest of the book. Bulk-charge models are also discussed, and the last part of the chapter introduces topics from small-device theory, such as scaling and hot electrons.

Chapter 2 is an overview of silicon fabrication and topics relevant to a CMOS process flow. Basic ideas in lithography and pattern transfer are covered, as are items such as design rules, FET sizing, isolation, and latch-up. This chapter can be skipped in a first reading, but it is important to understanding some problems that are specific to layout and fabrication issues. It is not meant to replace a dedicated course in the subject.

Circuit design starts in **Chapter 3**, which is a detailed analysis of the static CMOS inverter. The study is used to set the stage for all of the remaining chapters by defining important DC quantities, transient times, and introducing CMOS circuit analysis techniques. **Chapter 4** concentrates on a detailed study of the electrical characteristics of FETs when used as voltage-controlled electronic switches. In particular, the treatment is structured to emphasize the strong and weak points of nFETs and pFETs, and how both are used to create logic networks. This feeds into **Chapter 5**, which is devoted entirely to static logic gates. This includes fully complementary designs in addition to variants such as pseudo-nMOS circuits and novel XOR/XNOR networks. **Chapter 6** on transmission gate logic completes this part of the book.

Dynamic circuit concepts are introduced in **Chapter 7**. This chapter includes topics such as charge sharing and charge leakage in various types of CMOS circuit arrangements. RC modelling is introduced, and the Elmore formulas for the time constant of an RC ladder is derived. Clocks are introduced and used in various types of clocked static and dynamic circuits. Dynamic logic families are presented in **Chapter 8**. The discussion includes detailed treatments of precharge/evaluate ripple logic, domino logic cascades, self-resetting logic gates, single-phase circuits and others. I have tried to present the material in an order that demonstrates how the techniques were developed to solve specific problems. **Chapter 9** deals with differential dual-rail logic families such as CVSL and CPL with short overviews of related design styles.

The material in **Chapter 10** is concerned with selected topics in chip design, such as interconnect modelling and delays, crosstalk, ESD-protected input circuits, and the effects of transmission lines on output drivers. The level of the presentation in this chapter is reasonably high, but the topics are complex enough so that the discussions only graze the surface. It would take another volume (at least) to do justice to these problems. As such, the chapter was included to serve as an introduction for other courses or readings.

Use as a Text

There is more than enough material in the book for a 1-semester or 2-quarter sequence at the senior undergraduate or the first-year graduate level. The text itself is structured around a first-year graduate course entitled *Digital MOS Integrated Circuits* that is taught at Georgia Tech every year. The course culminates with each student completing an individual design project.

My objectives in developing the course material are two-fold. First, I want the students to be able to read relevant articles in the *IEEE Journal of Solid-State Circuits* with a reasonable level of comprehension by the end of the course. The second objective is more pragmatic. I attempt to structure the content and depth of the presentation to the point where the students can answer all of the questions posed in their job interviews and plant visits, and secure positions as chip designers after graduation. Moreover, I try to merge basics with current design techniques so that they can function in their positions with only a minimum amount of start-up time.

Problem sets have been provided at the end of every chapter (except Chapter 2). The questions are based on the material emphasized in the chapter, and most of them are calculational in nature. Process parameters have been provided, but these can easily be replaced by different sets that might be of special interest. Most of the problems have appeared on my homeworks or exams; others are questions that I wrote, but never got around to using for one reason or another. I have tried to include a reasonable number of problems without getting excessive. Students that can follow the level of detail used in the book should not have many problems applying the material. SPICE simulations add a lot to understanding, and should be performed whenever possible.

Apologies

No effort was made to include a detailed list of references in the final version of the book. I initially set out to compile a comprehensive bibliography. However, after several graduate students performed on-line literature searches that yielded results far more complete than my list, I decided to include only a minimal set here. The references that were chosen are books and a few papers whose contents are directly referenced in the writing. The task is thus left to the interested reader.

I have tried very hard to eliminate the errors in the book, but realize that many will slip through. After completing six readings of the final manuscript, I think that I caught most of the major errors and hope that the remaining ones are relatively minor in nature. I apologize in advance for those I missed.

Acknowledgments

Many thanks are due to Carl Harris of Kluwer who has shown amazing patience in waiting for this project to be completed. He never seemed to lose hope, even when I was quite ill (and crabby) for several months and unable to do much. Of course, those who know Carl will agree with me that he is a true gentlemen with exceptional qualities. And a real nice guy.

Dr. Roger P. Webb, Chair of the School of Electrical & Computer Engineering at Georgia Tech, has always supported my efforts in writing, and has my never ending thanks. Dr. William (Bill) Sayle, Vice-Chair for ECE Undergraduate Affairs, has also helped me more times than I can count during the many years we have known each other. I am grateful to my colleagues that have taken the time to discuss technical items with me. On the current project, this includes Dr. Glenn S. Smith, Dr. Andrew F. Peterson, and Dr. David R. Hertling in particular.

I am grateful to the reviewers that took the time to weed through early versions of the manuscript that were full of typos, missing figures, and incomplete sections to give me their comments. Feedback from the many students and former students that have suffered through the course have helped shape the contents and presentation.

Finally, I would like to thank my wife Melba and my daughters Valerie and Christine that have put up with dad sitting in front of the computer for hours and hours and hours. Their love has kept me going through this project and life in general!

John P. Uyemura
Smyrna, Georgia

Table of Contents

Chapter 1

Physics and Modelling of MOSFETs 1

Chapter 2

Fabrication and Layout of CMOS Integrated Circuits 61

Chapter 3

The CMOS Inverter: Analysis and Design 103

Chapter 4

Switching Properties of MOSFETs 155

Chapter 6

Transmission Gate Logic Circuits 259

Chapter 7

Dynamic Logic Circuit Concepts 287

Chapter 8

CMOS Dynamic Logic Families 349

CMOS LOGIC CIRCUIT DESIGN

Chapter 1

Physics and Modelling of MOSFETs

MOSFETs (metal-oxide-semiconductor field-effect transistors) are the switching devices used in CMOS integrated circuits. In this chapter, we will examine the current flow through a MOSFET by analyzing is the path that the charge carriers follow. This results in an equation set that will be used for the entire book. Some advanced VLSI effects are also discussed in the second half of the chapter.

1.1 Basic MOSFET Characteristics

The circuit symbol for an n-channel MOSFET (**nFET** or **nMOS**) is shown in Figure 1.1(a). The MOSFET is a 4-terminal device with the terminals named the **gate**, **source**, **drain**, and **bulk**. The device voltages are shown in Figure 1.1(b). In general, the gate acts as the control electrode. The value of the gate-source voltage V_{GSn} is used to control the drain current I_{Dn} that flows through the device from drain to source. The actual value of I_{Dn} is determined by both V_{GSn} and the drain-

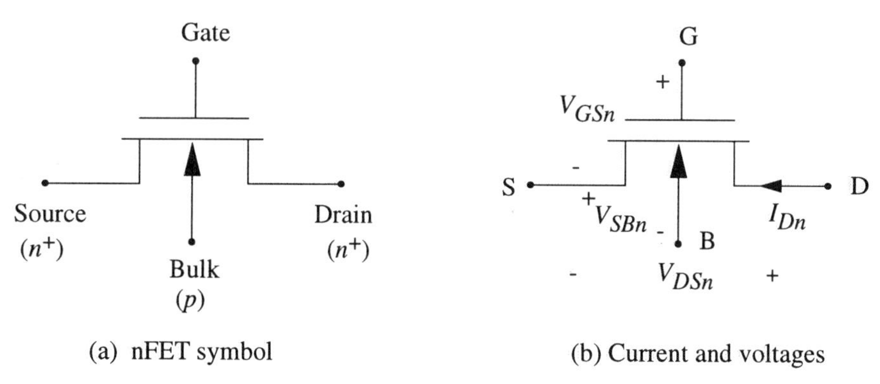

Figure 1.1 n-channel MOSFET symbol

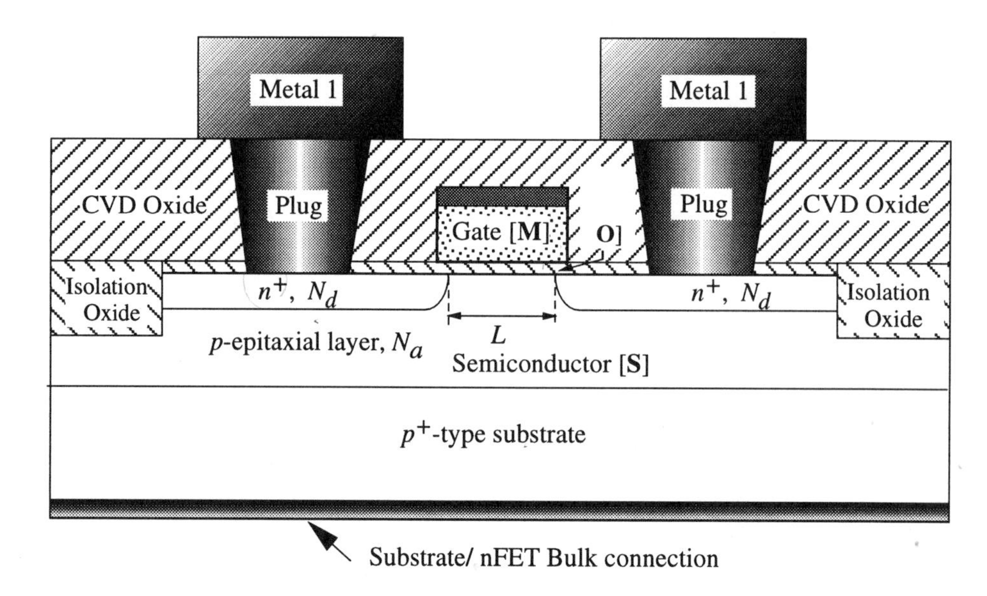

Figure 1.2 Cross-sectional view of a typical nFET

source voltage V_{DSn}. The source-bulk voltage V_{SBn} also affects the current flow to a lesser degree.

Figure 1.2 shows a typical nFET that will be used for the analysis. The central region of the device consists of a **metal-oxide-semiconductor** (**MOS**) subsystem made up of a conducting region called the **gate** [M], on top of an insulating silicon dioxide layer [O] shown as a cross-hatched region directly underneath the gate, and a p-type silicon [S] epitaxial layer on top of a p^+-substrate. The existence of this capacitor substructure between the gate and the semiconductor is implied by the schematic symbol. The *I-V* characteristics of the transistor result from the physics of the MOS system when coupled to the n^+ regions on the left and right sides. The n^+ regions themselves constitute the drain and source terminals of the MOSFET, while the bulk electrode corresponds to the electrical connection made to the p-type substrate. The distance between the two n^+ regions defines the **channel length** L of the MOSFET. As will be seen in the discussion, the channel length is one of the critical dimensions that establishes the electrical characteristics of the device.

A top view of the nFET is shown in Figure 1.3. This drawing defines the **channel width** W for the FET, and is the width of the region that supports current flow between the two n^+ regions. The ratio (W/L) of the channel width to the channel length is called the **aspect ratio**, and is the important circuit design parameter. Note that the top view shows the length L' as the visual distance between the two n^+ regions. This is called the **drawn channel length** and is larger than the electrical channel length L.

The *I-V* characteristics of a MOSFET are referenced to the **threshold voltage** V_T of the device; the actual value V_T for a particular device is set in the fabrication parameters. CMOS designs are based on **enhancement-mode** (E-mode) transistors where the gate voltage is used to enhance the conduction between the drain and source. By definition, an n-channel E-mode MOSFET has a positive threshold voltage $V_{Tn} > 0$, with a typical value ranging from about 0.5 to 0.9 volts. The value of the threshold voltage V_{Tn} is especially important to high performance circuit design.

In an ideal n-channel MOSFET, setting the gate-source voltage to a value $V_{GSn} < V_{Tn}$ places the transistor into **cutoff** where (ideally) the current flow is zero: $I_{Dn} = 0$; this is shown in Figure 1.4(a). Increasing the gate-source voltage to a value where $V_{GSn} > V_{Tn}$ allows the transistor to conduct cur-

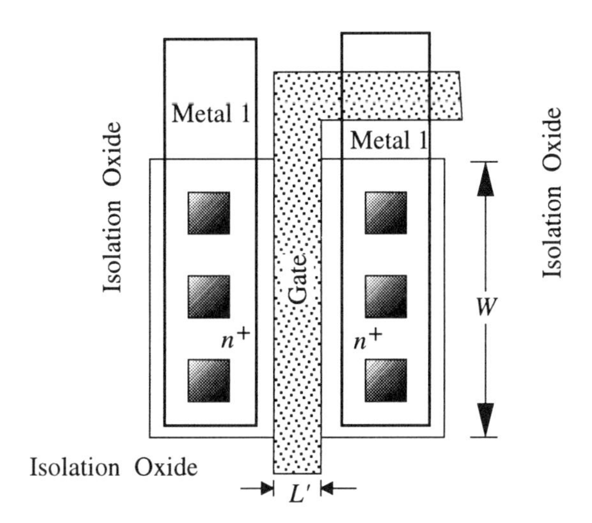

Figure 1.3 Top view of an nFET

rent I_{Dn}; this defines the **active** mode of operation as illustrated in Figure 1.4(b). Thus, the value of V_{GSn} relative to V_{Tn} determines if the transistor is **ON** (active) or **OFF** (no current flowing). The actual value of the current I_{Dn} depends on the voltages applied to the device.

1.1.1 The MOS Threshold Voltage

Conduction from the drain to the source in a MOSFET is possible because the central MOS structure has the characteristics of a simple capacitor. Figure 1.5(a) shows the gate-insulator-semiconductor system that acts as a capacitor. The top plate of the capacitor is shown as a two-layer conducting region as is typical in the state-of-the art. The bottom layer is **polycrystalline silicon**, which is usually called **polysilicon** or simply "**poly**." Poly is used because it provides good coverage and adhesion, and can be doped to either polarity. It does, however, have a relatively high resistivity, so that a refractory[1] metal layer is deposited on top; the drawing shows titanium (Ti), but other refractory metals such as platinum (Pt) can be used. The p-type semiconductor substrate acts as the bottom plate of the capacitor. The unique aspects of the MOS system arises from the fact that an electric field can penetrate a small distance into a semiconductor, thus altering the charge distri-

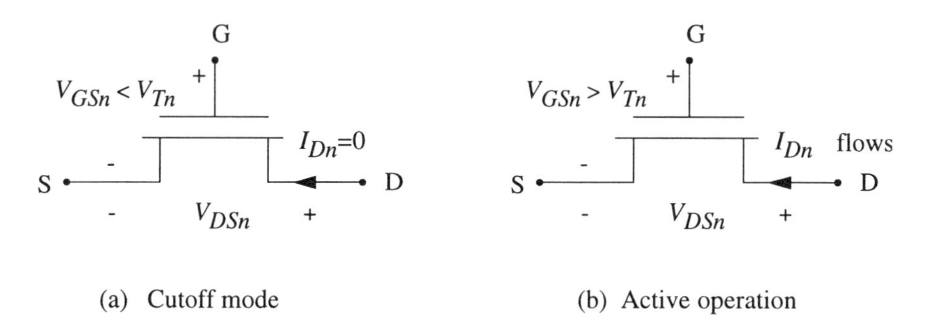

Figure 1.4 General behavior of an nFET with applied voltages

[1] "a refractory metal" is a metal with a high melting temperature.

bution at the surface.

The insulator between the top and bottom plates is silicon dioxide, SiO_2, which is generically known as quartz glass. Denoting the thickness of the gate oxide by x_{ox} (in units of cm) the oxide capacitance per unit area is given by the parallel plate formula

$$C_{ox} = \frac{\varepsilon_{ox}}{x_{ox}} \ F/cm^2 \tag{1. 1}$$

where the oxide permittivity is $\varepsilon_{ox} \approx (3.9)\varepsilon_o$ F/cm when silicon dioxide is used as the gate insulator. In this expression, ε_o is the permittivity of free space with a value of $\varepsilon_o \approx 8.854 \times 10^{-14}$ F/cm. Current technologies have oxide thicknesses x_{ox} less than about $100\mathring{A} = 0.01$ μm, giving a value for C_{ox} on the order of 10^{-7} F/cm^2 or greater. The most aggressive process lines have oxides as thin as $50\mathring{A}$. Thin oxides are desirable because they yield increased capacitance, which will in turn enhances the conduction through a MOSFET.

To understand the origin and characteristics of the threshold voltage, let us analyze the basic MOS structure in Figure 1.5. The charge carrier population at the semiconductor surface is controlled by the gate voltage V_G. If a positive voltage is applied to the gate electrode, negative charge is induced in the semiconductor region underneath the oxide. This is due to the penetration of the electric field into the silicon, and is termed the **field-effect**: the charge densities are controlled by the external voltage through the electric field. Applying KVL to the circuit,[2] we may write that

$$V_G = V_{ox} + \phi_S \tag{1. 2}$$

where V_{ox} is the voltage across the oxide, and ϕ_S is the **surface potential**, i.e., the voltage at the surface of the silicon; the behavior of the voltage is shown in Figure 1.5(b). This simple expression shows that increasing the gate voltage V_G increases the surface potential ϕ_S, thus giving a stronger electric field in the semiconductor.

The **surface charge density** Q_S C/cm^2 at the surface of the semiconductor represents the total charge seen looking downward from the oxide into the p-type bulk. For small values of V_G, the

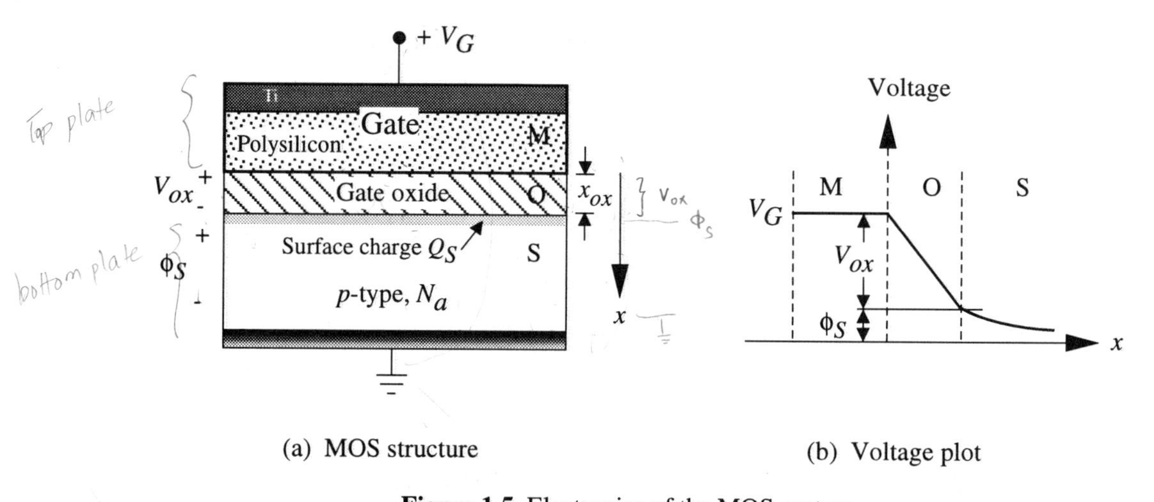

(a) MOS structure

(b) Voltage plot

Figure 1.5 Electronics of the MOS system

[2] KVL is short for Kirchhoff's Voltage Law.

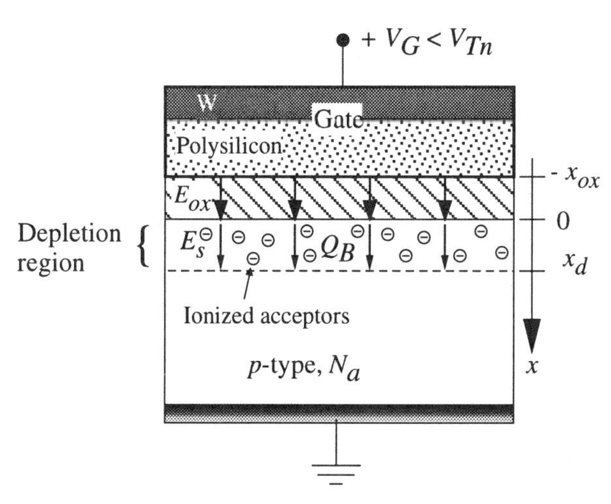

Figure 1.6 Depletion in the MOS structure

field creates a depletion region that consists of negative space charge to support the electric field. This mode of operation is called **depletion**, and has the characteristics shown in Figure 1.6. The depletion charge is usually referred to as the **bulk charge**, and is due to ionized acceptor atoms that have accepted a free electron into their electronic shell structure. The bulk charge density is given by

$$Q_B = -\sqrt{2q\varepsilon_{Si}N_a\phi_S} \tag{1.3}$$

with units of C/cm^2. In this equation, $\varepsilon_{Si} \approx (11.8)\varepsilon_o$ is the permittivity of silicon, N_a is the acceptor doping density in the substrate, and $q = 1.6 \times 10^{-19}$ C is the fundamental charge unit. In this mode of operation, the surface charge is made up entirely of bulk charge with

$$Q_S \approx Q_B. \tag{1.4}$$

Since bulk charge consists of ionized acceptor atoms, it is immobile.

Increasing the gate voltage to a value $V_G = V_{Tn}$, known as the **threshold voltage**, initiates the formation of a thin electron **inversion layer** with a surface charge density Q_n C/cm^2 at the silicon surface. Increasing the gate voltage to a value $V_G > V_{Tn}$ gives a buildup of the inversion charge, the total surface charge density is given by

$$Q_S = Q_B + Q_n. \tag{1.5}$$

Inversion charge is due to mobile electrons that are free to move in a direction parallel to the surface. This mode of operation is called **inversion**, and is characterized by the charges shown in Figure 1.7. It can be shown that the value of the surface potential ϕ_S needed to form this layer is given by

$$\phi_S \approx 2|\phi_F| = 2\left(\frac{kT}{q}\right)\ln\left(\frac{N_a}{n_i}\right) \tag{1.6}$$

where ϕ_F is called the **bulk Fermi potential**. The factor (kT/q) is the **thermal voltage**, and n_i is the **intrinsic carrier concentration**. At room temperature ($T = 300°K$), the thermal voltage is approximated by $(kT/q) \approx 0.026v$, and the intrinsic density in silicon is about $n_i \approx 1.45 \times 10^{10}$ cm^{-3}. Note that the value of the surface potential needed to invert the surface depends on the substrate doping

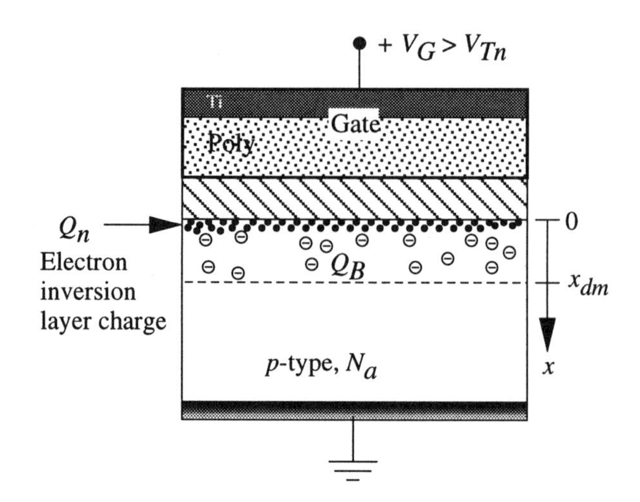

Figure 1.7 Inversion mode of operation in an MOS system

N_a. In a typical bulk CMOS process, $N_a \approx 10^{15}\ cm^{-3}$, which gives a value of $2|\phi_F| \approx 0.58$ volts.

The threshold voltage can be estimated at this point using the KVL equation and noting that at the onset of the inversion phenomenon, the inversion layer has just started to form. When $V_G = V_{Tn}$, $Q_n \approx 0$, so that $Q_S \approx Q_B$ is the total charge at the surface. Using the capacitive relation $Q=CV$ allows us to write the oxide voltage as[3]

$$V_{ox} = \frac{|Q_B|}{C_{ox}} = \frac{\sqrt{2q\varepsilon_{Si}N_a(2|\phi_F|)}}{C_{ox}}. \tag{1.7}$$

The Kirchhoff equation thus gives

$$V_{Tn}^{Ideal} = 2|\phi_F| + \frac{1}{C_{ox}}\sqrt{2q\varepsilon_{Si}N_a(2|\phi_F|)}. \tag{1.8}$$

for the *ideal* threshold voltage, which assumes that the MOS capacitor is a perfect insulator and ignores the fact that the gate and substrate are generally made out of different materials.

This equation must be modified before it can be applied to a realistic MOS structure which has (a) trapped charge within the oxide that alters the electric field, and (b) differences in the electrical characteristics of the gate and substrate materials. To account for these two effects, we add a term

$$V_{FB} = (\Phi_G - \Phi_S) - \frac{1}{C_{ox}}(Q_f + Q_{ox}) \tag{1.9}$$

which is called the **flatband voltage**[4]. In this expression, $(\Phi_G - \Phi_S)$ is the work function difference between the gate (G) and substrate (S), Q_f is the fixed surface charge density at the oxide-silicon interface, and Q_{ox} represents trapped charge within the oxide; both charge quantities have units of C/cm^2.

The trapped oxide charge term Q_{ox} is due mostly to mobile alkalai impurity ions Na^+ and K^+

---

[3] Note that both Q and C in this equation have units of '"per cm^2".

[4] The name *flatband voltage* is due to the fact that setting $V_G = V_{FB}$ gives flat energy bands in both gate and substrate.

that are trapped in the oxide. Denoting the volume charge density of the charged ions by ρ_{ox} in units of C/cm^3, Q_{ox} is found from

$$Q_{ox} = \int \frac{x}{x_{ox}} \rho_{ox}(x)\, dx \tag{1.10}$$

where the integral is performed over the extent of the oxide. Since these charges can move under the influence of an applied electric field, they can yield devices with unstable threshold voltages. Modern processing techniques generally reduce the effect of the trapped charge to negligible levels by performing the oxidation in a chlorinated atmosphere. In the case of the alkalai contaminants, this produces neutral NaCl and KCl salts that do not affect the electrical operation. The fixed charge Q_f, on the other hand, is due in part to the change in composition from silicon to silicon dioxide, and cannot be eliminated; thermal annealing can be used to minimize the value of Q_f.

The gate material used for basic MOSFETs is polycrystal silicon (**poly**), and the value of the work function difference $(\Phi_G - \Phi_S)$ depends on the doping of the gate relative to the substrate. Gates can be doped either n-type (with $N_{d,poly}$) or p-type (with $N_{a,poly}$). For an n-poly gate with the p-type substrate, the value of $(\Phi_G - \Phi_S)$ can be approximated by using

$$(\Phi_G - \Phi_S) \approx -\left(\frac{kT}{q}\right) \ln\left(\frac{N_a N_{d,\,poly}}{n_i^2}\right) \tag{1.11}$$

which results in a negative value. In the case of a p-poly gate and a p-type substrate, the calculation gives

$$(\Phi_G - \Phi_S) \approx \left(\frac{kT}{q}\right) \ln\left(\frac{N_{a,\,poly}}{N_a}\right) \tag{1.12}$$

In both equations, N_a is the background acceptor substrate doping[5].

Most advanced processes have gates that consist of polysilicon with a top layer of a high-conductivity refractory metal, such as Ti (titanium), W (tungsten), or a process-specific mixture of poly silicon and metal as was discussed earlier. In both cases, the gate work function is set by the lower polysilicon region directly over the gate oxide, and the upper refractory metal layer does not appreciably change the value of $(\Phi_G - \Phi_S)$.

Incorporating the flatband voltage contributions into the threshold voltage expression gives

$$V'_{Tn} = V_{FB} + 2|\phi_F| + \frac{1}{C_{ox}} \sqrt{2q\varepsilon_{Si} N_a (2|\phi_F|)}\ . \tag{1.13}$$

However, under normal processing conditions the flatband voltage V_{FB} is negative and usually yields a negative threshold voltage $V_{Tn}<0$. For CMOS switching circuits that use a positive power supply, a positive threshold voltage is needed. This is accomplished by performing a **threshold adjustment ion implant** with a dose D_I giving the number of implanted ions/cm^2 which modifies the equation to

$$V_{Tn} = V_{FB} + 2|\phi_F| + \frac{1}{C_{ox}} \sqrt{2q\varepsilon_{Si} N_a (2|\phi_F|)} \pm \frac{qD_I}{C_{ox}} \tag{1.14}$$

for the working value of the threshold voltage. Implanting acceptor ions into the substrate is equivalent to introducing additional bulk charge at the surface; the implant thus induces a positive shift

[5] These equations are the subject of Problem [1-2].

in V_{Tn} so that the " + " sign is applicable, with typical working voltages around $V_{Tn} \approx 0.7$ volts. The term (qD_I/C_{ox}) models the effect of the implanted ions as a charge sheet, and ignores the actual distribution into the substrate. If a donor ion implant is used, the threshold voltage is made more negative and the minus sign " - " must be used.

Once the gate voltage exceeds the threshold voltage, the electron inversion charge density may be approximated by

$$Q_n = -C_{ox}(V_G - V_{Tn}) \tag{1. 15}$$

since $(V_G\text{-}V_{Tn})$ represents the net voltage over that needed to create the inversion layer. An obvious result of this analysis is that the amount of mobile electron charge can be increased by increasing the gate voltage.

Example 1.1 Threshold Voltage Calculation

Consider an n-channel MOS system that is characterized by x_{ox}=100Å and N_a=10^{15} cm^{-3}. An n-type poly gate is used with $N_{d,poly}$ =$10^{19}cm^{-3}$. The fixed oxide charge is approximated as Q_f=$q(10^{11})$ C/cm^2 and is the dominant oxide charge term, and the acceptor ion implant dose is assumed to be $D_I \approx 2\times 10^{12}$ $cm.^{-2}$.

To determine the threshold voltage, we will first calculate the value of C_{ox} from

$$C_{ox} = \frac{\varepsilon_{ox}}{x_{ox}} = \frac{(3.9)\ (8.854\times10^{-14})}{100\times10^{-8}} \tag{1. 16}$$

which gives $C_{ox} \approx 3.45 \times 10^{-7}$ $F/cm.$ or $C_{ox} \approx 3.45$ $fF/\mu m$ where $1fF$ (femtofarad) is 10^{-15} F.

Now we compute each term in the expression. The flatband voltage is

$$V_{FB} = -(0.026)\ \ln\left(\frac{10^{15}\cdot 10^{19}}{(1.45\times10^{10})^2}\right) - \frac{(1.6\times10^{-19})\ (10^{11})}{3.45\times10^{-7}} \approx -0.865V \tag{1. 17}$$

and the surface potential is

$$2|\phi_F| \approx (0.026)\ \ln\left(\frac{10^{15}}{1.45\times10^{10}}\right) \approx 0.579V \tag{1. 18}$$

The bulk charge term contributes a value of

$$\frac{1}{C_{ox}}\sqrt{2q\varepsilon_{Si}N_a(2|\phi_F|)} = \frac{\sqrt{2\,(1.6\times10^{-19})\ (11.8)\ (8.854\times10^{-14})\ (10^{15})\ (0.579)}}{3.45\times10^{-7}} \approx 0.040V \tag{1. 19}$$

Finally, the ion implantation step increases the threshold voltage by an amount

$$\frac{qD_I}{C_{ox}} = \frac{(1.6\times10^{-19})\ (2\times10^{12})}{3.45\times10^{-7}} \approx 0.928V \tag{1. 20}$$

Combining terms gives the threshold voltage as

$$V_{T0} \approx -0.865 + 0.579 + 0.040 + 0.928 \tag{1.21}$$

or $V_{Tn} \approx 0.682V$. Note that the ion implant step is required to produce a positive threshold voltage $V_{Tn} > 0$ in this example. The threshold adjustment ion implant dose can be varied to give different working values of the threshold voltage.

1.1.2 Body Bias

Let us now examine the threshold voltage of a MOSFET. Although this is approximately the same as the value for the MOS capacitor structure, the application of voltages to the n^+ source and drain regions requires that we allow for modifications in the expression.

Consider the case where we bias the transistor with drain and source voltages as shown in Figure 1.8. Since the p-type bulk is grounded, this arrangement results in the application of a source-bulk voltage V_{SBn}, which induces the **body-bias effect** where the threshold voltage V_{Tn} is increased. This occurs because V_{SBn} adds reverse-bias across the p-substrate/n-channel boundary, which in turn increases the bulk depletion charge. The effect is identical to increasing the depletion charge in a pn junction by applying a reverse-bias voltage. With the source-bulk voltage V_{SBn}, the bulk charge increases to a value given by

$$\frac{1}{C_{ox}}\sqrt{2q\varepsilon_{Si}N_a(2|\phi_F| + V_{SBn})} \tag{1.22}$$

since the additional voltage increases the potential at the surface, effectively adding a reverse bias to the depletion region. The threshold voltage is then given by

$$V_{Tn} = V_{FB} + 2|\phi_F| + \frac{1}{C_{ox}}\sqrt{2q\varepsilon_{Si}N_a(2|\phi_F| + V_{SBn})} \pm \frac{qD_I}{C_{ox}} \tag{1.23}$$

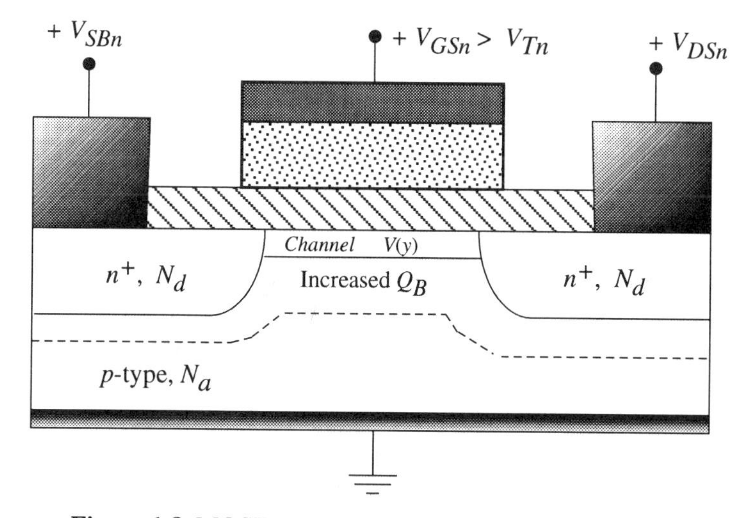

Figure 1.8 MOSFET biased into inversion

Let us denote the **zero-body bias threshold voltage** by V_{T0n}. Applying V_{SBs} increases the threshold voltage by an amount ΔV_{Tn}=(V_{Tn}-V_{T0n}) with

$$\Delta V_{Tn} = \gamma\left(\sqrt{2|\phi_F| + V_{SBn}} - \sqrt{2|\phi_F|}\right) \tag{1. 24}$$

where we have introduced

$$\gamma = \frac{1}{C_{ox}}\sqrt{2q\varepsilon_{Si}N_a} \tag{1. 25}$$

as the **body-bias factor** with units of $V^{1/2}$. For the general case, we write the threshold voltage as

$$V_{Tn} = V_{T0n} + \gamma\left(\sqrt{2|\phi_F| + V_{SBn}} - \sqrt{2|\phi_F|}\right) \tag{1. 26}$$

which evaluates to V_{T0n} when V_{SBn}=0. This has the square-root dependence illustrated in the plot of Figure 1.9. In practice, threshold voltage values are usually understood to be V_{T0n} with the zero-bias value V_{T0n} as the lowest value.

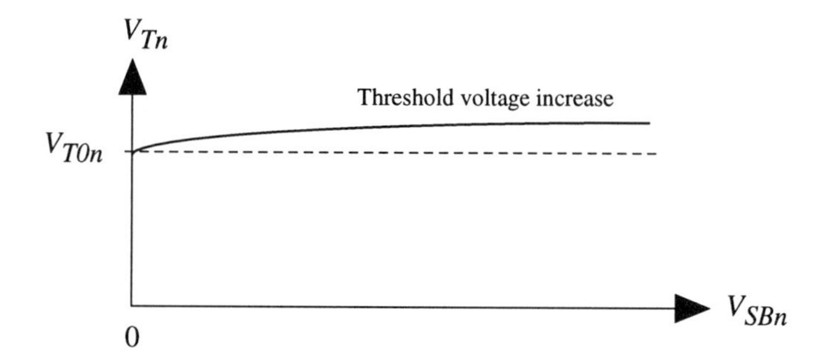

Figure 1.9 Increase of threshold voltage due to body bias

Example 1.2 Body-Bias Coefficient

The process parameters in Example 1.1 give a body-bias factor of

$$\gamma = \frac{\sqrt{2\,(1.6\times10^{-19})\,(11.8)\,(8.854\times10^{-14})\,(10^{15})}}{3.45\times10^{-7}} \approx 0.053V^{1/2} \tag{1. 27}$$

so that

$$V_{Tn} \approx 0.682 + 0.053\left(\sqrt{0.579 + V_{SB}} - \sqrt{0.579}\right) \tag{1. 28}$$

The body bias coefficient γ in this example is relatively small because of the large value of C_{ox}, which is due to the thin (100 Å) gate oxide.

1.2 Current-Voltage Characteristics

The MOSFET *I-V* characteristics can be extracted by modelling the characteristics of the charge as a function of the gate-source voltage V_{GSn}. Consider first the case of cutoff which occurs when

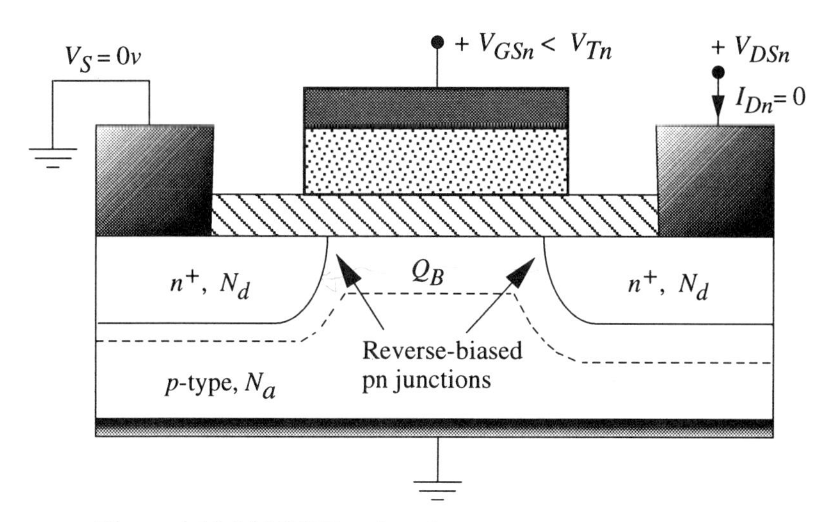

Figure 1.10 MOSFET pn junctions

$V_{GSn} < V_{Tn}$; this is shown in Figure 1.10. Since V_{GSn} is not sufficient to induce an electron inversion layer, only immobile bulk charge Q_B exists under the gate. The drain and source are separated by two pn junctions, one of which has zero-bias applied (the source) while the other has a reverse-bias across it. This blocks the flow of current, giving $I_{Dn} \approx 0$.

Active operation requires that $V_{GSn} \geq V_{Tn}$ be applied to the gate. This creates an electron inversion layer beneath the oxide, which in turn forms the FET conduction channel from drain to source. Since we have already characterized the electron charge in a simple MOS structure, we may modify our analysis to include the FET parameters, and compute I_{Dn} as a function of V_{GSn} and V_{DSn}. Modelling can be performed at various levels with the general tradeoff being complexity versus accuracy.

The basic analytic models are obtained using charge control arguments within the **gradual-channel analysis** below that makes some basic assumptions on the mechanism of current flow. Consider the device cross-section shown in Figure 1.11. To induce current flow, two conditions are needed. First, $V_{GSn} \geq V_{Tn}$ is required to create the channel region underneath the oxide. Second, a drain-to-source voltage V_{DSn} must be applied to produce the channel electric field E. This field forces electrons to move from the source to the drain, thereby giving drift current I_{Dn} in the opposite direction[6], i.e., the current flows into the drain and out of the source.

The electron inversion charge Q_n (in units of C/cm^2) in the channel is given by a capacitor relation of the form

$$Q_n = -C_{ox}[V_{GSn} - V_{Tn} - V(y)] \tag{1. 29}$$

where $V(y)$ represents the voltage in the channel due to V_{DSn}. The origin of the channel voltage $V(y)$ is easily understood by noting that the drain-source voltage V_{DSn} creates an electric field in the channel region, giving the electric potential function $V(y)$ such that

$$E = -\frac{dV}{dy}. \tag{1. 30}$$

[6] Remember that **conventional current** flows in the direction of positive charge motion, and is thus opposite to the direction that electrons move.

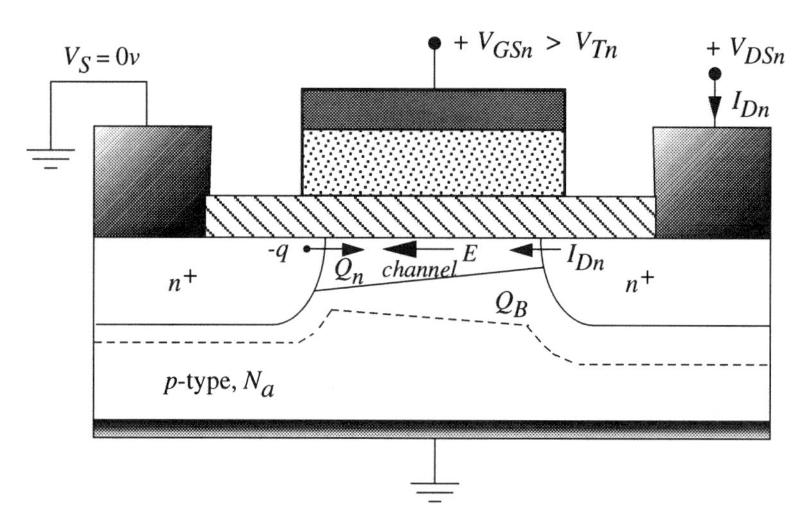

Figure 1.11 Current flow path in an nMOSFET

The channel potential has boundary conditions of

$$\begin{aligned} V(0) &= V_S = 0 \\ V(L) &= V_{DSn} \end{aligned} \tag{1. 31}$$

corresponding to the values we have chosen at the at the source and drain side, respectively. The factor $[V_{GSn}\text{-}V_{Tn}\text{-}V(y)]$ in Q_n thus gives the net effective voltage across the MOS structure at the point y , i.e., the value of the voltage that supports the electron inversion layer. The negative sign is required because the channel consists of negatively-charged electrons, so that $Q_n<0$.

To obtain the I-V equations for the MOSFET, we note that the channel region acts as a nonlinear resistor between the source and drain. The channel geometry is detailed in Figure 1.12. Consider the differential segment dy of the channel. Since this element has a simple rectangular shape with the current flow length of dy , the resistance is

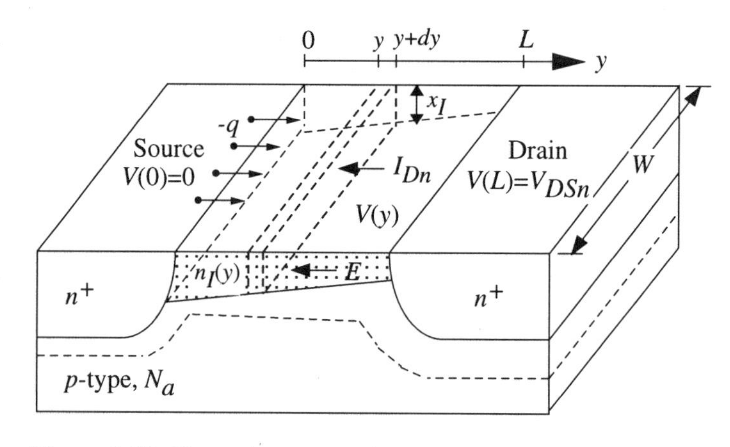

Figure 1.12 Channel geometry for current flow analysis

$$dR = \frac{dy}{\sigma A_I} \tag{1.32}$$

where σ is the conductivity of the region, and A_I is the cross-sectional area (perpendicular to the direction of current flow. Now note that the width of the channel is W, which allows us to write the areas as $A_I = Wx_I$, with x_I being the thickness of the channel inversion layer at that point. Also, the conductivity is given by $\sigma = q\mu_n n_I$ where μ_n is the **electron surface mobility** in units of $cm^2/V\text{-}sec$, and n_I is the electron density in the channel in units of cm^{-3}. Combining these relations allows us to write the denominator as

$$\begin{aligned} \sigma A_I &= q\mu_n n_c W x_c \\ &= -\mu_n W Q_n \end{aligned} \tag{1.33}$$

where the second line follows by noting the definition of the inversion charge density Q_n is equivalent to $(-qn_I x_I)$ since that is the electron charge density in units of C/cm^2.

Now, note that the current through the segment is I_{Dn}. The voltage dV across a differential segment dy of the channel is given by $dV = I_{Dn}\, dR$ or

$$dV = -\frac{I_{Dn} dy}{\mu_n W Q_n} \tag{1.34}$$

where the negative sign is required because the current is flowing in the $-y$ direction. Substituting for $Q_n(V)$ yields

$$I_{Dn} dy = \mu_n C_{ox} W [V_{GSn} - V_{Tn} - V(y)]\, dV \tag{1.35}$$

Rearranging and integrating y from $y=0$ to $y=L$ gives the general expression

$$I_{Dn} = k'_n \left(\frac{W}{L}\right) \int_{V=0}^{V_{DSn}} [V_{GSn} - V_{Tn} - V(y)]\, dV \tag{1.36}$$

We have introduced the nMOS **process transconductance**

$$k'_n = \mu_n C_{ox} \tag{1.37}$$

which has units of A/V^2, and is set by the processing parameters. The device geometry is specified by the **channel width** W and the **channel length** L; the **aspect ratio** (W/L) is the important geometrical factor that determines the current. Since the aspect ratio is set by the device layout, it is the easiest parameter to control for circuit design. The **device transconductance**

$$\beta_n = k'_n \left(\frac{W}{L}\right) \tag{1.38}$$

is used to characterize a specific device. The basic MOSFET device equations obtained from this analysis are discussed below.

It is important to note that analyzing the MOSFET by starting with the concept of a differential resistance assumes that current flow through a MOSFET is purely drift in nature, i.e., that the charge motion is induced solely by the electric field. Diffusion effects due to concentration gradients of the form (dn/dy) are neglected in the analysis. This approach was named the *gradual channel approximation* by Shockley since it assumes that the gradients are small. While it remains a useful vehicle for understanding conduction through a field-effect transistor, the equations derived below are only valid in devices with long channel lengths. This point is examined in more detail in the latter sections of this chapter.

1.2.1 Square-Law Model

The simplest description of current flow through a MOSFET is obtained by assuming that V_{Tn} is a constant in the channel. The integral (1.36) may then be evaluated to give

$$I_{Dn} = k'_n\left(\frac{W}{L}\right)\left[(V_{GSn} - V_{Tn})V_{DSn} - \frac{1}{2}V_{DSn}^2\right]$$
$$= \frac{\beta_n}{2}\left[2(V_{GSn} - V_{Tn})V_{DSn} - V_{DSn}^2\right] \tag{1.39}$$

which describes what we will call **non-saturated** current flow. Given a gate-source voltage V_{GSn}, this predicts a non-linear increase in current flow with increasing V_{DSn}. The peak current occurs when

$$\frac{\partial I_{Dn}}{\partial V_{DSn}} = \frac{\beta_n}{2}[(V_{GSn} - V_{Tn}) - V_{DSn}] = 0 \tag{1.40}$$

This value of V_{DSn} defines the **saturation voltage**

$$V_{sat} = V_{GSn} - V_{Tn} \tag{1.41}$$

such that eqn. (1.39) is valid for $V_{DSn} \leq V_{sat}$. Figure 1.13 is a plot of I_{Dn} as a function of V_{DSn} for a given gate-source voltage V_{GSn} applied to the device. Note that we have only used the parabolic function for drain-source voltages that satisfy $V_{DSn} \leq V_{sat}$. Beyond V_{sat}, the equation would predict a decrease in current, which is not observed in physical devices. The simplest approximation to make for the current above V_{sat} is to simply extend it at the same value as shown.

The physical significance of the saturation voltage is shown in Figure 1.14. When $V_{DSn} = V_{sat}$, the channel is "pinched off" at the drain side of the transistor. This can be verified mathematically by noting that the saturation condition corresponds to a channel voltage of $V(y=L)=V_{sat}$, so that the inversion charge in eqn. (1.29) evaluates to $Q_n(y=L)=0$. At this point the channel is viewed as being "compressed to its minimum thickness." The value of the current at the saturation voltage is

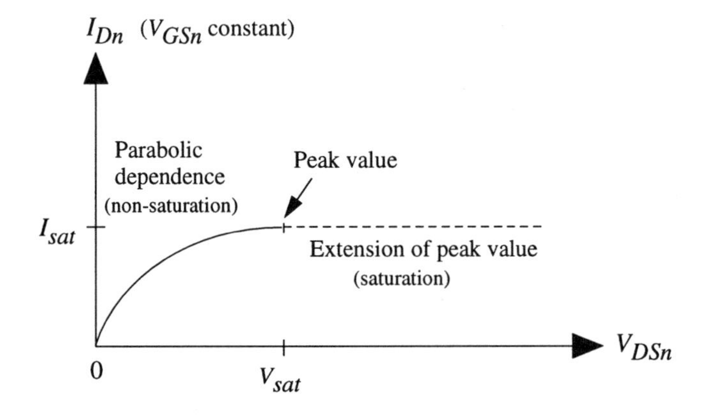

Figure 1.13 Basic MOSFET *I-V* relation for given V_{GS}

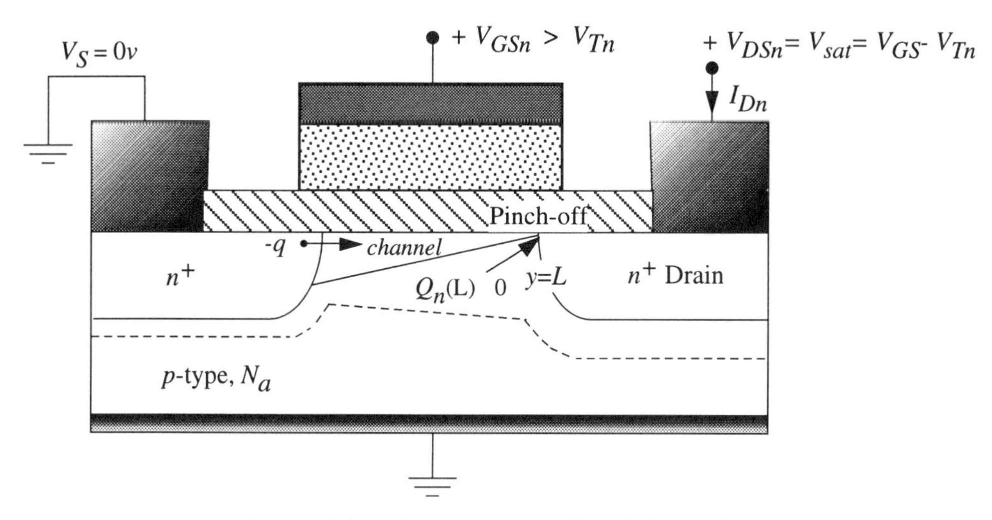

Figure 1.14 Channel pinch-off in a MOSFET

$$
\begin{aligned}
I_{Dn}\big|_{V_{DSn} = V_{sat}} &= \frac{\beta_n}{2}\left[2(V_{GSn} - V_{Tn})V_{sat} - V_{sat}^2\right] \\
&= \frac{\beta_n}{2}\left[2(V_{GSn} - V_{Tn})(V_{GSn} - V_{Tn}) - (V_{GSn} - V_{Tn})^2\right] \\
&= \frac{\beta_n}{2}(V_{GSn} - V_{Tn})^2
\end{aligned}
\tag{1.42}
$$

by direct substitution.

When the drain-source voltage is increased to $V_{DSn} \geq V_{sat}$, the device conducts in the **saturated** mode where the current flow has only a weak dependence on the drain-source voltage. As V_{DSn} increases, the effective length of the channel L_{sat} decreases as shown in Fig. 1.15; this phenomenon is called **channel-length modulation**. Since the drain current is proportional to $(1/L)$, channel-length modulation tends to increase the saturated current flow. The saturated current can be approximated by using the maximum value of the non-saturated current with an effective channel length. A simple expression for the saturated current is given by adding a factor to the peak current by writing

$$
I_{Dn} = \frac{\beta_n}{2}(V_{GSn} - V_{Tn})^2[1 + \lambda(V_{DSn} - V_{sat})]
\tag{1.43}
$$

where λ (with units of V^{-1}) is called the **channel-length modulation** parameter. This expression, which is valid for $V_{DSn} \geq V_{sat}$, is purely empirical, as there is no physical basis for the linear increase in current described by this expression. It does, however, remain a reasonable model for basic calculations. Channel-length modulation effects are important in analog networks. However, we will usually approximate $\lambda \approx 0$ for simplicity when analyzing circuits using square law models in hand calculations. The effects of channel length modulation should be included in a computer simulation.

Figure 1.16 illustrates the family of curves generated by the square law model without channel-length modulation effects included. Each curve corresponds to a different value of V_{GSn}. The border between saturation and non-saturation is approximately parabolic as shown by writing

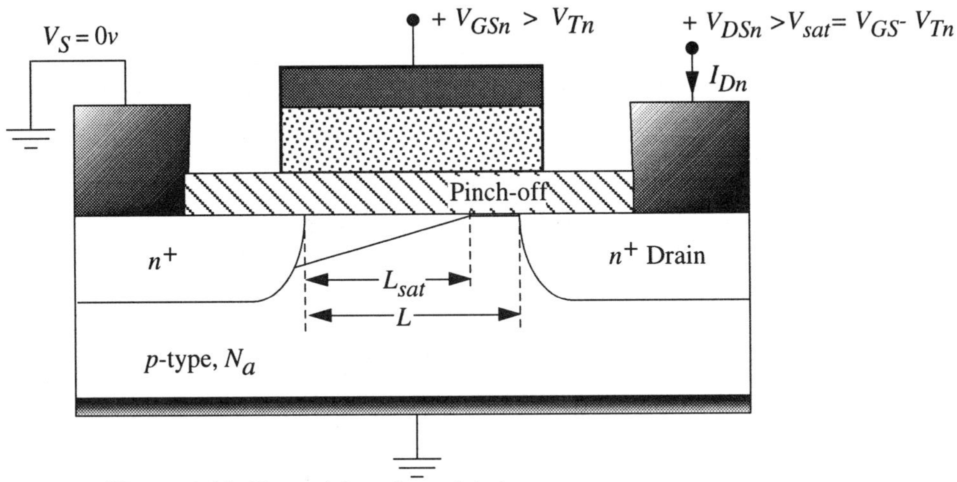

Figure 1.15 Channel-length modulation

$$
\begin{aligned}
I_{Dn}\big|_{Border} &= \frac{\beta_n}{2}\left[2\,(V_{GSn} - V_{Tn})\,V_{sat} - V_{sat}^2\right] \\
&= \frac{\beta_n}{2}V_{sat}^2
\end{aligned}
\qquad (1.44)
$$

Also note that the saturation voltage depends on the applied gate-source voltage. If we choose to include channel-length modulation effects, we arrive at the set of curves shown in Figure 1.17. This type of behavior assumes that the current increases in non-saturation, and then increases slightly once the device is saturated with $V_{DSn} \geq V_{sat}$.

Another useful characterization arises from examining the saturation current in more detail.

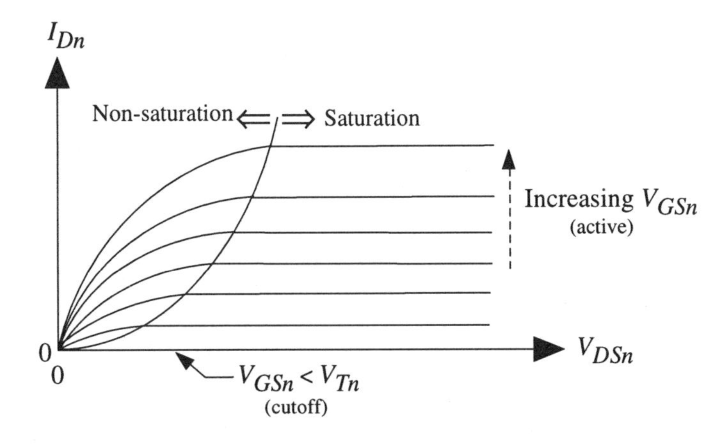

Figure 1.16 Basic MOSFET *I-V* relations

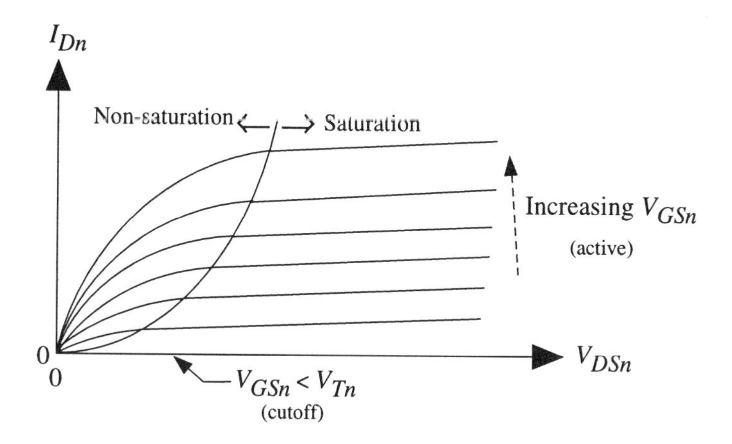

Figure 1.17 MOSFET *I-V* characteristic with channel-length modulation effects

Taking the square root of both sides gives

$$\sqrt{I_{Dn}} = \sqrt{\frac{\beta_n}{2}(V_{GSn} - V_{Tn})^2 [1 + \lambda (V_{DSn} - V_{sat})]} \tag{1. 45}$$

or, with $\lambda \approx 0$,

$$\sqrt{I_{Dn}} \approx \sqrt{\frac{\beta_n}{2}}(V_{GSn} - V_{Tn}) \tag{1. 46}$$

which is a linear plot. This is the **transfer curve** shown in Figure 1.18 which gives $\sqrt{I_{Dn}}$ as a function of V_{GSn} for a saturated MOSFET. The threshold phenomena at $V_{GSn}=V_{Tn}$ is evident from the figure. A saturated MOSFET is particularly useful for measuring the threshold voltage of transistors. There is, however, some distinction between the threshold voltage V_{Tn} of the MOS system and the value useful to a circuit designer.

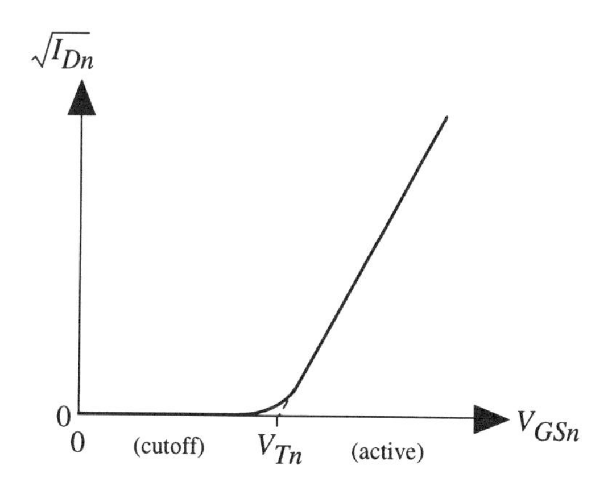

Figure 1.18 Saturation characteristics of an nMOSFET

To understand this comment, let us assume that we have a saturated MOSFET and measure the dependence $\sqrt{I_{Dn}}$ as a function of V_{GSn}. The extrapolated intersection of the line with the voltage axis can be interpreted as the threshold voltage for the device. Although this may seem to provide the necessary information, we find that in a realistic device, the current never really drops to a value $I_{Dn}=0$; this is due to leakage components discussed later. We can overcome this problem by making the distinction between a FET being "on" or "off" (i.e., active or in cutoff) using a small reference current I_{ref} such that when $I_{Dn}= I_{ref}$ when the device is at the edge of conduction. The complication in this approach is seen by noting that

$$I_{Dn} \sim \frac{\beta_n}{2} = \frac{k'_n}{2}\left(\frac{W}{L}\right)_n \tag{1.47}$$

This implies that devices with different aspect ratios $(W/L)_n$ will exhibit different values of "on voltage" when I_{ref} is used. In addition, other physical effects (as discussed later in the chapter) change the threshold voltage in small devices.

Square-law MOSFET models are usually chosen for circuit analysis due to their simplicity. Since this approach ignores some fundamental device physics, errors are automatically introduced into the analysis. This is not a problem so long as the equations are only used for general calculations. Crucial results must always be checked using computer simulations. This philosophy will be adopted here, and the square-law equations will be used extensively when analyzing a circuit.

1.2.2 Bulk-Charge Model

A more accurate equation set is obtained by noting that the channel voltage $V(y)$ is underneath the oxide and increases the effective bias on the gate-induced bulk charge. This increases the bulk charge term (1.18) in the threshold voltage equation to a value of

$$\frac{1}{C_{ox}}\sqrt{2q\varepsilon_{Si}N_a(2|\phi_F| + V)}\,, \tag{1.48}$$

where we assume for simplicity that $V_{SBn}=0$. Since V_{Tn} is now a function of the channel voltage V integrating equation (1.36) gives

$$I_{Dn} = \frac{\beta_n}{2}\left(2\left(V_{GSn} - V_{FB} - 2|\phi_F| \mp \frac{qD_I}{C_{ox}}\right)V_{DSn} - \frac{2\gamma}{3}\left((2|\phi_F| + V_{DSn})^{3/2} - (2|\phi_F|)^{3/2}\right) - V_{DSn}^2\right) \tag{1.49}$$

as the non-saturated drain current. V_{T0n} is still termed "the" threshold voltage, and physically represents the gate voltage needed induce surface inversion at the source end of the MOSFET. The device enters saturation at a drain-source voltage of V_{sat} corresponding to the value where I_{Dn} is a maximum. Explicitly,

$$V_{sat} = \beta_n\left(\left[V_{GSn} - V_{FB} - 2|\phi_F| \mp \frac{qD}{C_{ox}}\right]V_{DSn} - \frac{1}{2}V_{DSn}^2 - \frac{2\gamma}{3}\left[(2|\phi_F| + V_{DSn})^{3/2} - (2|\phi_F|)^{3/2}\right]\right) \tag{1.50}$$

in this model. The value of I_{Dn} evaluated at this voltage is the first order approximation to the saturation current.

Since we have included the variation of the threshold voltage along the channel, the bulk-charge model yields results that are inherently more accurate than predicted by the square-law equations. A detailed comparison between the two shows that the square-law model overestimates both the saturation current and the saturation voltage. Owing to this observation, the bulk-charge model is often used as the basis of many advanced SPICE models. However, for calculator-based circuit estimates, the increased complexity offsets the gain in precision. Because of this reason, square-law equations are the most common for manual circuit analysis.

1.2.3 The Role of Simple Device Models

Both the square-law and bulk-charge models derived above are oversimplifications of the actual physics in a modern transistor. One reason is that they are both derived using the gradual-channel approximation, which is only valid for long channel lengths, typically requiring values greater than about 20μm. With modern devices approaching values of $L \leq 0.20$ μm, the simple equations ignore many short-device effects that arise in the 2-dimensional electrostatics and current flow. These and other important items are discussed later in this chapter.

One must always be careful not to depend on closed-form analytic equations. This is particularly true for semiconductor devices such as MOSFETs. Although their accuracy is always limited, analytic models remain extremely useful for circuit design, even if sub-micron size devices are employed. This is because the equations provide reasonable *estimates* of the *behavior* of the current-voltage characteristics that can be used for obtaining first-cut designs. Once the circuit is created, it can be simulated, analyzed, and re-designed using computer tools. When applied in a careful manner, this approach allows the engineer to perform intelligent design, not just produce a trial-and-error set of SPICE runs.

1.3 p-Channel MOSFETs

Perhaps the most important circuit design aspect of CMOS is the use of both n-channel and p-channel transistors in complementary arrangements. Adhering to this technique guarantees the desired properties of minimum DC power dissipation and rail-to-rail output logic swings.

The operational physics of a p-channel MOSFET (**pFET** or **pMOS**) is the complement of that used to describe the operation of an n-channel device. This means that we should

- change n-type to p-type regions, and
- change p-type to n-type regions.

Then, we

- reverse the roll of electrons and holes,
- reverse the polarities of all voltages, and
- reverse the direction of current flow.

Rather than duplicating the derivation of the MOSFET equations, we will just write down the important equations using these observations. Square-law models will be used throughout for simplicity, but the analogy can be extended to more complex models if desired.

A p-channel MOSFET must be constructed in an n-type background region. If a p-substrate is used as a starting point, then an **n-well** must be provided at all pMOS locations.[7] Figure 1.19 illustrates a pMOS transistor in this type of process; note that the pMOS bulk (the n-well) is connected to the highest voltage in the system, typically the power supply voltage V_{DD}. This is required to insure that the pn junctions do not become forward biased. The channel length L for the pFET is defined as the distance between the two p^+ regions. A top view of the pFET is shown in Figure 1.20. The channel width W defines the width of the current flow path, and the aspect ratio is (W/L). L' is again used to represent the drawn channel length. The power supply connection (V_{DD}) to the n-well is shown explicitly in the layout; this bias is critical to the operation of the device, and the

[7] The concept of an n-well is discussed in Chapter 2.

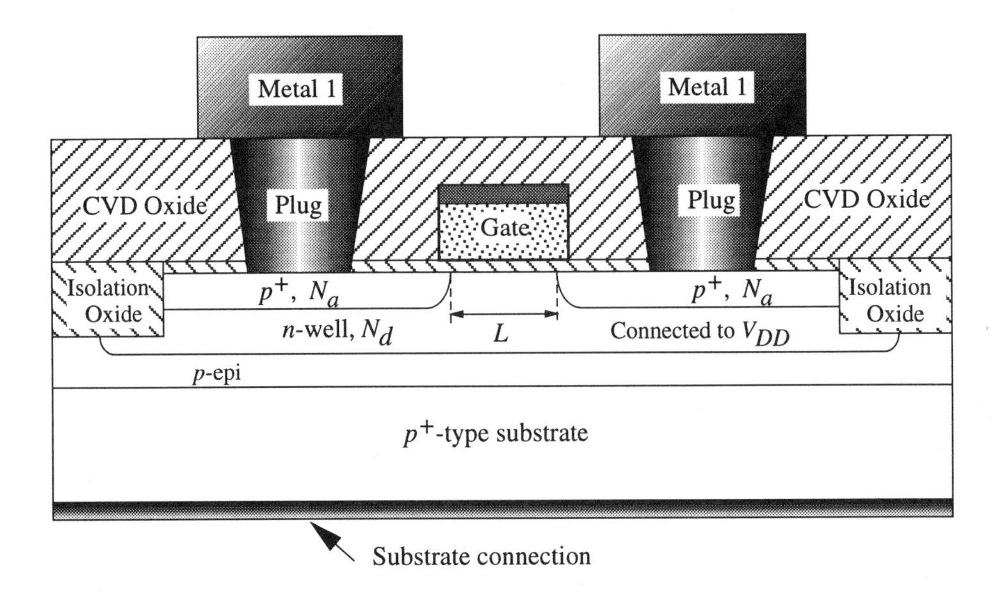

Figure 1.19 Cross-sectional view of a p-channel MOSFET in an n-well

contacts are usually applied quite liberally whenever space permits.

The basic circuit symbol for a pFET is shown in Figure 1.21(a). When we show the MOSFET as a 4-terminal device, the only difference between the nFET and pFET symbols is in the direction of the arrow on the bulk terminal. The arrow itself is used to acknowledge the polarity of the pn junction between the bulk and the channel.

In a pFET, current flow is due to the motion of positively charged holes. This then stipulates that the source side must be the p^+ region at the higher voltage, which is opposite to the naming of the terminals in an nFET. Conduction in a pFET is achieved by making the gate sufficiently negative to attract minority carrier holes to the silicon surface. Current flow is thus controlled by V_{SGp} and

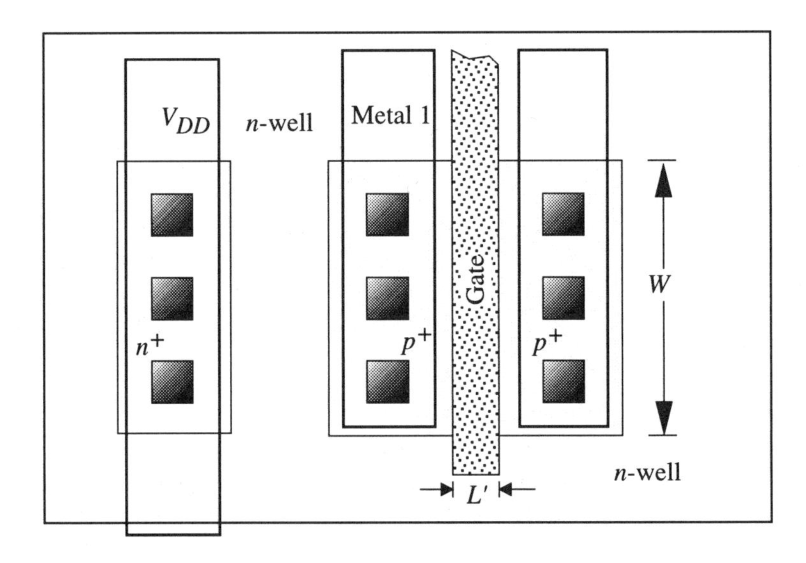

Figure 1.20 Top-view of a p-channel MOSFET

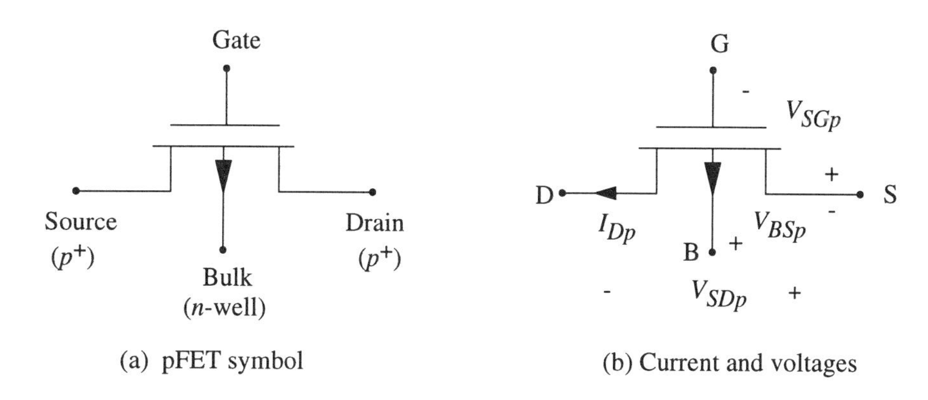

Figure 1.21 pFET circuit symbol and voltage and current definitions

V_{SDp}, and I_{Dp} flows out of the drain electrode as shown in Figure 1.21(b). This choice will result in having current flow equations that have the same form as those for an nFET.

Threshold voltages for field-effect transistors are, by convention, referenced to the value of the gate-to-source voltage. Since the pFET uses positive charge carriers, this implies that the threshold voltage should be opposite in sign to that of a comparable nFET. In our discussion, we will denote the pFET threshold voltage by $V_{Tp} < 0$, but will often use the absolute value $|V_{Tp}|$ in our calculations. In terms of the basic device parameters, the zero-body bias value is given by

$$V'_{T0p} = V_{FB,p} - 2\phi_{F,p} - \frac{1}{C_{ox}}\sqrt{2q\varepsilon_{Si}N_d(2\phi_{F,p})} < 0 \quad (1.51)$$

where N_d is the donor doping in the n-well,

$$2\phi_{F,p} = 2\left(\frac{kT}{q}\right)\ln\left(\frac{N_d}{n_i}\right) \quad (1.52)$$

is the bulk Fermi potential for the n-well, and $V_{FB,p}$ is the flatband voltage. As with the nFET, the working value of the threshold voltage is adjusted using a dedicated ion implantation step. Body-bias effects are described by writing the threshold voltage in the form

$$V_{Tp} = V_{T0p} - \gamma_p\left(\sqrt{2\phi_{F,p} + V_{BSp}} - \sqrt{2\phi_{F,p}}\right) \quad (1.53)$$

where V_{BSp} is the body-bias voltage. Since $V_{T0p} < 0$, body-bias makes V_{Tp} more negative, i.e., applying a bulk-source voltage V_{BSp} increases $|V_{Tp}|$. Note that the n-well of the pFET acts as the bulk electrode. In a CMOS network, this is connected to the highest positive voltage in the circuit, which is usually the power supply V_{DD}. The value of V_{BSp} is then referenced to V_{DD} by

$$V_{BSp} = V_{DD} - V_{Sp} \quad (1.54)$$

where V_{Sp} is the source voltage.

Now that we have examined the basic structure of the pFET, we may deduce the conduction characteristics by using a direct analogy with the n-channel MOSFET. Cutoff occurs when $V_{SGp} < |V_{Tp}|$, since this indicate that the source-gate voltage is not sufficient to support the formation of a hole inversion layer. As expected, cutoff is characterized by $I_{Dp} \approx 0$, with only leakage currents flowing in the device.

Active operation requires a source-gate voltage of $V_{SGp} > |V_{Tp}|$. Saturation occurs at the point

$$V_{sat} = (V_{SGp} - |V_{Tp}|) \ . \tag{1.55}$$

If $V_{SDp} \leq V_{sat}$, then the device is non-saturated with a current of

$$I_{Dp} = \frac{\beta_p}{2}\left[2(V_{SGp} - |V_{Tp}|)V_{SDp} - V_{SDp}^2\right] \tag{1.56}$$

In this equation,

$$\beta_p = k'_p\left(\frac{W}{L}\right)_p \tag{1.57}$$

is the pFET device transconductance with units of A/V^2, and

$$k'_p = \mu_p C_{ox} \tag{1.58}$$

is the p-channel process transconductance with μ_p the hole mobility in units of $cm^2/V\text{-}s$. The aspect ratio $(W/L)_p$ is the primary circuit design parameter.

When $V_{SDp} \geq V_{sat}$, the transistor is in saturation with

$$I_{Dp} = \frac{\beta_p}{2}(V_{SGp} - |V_{Tp}|)^2\left[1 + \lambda(V_{SDp} - V_{sat})\right] \tag{1.59}$$

As in the case of nFETs, we will usually ignore channel-length modulation in digital circuits by setting λ=0 and using the simpler expression

$$I_{Dp} \approx \frac{\beta_p}{2}(V_{SGp} - |V_{Tp}|)^2 \tag{1.60}$$

in hand calculations.

The complementary aspects of nMOS and pMOS transistors are exploited in many CMOS circuit design techniques. However, it should be noted that since electrons move faster than holes (which are vacant electron states), the electron mobility is always larger than the hole mobility, i.e., $\mu_n > \mu_p$ (for equal background doping). This implies that, in general,

$$k'_n > k'_p , \tag{1.61}$$

with $k'_n \approx (2.5)\, k'_p$ being typical. The difference between conduction levels due to different mobilities can significantly influence the circuit design choices. For example, this implies that nMOS transistors should be chosen over pMOS devices if the device speed is critical. Tradeoffs of this type will be discussed in great detail in the remaining chapters of the book.

1.4 MOSFET Modelling

An accurate electronic characterization of any devices requires both the *I-V* relationship and a model for all of the important parasitic elements that exist due to the physical structure and operation. This is particularly important for CMOS VLSI, since the parasitic components are often the limiting factor in the circuit performance.

In this section, we will develop the equivalent circuit for the n-channel MOSFET shown in Figure 1.22. This consists of a linear resistor R_n, drain and source capacitors C_D and C_S, respectively, a voltage-controlled switch that uses the gate electrode as the control terminal, and reverse-bias diodes. Although this is highly simplified and only provides first-order numerical estimates at best, it can be used as the basis for design and optimization analyses. Moreover, it is very useful as an aid in understanding most CMOS circuits regardless of their complexity.

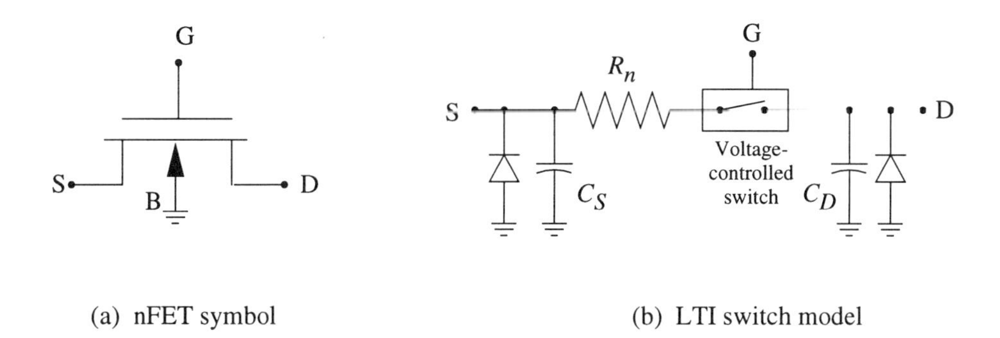

Figure 1.22 Simplified MOSFET equivalent circuit

1.4.1 Drain-Source Resistance

The *I-V* relations for an nFET can be used to define a drain-source resistance of

$$R_n = \frac{V_{DSn}}{I_{Dn}} \tag{1.62}$$

as implied by Figure 1.23(a). Using the square-law equations yields a voltage-dependent resistance r_n of

$$r_n = \frac{2}{\beta_n [2(V_{GSn} - V_{Tn}) - V_{DSn}]} \tag{1.63}$$

if the device is non-saturated, and

$$r_n = \frac{2V_{DSn}}{\beta_n (V_{GSn} - V_{Tn})^2} \tag{1.64}$$

if the transistor is saturated. Obviously, both expressions yield a resistance r_n that varies with V_{DSn}, i.e., r_n is nonlinear. A nonlinear resistance is of limited use in first-level circuit design since it gen-

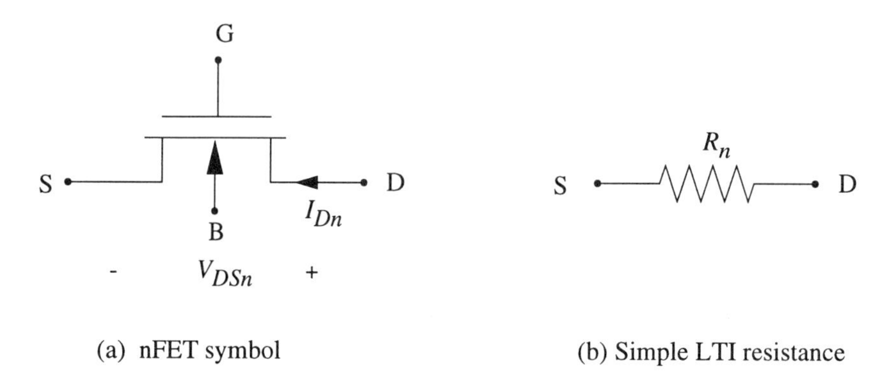

Figure 1.23 Drain-source resistance model

erally introduces a level of complexity that is best left to a computer simulation. It is, however, worthwhile to develop a simple expression for R_n that is a constant, but which reflects the important characteristics of the device. This allows us to model the drain-source conduction (using care) as shown in Figure 1.23(b).

A common approach is to define the linear, time-invariant (LTI) FET resistance by

$$R_n = \frac{1}{\beta_n (V_{Ref} - V_{Tn})} \tag{1.65}$$

where

$$\beta_n = k'_n \left(\frac{W}{L}\right)_n \tag{1.66}$$

provides the important dependence on $(W/L)_n$. In particular, this shows that

$$R_n \propto \frac{1}{\left(\frac{W}{L}\right)_n}, \tag{1.67}$$

i.e., the resistance is inversely proportional to the aspect ratio. V_{Ref} is a reference voltage that is chosen to normalize the value. A particularly simple choice is to take $V_{Ref} = V_{DD}$, with V_{DD} being the power supply voltage applied to the circuit. The value

$$\begin{aligned} R_n &= \frac{1}{\beta_n (V_{DD} - V_{Tn})} \\ &= \frac{1}{k'_n \left(\frac{W}{L}\right)(V_{DD} - V_{Tn})} \end{aligned} \tag{1.68}$$

is consistent with the transient behavior of the CMOS inverter circuit analyzed in Chapter 3, and is thus a popular choice for the more general model.

1.4.2 MOSFET Capacitances

MOSFETs exhibit a number of parasitic capacitances that greatly affect the circuit performance. The capacitances originate from the central MOS gate structure, the characteristics of the channel charge, and the pn junction depletion regions. MOS contributions are constants, but the channel and depletion capacitances are both nonlinear and vary with the applied voltage.

Parasitic capacitances give the fundamental limitation on the switching speed. Although computer simulations are usually required for an accurate analysis, the analytic estimates below suffice for first-order calculations. MOSFET capacitances may be divided into two major groups as shown in Figure 1.24. The first set is due to the MOS system itself through the gate oxide capacitance C_{ox}, giving rise to components such as C_{GS} and C_{GD}. The second set arises from the depletion capacitance of the drain-bulk and source-bulk pn junctions. These are denoted by C_{DB} and C_{SB} in the drawing.

MOS-Based Capacitances

The basic MOS capacitance is due to the physical separation of the gate conductor and the semiconductor by the gate oxide, which has a thickness x_{ox}. The gate regions of a FET is shown in Figure 1.25. We characterize the capacitance per unit area by the basic formula

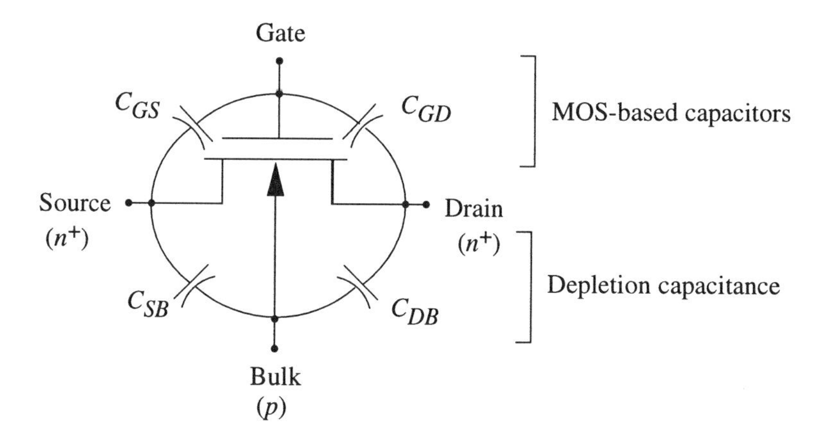

Figure 1.24 FET capacitance model

$$C_{ox} = \frac{\varepsilon_{ox}}{x_{ox}} \tag{1. 69}$$

with units of F/cm^2 as in the initial discussion of the field effect. The oxide capacitance is an intrinsic part of the transistor structure and the value of C_{ox} is important to the MOSFET current equations through k'. It does, however, introduce parasitic capacitance that slows down the transient response of digital switching circuits.

To account for the MOS structure, we first introduce the concept of the **gate capacitance** C_G. By definition, this is the **input capacitance** seen looking to the gate as shown in Figure 1.26. Examining the perspective drawing gives

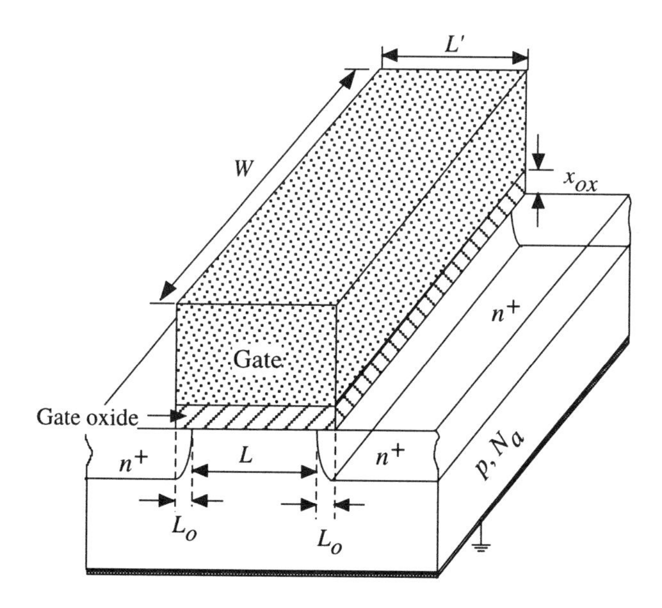

Figure 1.25 Perspective view of a MOSFET

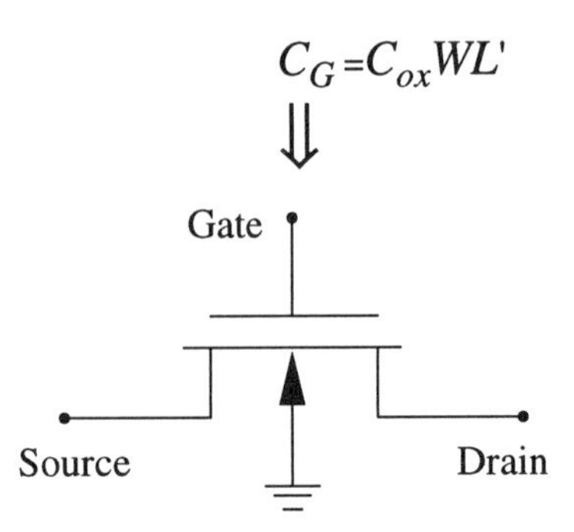

Figure 1.26 Gate capacitance definition

$$C_G = C_{ox}WL' \tag{1.70}$$

where we have used the total gate area of WL'. This simple equation ignores the presence of fringing electric fields which may be significant in the device. The drawn channel length L' that appears in this formula is related to "the" (electrical) channel length L by the relation

$$L' = L + 2L_o \tag{1.71}$$

with L_o known as the **gate overlap** distance. Gate overlap is due to lateral doping effects during the processing, and is discussed in more detail in Chapter 2. Although C_G itself is one of the most useful parameters to characterize the input capacitance in hand calculations, it may be broken down into components by writing

$$C_G = C_g + 2C_{ol} \tag{1.72}$$

where

$$C_g = C_{ox}WL \tag{1.73}$$

is the "central" gate capacitance due to the electrical channel length L only, and

$$C_{ol} = C_oW \tag{1.74}$$

is the **overlap capacitance** with

$$C_o = C_{ox}L_o \tag{1.75}$$

being the overlap capacitance per FET width in units of F/cm. In circuit simulations, this separation allows us to include fringing effects by modifying the value of C_o. Overlap exists on both the drain and source sides of the transistor and should be included in all calculations. In hand estimates, however, it is usually sufficient to simply use the total gate capacitance C_G with a reasonable value of C_{ox} to obtain reasonable values.

Now let us examine the concept of gate-channel capacitances that are represented by the gate-source and gate-drain equivalents C_{GS} and C_{GD}, respectively. The origin of these parasitics is shown in Figure 1.27: they represent the coupling of the gate electrode to the channel. If the device is in cutoff, then no channel exists; in this case, we use the gate-bulk capacitance C_{GB} to represent the charge effects. Since the channel depends upon the applied voltages, all three contributions are nonlinear in nature, i.e., they are functions of the voltages. In general, capacitors of this type are

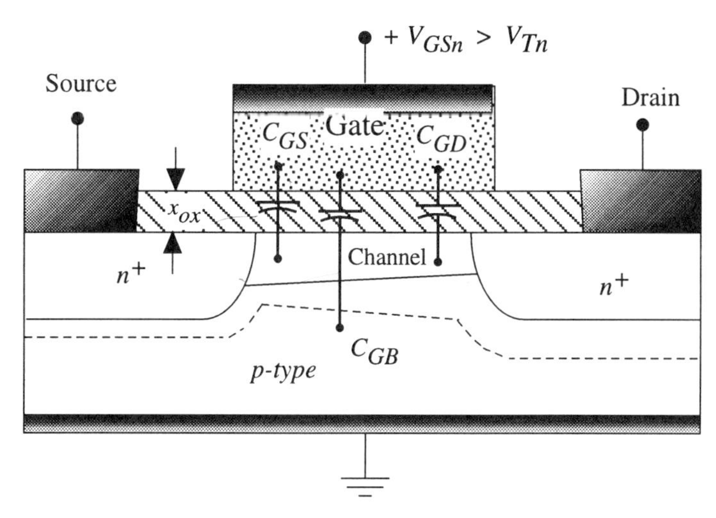

Figure 1.27 Gate-channel capacitances in a MOSFET

defined by calculating the rate of change of the charge with a given voltage. Using the gate charge Q_G, the appropriate derivatives for C_{GS} and C_{GD} are

$$C_{GS} = -\left(\frac{\partial Q_G}{\partial V_S}\right)$$
$$C_{GD} = -\left(\frac{\partial Q_G}{\partial V_D}\right) \tag{1. 76}$$

where V_D and V_S are the drain and source voltages, respectively, as measured with respect to the gate; a similar relation can be written for C_{GB}. These relations demonstrate that both depend upon the assumed nature of the channel, which is found to be much more complicated than in the simple models used here.

Analyzing the behavior of the three capacitors yields results that are similar to those shown in Figure 1.28. This portrays the values as normalized to the total gate capacitance C_G as functions of the operational regions of the MOSFET through V_{GSn}. The major contributions can be estimated for each region as follows:

Cutoff: $C_{GB} = C_G$

Saturation: $C_{GS}=(2/3)C_G$

Non-saturation: $C_{GS}= (1/2)C_G$ and $C_{GD}= (1/2)\ C_G$

subject to variations throughout the regions. Circuit simulation programs are capable of accounting for the nonlinear functional dependences. For the purposes of hand estimates in circuit design, however, we prefer to use simple linear approximations that will still provide some insight into the performance of the circuit. The easiest to use and remember are obtained by just splitting the total gate capacitance in half to write

$$C_{GD} \approx \frac{1}{2}C_G \ , \qquad C_{GS} \approx \frac{1}{2}C_G \ , \tag{1. 77}$$

i.e., the gate capacitance is viewed as being equally split between the drain and source. The gate-bulk capacitance appears as the total gate capacitance C_G where it is used as the total input value.

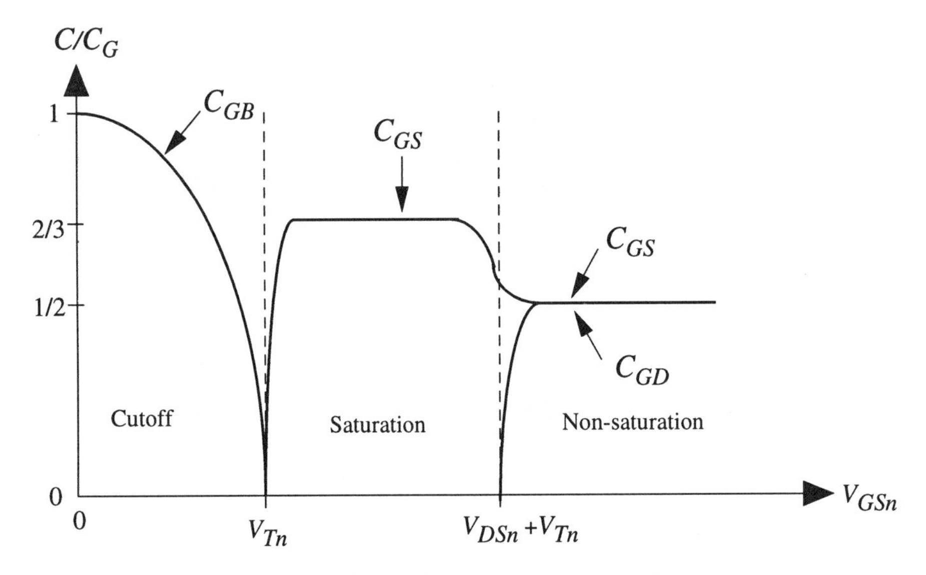

Figure 1.28 Gate-channel capacitances as functions of drain-source voltage

Although these are incorrect from the physical viewpoint, they are sufficient as first estimates in the initial stages of the design process. Note in particular that all of the MOS-base capacitances increase with the channel width *W*.

Depletion Capacitance

Depletion capacitance originates from the depletion region that consists of ionized dopants in the vicinity a pn junction as illustrated in Figure 1.29. Depletion capacitance is also a nonlinear parasitic that is defined by

$$C_j = \left(\frac{\partial Q_d}{\partial V}\right) \tag{1. 78}$$

with Q_d the depletion charge density in units of C/cm^2 and V the applied voltage. Assuming a step

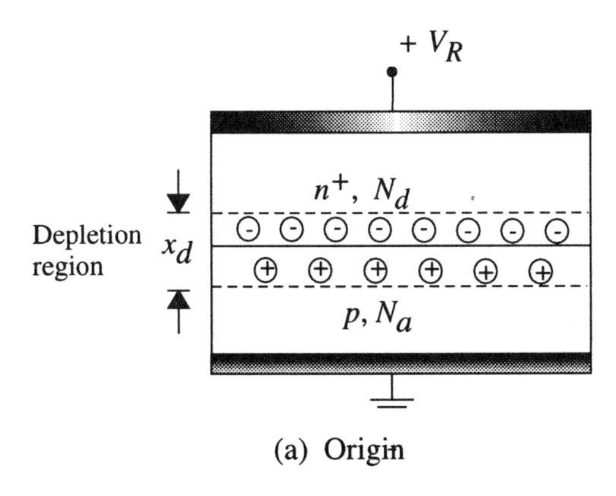

Figure 1.29 Depletion charge in a pn junction

(or, abrupt) doping profile with constant doping densities of N_a and N_d on the p-side and n-side, respectively, the derivative gives the zero-bias capacitance per unit area as

$$C_{j0} = \frac{\varepsilon_{Si}}{x_{d0}} \tag{1.79}$$

where

$$x_{d0} = \sqrt{\frac{2\varepsilon_{Si}\phi_o}{q}\left(\frac{1}{N_a} + \frac{1}{N_d}\right)} \tag{1.80}$$

is the zero-bias depletion width. In this equation,

$$\phi_o = \left(\frac{kT}{q}\right)\ln\left(\frac{N_a N_d}{n_i^2}\right) \tag{1.81}$$

is the **built-in voltage** that is set by the processing. Depletion widths increase with an applied reverse-bias voltage V_R according to the square-root behavior

$$x_d(V_R) = x_{d0}\sqrt{1 + \frac{V_R}{\phi_o}} \tag{1.82}$$

The junction capacitance C_j (per unit area) is then a nonlinear function of V_R with

$$C_j(V_R) = \frac{\varepsilon_{Si}}{x_d(V_R)} = \frac{C_{j0}}{\sqrt{1 + \frac{V_R}{\phi_o}}} \tag{1.83}$$

which shows that the junction capacitance C_j decreases as the reverse-bias voltage V_R increases. The general dependence is illustrated in Figure 1.30(a). The actual depletion capacitance C_d in farads is obtained by multiplying C_j by the area A_j of the junction. This gives

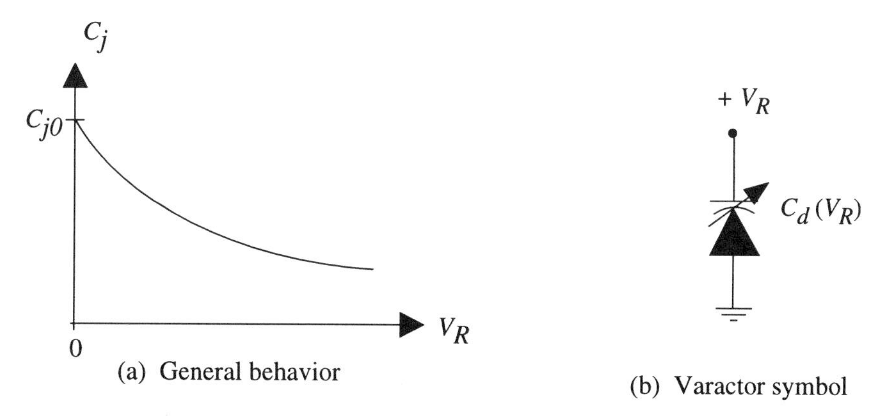

Figure 1.30 Depletion capacitance characteristics

$$C_d = C_j A_j \tag{1.84}$$

which may be calculated for each depletion region. In discrete electronics, the depletion capacitance is used to make a varactor (variable reactance) diode with the symbol shown in Figure 1.30(b). A varactor is used as a capacitor with a value that changes with the reverse voltage.

MOSFET depletion capacitances are found at the drain and source regions. A typical n^+p nFET region is illustrated in Figure 1.31. The region of interest has dimensions corresponding to the channel width W, the junction depth x_j, and the lateral extent labelled as X in the drawing. Due to the 3-dimensional features of the region, we divide up the calculation into the **bottom** and **sidewall** regions as shown. This is a natural grouping since the sidewall acceptor doping $N_{a,sw}$ is usually larger than the bottom doping N_a. The difference is due to field implants which are used for device isolation[8], or increased surface doping levels from a ion implantation step which adjusts the threshold voltage.

Figure 1.32 shows how the 3-dimensional structure is split into the two types of contributions; the total capacitance is then obtained by adding. Consider first the bottom capacitance. The zero-bias capacitance per unit area is given by C_{j0}. Since the area of the bottom is $A=WX$, the zero-biased value of the bottom contribution is

$$C_{bot} = C_{j0} WX \tag{1.85}$$

The sidewall capacitance C_{side} is calculated in a different manner. We will denote the zero-bias sidewall capacitance per unit area by C_{j0sw}. However, since each of the sidewalls regions has a depth x_j, we will define

$$C_{jsw} = C_{j0sw} x_j \tag{1.86}$$

as the **sidewall capacitance per unit perimeter** in units of F/cm. The total sidewall capacitance is given by

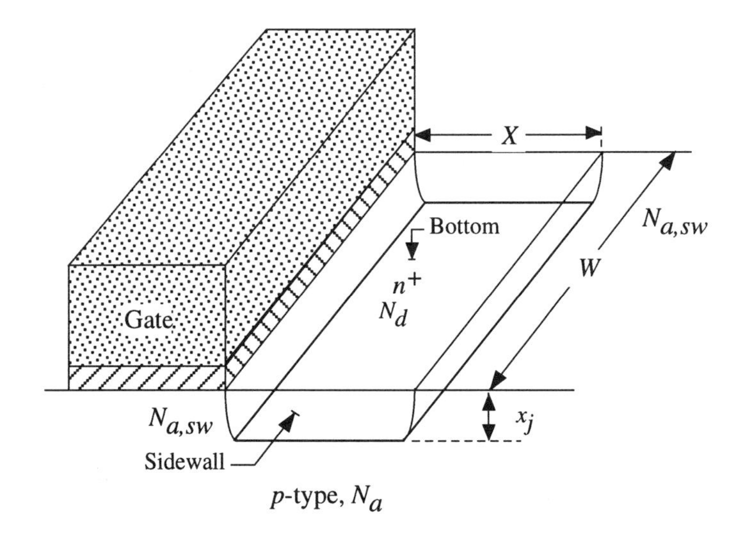

Figure 1.31 FET junction geometry for capacitance calculations

[8] Device isolation is discussed in detail in Section 2.4 of the next chapter.

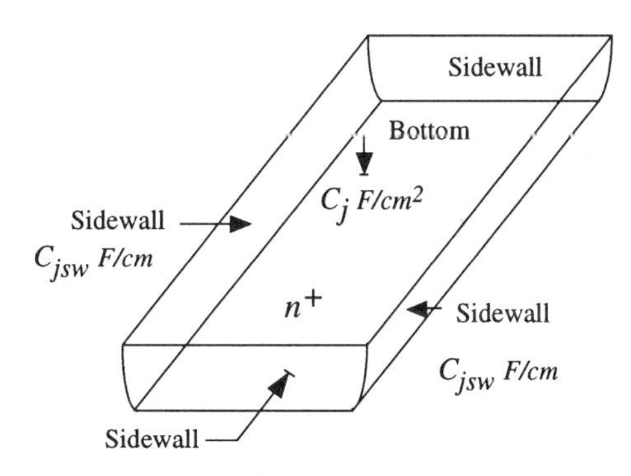

(a) 3-dimensional geometry

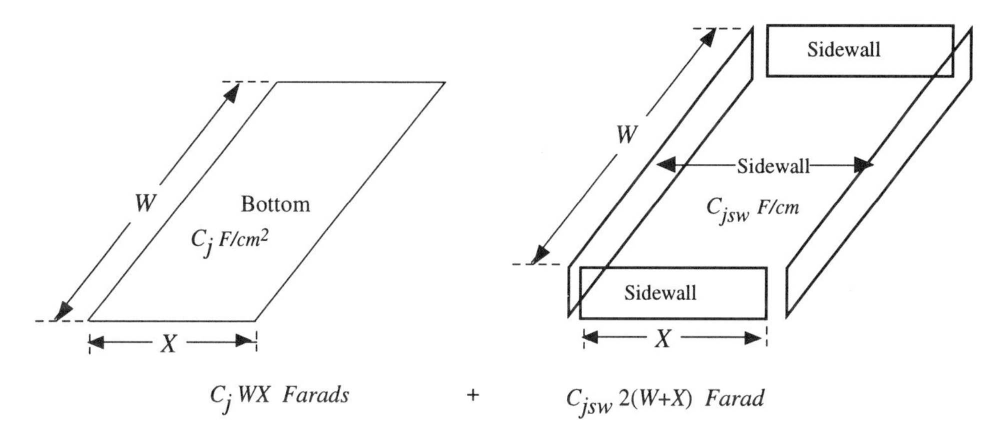

(b) Separation of bottom and sidewall contributions

Figure 1.32 Sidewall geometry

$$C_{side} = C_{jsw}P \tag{1.87}$$

where P is the total **perimeter length** around the n^+ region in units of centimeters. The value of P is obtained from the layout geometry for each transistor; in the present case,

$$P = 2(W + X) \tag{1.88}$$

This approach is also more useful in practice, since the perimeter length P can be determined directly from the layout. The total zero-biased depletion capacitance of the n^+ region is then given by

$$\begin{aligned} C_n &= C_{bot} + C_{side} \\ &= C_{j0}WX + C_{jsw}2(W + X) \end{aligned} \tag{1.89}$$

It is important to note that the depletion capacitance increases with the channel width W.

An example of this approach is shown in the FET layout in Figure 1.33 where the source and drain regions have different dimensions. The source is described by area and perimeter values of

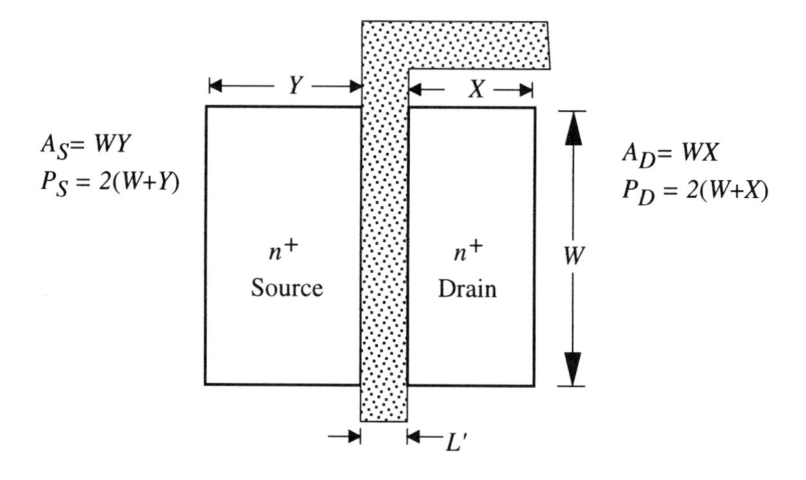

Figure 1.33 Example layout for capacitance calculation

$$A_S = WY$$
$$P_S = 2(W+Y) \tag{1.90}$$

so that the zero-bias source-bulk capacitance is

$$C_{SB0} = C_{j0}WY + 2C_{jsw}(W+Y) \tag{1.91}$$

Similarly, the drain geometry is described by

$$A_D = WX$$
$$P_D = 2(W+X) \tag{1.92}$$

which gives

$$C_{DB0} = C_{j0}WX + 2C_{jsw}(W+X) \tag{1.93}$$

as the zero-bias drain-bulk capacitance. Note that every term contains the channel width W. It is worth mentioning that we have ignored the gate overlap L_o in our calculations (it cannot be seen in he top-view drawing), but it could easily have been included by altering the values of X and Y to $(X+L_o)$ and $(Y+L_o)$.

Zero-bias depletion capacitances are often used as first estimates in analyzing the performance of a CMOS circuit. However, since depletion capacitance is nonlinear and decreases with an applied reverse voltage, the zero-bias contributions overestimate the values when used in a digital circuit analysis since the voltages vary over a wide range.[9] As an example, the drain-bulk capacitance has the form

[9] In an analog circuit, the capacitance may be calculated around a bias voltage, which is not possible in a large-signal digital logic network.

$$C_{DB}(V_D) = \frac{C_{j0}WX}{\sqrt{1+\frac{V_D}{\phi_o}}} + \frac{2C_{jsw}(W+X)}{\sqrt{1+\frac{V_D}{\phi_{osw}}}} \tag{1.94}$$

where V_D is the drain voltage (with respect to the bulk), and ϕ_{osw} is the built-in voltage for the sidewall doping level. Including even a simple square root dependence of this type in circuit equations makes the analysis intractable, so it is generally avoided.

To include depletion capacitance contributions in our hand analyses, we will often model a nonlinear capacitance using a simpler linear time-invariant (LTI) element. This can be accomplished by using an average capacitance C_{av} that is computed as a weighted value over a range of voltages as defined in

$$C_{av} = \frac{A}{(V_2 - V_1)} \int_{V_1}^{V_2} C_j(V_R)\, dV_R \tag{1.95}$$

where V_2 and V_1 are the high and low values, and A is the area. This yields expressions of the form

$$C_{av} = K(V_1, V_2)\, C_{j0} A \tag{1.96}$$

where C_{av} is by definition a constant. In general, the bottom junction and the sidewall junction have different doping profiles, and must be analyzed separately.

Linearly-graded doping profile models are sometimes more accurate for describing a real process than is possible using a simple step-profile. In this case, the voltage-dependent capacitance assumes the form

$$C_j(V_R) = \frac{C_{j0g}}{\left(1 + \frac{V_R}{\phi_{o,g}}\right)^{1/3}} \tag{1.97}$$

where C_{j0g} is the zero-bias capacitance per unit area. A general model for voltage-dependent depletion capacitance is based on the expression

$$C_j(V_R) = \frac{C_{j0m}}{\left(1 + \frac{V_R}{\phi_{o,m}}\right)^{m}} \tag{1.98}$$

where m is called the **grading coefficient** such that $m<1$. Use of these types of graded junctions are usually restricted to computer simulations due to the increased complexity of the equations. SPICE provides for various doping profiles via the grading parameter m_j and m_{jsw}. Step profile junctions have $m=0.5$ corresponding to the square root dependence, while a linearly graded junction is described by a cubed root dependence with $m=0.33$. Other values of m usually correspond to empirical values obtained by curve fitting, and are very common in circuit simulation models.

In a simple hand analysis, we often approximate the bottom junction as being step-like while the sidewalls are taken to be linearly graded. The voltage dependence then assumes the form

$$C_{dep}(V_R) = \frac{C_{j0}A}{\left(1 + \frac{V_R}{\phi_o}\right)^{1/2}} + \frac{C_{jsw}P}{\left(1 + \frac{V_R}{\phi_{osw}}\right)^{1/3}} \tag{1.99}$$

In this case, the weighted average value must be calculated using two separate integrals via

$$C_{av} = \frac{1}{(V_2 - V_1)} \left\{ \int_{V_1}^{V_2} \frac{C_{j0}A}{\left(1 + \frac{V_R}{\phi_o}\right)^{1/2}} dV_R + \int_{V_1}^{V_2} \frac{C_{jsw}P}{\left(1 + \frac{V_R}{\phi_{osw}}\right)^{1/3}} dV_R \right\} \tag{1. 100}$$

$$= K_{1/2}(V_1, V_2)\, C_{j0}A + K_{1/3}(V_1, V_2)\, C_{jsw}P$$

where $K_m(V_1,V_2)$ is determined by the grading parameter m for each term. Explicitly,

$$K_m(V_1, V_2) = \frac{1}{(V_2 - V_1)} \int_{V_1}^{V_2} \frac{1}{\left(1 + \frac{V_R}{\phi_o}\right)^m} dV_R \tag{1. 101}$$

which integrates to

$$K_m(V_1, V_2) = \frac{\phi_o}{(-m+1)(V_2 - V_1)} \left[\left(1 + \frac{V_2}{\phi_o}\right)^{(-m+1)} - \left(1 + \frac{V_1}{\phi_o}\right)^{(-m+1)} \right]. \tag{1. 102}$$

For the special case of abrupt (m=1/2) and linearly graded (m=1/3) junctions, we have

$$\begin{aligned} K_{1/2}(V_1, V_2) &= \frac{2\phi_o}{(V_2 - V_1)} \left[\left(1 + \frac{V_2}{\phi_o}\right)^{1/2} - \left(1 + \frac{V_1}{\phi_o}\right)^{1/2} \right] \\ K_{1/3}(V_1, V_2) &= \frac{3\phi_o}{2(V_2 - V_1)} \left[\left(1 + \frac{V_2}{\phi_o}\right)^{2/3} - \left(1 + \frac{V_1}{\phi_o}\right)^{2/3} \right] \end{aligned} \tag{1. 103}$$

which are both less than unity (i.e., $K<1$) as required to make the average capacitances less then their maximum values at zero-bias. It is important to remember that the value of the built-in voltage will generally be different for the bottom and the sidewall terms. The application of these expressions of circuit design will be examined in more detail in Chapter 3 for the case of an inverter gate.

Device Capacitance Model

The results above may be used to assign values to the capacitances that are used in basic calculations as shown in Figure 1.34. This model adds the source contributions in (a) together to write "the" source capacitance as

$$C_S = C_{GS} + C_{SB} \tag{1. 104}$$

with C_S a constant. Numerically, we would use the LTI average of the depletion capacitance. Similarly, the drain capacitance C_D is estimated as

$$C_D = C_{GD} + C_{DB} \tag{1. 105}$$

using the values calculated in the discussion above.

This simple model has the virtue of being easy to apply in practice. In digital CMOS circuits, we will estimate the total capacitance at every node by simply adding all contributions that are required to change voltage during a switching event. This automatically introduces errors into the calculations. However, the approach is easy to use and generally provides numerical results that are reasonable estimates of the actual circuit performance. Since we expect that every circuit will eventually be simulated on a computer anyway, this type of analytic modelling is adequate for a first

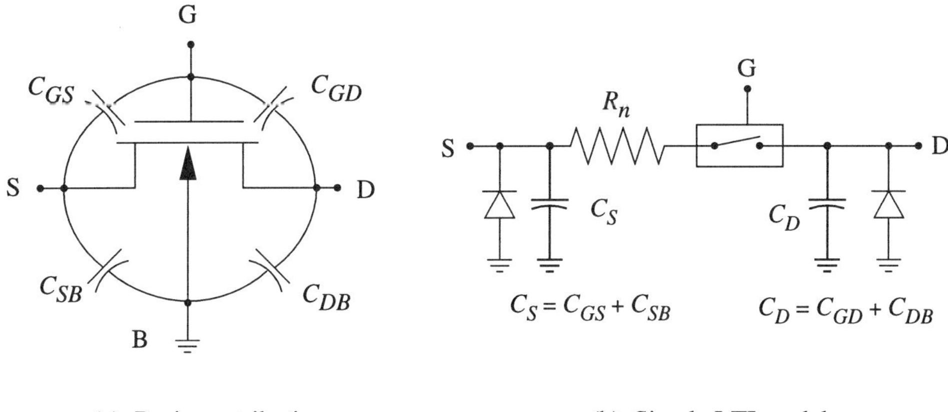

(a) Basic contributions (b) Simple LTI model

Figure 1.34 Simplified linear FET model for timing estimates

estimate on the performance during the design phases.

1.4.3 Junction Leakage Currents

In addition to the parasitic depletion capacitances, the pn junctions that are formed by the drain-bulk and source-bulk interfaces in a MOSFET introduce a component of leakage current that is often important in high-performance circuit design.

Consider the pn junction shown in Figure 1.35(a) which shows a pn junction with a reverse bias V_R applied. For a general forward bias voltage $V = -V_R$, the current is given by

$$I = I_o(e^{V/(kT/q)} - 1) + I_{dep} \tag{1. 106}$$

where I_o is the **saturation current**, and I_{dep} is the current due to recombination or generation events in the depletion region. In a MOSFET, the drain-bulk and source-bulk regions are always biased with a reverse voltage V_R=-V. The reverse leakage current I_R=-I through the junction is then

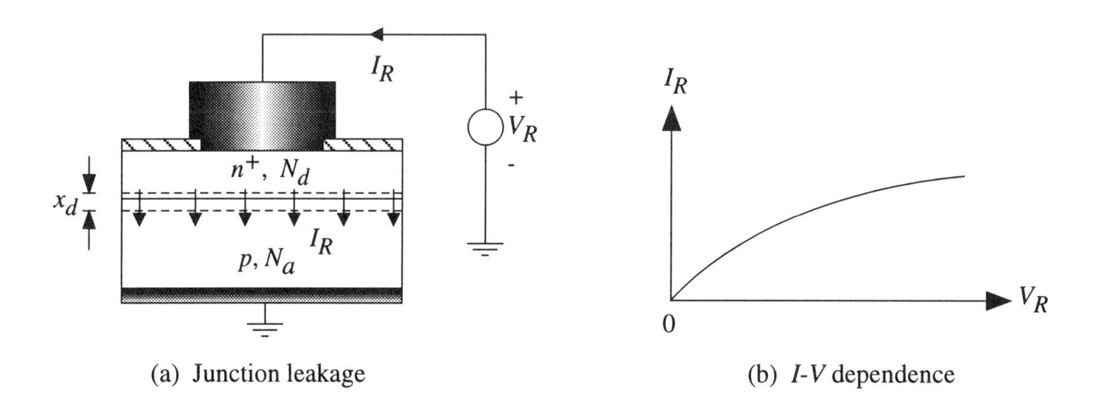

(a) Junction leakage (b) *I-V* dependence

Figure 1.35 Reverse-bias pn junction leakage current

$$I_R = I_o + I_{gen} \approx I_{gen} \tag{1.107}$$

where I_{gen} is the **generation current** that dominates the reverse current flow in a silicon diode at normal operating temperatures. Analyzing the generation current in a step-profile junction gives

$$I_{gen} \approx I_{go}\left[\sqrt{1+\frac{V_R}{\phi_o}} - 1\right] \tag{1.108}$$

where

$$I_{go} = \frac{qAn_i x_{d0}}{2\tau_o} \tag{1.109}$$

with A the area of the junction, and τ_o the effective carrier lifetime. The reverse current I_R is plotted as a function of the reverse voltage V_R in Figure 1.35(b). Note that I_R increases with the reverse bias voltage; this is opposite to the behavior of the junction capacitance, which decreases with increasing V_R.

The origin of the leakage current in a MOSFET is illustrated in Figure 1.36. Since the drain and source n^+ regions are always at a voltage greater than or equal to 0*v*, the grounding of the p-type bulk region implies that they will always exhibit leakage flows regardless of the state of conduction of the transistor itself.

A MOSFET switching model that includes both the capacitances and the junction leakage currents is shown in Figure 1.37. This models the leakage current as a controlled current source with a value

$$I_R = I_{gen} \approx I_{go}\left[\sqrt{1+\frac{V_R}{\phi_o}} - 1\right] \tag{1.110}$$

at each junction. Although this is highly simplified, it is generally sufficient for initial design estimates. It has the important characteristics that the leakage current is proportional to the junction area A, and increases with the reverse bias V_R.

For a more general doping profile, we may replace the voltage dependence with

$$I_{gen} \approx I_{gom}\left[\left(1+\frac{V_R}{\phi_o}\right)^m - 1\right] \tag{1.111}$$

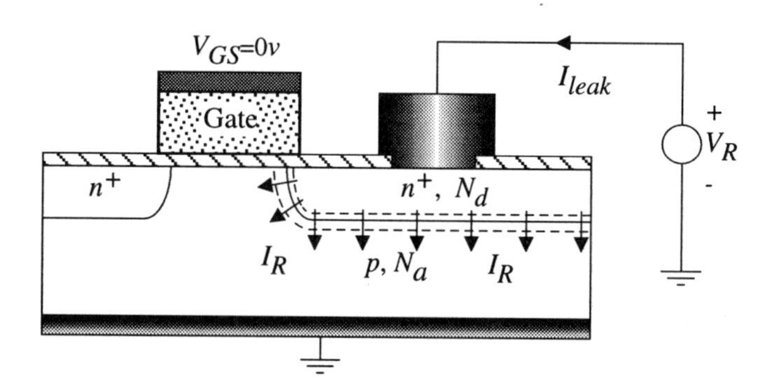

Figure 1.36 MOSFET reverse-bias junction leakage

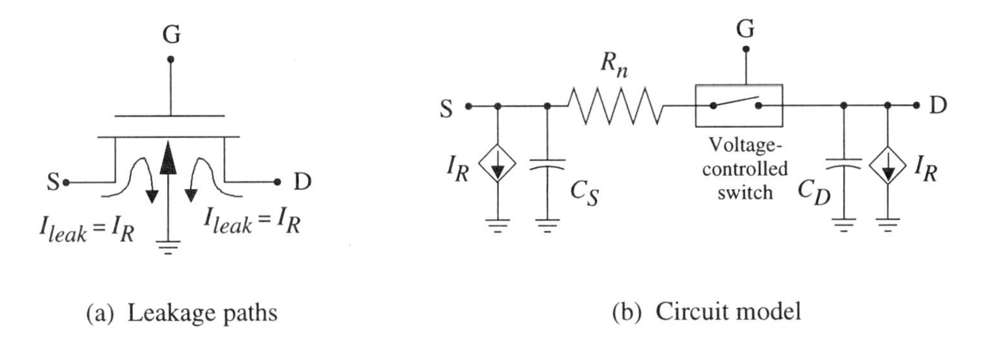

(a) Leakage paths (b) Circuit model

Figure 1.37 nFET junction leakage modelling

where m is the grading coefficient, but this adds little to the analysis since the equations are only estimates anyway. As we shall see later, the leakage current is a limiting factor in many advanced logic circuits such as the dynamic random-access memory (DRAM).

1.4.4 Applications to Circuit Design

The MOSFET models developed in this section are very useful for analyzing CMOS digital circuits. They provide the link between fundamental semiconductor parameters and the current-voltage relationships, and also contain the critical geometrical information such as the device aspect ratio (W/L). Combined with the I-V equations developed earlier in the chapter, we have a sound basis for analyzing the important characteristics of most digital CMOS circuits. More importantly, it is possible to use the analysis to develop a design philosophy that links the circuit performance to the device characteristics.

The following sections in this chapter deal with more advanced FET modelling concepts, and are not required in a first reading of the book. As such, the reader may wish to turn to Chapter 2 which deals with some of the details of silicon chip fabrication and CMOS layout, or jump directly to Chapter 3 where the CMOS circuit discussion begins.

1.5 Geometric Scaling Theory

The device I-V equations discussed thus far are only valid for "big" devices as they ignore many effects that arise when the channel length is reduced to values below about 20 μm. Although we will continue to stress the limited usefulness of analytic models in circuit design, it is important that the chip designer understand the physics of small devices and how they affect circuit operation.

Scaling theory deals with the question of how the device characteristics are changed as the dimensions of the transistor are reduced in an idealized well-defined manner. As a starting point, consider a MOSFET with a channel width W and a channel length L such that the channel area is $A=WL$ as shown in Figure 1.38(a). We introduce the concept of a **scaling factor** $S>1$ such that a new scaled device is created with reduced dimensions W' and L' where

$$W' = \frac{W}{S}, \; L' = \frac{L}{S} \tag{1. 112}$$

corresponding to, say, an improvement in the lithographic resolution. Since the current is proportional to the aspect ratio, and

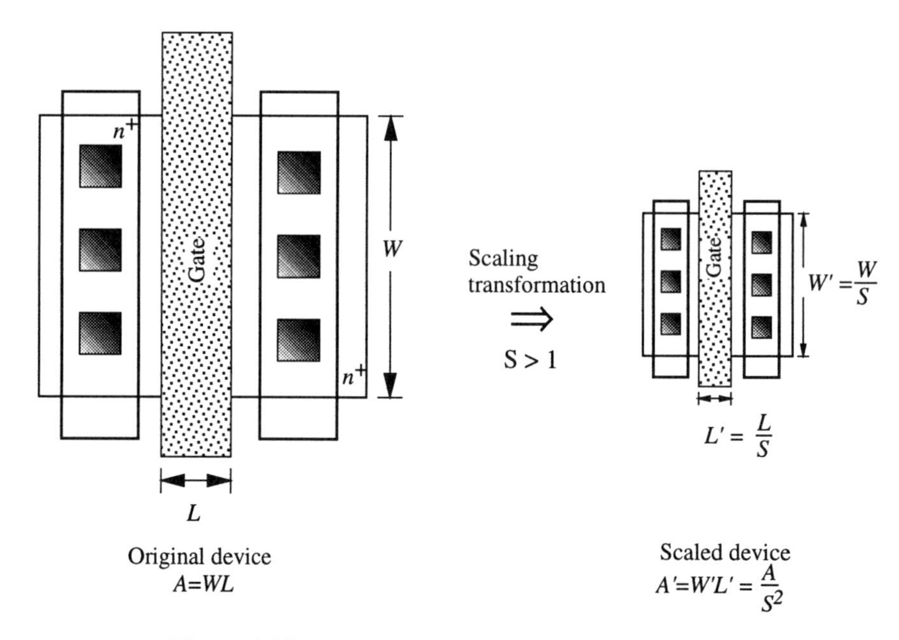

Figure 1.38 MOSFET scaling

$$\left(\frac{W}{L}\right) = \left(\frac{W'}{L'}\right) \tag{1. 113}$$

shows that the aspect ratio is invariant, this type of scaling results in a reduced channel area $A' = W'L'$ of

$$A' = \frac{A}{S^2} \ . \tag{1. 114}$$

The scaled device is portrayed in Figure 1.38(b). The primary objective of reducing the real estate area needed for a transistor is thus accomplished.

Even though the scaled device has a smaller area, it must be remembered that the MOSFET is an electrical field-effect structure. By reducing the dimensions of the channel region, the internal device physics is altered, and we expect that the current flow characteristics will change. Moreover, improvements in other aspects of the processing will further effect the operation of the device, so that is important to delve deeper into the problem.

To be consistent with the surface scaling, we should also examine the effects on the vertical features. The problem is portrayed in Figure 1.39. In (a), the most important vertical dimensions of the original device are the oxide thickness x_{ox} and the junction depth x_j. In Figure 1.39(b), we have reduced both by the same scale factor S used on the surface dimensions. The most obvious effect is that the electrical characteristics of the device will be altered. To see this, consider first the process transconductance $k = \mu_n C_{ox}$ (where we will omit the prime to avoid confusion with the notation used for the scaled device). A large value of k is desirable, and this is achieved by increasing the oxide capacitance C_{ox}. This, in turn, requires that we reduce the oxide thickness x_{ox}. If we subject this to the same scaling factor and write

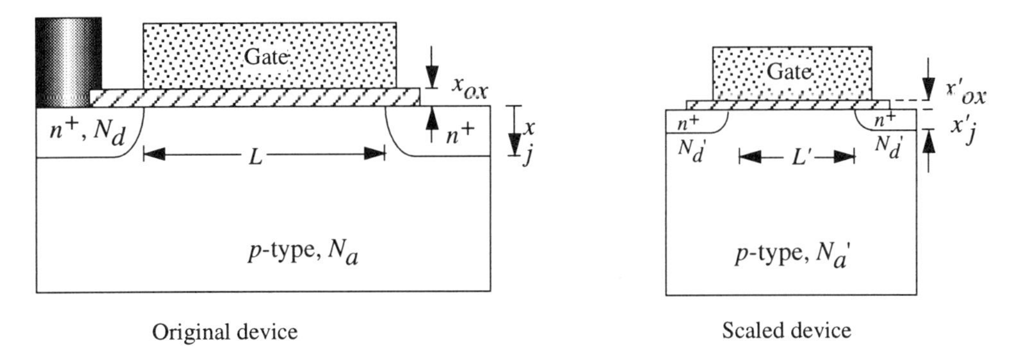

Figure 1.39 Vertical scaling in a MOSFET

$$x'_{ox} = \frac{x_{ox}}{S}. \tag{1. 115}$$

as shown, then the oxide capacitance in the scaled device becomes

$$C'_{ox} = \frac{\varepsilon_{ox}}{x'_{ox}} = SC_{ox} \tag{1. 116}$$

This gives the new scaled process transconductance as being increased to

$$k' = Sk. \tag{1. 117}$$

Similarly, the device transconductance scales according to

$$\beta' = S\beta \tag{1. 118}$$

since the aspect ratio is invariant. This approach defines **isotropic 3-dimensional scaling** where the same scaling factor S has been used for both vertical and lateral dimensions. We also note at this point that the field-effect is intrinsically related to the doping densities N_a and N_d in the transistor. This observation implies that the scaled device may require modifications in these values to new levels N'_a and N'_d in order to maintain the operational characteristics.

As seen by the simple analysis above, geometrical scaling is straightforward. However, the physics of the field effect is more complicated since it is based in the charge-field relationship contained in the Poisson equation

$$\nabla^2\phi = -\frac{\rho}{\varepsilon} \tag{1. 119}$$

where ϕ is the electrostatic potential, ρ is the volume charge density in units of C/cm^3, and ε is the permittivity. The connection between geometrical scaling and the field-effect is due to the fact that the Laplacian expands to

$$\nabla^2\phi = \frac{\partial^2\phi}{\partial x^2} + \frac{\partial^2\phi}{\partial y^2} + \frac{\partial^2\phi}{\partial z^2} \tag{1. 120}$$

with $\phi=\phi(x,y,z)$, where (x,y,z) are the spatial coordinates. If we scale the dimensions by means of

$$(x', y', z') = \left(\frac{x}{S}, \frac{y}{S}, \frac{z}{S}\right), \tag{1. 121}$$

the Laplacian operator in the scaled coordinate system is given by

$$\nabla'^2 = S^2\nabla^2, \tag{1. 122}$$

which modifies the Poisson equation accordingly.

Let us consider a practical application of scaling theory before proceeding further. Suppose that we have a functional circuit with a complete mask set that works well in a given process. Our task is to adapt the circuit to a new processing line that has smaller linewidths and a thinner gate oxide. The simplest approach would be to subject all of the masks to the same shrink, which obviously gives a smaller circuit. However, this does not imply that the reduced circuit will have the same *electrical* performance. The only way that we might expect similar electrical characteristics is if the *form* of the *I-V* equation is invariant under the transformation.

In terms of the Poisson equation, this means that the scaled potential $\phi'=\phi'(x',y',z')$ in the new coordinate system should satisfy an equation of the form

$$\nabla'^2\phi' = -\frac{\rho'}{\varepsilon}, \tag{1. 123}$$

i.e., with the same *form* as the original equation. Invoking the geometrical scaling factor S automatically affects the charge field relation through the Laplacian operator. To preserve the form of the equation, we must choose how the electrostatic potential behaves in the scaling process.

1.5.1 Full-Voltage Scaling

Let us assume that the potential is also scaled by the same factor S such that

$$\phi' = \frac{\phi}{S}. \tag{1. 124}$$

This requires that the charge density ρ be increased according to

$$\rho' = S\rho \tag{1. 125}$$

if we are to arrive at the scaled equation (1.123).

Now let us apply this analysis to the square law equations of a MOSFET. Consider the original device shown in Figure 1.40(a). Our program is to examine how the current and voltages will transform when the device is scaled by the factor S. Let us start with the non-saturated current flow equation

$$I_D = \frac{\beta}{2}\left[2\left(V_{GS} - V_T\right)V_{DS} - V_{DS}\right] \tag{1. 126}$$

for the original (unscaled) device. After we apply the scaling factor, β is increased to $S\beta$, while the applied voltages are assumed to be reduced according to

$$V'_{DS} = \frac{V_{DS}}{S}, \qquad V'_{GS} = \frac{V_{GS}}{S}. \tag{1. 127}$$

The threshold voltage V_T is set by the processing, and is cannot be changed by the circuit designer. However, we will assume that the threshold adjustment ion implant is capable of providing a scaled threshold voltage of

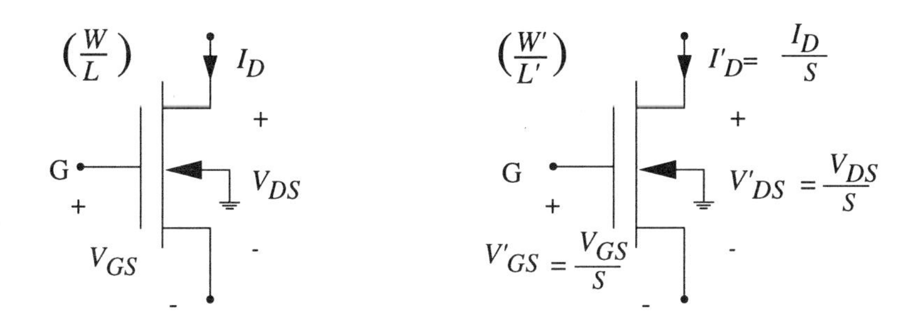

(a) Original device (b) Scaled device

Figure 1.40 Full voltage scaling in a MOSFET

$$V'_T = \frac{V_T}{S}. \tag{1. 128}$$

The scaled device is then described by the current versus voltage equation

$$\begin{aligned} I'_D &= \frac{\beta'}{2}\left[2(V'_{GS} - V'_T)V'_{DS} - V'^2_{DS}\right] \\ &= \frac{S\beta}{2}\left[2\left(\frac{V_{GS}}{S} - \frac{V_T}{S}\right)\frac{V_{DS}}{S} - \frac{V^2_{DS}}{S^2}\right] \\ &= \frac{I_D}{S} \end{aligned} \tag{1. 129}$$

so that the scaled current is simply the original current divided by the scale factor S. The saturated current

$$I_D' = \frac{\beta'}{2}(V_{GS}' - V_T')^2 = \frac{I_D}{S} \tag{1. 130}$$

also scales in the same manner. This illustrates that the scaled circuit should still behave in the same manner as the original circuit, although the actual values of the critical circuit benchmarks will be different. These values are summarized in the drawing of Figure 1.40(b).

An interesting aspect of full-voltage scaling is seen by comparing the power dissipation of a transistor. In the original device, the power is

$$P = I_D V_{DS}, \tag{1. 131}$$

while the scaled device has

$$P' = I'_D V'_{DS} = \frac{I_D}{S}\frac{V_{DS}}{S} = \frac{P}{S^2} \qquad (1.\ 132)$$

Thus, both the chip area and the power dissipation are reduced by the same factor, so that the power dissipation per unit area remains constant.

Example 1.3

Let us look at the effects of a simple scaling with S=2. This would imply that the channel dimensions are reduced from W and L to new values

$$L' = \frac{L}{2}, \qquad W' = \frac{W}{2} \qquad (1.\ 133)$$

The original area A=WL is then reduced to

$$A' = \frac{A}{4} \qquad (1.\ 134)$$

as portrayed schematically in the drawing below (Figure 1.41). The reduction in surface area is

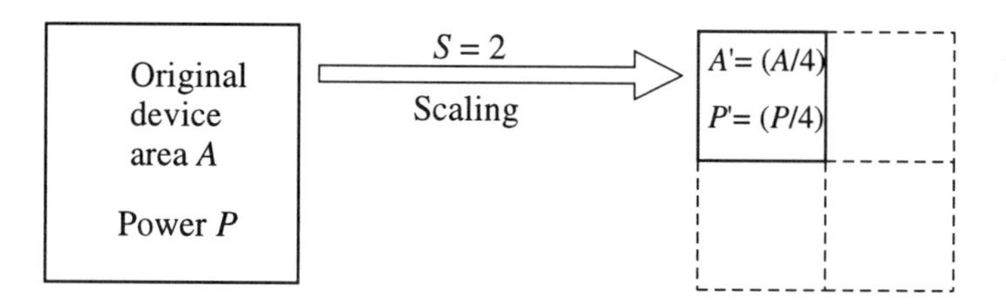

Figure 1.41 Example of basic scaling theory

obvious. However, note that full voltage scaling also reduces the power of the scaled device to one-fourth that of the original layout.

One more important point must be mentioned in the context of full-voltage scaling. The requirement that the charge density be increased means that the doping densities are scaled according to

$$N'_a = SN_a \qquad N'_d = SN_d \qquad (1.\ 135)$$

to preserve the form. Unfortunately, increasing the doping densities at a pn junction increases the junction capacitance, and reduces the reverse breakdown voltage. Both are undesirable in circuit design.

1.5.2 Constant-Voltage Scaling

In general, the circuit designer does not have the option of arbitrarily changing the power supply in a circuit, making full-scaling difficult to accomplish. Constant-voltage scaling deals with the effects when we maintain the same voltage levels, but scale the dimensions. For this case, we choose

$$\phi' = \phi \tag{1. 136}$$

which then requires that the charge densities scale according to

$$\rho' = S^2 \rho \tag{1. 137}$$

in order to preserve the form of the Poisson equation.

The problem is illustrated in Figure 1.42 where we want to take the device in (a), scale it by a factor S, and then determine the characteristics of the new transistor when the voltages are not changed. By inspection we may write the scaled value of the non-saturation current as

$$\begin{aligned} I'_D &= \frac{\beta'}{2}\left[2(V'_{GS} - V'_T)V'_{DS} - V'^2_{DS}\right] \\ &= S\frac{\beta}{2}\left[2(V_{GS} - V_T)V_{DS} - V^2_{DS}\right] \\ &= SI_D \end{aligned} \tag{1. 138}$$

which shows that the current is increased by S. Similarly, the saturated current scales according to

$$\begin{aligned} I'_D &= \frac{\beta'}{2}(V'_{GS} - V'_T)^2 \\ &= SI_D \end{aligned} \tag{1. 139}$$

so that it has the same increase. These results are summarized in Figure 1.42(b). The power dissipation scales according to

$$\begin{aligned} P' &= I'_D V'_{DS} \\ &= SI_D V_{DS} \\ &= SP \end{aligned} \tag{1. 140}$$

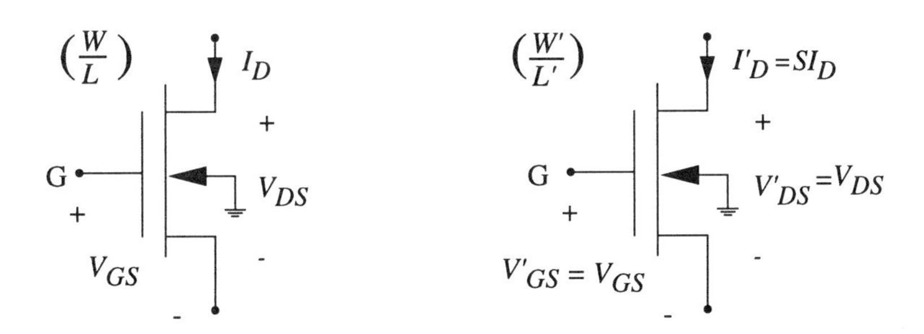

(a) Original device (b) Scaled device

Figure 1.42 Constant-voltage scaling in a MOSFET

showing that the power dissipation increases by a factor of S. Another problem is the fact that the doping densities must be increased according to

$$N'_a = S^2 N_a \qquad N'_d = S^2 N_d \tag{1. 141}$$

in order to preserve the form of the Poisson equation. This lowers the reverse breakdown voltage of the junction and can lead to unwanted effects at the chip level.

1.5.3 Second-Order Scaling Effects

The first-order scaling discussed above deals with direct transformations of the MOSFET dimensions, doping levels, voltages, and currents. Second order effects arise when the interdependence of device characteristics on these parameters is examined.

The mobility provides an example of a second order effect. In general, μ is a function of background doping N such that

$$\left(\frac{\partial \mu}{\partial N}\right) < 0 \tag{1. 142}$$

i.e., the mobility decreases with increased doping. Since scaling theory uses an upward scaling to $N'=SN$ or $N'=S^2N$, increased impurity scattering gives that

$$\mu' < \mu \tag{1. 143}$$

When this is included in the scaling equations, the transconductance no longer exhibits linear scaling.

The effects of ion implants also change when the vertical dimensions are scaled. An example of this is the term (qD_I/C_{ox}) in V_T. This assumes that the implanted layer is thin enough to be modeled as a sheet. However, if the drain and source junction depths are scaled according to

$$x'_j = \frac{x_j}{S} \tag{1. 144}$$

then the ion implant penetration depth[10] may be comparable with x_j'. A similar comment applies to the charge terms

$$-\frac{1}{C_{ox}}(Q_f + Q_{ox}) \tag{1. 145}$$

in the flatband voltage as x_{ox} is scaled. These and other considerations often dictate the need for a more accurate analysis.

Another point that should be raised is that, to be consistent, we should also scale the thickness of all other layers in the device. This includes the gate material (e.g., poly) and the metal interconnects in particular. Although these changes do not appear to have any direct affect on the device characteristics, they will affect the performance of a circuit since they will alter the values of various parasitics.

1.5.4 Applications of Scaling Theory

Reducing the device dimensions allows higher density logic integration. This philosophy has been the primary motivating factor for developing sub-micron size structures. One price paid for this

[10] This is the projected range R_p discussed in the next chapter.

improved technology is an increase in the complexity of the device physics. Effects which are negligible in "large" MOSFETs become extremely important when the transistor dimensions are reduced. Scaling theory provides a general guide to making MOSFETs smaller. Even though it is not possible (or desirable) to follow every aspect of the theory, it remains a useful metric for measuring progress in device physics. It is also of some use in predicting how circuits will behave when they are scaled to smaller dimensions, and scaling theory is commonly used for this purpose.

1.6 Small-Device Effects

Scaling theory is idealized, and ignores many of the small-device effects that govern the performance of MOSFETs found in modern integrated circuits. It is often desirable to adhere to the "big-device" models for simplicity, but modify the parameters to account for the more important changes in the transistor parameters. Our study will divide the effects into two categories. The first group deals with changes in the device parameters that are found when the approximations used to derive the parameters are no longer valid in small structures; these are generally concerned with the value of the threshold voltage. The second category concentrates on changes in the electron mobility and how they affect the current flow through the channel.

1.6.1 Threshold Voltage Modifications

The basic threshold voltage expression

$$V_{Tn} = V_{FB} + 2|\phi_F| + \frac{1}{C_{ox}}\sqrt{2q\varepsilon_{Si}N_a(2|\phi_F|)} \tag{1.146}$$

contains an implicit approximation that breaks down when applied to a MOSFET with small dimensions. This arises because the third term representing the bulk charge is based upon the relation

$$|Q_B| = C_{ox}V_{ox} \tag{1.147}$$

which uses charge per unit area and capacitance per unit area, i.e., Q_B has units of C/cm^2 and C_{ox} has units of F/cm^2. In order to obtain a true charge-voltage relation, we must multiply this by an area, which is implicitly assumed to be the area of the channel, i.e., WL. In other words, the full expression should read

$$V_{Tn} = V_{FB} + 2|\phi_F| + \frac{1}{C_{ox}}\sqrt{2q\varepsilon_{Si}N_a(2|\phi_F|)}\left(\frac{WL}{WL}\right) \tag{1.148}$$

which shows that the bulk charge term in the threshold voltage is actually supporting the entire depletion charge underneath the gate.

While this is a good approximation for big transistors, it ignores two effects that may arise in a small-geometry transistor. First, the drain-bulk and source-bulk pn junctions induce the formation of a depletion charge that is ignored by this equation; the junction-induced charge becomes more noticeable in short channel length devices. Second, the overhang of the gate over the field regions implies that the gate voltage will induce depletion charge outside of the gate area. This additional charge can be important in small channel width devices. Both can be analyzed by performing a more detailed examination of the electric fields and charge distributions.

Short-Channel Effect

Short-channel effects are important when the channel length L is small, typically when L is reduced below about 2μm or so. The main result is that the threshold voltage is reduced below the value

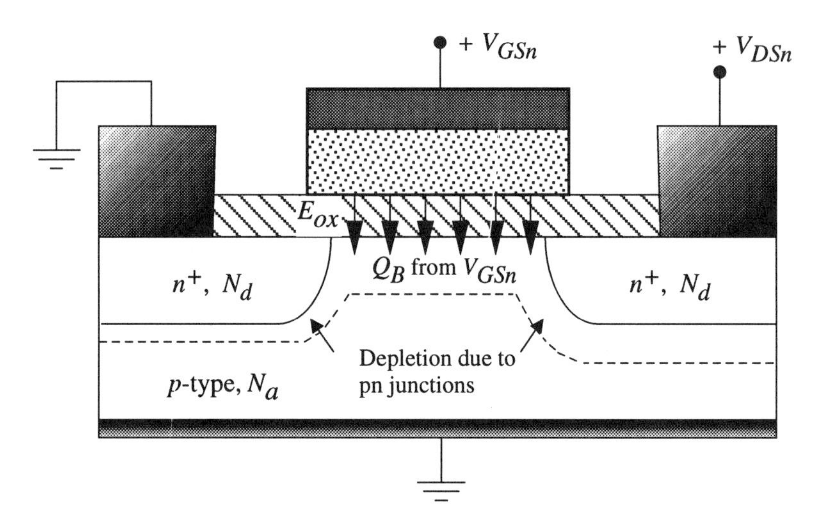

Figure 1.43 Bulk charge in a MOSFET

found using the long-channel analysis. A cross-section of the channel region is shown in Figure 1.43. The origin of the short channel reduction of the threshold voltage is due to the fact that the gate voltage (and hence, the threshold voltage) does not support all of the bulk charge with an area of WL underneath the gate. Rather, some of the depletion charge at the drain and source sides of the channel is automatically induced by the ionized dopants that form the pn junctions. This says that the bulk charge term in V_T overestimates the size of this contribution.

A simple model for calculating the threshold voltage reduction is shown in Figure 1.44. This approximates the geometry of the gate-supported bulk charge to have a cross-sectional profile that is trapezoidal in shape. The upper edge of the trapezoid has a length of L, while the lower edge length is denoted by L_1. To calculate the reduction in the threshold voltage, we must find the reduc-

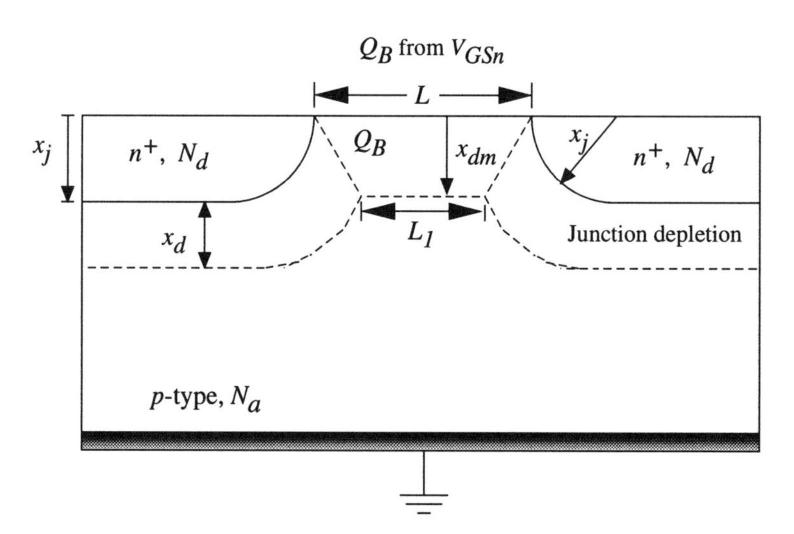

Figure 1.44 Bulk-charge distribution by origin

tion in bulk charge from the simpler rectangular geometry.

Figure 1.45(a) provides an expanded view of the problem. The value of L_1 is now given by

$$L_1 = L - 2(\Delta L) . \qquad (1.149)$$

where (ΔL) is the same on both sides. To proceed in the calculation, we will assume that the depletion thickness x_d of the pn junction is about the same as the maximum MOS depletion depth

$$x_{dm} = \sqrt{\frac{2\varepsilon_{Si}(2|\phi_F|)}{qN_a}} \qquad (1.150)$$

as shown in the drawing. This allows us to construct the simple right triangle shown in Figure 1.45(b). The value of (ΔL) can be calculated by using the Pythagorean theorem to write

$$(x_j + x_{dm})^2 = x_{dm}^2 + (x_j + \Delta L)^2 \qquad (1.151)$$

where x_j is the junction depth. Solving the quadratic equation gives

$$\Delta L = -x_j + \sqrt{x^2{}_j + 2x_j x_{dm}} \qquad (1.152)$$

for one-half of the difference between the top and bottom of the trapezoid.

To incorporate this result into the threshold voltage expression, we first write

$$\begin{aligned} \frac{WL}{WL} \to \frac{L-(\Delta L)}{L} &= 1 - \frac{(\Delta L)}{L} \\ &= 1 - \frac{x_j}{L}\left[\sqrt{1 + \frac{2x_{dm}}{x_j}} - 1\right] \\ &= f \end{aligned} \qquad (1.153)$$

where we see that $f<1$. Then, the threshold voltage that accounts for short channel effects is given by

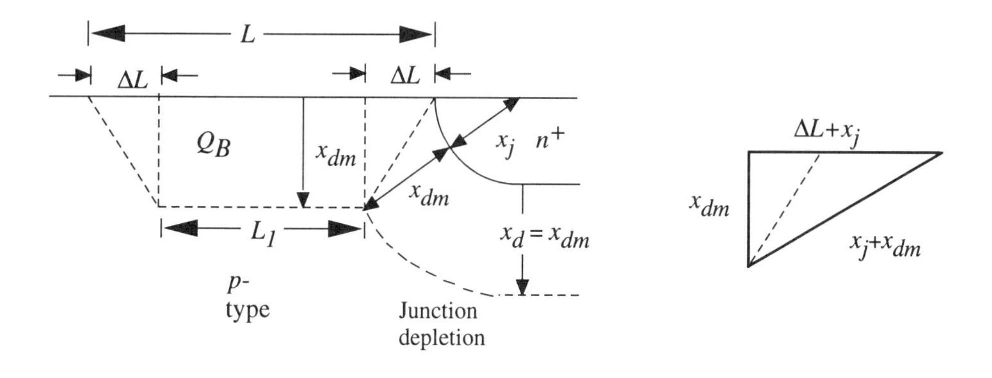

(a) Details of the geometry (b) Triangle sides

Figure 1.45 Bulk-charge geometry for short-channel effect calculation

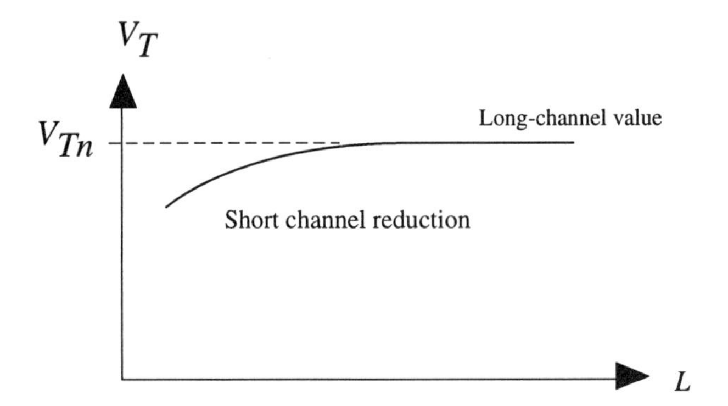

Figure 1.46 Threshold voltage dependence due to short-channel effect

$$(V_{Tn})_{SCE} = V_{FB} + 2|\phi_F| + \frac{1}{C_{ox}}\sqrt{2q\varepsilon_{Si}N_a(2|\phi_F|)}\, f\,. \tag{1. 154}$$

The change in the threshold voltage is

$$(\Delta V_{Tn})_{SCE} = -\frac{x_j|Q_B|}{L\,C_{ox}}\left[\sqrt{1+\frac{2x_{dm}}{x_j}}-1\right] \tag{1. 155}$$

such that

$$(V_{Tn})_{SCE} = V_{Tn} + (\Delta V_{Tn})_{SCE} \tag{1. 156}$$

where V_{Tn} is the long-channel value. This shows the reduction from the long-channel value. Although the geometry is only a first-order approximation, it is sometimes used as the basis for defining a short channel effect as occurring when $L \sim x_j$. The general dependence of the threshold voltage as a function of L is shown in Figure 1.46. In practice, the actual shift will be different than that estimated using these equations because of the approximations involved. However, the decrease of the threshold voltage with decreasing channel length is a general dependence in a FET.

Narrow Width Effects

Narrow width effects occur when the channel width W is decreased to small values. With regards to the threshold voltage, we will see that narrow-width MOSFETs exhibit threshold voltages that are larger than that predicted with the gradual channel approximation. The amount of increase $(\Delta V_{Tn})_{NWE} > 0$ of the threshold voltage is due to bulk charge outside of the gate region that is ignored in a simple analysis. Figure 1.47 illustrates the basic problem: fringing electric fields deplete the silicon beyond the gate region as defined by the channel width W. Since the corresponding bulk charge was ignored, our expression for V_{Tn} will underestimate the contribution due to Q_B.

The shape of the depletion edge must be known to calculate the total amount of bulk charge in the region. Figure 1.48 shows a cross-sectional view of the geometry. In general, we may write the total bulk charge in Coulombs using

$$|Q_B|A_c = qN_a x_{dm} A_C \tag{1. 157}$$

where A_C is the total area of the region. Explicitly,

$$A_C = x_{dm}W + 2A_{NWE}, \tag{1. 158}$$

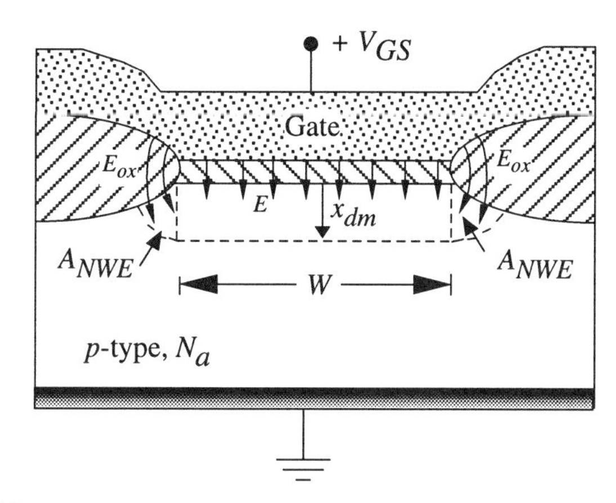

Figure 1.47 Bulk charge distribution for narrow-width effect calculation

with $2A_{NWE}$ as the total area of the charge that causes the NWE correction. The basic threshold voltage expression now assumes the form

$$V_{Tn} = V_{FB} + 2|\phi_F| + \frac{1}{C_{ox}}\sqrt{2q\varepsilon_{Si}N_a(2|\phi_F|)}\,g \tag{1. 159}$$

where

$$g = 1 + \frac{A_{NWE}}{x_{dm}W} > 1 \tag{1. 160}$$

is the NWE form factor. Since it is not possible to accurately determine the boundary without a detailed numerical calculation, the value of A_{NWE} is estimated by assuming a geometrical shape for the depletion edge. A simple choice is to use a circular boundary that has a radius x_{dm}. Since the area for both contributions is

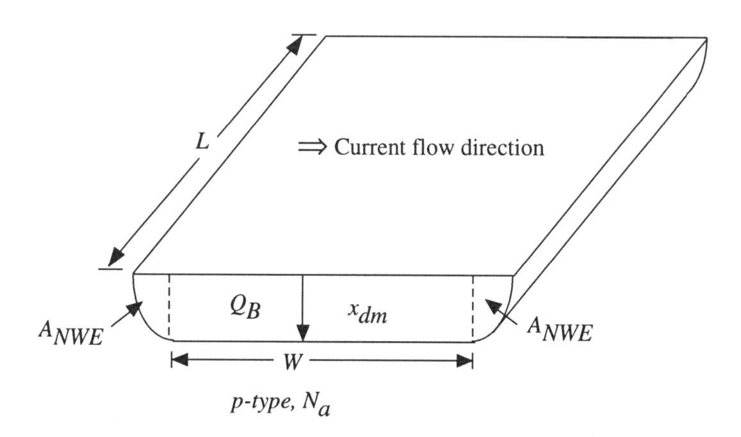

Figure 1.48 Bulk charge geometry for narrow-width calculations

$$2\left(\frac{\pi x_{dm}^{\ 2}}{4}\right) \tag{1. 161}$$

the form factor is given by

$$g \approx 1 + \frac{\pi x_{dm}}{2W}. \tag{1. 162}$$

Another approach is to use an empirical factor κ and write

$$g \approx 1 + \kappa\left(\frac{x_{dm}}{W}\right). \tag{1. 163}$$

This can be used to define a narrow-width effect as occurring when W $\approx x_{dm}$. Regardless of the details, the analysis gives the threshold voltage as having the form

$$(V_{Tn})_{NWE} = V_{Tn} + (\Delta V_{Tn})_{NWE} \tag{1. 164}$$

with

$$(\Delta V_{Tn})_{NWE} = \frac{A_{NWE}\sqrt{2q\varepsilon_{Si}N_a(2|\phi_F|)}}{xC_{oxdm}W} > 0 \tag{1. 165}$$

being a positive number. The general dependence of the threshold voltage due to narrow-width effects is shown in Figure 1.49. As with short-channel effects, circuit simulation models employ different equations but most tend to be very similar to the results above.

1.6.2 Mobility Variations

Another important parameter that must be modified is the surface mobility μ_n. This appears in the transconductance parameter k_n', and is a sensitive function of the doping, temperature, and operating voltages. It is therefore important to examine the mobility in more detail. Let us first recall that the gradual-channel approximation was based upon modelling the channel as a small differential resistor with a value

$$dR = \frac{dy}{\sigma A_c} \tag{1. 166}$$

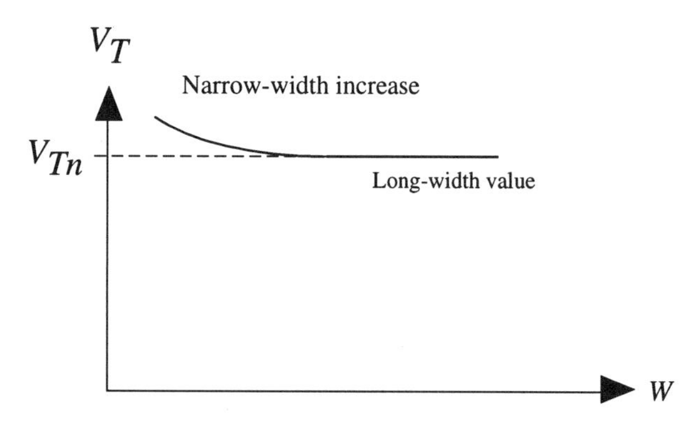

Figure 1.49 Narrow-width threshold voltage effect

where σ is the conductivity of the region as defined by Ohm's Law

$$\sigma = \frac{J}{E} \tag{1.167}$$

with J the current density in units of A/cm^2 and E the electric field intensity. Drift current is due to charge moving under the influence of an electric field. To obtain an expression involving the carrier density, we note that J can also be expressed as

$$J = \rho_v v_e \tag{1.168}$$

with ρ_v the volume charge density[11] in units of coulombs per cubic centimeter and v the velocity of the charge.

In performing the gradual-channel analysis of a MOSFET, the conductivity was expressed by

$$\sigma = q\mu_n n_I \tag{1.169}$$

It is seen that this assumes that the electron velocity is proportional to the electric field such that

$$v_e = \mu_n E \tag{1.170}$$

with the mobility μ_n acting as the proportionality constant. This ignores two main features of the electron motion. First, the gate-source voltage V_{GS} that induces the field effect will alter the local electric field. This can be treated using the empirical expression

$$\mu_n = \frac{\mu_n}{[1 + \theta(V_{GS} - V_T)]} \tag{1.171}$$

where μ_n is the "normal" surface mobility, and θ is an empirical constant with units of V^{-1}. This effectively reduces the value of the process transconductance k'_n when the MOSFET is used in a circuit.

The second feature is more complicated. It is well known that the velocity-field dependence for electrons in silicon is non-linear with the characteristics portrayed in Figure 1.50. The linear relationship between the mobility and the electric field is only valid for small field intensities. As the strength of the electric field is increased, the v-E curve bends and it is no longer possible to use a constant value for the mobility for either electrons or holes. Instead, we write that $\mu=\mu(E)$ and refer to this as a **nonlinear mobility**. As the electric field strength is increased further, the curves eventually hit a value $v = v_s$ which is called the **saturation velocity**. This is the maximum drift speed that a particle can attain in the lattice; this limit is due to the fact that collisions remove energy from the particle even as the electric field interaction attempts to increase it. As indicated in the drawing, the saturation velocity in silicon has a numerical value of about $v_s = 10^7$ *cm/sec* at room temperature.

The nonlinear velocity-field relation has a number of significant consequences when applied to modern short-channel processes. The reason for this can be seen with a simple calculation. Suppose that we consider a fairly large FET that has an electrical channel length of L=0.5μm. With an applied drain-source voltage of V_{DS}= 2v, the strength of the channel electric field can be estimated as

$$E \approx \frac{V_{DS}}{L} = \frac{2}{5\times10^{-5}} = 4\times10^4 \quad V/cm \tag{1.172}$$

[11] Do not confuse this with the resistivity $\rho = (1/\sigma)$.

which places us in the nonlinear region. Since processes continually shrink the size of a FET, it is increasingly more difficult to ignore the problems that arise if we neglect the nonlinear effects. Most of the accurate modelling for circuit analysis is accomplished by CAD models. There is, however, much to be said for circuit designers that can understand and anticipate complicated device effects in their work.

The concept of an **electron temperature** is useful for classifying the velocity regions. If the excitation were due solely to thermal means, then v would increase with the temperature T; this view provides an analogy for describing field-aided transport. Analytically we may equate the particle kinetic energy to the thermal energy by writing

$$\frac{1}{2}m^* v^2 = \frac{3}{2}k_B T \tag{1. 173}$$

where m^* is the (effective) mass of the particle, and k_B is Boltzmann's constant from thermodynamics. Using this relationship allows us to relate the speed to an equivalent temperature. For low electric fields, v is small defining the **cold electron** region. As the velocity increases and the $v(E)$ curve goes nonlinear, we enter the **warm electron** region. Finally, when the velocity reaches the saturation value v_s, we are in the **hot electron** region where the transport is complicated by high-field effects.

1.6.3 Hot Electrons

Hot electron effects have been observed in MOSFETs, and are particularly important in modern devices where the channel lengths are smaller than 1 micron. Standard transistors can be degraded by tunnelling effects where highly energetic particles can leave the silicon and enter the gate oxide. Trapped electrons increase the oxide charge Q_{ox}, leading to instability of the threshold voltage. Long-term reliability problems may result from this mechanism. In addition, hot electrons may induce leakage gate currents I_g and excessive substrate currents I_s. Although a detailed discussion of hot electron effects in FETs is beyond the scope of the present treatment, it is worthwhile to mention one structure that is commonly found in modern integrated circuits.

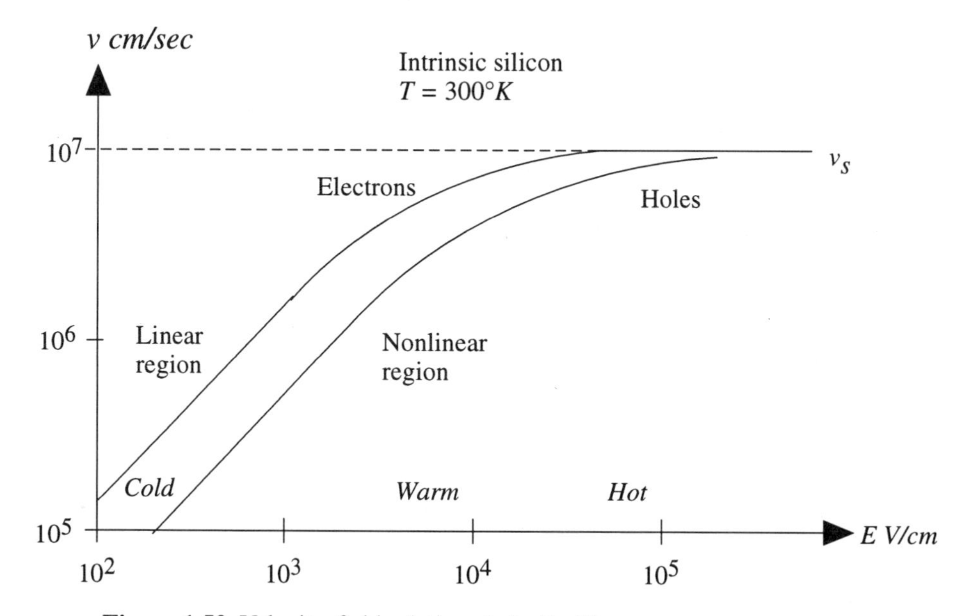

Figure 1.50 Velocity-field relations in bulk silicon

The LDD MOSFET

Reliable design of small MOSFETs usually requires that hot electron effects be minimized. This is accomplished by reducing the magnitude of the electric field that acts on the channel charge carriers. This is particularly important in short-channel devices, since the drain-source voltage V_{DS} must be dropped along the current flow path. The gradual channel analysis shows that the problem is much worse than this on the drain side of the transistor where the electric field is a maximum. One obvious solution is to reduce the operating voltage, but even dropping the value of power supply V_{DD} will result in hot electron effects in short-channel devices. To overcome these adverse effect, device engineers have developed structures that are designed to minimize hot electron effects. Various approaches have been presented in the literature. In the present discussion we will examine using lightly-doped drains (LDD) structures to deal with hot electrons.

An LDD structure is shown in shown in Figure 1.51. The drain and source regions have been modified by inserting lightly doped n regions between the channel and the low resistance n^+ areas. This reduces the effective built-in electric field on the drain side of the channel where the hot electron tunnelling probability is the largest (it has no direct effect on the applied field component). To understand this, let us denote the n^+ donor doping by N_d and approximate the n^+-p boundary as a step junction. The maximum value of the built-in electric field occurs at the junction with

$$E_{max} = \frac{qN_d x_d}{\varepsilon_{Si}} \tag{1. 174}$$

where x_d is the depletion width. Setting N_d to be 1 or 2 orders of magnitude smaller than the doping in the n^+ regions decreases the tunneling probability at the drain. Both I_g and I_s are reduced accordingly. The price paid for this design is the fact that the drain and source resistances are larger and the processing is more complicated since it requires another masking step. (The increase in drain and source resistance actually reduces the channel electric field even further, but they degrade the transient response.)

1.7 Small Device Model

The physics of the electron transport in a small-geometry MOSFET is markedly different from that

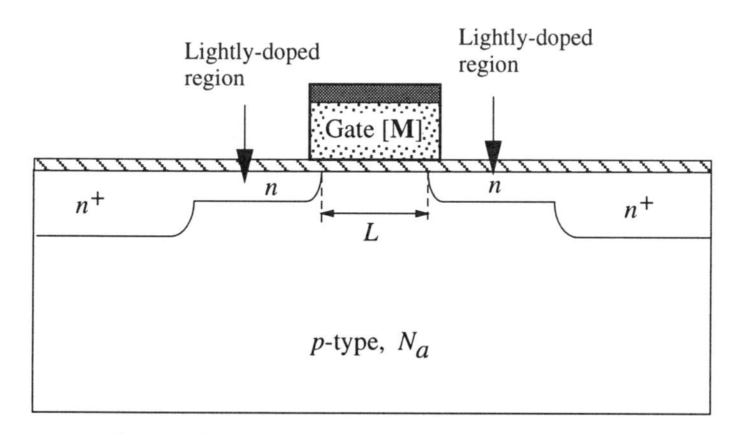

Figure 1.51 Lightly-doped drain (LDD) structure

observed in long-channel structures. Although the analysis can be quite complicated, we can include some of the differences by introducing correction into the simpler device equations. This provides improved accuracy without resorting to excessively complicated device models in our basic design calculations. Of course, the analytic complexity of a model is not an issue in a circuit simulator, so computer models should be as precise as possible.

In this section we will examine an analytic model for small MOSFETs that accounts for velocity saturation. This has been included to illustrate the problems that are involved in describing small transistors, and was chosen as being representative of work in the area. Several other approaches can be found in the literature.

The MOSFET models discussed thus far in this chapter are both based on low-electric field charge transport where the electron drift velocity v_e is proportional to the electric field. In a short-channel MOSFET, the field intensity can easily exceed 10^3 [*V/cm*], driving the transport into the nonlinear regime. A simple model for the velocity is obtained by writing

$$v_n = \frac{\mu_n |E|}{1 + \frac{\mu_n |E|}{v_s}} \tag{1. 175}$$

where $v_s \approx 10^7$ [*cm/sec*] is the saturation velocity for electrons in silicon.[12] Let us define the **critical electric field** E_c by

$$E_c = \frac{v_s}{\mu_n}, \tag{1. 176}$$

with μ_n as the linear mobility. As shown in Figure 1.52, the critical electric field is obtained by extrapolating the linear dependence until it intersects the saturation velocity line. This allows us to use a field-dependent velocity expression in the form

$$v = \frac{v_s (|E|/E_c)}{1 + (|E|/E_c)} \tag{1. 177}$$

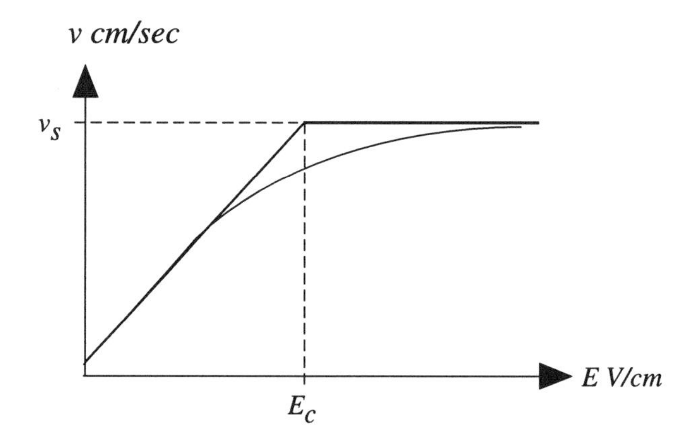

Figure 1.52 Piecewise linear modelling of velocity-field curve

[12] The numerical value is for a temperature of 300° *K*.

which defines a nonlinear mobility function as

$$\mu_{NL} = \frac{\mu_n}{1 + (|E|/E_c)} \quad (1.178)$$

such that

$$v_n = \mu_{NL} E \quad (1.179)$$

is valid.

To use this in describing a MOSFET, we start with the drift current expression

$$\begin{aligned} I_{Dn} &= -WQ_n v_n \\ &= WC_{ox}(V_{GS} - V_{Tn} - \acute{V}) \left[\frac{\mu_n \left(\frac{dV}{dy} \right)}{1 + \frac{\mu_n}{v_s} \left(\frac{dV}{dy} \right)} \right] \end{aligned} \quad (1.180)$$

which is obtained directly by substituting the nonlinear mobility into the GCA expression and using

$$|E| = \left(\frac{dV}{dy} \right) \quad (1.181)$$

with $V(y)$ the channel voltage. The expressions may be written in terms of the critical electric field if desired. Grouping terms and integrating from $y=0$ to $y=L$ with the boundary conditions $V(0)=0$ and $V(L)=V_{DS}$ gives the non-saturated current in the form

$$I_{Dn} = \beta_n \frac{[2(V_{GS} - V_{Tn})V_{DS} - V^2_{DS}]}{\left(1 + \frac{\mu_n}{v_s} V_{DS} \right)} \quad (1.182)$$

where

$$\beta_n = \mu_n C_{ox} \left(\frac{W}{L} \right) \quad (1.183)$$

is the usual device transconductance. It is seen that the main effect is the decrease of the current with the V_{DS} term in the denominator, corresponding to the electrons attaining the saturation velocity. The saturation voltage found by calculating the maximum current point is found to be

$$V_{sat} = \frac{Lv_s}{\mu_n} \left[\sqrt{1 + 2\frac{\mu_n}{Lv_s}(V_{GS} - V_{Tn})} - 1 \right] \quad (1.184)$$

This in turn gives the saturation current

$$I_{Dn} = WC_{ox} v_s (V_{GS} - V_T) \quad (1.185)$$

for $V_{DS} \geq V_{sat}$. Note that this is a linear function of the gate-source voltage V_{GS}, which is distinctly different from the quadratic relation found for the linear mobility regime. Although the model is highly simplified, this functional dependence is observed in short-channel MOSFETs.

1.8 MOSFET Modelling in SPICE

SPICE[13] and its relatives are particularly useful in complex chip design and analysis. The accuracy of a SPICE simulation depends on the accuracy of the device models. Increased precision generally requires better modelling and longer run times, with much effort being devoted towards the development of better models. This has proven to be more difficult as the device dimensions are reduced because of the complexity of submicron phenomena. In this section, we will review the most basic SPICE models for a MOSFET. The interested reader is directed to the books listed in the Reference section for more detailed discussion.

1.8.1 Basic MOSFET Model

To describe a MOSFET in SPICE, we must number the nodes and provide the appropriate parameters. Any MOSFET is represented in the element list using the syntax

Mname ND NG NS NB ModName *<parameters>*

where Mname is the name of the specific MOSFET (that must begin with the letter "M"), the connections are specified by the node numbers

ND = number of drain node

NG = number of gate node

NS = number of source node

NB = number of bulk node,

and ModName is the name of the model listing that provides the processing parameters. The *<parameters>* entry denotes is a set of parameters that provide specifics of the device geometry including

L is the channel length

W is the channel width

AD is the area of the drain n^+ or p^+

PD is the perimeter of the drain n^+ or p^+

AS is the area of the source n^+ or p^+

PS is the perimeter of the source n^+ or p^+

It is important to remember that strict MKS units are employed in SPICE so that all dimensions must be specified in units of meters; useful scaling factors are U=10^{-6} and P=10^{-12}. The channel length L and channel width W may be the actual electrical lengths that have been used in the analysis here. However, these are often replaced by the drawn values with processing effects included in the processing parameters as discussed previously.

Intricate details of the MOSFET are contained in the .MODEL statement, as with all active devices in SPICE. In general, this assumes the form

.MODEL ModName *<parameter list>*

where the information provided in *<parameter list>* depends upon the model being used. Models are selected by the specification

Level=N

in the *<parameter list>*, where N is predefined to reference a particular set of values and equations. In the original implementations of SPICE as released by the University of California, Berke-

[13] This is an acronym that stands for *Simulation Program with Integrated Circuit Emphasis*

ley, three levels of accuracy were provided for MOSFET models in standard SPICE. These were denoted by Level=1, Level=2, and Level=3. Level 1 uses the square-law equations of current flow, while Level 2 is based on the bulk charge equations. Level 3 is an empirical database that can extrapolate between stored values. Current SPICE programs generally provide for a wide variety of Level choices, each being applicable to a different type of device structure. Most of these are still based on the bulk-charge equations, but include more details about other effects. The BSIM model can be used to providing a smooth interface between a set of basic wafer measurements and accurate simulations.

A complete discussion of SPICE device modelling is beyond the scope of the present treatment. Excellent treatments can be found in the books by Foty, and Antognetti and Massobrio. Commercial user manuals are also useful in this regard. Figure 1.53 provides a Level=2 listing of the model parameters for a MOSIS[14] 2.0 micron process. Many of the parameters have a direct one-to-one correlation with the square-law and bulk-charge expressions.

It is worthwhile to discuss the manner in which SPICE treats parasitic capacitances. Since the program itself performs transient analyses by incrementing the time variable, nonlinear capacitances are easily included in the calculations. The zero-bias values are given by CJ, CJSW, CGBO, CGDO, and CGSO. The grading parameters MJ and MJSW can be adjusted to model the doping profile. Area and perimeter information is input into the device descriptions using AD, AS, PD, and PS in the device line. In general, the division of area among devices with common drain or source regions is arbitrary.

In a realistic parameter list, the values of the overhang capacitances CGDO, CGBO include fringing field effects. If one attempts to compute these values using, for example,

```
.MODEL CMOSN NMOS LEVEL=2 PHI=0.700000 TOX=3.8700E-08
+ XJ=0.200000U TPG=1  VTO=0.7513 DELTA=3.7170E+00
+ LD=1.3450E-07 KP=5.8052E-05  UO=650.6 UEXP=9.8560E-02
+  UCRIT=5.5460E+03 RSH=1.1950E+01  GAMMA=0.5675
+ NSUB=7.7230E+15 NFS=9.1070E+10 VMAX=5.0410E+04
+ LAMBDA=3.6540E-02 CGDO=1.8002E-10 CGSO=1.8002E-10
+ CGBO=3.4582E-10 CJ=1.27E-04 MJ=0.950 CJSW=5.05E-10
+ MJSW=0.286 PB=0.86
* Weff = Wdrawn - Delta_W
* The suggested Delta_W is 2.0000E-09
.MODEL CMOSP PMOS LEVEL=2 PHI=0.700000 TOX=3.8700E-08
+ XJ=0.200000U TPG=-1 VTO=-0.9784 DELTA=2.9080E+00
+ LD=5.8670E-08 KP=1.6828E-05 UO=188.6 UEXP=2.4140E-01
+ UCRIT=1.1560E+05 RSH=9.0910E-02  GAMMA=0.6412
+ NSUB=9.8610E+15 NFS=1.4700E+11 VMAX=9.9990E+05
+ LAMBDA=4.2450E-02 CGDO=1.8526E-10 CGSO=1.8526E-10
+ CGBO=4.2819E-10 CJ=3.39E-04 MJ=0.581 CJSW=2.63E-10
+ MJSW=0.308 PB=0.90
* Weff = Wdrawn - Delta_W
* The suggested Delta_W is -4.9740E-07
```

Figure 1.53 LEVEL 2 MOSFET .model statements

[14] MOSIS stands for "MOS Implementation Service." The most recent parameter sets may be downloaded from their web site.

$$C_o = C_{ox}L_o \quad (1.186)$$

with

$$C_{ox} = \frac{\varepsilon_{ox}}{x_{ox}} \quad (1.187)$$

the computed and listed values should be different since the simple analytic approach ignores the fringing fields.

These comment have particular relevance to the analytic models derived in this chapter. As we will see later, the use of simple FET models based on LTI elements introduces errors into the hand calculations. These discrepancies may be blatantly obvious when the values are compared to those obtained from a computer simulation. However, in the circuit design philosophy used throughout the book, the hand estimates serve only as a basis for first-level design. Although the numbers are only estimates, the values are generally reasonable. Circuits should always be simulated as a check on the final characteristics.

1.9 Problems

[1-1] Consider a MOS structure that is made with the following characteristics:

$x_{ox} = 200\text{Å}$, $N_a = 10^{15} cm^{-3}$, n-poly gate with $N_d = 2 \times 10^{19} cm^{-3}$, $Q_f = q10^{10} C/cm^{-2} >> Q_{ox}$.

(a) Calculate the value of C_{ox} in units of F/cm^2 and $fF/\mu m^2$.

(b) Calculate the value of the flatband voltage.

(c) Find the value of the threshold voltage before a threshold adjustment ion implant.

(d) Now find the value of the acceptor implant dose D_I needed to set V_{T0} to 0.7*v*.

[1-2] The work function of an n-type sample of silicon that is doped with a donor density N_d can be written in the form

$$\Phi = \Delta\phi - \left(\frac{kT}{q}\right)\ln\left(\frac{N_d}{n_i}\right) \quad (1.188)$$

where $\Delta\phi$ is a constant. In a similar manner, the work function for a p-type sample with an acceptor doping of N_a is given by

$$\Phi = \Delta\phi + \left(\frac{kT}{q}\right)\ln\left(\frac{N_a}{n_i}\right) \quad (1.189)$$

where $\Delta\phi$ is the same constant as for the n-type sample. These formulas are used to approximate the work functions of both crystal and polycrystal silicon, and will be used to derive equations (1.11) and (1.12) in the text.

Consider a n-channel MOSFET that is built in a p-type substrate with a doping density of N_a.

(a) Find a general equation for ($\Phi_G - \Phi_S$) if an n-type gate with doping $N_{d,poly}$ is used. Then calculate the numerical value of ($\Phi_G - \Phi_S$) for the case where $N_a = 10^{15}\ cm^{-3}$ and $N_{d,poly} = 10^{19}\ cm^{-3}$.

(b) Suppose instead that the gate is p-type polysilicon with an acceptor doping of $N_{a,poly}$ .Find a general expression value for ($\Phi_G - \Phi_S$) for this case. Then find the numerical value of ($\Phi_G - \Phi_S$) for the case where $N_a = 10^{15}\ cm^{-3}$ and $N_{1,poly} = 10^{17}\ cm^{-3}$.

[1-3] Consider an nFET process that uses an n-type poly gate. The important processing parameters are as follows: $x_{ox} = 100\text{Å}$, $N_a = 10^{15}\ cm^{-3}$, $\mu_n = 580\ cm^{-2}/V\text{-}sec$, $V_{T0n} = +0.7$v.

(a) Calculate the value of the oxide capacitance per unit area C_{ox}. Place your answer in units of

F/cm^2 and $fF/\mu m^2$.

(b) Find the value of k'_n in units of $\mu A/V^2$.

(c) Find the value of the body-bias coefficient γ. Then plot V_{Tn} as a function of the source-bulk voltage V_{SB} in the range $V_{SB} = 0$ to 5 volts.

(d) Suppose that an nFET is made with $W= 10\mu m$ and $L=1\ \mu m$. Voltages of $V_{GSn}=2.5v$, $V_{DSn}=2v$, and $V_{SB}=1.25v$ are applied. Calculate the drain current using the square law equations.

(e) Repeat the calculation in (d) if the drain-source voltage is changed to a value of $V_{DSn}=4v$.

[1-4] A CMOS process specifies that $\gamma=0.071V^{1/2}$ and $2|\phi_F|=0.58v$ for an nFET. The mobilities are given by $\mu_n=560\ cm^2/V\text{-}sec$ and $\mu_p=220\ cm^2/V\text{-}sec$.

(a) Calculate the thickness x_{ox} of the gate oxide and then find C_{ox} .

(b) Calculate the process transconductance factors for both nFETs and pFETs.

[1-5] Suppose that we have the following information about an nFET: $x_{ox}= 100Å$, $N_a=8\times10^{14}cm^{-3}$, $V_{TOn}= +0.70v$, $\mu_n=580\ cm^2/V\text{-}sec$.

(a) Calculate the process transconductance.

(b) Calculate the body-bias coefficient.

(c) An nFET is made with $W=10\mu m$ and $L=1\mu m$. Calculate the drain current if the voltages are set to $V_{GSn}= 2v$, $V_{DSn}= 1v$, and $V_{SBn}= +1v$.

(d) Repeat the calculation for the current when the voltages are specified as $V_{GSn}= 2v$, $V_{DSn}= 3v$, and $V_{SBn}= 0v$.

[1-6] An nFET process is described by the parameters $k'_n = 110\mu A/V^2$ and $V_{TOn}= +0.70v$. The voltages are measured to be $V_{GSn}= 2v$ and $V_{SBn}= 0v$.

(a) A FET with an aspect ratio of 4 has a drain current of $340\mu A$ flowing through it. Find the drain-source voltage V_{DSn}.

(b) A different FET is biased with $V_{DSn}= 2v$ and $V_{SBn}= 0v$, and the current is measured as $440\mu A$. Find the gate-source voltage V_{GSn} if the aspect ratio is known to be 8.

[1-7] A step-profile pn junction has p- and n-side doping densities of $N_a=8\times10^{14}cm^{-3}$ and $N_d=7\times10^{19}cm^{-3}$, respectively.

(a) Find the built-in voltage ϕ_o.

(b) Find x_{dO}.

(c) Calculate C_{jO}.

(d) Junction junction area is $14\mu m^2$. Plot the capacitance $C(V_R)$ for V_R from 0 to 5 volts.

[1-8] All dimensions shown in the device layout of Figure P1.1 are in units of microns. The process is characterized by $L_o=0.1\mu m$, $C_j=0.17\ fF/\mu m^2$, $C_{jsw}=0.64\ fF/\mu m$, $x_{ox}= 120Å$, and $\mu_n=580\ cm^2/V\text{-}sec$. Construct the RC equivalent circuit for the nFET. Use zero-bias values for the depletion capacitances.

[1-9] Use SPICE to obtain the I_{Dn} vs. V_{DSn} family of curves for the voltage range [0,5] for nFET described by the .MODEL state in Figure 1.53. Step V_{GSn} over the same range.

1.10 References

MOSFETs are discussed in a large number of books and an even larger number of journal articles. All of the books listed below contain excellent discussions of MOSFET characteristics. A few select journal articles have also been listed to aid the reader in performing a literature search that starts with some original papers.

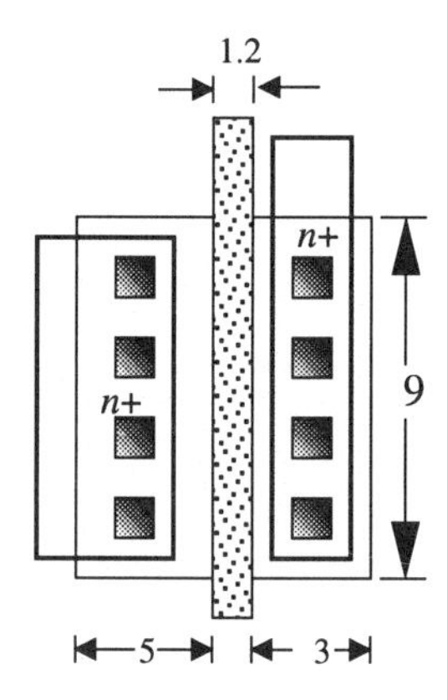

Figure P1.1

[1] L.A. Akers and J.J. Sanchez, "Threshold Voltage Models of Short, Narrow and Small Geometry MOSFET's: A Review", *Solid-State Electronics*, vol. 25, pp. 621-641, July, 1982.

[2] P. Antognetti and G. Massobrio (eds.), **Semiconductor Device Modelling with SPICE**, McGraw-Hill, New York, 1988.

[3] J.R. Brews, W. Fichtner, E.H. Nicollian and S.M. Sze, "Generalized guide for MOSFET Miniaturization", *IEEE Electron Device Letters*, vol. EDL-1, pp. 2-4, 1980.

[4] J.Y. Chi and R.P. Holstrom, "Constant voltage scaling of FET's for high frequency and high power applications", *Solid-State Electronics*, vol. 26, pp. 667-670, July, 1983.

[5] P.E. Cottrell, R.R. Troutman, and T.H. Ning, "Hot-Electron Emission in N-Channel IGFET's", *IEEE Trans. Electron Devices*, vol. ED-26, pp. 520-532, April, 1979.

[6] M.J. Deen and Z.P. Zuo, "Edge Effects in Narrow-Width MOSFET's", *IEEE Trans. Electron Devices*, vol. 38, pp. 1815-1819, August, 1991.

[7] R.H. Dennard, et. al, "Design of ion-implanted MOSFETs with very small physical dimensions", *IEEE J. Solid-State Circuits*, vol. SC-9, pp. 256-268, October, 1974.

[8] D. A. Divekar, **FET Modeling for Circuit Simulation**, Kluwer Academic Publishers, Boston, 1988.

[9] D. Foty, **MOSFET Modelling with SPICE**, Prentice-Hall, Upper Saddle River, NJ, 1997.

[10] G. Krieger, R. Sikora, P.P. Ceuvas, and M.N. Misheloff, "Moderately Doped NMOS (M-LDD)-Hot Electron and Current Drive Optimization", *IEEE Trans. Electron Devices*, vol. 38, pp. 121-127, January, 1991.

[11] Robert F. Pierret, **Semiconductor Device Fundamentals**, Addison-Wesley, Reading, MA, 1996.

[12] B.J. Sheu, D.L. Scharfetter, P-K. Ko, M-C Jeng, "BSIM: Berkeley Short-Channel IGFET Model for MOS Transistors", *IEEE J. Solid-State Circuits*, vol. SC-22, No. 4, pp. 558-565, August, 1987.

[13] S. Sze, **Physics of Semiconductor Devices**, 2nd ed., John Wiley & Sons, New York, 1981.

[14] K-Y. Toh, P-K. Ko, and R.G. Meyer, "An Engineering Model for Short-Channel MOS Devices"', *IEEE J. Solid-State Circuits*, vol. 23, No. 4, pp. 950-957, August, 1988.

[15] E.S. Yang, **Microelectronic Devices**, McGraw-Hill, New York, 1988.

[16] Y. P. Tsividis, **Operation and Modeling of The MOS Transistor**, McGraw-Hill, New York, 1987.

Chapter 2

Fabrication and Layout of CMOS Integrated Circuits

The design of CMOS integrated circuits is highly dependent upon the fabrication steps and resulting electrical properties. In this chapter, we will examine the basic features of the processing line that are crucial to formulating a circuit design philosophy and style.

2.1 Overview of Integrated Circuit Processing

In the basic sense, an integrated circuit is a set of patterned material layers that combine to form a 3-dimensional physical structure that are the electronic devices and interconnects. A typical CMOS integrated circuit will consist of many individual layers such as polycrystalline silicon (poly), silicon dioxide (quartz glass), and metal conductors. Each layer is defined by its own distinct pattern made up of geometrical objects that are strategically placed relative to other layers to form transistors and the needed interconnect lines that define the circuit itself. The processing sequence consists of the physical steps that must be performed in order to create the patterned layers on a silicon substrate.

In this section we will characterize the fabrication steps that are the most important to understanding CMOS integrated circuit design. The treatment here is necessarily short, but the subject is well presented in the open literature for those desiring a deeper understanding.

2.1.1 Oxides

Silicon dioxide (SiO_2) is used extensively in integrated circuits because it is easy to grow or deposit, and has excellent insulating characteristics.[1] It is used as the gate insulator in a MOSFET, and provides insulation between conducting layers. There are two ways that oxides are created, thermal growth and CVD (chemical vapor deposition). Both are summarized below.

---

[1] The generic name for silicon dioxide is quartz glass.

Thermal Oxides

Thermally-grown oxides use silicon atoms from the substrate in the reaction

$$Si + O_2 \rightarrow SiO_2$$

which creates silicon dioxide at an elevated growth temperature, typically from about 900°C to 1100°C. This is called dry oxidation to distinguish it from a wet oxidation process that obtains the oxygen from steam via

$$Si + 2H_2O \rightarrow SiO_2 + 2H_2$$

in about the same temperature range. In general, dry oxidation produces a better insulator but is characterized by relatively slow growth rate; steam oxidation is much faster, but the resulting oxides are of lower quality. One important characteristic of the thermal oxidation process is that silicon is consumed during the oxide growth.

It is important to understand the basics of the oxidation process as it has a direct effect on the manner in which we design chips. An analysis of oxidation process shows that the oxide thickness x_{ox} can be approximated by the quadratic equation

$$x_{ox}^2(t) + Ax_{ox}(t) = Bt \quad (2.1)$$

where A and B are coefficients that depend upon the temperature, crystal orientation, and gas mixture. Solving gives the time dependence as

$$x_{ox}(t) = \frac{A}{2}\left[\sqrt{1+\frac{4Bt}{A^2}} - 1\right] \quad (2.2)$$

which is plotted in Figure 2.1. During the initial phases of the oxide growth process, t is small and the thickness increases linearly with time:

$$x_{ox}(t) \approx \frac{B}{A}t \; . \quad (2.3)$$

As illustrated in Figure 2.2(a), the rapid initial growth is due to the fact that silicon atoms are readily available at the surface for consumption in the reaction. The ratio (B/A) is known as the linear rate constant. For large times t,

$$x_{ox}(t) \approx \sqrt{Bt}\,. \quad (2.4)$$

This slower growth rate is due to the fact that the oxygen molecules must diffuse through existing

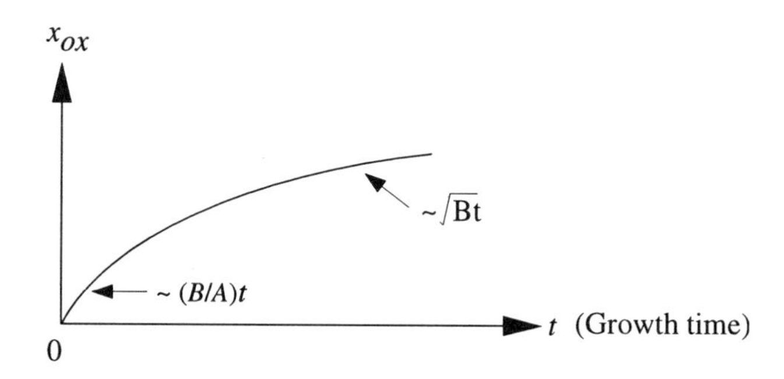

Figure 2.1 Oxide thickness as a function of time

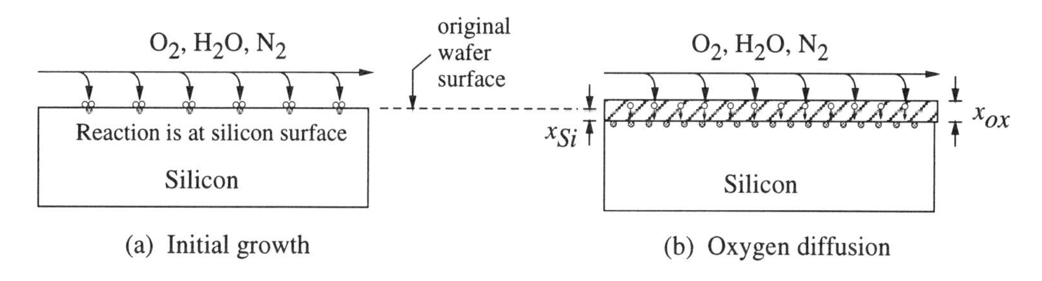

Figure 2.2 Visualization of the oxide growth process

oxide before reaching the silicon atoms at the substrate surface as shown in Figure 2.2(b). The parameter B by itself is called the parabolic rate constant.

The results bring out two interesting points. First, thick oxides require a long growth period, making them costly to include in the fabrication flow. However, since silicon is consumed during the thermal oxidation process, this allows us to create a recessed insulating glass layer that is useful for electrically isolating tightly packed transistors. The recession of the silicon surface is shown in Figure 2.2 by comparing (a) and (b). Growing an oxide with a thickness x_{ox} consumes a layer of silicon of thickness x_{Si}. Let us denote the density of silicon atoms by $N_{Si} \approx 5 \times 10^{22}\ cm^{-3}$, and the density of silicon dioxide molecules by $N_{ox} \approx 2.3 \times 10^{22} cm^{-3}$. Since it takes one atom of silicon to create one molecule of silicon dioxide, we may equate

$$N_{Si} x_{Si} = N_{ox} x_{ox} . \tag{2.5}$$

Rearranging and substituting values gives

$$x_{Si} \approx 0.46 x_{ox} , \tag{2.6}$$

illustrating that the recession is about 46% of the oxide thickness. As an example, this says that growing a $1\mu m$-thick oxide consumes $0.46\mu m$ of silicon. This observation forms the basis for the local oxidation of silicon (LOCOS) technique discussed later.

CVD Oxides

Chemical vapor deposition (CVD) oxides are created using reactions such as the one below that combines silane (SiH_4) with oxygen:

$$SiH_4 + 2O_2 \rightarrow SiO_2 + 2H_2O$$

which is valid for temperatures below 1000°C. Reactions of this type create SiO_2 molecules that are then deposited on top of the wafer. The important characteristic of a CVD oxide is that it does not use silicon atoms from the substrate, so it can be deposited on top of any existing layer. CVD oxides are used extensively as insulating dielectrics between conducting layers such as metals above the surface of the silicon.

2.1.2 Polysilicon

Modern MOS technology makes heavy use of polycrystalline silicon (which is called **polysilicon** or simply **poly**) as a deposited conducting layer on top of oxides. A simple reaction to produce poly is the pyrolysis of silane

$$SiH_4 \rightarrow Si + 2H_2$$

which gives varying characteristics as the temperature is varied. Polysilicon is of particular importance in CMOS since MOSFET gates generally consist of a deposited poly layer with a refractory metal (such as W or Pt) either on top of it, or mixed in during the deposition process. This combination is called a **silicide**.

Polycrystal silicon gains it name because it consists of many small regions crystal, called crystallites, instead of having a single crystal structure throughout (such as in a silicon wafer). This state is achieved by depositing silicon over an amorphous material such as silicon dioxide. Silicon atoms attempt to form crystals, but do not have a well-defined base to grow on. This results in the formation of the local crystal regions called **crystallites**. Polysilicon is used because it provides excellent coverage, has good adhesion properties to the silicon dioxide surface, and can be subjected to high temperature processing steps. One of the drawbacks of the material is that even heavily doped poly has a substantial resistance to current flow. Adding a refractory metal overcomes this problem, and allows poly lines to be used as interconnect wiring.

2.1.3 Doping and Ion Implantation

High-density VLSI circuits use ion implantation to create **doped** n and p regions in the silicon substrate. A doped region is simply a section of the silicon into which impurity atoms have been purposely added to alter the electrical properties. Arsenic (As) and phosphorus (P) are used for n-type regions in which there is an enhanced free electron concentration. Boron (B) is used to create p-type regions where the equilibrium density of positively charged holes is greater than that of the electrons. In VLSI fabrication, doped regions are most often created using the technique of ion **implantation**.

In the ion implantation process, dopant ions are accelerated to high energies and literally smashed into the silicon wafer as shown in Figure 2.3. Collisions between the ions and the silicon atoms and electrons eventually bring the ions to rest within the wafer. This stopping mechanism creates a lot of damage to the crystal that must be repaired; in addition, the ions must find their way to normal silicon lattice sites in order to act like substitutional impurities. Both objectives are achieved by heating the wafer in an **annealing step** where thermal energy is used to induce diffusion of the atoms.

A simple analytic approximation for the ion implanted doping density is given by the Gaussian expression

$$N_{ion}(x) = N_p \exp\left[-\left(\frac{x - R_p}{2(\Delta R_p)}\right)^2\right] \tag{2.7}$$

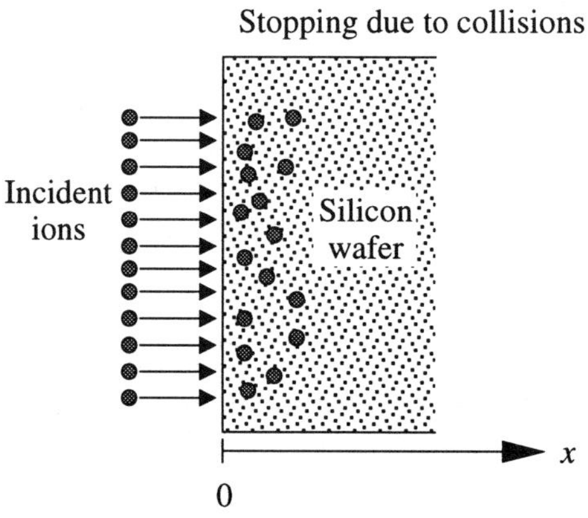

Figure 2.3 The ion stopping process

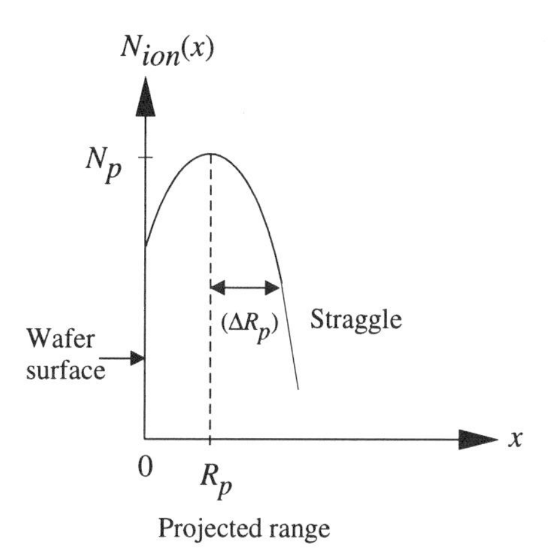

Figure 2.4 Gaussian model for ion implant doping profile

which is plotted in Figure 2.4. R_p is known as the **projected range** and is the average stopping distance in the silicon. ΔR_p is called the **straggle** and represents the spread in the distribution, while N_p is the peak density. The implant dose D_I is the implant density per unit area, and is calculated from

$$D_I = \int_{-\infty}^{\infty} N(x)\,dx = \sqrt{2\pi}\,(\Delta R_p)\,N_p \tag{2. 8}$$

This allows us to write the implant profile as

$$N_{ion}(x) = \frac{D_I}{\sqrt{2\pi}\,(\Delta R_p)} \exp\left[-\left(\frac{x-R_p}{2(\Delta R_p)}\right)^2\right] \tag{2. 9}$$

which shows the general relationship among the parameters. Annealing induces diffusive motion of the dopants and has two main effects:

- The effective value of the straggle ΔR_p is increased, and
- The peak value of the implant is decreased.

A **RTA** (rapid thermal anneal) process is usually employed to minimize these effects.

In a practical setup, the dose D_I of the implant can be measured by noting that the ion beam consists of moving charges that gives rise to a measurable electrical current i_{ion}. Assuming that the current is a constant in time, the total ion charge Q_{tot} imparted to the wafer during an implantation time t_{tot} is

$$Q_{ion} = \int_0^{t_{ion}} i_{ion}\,dt = i_{ion} t_{ion}\,. \tag{2. 10}$$

The dose is then related to the total ion charge by

$$D_I = \frac{Q_{ion}}{sqA_{ion}} \quad (2.11)$$

where s is the charge state (1 or 2) of the ion, q is the basic charge unit, and A_{ion} is the area of the implanted region.

Realistic ion implant profiles are much more complicated than can be modelled by the simple Gaussian form. The distribution is not symmetric around R_p, and there is a substantial difference in the point-by-point shape. Pearson distributions provide a much more accurate description at the expense of increased complexity. In this approach, we introduce a normalized distribution function $f(x)$ that satisfies

$$\int_{-\infty}^{\infty} f(x)\, dx = 1 \quad (2.12)$$

and has a gradient defined by

$$\frac{df}{d\xi} = \frac{(\xi - a)\, f(\xi)}{b_0 + b_1\xi + b_2\xi^2} \quad (2.13)$$

where $\xi = x - R_p$, and b_0, b_1, and b_2 are constants determined by details of the specific implant being studied. Various implant parameters are related to the distribution function by moments. The first moment gives the projected range

$$R_p = \int_{-\infty}^{\infty} xf(x)\, dx \quad (2.14)$$

while the second moment is the square of the straggle:

$$\Delta R_p^{\ 2} = \int_{-\infty}^{\infty} (x - R_p)^2 f(x)\, dx \quad (2.15)$$

The third moment defines the skewness

$$\gamma = \frac{1}{\Delta R_p^{\ 3}} \int_{-\infty}^{\infty} (x - R_p)^3 f(x)\, dx \quad (2.16)$$

while the fourth moment is the kurtosis

$$\beta = \frac{1}{\Delta R_p^{\ 4}} \int_{-\infty}^{\infty} (x - R_p)^4 f(x)\, dx \quad (2.17)$$

with γ and β used to specify the details of the shape of the curve itself. The four constants in the gradient definition are related to the moments by

$$a = b_1 = -\frac{\gamma\left(\Delta R_p^{\ 2}\right)(\beta+3)}{A}$$

$$b_0 = \frac{\left(\Delta R_p^{\ 4}\right)(3\gamma^2 - 4\beta)}{A} \qquad (2.\ 18)$$

$$b_2 = \frac{(6 + 3\gamma^2 - 2\beta)}{A}$$

where

$$A = 10\beta - 12\gamma^2 - 18 \qquad (2.\ 19)$$

A **Pearson Type IV** distribution is defined by the coefficient range

$$0 < \frac{b^2_{\ 1}}{4b_0 b_2} < 1 \qquad (2.\ 20)$$

and is particularly useful for describing ion implant profiles. The profile is of the form

$$N_{ion}(x) = N_p \exp\left[u(x)\right] \qquad (2.\ 21)$$

where

$$u(x) = \frac{1}{2b_2}\ln\left[b_2(x-R_p)^2 + b_1(x-R_p) + b_o\right] - \frac{(b_1/b_2) + 2a}{\sqrt{4b_2b_0 - b^2_{\ 1}}}\operatorname{atan}\left[\frac{2b_2(x-R_p) + b_1}{\sqrt{4b_2b_0 - b^2_{\ 1}}}\right] \qquad (2.\ 22)$$

Programs such as SUPREM provide for modelling of the process using this type of distribution.

2.1.4 Metal Layers

Most interconnects are created using a patterned layer of metals or metal alloys and compounds. In modern processing, 4 to 7 or more separate metal interconnect layers have become commonplace. In the early days of MOS processing, aluminum (Al) was used exclusively for FET gates and interconnects. It was the metal of choice because it was easy to evaporate and deposit on the wafer and exhibited good adhesion to the surfaces. One of the drawbacks was its relatively low melting temperature, which all but eliminated high-temperature processing steps after the first Al deposition. Although Al is still widely used in many steps of IC processing, other metals have entered the picture. Refractory metals such as platinum and tungsten are deposited on top of polysilicon to create silicides that have a low sheet resistance. Tungsten itself is used for "plugs" between metal layers. And, more recently, work by IBM has led to the use of copper as a viable interconnect material.

At the circuit design level, our interest in the choice of metals is usually directed to the values of parasitic resistance. The details of multi-level layering (along with the dielectric characteristics) give us the self and coupling capacitances that affect our sensitive system-level designs. Because of these observations, we will not go into any details on the metallization process. The interested reader is directed towards books that are deal solely with silicon processing.

2.2 Photolithography

Patterning of the material layers is achieved using the process of **photolithography**. This starts with the design of each layer using a CAD tool called a **layout editor**, which is a graphics package where different material patterns are represented by different colors on the screen. The output of the layout editor is sent to a mask-making apparatus where the information is used to create a **mask** for each layer. Physically, the mask is a plate of high quality glass called a **reticle** that has the pattern defined by chromium features. The chromium layer blocks light transmission through the reticle in selected areas. The reticle is typically around a 10× representation of the actual pattern size.

The actual transfer of the pattern to the layer is obtained in the sequence of steps. Suppose that we want to created a patterned polysilicon gate. The first step is to deposit the polysilicon uniformly over the entire surface of the wafer as shown in Figure 2.5(a). Then we coat the wafer with a thin layer of a light sensitive organic polymer (i.e., a plastic) that is called **photoresist** (or simply **resist** for short). The resist is initially in a liquid state that is applied to a spinning wafer to provide a uniform coating; the wafer is then baked in an oven to dehydrate it and leave a solid layer as shown in Figure 2.5(b); this is called a soft-bake step. The important characteristic of the resist is that it is sensitive to ultraviolet (UV) light.

Pattern transfer is achieved using the **step-and-repeat** printing approach shown in Figure 2.6. The photoresist is exposed using ultraviolet (UV) light that is passed through the reticle creating an optical image of the reticle pattern. Since regions of chromium block the passage of light, we may view the system as projecting the shadow of the mask onto the surface of the resist. High-quality optical focussing is used to provide the micron-size resolution needed in VLSI. The shadow pattern created by the reticle image determines which areas of the photoresist are exposed to the light. The reticle represents one die, so that the process is repeated until every die site on the wafer is exposed.[2]

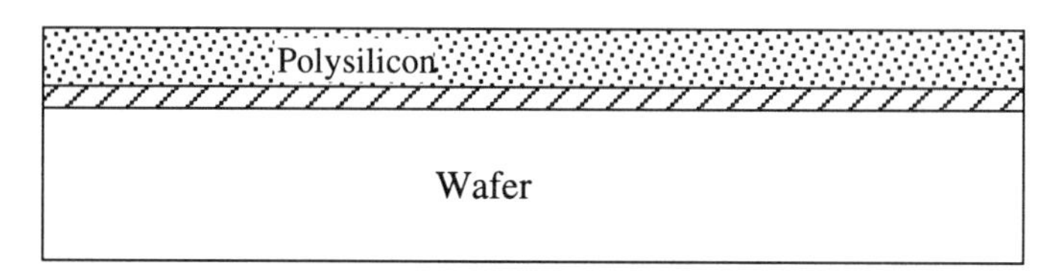

(a) After deposition of the polysilicon layer

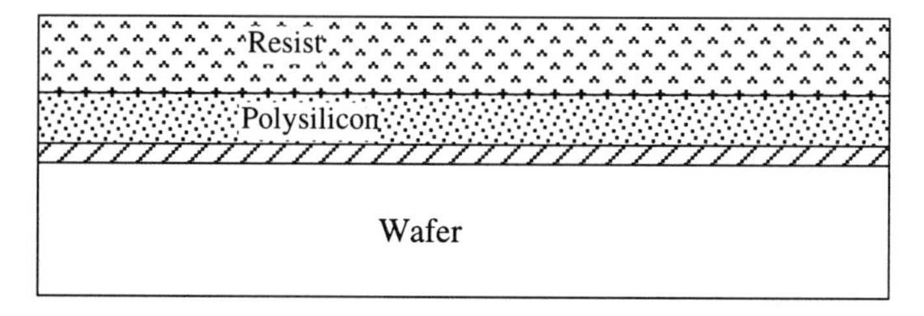

(b) Photoresist layer applied

Figure 2.5 Initial steps in the lithographic sequence

[2] The apparatus itself is called a stepper.

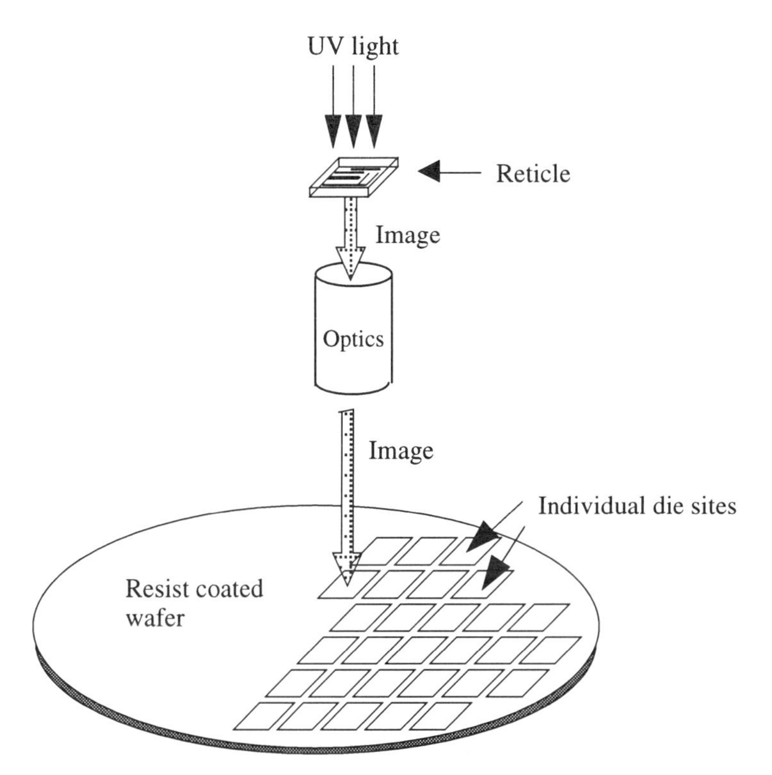

Figure 2.6 Wafer printing with a stepper

After the exposure step, the photoresist is subjected to the development step; this is similar to developing a roll of ordinary photographic film. The effect of the exposure depends upon the type of photoresist being used. If we use **positive photoresist**, then the illuminated regions become soluble when the wafer is placed in a developing solution (similar to the process used to develop everyday photographic film), while areas that were shielded from the light are hardened. A **negative photoresist** acts in the exact opposite manner, with illuminated regions becoming hardened and shielded areas being soluble during the developing process.

Regardless of the type of photoresist used, we will obtain the situation portrayed in Figure 2.7(a) in which some of the resist is hardened while the remaining regions are soluble. After development, we are left with sections of hardened resist that correspond to the desired polysilicon pattern as in Figure 2.7(b). The actual patterning of the poly layer is in the **etching** step that follows. Typically, we use a **reactive ion etch** (RIE) where the wafer is subjected to a gaseous mixture that removes bare polysilicon.[3] Regions that are protected by hardened photoresist are not affected much by the etching, so that the resist pattern determines the shape of the underlying poly layer; this is illustrated in Figure 2.7(c). Finally, the resist is removed leaving the desired result: a polysilicon layer that has the same patterning as the reticle.

Photolithography is the key to creating structures for VLSI circuits. The dimensions of a transistor and associated interconnected are determined by the resolution of the lithographic system. Every layer in an IC must be patterned, so that the lithographic sequence is repeated many times in the creation of a single circuit.

Although the circuit designer is not directly concerned with the printing process, the reticle patterns are a direct consequence of the design process. High-performance CMOS circuits require very

---

[3] RIE employs inert gases such as Ar or K that are ionized and combined with reactants to perform the etch

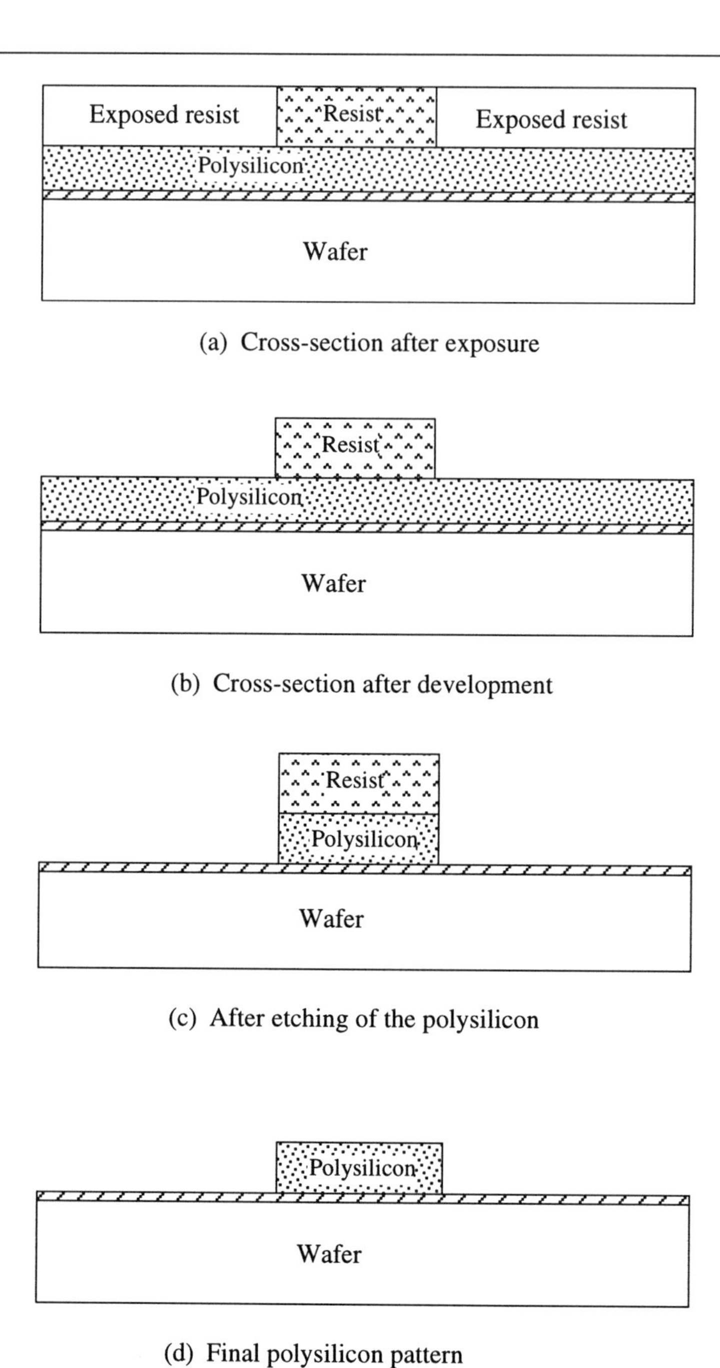

(a) Cross-section after exposure

(b) Cross-section after development

(c) After etching of the polysilicon

(d) Final polysilicon pattern

Figure 2.7 Lithographic definition of a polysilicon feature

careful device sizing and layout. Transistor characteristics, parasitic values, and electrical characteristics all vary with the geometry. Owing to this fact, a skilled CMOS designer is always aware of the physical limitations in the fabrication process.

2.3 The Self-Aligned MOSFET

The ability to create a small MOSFET is paramount to designing VLSI networks. **Self-aligned MOSFETs** provide working transistors using a minimum number of masking steps, giving both simplicity and a reliable manufacturing process.

The basic sequence for building a self-aligned n-channel MOSFET starts with a p-type silicon surface as shown in Figure 2.8(a); note that the background acceptor doping is shown as N_a. Next, a gate oxide is grown on top of the silicon to a thickness of x_{ox} [see Figure 2.8(b)]. This is the most critical step in the process, since the value of the gate oxide is crucial to the operation of the transistor. Next, a layer of polysilicon is deposited on top of the oxide [Figure 2.8(c)], and then the poly layer is patterned to define the gate region of the transistor [Figure 2.8(d)].

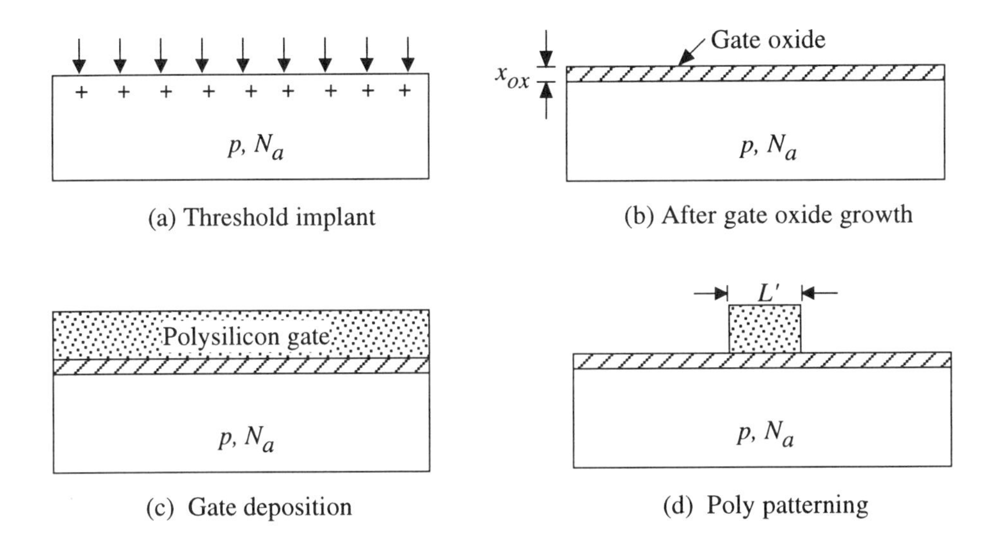

Figure 2.8 Initial steps for creating a self-aligned MOSFET

The self-aligned MOSFET gets its name from the fact that the drain and source regions are automatically aligned to the location of the gate, eliminating misalignment problems that might occur if we used a separate masking step. In the present example, the gate acts as a mask for a donor ion implant as shown in Figure 2.9(a); ions only penetrate the wafer in regions where the poly has been removed. An annealing step activates the implant, yielding the n^+ regions shown in

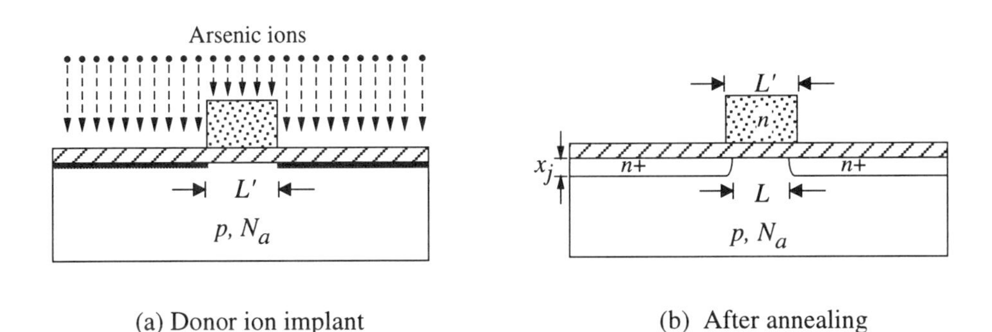

Figure 2.9 Implantation of drain and source regions

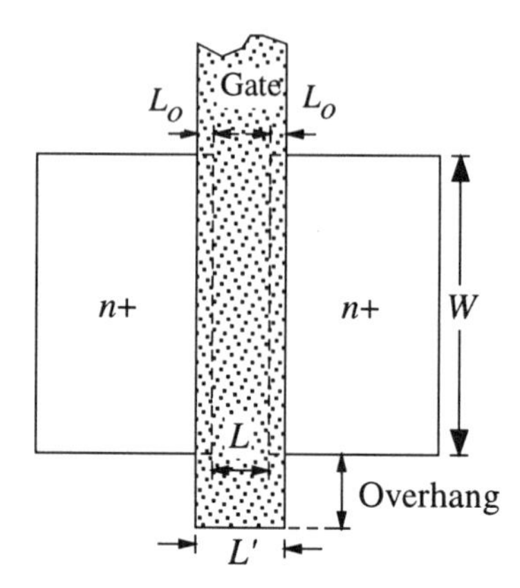

Figure 2.10 Top-view geometry of the self-aligned FET

(b). Note that lateral diffusion effects give an electrical channel length L that is smaller than the "drawn" channel length L' defined by the patterning of the polysilicon layer. The top view drawing of the FET provided in Figure 2.10 clarifies the geometry. Lateral doping effects induce the **gate overlap** distance L_o; the value of overlap is the same on both sides of the transistor. We may thus write that

$$L' = L + 2L_o \tag{2. 23}$$

so that the electrical channel length is given by

$$L = L' - 2L_o . \tag{2. 24}$$

This is the value that must be used in the aspect ratio and equations of current flow, since it represents the length of the path that electrons must traverse between the source and drain regions. We note that L is sometimes referred to as the **effective channel length** L_{eff}, and the overlap is related to the **lateral diffusion length** $L_D = L_o$; for example, SPICE uses this terminology. The distinction between the drawn and electrical length can be accounted for in these programs by allowing the user to input the drawn length, and including the value of L_o in the modelling data. Regardless of the convention and terminology used, it is always important to distinguish between the two channel length values.

2.3.1 The LDD MOSFET

Lightly-doped drain (LDD) FETs are used to decrease the channel electric field and reduce hot electron effects as discussed in Chapter 1. To understand the idea, consider the electric field E at the drain side of a MOSFET. This can be written as the sum of two contributions in the form

$$E = E_{app} + E_{pn} \tag{2. 25}$$

where E_{app} is from the applied voltages, and E_{pn} is due to the built-in junction field. If we model the pn junction as having a step profile for simplicity, then the analysis shows that the maximum electric field occurs at the junction and is given by

$$E_{max} = \frac{qN_d x_n}{\varepsilon_{Si}} = \frac{qN_a x_p}{\varepsilon_{Si}} \tag{2. 26}$$

where N_a and N_d are the acceptor and donor doping densities, respectively, and x_n and x_p are the respective depletion widths. The most important observation is that the built-in field is proportional to the doping density. The LDD structure is designed to decrease the pn junction field by reducing the doping there.

A typical fabrication sequence for an LDD MOSFET is shown in Figure 2.11. This particular approach has the advantage that it allows the LDD structure to be achieved without adding another masking step. The starting point in Figure 2.11(a) corresponds to the standard self-aligned sequence directly after the poly gate has been patterned and the light n-implant has been performed. The implant dose is reduced from the normal level, and leaves the n^- regions shown. These constitute the "lightly-doped drain" sections. Figure 2.11(b) depicts the structure after a poly oxidation step that results in an oxide layer all around the gate. An etchback of the poly oxide leaves the sidewall oxide as shown in Figure 2.11(c); the oxide sidewall act as spacers for a high dose implant, which creates the low-resistance n^+ regions. Since LDD structures can be created without an additional masking step, their presence is often transparent to the designer.

2.4 Isolation and Wells

A CMOS integrated circuit consists of MOSFETs that are wired together by conducting lines called interconnect. As discussed in Chapter 1, there are two types of MOSFETs used in CMOS: the n-channel MOSFET and the p-channel MOSFET. At the chip level, the two are electrical complements of each other with the n-type and p-type regions reversed.

There are two important items that must be examined before we progress into the details of the CMOS process flow. First, a silicon integrated circuit is fabricated on a substrate wafer that has

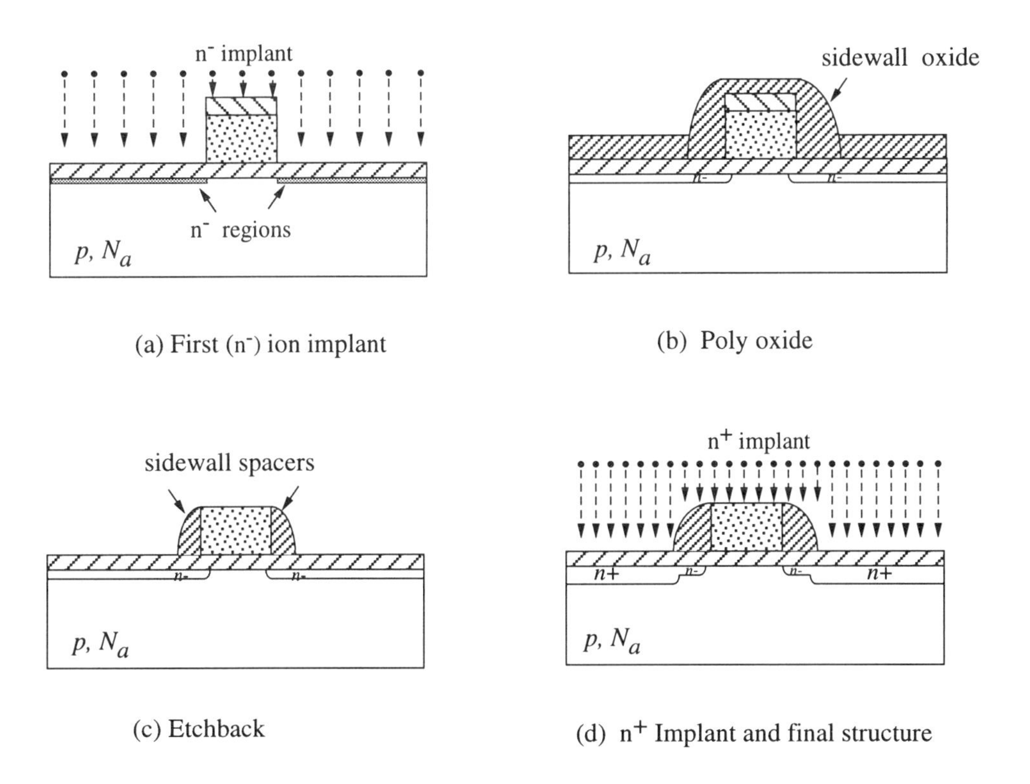

Figure 2.11 Fabrication sequence for an LDDD MOSFET

been doped to a given polarity (p-type or n-type) during the ingot growth. At the device level, nFETs require a p-type background, while pFETs are built in an n-type background. To create a complementary circuit that uses both types of transistors, we must provide an opposite polarity **well** in the process. This means that if a p-type substrate is used as the starting point, then nFETs can be fabricated directly in the substrate, but pFETs must reside in n-well regions that are added in a separate masking step.

The second important point that we need to consider is the fact that VLSI is based on the ability to achieve large transistor packing densities. For example, current commercial designs typically employ between 5 to 10 million MOSFETs or more on a single die. When the transistors are fabricated on the substrate, they must be electrically isolated from each other unless the circuit requires a connection. Isolation techniques are concerned with achieving this goal using a reasonably simple process that does not waste too much surface area.

As an example of the problem, let us examine the 3-transistor layout shown in Figure 2.12. The main idea is that we want to insure that there are no unwanted conduction paths between any two FETs. In particular, FET 1 and FET 2 are assumed to be isolated from each other, as are FET 2 and FET 3, using regions that are labelled "Isolation Oxide" which may be viewed as layers of glass that separate the individual transistors. If we want to establish an electrical connection between two devices, it must be done by means of a conducting layer. In the drawing, we have used a Metal1 layer to connect FET 1 to FET 3. Isolation is critical to the layout designer. It allows one to assume that adjacent transistors do not "talk" to each other unless an electrical connection is purposely added.

2.4.1 LOCOS

The most commonly used isolation technique is based on the **local oxidation of silicon (LOCOS)**. This creates (relatively) thick silicon dioxide (quartz glass) insulating regions that surround every MOSFET. In standard terminology, we defined the substrate-level surface sections of the die as being either **Active** or **Field** such that

$$\text{Die} = \text{Active} + \text{Field}.$$

Active areas are flat regions on the silicon where MOSFETs reside; any region that is not Active is defined as Field.

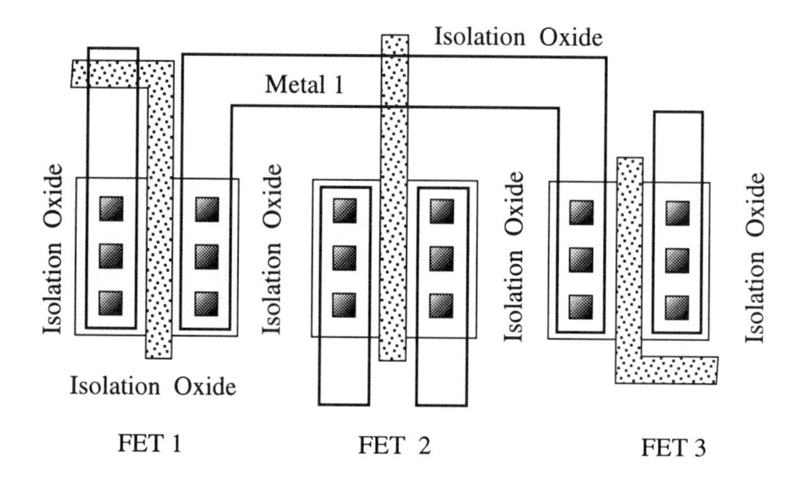

Figure 2.12 Use of isolation oxides to electrically separate FETs

Recall that the two reactions used to grow a thermal oxide layer on the silicon substrate are

$$Si + O_2 \rightarrow SiO_2$$

$$Si + 2H_2O \rightarrow SiO_2 + 2H_2 .$$

Both processes use silicon atoms from the substrate to create the SiO_2 layer with

$$x_{Si} \approx 0.46 x_{ox}$$

as derived earlier. LOCOS uses this fact to grow a recessed isolation oxide only in local field regions. LOCOS uses silicon nitride (Si_3N_4) to inhibit the growth of thermal oxides in Active areas of the die. The process is shown in Figure 2.13. The drawing in Figure 2.13(a) defines the starting point we have pre-determined the location of two adjacent nFETs. The drawing in (b) shows wafer after the initial first steps are completed. The substrate has been covered with a CVD silicon nitride layer that has been deposited on top of a thin thermal oxide layer. The underlying SiO_2 layer is called a **stress-relief oxide**, and is used to absorb the mechanical stress between the surfaces of the silicon wafer and the silicon nitride. If the stress-relief oxide is omitted, then the nitride[4] layer has a tendency to crack. An active device region is defined as a surface area that remains flat; Figure

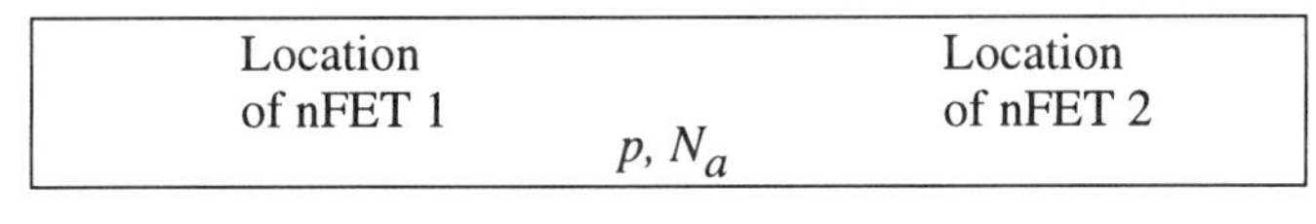

(a) Starting substrate

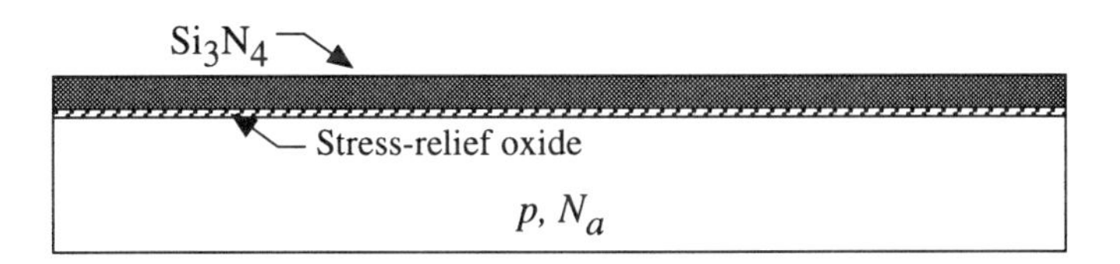

(b) Stress-relief oxide and nitride deposition

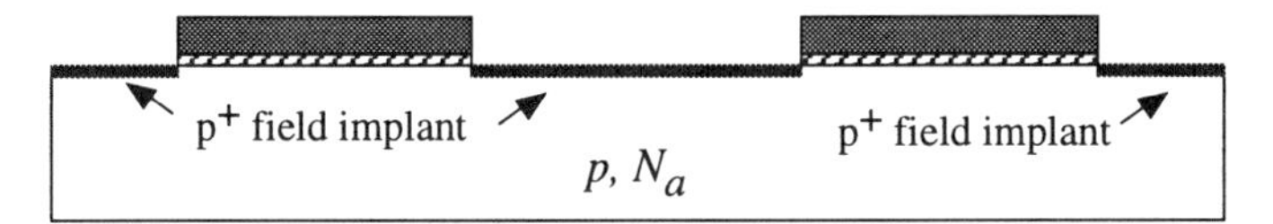

(c) Active area patterning

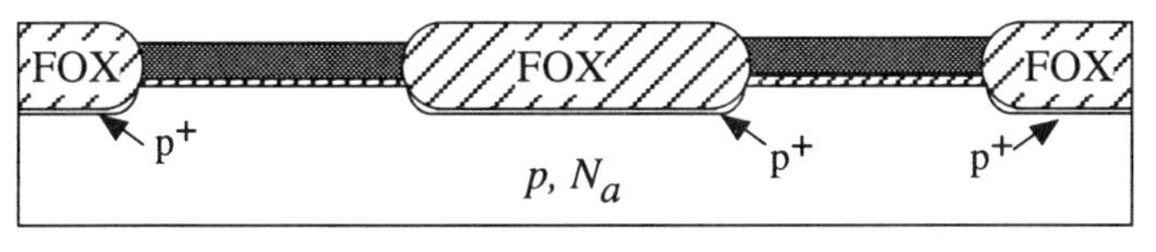

(d) Local oxidation

Figure 2.13 Steps in the LOCOS isolation process

[4] We use the term "nitride" to mean silicon nitride where there is no possibility of confusion.

2.13(c) shows the wafer after the nitride/oxide layer has been patterned such that the nitride remains in FET regions. Once the nitride has been patterned, boron ions are implanted into the exposed regions of the silicon wafer; this is called a **field implant** and is discussed in more detail below. The next step is the actual growth of the isolating **field oxide** (FOX). The surface of the wafer is subjected to an oxygen rich gas flow, producing thermal oxide in regions where the silicon is exposed, but protecting Active areas covered by the nitride. Due to the recession of the silicon surface, adjacent active areas are separated from each other by a recessed glass insulating region as shown in Figure 2.13(d). An alternate name for the field oxide is the **recessed oxide** (ROX) since it is indeed below the original surface of the silicon. The recession of the glass layer provides the desired electrical isolation between adjacent devices since it acts like a "wall of insulating glass" between them.

Growing the field oxide gives rise to the formation of an interface between the thick field oxide and the stress-relief oxide that is called the **bird's beak** region because of its shape. This is shown in Figure 2.14. Bird's beaking occurs because oxygen diffuses underneath the edges of the nitride, allowing oxide growth there. In creating this transition region, the edges of the nitride are lifted, reducing the size of the flat Active area. This phenomena is called **encroachment** (of the Active area) and reduces the integration density.

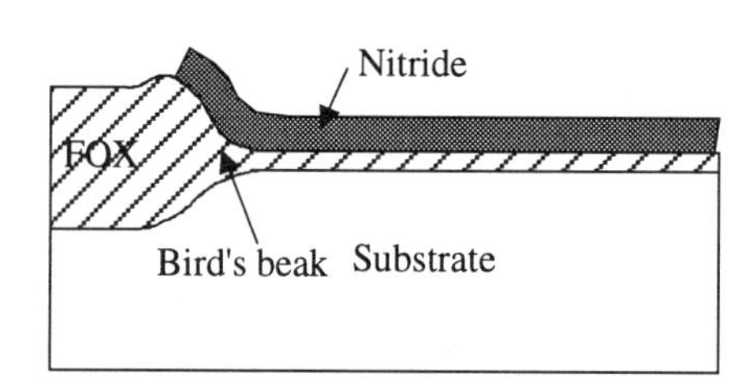

Figure 2.14 Bird's beak at nitride edge

Now let us examine the transistor isolation in more detail. Two adjacent FETs are portrayed in Figure 2.15. Since the FOX is recessed into the wafer surface, the glass acts to block direct electrical conduction between the two transistors. The drawing also shows the use of a metal interconnect line that is routed over the central field region. This creates a parasitic MOS capacitor structure that is characterized by the field oxide capacitance per unit area

$$C_{FOX} = \frac{\varepsilon_{ox}}{X_{FOX}} \tag{2.27}$$

which is much smaller than C_{ox}. The field threshold voltage V_{TF} will be larger than the FET thresh-

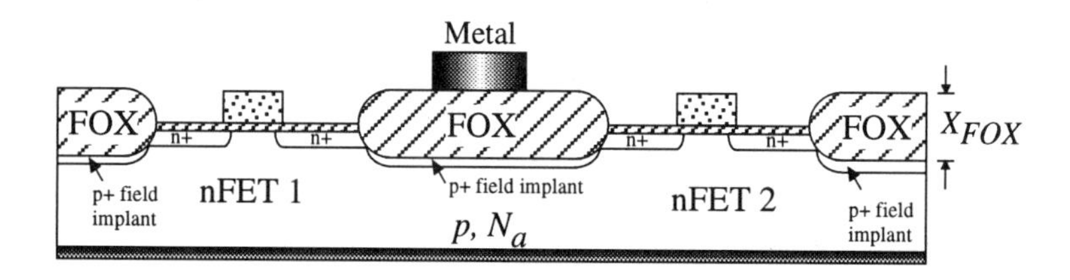

Figure 2.15 Side-view of adjacent nFETs isolated by a field oxide

old voltage V_{TOn}, but it is important to insure that the voltage on the interconnect is always less than V_{TF}. If the voltage exceeds this value, an inversion layer will form under the field oxide, and the isolation scheme fails. The field implant is used to adjust the field threshold voltage V_{TF} to a value that is much larger than normal operating voltages. It is also used for special input circuits that protect the internal circuitry from electrostatic discharge (ESD) problems; this is discussed in more detail in Chapter 10.

2.4.2 Improved LOCOS Process

Another problem with the simple process described above is that there is a difference in the height of Active and Field regions. This can be overcome by using an additional etching step as summarized in Figure 2.16. As seen in Figure 2.16(b), field regions of the silicon wafer are etched according to the patterning of the nitride layer. The depth is chosen to give a smooth surface after the field oxide is grown (or deposited, depending upon the process flow) as in (c). The final result helps maintain a flat surface as additional layers are added. CMP (chemical-mechanical polishing) techniques are very useful in this type of technology. Planarization has become increasingly important as the number of interconnect layers has increased.

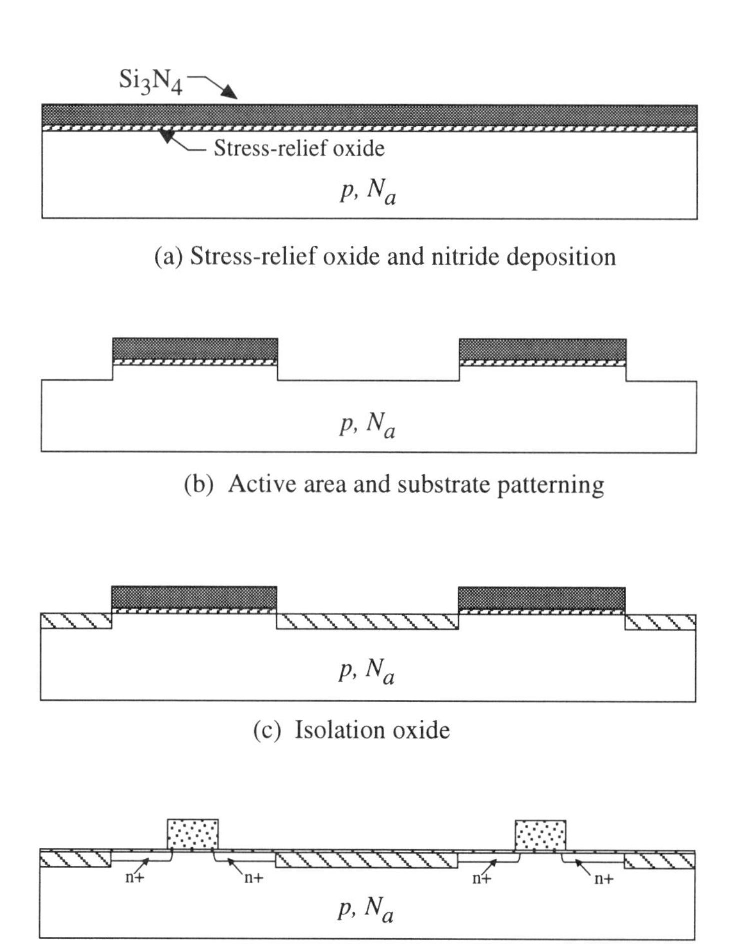

Figure 2.16 Planarized isolation technique

2.4.3 Trench Isolation

An alternate to LOCOS uses oxide-coated trenches that have been etched into the silicon wafer to isolate devices. Although the technology needed to perform trench isolation is more difficult than that used for LOCOS, the trenches require less surface area and thus allow for a higher integration density. In addition, some dynamic random-access memory (DRAM) cells use the trenches to create charge storage capacitors.

Conceptually, trench isolation is very straightforward to study. Reactive ion etching (RIE) is used to define vertical-walled trenches in the silicon substrate as shown in Figure 2.17(a). Oxygen is passed over the wafer, growing an oxide layer on the walls and bottom. Polysilicon is then used to fill-up the trenches, and a final oxide growth and addition of the FETs gives the structure shown in Figure 2.17(d). The isolation action is obvious: lateral current flows (i.e., parallel to the surface) are blocked by the insulating trenches. Overall, this yields a more planar surface that makes it easier to add above-wafer interconnect layers.

2.5 The CMOS Process Flow

Let us now follow the basic sequence that is needed to fabricate nFETs and pFETs in a p-type wafer. This is called an n-well process, since an n-region must be introduced to accommodate the pFETs. The cross-sectional views of the wafer during the steps described in the next few para-

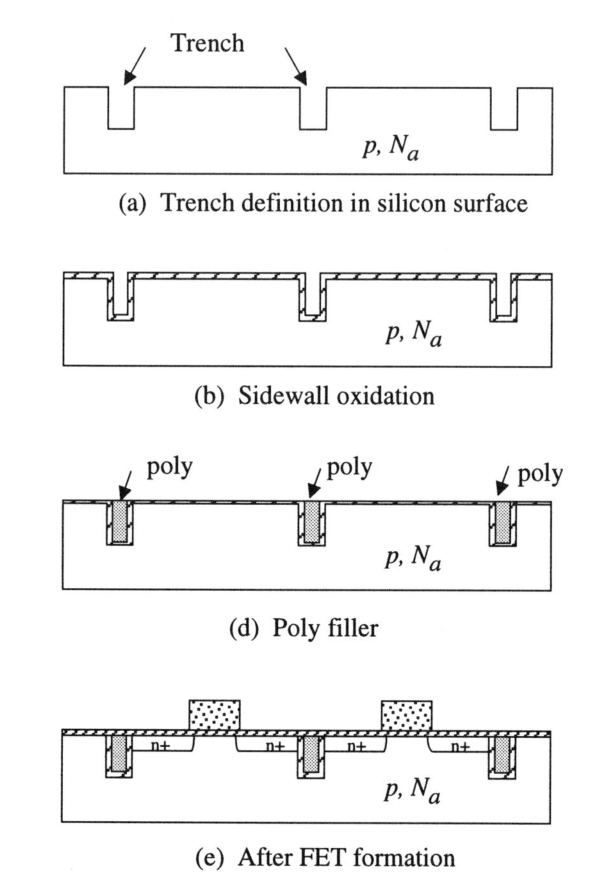

Figure 2.17 Basic steps in trench isolation

graphs are shown in the drawings of Figures 2.18 to 2.22. It is important to note that the drawings are not to scale since the thicknesses have been exaggerated for clarity.

The starting point in our example process is a heavily doped p^+ wafer that we will generally call the substrate. A thin p^- silicon epitaxial layer is grown on top of the wafer to provide a well-defined background for the transistors; this results in the cross-sectional view with the general structure depicted in Figure 2.18(a). Typically, the boron (acceptor) doping of the epitaxial layer is less than about $N_a = 10^{15}\ cm^3$, and the epi layer is only a few microns thick. In the remaining views, only the epitaxial layer will be shown; the wafer itself will be omitted in an attempt to preserve at least some of the scaling. Since nFETs have a p-type bulk, they can be created in the epitaxial layer. A pFET, on the other hand, requires an n-type bulk, so that we provide an n-well in the p-epitaxial layer for these transistors. The n-well is created using a deep ion implant that is diffused deeper into the substrate, resulting in the cross-section shown in Figure 2.18(b). Once this has been accomplished, the threshold voltage ion implant adjustments must be performed for both nFETs and pFETs. These are shown in Figures 2.18(c) and (d). Creating a positive nFET threshold voltage V_{Tn} requires a boron (p-type) with a dose D_I determined by the oxide thickness, doping levels, and other physical parameters. The working value of the pFET threshold voltage $V_{Tp} < 0$ is also established by a ion implantation step; a donor implant makes the value more negative while an acceptor implant makes the value less negative. The next step is the creation of the dielectric isolation regions. This is summarized by the steps shown in Figure 2.18(e) and (f) for a LOCOS process, and results in the characteristics discussed in the previous section.

The next group of processing steps are used to form the transistors themselves. Access to the bare silicon surface [Figure 2.19(a)] is achieved by stripping the nitride and stress-relief oxide layers. This allows the careful growth of the gate oxide layer in which x_{ox} is established as in Figure 2.19(b). The gate oxide establishes the value of the oxide capacitance per unit area

$$C_{ox} = \frac{\varepsilon_{ox}}{x_{ox}} \tag{2. 28}$$

and is considered one of the most critical steps in the CMOS process flow. The next step is to deposit the gate polysilicon layer, which is then patterned by lithography according to the location of the transistor gates; this results in the structure shown in Figure 2.19(c).

The transistors themselves are formed by ion implants using the self-aligned scheme. pFETs are created using a p-type boron implant in which nFET locations are blocked with photoresist; the resulting cross-section is portrayed in Figure 2.19(d). Similarly, nFETs require an n-type implant for drain and source regions. To accomplish this, pFET locations are blocked with resist while the n-implant is performed, leaving the structure shown in Figure 2.19(e). At this point, both FET polarities are established.

Figure 2.20 illustrates the "above-wafer" steps in the processing sequence. In (a), the surface has been coated with a CVD oxide that acts to electrically insulate the device from overlaying conductors. In advanced processes, this is followed by steps that planarize the surface for the next material layer. Contact cuts are etched into the oxide where needed, and then filled with a metal "plug" such as tungsten [Figure 2.20(b)] . The first layer of metal interconnect, denoted generically as "Metal1" is deposited and patterned as implied by the view in Figure 2.20(c).This is repeated for each subsequent interconnect level. Figure 2.21 illustrates the cross-sectional view after the deposition of the second metal (denoted as metal2 in the drawing); in this case, the contacts are called "vias."

State-of-the-art CMOS processing has become quite complex, with many interconnect layers used to ease the layout and signal distribution problems. Figure 2.22 shows the layers in a hypothetical 4-metal process.The numerical values indicate thicknesses in units of microns, and are typical of those found in a modern process. The materials themselves vary with the manufacturer. Various compounds are used at interface layers, and the aluminum is often the material of choice for the

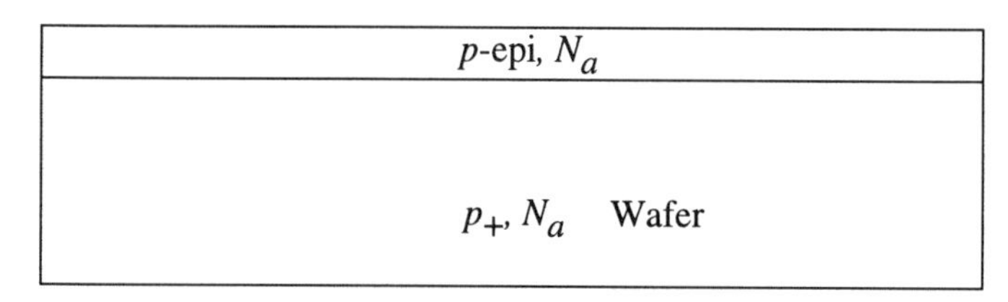

(a) Epitaxial layer growth

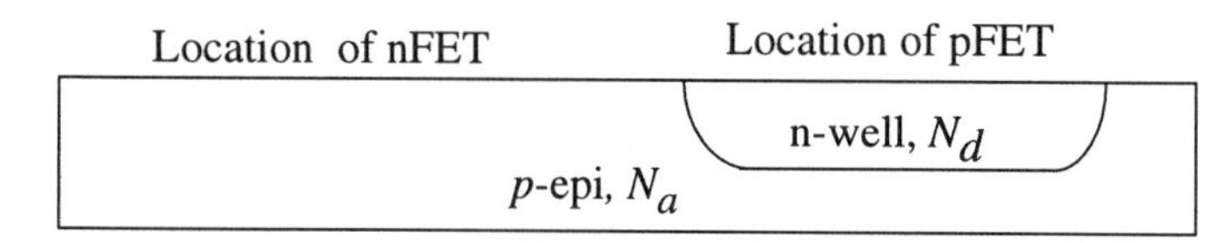

(b) After n-well creation

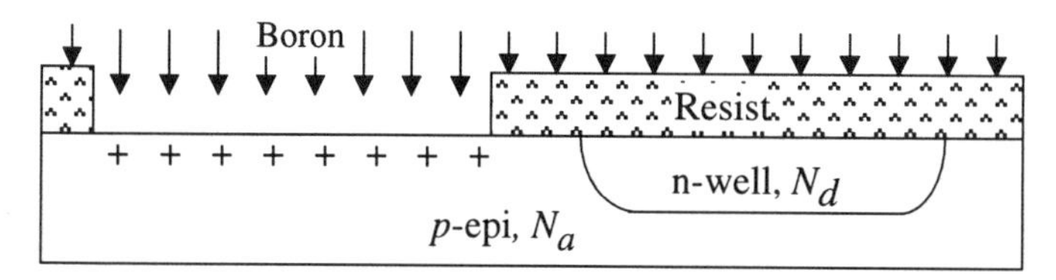

(c) nFET threshold implant

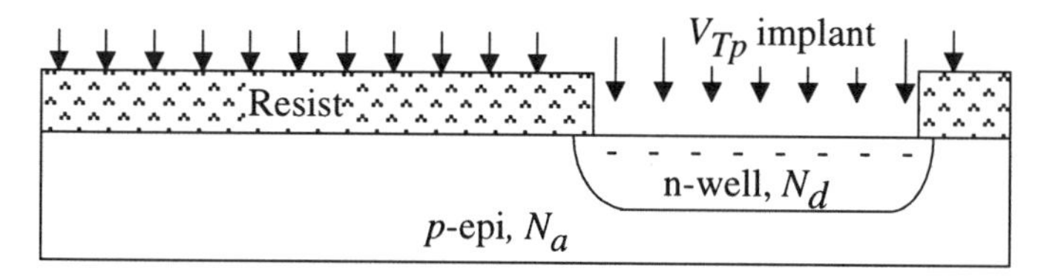

(d) pFET threshold implant

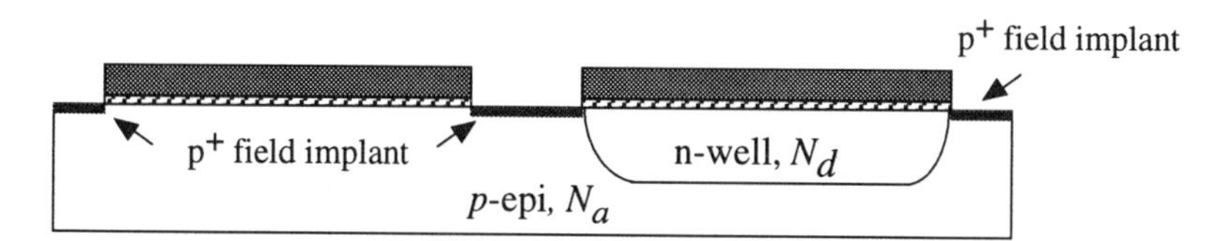

(e) Active area definition and field implant

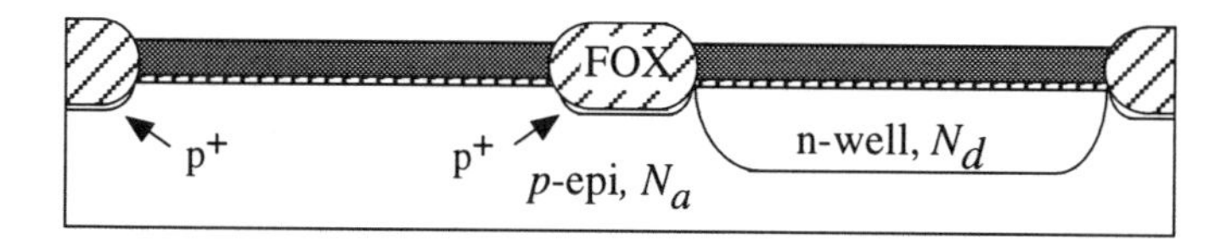

(f) Local oxidation

Figure 2.18 Initial steps in a bulk n-well CMOS process

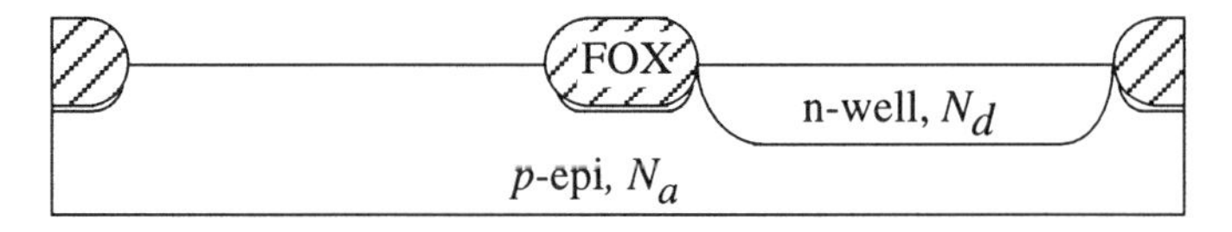

(a) Etch to bare silicon

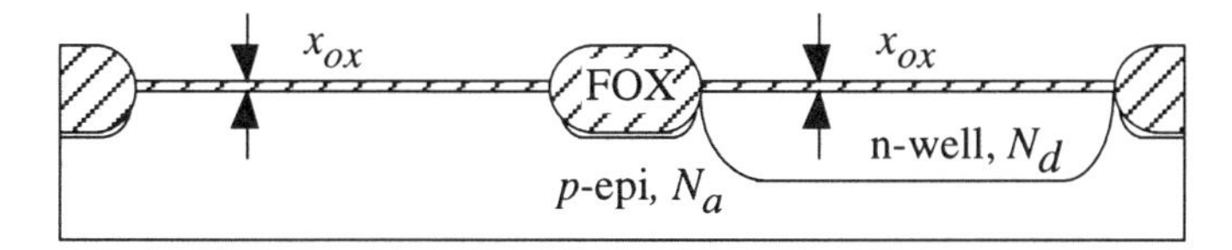

(b) Gate oxide growth

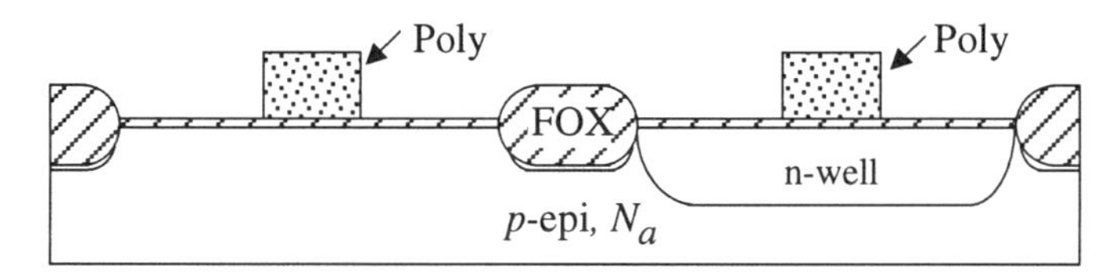

(c) After poly deposition and patterning

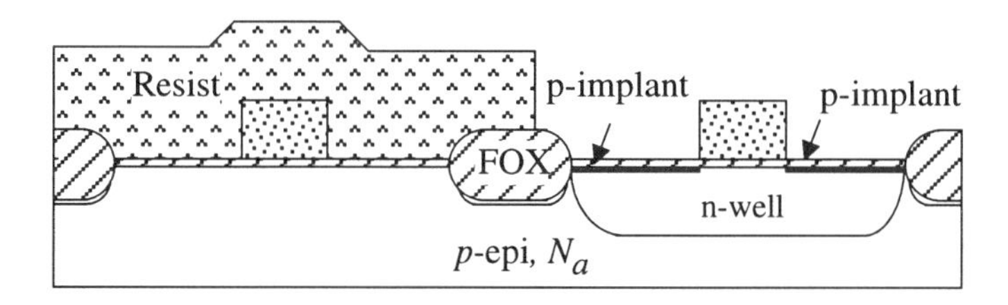

(d) pFET p^+ drain/source implant

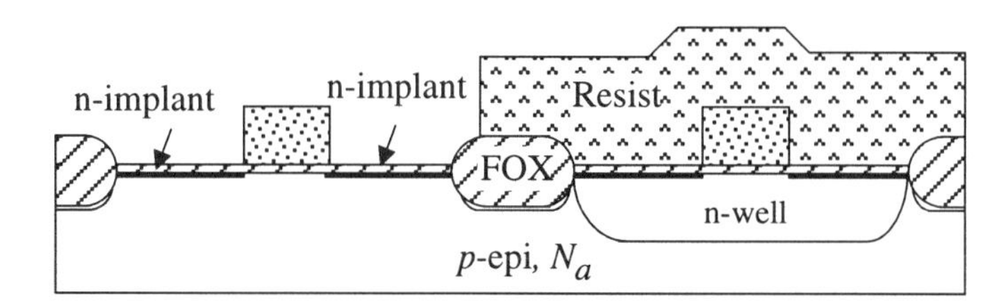

(h) nFET n^+ drain/source implant

Figure 2.19 Formation of MOSFETs in the CMOS process

high level layer(s). Although the structures vary, the circuit designer is generally not overly concerned with specifics such as the materials, since it is the electrical characteristics that are of prime importance.

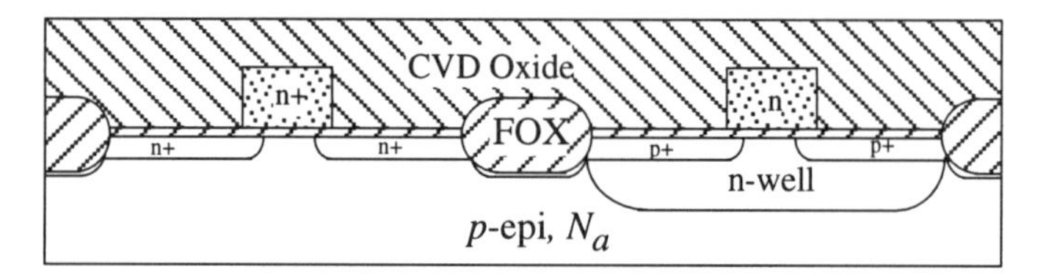

(a) CVD oxide and planarization

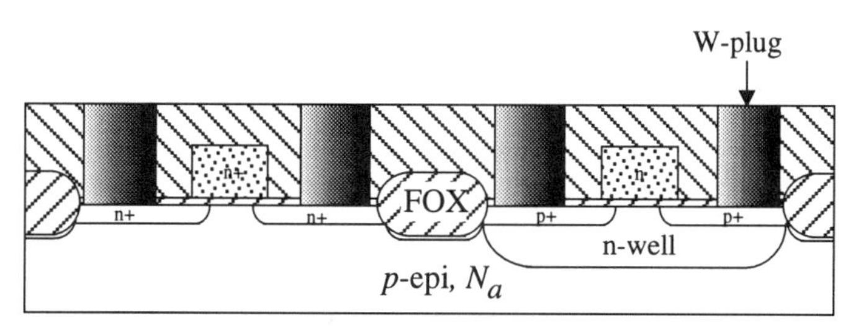

(b) Active contacts and tungsten plugs

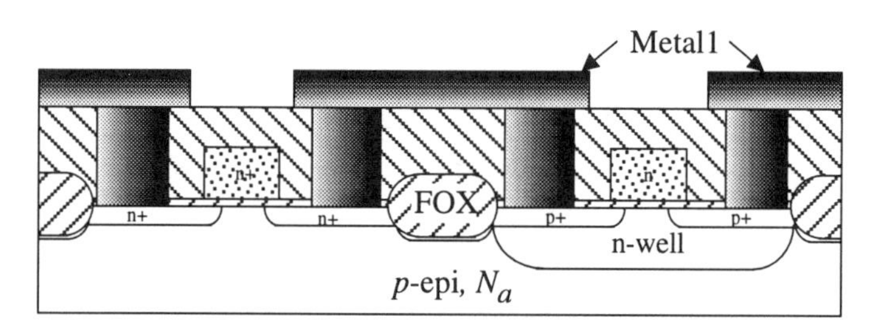

(c) Metal1 Deposition and patterning

Figure 2.20 First metallization step

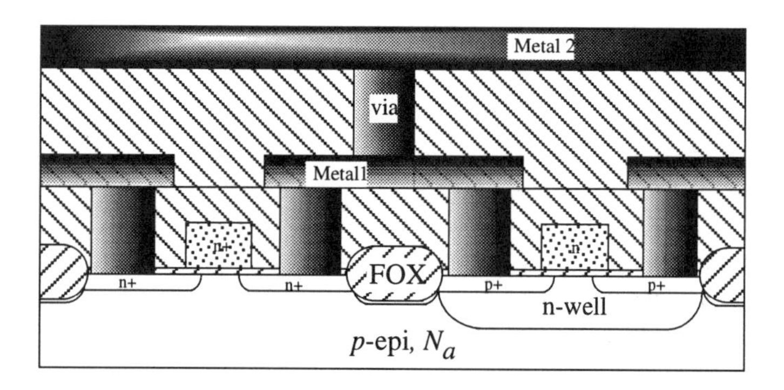

Figure 2.21 Cross-sectional view after second metal deposition

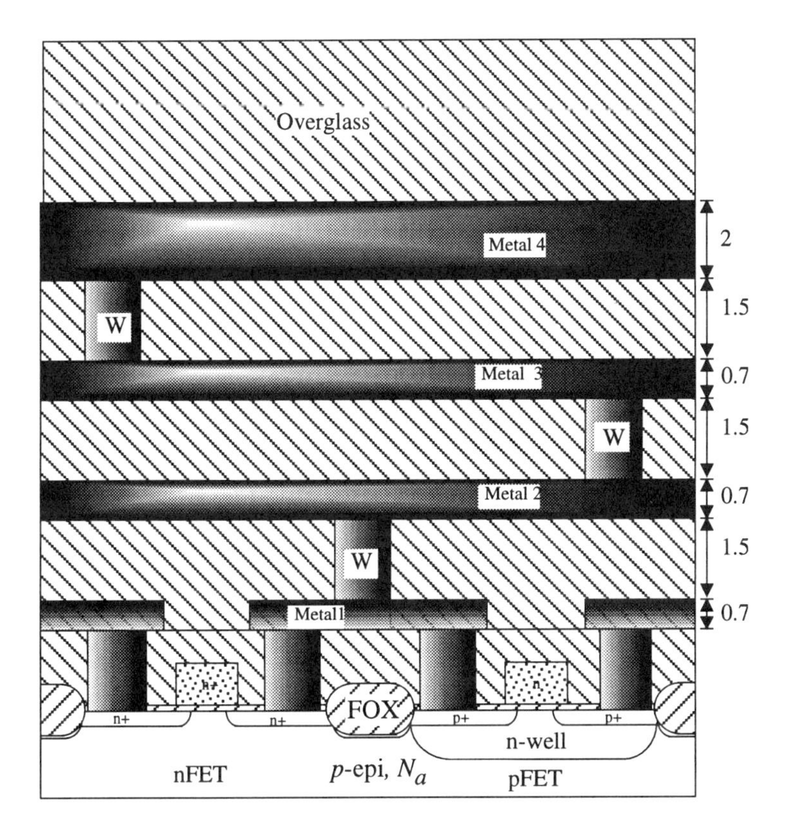

Figure 2.22 Visualization of 4-metal interconnect structure

2.5.1 Silicide Structures

Polysilicon is used for the gate material because the material has excellent coverage, adheres well to the silicon dioxide, and can be doped. Unfortunately, even heavily doped poly has a relatively high resistivity which limits it use as an interconnect. This problem is solved by using a high-temperature (refractory) metal such as titanium as a "coating" on the top, creating what is called a **silicide**.

The sequence in Figure 2.23 shows how a silicide can be created in the basic process flow. After the drain and source implants have been completed, a layer of titanium is patterned on top of the transistors; this yields titanium silicide $TiSi_2$ on both the poly gate and the drain/source regions as shown in Figure 2.23(a). Next, a CVD insulating oxide layer is applied [Figure 2.23(b)]. Contact cuts and tungsten plugs complete the sequence, and results in the structure shown in Figure 2.23(c). In this example, the tungsten plugs are used to contact the silicide drain and source regions. Silicides appear in many different forms and structures, but all have the same objective: reduce the resistance of the polysilicon line.

2.5.2 Other Bulk Technologies

In addition to n-well CMOS, two other bulk processes can be created: p-well and twin-well (also known as twin-tub). These are illustrated in Figure 2.24. The p-well process in 2.24(a) starts with an n-type bulk wafer. pFETs can be placed in the substrate, but a p-well must be added to accom-

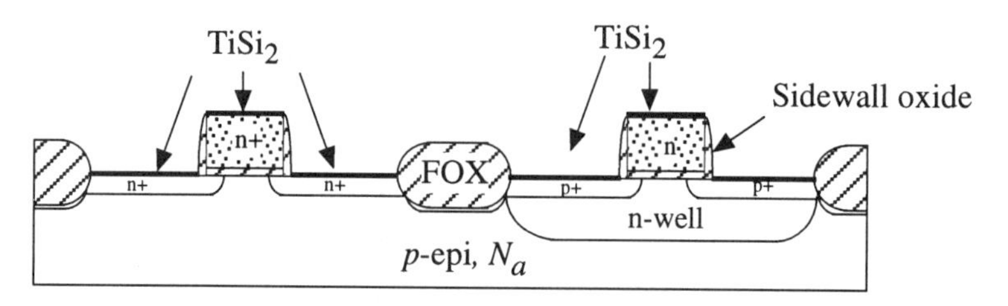

(a) Titanium deposition

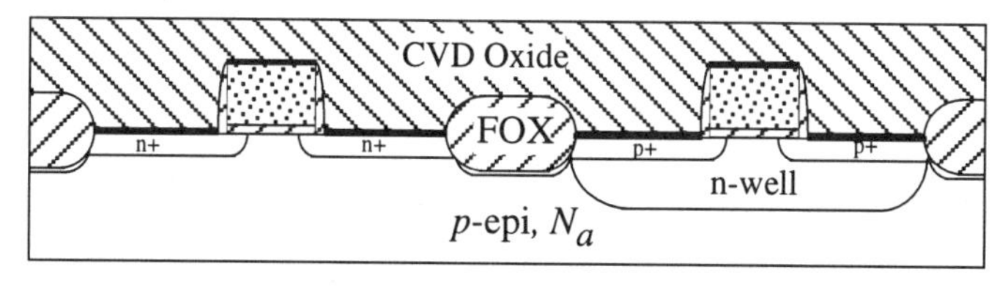

(b) CVD oxide and planarization

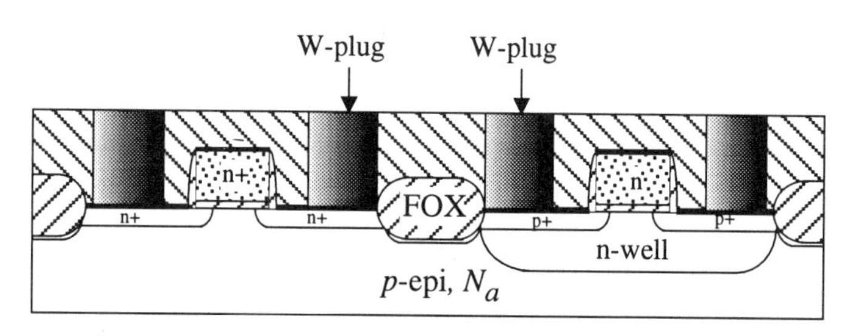

(c) Active contacts and tungsten plugs

Figure 2.23 Silicides and plug creation in a CMOS process

modate nFETs. In a twin-tub process, a high-resistivity epitaxial layer is grown on top of the wafer, and the separate n-well and p-well regions are introduced for pFETs and nFETs, respectively. This approach is shown in Figure 2.24(b). Another technology called silicon-on-insulator (SOI) has undergone many reincarnations over the past 20+ years. The most recent variation uses an oxide that has been grown on the wafer, with device regions created by selective epitaxial growth above the oxide.

Variations in the technology are done for many reasons: economics, speed, radiation-hardening, and others. For our purposes, we will continue to use the n-well process as being typical in the circuit design process. This provides a good foundation for learning to design in any technology. One must, of course, pay attention to the values of critical parameters whenever switching to a new process. Effects that are negligible in one generation may be critical in the next!

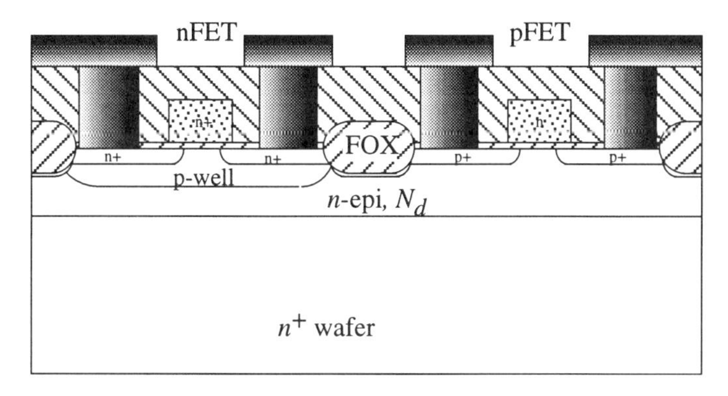

(a) P-well technology

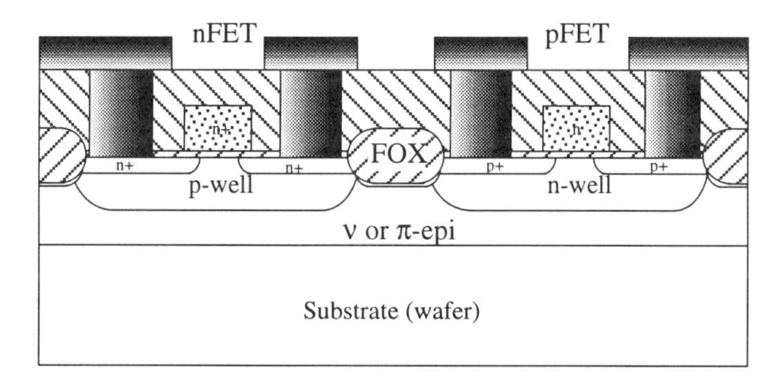

(b) Twin-well technology

Figure 2.24 Alternate bulk CMOS technologies

2.6 Mask Design and Layout

Chip design centers around two main tasks:

- Translate the necessary logic function into equivalent electronic circuits, and,
- Create fast switching networks.

As we will see in later chapters,[5] synthesizing logic operations is accomplished by proper placement and connection of the MOSFETs, i.e., the circuit topology. Switching performance, on the other hand, is more difficult to control as it depends upon the sizes of the transistors, the characteristics of the circuit connections, and the parasitic resistance and capacitance in the circuit.

Physical design deals with specifying the exact size and location of every geometrical shape on every material layer of the chip. At the design level, this is accomplished by designing every mask that is needed to fabricate the 3-dimensional structure. This is done by using a CAD drawing tool known as a **layout editor** that allows the engineer to specify the pattern of every lithographic step in the process flow. The physical design step gives important characteristics such as the transistor

[5] Circuit and logic design starts in Chapter 3, and is the subject of the remainder of the book.

packing density and the electrical transmission properties of the interconnect "wires" that are possible in a given fabrication process line.

Every patterning step in the process flow requires a separate mask. The masking steps needed in the basic n-well CMOS process that was described above are as follows.

1. **Nwell**: the n-well mask
2. **Active**: regions where FETs will be placed
3. **Poly**: the polysilicon gate pattern
4. **Pselect:** regions where the p-type ion implant will form p^+ regions
5. **Nselect**: regions where the n-type implant will form n^+ regions
6. **Poly contact:** cuts in the oxide that provide Metal1 contacts to the poly
7. **Active contact**: cuts in the oxide that provide connectrions from Metal1 to n^+ or p^+
8. **Metal1**: pattern for the first layer of metal
9. **Via**: oxide cut for Metal1 to Metal2 connections
10. **Metal2**: pattern for the second layer of metal.

More complicated processes include other layers; for example, additional metal layers are required in advanced circuit designs. However, the general comments here remain valid. It is worthwhile to point out that every conducting layer is separated from the next layer (both above and below) by an insulating oxide. The presence of oxide layers is not denoted explicitly in the mask set listing; they are, however, implied by the contact cut masks such as Poly contact and Via.

The alert reader may have noticed that the mask set listed above does not explicitly list separate masks for the nFET and pFET threshold adjustment implants. This is because the necessary masks are **derived** from others in the group. For example, a mask **Nthresh** used to define the nFET ion implant step can be constructed from the logical expression

Nthresh = (**Active**) AND (**Nselect**)

as the overlap of the two defines all nFET regions that require the implant. Similarly, the threshold implant pattern for pFETs can be obtained from

Pthresh = (**Active**) AND (**Pselect**)

using the same reasoning.

In the absence of a layout editor, it is useful to visualize the effect of each mask by means of a set of drawings that show the surface geometry after each step. In Figure 2.25(a), the first mask Nwell defines the locations of the nwells needed for pFETs. The first masking layer is also used to provide **registration marks** (or alignment marks/targets) on the wafer that are used to align several of the masks that follow. The next mask in the sequence defines the active areas as in Figure 2.25(b). The Active mask definition provides flat areas that are used for MOSFETs and substrate or nwell bias contacts. After this is completed, the gate oxide is grown and the polysilicon gate deposition takes place. The poly layer is then patterned using the Poly mask as shown in Figure 2.25(c). In most processes, the poly layer is n-doped *in situ* (meaning that it is doped while being grown) and has a refractory metal layer on top; this allows poly lines to be used as short-run interconnects.

The next masking layers in the sequence are illustrated in Figure 2.26. The drawing in (a) shows the Pselect mask (or Nblock) that defines the silicon regions that are exposed to the boron p-type ion implant. This creates p+ regions in the silicon that are used for both pFETs and low-resistance substrate contacts as shown. The p-implant is followed by the n-type ion implant that has a pattern defined by the Nselect (or Pblock) mask. Nselect gives nFET drain and source regions and n-well contact regions (for applying the power supply voltage). After the FETs are created, the wafer is covered with oxide. The next mask is the Active contact, which is used to define where the oxide cuts are made for connections to the metal1 lines. The addition of these contact cuts is shown in Figure 2.26(c). After the plug-material is deposited into the contact cuts, the first layer of metal is

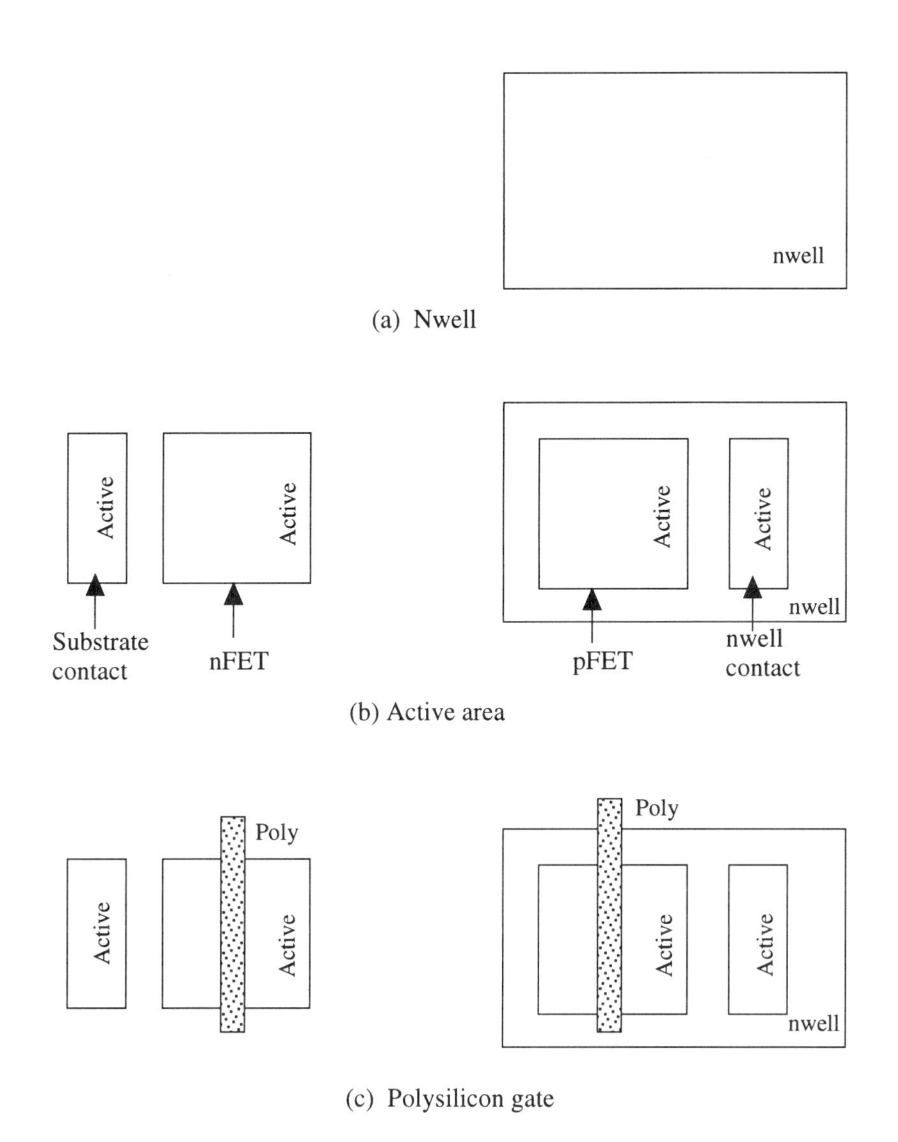

Figure 2.25 Basic masking steps used in defining FETs

deposited over the surface. The final component of the basic mask set is Metal1 which defines the patterns of the Metal1 interconnect layer. This is shown by the drawing in Figure 2.27 where the outlines of the Metal1 features are shown by heavy lines. It should be noted that Metal1 lines are used as an interconnect over the entire chip. The material can be electrically connected to Active areas (n^+ or p^+), polysilicon lines or to higher level metal lines.

The sequence defined by the drawings above only provide examples of the Active contact mask. Figure 2.28 shows the three types of contacts listed in the process flow (Active, Poly, and Via) to more clearly illustrate the characteristics. An Active contact is used as an electrical connection between the drain/source regions of a FET and the Metal1 interconnect layer. A Poly contact is used to connect a Poly line with a Metal1 line. Finally, a Via gives an electrical connection between Metal1 and Metal2 lines. The drawing also illustrates that Metal1 and Metal2 can overlap without shorting. The same statement holds for Poly to both Metal1 and Metal2.

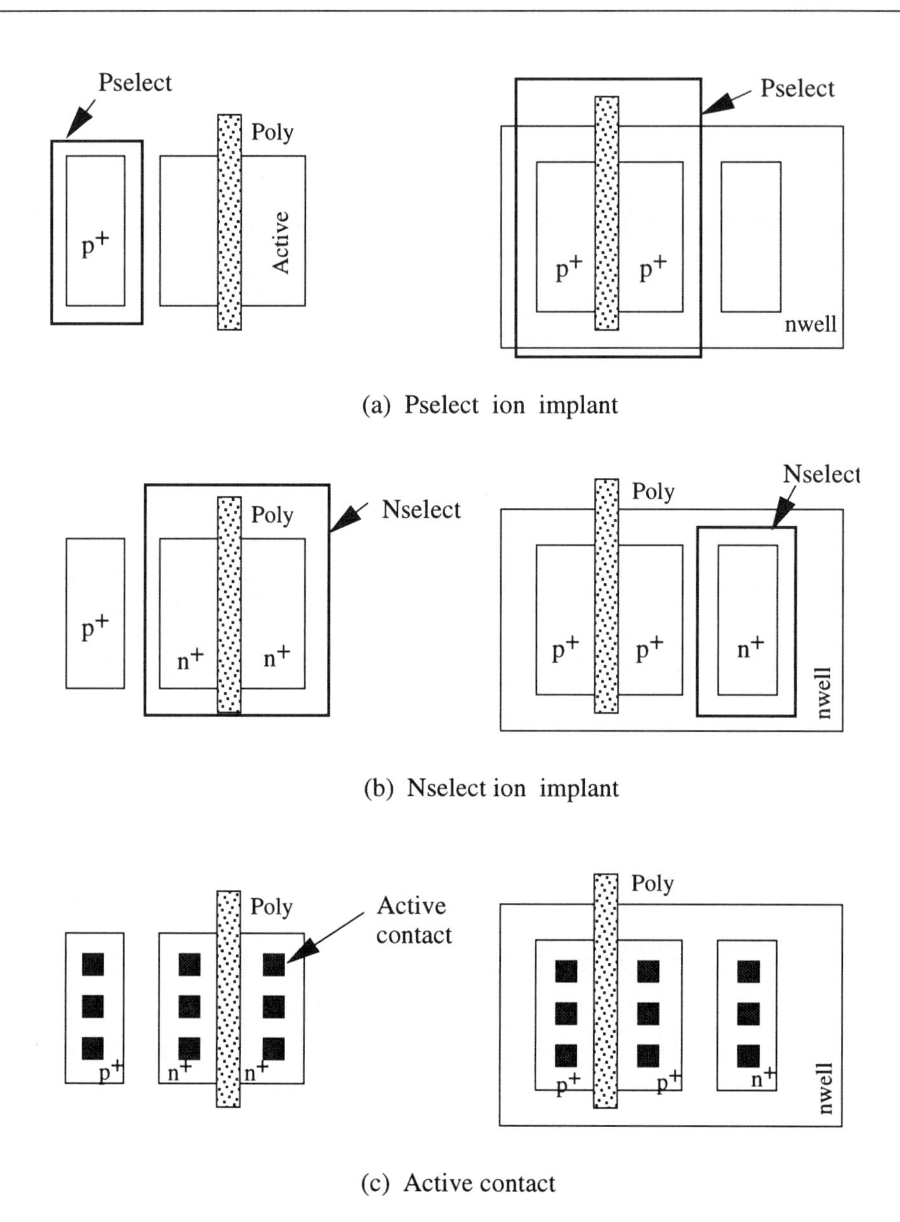

(a) Pselect ion implant

(b) Nselect ion implant

(c) Active contact

Figure 2.26 Remaining masking steps used to define FETs

2.6.1 MOSFET Dimensions

The values of the channel width W and channel length L combine to give the aspect ratio (W/L), which is (as we will see) the primary design parameter in CMOS integrated circuits. The process flow above allows us to see the relationship between the masks and the physical dimensions of the final device. These are shown in Figure 2.29. Consider first the channel length L. As discussed previously, we may write

$$L = L' - 2L_o \tag{2. 29}$$

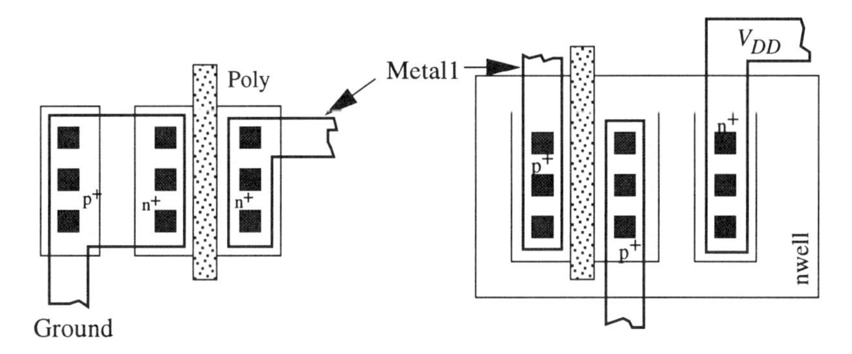

Figure 2.27 Metal1 patterning

where L' is the drawn channel length and L_o is the overlap due to lateral doping effects. In general, the drawn channel length L' is established by the minimum linewidth of the polysilicon gate, while L_o is a result of the processing recipe. We therefore conclude that the resolution of the polysilicon gate mask determines the minimum channel length of a transistor. Often one refers to the characteristics of a particular process by referring to the value of L, the electrical channel length. For example, a "0.35 micron process" usually implies that $L=0.35\mu m$, from which one might extrapolate that the poly linewidth is around $L' \approx 0.40\mu m$ for a simple FET process. Alternately, a "0.35 micron process" might specify the lithographic resolution for the gate, i.e., the drawn channel length L'.

The channel width W is set by the dimensions of the Active mask, since this defines the width of the region that will accept the Nselect or Pselect implants into the silicon. The complicating factor in LOCOS isolation is that encroachment decreases the size of the active area from the original Active mask definition. Defining the encroachment amount as ΔW, then channel width may be expressed as

$$W = W' - 2(\Delta W) \tag{2.30}$$

where W' is the drawn width of the Active mask. The aspect ratio of a transistor is thus given in terms of the drawn dimensions by

$$\left(\frac{W}{L}\right) = \frac{W' - 2(\Delta W)}{L' - 2L_o} \tag{2.31}$$

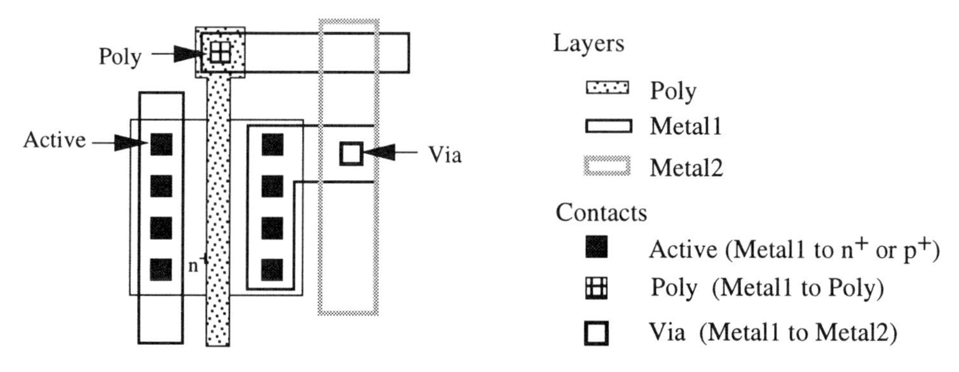

Figure 2.28 Contacts, vias, and interconnect layers

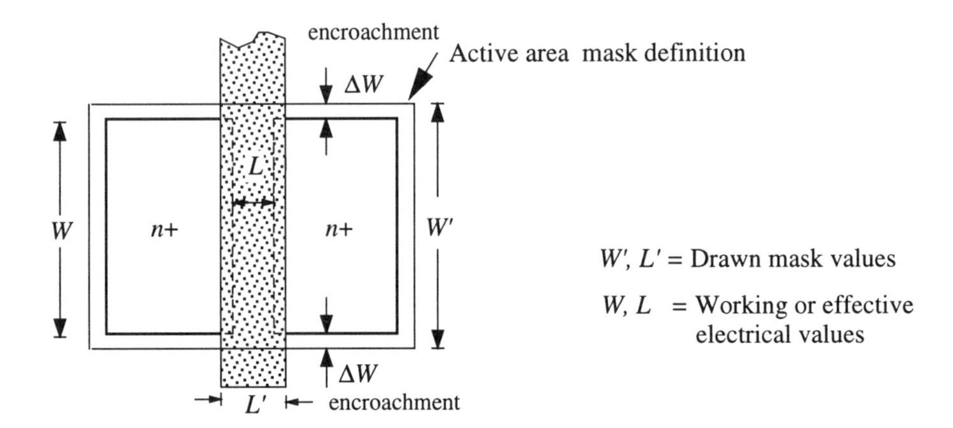

Figure 2.29 Geometrical definitions for a MOSFET

It is this value that must be used in the device equations, as it represents the electrical dimensions seen by the current flow lines.

In practice, one often uses the drawn values W' and L' as extracted from the layout as the basic input parameters into a circuit simulation program. The electrical values are then calculated in the simulation code using addition input from the processing values. This allows one to work directly with the layout without having to visualize the differences between drawn and effective (electrical) values. For example, a SPICE simulation may use a statement such as

```
Minput 20 5 0 0 NFET W=10U L=0.5U
```

where $10\mu m$ and $0.5\mu m$ are the *drawn* values. The electrical values used in the program are calculated from data listed in the model statement

```
.MODEL NFET <parameters>
```

in the parameter listing. The model itself defines the relationship between the drawn and effective values. SPICE-related literature often refers to the effective values L_{eff} and W_{eff}, which are in fact the true electrical lengths that are defined by the device geometry, not the layout masks. An example can be seen in the .MODEL example listing provided at the end of Chapter 1.

In this book, we will continue to use the notation W and L to denote the electrical values, and W' and L' as the drawn values. This convention for W and L is consistent with that used in device physics, and also keeps our equations simple by not having to add so many subscripts!

2.6.2 Design Rules

Design rules are a listing of critical geometrical size and spacing constraints that must be observed when designing a lithographic mask pattern. Each rule originates from limitations imposed by items such as the lithographic process, equipment characteristics, and physical considerations. Every process flow is characterized by a distinct set of design rules that are derived from the relevant limitations that arise in the manufacturing equipment, physical properties of the materials, or critical circuit parameters. Failure to adhere to every rule can mean that the mask set will not result in a functional chip.

2.6.3 Types of Design Rules

Although a design rule set may be quite long and involved, most rules can be grouped into a few major classes. These are summarized below.

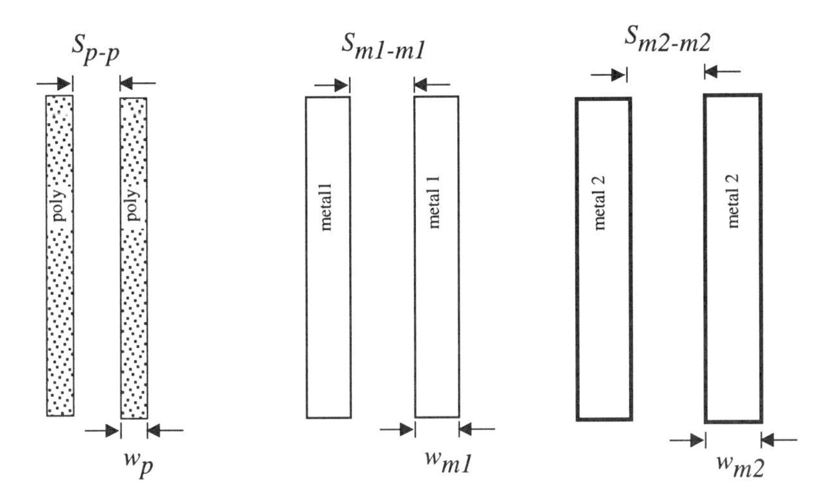

Figure 2.30 Linewidth and spacing rules

Minimum Width and Spacing

The first group deals with the minimum width for defining a line on a layer, and the minimum spacing to an adjacent line on the same layer. Figure 2.30 shows these values for lines created on layers of polysilicon, metal 1, and metal 2. Minimum width specifications are shown as w_p, w_{m1}, and w_{m2} respectively. Minimum spacing distances are denoted by $S_{p\text{-}p}$, $S_{m1\text{-}m1}$, and $S_{m2\text{-}m2}$ in the same order. Similar specifications apply to every masking layer in the process, including nwell, active, nselect, pselect, and so on.

Exact Size and Surround

An exact size rule dictates the dimensions of a particular object on the mask. In CMOS processing, exact size rules arise in specifying the dimensions of oxide cuts that are used to provide access between two conducting regions. These are usually square, or close to square, and are due to considerations in the equipment characteristics. An example is shown in Figure 2.31(a) where the contact is specified to have dimensions x_c by y_c. The spacing $S_{c\text{-}c}$ is also indicated in the drawing.

A surround rule is required when a feature is to be placed in a region that has already been pat-

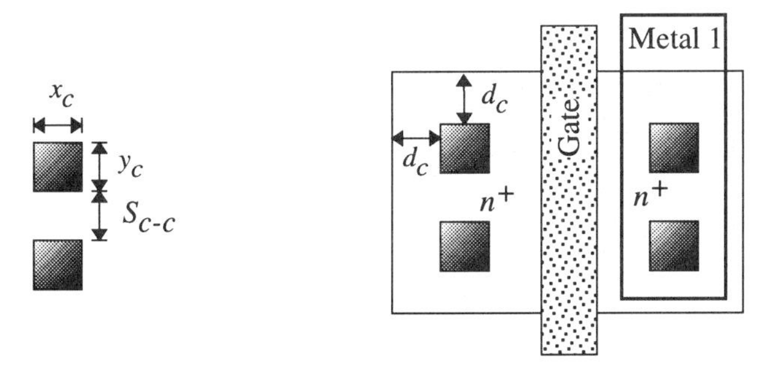

(a) Exact size rule (b) Surround rule example

Figure 2.31 Exact size and surround rule examples

terned by a previous masking step. Consider the MOSFET shown in Figure 2.31(b). The shape of the n^+ region is determined by the ACTIVE and NSELECT masks; the active contact must reside within the borders of the region. The surround spacing d_c is used to compensate for small registration errors in the alignment during the exposure and insure a working contact.

MOSFET Rules

Self-aligned MOSFETs require an additional set of rules to insure that the devices will operate. These are generally required to compensate for any misalignment between a mask and the features already on the die.

Gate overhang is shown in Figure 2.32 as D_{GO}, and shows the distance that the gate must extend beyond the Active area. Recall that the self-aligned MOSFET uses the gate as a mask to the n^+ or p^+ ion implant. The overhang distance insures that the doped regions formed by the implant are physically separated even if the poly gate mask is not perfectly aligned to the existing active area region. An extension distance D_E must be used when the active border changes as seen in the upper left side of the device. This is included for the same reason as D_{GO}, namely, to allow for registration errors (in the horizontal direction) of the poly gate mask.

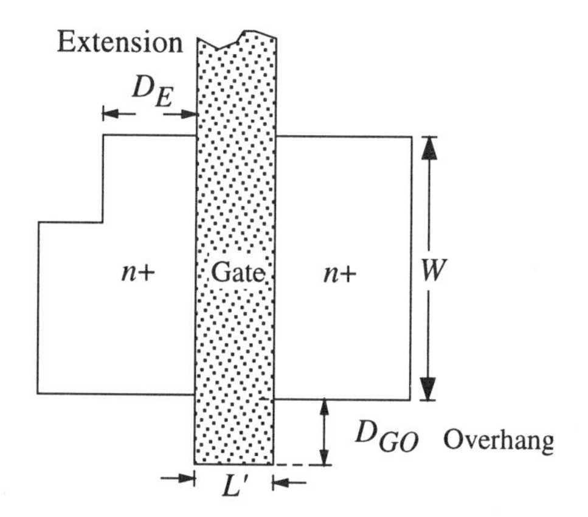

Figure 2.32 Two basic MOSFET design rules

Active Contacts

One important point that arises from the exact size specification of active contacts is the use of multiple contacts between metal and n^+ or p^+ silicon regions. This is particularly important in MOSFETs for the reduction of parasitic resistance and the proper operation of the device.

Consider the FET layout shown in Figure 2.33(a). Both the source and drain regions use multiple active contacts with the design rules specifying the size and spacing of the contacts. The cross-sectional view along the line *X-X'* in Figure 2.33(b) shows the details of the connections. One reason for using many contacts is to reduce the effective **contact resistance** between the metal and the semiconductor. Each contact point is characterized by a resistance R_c. If we use *m*-contacts, then the individual contributions are all in parallel as shown in Figure 2.33(c). This reduces the effective value to

$$R_{c,\,eff} = \frac{R_c}{m} \tag{2. 32}$$

which will in turn help the circuit to switch faster.

The second reason that multiple contacts are used is to insure that the current flow between the

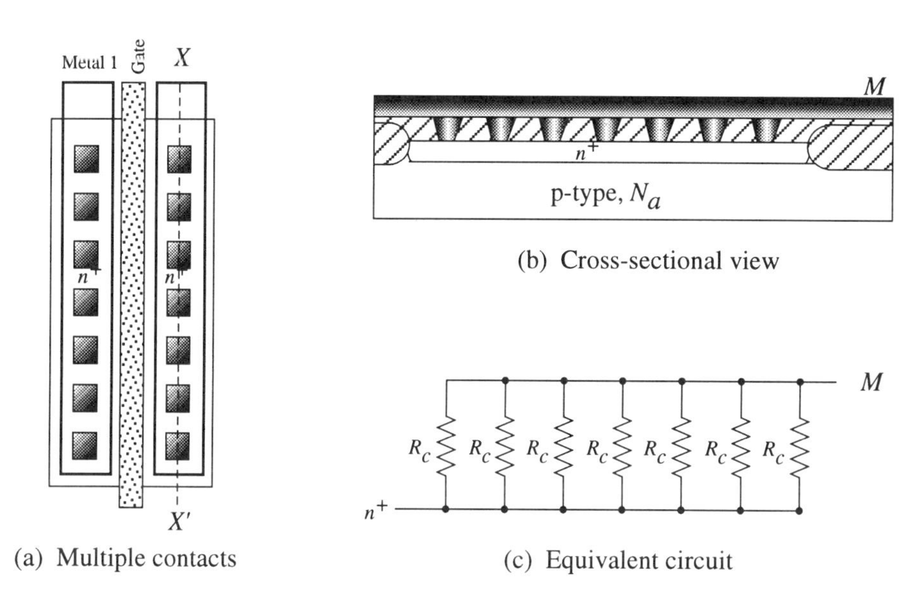

(a) Multiple contacts

(b) Cross-sectional view

(c) Equivalent circuit

Figure 2.33 Use of multiple FET contacts

drain and source is spread out over the entire electrical width W of the FET. This is portrayed in the two drawings of Figure 2.34. In figure (a), only one contact is used for each side of the device. From basic electrostatics, the current flow is along electric field lines that must satisfy all boundary conditions. Since the field lines originate and terminate at the contacts, the flow is concentrated in the region directly between the two contacts. This implies that the current flow density is lower as we move away from the location of the contacts. In terms of MOSFETs, this says that the electrical channel width will be smaller than the geometrical channel width W. This, of course, is undesirable as it negates the design implied by the layout. The use of multiple contacts as shown in drawing (b) overcomes this problem by spreading out the current flow lines over the entire extent of the device. These arguments illustrate that multiple contacts are necessary whenever the design rule set specifies exact sizes for oxide cuts that access active regions.

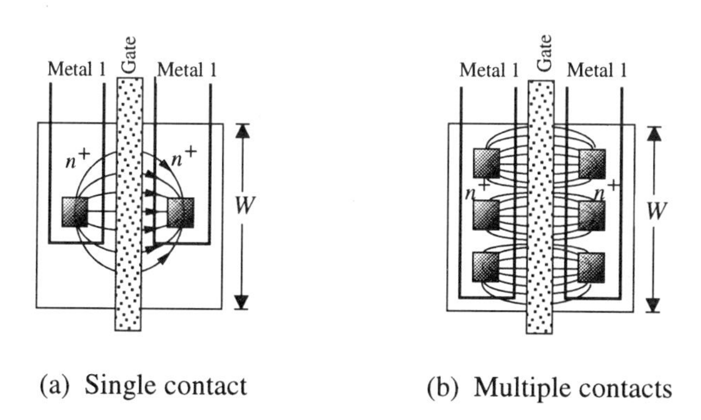

(a) Single contact

(b) Multiple contacts

Figure 2.34 Current flow paths between contacts

2.6.4 General Comments

Design rule listings are process-specific. Numerical values for every required mask dimension are derived by considering details such as the capabilities of the lithographic imaging equipment, and physical effects such as pn junction depletion widths and coupling parasitics. Since they are concerned with minimum sizes, the design rule set acts as the limiting factor for circuit integration. A critical observation is the fact that the size of a MOSFET is almost negligible compared to the area consumed by interconnect lines and other wiring. The means that the area of a CMOS layout tends to be limited by the interconnect itself.

CMOS design is directly related to layout, but in modern circuit engineering, the need to master layout varies with the scope of the position. Some chip designers "push polygons" on a daily basis, while others let layout technicians provide amazingly compact solutions that can be analyzed and simulated. The philosophy as to whether a chip designer should or should not perform the layout seems to vary with the company, or even depend upon the "culture" of a particular group within a company. If you are a student and just learning the subject, it is generally accepted that "the more you know, the better off you are."[6]

2.7 Latch-Up

The structure of a bulk CMOS process introduces a problem known as latch-up in which the circuits fail to operate and the chip draws excessive power supply current. In practice, this may arise in two different situations:

- The chip is operating normally, and then goes into the latch-up state. The only way to restore normal conditions is to disconnect the power supply, and then reapply. The chip may go into latch-up again.
- The chip goes into latch-up immediately upon application of the power supply.

In the worst case scenario, the chip will be destroyed by heat. In the early days of CMOS development, latch-up was a major problem that slowed the growth of the technology. Although the factors that induce the condition are now understood, there are times when it can still be a problem.

The origins of latch-up can be shown using the drawing in Figure 2.35(a). This identifies four distinct layers and the current flow path from the power supply voltage (V_{DD}) to ground associated with the latch-up condition. Under normal operating conditions, current along this path would only consist of leakage components. However, the nature of the layering scheme gives rise to parasitic bipolar transistors as shown in Figure 2.35(b). As discussed below, the bipolar transistors form a feedback loop that may induce latch-up.

The left drawing in Figure 2.36 shows how the layers 1 through 4 in the CMOS structure can be viewed as creating a 4-layer device with the pnpn layering. In power electronics, this is called a **silicon-controlled rectifier** (SCR) where it is used as a switching device between the top (p) and bottom (n) regions. From the qualitative viewpoint, the 4-layer structure blocks current flow from the power supply to ground due to the presence of reverse-biased pn junctions in the path. However, if one of the internal regions (n-well or p-epi) can be electrically shorted, then we would be left with a forward biased pn junction from the top to the bottom. This would give I_{DD} exponential characteristics of the form

$$I_{SS} \approx I_S \exp\left(\frac{V_{DD}}{kT/q}\right) \tag{2. 33}$$

indicating a large current flow. To describe this type of behavior, we will construct an equivalent

[6] So you should sign up for the next VLSI systems course that is offered!

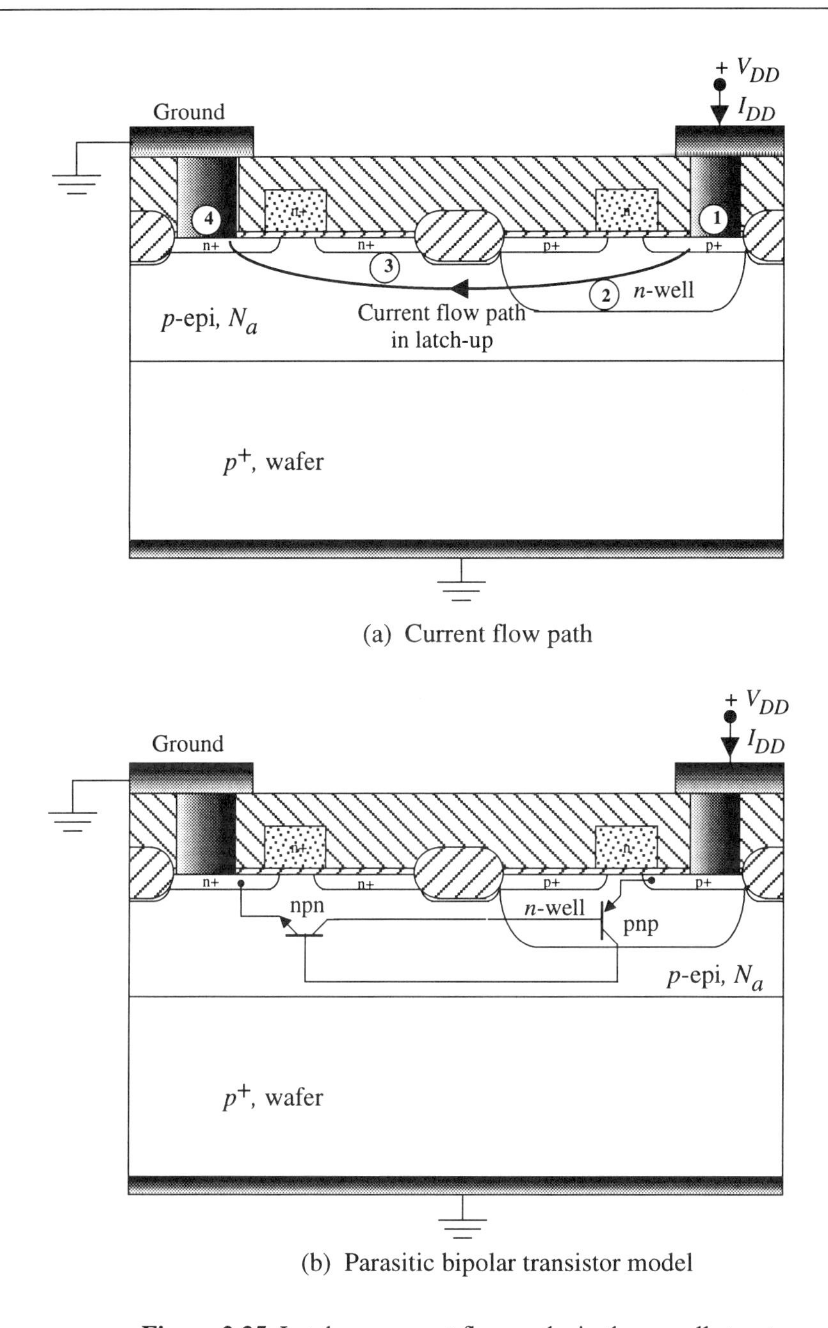

Figure 2.35 Latch-up current flow paths in the n-well structure

pair of npn and pnp bipolar transistors from the structure as shown on the right side of Figure 2.36. This shows that the base of the pnp also acts as the collector of the npn transistor. Similarly, the collector of the pnp is electrically the same as the base of the npn. The pnp-npn pair thus form a set of coupled bipolar transistors where the current flow through one affects the conduction characteristics of the other.

The effect of the parasitic bipolar transistors can be understood using the equivalent circuit shown in Figure 2.37(a). Resistors have been added to represent the parasitic effects of the silicon regions. This circuit diagram illustrates the fact that the two transistors are connected in a manner that creates a feedback loop between the them; this is the middle window in the circuit schematic.

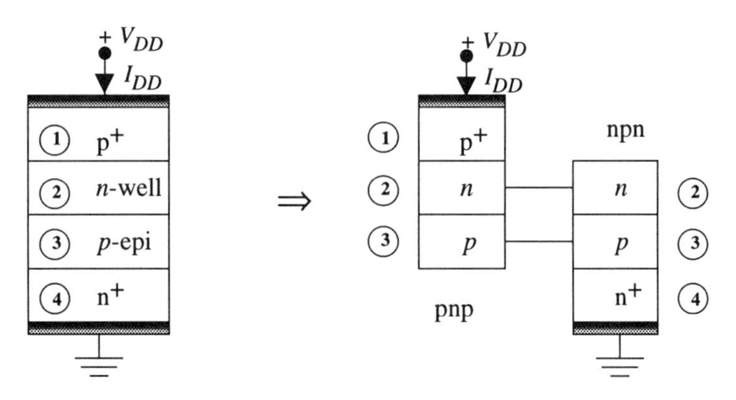

Figure 2.36 4-layer modelling of the latchup network

Consider the current I_1 on the right-hand side of the circuit. Under normal conditions, this will be leakage current. However, it creates a voltage V_1 that acts as an emitter-base voltage $V_1 = V_{EB1}$ across the pnp transistor, enhancing the flow of I_{Ep} and I_{Cp} in the device. Most of this current contributes to I_2 which establishes the voltage $V_2 = V_{BE2}$ across the base-emitter junction of the npn transistor. This in turn enhances the collector current I_{Cn}, which increases I_1 and completes the feedback loop. If both transistors are conducting, then the structure may induce the latchup condition.

A plot of I_{DD} as a function of V_{DD} is shown in Figure 2.37(b). For small values of V_{DD}, the current is restricted to small leakage levels and there is no problem with the structure. However, if V_{DD} is increased to a value V_{BO}, known as the **break-over voltage**, the blocking characteristics of the 4-

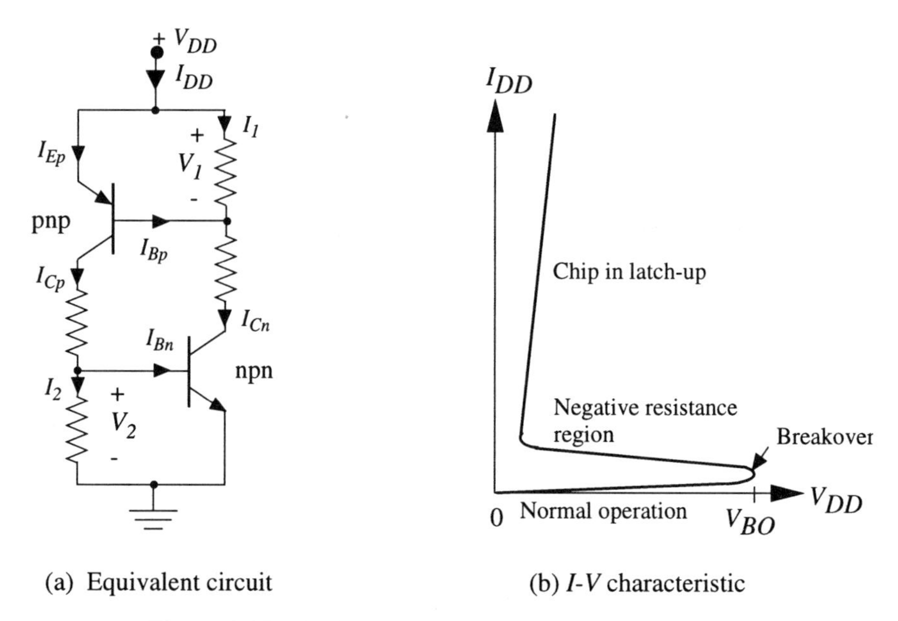

(a) Equivalent circuit (b) *I-V* characteristic

Figure 2.37 Bipolar transistor modelling of latch-up

layered pnpn device break down, allowing I_{DD} to increase as the voltage V_{DD} drops. Since this has a negative slope, it is called the **negative conductance region** of the *I-V* curve. The voltage falls until the device "catches" the exponential curve, which describes the high current flow levels from the power supply to ground. The chip is classified as being in latch-up in this region.

An analysis of the circuit shows that the critical condition for latch up is when the sum of the common base current gains is equal to 1, i.e.,

$$\alpha_{npn} + \alpha_{pnp} = 1 \tag{2. 34}$$

where

$$\alpha_{npn} = \frac{I_{Cn}}{I_{En}} , \qquad \alpha_{pnp} = \frac{I_{Cp}}{I_{Ep}} \tag{2. 35}$$

Recall that the forward alpha α_F of a bipolar transistor depends on the emitter current as illustrated by Figure 2.38. For small values of I_E, α_F increases as the forward injection builds. It then levels out for a range of currents, and finally exhibits a roll-off at high currents due to high-level injection effects. For the present situation, we note that the current levels are initially restricted to small leakage currents, which in turn gives small α_F values. However, since the current flow is enhanced by the feedback mechanism, both α_{npn} and α_{pnp} increase. This illustrates how the common-base current gain condition in equation (2.34) can be reached: the feedback loop enhances both the npn and the pnp current, which in turn increases the gain.

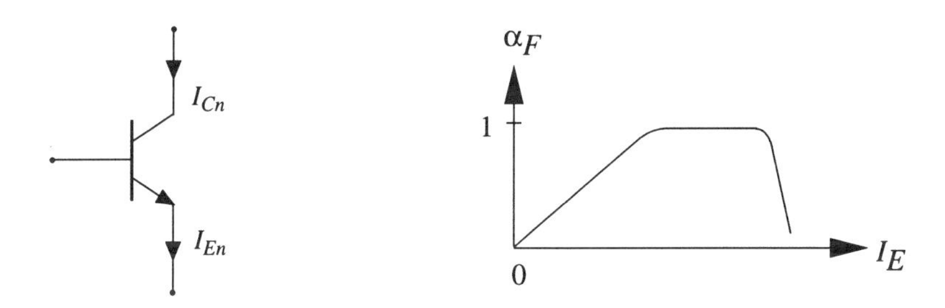

Figure 2.38 Common-base current gain in a bipolar junction transistor

2.7.1 Latch up Prevention

Latch up prevention is accomplished by designing the structures in a manner that acts against the formation of the feedback network. Various techniques have been developed and are usually included in the design rule specifications.

Consider, for example, the cross-sectional view in Figure 2.39 where we have added resistors to represent the path seen by the current flowing through the semiconductive regions. One way to break the feedback loop between the npn and the pnp transistors is to insure that the effective resistance r_x is very large. This can be achieved by using trench isolation (instead of LOCOS) to provide a glass block between the two parasitic devices. Another effective deterrent is to insure that the substrate resistance r_{sub} is very small, since this would effectively break the path between the collector of the npn and the base of the npn transistors. This implies that the wafer doping should be heavily p-type, as in our earlier example of a CMOS process flow.

Other rules to prevent latchup can be summarized by the following statements that apply to an

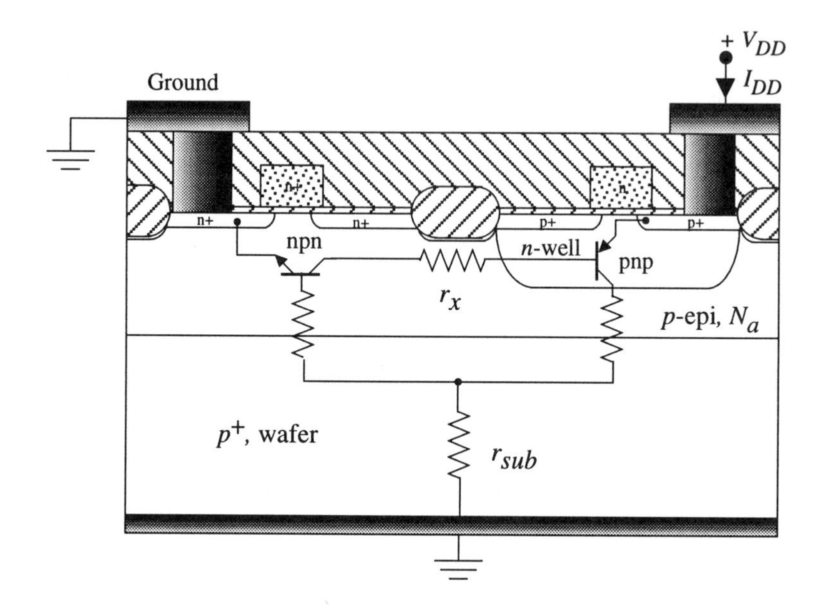

Figure 2.39 Semiconductor parasitics in the latch-up paths

n-well process. Each is portrayed in the various aspects of Figure 2.40.

- Apply ground-to-substrate contacts whenever possible;
- Add V_{DD}-to-n well contacts whenever possible.

These help eliminate up the voltage drops that might bias the bipolar transistors into the active operational region.

- Obey all design rule spacings, especially those that affect the formation of the parasitic BJTs.

This one is aimed at reducing the gain of the parasitic bipolar transistors by making the BJT base

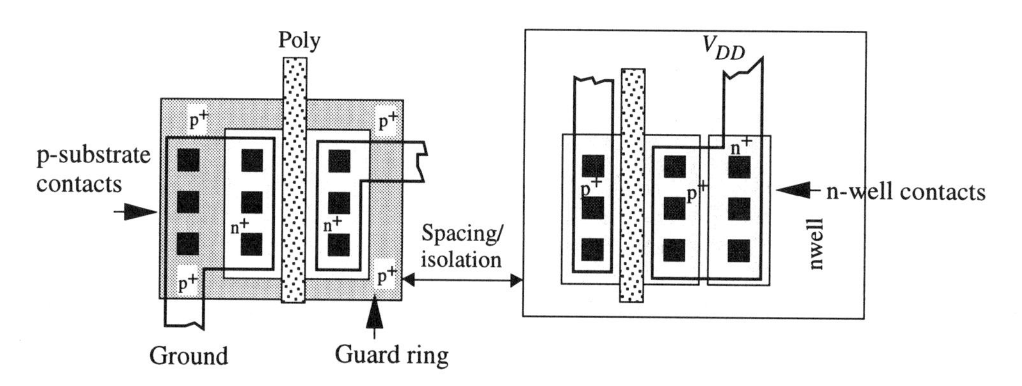

Figure 2.40 Layout for latchup prevention

widths large.

- Use **guard rings** around devices or groups of same-polarity devices.

A guard ring is a doped region that surrounds the MOSFET(s) and is biased by the power supply (if it is an n-type ring) or ground (for the case of a p type ring). The physical extent of the guard ring increases the BJT base widths, while the bias helps maintain well-defined potentials. Rings help avoid latchup, but do consume chip real estate.

2.8 Defects and Yield Considerations

High-density chip designs consist of a few tens of millions of MOSFETs. This has become so commonplace that the technical achievement of silicon processing is generally overlooked. Consider the implications of having a "good die". This means that, so far as the testing process has shown, every circuit in the die operates as it should. In other words, every important feature of every transistor and the interconnect wiring has the correct behavior and, therefore, the correct structure.

In the real world of semiconductor manufacturing, we are continually faced with the fact that only a percentage of the die are functional. This is expressed by the **yield** Y of a process such that

$$Y = \frac{\text{Number of Good Die}}{\text{Number of Total Die}} \times 100\% \tag{2. 36}$$

Obviously, a high yield is required to have a profitable design. The study of **yield enhancement** centers on the problem of achieving this goal. While most of the problems originate in the fabrication process and are the responsibility of those involved in the process flow definition and design, some factors have a direct effect on the chip and circuit designer. An example is in the formulation of a design rule set, since these are derived to insure functional chips that can be manufactured within the limits of the equipment. The second most important concept is that of the area A of an individual die.

Silicon wafers cannot be manufactured without random defects on the surface. These are usually specified using the defect density D that has informal units of # of defects per cm^2. In a highly simplified analysis, we may state that the presence of a defect within a die boundary will lead to a non-functional circuit. In order to maintain a reasonable yield, this implies that the die area A should be kept as small as possible. The concept can be expressed by the simple equation

$$Y \sim e^{-\sqrt{DA}} \times 100\ \% \tag{2. 37}$$

which is based on an empirical model and is drawn in Figure 2.41. The reasoning for this dependence can be understood using the drawings in Figure 2.42. Suppose that a given wafer is used for two chip designs with the area A_1 of design 1 larger than the area A_2 of design 2: $A_1 > A_2$. In Figure

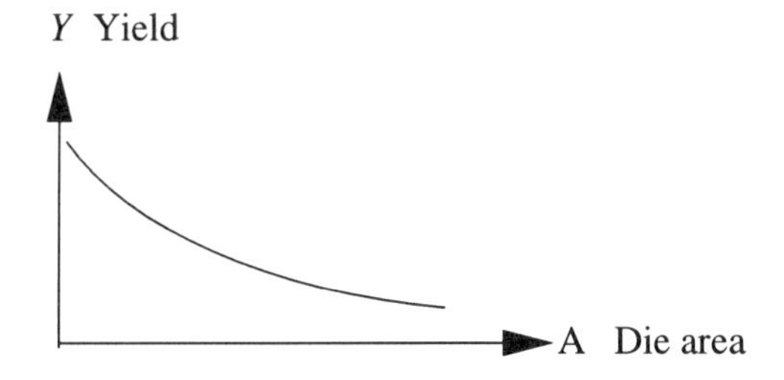

Figure 2.41 Yield as a function of die area

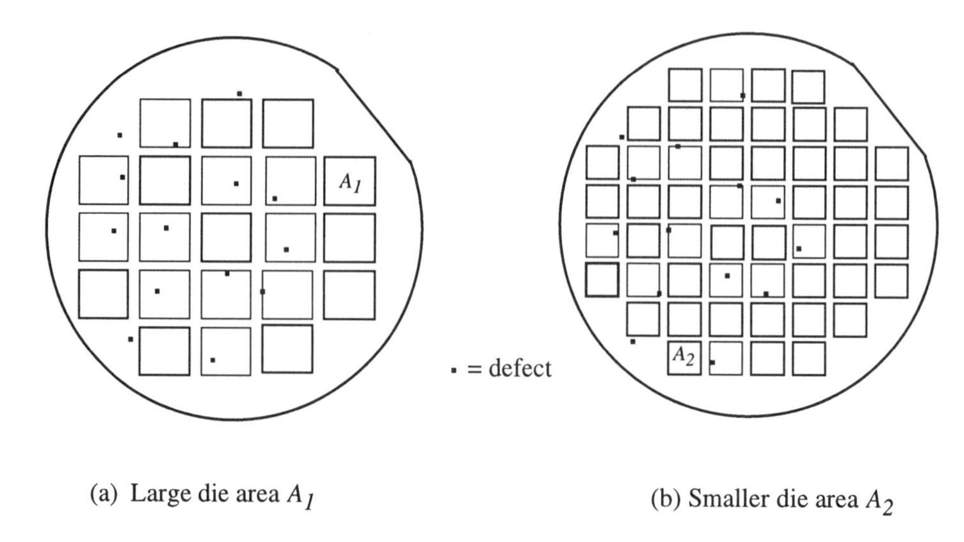

(a) Large die area A_1

(b) Smaller die area A_2

Figure 2.42 Dependence of die size on yield

2.42(a), the large die area implies that there is a high probability of overlaying a defect, which reduces the yield. However, the smaller area design in Figure 2.42(b) reduces the probability that a die boundary will surround a defect, thus increasing the yield.

This problem highlights one of the main aspects of modern CMOS VLSI designs. To maximize the yield, we want to use a small die area. This in turn requires that we have

- the ability to achieve small lithographic features,
- compact layout of the circuits, and
- efficient algorithms that allow us to compact the maximum amount of logic into as small an area as possible.

To the chip designer, the latter two problems are of paramount importance. The physical design is the manifestation of a logic network, but every device, wire, and connection requires surface area. Even though two circuits may give the same logical output, the internal characteristics dictate how large the circuit will be and the factors that limit the response. This is one of the main themes we will follow throughout this book.

2.8.1 Other Failure Modes

Several failure modes exist in MOS integrated circuits. These originate in the fabrication sequence, and cannot be eliminated completely. Short-term (immediate) problems include line breaks and metallization failures, lithographic problems, and short circuits. Long-term effects are more difficult to characterize and may require an extensive analysis of the problem. Gate oxide shorts are unique to MOSFETs, and deserve a more detailed explanation.

Gate Oxide Shorts

A gate oxide short (GOS) nullifies the field effect, and thus renders a MOSFET non-functional. Qualitatively, a GOS is due to the failure of the oxide to act as an insulator between the gate and the substrate. The origin of this type of defect is shown in Figure 2.43. During the initial stages of the growth of the gate oxide, it is possible that a defect or surface non-uniformity inhibits the local growth rate of the SiO_2 layer as shown in (a); this yields a pinhole in the oxide. Depositing the poly over a pinhole yields the MOSFET shown in (b). Assuming that the poly gate is doped n-type, then

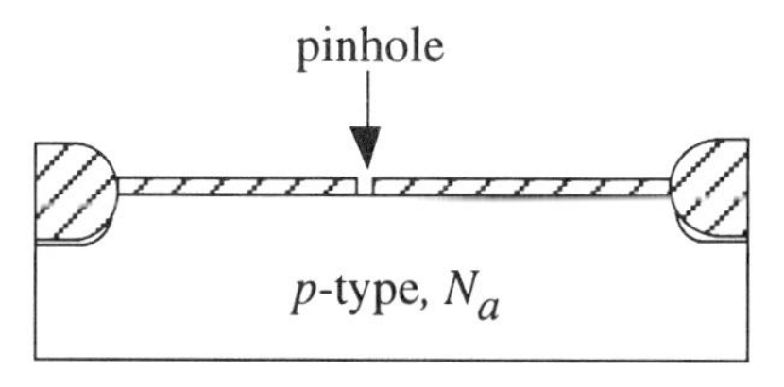

(a) Gate oxide growth

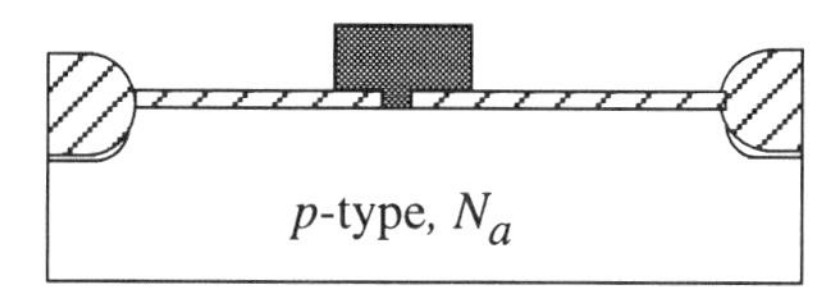

(b) Gate poly deposition

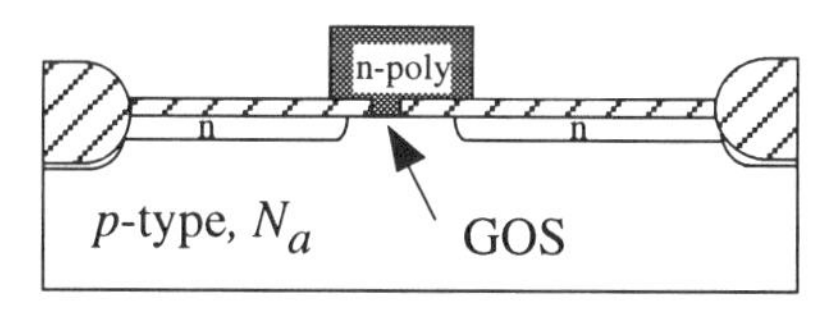

(c) After drain/source implant

Figure 2.43 Gate-oxide shorts in MOSFETs

the GOS is electrically equivalent to a pn junction, and can be modelled using an equivalent circuit where the gate and substrate are connected by a reverse-biased diode. The prohibits the formation of the drain-source channel inversion layer, and the FET will not function as intended.

Consider the formation of the oxide itself. If the oxidation process is allowed to continue (yielding a thick oxide), then it is highly probable that the pinhole will "fill up" and not be a problem. However, the probability of a having GOS increases as thinner oxides are used. With modern devices having gate oxide thicknesses x_{ox} less than 70-80 Å, this type of failure mechanism becomes increasingly important. In static logic circuits (discussed in the next three chapters), it is possible to use a special testing technique known as "I_{DDQ}-Approach" which is particularly adept at finding clusters of GOS failures. The interested reader is directed to the literature for more information on this subject.

2.9 Chapter Summary

High-speed CMOS circuit design is intimately related to the structures that can be fabricated on a silicon substrate and then be transferred to a manufacturing line for mass production. Many of the limitations found in modern chip design are related to the fabrication process.

As we have seen in our short discussion, the electrical characteristics of FETs are established by a fairly complex interplay of parameters and dependences in the process flow. Experienced designers always examine how these values affect the performance of a logic network, and try to work with a given set of electrical characteristics to achieve their goals. We will adhere to this philosophy through the remaining chapters of this book.

2.10 References

The circuit designer usually views the fabrication technology as the basis for layout. There are many excellent books on the subject. A few select titles are listed below, but additional literature can easily be found in a library or using the resources of the world-wide web.

[1] S. Campbell, **The Science and Engineering of Microelectronic Fabrication**, Oxford University Press, New York, 1996.

[2] C.Y. Chang and S.M. Sze, **ULSI Technology**, McGraw-Hill Book Company, New York, 1996.

[3] B. Ciciani (ed.), **Manufacturing Yield Evaluation of Vlsi/Wsi Systems**, IEEE Computer Society, 1995.

[4] J-P. Colinge, **Silicon-on-Insulator Technology**, Kluwer Academic Publishers, Boston, 1990.

[5] D. De Cogan, **Design and Technology of Integrated Circuits**, John Wiley & Sons, New York, 1990.

[6] S.K. Ghandhi, **VLSI Fabrication Principles**, 2nd ed., John Wiley & Sons, New York, 1994.

[7] R. K. Gulati and C.F. Hawkins (eds.), **IDDQ Testing of VLSI Circuits**, Kluwer Academic Publishers, Boston, 1993.

[8]

[9] N. J. Jha and S. Kundu, **Testing and Reliable Design of CMOS Circuits**, Kluwer Academic Publishers, Boston, 1990.

[10] M. Madou, **Fundamentals of Microfabrication**, CRC Press, 1997.

[11] R. Rajsuman, **Iddq Testing for Cmos Vlsi** , The Artech House, 1995.

[12] M. Sarrafzadeh and C.K. Wong, **An Introduction to VLSI Physical Design**, McGraw-Hill Book Company, New York, 1996.

[13] N. Sherwani, **Algorithms for VLSI Physical Design Automation**, Kluwer Academic Publishers, Norwell, MA, 1993.

[14] S.M. Sze, **VLSI Technology**, 2nd ed., McGraw-Hill Book Company, New York, 1988

[15] R. Troutman, **Latchup in CMOS Technology**, Kluwer Academic Publishers, Boston, 1986.

[16] J. P. Uyemura, **Physical Design of CMOS Integrated Circuits Using L-Edit**, PWS Publishers, Boston, 1995.

Chapter 3

The CMOS Inverter: Analysis and Design

One of the basic functions in digital logic is the NOT operation. A CMOS inverter circuit provides this operation in a straightforward manner. The inverter is quite simple and is built using an nFET-pFET pair that share a common gate. The circuit gives a large output voltage swing and only dissipates significant power when the input is switched; these are two important properties of static CMOS logic circuits. This chapter provides a detailed examination of a CMOS inverter and sets the foundations for most higher-level CMOS design styles in the rest of the book.

3.1 Basic Circuit and DC Operation

Figure 3.1(a) shows the logic symbol for an inverter. Given a Boolean input variable A, then NOT(A) = $\overline{A}$ takes a value of A=0 and produces $\overline{A}$=1, and vice versa. A CMOS inverter circuit is shown in Figure 3.1(b). It consists of two opposite polarity MOSFETs Mn (the nFET) and Mp (the pFET) with their gates connected together at the input; the applied voltage is denoted by V_{in}. An nFET-pFET group with a common gate is called a **complementary pair**, which gives us the "C" in "CMOS." As we will see in later chapters, the complementary pair forms the basis for CMOS logic circuits. The inverter output voltage V_{out} is taken from the common drain terminals. The transistors are connected in a manner that ensures that only one of the MOSFETs conducts when the input is stable at a low or high voltage; this is due to the use of the complementary arrangement.

The inverter circuit operation can be understood by examining the relationship between V_{in} and the gate-source voltages of the FETs. Figure 3.2 shows the device voltages using the simplified MOSFET symbols; these are used because there are no body-bias effects in the circuit, so that the bulk voltages do not affect the operation. From the drawing we see that

$$\begin{aligned} V_{GSn} &= V_{in} \\ V_{SGp} &= V_{DD} - V_{in} \end{aligned} \tag{3.1}$$

where V_{in} is assumed to be in the voltage range $[0, V_{DD}]$ with V_{DD} the power supply. Let us apply a

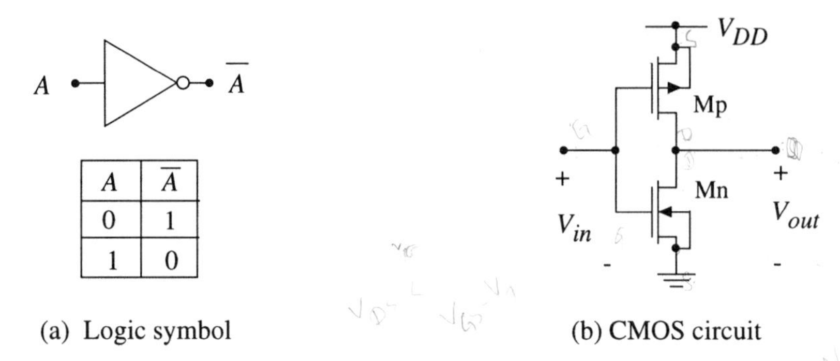

(a) Logic symbol (b) CMOS circuit

Figure 3.1 Basic inverter symbol and circuit

high input voltage of $V_{in} = V_{DD}$ so that

$$\begin{aligned} V_{GSn} &= V_{DD} \\ V_{SGp} &= 0 \end{aligned} \qquad (3.\ 2)$$

In this case, the p-channel MOSFET Mp is in cutoff while the n-channel MOSFET Mn is conducting in the non-saturated mode. Mn then provides a current path to ground, resulting in an output voltage of

$$\min(V_{out}) = V_{OL} = 0 \qquad (3.\ 3)$$

where V_{OL} is called the **output low voltage**, and represents the smallest voltage available at the output. Conversely, a low input voltage of $V_{in} = 0$ results in

$$\begin{aligned} V_{GSn} &= 0 \\ V_{SGp} &= V_{DD} \end{aligned} \qquad (3.\ 4)$$

which shows that Mn is in cutoff while Mp conducts in the non-saturated mode. The pFET Mp then provides a conductive path to the power supply and gives

$$\max(V_{out}) = V_{OH} = V_{DD} \qquad (3.\ 5)$$

which defines the **output high voltage** V_{OH} of the circuit; V_{OH} is the largest value of V_{out}. Because

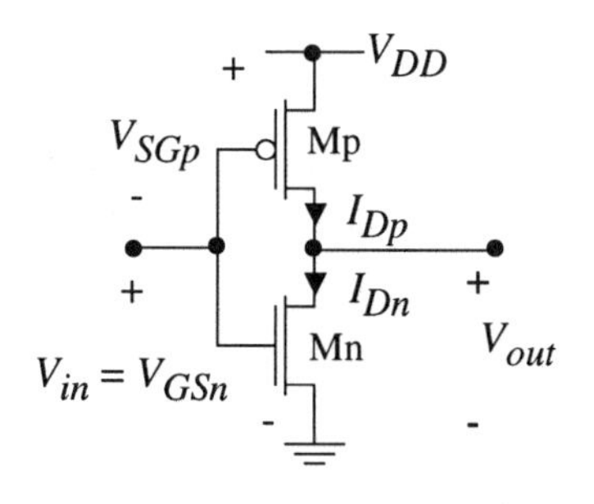

Figure 3.2 Input and output voltages

of the placement and operation of each MOSFET, Mn is sometimes called a *pull-down* transistor, while Mp is termed a *pull-up* device.

We may summarize the behavior of the simple two-transistor inverter circuit by writing that

$$V_{in} = 0v \Rightarrow V_{out} = V_{DD}$$

and

$$V_{in} = V_{DD} \Rightarrow V_{out} = 0v$$

To create the NOT operation, we can use these statements to associate Boolean values of 0 and 1 with the **ideal** voltage levels

$$A = 0 : V = 0v$$

$$A = 1 : V = V_{DD}$$

This defines the **positive logic** convention used in CMOS digital circuits where small voltages are associated with a logic 0, while large positive voltages define a logic 1 state. While these values are useful as references, they do not tell us what the output voltage is for arbitrary values of V_{in}.

The DC input-output characteristics are portrayed graphically using the **Voltage-Transfer Curve** (or VTC) shown in Figure 3.4.[1] This is simply a plot of V_{out} as a function of V_{in}. The inversion operation is seen directly from the curve: when V_{in} is small, V_{out} is large, and vice-versa.[2] Qualitatively, the sharpness of the transition is a measure of how well the circuit is able to perform digital operations. All of the important DC circuit characteristics can be extracted from the VTC. This plot allows us to extend the logic 0 and logic 1 voltage definitions to a *range of voltages* for each logic level. Moreover, the limits used to define logic 0 and logic 1 values are different for the input and output terminals. The VTC itself gives a set of **critical voltages** to work with in creating the defined ranges. The important values are the output voltages V_{OH} and V_{OL}, the input voltages V_{IH} and V_{IL}, and the inverter threshold voltage V_I. Each is analyzed in detail below.

In our calculations we will assume that the MOSFETs have known device transconductance values of

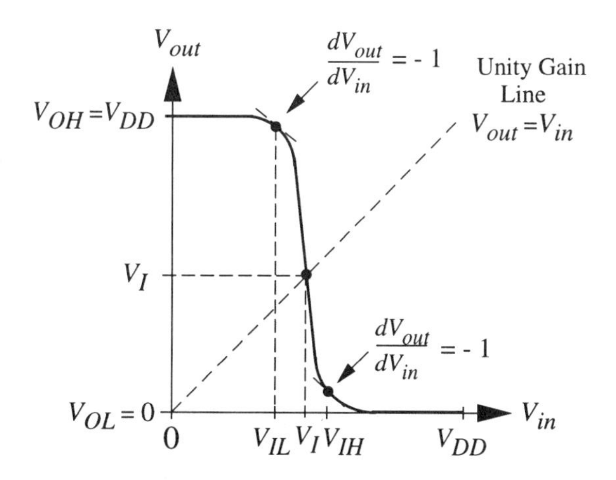

Figure 3.3 Critical VTC voltages

[1] This is also known as the voltage-transfer *characteristic*

[2] The DC VTC assumes that transient effects have decayed away.

$$\beta_n = k'_n\left(\frac{W}{L}\right)_n \quad \text{and} \quad \beta_p = k'_p\left(\frac{W}{L}\right)_p \tag{3.6}$$

This means that the layout has been completed and the aspect ratios $(W/L)_n$ and $(W/L)_p$ are known quantities. We will investigate the design of the circuit later.

3.1.1 DC Characteristics

The critical voltages are established by the MOSFET parameters and characterize the DC response of the circuit. These are directly related to the method we use to define logic 0 and logic 1 levels. In CMOS circuit design, only the device aspect ratios $(W/L)_n$ (for the nFET) and $(W/L)_p$ (for the pFET) can be adjusted in the design phase. The other electrical parameters such as k' and V_T are a result of the fabrication and cannot be changed.

The critical voltages may be computed by setting the input voltage to the desired value and then equating the drain currents $I_{Dn}=I_{Dp}$ at the output node. We will use the square-law MOSFET current flow equations to obtain closed form expressions in our calculations. Channel-length modulation effects are ignored for simplicity, but are usually included in computer simulations.

Output-High Voltage

As discussed above, the output-high voltage V_{OH} is the largest value of V_{out}. It may be calculated by applying an input voltage $V_{in} < V_{Tn}$ which insures that the nFET is in cutoff while the pFET is biased into the active region. Ideally, the simplified MOSFET equations give $I_{Dp}=0$, which implies that the source-drain voltage $V_{SDp}=0v$. Since the source of the pFET is connected to the power supply voltage V_{DD}, KVL gives

$$\begin{aligned} V_{out} &= V_{DD} - V_{SDp} \\ &= V_{DD} \\ &= V_{OH} \end{aligned} \tag{3.7}$$

in agreement with our earlier statement. In realistic circuits, leakage currents (which are ignored in our simple square-law model) reduce the value slightly.

Output-Low Voltage

The output-low voltage V_{OL} represents the smallest value of V_{out} from the circuit. Setting the input voltage to a value $V_{in} = V_{DD} > (V_{DD} - |V_{Tp}|)$ places Mp in cutoff and defines the condition needed to calculate the value of $V_{out}=V_{OL}$. Since Mn is biased active but has $I_{Dn} = 0$, the drain-source voltage across the nMOSFET is $V_{DSn}=0v$. At this point, the inverter output is given by

$$V_{out} = V_{DSn} = 0 = V_{OL} \tag{3.8}$$

Leakage currents increase the actual value of V_{OL} slightly.

An important property of CMOS is that the output logic swing V_L is given by

$$\begin{aligned} V_L &= V_{OH} - V_{OL} \\ &= V_{DD} \end{aligned} \tag{3.9}$$

This shows that the CMOS inverter exhibits a **full-rail** output voltage swing, i.e., the entire power supply range. This helps provide well-defined logic 0 and logic 1 voltages.

Input Low Voltage

The input-low voltage V_{IL} represents the largest value of V_{in} that can be interpreted as a logic 0 input. This can be seen from the voltage-transfer curve in Figure 3.3. If the input voltage satisfies $V_{in}<V_{IL}$, then the output voltage V_{out} is either at V_{DD}, or close to it, indicating that the output can be interpreted as a logic 1. If V_{in} is increased above V_{IL} the circuit moves into the transition region. Using stability arguments, we define V_{IL} as the point where the slope of the VTC has a value of -1, i.e.,

$$\frac{dV_{out}}{dV_{in}} = -1 \tag{3. 10}$$

To calculate V_{IL} we note that at this point, the nFET Mn is saturated while the pFET Mp is conducting in the non-saturated mode. Equating currents $I_{Dn}= I_{Dp}$ gives

$$\frac{\beta_n}{2}(V_{in} - V_{Tn})^2 = \frac{\beta_p}{2}\left[2(V_{DD}-V_{in} - |V_{Tp}|)(V_{DD}-V_{out}) - (V_{DD}-V_{out})^2\right] . \tag{3. 11}$$

The derivative condition is applied by first writing the functional relationship

$$I_{Dn}(V_{in}) = I_{Dp}(V_{in}, V_{out}) \tag{3. 12}$$

Taking differentials of both sides gives

$$\frac{dI_{Dn}}{dV_{in}}dV_{in} = \frac{\partial I_{Dp}}{\partial V_{in}}dV_{in} + \frac{\partial I_{Dp}}{\partial V_{out}}dV_{out} \tag{3. 13}$$

Rearranging as

$$\frac{dV_{out}}{dV_{in}} = \frac{\dfrac{dI_{Dn}}{dV_{in}} - \dfrac{\partial I_{Dp}}{\partial V_{in}}}{\dfrac{\partial I_{Dp}}{\partial V_{out}}} = -1 \tag{3. 14}$$

shows that the derivative condition may be found using the equations for current flow. Substituting and calculating the derivatives yields

$$V_{in}\left(1 + \frac{\beta_n}{\beta_p}\right) = 2V_{out} - V_{DD} - |V_{Tp}| + \frac{\beta_n}{\beta_p}V_{Tn} . \tag{3. 15}$$

Solving this simultaneously with eqn. (3.11) gives the value of $V_{in}=V_{IL}$. We note that two equations are necessary because there are really two unknowns, V_{IL} and the value of V_{out} with this input.

Input High Voltage

The input-high voltage V_{IH} is the smallest value of V_{in} that can be interpreted as a logic 1 level. This interpretation is verified from the VTC plot where it is seen that an input voltage of $V_{in} \geq V_{IH}$ gives an output voltage that is either 0*v*, or close to it. To calculate V_{IH}, we use the current flow equations and the unity slope condition as in finding V_{IL}. Now, however, Mn is non-saturated while Mp is saturated so that equating currents gives

$$\frac{\beta_n}{2}\left[2(V_{in}-V_{Tn})V_{out}-V_{out}^2\right] = \frac{\beta_p}{2}(V_{DD}-V_{in}-|V_{Tp}|)^2 . \qquad (3.16)$$

Since the functional relationship for this case is seen to be

$$I_{Dn}(V_{in}, V_{out}) = I_{Dp}(V_{in}) \qquad (3.17)$$

taking differentials of both sides gives the slope requirement as

$$\frac{dV_{out}}{dV_{in}} = \frac{\dfrac{dI_{Dp}}{dV_{in}} - \dfrac{\partial I_{Dn}}{\partial V_{in}}}{\dfrac{\partial I_{Dn}}{\partial V_{out}}} = -1 \qquad (3.18)$$

Substituting the current equations and differentiating gives the second equation

$$V_{in}\left(1+\frac{\beta_p}{\beta_n}\right) = 2V_{out} + V_{Tn} + \frac{\beta_p}{\beta_n}(V_{DD}-|V_{Tp}|) \qquad (3.19)$$

which must be solved with eqn. (3.16) to find $V_{in}=V_{IH}$. As with the V_{IL} analysis, we need two equations to solve for the input and output voltages.

The Inverter Threshold (Midpoint) Voltage

The voltage V_I is called the **inverter gate threshold voltage**, and is defined by the point where the voltage transfer curve intersects the unity gain line defined by $V_{out}=V_{in}$. V_I is the midpoint between the borders of the logic 0 and logic 1 input voltages V_{IL} and V_{IH}, and is a very useful parameter that characterizes the entire VTC. This gives rise to the alternate notation and terminology in the literature for this voltage as V_M, **the midpoint voltage**. In our discussion of more complex static logic gates, we will use V_I and V_M interchangeably. The value of V_I can be found by first noting that, by definition,

$$V_{out} = V_{in} = V_I \qquad (3.20)$$

so that

$$\begin{aligned} V_{GSn} &= V_I = V_{DSn} \\ V_{SGp} &= V_{DD} - V_I = V_{SDp} \end{aligned} \qquad (3.21)$$

holds for the MOSFET voltages. These equations show that both the nFET and the pFET are operating in the saturation region. Equating currents gives

$$\frac{\beta_n}{2}(V_I - V_{Tn})^2 = \frac{\beta_p}{2}(V_{DD}-V_I-|V_{Tp}|)^2 \qquad (3.22)$$

so that

$$V_I = \frac{V_{DD}-|V_{Tp}| + \sqrt{\frac{\beta_n}{\beta_p}}V_{Tn}}{1+\sqrt{\frac{\beta_n}{\beta_p}}} = V_M \qquad (3.23)$$

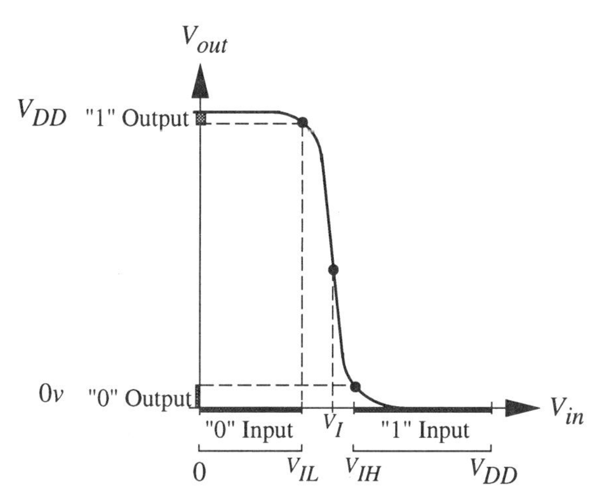

Figure 3.4 Interpretation of critical voltages

gives the desired result. Note that the device ratio (β_n/β_p) is the important quantity that determines the value of V_I for a given circuit.

Interpretation of the Critical Voltages

The most important aspects of the DC calculation are summarized by the VTC in Figure 3.4. Consider the input voltage V_{in} as it is increased from 0*v*. For input voltages in the range

$$0 \le V_{in} \le V_{IL} \tag{3. 24}$$

the output voltage is high at either a perfect logic 1 voltage $V_{out} = V_{DD}$, or very close to it. This then allows us to identify this range of voltages as logic 0 input values. When V_{in} is in the high range defined by

$$V_{IH} \le V_{in} \le V_{DD} \tag{3. 25}$$

then the output voltage is either at a perfect logic 0 value $V_{out} = 0v$ or very close to it. We thus identify these values of V_{in} as corresponding to logic 1 inputs. The inverter threshold voltage V_I always has the characteristic that

$$V_{IL} \le V_I \le V_{IH} \tag{3. 26}$$

We may thus interpret this as the midpoint voltage $V_M = V_I$ such that

$$V_{in} < V_M \Rightarrow \text{Input is probably a logic 0}$$

and,

$$V_{in} > V_M \Rightarrow \text{Input is probably a logic 1}$$

with V_{IL} and V_{IH} providing more precise limits.

3.1.2 Noise Margins

The meaning of the critical input and output voltages gains greater significance when coupled to the concept of **noise margins**. Consider the situation in Figure 3.5(a) where the input of an inverter is close to a neighboring interconnect line. A parasitic coupling capacitance C_c exists between the two, so that applying a voltage pulse to one line will cause a change in the voltage in the other. Sup-

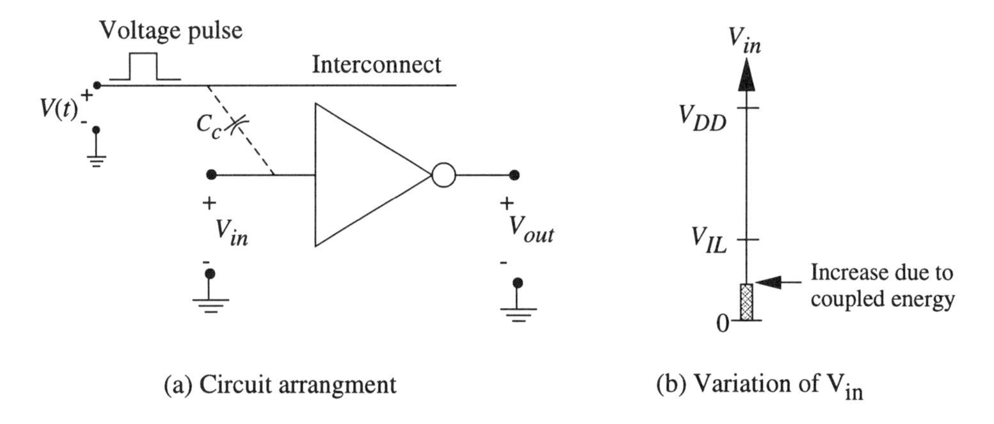

(a) Circuit arrangment

(b) Variation of V_{in}

Figure 3.5 Stray coupling at the input of a gate

pose that the input to the inverter is initially at 0*v*. With this type of electric coupling, the value of V_{in} can jump to a voltage $V_{in} > 0$ as shown in Figure 3.5(b). If the increase is large, then an incorrect switching event may occur. However, so long as the input voltage remains below V_{IL}, then the input will still be correctly interpreted as a logic 0 voltage. Noise margins provide a quantitative measure of how resistant a circuit is to false switches.

Figure 3.6 provides a graphical means to define and interpret noise margins. Consider first the logic 1 voltages. In general, we define the voltage noise margin for logic 1 (high) voltages as

$$VNM_H = V_{OH} - V_{IH} \tag{3.27}$$

Since the CMOS inverter has $V_{OH}=V_{DD}$, this is

$$VNM_H = V_{DD} - V_{IH} \tag{3.28}$$

Similarly, the voltage noise margin for logic 0 (low) voltage is given by

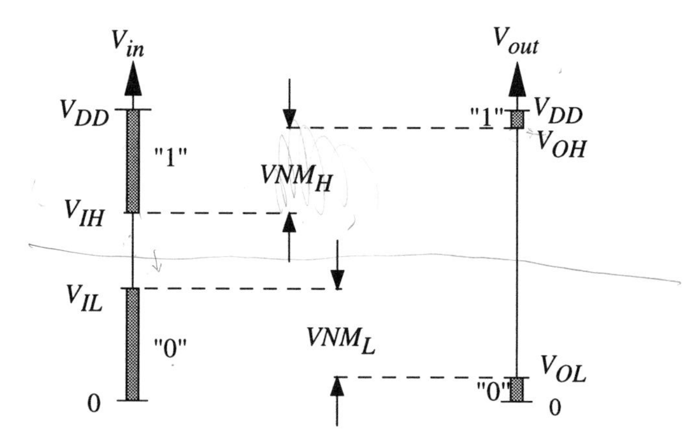

Figure 3.6 Definition of the voltage noise margins

$$\begin{aligned} VNM_L &= V_{IL} - V_{OL} \\ &= V_{IL} \end{aligned} \tag{3. 29}$$

where the second line is obtained by noting that the CMOS inverter has a value of $V_{OL} = 0v$. For a functional digital circuit, we must have $V_{NML} > 0$ and $V_{NMH} > 0$.

Noise margins are particularly important when designing low voltage circuits, i.e., those with small power supply values. This is because as V_{DD} shrinks, the definitions of logic 0 and 1 voltage ranges also shrink, and the gates are more susceptible to spurious signals from neighboring lines.

Example 3-1 CMOS Inverter Characteristics

To illustrate the calculations, let us assume a circuit designed with $(W/L)_n$=8 and $(W/L)_p$=12 which is fabricated in a process where

$$k_n' = 100\mu A/V^2,\ V_{TOn} = 0.75v,\ k_p' = 40\mu A/V^2,\ V_{TOp} = -0.8v$$

Using the aspect ratios values gives $\beta_n = 800\mu A/V^2$ and $\beta_p = 480\mu A/V^2$, so that $(\beta_n/\beta_p) = 1.67 > 1$. We will assume a power supply of $V_{DD} = 3.3v$ for our calculations.

First note that the output voltages are given by $V_{OH} = 3.3v$ and $V_{OL} = 0v$ due to the complementary structuring. To compute V_{IL} we must solve the two equations

$$\begin{aligned} 1.67\,(V_{IL} - 0.75)^2 &= 2\,(2.5 - V_{IL})\,(3.3 - V_{out}) - (3.3 - V_{out})^2 \\ V_{IL}\,(1 + 1.67) &= 2V_{out} + (1.67)\,(0.75) - 2.5 \end{aligned} \tag{3. 30}$$

The algebra gives

$$V_{IL} \approx 1.26v \tag{3. 31}$$

at an output voltage of

$$V_{out} \approx 3.11v \tag{3. 32}$$

Similarly, V_{IH} is obtained by solving

$$\begin{aligned} \left(\frac{1}{1.67}\right)(2.5 - V_{IH})^2 &= 2\,(V_{IH} - 0.75)\,V_{out} - V_{out}^2 \\ V_{IH}\left(1 + \frac{1}{1.67}\right) &= 2V_{out} + \frac{1}{1.67}\,(2.5) + 0.75 \end{aligned} \tag{3. 33}$$

yielding

$$V_{IH} \approx 1.70v \tag{3. 34}$$

with

$$V_{out} \approx 0.233v \tag{3. 35}$$

at this point.

Finally, the inverter gate threshold voltage is determined by

$$V_I = \frac{3.3 - 0.8 + \sqrt{1.67}\,(0.75)}{1 + \sqrt{1.67}} \approx 1.51v \tag{3. 36}$$

which completes the calculation of the critical voltages for the circuit. Note that

$$V_{IL} < V_I < V_{IH} \tag{3. 37}$$

is satisfied and provides a check on our numerical results.

3.1.3 Layout Considerations

The MOSFET aspect ratios $(W/L)_n$ and $(W/L)_p$ determine the values of the critical input voltages V_{IL}, V_{IH}, and V_I. The size of the transistors is set by the layout masking, which shows the important link between the fabrication sequence and the resulting electrical characteristics.

Since the inverter consists of only two FETs, the layout of the circuit is relatively simple. Figure 3.7 shows a basic approach in which the two FETs are oriented in the same manner as the circuit schematic. The polysilicon gate with drawn channel length L' serves as the input connection, while the output is taken from a metal line on the right side of the circuit. In this example the FETs have been chosen to have equal channel widths with $W_n = W_p$, so that $(W/L)_n = (W/L)_p$. The dimensions of the drain sections are shown as $(W_p \times Y)$ for the pFET p^+ region, and $(W_n \times X)$ for the nFET n^+ region. These values do not affect the DC operation, but enter into the calculation of the switching transients when finding the parasitic capacitance contributions that affect the circuit.

Another layout style is shown in Figure 3.8. This one uses MOSFET that are oriented horizontally, which allows a simple vertical polysilicon line to be used as the gate for both of the transistors. This example employs different aspect ratios for the two transistors, with $W_p > W_n$. This is sometimes done to compensate for the fact that $k_n' > k_p'$. One important point to note from the DC analysis is that the critical voltages in the VTC depend only upon the ratio

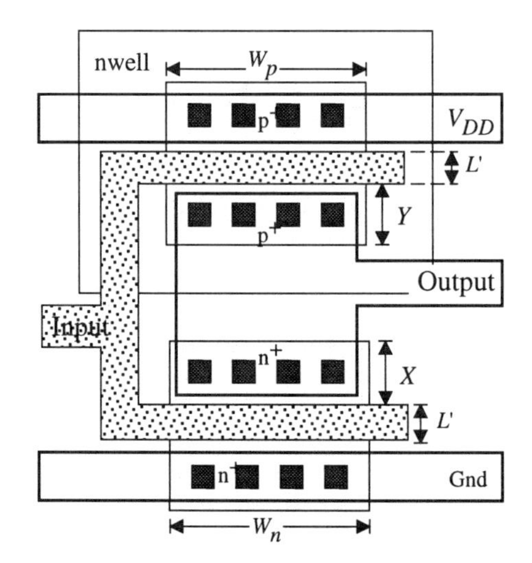

Figure 3.7 Inverter layout example

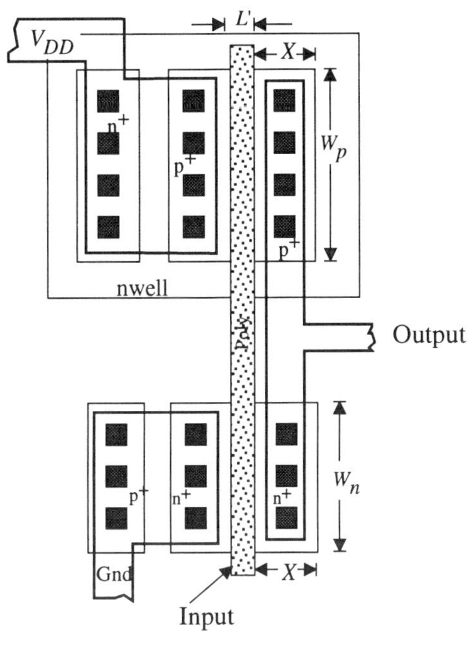

Figure 3.8 Alternate layout approach

$$\frac{\beta_n}{\beta_p} = \frac{k_n'\left(\frac{W}{L}\right)_n}{k_p'\left(\frac{W}{L}\right)_p} \tag{3. 38}$$

of the device transconductance factors.[3] As we will see in the next section, the value of β_p and β_n determine the transient rise and fall times, respectively.

3.2 Inverter Switching Characteristics

Transient switching times are used to calculate data throughput rates and are also important in system timing. Switching times are determined by two circuit properties: transistor current flow levels and parasitic capacitances. Both are set by the chip design parameters, and are sensitive to the transistor aspect ratios, layout geometry, and logic routing.

To model the basic problem, we introduce the **output capacitance** C_{out} shown in Figure 3.9; this represents the total capacitance at the output node, and consists of contributions from the MOSFETs and the external network. For our analytic calculations, C_{out} is assumed to be a linear, time-invariant (LTI) quantity. This allows us to obtain closed-form expressions for the important switching times that characterize the inverter. Moreover, we will be able to clarify the design issues that affect CMOS designs in general. It is important to note, however, that C_{out} has both linear and non-linear (voltage-dependent) terms that we will need to deal with later.

[3] This conclusion is based upon the fact that every equation has both β_n and β_p in it, so that the individual values are not important.

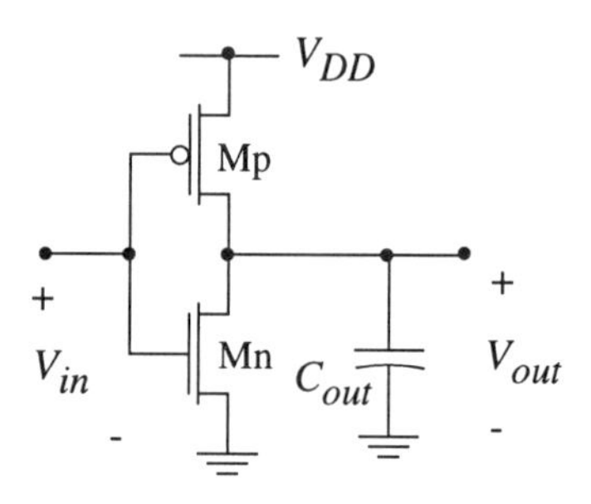

Figure 3.9 Output capacitance

3.2.1 Switching Intervals

Switching performance of CMOS digital circuits are characterized by the time intervals required to charge and discharge capacitors at output nodes. CMOS inverters use transistors to provide current flow paths between the power supply (Mp) and ground (Mn). All switching times are thus set by the current levels and the value of C_{out}. Figure 3.10 shows the inverter input and output voltages as functions of time. The input waveform $V_{in}(t)$ has been taken to have idealized step characteristics; this choice simplifies the calculations and also provides a standard reference. When the input voltage is low with V_{in}=0v, the output voltage is high at a value of $V_{out} = V_{DD}$. This corresponds to the case where the nFET is OFF, while the pFET is ON and provides the connection to the power supply. Changing the input voltage to high value $V_{in} = V_{DD}$ reverses this; now the nFET is active while the pFET is in cutoff. The capacitor C_{out} discharges to 0v through Mn, and the output voltage decays to a final value of V_{in} = 0v as shown. The switching time associated with this decay is the output high-to-low time t_{HL}. If the input voltage is returned to a low voltage V_{in} = 0v, the nFET is

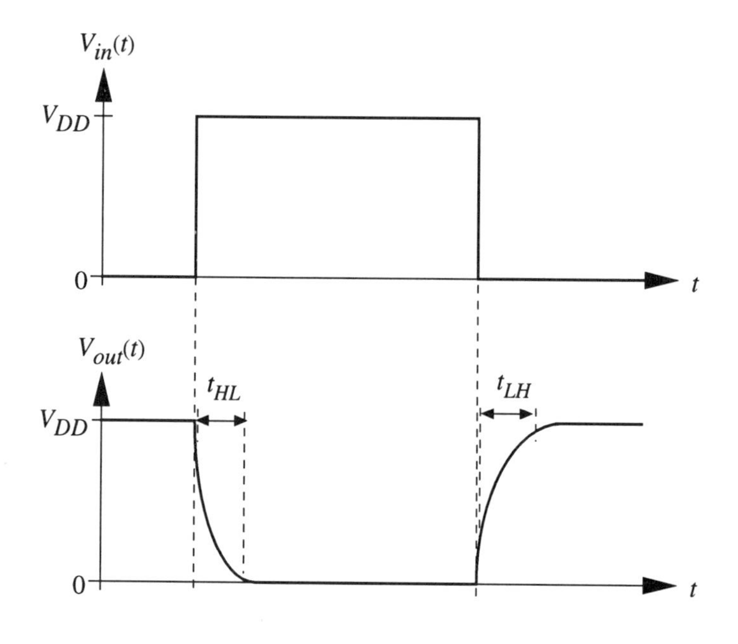

Figure 3.10 Switching time definitions

driven into cutoff while the pFET reconnects C_{out} to the power supply. This allows C_{out} to charge to a final voltage of $V_{out}=V_{DD}$ in a characteristic time t_{LH}, the output low-to-high time.

The importance of the switching times t_{HL} and t_{LH} is obvious: they represent the times required for the output to stabilize to a final value in response to changes of the input voltage. These are limiting factors in the performance of a digital CMOS logic circuit. Owing to their importance in circuit design, we will analyze each in the subsections below. It is important to keep in mind that the two quantities can be treated as being distinct, and that each depends on the behavior of only one FET.[4]

3.2.2 High-to-Low Time

The output high-to-low time is calculated using the subcircuit in Figure 3.11(a). It represents the time interval needed for the output capacitor to discharge through the n-channel MOSFET Mn when Mp is in cutoff. t_{HL} is also referred to as the **fall time** t_f for the circuit since it gives the time needed for the output to decay from a well-defined logic 1 state to a well-defined logic 0 state.

The discharge is described by the capacitor equation

$$I_{Dn} = -C_{out}\frac{dV_{out}}{dt} \tag{3.39}$$

where we will assume an initial condition of $V_{out}(t=0)=V_{DD}$; the minus sign is required because the current is leaving the positive terminal. At the beginning of the discharge, Mn is saturated, so that

$$\frac{\beta_n}{2}(V_{DD}-V_{Tn})^2 = -C_{out}\frac{dV_{out}}{dt} \tag{3.40}$$

describes the initial discharge. Integrating gives a linear decay in time be means of the function

$$V_{out}(t) = V_{DD}-\frac{\beta_n(V_{DD}-V_{Tn})^2 t}{2C_{out}} \tag{3.41}$$

This is valid until a time t_o when the output voltage drops to $V_{out} = (V_{DD} - V_{Tn})$ when the MOSFET enters the non-saturated conduction region; this is indicated in Figure 3.11(b). The value of t_o is found by setting

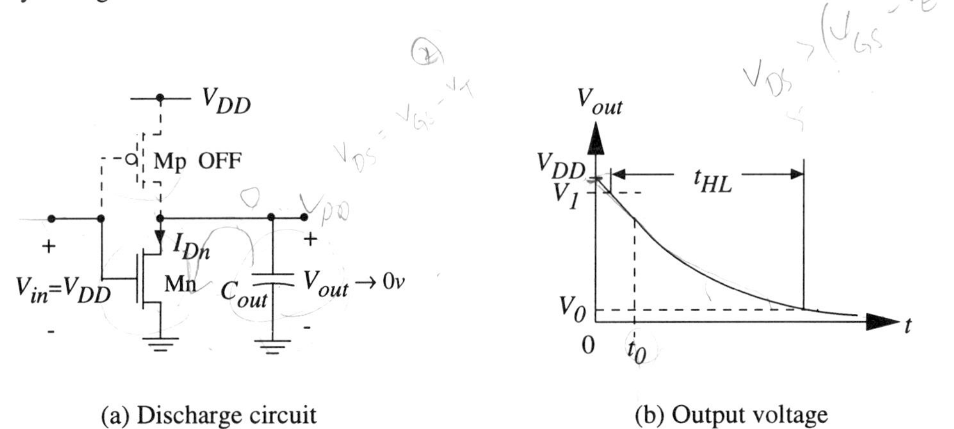

(a) Discharge circuit (b) Output voltage

Figure 3.11 Subcircuit for fall-time calculation

[4] These calculations assume the step input voltage.

$$V_{out}(t_o) = V_{DD} - \frac{\beta_n (V_{DD} - V_{Tn})^2 t_o}{2C_{out}} \qquad (3.42)$$
$$= V_{DD} - V_{Tn}$$

so that

$$t_o = \frac{2C_{out} V_{Tn}}{\beta_n (V_{DD} - V_{Tn})^2} \qquad (3.43)$$

For times $t \geq t_o$, the differential equation describing the discharge is

$$\frac{\beta_n}{2}\left[2(V_{DD} - V_{Tn}) V_{out} - V_{out}^2\right] = -C_{out}\frac{dV_{out}}{dt} \qquad (3.44)$$

because the nFET is non-saturated. This integrates to

$$V_{out}(t) = (V_{DD} - V_{Tn})\left[\frac{2e^{-(t-t_o)/\tau_n}}{1 + e^{-(t-t_o)/\tau_n}}\right] \qquad (3.45)$$

where

$$\tau_n = \frac{C_{out}}{\beta_n (V_{DD} - V_{Tn})} \qquad (3.46)$$

is the **time constant** for the discharge circuit.

The value of t_{HL} is usually defined between the 10% and 90% voltages V_0 and V_1, respectively, with

$$V_0 = 0.1 V_{DD}$$
$$V_1 = 0.9 V_{DD} \qquad (3.47)$$

for a full-rail output CMOS circuit. This can be computed by using the integrals

$$t_{HL} = C_{out} \int_{(V_{DD} - V_{Tn})}^{V_1} \frac{dV_{out}}{I_{Dn(sat)}} + C_{out} \int_{V_0}^{(V_{DD} - V_{Tn})} \frac{dV_{out}}{I_{Dn(non\text{-}sat)}}, \qquad (3.48)$$

or by determining the required time intervals from the equations above. Either approach gives the result

$$t_{HL} = s_n \tau_n \qquad (3.49)$$

where

$$s_n = \frac{2(V_{Tn} - V_0)}{(V_{DD} - V_{Tn})} + \ln\left(\frac{2(V_{DD} - V_{Tn})}{V_0} - 1\right) \qquad (3.50)$$

is a voltage-dependent scaling multiplier. The first term in s_n represents the time when Mn is saturated, while the second term is due to non-saturated conduction.

As a final point, note that the definition of the time constant τ_n allows us to write that $\tau_n = R_n C_{out}$ where

$$R_n = \frac{1}{\beta_n (V_{DD} - V_{Tn})} \tag{3.51}$$

represents an equivalent LTI value for the drain-to-source resistance. This analysis provides a simple rule-of-thumb for many circuit performance estimates and is more realistic than the best-case value discussed in the simple RC equivalent circuit model. However, the concept of the MOSFET resistance must be used with care, since the MOSFET is inherently a non-linear device while resistances are usually assumed to be linear.

3.2.3 Low-to-High Time

The low-to-high time t_{LH}, also known as the rise time t_r, is found in the same manner. During this time interval, Mn is in cutoff while Mp is conducting from the power supply. As shown in Figure 3.12(a), t_{LH} is the time required to charge C_{out} through Mp. It is also referred to as the charge time t_{ch} in the literature.

Charging is described by the general equation

$$I_{Dp} = C_{out} \frac{dV_{out}}{dt} \tag{3.52}$$

with the initial condition $V_{out}(t{=}0) = 0v$. When the charging starts, Mp is saturated and the integration gives

$$V_{out}(t) = \frac{\beta_p (V_{DD} - |V_{Tp}|)^2 t}{2C_{out}} \tag{3.53}$$

This is valid until a time

$$t_1 = \frac{2C_{out}|V_{Tp}|}{\beta_p (V_{DD} - |V_{Tp}|)^2} \tag{3.54}$$

where $V_{out}(t_1)=|V_{Tp}|$. This point is shown in the graph of Figure 3.12(b). For times $t \geq t_1$, Mp is non-saturated and the output voltage is described by

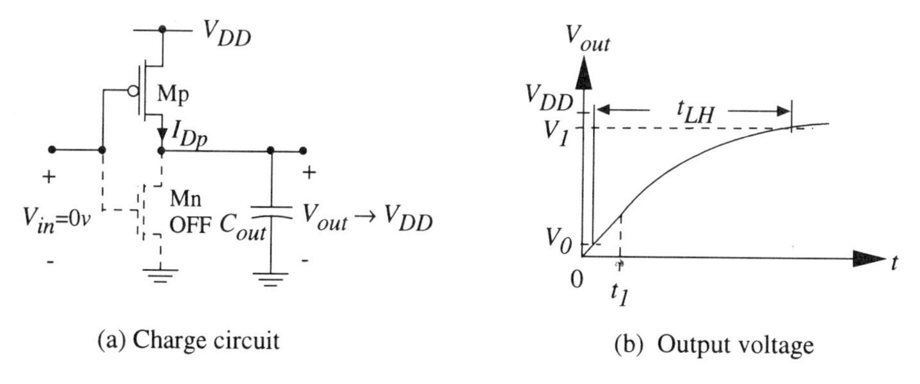

(a) Charge circuit (b) Output voltage

Figure 3.12 Subcircuit for rise-time calculation

$$V_{out}(t) = V_{DD} - (V_{DD} - |V_{Tp}|)\left[\frac{2e^{-(t-t_1)/\tau_p}}{1+e^{-(t-t_1)/\tau_p}}\right] \tag{3.55}$$

where the charging time constant is

$$\tau_p = \frac{C_{out}}{\beta_p(V_{DD} - |V_{Tp}|)} \tag{3.56}$$

Defining t_{LH} as the time to charge C_{out} from V_0 (the 10% point) to V_1 (the 90% point) gives

$$t_{LH} = s_p \tau_p \tag{3.57}$$

with

$$s_p = \frac{2(|V_{Tp}| - V_0)}{(V_{DD} - |V_{Tp}|)} + \ln\left(\frac{2(V_{DD} - |V_{Tp}|)}{V_0} - 1\right) \tag{3.58}$$

as the multiplier for this time interval. Note that t_{LH} has the same form as the fall time t_{HL}, except that pMOSFET parameters appear instead of the nFET quantities. This is expected from the complementary symmetry of the circuit. In fact, a moment's reflection will convince the reader that the results of this calculation could have been written down by simply modifying the nFET equations!

A pMOS resistance may be approximated by

$$R_p = \frac{1}{\beta_p(V_{DD} - |V_{Tp}|)} \tag{3.59}$$

such that $\tau_p = R_p C_{out}$ gives the charging time constant. Note that both R_p and R_n are inversely proportional to (W/L); increasing the aspect ratio thus decreases the equivalent resistance. This analogy is quite useful to remember, and gives an excellent rule-of-thumb. However, we once again remind the reader that a field-effect transistor is inherently a nonlinear device, so that equivalent linear resistances such as R_p and R_n must be used with care.

3.2.4 Maximum Switching Frequency

The sum of the transient times $(t_{HL}+t_{LH})$ represents the minimum time needed for a gate to undergo a complete switching cycle, i.e, for the output to change from a logic 1 to a logic 0 voltage, and then back up to a logic 1 value. We may use this to define the **maximum switching frequency** by

$$\begin{aligned} f_{max} &= \frac{1}{t_{HL} + t_{LH}} \\ &= \frac{1}{s_n \tau_n + s_p \tau_p} \end{aligned} \tag{3.60}$$

This represents the maximum rate of data transfer for the gate. In system design, the working value of f_{max} is set by the slowest gate or datapath element in the network. Figure 3.13 illustrates the importance of f_{max} for the inverter.[5] For signal frequencies $f < f_{max}$ as in Figure 3.13(a), the output has sufficient time to react to changes in the inputs and exhibits proper form. However, if we increase the signal frequency to $f > f_{max}$, the circuit does not have enough time to complete the charge or discharge event. This gives an output signal that has a limited amplitude that may cause a

[5] The results of this discussion may be applied to the more complex logic gates discussed in later chapters.

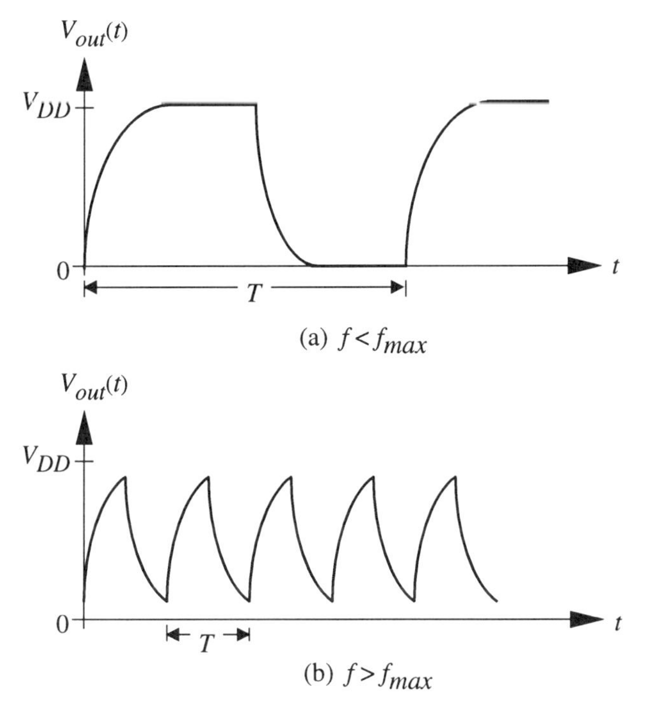

Figure 3.13 Output voltage waveforms

logic error.

3.2.5 Transient Effects on the VTC

Although the VTC is defined to be the DC transfer curve, it is useful for illustrating the transient switching effects by generating the family of curves as shown in Figure 3.14. This corresponds to merging the transient behavior manifest in $V_{out}(t)$ and $V_{in}(t)$ by eliminating the time t as a variable.

For low switching frequencies we obtain the "usual" plot that gives the DC behavior. However,

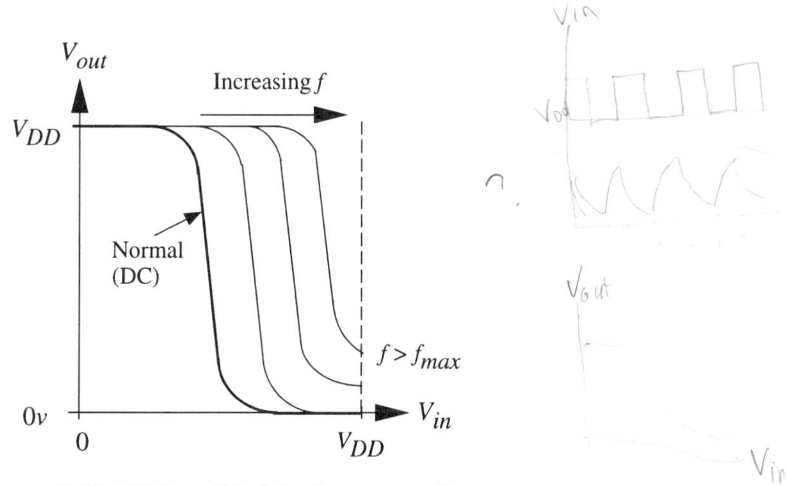

Figure 3.14 VTC modified for frequency effects

as the signal frequency is increased, the behavior of V_{out} as a function of V_{in} shows that there is a change in the response of the network. When the switching frequency exceeds f_{max}, then the circuit cannot respond to the quickly changing input. For this case, the output voltage never reaches a value of 0v. This is the same type of information as that contained in Figure 3.13, but in a slightly different form. Often times one concentrates only on the transient switching characteristics, so that the DC characteristics are of limited use. However, they still provide useful information for stable input states so long as care is taken when applied to a high-speed network.[6]

3.2.6 RC Modelling

A simple RC network model can be used to obtain first-order estimates of the switching times. In Figure 3.15(a) the MOSFETs are replaced by resistor-switch sub-networks. In the model, a cutoff transistor is represented by an open circuit, while an active MOSFET is represented by a closed switch in series with a parasitic drain-source resistance. The nFET equivalent resistance will be denoted by R_n, while R_p represents the pFET equivalent resistance. In this model, the logic circuit is based on the charging and discharging of C_{out} through the appropriate resistors. The operation of the switches is summarized in Figure 3.15(b), and is based on the behavior of MOSFETs under the same situation. When the input is at G=0, switch SWp is closed while SWn is open; a high input G=1 gives the opposite situation, with SWp open and SWn closed.

Consider first the charging circuit. This corresponds to having a low input voltage $V_{in} \approx 0v$, so that Mp is ON and Mn is OFF. Assuming the worst-case situation where $V_{out}(t=0) = 0v$, the voltage buildup is given by

$$V_{out}(t) = V_{DD}(1 - e^{-t/\tau_p}) \tag{3.61}$$

where $\tau_p = R_p C_{out}$ is the time constant. Since the MOSFET is a nonlinear device, R_p can only be approximated. The best-case value[7] of the resistance is where Mp is assumed to be saturated. With a drain-source voltage of V_{DD}, the pMOS resistance is approximated by

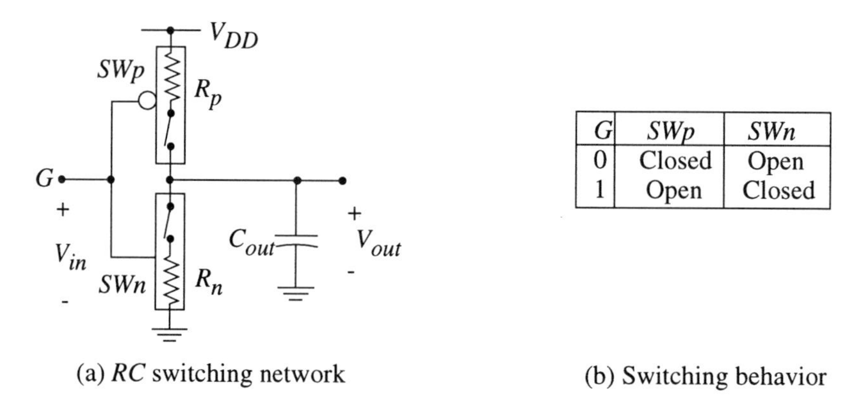

G	SWp	SWn
0	Closed	Open
1	Open	Closed

(a) *RC* switching network (b) Switching behavior

Figure 3.15 RC switch model of the CMOS inverter

[6] In fact, one enlightened engineer once told the author that DC curves are "virtually useless," a comment that indicated a rather myopic view of a very broad field where there are countless types of circuits and problems!

[7] In this case, "best case" means smallest, since this will give the fast voltage change.

$$R_p = \frac{V_{SD}}{I_{D,sat}} \\ = \frac{2V_{DD}}{\beta_p (V_{DD} - |V_{Tp}|)^2} \tag{3. 62}$$

for an order-of-magnitude estimate. In practice, it is more common to use the results of the current flow analysis which gave

$$R_p = \frac{1}{\beta_p (V_{DD} - |V_{Tp}|)} \tag{3. 63}$$

as the equivalent resistance.

The discharge event may be computed in a similar manner. With a high input $V_{in} > (V_{DD}-|V_{Tp}|)$, Mn is conducting while Mp is OFF. Within the RC model, the output voltage is approximated by

$$V_{out}(t) = V_{DD} e^{-t/\tau_n} \tag{3. 64}$$

where we have assumed an initial condition of $V_{out}(0)=V_{DD}$. The discharge time constant through Mn is given by $\tau_n = R_n C_{out}$ such that

$$R_n = \frac{V_{DD}}{I_{D,sat}} = \frac{2V_{DD}}{\beta_n (V_{DD} - V_{Tn})^2} \tag{3. 65}$$

is the best-case value of the nMOS drain-source resistance. As with the charging time calculation, it is more common to use the expression

$$R_n = \frac{1}{\beta_n (V_{DD} - V_{Tn})} \tag{3. 66}$$

for the nFET resistance since this is based on the more rigorous analysis.

Exponential models provide first-order approximations for estimating the gate delays. Simplified networks based on RC time constants are useful for evaluating complex high-performance designs and also provide valuable insight into the operation. Logic simulation tools are often based on switching networks of this type. Once the decision has been made to invoke exponential approximations, switching times can be computed by defining the starting and ending points of the voltages.

Consider first the high-to-low time t_{HL} which is defined as the time needed for the output voltage to fall from $0.9V_{DD}$ and $0.1V_{DD}$. With the exponential decay, the time t_x needed for the voltage to fall from V_{DD} to an arbitrary value V_x is

$$t_x = \tau_n \ln\left[\frac{V_{DD}}{V_x}\right] \tag{3. 67}$$

Thus, t_{HL} can be estimated using

$$\begin{aligned} t_{HL} &= t_{0.1} - t_{0.9} \\ &= \tau_n \ln\left[\frac{V_{DD}}{0.1V_{DD}}\right] - \tau_n \ln\left[\frac{V_{DD}}{0.9V_{DD}}\right] \\ &= \tau_n \ln [9] \end{aligned} \tag{3. 68}$$

or

$$t_{HL} \approx 2.2\tau_n \tag{3. 69}$$

Similarly, the low-to-high time t_{LH} can be estimated as

$$t_{LH} \approx 2.2\tau_p \tag{3. 70}$$

using the same type of analysis on the exponential charging expression. The maximum switching frequency in this approximation is given by

$$f_{max} \approx \frac{0.45}{(\tau_n + \tau_p)} \tag{3. 71}$$

In simplistic terms, the exponential approximation replaces the scaling factors s_n and s_p by the constant $\ln(9) \approx 2.2$. However, note that s_n and s_p depend upon the power supply and the threshold voltages, while the ln(9) factor arises solely from the form of the equations. The two results are therefore distinct, with the more rigorous analysis yielding a higher level of accuracy.

Individual circuits can be more accurately characterized by including device conduction properties and employing the full analysis in conjunction with the results of a computer simulation. Although this requires substantially more work, gate-level optimization is important for custom circuits, ASIC cell design, and transistor network arrays.

3.2.7 Propagation Delay

Logic delay through a gate is conveniently described by the propagation delay time t_P. Physically we interpret t_P as the average time needed for the output to respond to a change in the input logic state. By definition,

$$t_P = \frac{(t_{PHL} + t_{PLH})}{2} \tag{3. 72}$$

where t_{PHL} and t_{PLH} represent the propagation delays for a high-to-low, and a low-to-high transition, respectively. Let us define the 50% voltage points as $V_{1/2} = 0.5V_{DD}$. Then, t_{PHL} and t_{PLH} are defined by the time intervals between the input and output voltages as shown in Figure 3.16.

The high-to-low propagation delay represents the time needed for the output to fall from V_{DD} to V_I; to simplify the calculations, we usually approximate $V_I \approx (V_{DD}/2)$. This yields the general expression

$$t_{PHL} = C_{out} \int_{(V_{DD}-V_{Tn})}^{V_{DD}} \frac{dV_{out}}{I_{Dn\,(sat)}} + C_{out} \int_{V_{DD}/2}^{(V_{DD}-V_{Tn})} \frac{dV_{out}}{I_{Dn\,(non\text{-}sat)}} \tag{3. 73}$$

which defines the basic integrals. Evaluating yields

$$t_{PHL} = s_n'\tau_n \tag{3. 74}$$

where $\tau_n = R_n C_{out}$ is the time constant, and the new scaling factor is

$$s_n' = \left[\frac{2V_{Tn}}{(V_{DD} - V_{Tn})} + \ln\left(\frac{4(V_{DD} - V_{Tn})}{V_{DD}} - 1\right)\right] \tag{3. 75}$$

The value of t_{PLH} is computed in the same manner with

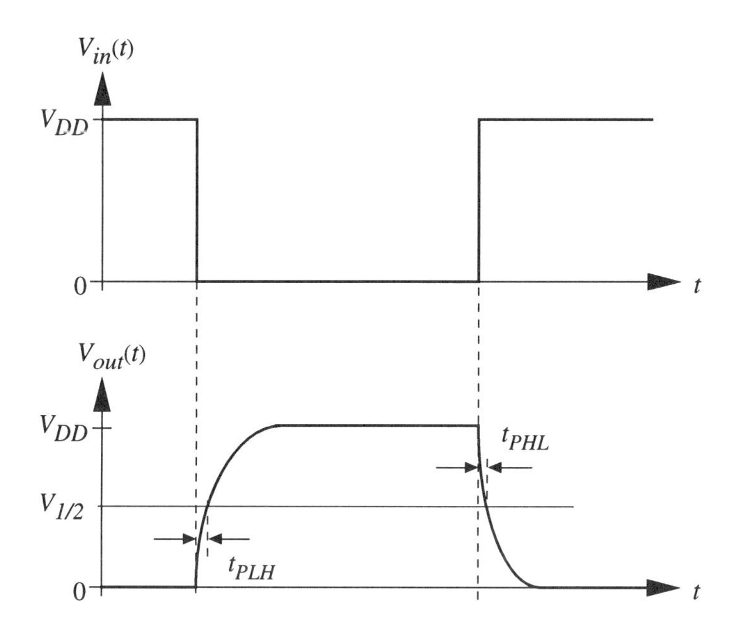

Figure 3.16 Propagation delay times with a step-input voltage

$$t_{PLH} = s_p{}'\tau_p \tag{3.76}$$

where

$$s_p{}' = \left[\frac{2|V_{Tp}|}{(V_{DD}-|V_{Tp}|)} + \ln\left(\frac{4(V_{DD}-|V_{Tp}|)}{V_{DD}} - 1\right)\right] \tag{3.77}$$

provides the scaling factor. Combining terms thus gives

$$t_P = \frac{1}{2}(s_n{}'\tau_n + s_p{}'\tau_p) \tag{3.78}$$

as the total propagation delay. The important factors are once again seen to be the values of the times constants τ_n and τ_p. If one of the times t_{PHL} or t_{PLH} dominates the other, then we often use it for the value of t_P instead of the average value since it is more realistic.

RC Model

The RC model may be used to obtain simpler estimates for the propagation delay. Following the analysis of the RC model in computing t_{HL} allows us to compute the value of t_{PHL} using

$$\begin{aligned} t_{PHL} &= t_{0.5} \\ &= \tau_n \ln\left[\frac{V_{DD}}{0.5V_{DD}}\right] \\ &= \tau_n \ln(2) \end{aligned} \tag{3.79}$$

Similarly,

$$t_{PLH} = \tau_p \ln(2) \tag{3. 80}$$

so using ln(2)≈0.693 gives

$$\begin{aligned} t_P &= \frac{(t_{PHL} + t_{PLH})}{2} \\ &\approx 0.347\,(\tau_n + \tau_p) \end{aligned} \tag{3. 81}$$

as a first estimate. We note once again that the exponential approximation is based on symmetrical rise and fall times, and that the multiplier of ln(2)≈0.693 is due solely to the shape of the curve, not the circuit or device characteristics.

3.2.8 Use of the Step-Input Waveform

The results in this section are based on the assumption that the input waveform has step-like characteristics. This, of course, does not correspond to the real world, but is only a idealized approximation. We can understand this comment by looking at the two cascaded inverters shown in Figure 3.17(a). The input voltage $V_{in}(t)$ to the second inverter is the same as the output of the first inverter,

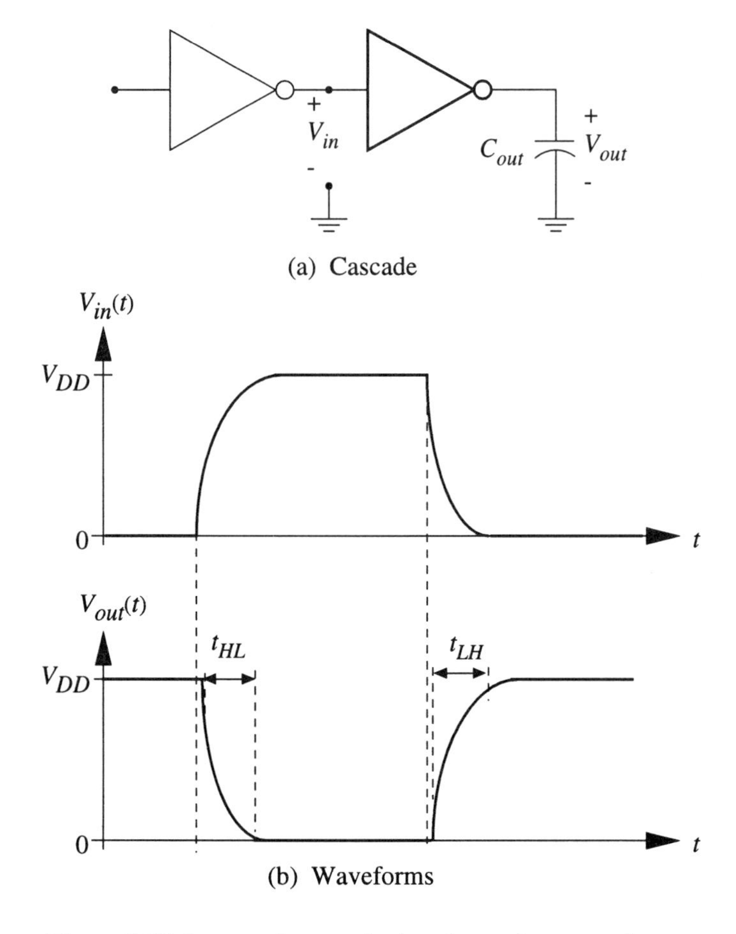

Figure 3.17 Input and output logic voltages in a cascade

so it cannot have step-like characteristics. Instead, the waveform will be more like that illustrated in the plots of Figure 3.17(b). Although it is possible to analyze the response of the circuit using more realistic (and messy) equations for $V_{in}(t)$, the exercise is really academic in nature due to the nature of the approximations contained in the device equations and parasitic elements. When a precise calculation is needed, it should be obtained using a circuit simulator.

With these statements, the reader may be wondering why the step-input analysis was even introduced in the first place. The answer is that, despite their limitations, the results provide reasonable first order estimates of the actual circuit response in a real circuit. More importantly, the equations provide us with a basis for doing circuit design that "tracks" changes in the design iterations. For example, if we change the aspect ratio of one transistor in the inverter and then analyze the new circuit using the simple equations, the new calculated values will change in a manner that will be very close to the differences observed when simulating the old and new designs. This characteristic applies to all of the circuits analyzed in this book, not just the simple inverter.

The propagation time t_p is the exception to this statement. As shown in Figure 3.18, the finite ramps on the input voltage $V_{in}(t)$ are important for determining the values of t_{PHL} and t_{PLH} as they should be calculated between 50% points of V_{in} and V_{out}. This means that we expect a larger error when computing the value of t_p using the step-input waveform. In spite of this observation, it still remains useful for first estimates of the delay time.

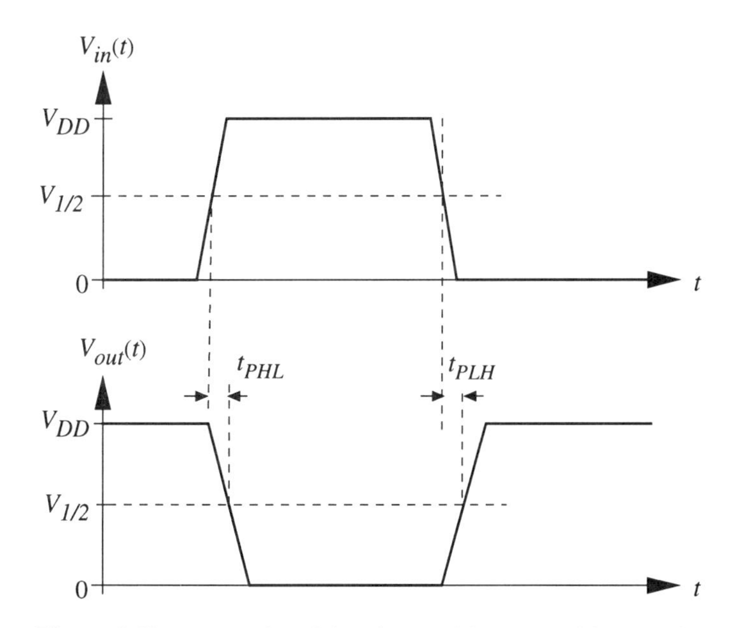

Figure 3.18 Propagation delay times with a ramped input voltage

3.3 Output Capacitance

A quick examination of the calculations above shows that all of the transient times are proportional to the output capacitance C_{out}. A major problem with the approach is the assumption that C_{out} is a linear, time-invariant (LTI) element. This arises because C_{out} contains MOSFET contributions, and the gate-channel and depletion contributions in a FET are nonlinear functions of the voltages. All is not lost, however. We may still use the formulas by defining C_{out} to be an *average value* over the voltage range so long as we exercise caution in interpreting the final results. This means that the analytic approach can be used for the initial design of the circuit and the performance estimates. Increased accuracy can be obtained by a computer simulation when it is needed. In this section we will examine the contributions to C_{out} and illustrate the averaging process.

Figure 3.19 shows the main contributions to C_{out}. Only those capacitors that are directly driven by the output node and also experience a change in voltage during a switching event have been included. The effective value of C_{out} can be obtained by examining the load presented to the output node during a switching event. Consider, for example, the case where V_{in} is initially high, and then falls to a value of V_{in}=0 at time t=0. All of the capacitors shown in the drawing change voltage as C_{out} charges from 0 volts to V_{DD}. Thus, we estimate the output capacitance by

$$\begin{aligned} C_{out} &= [(C_{GDn} + C_{GDp}) + (C_{DBn} + C_{DBp})] + [C_{line} + C_{FO}] \\ &= C_{int} + C_L \end{aligned} \tag{3. 82}$$

which is equivalent to having all of the contributions in parallel.[8] We have split the output capacitance into the internal FET contributions C_{int} and the external load C_L as defined by the square brackets. The gate-drain contributions C_{GD} are due to the gate coupling capacitances, while the drain-bulk terms C_{DB} are nonlinear depletion capacitances. All FET contributions are calculated from the transistor geometries, and the circuit layout provides the necessary dimensions. C_{line} is the interconnect line contribution (which is also sensitive to layout), and C_{FO} represents the fan-out capacitance representing the input values seen looking into the next stage(s).

To obtain an average value for C_{out}, we will approximate the nonlinear terms using simple formulas. First, the gate capacitances for Mn and Mp are given by the respective formulas

$$\begin{aligned} C_{Gn} &= C_{ox} W_n L'_n \\ C_{Gp} &= C_{ox} W_p L'_p \end{aligned} \tag{3. 83}$$

where L'_n and L'_p are the drawn channel lengths. Then, a simple approximation for the gate-drain capacitance for each transistor is to write

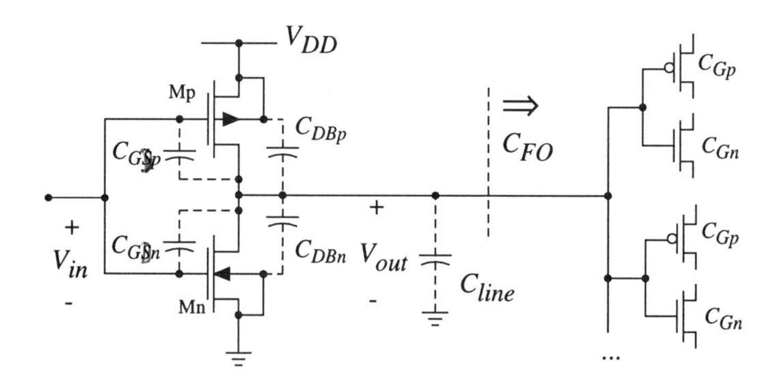

Figure 3.19 Capacitance contributions to C_{out}

[8] Note that C_{DBp} actually discharges in this case. This implies that the formula for C_{out} gives pessimistic results for the switching times.

$$C_{GDn} \approx \frac{1}{2} C_{Gn} \quad \text{and} \quad C_{GDp} \approx \frac{1}{2} C_{Gp} \tag{3.84}$$

for these terms.

Consider next the depletion capacitances C_{DBn} and C_{DBp}. Since the output node voltage reverse biases both drain-bulk junctions, these terms vary during a switching event. An average value is obtained by defining the LTI average value

$$C_{av} = \frac{1}{(V_{OH} - V_{OL})} \int_{V_{OL}}^{V_{OH}} \frac{C_{j0} A}{\left(1 + \frac{V_R}{\phi_o}\right)^m} \tag{3.85}$$

where A is the junction area and m is the grading coefficient. When we apply this formula to a MOSFET junction, both bottom and sidewall contributions must be included. From Chapter 1 we have that for general m,

$$C_{av} = K_m (V_{OL}, V_{OH}) \, C_{j0} A \tag{3.86}$$

where

$$K_m (V_{OL}, V_{OH}) = \frac{\phi_o}{(-m+1)(V_{OH} - V_{OL})} \left[\left(1 + \frac{V_{OH}}{\phi_o}\right)^{(-m+1)} - \left(1 + \frac{V_{OL}}{\phi_o}\right)^{(-m+1)} \right] \tag{3.87}$$

Note that, for the CMOS circuit, $V_{OL} = 0v$ and $V_{OH} = V_{DD}$; the lower limit of 0 helps to simplify the expressions.

To calculate the average capacitance on the bottom we will assume that $m = 1/2$ corresponding to a step-like doping profile. In this case, the expression yields

$$C_{av,bot} = K_{1/2} (0, V_{DD}) \, C_{j0} A_{bot} \tag{3.88}$$

where

$$K_{1/2} (0, V_{DD}) = \frac{2\phi_o}{V_{DD}} \left[\left(1 + \frac{V_{DD}}{\phi_o}\right)^{1/2} - 1 \right] \tag{3.89}$$

is a voltage-averaging factor. Note that $K_{1/2}(0, V_{DD}) < 1$, since the zero-bias capacitance C_{j0} represents the maximum value.

Sidewall doping profiles can often be modelled using $m=1/3$, which is a linearly graded junction. The formula gives

$$C_{av,sw} = K_{1/3} (0, V_{DD}) \, C_{jsw} P \tag{3.90}$$

where C_{jsw} is the zero-bias perimeter capacitance per unit length, P is the perimeter length, and

$$K_{1/3} (0, V_{DD}) = \frac{3\phi_o}{2V_{DD}} \left[\left(1 + \frac{V_{DD}}{\phi_{osw}}\right)^{2/3} - 1 \right] \tag{3.91}$$

is the averaging factor. Note that the sidewall generally has a different built-in potential $\phi_{osw} > \phi_o$ due to higher doping levels.

Combining the bottom and sidewall contributions gives the LTI average values for this case as

$$C_{DBn} = K_{1/2}(0, V_{DD})\, C_{j0n} A_{Dn} + K_{1/3}(0, V_{DD})\, C_{jswn} P_{Dn}$$
$$C_{DBp} = K_{1/2}(0, V_{DD})\, C_{j0p} A_{Dp} + K_{1/3}(0, V_{DD})\, C_{jswp} P_{Dp} \qquad (3.92)$$

where A_{Dn} and P_{Dn} are the bottom area and perimeter for the nFET drain regions, respectively, and A_{Dp} and P_{Dp} are the needed quantities for the pFET drain. Also, the values of C_{j0} and C_{jsw} are different for the nFET and the pFET, as implied by the notation. It is important to note the dependence on layout geometry of the n^+ and p^+ regions.

The line capacitance C_{line} is due to the interconnect wiring. For a simple straight-line geometry, an interconnect that has a distance D and width w gives

$$C_{line} = \frac{\varepsilon_{ox}}{X_{int}} Dw \qquad (3.93)$$

where X_{int} is the thickness of the oxide between the line and the substrate. If another insulator such as silicon nitride is used, then the permittivity must be changed accordingly. Also, this formula ignores both fringing fields and coupling capacitance with neighboring lines.

Finally, the fan-out capacitance C_{FO} represents the capacitance seen looking in the next stage(s) in the chain. If we assume that the output of the inverter is connected to a complementary pair consisting of an nFET and a pFET, then this is simply the sum of gate capacitances in the form

$$C_{FO} = FO \cdot (C_{Gn} + C_{Gp}) \qquad (3.94)$$

where FO is the fan-out. The drawing in Figure 3.19 explicitly shows the case with FO=2 (with the possibility for additional loads denoted by the ellipses ...).

Example 3-2

Consider the inverter layout in Figure 3.20; all dimensions are in units of microns μm. We will go through the complete calculation of the inverter characteristics with the information provided. Many of the calculational techniques can be applied to any MOS circuit.

Basic FET Dimensions

Since the drawn channel is shown as L'=1.2μm and the overlap distance is L_o=0.1μm , the electrical channel length is

$$L = L' - L_o = 1\mu m \qquad (3.95)$$

The aspect ratios are then

$$\left(\frac{W}{L}\right)_n = \left(\frac{5}{1}\right) , \quad \left(\frac{W}{L}\right)_p = \left(\frac{7}{1}\right) \qquad (3.96)$$

and the device transconductance values are calculated as

$$\beta_n = k'_n\left(\frac{W}{L}\right)_n = 500\mu A/V^2$$
$$\beta_p = k'_p\left(\frac{W}{L}\right)_p = 266\mu A/V^2 \qquad (3.97)$$

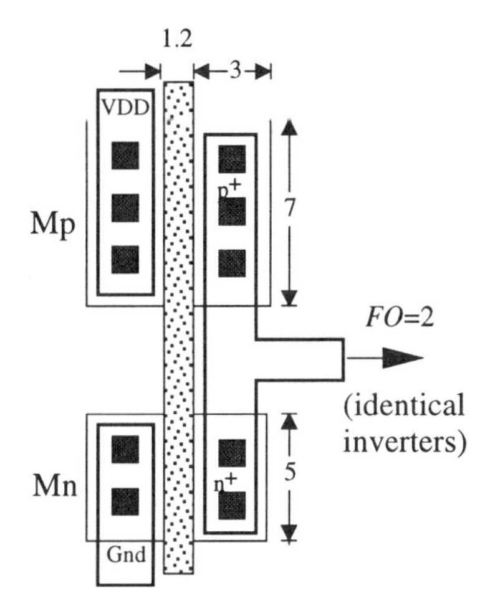

All dimensions in units of μm

Oxide thickness: x_{ox}= 200Å
Gate overlap: L_o−0.10μm
Power supply: V_{DD} =3.3v

nFET Parameters
V_{Tn} = + 0.74 v, k_n' = 100 $\mu A/V^2$
C_{j0} = 2.82 × 10^{-8} F/cm^2, ϕ_o=.90v, m_j=0.5
C_{jsw} = 4.62 × 10^{-12} F/cm, ϕ_{osw}=.95v, m_{jsw}=0.33

pFET Parameters
V_{Tp} = - 0.90 v, k_p' = 38 $\mu A/V^2$
C_{j0} = 4.85 × 10^{-8} F/cm^2, ϕ_o=.92v, m_j=0.5
C_{jsw} = 1.95 × 10^{-12} F/cm, ϕ_{osw}=.97v, m_{jsw}=0.33

Figure 3.20 Layout used in Example 3-2

Inverter Threshold Voltage

This may be calculated directly using

$$V_I = \frac{V_{DD} - |V_{Tp}| + \sqrt{\frac{\beta_n}{\beta_p}}V_{Tn}}{1 + \sqrt{\frac{\beta_n}{\beta_p}}} = \frac{3.3 - 0.9 + \sqrt{\frac{500}{266}}0.74}{1 + \sqrt{\frac{500}{266}}} = 1.44v \tag{3. 98}$$

which is less than one-half of the power supply voltage as expected. The other critical VTC voltages can be computed if needed.

Gate Capacitances

The gate oxide thickness is specified to be x_{ox}=200$\AA$ so

$$C_{ox} = \frac{\varepsilon_{ox}}{x_{ox}} = \frac{(3.9)\ (8.854\times10^{-14})}{200\times10^{-8}} = 1.727\times10^{-7} F/cm^2 \tag{3. 99}$$

or

$$C_{ox} = 1.727\, fF/\mu m^2 \qquad (3.100)$$

where the units of the final answer have been converted to $fF/\mu m^2$ for convenience in hand calculations.

The total gate capacitance of the pFET is

$$\begin{aligned} C_{Gp} &= C_{ox} W_p L' \\ &= (1.727)\,(7)\,(1.2) \\ &= 14.507\, fF \end{aligned} \qquad (3.101)$$

while the nFET gate capacitance is

$$\begin{aligned} C_{Gn} &= C_{ox} W_n L' \\ &= (1.727)\,(5)\,(1.2) \\ &= 10.362\, fF \end{aligned} \qquad (3.102)$$

These may be used to estimate the gate-drain capacitances as

$$\begin{aligned} C_{GDp} &= \frac{1}{2} C_{Gp} = 7.254\, fF \\ C_{GDn} &= \frac{1}{2} C_{Gn} = 3.627\, fF \end{aligned} \qquad (3.103)$$

Depletion Capacitances

We will start with the drain capacitance of the pFET. Consider first the zero-bias values. These are given by

$$\begin{aligned} C_{bot,\,p} &= C_{j0} A_{Dp} = (0.485)\,(3.1)\,(7) = 10.525\, fF \\ C_{sw,\,p} &= C_{jsw} P_{Dp} = (0.195)\,(2)\,(3.1 + 7) = 3.939\, fF \end{aligned} \qquad (3.104)$$

where we have included the gate overlap distance L_o in computing the area and perimeter, and we have made the unit conversions to C_{j0}=$0.485 fF/\mu m^2$ and C_{jsw} =$0.195 fF/\mu m$ for convenience. To obtain average values, we first calculate the averaging factors

$$\begin{aligned} K_{1/2}(0, 3.3) &= \frac{2\,(0.92)}{(3.3)} \left[\left(1 + \frac{3.3}{0.92} \right)^{1/2} - 1 \right] = 0.637 \\ K_{1/3}(0, 3.3) &= \frac{3\,(0.97)}{2\,(3.3)} \left[\left(1 + \frac{3.3}{0.97} \right)^{2/3} - 1 \right] = 0.813 \end{aligned} \qquad (3.105)$$

so that the average value of the drain-bulk capacitance is

$$\begin{aligned} C_{DBp} &= K_{1/2}(0, 3.3)\, C_{bot} + K_{1/3}(0, 3.3)\, C_{sw} \\ &= (0.637)\,(10.525) + (0.813)\,(3.939) \\ &= 9.907\, fF \end{aligned} \qquad (3.106)$$

which completes the calculations.

The value of C_{DBn} is computed in the same manner. First we convert the capacitances to C_{j0}=$0.282 fF/\mu m^2$ and C_{jsw} =$0.462 fF/\mu m$ to find the zero-bias values

$$C_{bot,n} = C_{j0}A_{Dn} = (0.282)\,(3.1)\,(5) = 4.371\,fF$$
$$C_{sw,n} = C_{jsw}P_{Dn} = (0.462)\,(2)\,(3.1+5) = 7.484\,fF \tag{3. 107}$$

The averaging factors are

$$K_{1/2}(0, 3.3) = \frac{2\,(0.90)}{(3.3)}\left[\left(1+\frac{3.3}{0.90}\right)^{1/2}-1\right] = 0.633$$
$$K_{1/3}(0, 3.3) = \frac{3\,(0.95)}{2\,(3.3)}\left[\left(1+\frac{3.3}{0.95}\right)^{2/3}-1\right] = 0.741 \tag{3. 108}$$

so that the average capacitance is

$$C_{DBn} = (0.633)\,(4.371) + (0.741)\,(7.484) = 8.312\,fF \tag{3. 109}$$

which completes the depletion capacitance calculations.

FET Contribution to Output Capacitance

The portion of the output capacitance due to the FETs is the internal capacitance

$$C_{int} = C_{GDp} + C_{GDn} + C_{DBp} + C_{DBn} \tag{3. 110}$$

which is summed to give

$$C_{int} = 7.253 + 5.181 + 9.907 + 8.312 = 30.653\,fF \tag{3. 111}$$

It is important to remember that this is an approximate LTI value.

Fan-Out Capacitance

The drawing shows a fan out into two identical inverter stages. The fan-out capacitance is then computed from

$$C_{FO} = 2\,(C_{Gn} + C_{Gp}) = 2\,(10.362 + 14.507) = 49.738\,fF \tag{3. 112}$$

for this circuit.

Interconnect

The drawing does not provide any information on the interconnect line capacitance, so we could just ignore that contribution and write $C_{line} = 0$. However, it is useful to estimate some contribution for completeness.

Let us assume for the moment that the metal interconnect runs over a field oxide that has a thickness of 0.6 μm. The interconnect capacitance per unit area would be

$$C_{int} = \frac{\varepsilon_{ox}}{x_{int}} = \frac{(3.9)\,(8.854\times10^{-14})}{0.6\times10^{-4}} = 5.76\times10^{-9}\,F/cm^2 \tag{3. 113}$$

The metal line is approximately 2μm wide so we will guess a length of 50μm as being reasonable.

This gives

$$C_{line} = (5.76\times10^{-9})\ (2\times10^{-4})\ (50\times10^{-4}) = 5.755\,fF \tag{3.114}$$

which is a "ballpark" estimate, sufficient for the present purposes.

The external load capacitance C_L is then obtained by

$$\begin{aligned} C_L &= C_{FO} + C_{line} \\ &= 49.738 + 5.755 \\ &= 55.493\,fF \end{aligned} \tag{3.115}$$

Note that this increases if we change the loading on the circuit (or we have more details to calculate C_{line} with.)

Total Capacitance

Summing the contributions above gives C_{out} as

$$\begin{aligned} C_{out} &= C_{int} + C_L \\ &= 30.653 + 55.493 \\ &= 86.146\,fF \end{aligned} \tag{3.116}$$

as the total LTI capacitance at the output node.

Time Constants

We are now in a position to calculate the time constants for the output circuit. Consider first the nFET. The LTI resistance is

$$\begin{aligned} R_n &= \frac{1}{\beta_n (V_{DD} - V_{Tn})} \\ &= \frac{1}{500\times10^{-6}\,(3.3 - 0.74)} \\ &= 781.25\Omega \end{aligned} \tag{3.117}$$

so that the discharge time constant is

$$\begin{aligned} \tau_n &= R_n C_{out} \\ &= (781.25)\ (86.146\times10^{-15}) \\ &= 67.302\,ps \end{aligned} \tag{3.118}$$

Similarly, the pFET resistance is

$$\begin{aligned} R_p &= \frac{1}{\beta_p (V_{DD} - |V_{Tp}|)} \\ &= \frac{1}{266\times10^{-6}\,(3.3 - 0.9)} \\ &= 1566.42\Omega \end{aligned} \tag{3.119}$$

which leads to a time constant of

$$\begin{aligned} \tau_p &= R_p C_{out} \\ &= (1566.416)\,(86.146\times10^{-15}) \\ &= 134.94\,ps \end{aligned} \tag{3. 120}$$

for the charging event.

Switching Times

The switching times are calculated from

$$\begin{aligned} t_{HL} &= s_n \tau_n \\ t_{LH} &= s_p \tau_p \end{aligned} . \tag{3. 121}$$

The voltage multiplier for the high-to-low time is

$$\begin{aligned} s_n &= \frac{2\,(V_{Tn} - V_0)}{(V_{DD} - V_{Tn})} + \ln\left(\frac{2\,(V_{DD} - V_{Tn})}{V_0} - 1\right) \\ &= \frac{2\,(0.74 - 0.33)}{(3.3 - 0.74)} + \ln\left(\frac{2\,(3.3 - 0.74)}{0.33} - 1\right) \\ &= 2.995 \end{aligned} \tag{3. 122}$$

where we have use the 10% voltage $V_0 = 0.1V_{DD} = 0.33v$. Similarly, the pFET multiplier for the low-to-high time is

$$\begin{aligned} s_p &= \frac{2\,(|V_{Tp}| - V_0)}{(V_{DD} - |V_{Tp}|)} + \ln\left(\frac{2\,(V_{DD} - |V_{Tp}|)}{V_0} - 1\right) \\ &= \frac{2\,(0.90 - 0.33)}{(3.3 - 0.90)} + \ln\left(\frac{2\,(3.3 - 0.90)}{0.33} - 1\right) \\ &= 3.082 \end{aligned} \tag{3. 123}$$

The switching times are thus found to be

$$\begin{aligned} t_{HL} &= (2.995)\,(67.302\times10^{-12}) = 201.57\,ps \\ t_{LH} &= (3.082)\,(134.940\times10^{-12}) = 415.89\,ps \end{aligned} \tag{3. 124}$$

as our final result.

The maximum switching frequency is

$$f_{max} = \frac{1}{(t_{HL} + t_{LH})} = 1.62\,GHz \tag{3. 125}$$

and represents the upper limit for the single circuit.

3.4 Inverter Design

Static CMOS gates are straightforward to design. Since rail-to-rail output voltage levels are automatic from the topology of the circuit[9], the design is directed towards either shaping the voltage transfer curve or establishing the necessary transient switching times. This is accomplished by specifying the device parameters β_n and β_p where

$$\beta_n = k'_n\left(\frac{W}{L}\right)_n$$
$$\beta_p = k'_p\left(\frac{W}{L}\right)_p \quad . \tag{3. 126}$$

Choosing the aspect ratios establishes both the DC and the transient switching times. As we will see below, the ratio (β_n/β_p) is chosen to set the inverter midpoint voltage V_I, while the individual values of β_n and β_p are adjusted according to the needed switching times. The design procedure presented here provides the basis for the complex static logic gates examined in Chapter 5.

3.4.1 DC Design

Consider first the DC transfer characteristics. Recall that the inverter threshold voltage V_I is computed from the basic formula

$$V_I = \frac{V_{DD} - |V_{Tp}| + \sqrt{\frac{\beta_n}{\beta_p}}V_{Tn}}{1 + \sqrt{\frac{\beta_n}{\beta_p}}} . \tag{3. 127}$$

The value of V_I can be set by the designer by adjusting the nFET/pFET device transconductance ratio (β_n/β_p), since this determines the gate threshold voltage V_I. This ratio also establishes the critical input voltages V_{IL} and V_{IH}, but there are no simple closed form expressions to show the dependence.

The design equation can be obtained by simply rearranging the current equation to the form

$$\frac{\beta_n}{\beta_p} = \left(\frac{V_{DD} - V_I - |V_{Tp}|}{V_I - V_{Tn}}\right)^2 . \tag{3. 128}$$

This gives the ratio (β_n/β_p) for the desired value of V_I.

Symmetrical Inverter

A symmetrical CMOS inverter is defined to have

$$V_I = \frac{1}{2}V_{DD} \tag{3. 129}$$

i.e., the switch point is at one-half the power supply voltage; this is shown by the plot in Figure 3.21(a). Assuming that the process is polarity-symmetric with $V_{Tn} = |V_{Tp}| = V_T$, this condition can be achieved by designing the circuit with $\beta_n = \beta_p$ since this gives

[9] The topology is how the transistors are connected to form the network.

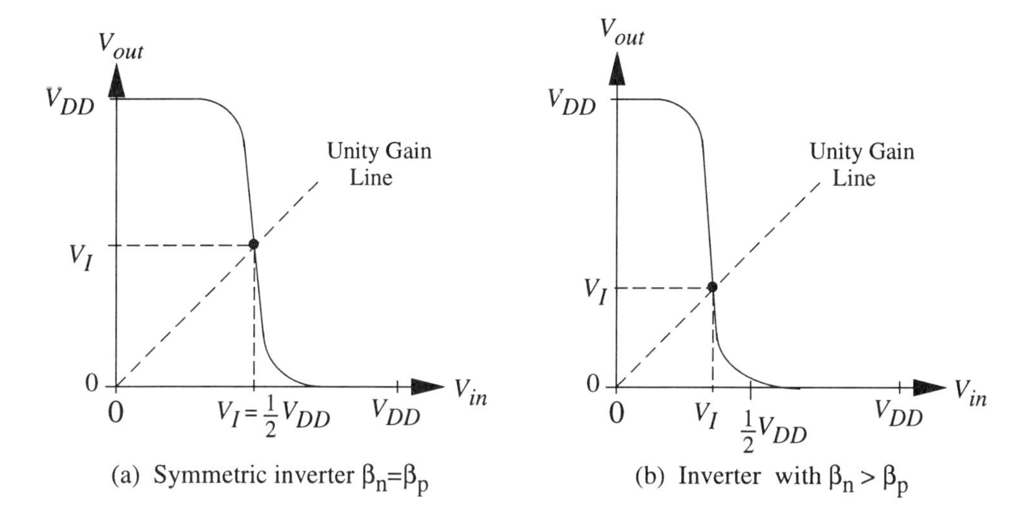

Figure 3.21 Examples of inverter designs

$$V_I = \frac{V_{DD} - |V_{Tp}| + \sqrt{1}V_{Tn}}{1 + \sqrt{1}} = \frac{1}{2}V_{DD} \tag{3. 130}$$

as desired. In terms of the aspect ratios, this requires that

$$\left(\frac{W}{L}\right)_p = \frac{k_n'}{k_p'}\left(\frac{W}{L}\right)_n \quad . \tag{3. 131}$$

Since $k'_n > k'_p$, this design requires that the pFET aspect ratio is larger than the nFET aspect ratio by a factor of $(k'_n/k'_p) \approx 2.5$.

A symmetrical inverter deigned in this manner has critical input voltages given by

$$\begin{aligned} V_{IL} &= \frac{1}{4}\left(V_{Tn} + \frac{3}{4}V_{DD}\right) \\ V_{IH} &= \frac{1}{4}\left(\frac{5}{2}V_{Tn} - V_{DD}\right) \end{aligned} \tag{3. 132}$$

as can be derived directly from the general expressions. Note that

$$V_{IL} + V_{IH} = V_{DD} \tag{3. 133}$$

is valid for this design. The noise margins are equal with

$$VNM_H = VNM_L = \frac{1}{4}\left(V_{Tn} + \frac{3}{4}V_{DD}\right) \tag{3. 134}$$

as verified by direct calculation. Moreover, it will be shown that the switching times are equal: $t_{HL}=t_{LH}$. The main price paid for this type of circuit is that the pFET is relatively large.

Equal Size MOSFETs

Next consider the case where we choose the aspect ratios to be the same: $(W/L)_n = (W/L)_p$. This choice gives a beta-ratio of

$$\frac{\beta_n}{\beta_p} = \frac{k_n'\left(\frac{W}{L}\right)_n}{k_p'\left(\frac{W}{L}\right)_p} = \frac{k_n'}{k_p'} \approx 2.5 \tag{3. 135}$$

i.e., it is set by the ratio of process transconductance values. The inverter threshold voltage is then

$$V_I = \frac{V_{DD} - |V_{Tp}| + \sqrt{\frac{k_n'}{k_p'}}V_{Tn}}{1 + \sqrt{\frac{k_n'}{k_p'}}} < \frac{1}{2}V_{DD} \tag{3. 136}$$

where the inequality is valid for $V_{Tn} \approx |V_{Tp}|$. This shifts the VTC to the left as shown in Figure 3.21(b). It is seen by inspection that, compared to the symmetric design, the value of VNM_L is decreased, while the value of VNM_H is larger.

General Sizing

In the general case, the DC design is controlled by the value of

$$\frac{\beta_n}{\beta_p} = \frac{k_n'\left(\frac{W}{L}\right)_n}{k_p'\left(\frac{W}{L}\right)_p} \tag{3. 137}$$

as discussed in the analysis section above. If the ratio of (β_n/β_p) is greater than 1, then V_I is moved to the left of $(V_{DD}/2)$, while the less-likely case where $(\beta_n/\beta_p)<1$ gives $V_I > (V_{DD}/2)$ as a result.[10] In practice, the actual values of the aspect ratio may be chosen by the layout designer in non-critical datapath circuits simply to fit in the allocated area.

Example 3-3 Example

Consider a process where V_{DD}=3.3v and the device parameters are

$$\begin{aligned} V_{Tn} &= 0.74v \ , \qquad k_n' = 100\mu A/v^2 \ , \\ V_{Tp} &= -0.90v \ , \qquad k_p' = 38\mu A/v^2 \end{aligned} \tag{3. 138}$$

Suppose that we want to design an inverter with V_I=1.5v . The ratio for the device transconductances is

$$\frac{\beta_n}{\beta_p} = \left(\frac{3.3 - 1.5 - 0.9}{1.5 - 0.74}\right)^2 = 1.40 \tag{3. 139}$$

so that the relative device sizes are

---

[10] The case for $(\beta_n/\beta_p)<1$ implies the use of large pFETs, which is usually avoided.

$$\frac{100\left(\frac{W}{L}\right)_n}{38\left(\frac{W}{L}\right)_p} = 1.4 \quad (3.140)$$

or,

$$\left(\frac{W}{L}\right)_p = 1.88\left(\frac{W}{L}\right)_n \quad (3.141)$$

This has the expected results that the pFET must be larger than the nFET.

3.4.2 Transient Design

Using the expressions

$$t_{HL} = s_n\tau_n$$
$$t_{LH} = s_p\tau_p \quad (3.142)$$

we see that the time constants τ_n and τ_p are the important factors, since the multipliers s_n and s_p are constants for a given technology. Transient design revolves around choosing the two aspect ratios $(W/L)_n$ and $(W/L)_p$ for the transistors. The nFET device transconductance β_n establishes the value of t_{HL}. Similarly, the pFET is responsible for charging C_{out} so that β_p is the factor that determines the low-to-high time t_{LH}

Relationship to DC Design

Let us first note that once $(W/L)_n$ and $(W/L)_p$ have been chosen, then are t_{HL} and t_{LH} determined. If we design an inverter with a symmetric VTC using $\beta_n = \beta_p$, then

$$t_{LH} = t_{HL} \quad (3.143)$$

for the situation where $V_{Tn} = |V_{Tp}|$. If instead the devices have equal aspect ratios so that $\beta_n > \beta_p$, then

$$t_{HL} < t_{LH} , \quad (3.144)$$

since the charging current passing through the pFET is less than the discharge current through the nFET. This shows that the DC design sets the general shape of the switching waveforms.

High-Performance Design

High-performance design is generally directed towards the problem of achieving small time delays. To study this problem, let us write the output capacitance as

$$C_{out} = C_{int} + C_L \quad (3.145)$$

where C_{int} is the internal MOSFET capacitance and C_L is the external load capacitance. In terms of the contributions shown in Figure 3.22 we have

$$C_{int} = (C_{GDn} + C_{GDp}) + (C_{DBn} + C_{DBp}) \quad (3.146)$$

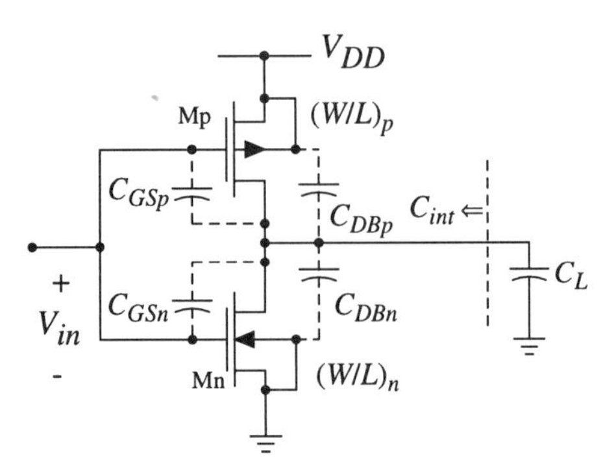

Figure 3.22 Internal and external capacitance terms

for the internal capacitance terms. This split is useful since the external load capacitance C_L is independent of the device aspect ratios, while the parasitic MOSFET contributions in C_{int} depend upon the device sizes. The time constants can thus be broken down into

$$\tau_n = \frac{C_{int}}{\beta_n(V_{DD} - V_{Tn})} + \frac{C_L}{\beta_n(V_{DD} - V_{Tn})} = \tau_n^{int} + \tau_n^L$$
$$\tau_p = \frac{C_{int}}{\beta_p(V_{DD} - |V_{Tp}|)} + \frac{C_L}{\beta_p(V_{DD} - |V_{Tp}|)} = \tau_p^{int} + \tau_p^L \tag{3. 147}$$

which helps to illustrate the design aspects. The switching times are then split into two distinct contributions

$$t_{HL} = s_n\tau_n^{int} + s_n\tau_n^L$$
$$t_{LH} = s_p\tau_p^{int} + s_p\tau_p^L \tag{3. 148}$$

Fast switching is achieved by having small time constants τ_n and τ_p. At first sight, it appears that using large values for β_n and β_p reduces both time constants. However, this is only true for the second terms that are proportional to C_L. The internal capacitance C_{int} is more complicated by the fact that both W_n and W_p enter into the calculations. This can be seen by writing C_{int} as

$$C_{int} = \frac{C_{ox}L'(W_n + W_p)}{2} + K_{1/2}C_{j0n}A_{Dn} + K_{1/3}C_{jswn}P_{Dn} + K_{1/2}C_{j0p}A_{Dp} + K_{1/3}C_{jswp}P_{Dp} \tag{3. 149}$$

where we have assumed that the drawn channel length L' is the same for both devices. In this equation, the depletion areas A_{Dn} and A_{Dp}, as well as the perimeter terms P_{Dn} and P_{Dp}, depend upon the channel widths W_n and W_p. Assuming that the nFET n^+ region is a rectangle with dimensions ($W_n \times X_n$), and that the pFET p^+ region has dimensions ($W_p \times X_p$) allows us to write

$$C_{int} = \frac{C_{ox}L(W_n + W_p)}{2} + K_{1/2}C_{j0n}W_nX_n + K_{1/3}C_{jswn}2(W_n + X_n) + K_{1/2}C_{j0p}W_pX_p + K_{1/3}C_{jswp}2(W_p + X_p) \tag{3. 150}$$

showing the explicit dependence on the values of W_n and W_p. For simplicity, let us write this as

$$C_{int} = C'_{nFET}W_n + C'_{swn}(W_n + X_n) + C'_{pFET}W_p + C'_{swp}(W_p + X_p) \quad (3.151)$$

where

$$\begin{aligned} C'_{nFET} &= \frac{1}{2}C_{ox}L + K_{1/2}C_{j0n}X_n \\ C'_{pFET} &= \frac{1}{2}C_{ox}L + K_{1/2}C_{j0p}X_p \end{aligned} \quad (3.152)$$

give the capacitances per unit channel width W, and

$$\begin{aligned} C'_{swn} &= 2K_{1/3}C_{jswn} \\ C'_{swp} &= 2K_{1/3}C_{jswp} \end{aligned} \quad (3.153)$$

are the sidewall multiplier factors in units of F/cm. Using this for the internal time constants gives

$$\begin{aligned} \tau_n^{int} &= \frac{C'_{nFET}W_n + C'_{swn}(W_n + X_n) + C'_{pFET}W_p + C'_{swp}(W_p + X_p)}{k'_n(W_n/L)(V_{DD} - V_{Tn})} \\ \tau_p^{int} &= \frac{C'_{nFET}W_n + C'_{swn}(W_n + X_n) + C'_{pFET}W_p + C'_{swp}(W_p + X_p)}{k'_p(W_p/L)(V_{DD} - |V_{Tp}|)} \end{aligned} \quad (3.154)$$

which shows the dependences on the channel widths. Since both W_n and W_p appear in the numerator, increasing both aspect ratios $(W/L)_n$ and $(W/L)_p$ by the same amount has a minimal effect on the internal time constants. In fact, if the sidewall terms containing X_n and X_p were absent, then both expressions would be invariant to changes in W_n and W_p.

Improving the performance of the circuit (which means reducing the transient times t_{HL} and t_{LH}) must be accomplished by reducing the time constants due to the external load capacitance C_L. Since these are given by

$$\begin{aligned} \tau_n^L &= \frac{C_L}{k'_n(W_n/L)(V_{DD} - V_{Tn})} \\ \tau_p^L &= \frac{C_L}{k'_p(W_p/L)(V_{DD} - |V_{Tp}|)} \end{aligned} \quad (3.155)$$

we see that using larger aspect ratios can speed up the switching. Physically, this is due to the fact that increasing $\beta = k'(W/L)$ allows for a larger current flow, thus giving faster charging or discharging. For this design approach to be effective, C_L must be the dominant contribution to C_{out}. If, on the other hand, $C_{int} \approx C_L$, then increasing the device sizes has a limited effect on the switching times.

Designing for Load Values

The discussion above shows that transient times can be written in the form

$$\begin{aligned} t_{HL} &= t_{nFET} + aC_L \\ t_{LH} &= t_{pFET} + bC_L \end{aligned} \quad (3.156)$$

where

$$t_{nFET} = s_n \tau_n^{int}$$
$$t_{pFET} = s_p \tau_p^{int} \quad (3.157)$$

are the time delays due to the internal FET capacitances, and

$$a = s_n R_n$$
$$b = s_p R_p \quad (3.158)$$

are the factors (in units of *sec/F*) that set the dependence on the value of the load capacitor. With this form, both t_{HL} and t_{LH} are linear functions of C_L. Plotting the switching times yields graphs that exhibit the features shown in Figure 3.23.

The dependences of $t_{HL}(C_L)$ and $t_{LH}(C_L)$ summarize the transient switching design. The value of the switching times with C_L=0 corresponds to the time needed to drive the internal parasitic capacitances, and are the shortest times possible for a given circuit. The characteristics for driving an external load are determined by the slopes a and b, both of which are inversely proportional to the appropriate FET aspect ratio. Changing (W/L) alters both the vertical (switching time) axis corresponding to the zero-load (C_L=0) transient times, and the slope of the response curve. This can be used to design the circuit to meet timing specifications.

3.5 Power Dissipation

Standard CMOS circuits draw a significant level of current from the power supply only during a switching event. This low power characteristic is due to the complementary behavior of the nMOS and pMOS transistors, and is one reason that the popularity of CMOS increased at such a rapid pace early in the 1990's. Figure 3.24 shows the basic inverter circuit consisting of two MOSFETs. The power supply current I_{DD} is the important factor when calculating power dissipation since

$$P = I_{DD} V_{DD} \quad (3.159)$$

gives the value in units of watts. However, we need to break up the contributions into DC and transient terms to be consistent with the operation of the circuit.

Consider the DC characteristic first; Figure 3.25 provides the important plots. When V_{in} is at a

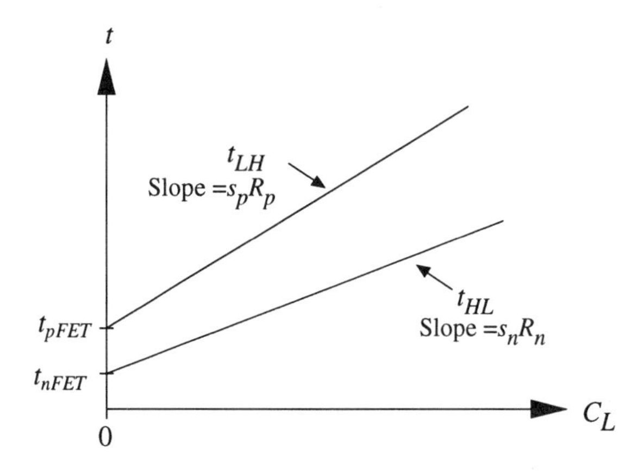

Figure 3.23 Switching times as functions of load capacitance

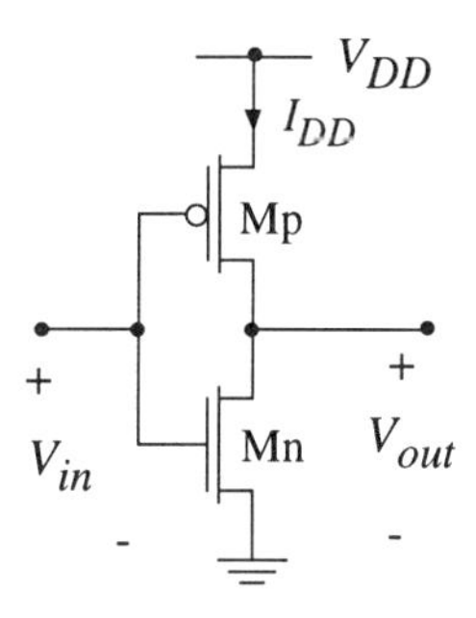

Figure 3.24 Power supply current

stable logic 0 or logic 1 voltage level, either the nFET or the pFET is in cutoff. In this case, there is no direct current flow path through the transistors between the power supply and ground. In a realistic circuit, however, a small amount of quiescent leakage current I_{DDQ} flows across the reverse-biased the drain-bulk regions. The quiescent DC power dissipation is then given by

$$P_{DQ} = I_{DDQ}V_{DD} \tag{3. 160}$$

with I_{DDQ} on the order of $10^{-13}A$ per gate. This is quite small, particularly when compared with bipolar or nMOS-only circuits. When the input voltage is switched between 0 and 1 logic voltages, both FETs conduct as discussed in the analysis of the VTC. The maximum power supply current I_{max} occurs when $V_{in}=V_I$ shown in the plot; this is verified by noting that both the nFET and the pFET are saturated at this point.

The transient power dissipation is more complicated. As seen in the drawing, direct power supply current is consumed when the inverter input voltage is switched. The maximum value I_{max} is set by the device geometries and power supply. However, the DC curves neglect the fact that the output capacitor C_{out} can store electric energy. If we apply a pulsed voltage to the input, then the complete cycle will lead to an additional component of power dissipation that increases with increasing signal frequency f. This can be seen in Figure 3.26. In drawing (a), the input voltage is at a value of V_{in}=0v, which places the nFET in cutoff and the pFET in conduction. C_{out} thus charges to a final voltage of $V_{out} = C_{out} = V_{DD}$. When the input voltage is increased to a high voltage $V_{in}=V_{DD}$ as in Figure 3.26(b), the situation is reversed: Mn is active and Mp is in cutoff. This allows C_{out} to discharge to a final voltage of V_{out}=0v. Since the power supply current is allowed to flow to ground, the sequence of events gives power dissipation when we consider the entire sequence. The amount of transient power dissipated obviously depends upon the rate at which we switch the input, i.e., the signal frequency f.

The **power-delay product** (*PDP*) is often introduced to compare the performance of competing digital technologies. It is defined by

$$PDP = P_{av}t_P \tag{3. 161}$$

where P_{av} is the average power dissipation over a switching cycle, and t_P is the propagation delay time. The *PDP* has units of *Watt-sec* = *Joules*, so that it is often interpreted as the average "energy per switch." Small *PDP* values are desirable, as this implies fast switching and small power dissipation. Since CMOS circuits are characterized by propagation delay times on the order of a nanosecond, they have *PDP* values on the order of picojoules (*pJ*).

To estimate the *PDP* for the CMOS inverter, we first note that the DC contribution

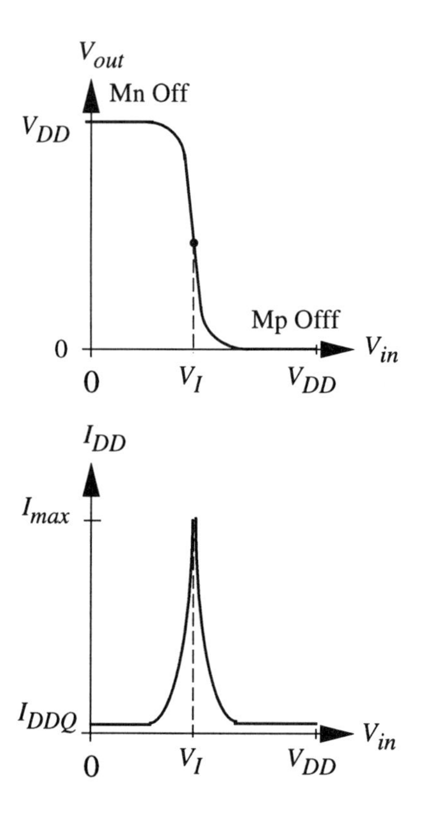

Figure 3.25 Current flow in an inverter circuit

$$PDP = P_{DQ}t_P \tag{3. 162}$$

due to quiescent leakage current is negligible. To find the dynamic contribution to the *PDP*, we first compute the average power dissipation over a cycle with period $T=(1/f)$ by writing

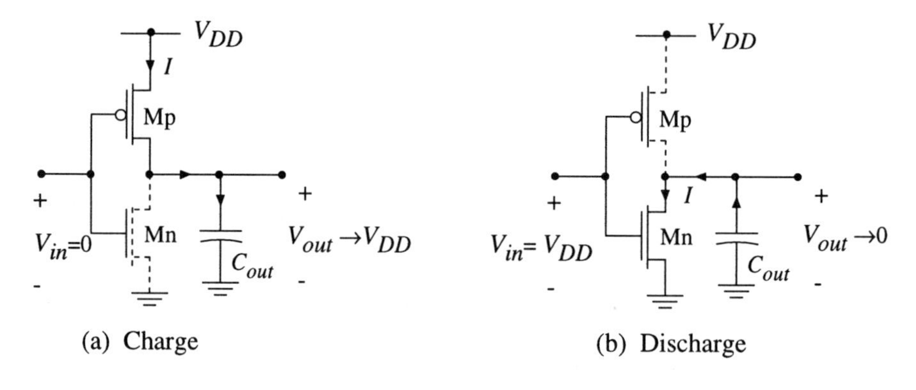

Figure 3.26 Dynamic power dissipation

$$P_{av} = \frac{1}{T}\int_0^T V_{DD} I(\tau)\, d\tau \tag{3. 163}$$

where the current is given by

$$I = \frac{dQ}{dt} = \frac{d}{dt}(C_{out}V) \tag{3. 164}$$

with Q the charge stored on the capacitor. Integrating and taking the maximum value of the voltage to be V_{DD} gives

$$P_{av} = \frac{V_{DD}Q}{T} = \frac{1}{T}C_{out}V_{DD}^2. \tag{3. 165}$$

for the average power.

To account for a switching event during T, we multiply this value by a factor of (1/2), which is equivalent to assuming the clock is high half of the time. Then, the dynamic contribution to the PDP is

$$PDP_{Dynamic} = \frac{1}{2}C_{out}V_{DD}^2\left(\frac{t_P}{T}\right) \tag{3. 166}$$

The factor

$$w_e = \frac{1}{2}C_{out}V_{DD}^2 \tag{3. 167}$$

is simply the energy stored in the capacitor with a voltage V_{DD}. Finally, we may write the actual signal frequency and the maximum signal frequency using

$$f = \frac{1}{T} \qquad f_{max} \approx \frac{1}{2t_P} \tag{3. 168}$$

where we have estimated f_{max} as being determined by twice the propagation delay time. Then we have

$$PDP_{Dynamic} \approx C_{out}V_{DD}^2\left(\frac{f}{f_{max}}\right) \tag{3. 169}$$

as a simple estimate for the dynamic contribution. The total *PDP* is the sum of the DC and transient terms. As the switching frequency increase, so does the *PDP*. This shows that the often-quoted *low-power property* of CMOS really only holds at low frequencies or when circuit are in stable states.

A common rule-of-thumb obtained from this expression is that the transient power dissipation is estimated by

$$P_{tran} = C_{out}V_{DD}^2 f \tag{3. 170}$$

which states that the power dissipation increases with signal frequency. This says that doubling the switching speed of a chip will double the heat dissipation. In other words, "fast chips get hot" is a manifestation of the laws of physics, not poor design.

Example 3-4 CMOS PDP Estimate

Consider an inverter with C_{out}=300fF which is operated with a 5v supply. For f=0.5 f_{max}, we compute

$$PDP_{Dynamic} \approx (300\times10^{-12})\ (25)\ (0.5) = 0.38\,pJ\ . \tag{3. 171}$$

This is a typical order of magnitude.

Power dissipation problems increase with the circuit density when measured in metrics such as the number of transistors per unit area. As the field of high-performance VLSI has advanced, power dissipation and heating have become major aspects of chip design. One reason is obvious: fast switching increases heating. Advances in packaging and novel circuit design techniques have been introduced to deal with this problem. From the viewpoint of circuits, low-power CMOS techniques have been published in the literature, and remain a constant point of interest in research. In this book, we have tried to maintain a general approach to CMOS, so that low-power circuits are not addressed in any detail. The interested reader should consult the journals or one of the fine books that have already been published on the subject.[11]

3.6 Driving Large Capacitive Loads

Consider the situation where we need to drive a large load capacitance C_L as shown in Figure 3.27. Using the results of Section 3.4.2 above, the switching times can be written as

$$\begin{aligned} t_{HL} &= s_n\tau_n^{int} + s_n\tau_n^L \\ t_{LH} &= s_p\tau_p^{int} + s_p\tau_p^L \end{aligned} \tag{3. 172}$$

where the load time constants are given by

$$\begin{aligned} \tau_n^L &= \frac{C_L}{k'_n (W/L)_n (V_{DD} - V_{Tn})} \\ \tau_p^L &= \frac{C_L}{k'_p (W/L)_p (V_{DD} - |V_{Tp}|)} \end{aligned} \quad . \tag{3. 173}$$

In order to maintain fast switching speeds (i.e., short t_{HL} and t_{LH}), the aspect ratios of both transistors must be large. While this solves the problem at the output, the gate will have a large input capacitance since

$$C_{in} = C_{ox}[\,(WL)_n + (WL)_p] \tag{3. 174}$$

is directly proportional to the device sizes. This in turn slows down the preceding gate unless it too is made using large transistors. The problem continues to replicate itself as we move away from the load until we reach a stage where the FETs are "normal" size. Although this requires additional circuitry, using a chain of cascaded gates can help maintain the switching speed of the network. This

[11] Note, however, that most treatments will assume a background at about the level of this entire book, not just the inverter analysis in this chapter.

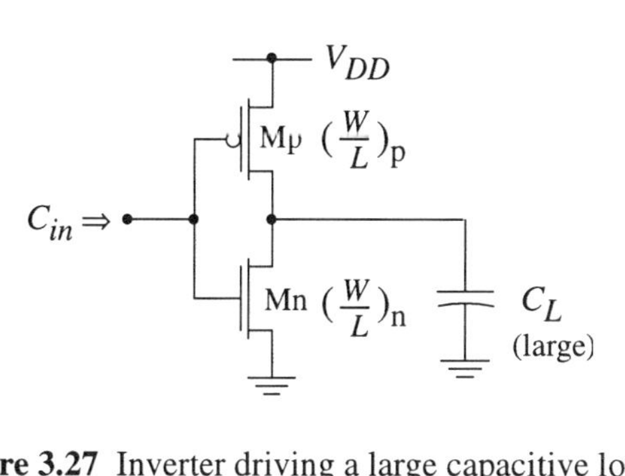

Figure 3.27 Inverter driving a large capacitive load

basic set of observations results in the rule that, when driving a large capacitive load, the switching speed can be made faster by *inserting* a properly designed chain of gates between the input and the load.

The problem of driving a large load capacitor is managed by using a chain of scaled inverters[12] where the FET sizes are increased from the input towards the load. Figure 3.28 illustrates the idea. In Figure 3.28(a), a single inverter would be forced to drive the load. Interestingly enough, the chain of drivers in Figure 3.28(b) can result in a shorter delay if the relative drive capacity of each stage is properly chosen. The chain consists of N inverters, with the input applied to Inverter 1 and the load capacitor C_L attached to the output of Inverter N. If we apply a voltage pulse to $V_{in}(t)$ as shown in Figure 3.29, then the output $V_{out}(t)$ will be delayed by a time t_D that cannot be eliminated.[13] Our program in this section will be to find (i) the number N of inverters, and (ii) the relative sizes of the FETs used in the various stages in a manner that minimizes the propagation delay through the chain.

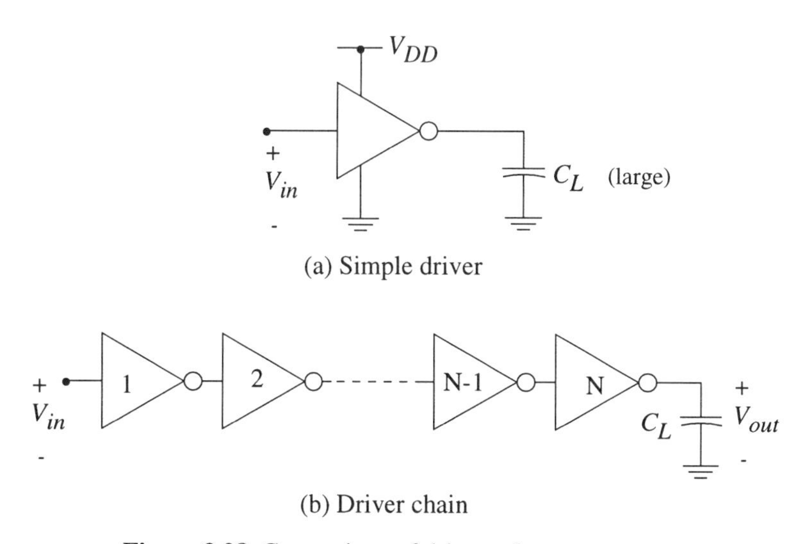

Figure 3.28 Comparison of driver schemes

[12] This approach is not limited to inverters, but can be applied to the more general static logic gates presented in Chapter 5.

[13] We have assumed a non-inverting chain without loss of generality.

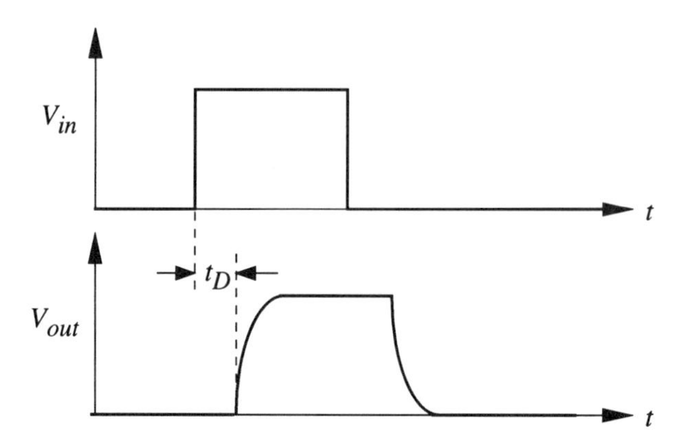

Figure 3.29 Time delay in a driver chain (with an even number of stages)

To attack the problem, we will choose Inverter 1 as a reference and denote the aspect ratio of this stage by $(W/L)_1$; this can be used to represent either $(W/L)_{n1}$ and $(W/L)_{p1}$ as needed, and is assumed to be a known value.[14] The remaining stages in the chain have aspect ratios that increase monotonically such that

$$\left(\frac{W}{L}\right)_1 < \left(\frac{W}{L}\right)_2 < \left(\frac{W}{L}\right)_3 < \ldots < \left(\frac{W}{L}\right)_N \tag{3. 175}$$

This implies that Inverter 1 has the smallest FETs, while Inverter N has the largest FETs. The relative size of the αth stage ($2 < \alpha < N$) with respect to the reference Inverter 1 is designated by the scaling factor $S_\alpha > 1$ such that

$$\left(\frac{W}{L}\right)_\alpha = S_\alpha \left(\frac{W}{L}\right)_1 \tag{3. 176}$$

The breakdown of the problem at this point is shown in Figure 3.30. Note that since every state is referenced to Stage 1, we are allowing for a distinct value of S_α for each stage.

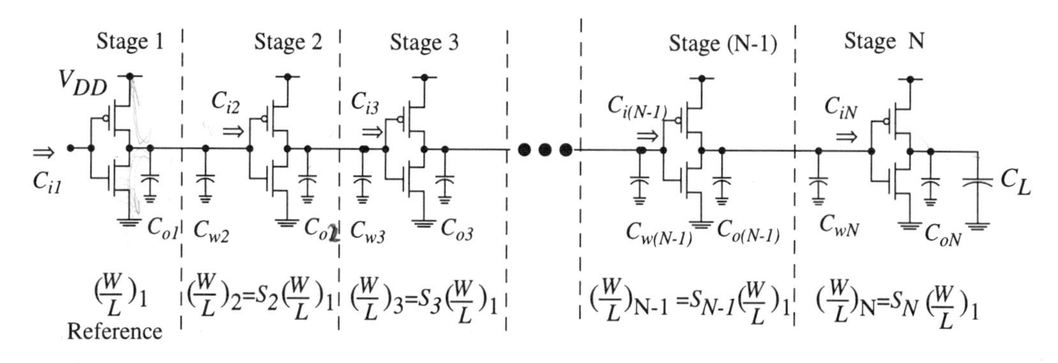

Figure 3.30 Scaling of FETs in the driver chain

[14] Alternately, we could perform the entire analysis using β_1 for Stage 1 as a reference.

To understand the effect of the scaling, recall that the parasitic resistance and capacitance of a MOSFET vary with the aspect ratio. Inverter 1 has an input capacitance C_{i1} given by

$$C_{i1} = C_{ox}[(WL)_{n1} + (WL)_{p1}] \tag{3.177}$$

and an output capacitance C_{o1} that is estimated by

$$C_{o1} = C_{GDn1} + C_{GDp1} + C_{SBn1} + C_{SBp1} \tag{3.178}$$

with each contribution dependent on the width of the transistors.

The drain-source resistance of either FET can be estimated by using the simple LTI expression

$$R_1 \approx \frac{1}{k\left(\frac{W}{L}\right)_1 (V_{DD} - V_T)} \tag{3.179}$$

with appropriate nFET or pFET parameters. The values of R_1, C_{i1} , and C_{o1} are used as the reference for stages further along the chain. The parasitics of the αth stage are summarized in Figure 3.31. Recall that the drain-source resistance of a MOSFET is inversely proportional to the aspect ratio, while the device capacitances are approximately proportional to the channel width. Since the aspect ratio of this stage is increased by a factor S_α relative to Inverter 1, we write

$$R_\alpha = \frac{R_1}{S_\alpha} \qquad C_{i\alpha} = S_\alpha C_{i1} \qquad C_{o\alpha} \approx S_\alpha C_{o1} \tag{3.180}$$

In addition, we have include the line "wiring" capacitance $C_{w\alpha}$ in the drawing to provide a bit more accuracy. This is associated with the input side of the inverter, and is assumed to scale according to

$$C_{w\alpha} \approx S_\alpha C_w \tag{3.181}$$

where C_{w1} is the input wiring capacitance of Invert

Now let us turn to the problem of calculating the delay through the inverter chain. As shown in Figure 3.31, the αth stage drives the $(\alpha+1)^{th}$ stage which consists of parasitic capacitors $C_{o\alpha}$, $C_{i(\alpha+1)}$, and $C_{w(\alpha+1)}$ in parallel with each other. The time constant for the αth stage can thus be written as

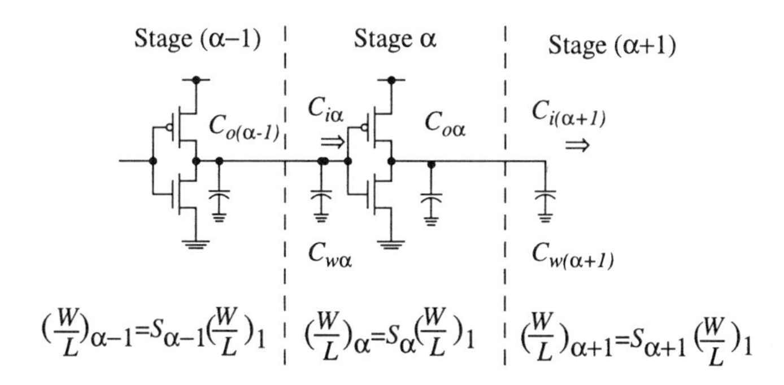

Figure 3.31 Circuitry for the αth stage in the chain

$$\tau_\alpha = R_\alpha [C_{o\alpha} + C_{i(\alpha+1)} + C_{w(\alpha+1)}]$$
$$= \frac{R_1}{S_\alpha}[S_\alpha C_{o1} + S_{\alpha+1}(C_{i1} + C_{w1})] \tag{3. 182}$$

The total time delay t_D through the chain can be estimated by summing over all N stages to give

$$t_D = \sum_{\alpha=1}^{N} R_1\left[C_{o1} + \frac{S_{\alpha+1}}{S_\alpha}(C_{i1} + C_{w1})\right] \tag{3. 183}$$

Note that the first term in the square brackets is independent of the scaling factor S_α.

To find the minimum delay time, let us differentiate t_D with respect to S_α and set the derivative to zero. Since a summation is involved, we must include all relevant terms in the calculation. These are obtained by looking at the $(\alpha-1)^{th}$ and α^{th} terms since both include S_α as shown by

$$\frac{\partial t_D}{\partial S_\alpha} = \frac{\partial}{\partial S_\alpha}\left(\ldots + \frac{S_\alpha}{S_{\alpha-1}}(C_{i1} + C_{w1}) + \frac{S_{\alpha+1}}{S_\alpha}(C_{i1} + C_{w1}) + \ldots\right) \tag{3. 184}$$

Performing the differentiating gives the **recursion relation**

$$\frac{S_\alpha}{S_{\alpha-1}} = \frac{S_{\alpha+1}}{S_\alpha} \tag{3. 185}$$

for the scaling factors. A moment's reflection shows that the only way this can be satisfied for any value of α is to have both sides equal to a constant K such that

$$\frac{S_{\alpha+1}}{S_\alpha} = K. \tag{3. 186}$$

To apply this, we will note that the problem has associated with it a pair of boundary conditions on S_α such that

$$S_1 = 1$$
$$S_{N+1} = \frac{C_L}{C_1} \tag{3. 187}$$

The first statement merely reflects the fact that we are using Inverter 1 as the reference. The second expression views the load capacitor C_L as the input capacitance into the "next stage" which would be numbered as $(N+1)$; as illustrated in Figure 3.32, this requires that

$$C_{N+1} = C_L = S_{N+1}C_1 \tag{3. 188}$$

to be consistent.

Now, let us form the product

$$\frac{S_2}{S_1}\frac{S_3}{S_2}\frac{S_4}{S_3}\cdots\frac{S_N}{S_{N-1}}\frac{S_{N+1}}{S_N} = \frac{S_{N+1}}{S_1} = K^N \tag{3. 189}$$

which gives

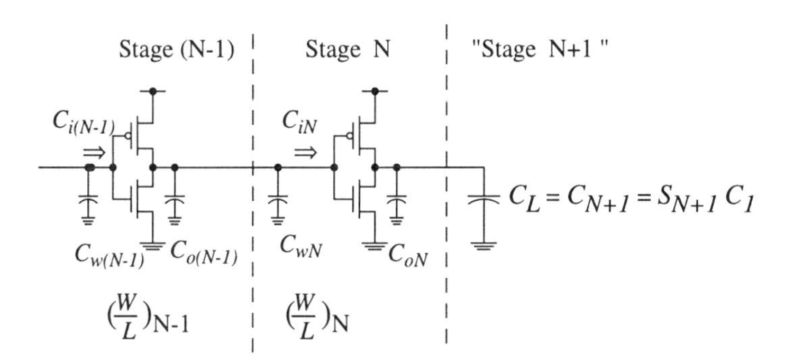

Figure 3.32 End boundary condition for the chain

$$\frac{C_L}{C_1} = K^N \tag{3. 190}$$

by applying the boundary conditions. Thus, the constant K is found as

$$K = \frac{S_{\alpha+1}}{S_\alpha} = \left(\frac{C_L}{C_1}\right)^{1/N} \tag{3. 191}$$

with C_L and C_1 being known values. Using this relationship gives the scaling factors as

$$\begin{aligned} S_1 &= 1 \\ S_2 &= K \\ S_3 &= K^2 \\ &\cdots \\ S_N &= K^{N-1} \end{aligned} \tag{3. 192}$$

for each stage. The total time delay associated with this choice is

$$\begin{aligned} t_D &= \sum_{\alpha=1}^{N} R_1 [C_{o1} + K(C_{i1} + C_{w1})] \\ &= NR_1 [C_{o1} + K(C_{i1} + C_{w1})] \end{aligned} \tag{3. 193}$$

This shows that the minimum delay through the chain is achieved by equalizing the delay through every stage.

To complete the analysis, we need to determine the number of stages N required in the chain. This is accomplished by differentiating

$$t_D = NR_1 \left[C_{o1} + \left(\frac{C_L}{C_1}\right)^{1/N} (C_{i1} + C_{w1}) \right] \tag{3. 194}$$

with respect to N. Taking the derivative gives

$$\frac{dt_D}{dN} = R_1\left[C_{o1} + \left(\frac{C_L}{C_1}\right)^{1/N}(C_{i1} + C_{w1})\right] + N(C_{i1} + C_{w1})\frac{d}{dN}\left(\frac{C_L}{C_1}\right)^{1/N} \tag{3. 195}$$

where we have evaluated the simple terms, but will take time to examine how to perform the last differentiation.[15] Let us first take the natural logarithm to write

$$\begin{aligned} \ln\left(\frac{C_L}{C_1}\right)^{1/N} &= \left(\frac{1}{N}\right)\ln\left(\frac{C_L}{C_1}\right) \\ &= f(N) \end{aligned} \tag{3. 196}$$

so that exponentiating both sides gives us the alternate expression

$$\left(\frac{C_L}{C_1}\right)^{1/N} = e^{f(N)} . \tag{3. 197}$$

Differentiating gives

$$\begin{aligned} \frac{d}{dN}\left(\frac{C_L}{C_1}\right)^{1/N} &= e^{f(N)}\frac{df}{dN} \\ &= \left(\frac{C_L}{C_1}\right)^{1/N}\left[-\frac{1}{N^2}\ln\left(\frac{C_L}{C_1}\right)\right] \end{aligned} \tag{3. 198}$$

Substituting and equating the derivative to zero then shows that the minimum value of N is obtained when

$$C_{o1} + \left(\frac{C_L}{C_1}\right)^{1/N}(C_{i1} + C_{w1})\left[1 - \frac{\ln(C_L/C_1)}{N}\right] = 0 \tag{3. 199}$$

is satisfied. If C_{ol} is small, this gives the classical result

$$N \approx \ln\left(\frac{C_L}{C_1}\right) \tag{3. 200}$$

In a practical situation, the nearest integer value would be chosen. Note that this implies that the equation for scaling constant K assumes the form

$$K^N = K^{\ln(C_L/C_1)} = \left(\frac{C_L}{C_1}\right) \tag{3. 201}$$

with the solution $K=e \approx 2.71$ (the Euler constant). In general, however, the value of the scaling constant K is larger than e when the other terms are included.

[15] This can be added to your mathematical "bag of tricks" for future reference!

Example 3-5

Suppose that we want to drive a load capacitor of C_L=50pF. Our "standard-size" inverter has an input capacitance of C_I=140fF. Using these values, we calculate the number of stages in our driver chain as

$$\hat{N} \approx \ln\left(\frac{50\times10^{-12}}{140\times10^{-15}}\right) = 5.88 \tag{3. 202}$$

which implies that we would need N=6 stages. Estimating K=3 then gives the scaling factors of

$$S_1 = 1 \qquad S_2 = 3 \qquad S_3 = 3^2 = 9$$
$$S_4 = 3^3 = 27 \qquad S_5 = 3^4 = 108 \qquad S_6 = 3^5 = 324 \tag{3. 203}$$

which would require exceptionally large FETs in the 5th and 6th stages. While aspect ratios of 100 would not be unacceptable in practice, it is still difficult to attain the optimization using the algorithm derived here.

A simpler approach to driving large capacitive loads can be illustrated by the drawing in Figure 3.33. Since the most important objective is to drive C_L, we can start with the output stage N and choose the aspect ratios $(W/L)_n$ and $(W/L)_p$ to meet specific t_{HL} and t_{LH} objectives. Once these are established, we can find the input capacitance C_x which can be used to design the (N-1)st stage for values $(W/L)_{nx}$ and $(W/L)_{px}$; using the same t_{HL} and t_{LH} values is equivalent to equalizing the delay through the two stages. The procedure is continued by using C_y as a basis for computing $(W/L)_{ny}$ and $(W/L)_{py}$ and so on, until the FET sizes reduce to the desired values. This is very similar to the more mathematical algorithm derived above, but concentrates on the actual value of the output switching times needed to achieve the design specifications.

Other approaches to chain scaling can be found in the literature. Most are based on similar models but propose other types of sizing schemes such as nonlinear or exponential increases from the input to the load.

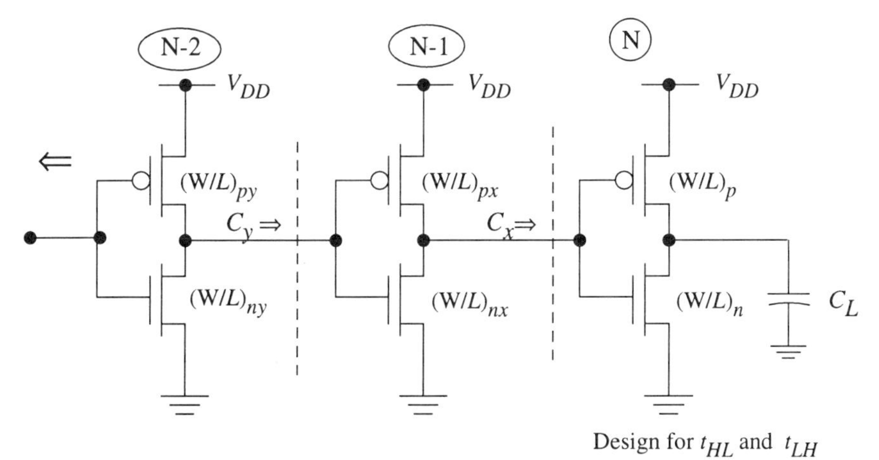

Figure 3.33 Simple approach to driver chain design

3.7 Problems

[3-1] Consider a CMOS inverter that is designed in a process with the following parameters: $k'_n = 100\ \mu A/V^2$, $k'_p = 40\ \mu A/V^2$, $V_{TOn} = +0.7v$, and $V_{TOp} = -0.8v$. The transistors have aspect ratios of $(W/L)_n = 10$ and $(W/L)_p = 15$, and the power supply is chosen to be $5v$.

(a) Calculate the value of the inverter midpoint voltage $V_M = V_I$.

(b) Calculate the values of V_{IL} and V_{IH}, and then find the voltage noise margins.

(c) Obtain the VTC using a SPICE .DC analysis.

[3-2] Rework Problem [3-1] for a power supply voltage of $3v$.

[3-3] Consider a CMOS inverter that is designed in process with parameters of $x_{ox} = 150$ Å, $\mu_n = 580 cm^2/V\text{-}sec$, $\mu_p = 235\ cm^2/V\text{-}sec$, $V_{TOn} = +0.75v$, and $V_{TOp} = -0.80v$. The transistors have aspect ratios of $(W/L)_n = 12$ and $(W/L)_p = 12$, and the power supply is chosen to be $3v$.

(a) Calculate the value of the inverter midpoint voltage $V_M = V_I$.

(b) Calculate the values of V_{IL} and V_{IH}, and then find the voltage noise margins.

(c) Obtain the VTC using a SPICE .DC analysis.

(d) Suppose that the output capacitance is estimated to be $C_{out} = 140 fF$. Calculate the values of t_{HL} and t_{LH} for the circuit.

(e) Construct the RC-equivalent switching model for the circuit, and then compute t_{HL} and t_{LH}. Compare your values with those found in part (d).

[3-4] A CMOS inverter is constructed using the process described in Problem [3-1].

(a) Design the circuit to have a value of $V_I = 0.4V_{DD}$ with $V_{DD} = 5v$.

(b) Suppose that we estimate the total output capacitance for a FO=2 circuit to be $C_{out} = 3C_{in}$. with $L' = 2\mu m$. Calculate the values of t_{HL} and t_{LH} for your design if the smallest permitted aspect ratio is 4. (Assume that the drawn and electrical channel lengths are approximately the same for this calculation.)

[3-5] The inverter of Problem [3-1] has a total output capacitance of $C_{out} = 120 fF$.

(a) Calculate the value of t_{HL}.

(b) Calculate the value of t_{LH} and f_{max}.

(c) Calculate the propagation delay t_p.

(d) Construct the RC-equivalent circuit for the inverter. Then use the simplified expressions to find the transient times. How do they compare with the values above?

(e) Perform a SPICE simulation of the transient characteristics. Compare your results to the hand estimates. [Use only the information provided in the problem statements, i.e., do not add any other parameters to the .Model listing].

[3-6] Consider a CMOS inverter that is designed in a process with the following parameters:

$$k'_n = 200\ \mu A/V^2,\ k'_p = 85\ \mu A/V^2,\ V_{TOn} = +0.75v, \text{ and } V_{TOp} = -0.7v.$$

The transistors have aspect ratios of $(W/L)_n = 8$ and $(W/L)_p = 12$, and the power supply is chosen to be $3.3v$.

(a) Calculate the value of the inverter midpoint voltage V_M.

(b) Calculate the values of V_{IL} and V_{IH}, and then find the voltage noise margins.

(c) Obtain the VTC using SPICE.

[3-7] A CMOS inverter that is designed in process that has a gate oxide thickness of 270 Å. In addition, the following parameters are known: $\mu_n = 580\ cm^2/V\text{-}sec$, $\mu_p = 220\ cm^2/V\text{-}sec$, $V_{TOn} = +0.75v$, and $V_{TOp} = -0.85v$. The transistors have aspect ratios of $(W/L)_n = (8/1)$ and $(W/L)_p = (16/1)$ based on an electrical channel length of $1\mu m$ and a gate overlap of L_o=$0.1\mu m$ (so that the drawn channel length is $1.20\mu m$). The power supply is chosen to be $5v$.

(a) Calculate the device transconductance values for both FETs.

(b) Calculate the value of the inverter threshold voltage V_I.

(c) Find the values of V_{IL} and V_{IH}.

(d) Calculate the input capacitance C_{in} seen looking into the inverter.

(e) Assume that the internal (FET) capacitance has a value of $C_{int}=0.7C_{in}$. Plot the values of t_{LH} and t_{HL} for the inverter driving external loads ranging from $C_L=0$ to $C_L=5C_{in}$.

[3-8] Consider the layout shown in Figure P3.1 and perform the calculations below. Use the process parameters given in Example 3-2 on page 129 unless the instructor provides an alternate list.

(a) Calculate the values of C_{GDn} and C_{GDp}.

(b) Find the voltage-average value of C_{DBn}.

(c) Find the voltage-average value of C_{DBp}.

(d) Estimate the switching times t_{LH} and t_{HL} when an external load of C_L=$100fF$ is attached to the output.

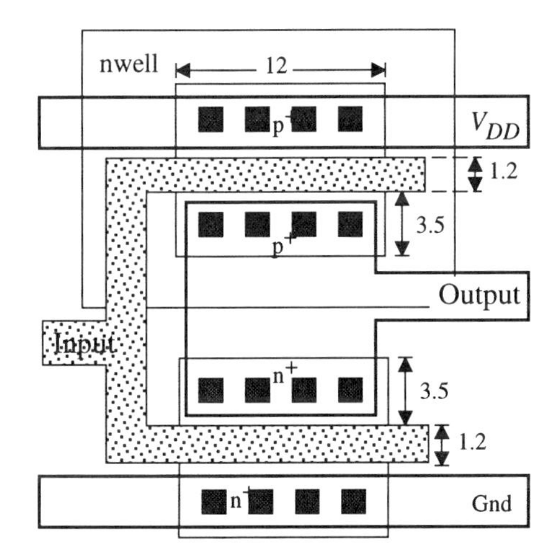

Figure P3.1

[3-9] Use the information in Problem [3-6] to perform a full SPICE simulation of the inverter shown in Figure P3.1. Be sure to convert the parameters to appropriate units. Obtain the VTC and simulate the transient response using a PULSE input; assume an external load of C_L=$100fF$.

[3-10] Construct the simple RC switching equivalent for the inverter described in Problem [3-6] and drawn in Figure P3.1. Assume an external load of C_L=$100fF$, and then estimate the switching times t_{LH} and t_{HL}.

[3-11] Consider a CMOS process that has x'_{ox}=$120Å$. We wish to design an inverter chain to drive a load capacitance of C_L=$25pF$. The first stage in the chain is an inverter that uses FETs with the same aspect ratio of $(W/L)_n=(W/L)_p=(10\mu m/1\mu m)$.

(a) Use the idealized analysis to find the number of stages needed in the chain and the scaling factors.

(b) The equal-sized FETs will give an asymmetric output signal from each inverter. Is there a way to produce a symmetrical output waveform from the chain if we have a symmetrical input waveform?

[3-12] Consider the circuit shown in Figure P3.2. Use the process parameters given in Figure 3.20 on page 129 to do the following.

(a) Calculate the value of the inverter midpoint voltage $V_M = V_I$.

(b) Calculate the values of V_{IL} and V_{IH}, and then find the voltage noise margins.

(c) Obtain the VTC using a SPICE .DC analysis.

(d) Suppose that the external load capacitance is $C_L = 125fF$. Calculate the values of t_{HL} and t_{LH} for the circuit.

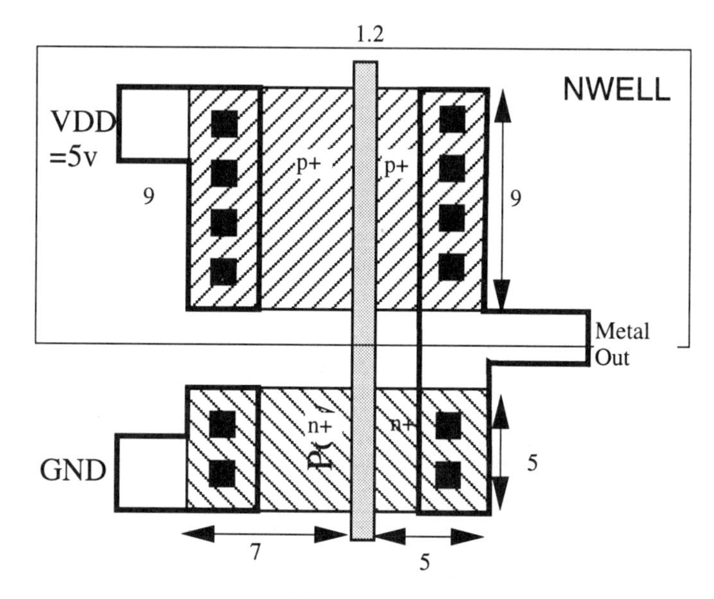

Figure P3.2

3.8 References

All of the books listed below examine the CMOS inverter from different perspectives.

[1] A. Bellaouar and M.I. Elmasry, **Low Power Digital VLSI Design**, Kluwer Academic Publishers, Norwell, MA, 1995.

[2] A.P. Chandrakasan and R.W. Broderson, **Low Power Digital CMOS Design**, Kluwer Academic Publishers, Norwell, MA, 1995.

[3] T.A. DeMassa and Z. Ciccone, **Digital Integrated Circuits**, John Wiley & Sons, New York, 1996.

[4] D.A. Hodges and H.J. Jackson, **Analysis and Design of Digital Integrated Circuits**, 2nd. ed., McGraw-Hill, New York, 1988.

[5] J.M. Rabaey, **Digital Integrated Circuits**, Prentice-Hall, Upper Saddle River, NJ, 1996.

[6] M. Shoji, **CMOS Digital Circuit Technology**, Prentice-Hall, Englewood Cliffs, NJ, 1988.

[7] N. Weste and K. Eshraghian, **CMOS VLSI Design**, 2nd ed., Addison-Wesley, Reading, MA, 1994.

[8] J.M. Rabaey and M. Pedram (eds.), **Low Power Design Methodologies**, Kluwer Academic Publishers, Norwell, MA, 1996.

[9] Gary K. Yeap, **Practical Low Power Digital VLSI Design**, Kluwer Academic Publishers, Boston, 1998.

Chapter 4

Switching Properties of MOSFETs

High-density logic design requires circuits that are compact and exhibit fast switching speeds. Regardless of the design style used to construct the circuits, the switching characteristics of MOSFETs constitute the fundamental limitation on performance. In this chapter, we will analyze the behavior of FETs when used as **pass transistors**, i.e., voltages are passed through the device from source to drain or vice versa, with the gate acting only as a control electrode. This characterization provides us with the details of how nFETs and pFETs can be used in logic design.

4.1 nFET Pass Transistors

Let us analyze the nFET pass transistor shown in Figure 4.1. The input has a voltage V_{in} and the output is connected to a capacitive load C_{out} that has a voltage V_{out} across it. The capacitor C_{out} represents the total capacitance at the output node and has several contributions. Placing a high voltage V_G=V_{DD} onto the gate biases the nFET into conduction, connecting the input and output nodes. To clarify the voltage transmission properties of the transistor when used in this arrangement, we will

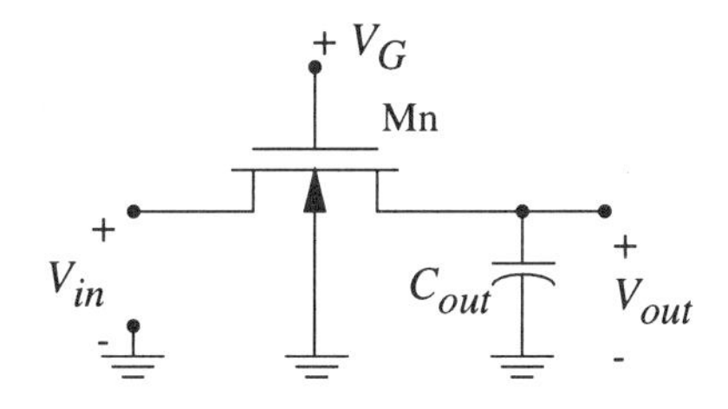

Figure 4.1. Basic nFET pass transistor

examine the two cases of transmitting logic 1 (high voltage) state, and transmitting a logic 0 (low voltage) state through the device using separate analyses.

4.1.1 Logic 1 Input

Transferring a logic 1 state through the nFET corresponds to setting $V_{in} = V_{DD}$ at the input, and then computing the voltage $V_{out}(t)$ at the output. Figure 4.1(a) shows the voltages for this case. Let us assume that C_{out} is initially uncharged so that at time $t = 0$, $V_{out}(0) = 0v$. In this case, the input side of the transistor is the drain, while the output side is the source terminal.[1] The device voltages are given by

$$\begin{aligned} V_{GSn} &= V_{DD} - V_{out}(t) \\ V_{DSn} &= V_{DD} - V_{out}(t) \end{aligned} \tag{4. 1}$$

Since $V_{DSn} > V_{sat} = (V_{GSn} - V_{Tn})$, the device conducts in the saturated mode giving

$$I = C_{out}\frac{dV_{out}}{dt} = \frac{\beta_n}{2}(V_{DD} - V_{out} - V_{Tn})^2 \tag{4. 2}$$

for the charging operation. This may be rearranged and integrated to read as

$$\int\frac{dV_{out}}{(V_{DD} - V_{out} - V_{Tn})^2} = \int\frac{\beta_n}{2C_{out}}dt \tag{4. 3}$$

Using the initial condition $V_{out}(0) = 0$ in the integral results in the expression

$$V_{out}(t) = (V_{DD} - V_{Tn})\left[\frac{t/2\tau_n}{1 + t/2\tau_n}\right] \tag{4. 4}$$

where

$$\tau_n = \frac{C_{out}}{\beta_n(V_{DD} - V_{Tn})} = R_n C_{out} \tag{4. 5}$$

is the time constant for the event.[2]

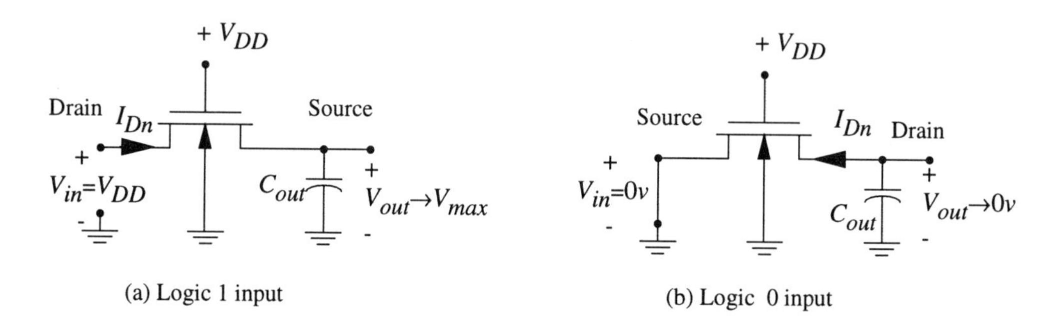

(a) Logic 1 input (b) Logic 0 input

Figure 4.2. nFET pass arrangements

[1] Recall that, for an nFET, the drain is defined as the n^+ side at the higher voltage.

[2] Note that the time constant does not refer to an exponential function, but a more complicated behavior.

The problems involved in using a single nMOS pass transistor are obvious. First, the charging of the output capacitor is described by time dependence that starts out linear as $(t/2\tau_n)$, and then slowly levels out. Physically, this is due to the fact that since $V_{out}(t)$ increases in time, both of the device bias voltages

$$V_{GSn} = V_{DD} - V_{out}(t) = V_{DSn} \tag{4.6}$$

decrease with time. A decreasing value of V_{GSn} reduces the channel charge density, while a smaller V_{DSn} indicates a reduction in the drain-source electric field. Both factors imply that it is difficult to pass a logic 1 voltage through the n-channel transistor. The second important point to note is that, in the limit where $t \to \infty$,

$$V_{out} \to (V_{DD} - V_{Tn}) = V_{max} \tag{4.7}$$

showing that the nFET cannot pass the power supply voltage V_{DD} from drain to source. This is called **a threshold voltage loss**. Although it may appear that this is due to a limit from the applied voltage at the drain, it is really due to the fact that V_{DD} is applied to the gate. In order to maintain conduction through the device, the gate-source voltage must have a minimum value of $\min(V_{GSn}) = V_{Tn}$ to maintain the inversion layer; applying KVL then requires that

$$\begin{aligned} \max(V_{out}) &= V_G - \min(V_{GSn}) \\ &= V_{DD} - V_{Tn} \end{aligned} \tag{4.8}$$

which is in agreement with the original equation. Also note that since $V_{SBn} = V_{out}$, body bias effects are present with

$$V_{Tn} = V_{T0n} + \gamma(\sqrt{2|\phi_F| + V_{out}} - \sqrt{2|\phi_F|}) \tag{4.9}$$

The maximum value of the output voltage $V_{out} = V_{max}$ is computed from solving the transcendental equation

$$\begin{aligned} V_{max} &= V_{DD} - [V_{T0n} + \gamma(\sqrt{2|\phi_F| + V_{max}} - \sqrt{2|\phi_F|})] \\ &= (V_{DD} - V_{T0n}) - \gamma(\sqrt{2|\phi_F| + V_{max}} - \sqrt{2|\phi_F|}) \end{aligned} \tag{4.10}$$

for the logic 1 voltage. The two characteristics of slow logic 1 transfers and the threshold voltage loss can be critical factors when using nFETs as pass transistors in high-speed circuit design.

Example 4-1

Consider an n-channel pass transistor that is connected as shown in Figure 4.2(a). We will assume device parameter value of V_{T0n}= 0.7v, $2|\phi_F|$=0.58v, γ= 0.053 $v^{1/2}$. With V_{DD}=3.3v applied to the gate, the maximum voltage that can be passed is calculated from

$$\begin{aligned} V_{max} &= (3.3 - 0.7) - 0.053(\sqrt{0.58 + V_{max}} - \sqrt{0.58}) \\ &= 2.6 - 0.053(\sqrt{0.58 + V_{max}} - \sqrt{0.58}) \end{aligned} \tag{4.11}$$

for V_{max}. Although this can be viewed as a quadratic equations, it is simpler to use an iterative numerical technique.

The algorithm is as follows.

1. Guess a value for V_{max}
2. Calculate the right-hand side (RHS) of the equation
3. If the RHS is equal to the original guess for V_{max}, then we have the solution
4. Otherwise, use the RHS as the new guess, and redo the calculation starting at step 2

Since the unknown is inside a square root, the convergence will be very rapid.The table below summarizes the calculations for the present example.

Guess	RHS
2.6	2.545
2.545	2.547
2.547	2.547

The solution is thus found to be $V_{max} \approx 2.55v$. In the example, the parameter values result from x_{ox}=200 Å and $N_a = 10^{15}\ cm^{-3}$. Since C_{ox} will be relatively large, the value of γ is small, which in turn makes the body-bias effects small in this example.

4.1.2 Logic 0 Input

Let us now consider the case where $V_{in} = 0v$ corresponding to a logic 0 input as shown previously in Figure 4.2(b). To be consistent with the logic 1 analysis, we assume that C_{out} is initially charge to a voltage $V_{out} = V_{max} = (V_{DD} - V_{Tn})$. Transferring a logic 0 voltage through the nFET corresponds to discharging C_{out}. In this case, the input side of the MOSFET is the source, while the output side is the drain. The transistor voltages are given by

$$\begin{aligned} V_{GSn} &= V_{DD} \\ V_{DSn} &= V_{out}(t) \end{aligned} \tag{4. 12}$$

Since the maximum value of $V_{out} = V_{max} = V_{sat}$ occurs at the beginning of the discharge (which is assumed to be at $t = 0$), $V_{DSn} \leq V_{sat}$ is always maintained, so that the nFET is always non-saturated. The discharge event is thus described by

$$I = -C_{out}\frac{dV_{out}}{dt} = \frac{\beta_n}{2}\left[2(V_{DD} - V_{Tn})V_{out} - V_{out}^2\right] \tag{4. 13}$$

where the minus sign is required because the current flow is out of the positive terminal. Rearranging yields the equation

$$\int\frac{dV_{out}}{2(V_{DD} - V_{Tn})V_{out} - V_{out}^2} = -\int\frac{\beta_n}{2C_{out}}dt \tag{4. 14}$$

Integrating and applying the initial condition gives the voltage decay in the form

$$V_{out}(t) = (V_{DD} - V_{Tn})\left[\frac{2e^{-t/\tau_n}}{1 + e^{-t/\tau_n}}\right] \tag{4. 15}$$

showing a rapid exponential decay. In the limit where $t \to \infty$, $V_{out} \to 0v$. Physically, this occurs because the nFET has a constant gate-source bias of $V_{GSn} = V_{DD}$, so that it remains active throughout the entire discharge. The most important point to note is that the nFET can pass the 0*v* input voltage without any problems, and that the discharge is fast.

4.1.3 Switching Times

The analyses above may be used to calculate t_{LH} and t_{HL} for the nFET. Consider first the low-to-high time t_{LH} between the voltage levels 0*v* and $V_1 = 0.9V_{max}$ which will serve as our limits.[3] Rearranging eqn. (4.4) into the form

$$t = 2\tau_n \left[\frac{1}{1 - \frac{V_{out}(t)}{V_{max}}} - 1 \right] \tag{4.16}$$

and substituting $V_{out} = 0.9V_{max}$ gives

$$t_{LH} = 18\tau_n \tag{4.17}$$

The high-to-low time t_{HL} between the voltage levels V_{max} and $V_0 = 0.1V_{max}$ can be computed by solving eqn. (4.15) for time *t* in the form

$$t = \tau_n \ln\left[\frac{2V_{max}}{V_{out}(t)} - 1\right] \tag{4.18}$$

Substituting $V_{out} = 0.1V_{max}$ yields the value of

$$t_{HL} = \ln(19)\,\tau_n \approx 2.94\tau_n \quad . \tag{4.19}$$

It is clear that the discharge (a logic 0 transmission) is much faster than the charging event (a logic 1 transmission) by a factor of

$$\frac{t_{LH}}{t_{HL}} \approx \frac{18}{2.94} \approx 6.11 . \tag{4.20}$$

When combined with the threshold voltage loss, we conclude that the nFET acts very well to pass a strong logic 0 voltage, but exhibits problems in both the DC and transient characteristics when attempting to pass a logic 1 voltage of V_{DD}.

The plots in Figure 4.3 illustrate the important aspects of the switching times. The input voltage $V_{in}(t)$ is taken to be a pulse that swings from 0*v* to V_{DD}. The slow rise time associated with the logic 1 transfer is important to the circuit designer as it represents a limiting aspect of using nFETs as logic switches. When the input voltage $V_{in}(t)$ falls from V_{DD} to 0*v*, the response is very rapid and does not present a problem. At the circuit level, however, we must always assume the worst-case situation. Thus, we always keep in mind that circuits that require a logic 1 voltage transmission through an nFET may cause a slowdown in the logic throughput.

4.1.4 Interpretation of the Results

To understand the significance and use of the above analyses, consider the general switching network shown in Figure 4.4. In this drawing, the output capacitance C_{out} has been broken down into contributions of

---

[3] The usual 10%-to-90% voltage limits have been replaced in these calculations for simplicity.

$$C_{out} = C_{MOS} + C_n + C_L \tag{4.21}$$

where C_{MOS} is the gate-drain or gate-source capacitance, C_n is the junction drain-bulk or source-bulk capacitance, and C_L represents the external load capacitance. Defining the internal FET capacitance by

$$C_{int} = C_{MOS} + C_n \tag{4.22}$$

allows us to write

$$C_{out} = C_{int} + C_L . \tag{4.23}$$

as we did for the inverter circuit analyzed in the previous chapter. The time constant can then be broken down into two distinct terms:

$$\begin{aligned} \tau_n &= \frac{(C_{int} + C_L)}{\beta_n (V_{DD} - V_{Tn})} \\ &= R_n C_{int} + R_n C_L \\ &= \tau_{n,\,int} + \tau_{n,\,L} \end{aligned} \tag{4.24}$$

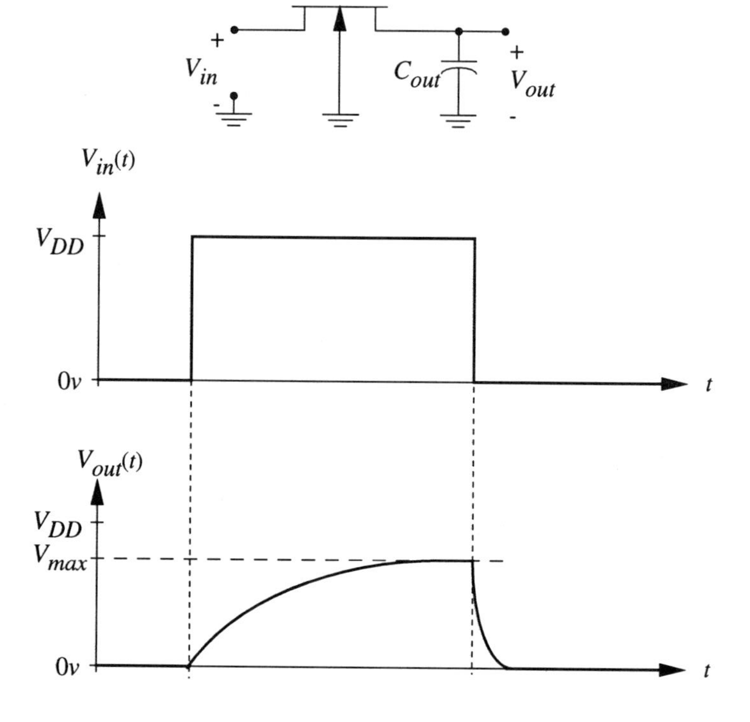

Figure 4.3. Pulse transmission through an nFET

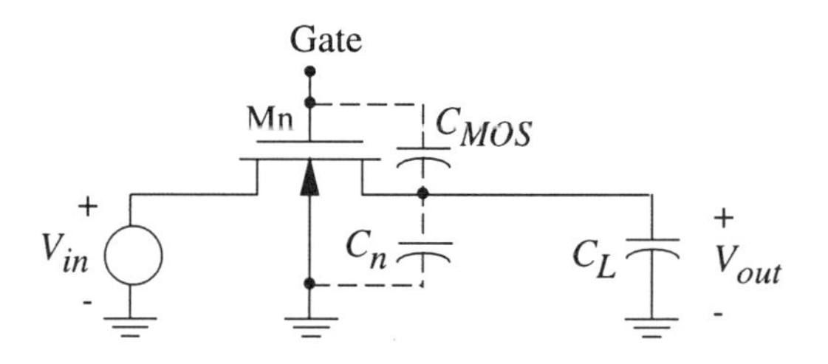

Figure 4.4. Output capacitance contributions

The first term represents the charging time needed for the internal parasitic capacitances, while the second term is the additional time needed to charge the external load capacitance.

The value of the internal device capacitance C_{int} is set by

- The processing parametersx_{ox}, k'_n, etc.,

and,

- The dimensions W and L of the MOSFET.

For an nFET FET of a given size, the internal time constant

$$\tau_{n,\,int} = \frac{C_{int}}{\beta_n (V_{DD} - V_{Tn})} \tag{4. 25}$$

represents the fastest switching that can be attained by the technology under ideal no-load conditions. While the no-load situation never occurs in a realistic circuit, it does provide a useful upper frequency limit for a circuit created in a given fabrication process. Once a load is connected to the circuit, then we must consider the additional delays due to the load through the term

$$\tau_{n,\,L} = \frac{C_L}{\beta_n (V_{DD} - V_{Tn})} \tag{4. 26}$$

which must be included in the total time constant. For the case $C_L >> C_{int}$, this would imply that we would have to use a large aspect ratio (W/L) to decrease the FET resistance and speed up the transfer. If C_L and C_{int}, are about the same value, then increasing the aspect ratio has less effect on shrinking the time constant.

4.1.5 Layout

The element values in the RC-equivalent network are determined from the layout. Figure 4.5(a) shows the geometry when viewed at the silicon level. To translate to the switching circuit shown in Figure 4.5(b), we simply apply the modelling formulas directly.

The FET resistance is given by

$$R = \frac{1}{k'_n \left(\frac{W}{L}\right)(V_{DD} - V_{Tn})} \tag{4. 27}$$

where $L=L'-2L_o$, with L_o the gate overlap (which is not shown in the drawing). The total gate capacitance is calculated from

$$C_G = C_{ox}WL' \tag{4.28}$$

so that we estimate

$$C_{GS} \approx \frac{1}{2}C_G \approx C_{GD} \tag{4.29}$$

for the gate-source and gate-drain contributions. The LTI values of the depletion capacitances are obtained from

$$\begin{aligned} C_{n,X} &= K_m(0, V_{max})\, C_{j0}W(X + L_o) + K_{mj}(0, V_{max})\, 2C_{jsw}(W + X + L_o) \\ C_{n,Y} &= K_m(0, V_{max})\, C_{j0}W(Y + L_o) + K_{mj}(0, V_{max})\, 2C_{jsw}(W + Y + L_o) \end{aligned} \tag{4.30}$$

where we have included the overlap distance L_o in our formulas. Note that the averaging has been performed over the voltage range from 0 to V_{max}; this is important on the right (load) side since the voltage there never exceeds V_{max}. However, if the left side is driven by an inverter, then V_{max} should be replaced by V_{DD} in the C_X equation to be consistent. In practice, the difference in K values is small, and we could use V_{max} in both for simplicity without introducing significant error.

To complete the calculation of the RC-equivalent circuit elements values, we add contributions to obtain

$$\begin{aligned} C_i &= C_{n,X} + C_{GS} \\ &= C_{n,X} + \frac{1}{2}C_G \end{aligned} \tag{4.31}$$

as the input capacitance, and

$$\begin{aligned} C_{out} &= C_{n,Y} + C_{GD} + C_L \\ &= C_{n,Y} + \frac{1}{2}C_G + C_L \end{aligned} \tag{4.32}$$

for the output capacitance. This completes the calculations for the values in the equivalent circuit. The most important point to keep in mind is that the numerical values depend explicitly on the layout dimensions.

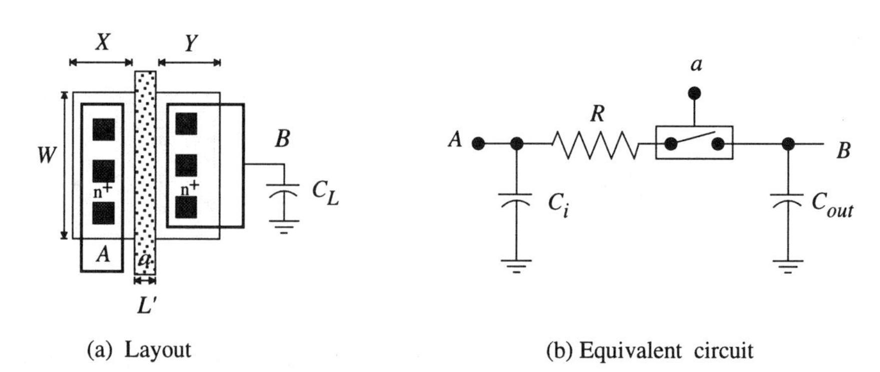

(a) Layout (b) Equivalent circuit

Figure 4.5. nFET layout and RC equivalent circuit

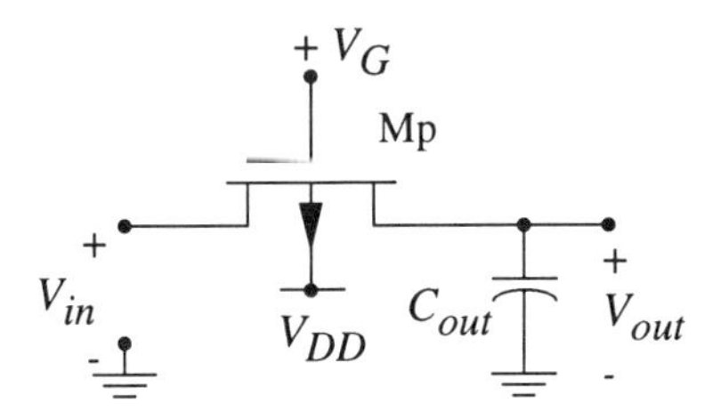

Figure 4.6. pFET pass transistor voltages

4.2 pMOS Transmission Characteristics

Figure 4.6 shows a pMOSFET used as a voltage-controlled switching device. In this section, we will study the pFET transmission characteristics for logic 1 and logic 0 input voltages. Since the pFET is the electrical complement of the nFET, we expect to find that it behaves in exactly the opposite manner when passing low and high voltages.

4.2.1 Logic 0 Input

Let us first examine the case where we transmit a logic 0 through a pFET. The problem is illustrated in Figure 4.7(a) where we have set $V_{in} = 0v$. Let us assume that $V_{out}(t{=}0) = V_{DD}$ (which will be consistent with the logic 1 analysis) so that the terminal voltages are given by

$$\begin{aligned} V_{SGp} &= V_{out}(t) \\ V_{SDp} &= V_{out}(t) \end{aligned} \tag{4. 33}$$

Since it is obvious that

$$V_{SDp} > V_{sat} = V_{out} - |V_{Tp}| \tag{4. 34}$$

the pFET conducts in the saturation mode with a current flow of

$$I = -C_{out}\frac{dV_{out}}{dt} = \frac{\beta_p}{2}(V_{out} - |V_{Tp}|)^2 \tag{4. 35}$$

This differential equation describes the discharge of the output capacitor C_{out} through the transistor. Solving the current flow equation by direct integration as above for the case of an nFET yields an

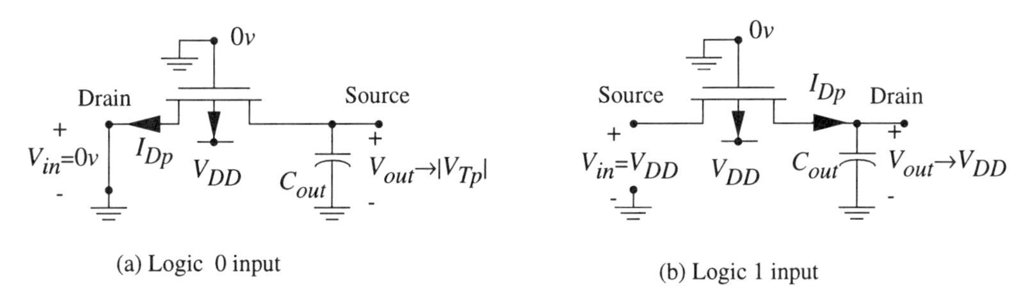

Figure 4.7. Logic 0 and logic 1 inputs

output voltage of the form

$$V_{out}(t) = |V_{Tp}| + \frac{(V_{DD} - |V_{Tp}|)}{1 + t/2\tau_p} \tag{4.36}$$

In this expression, we have introduced

$$\tau_p = \frac{C_{out}}{\beta_p (V_{DD} - |V_{Tp}|)} = R_p C_{out} \tag{4.37}$$

as the pFET time constant for this circuit. In this equation,

$$R_p = \frac{1}{\beta_p (V_{DD} - |V_{Tp}|)} \tag{4.38}$$

acts as an equivalent pFET resistance.

For large times t we see that

$$V_{out} \rightarrow |V_{Tp}| \tag{4.39}$$

indicating that is not possible to discharge the capacitor to 0v through the p-channel transistor. This is easily understood by noting that the pFET must maintain a source-gate voltage of

$$\min(V_{SGp}) = |V_{Tp}| \tag{4.40}$$

for the conducting channel to exist. Since the gate is grounded at 0v, applying KVL shows that the smallest voltage that can be supported by the pFET at the output node is $|V_{Tp}|$ corresponding the to minimum source-gate voltage.

Including body-bias effects gives the complete expression

$$V_{min} = \left|V_{T0p} - \gamma\left(\sqrt{2|\phi_F| + (V_{DD} - V_{max})} - \sqrt{2|\phi_F|}\right)\right| \tag{4.41}$$

which constitutes a self-iterating equation for V_{min}. This may be solved using the same approach as for the V_{max} equation that was found in the nFET logic 1 transfer analysis.[4] In general terms, we say that the pFET transistor can only pass a "weak" logic 0 since it can never reach 0v.

4.2.2 Logic 1 Input

The pass characteristics of the pFET are much better when we apply a high voltage to the input. To analyze the logic 1 pass characteristics of a pFET, we assume that the output voltage is initially at a value $V_{out}(t=0) = |V_{Tp}|$, since this is the lowest voltage that can be passed through the device. Figure 4.7(b) shows the designation of the drain and source for this case.By inspection, we see that

$$\begin{aligned} V_{SGp} &= V_{DD} \\ V_{SDp} &= V_{DD} - V_{out}(t) \end{aligned} \tag{4.42}$$

At time t=0, the voltages are given by

[4] Note that the values of γ and $2|\phi_F|$ for a pFET are distinct from the equivalent nFET values.

$$V_{sat} = V_{DD} - V_{out}(0) \\ = V_{DD} - |V_{Tp}| \tag{4.43}$$

which indicates that the pFET is at the border between saturation and non-saturation. Physically, we know that the capacitor will charge such that

$$\frac{dV_{out}}{dt} > 0 \tag{4.44}$$

i.e., $V_{out}(t)$ increase in time. Thus, for times t > 0, $V_{SDp} < V_{sat}$ and the pFET conducts in the non-saturated mode. This gives the charging equation as

$$I = C_{out}\frac{dV_{out}}{dt} = \frac{\beta_p}{2}\left[2(V_{DD} - |V_{Tp}|)(V_{DD} - V_{out}) - (V_{DD} - V_{out})^2\right] \tag{4.45}$$

which can be rearranged to the form

$$t = \frac{2C_{out}}{\beta_p}\int_{|V_{Tp}|}^{V_{out}(t)} \frac{dV_{out}}{\left[2(V_{DD} - |V_{Tp}|)(V_{DD} - V_{out}) - (V_{DD} - V_{out})^2\right]} \tag{4.46}$$

Integrating and applying the limits yields the output voltage as

$$V_{out}(t) = V_{DD} - (V_{DD} - |V_{Tp}|)\left[\frac{2e^{-t/\tau_p}}{1 + e^{-t/\tau_p}}\right] \tag{4.47}$$

The pFET time constant

$$\tau_p = \frac{C_{out}}{\beta_p(V_{DD} - |V_{Tp}|)} = R_p C_{out} \tag{4.48}$$

is again our reference time for this event, but it should be noted that it has a different meaning because of the difference in the equations. As the time $t \to \infty$, this equation shows that $V_{out} \to V_{DD}$. We thus say that a p-channel MOSFET can pass a "strong" logic 1 voltage without any problems.

4.2.3 Switching Times

The curves in Figure 4.8 shows the pulse transfer characteristics of the pMOS transistor. Once again we note that the pFET is the voltage complement of the nFET. This allows us to write by inspection the switching times as

$$t_{HL} = 18\tau_p \\ t_{LH} = \ln(19)\,\tau_p \approx 2.94\tau_p \tag{4.49}$$

which can be verified by direct calculation. The meaning of these expressions is clear: the pFET is very good at passing a logic 1 input (small t_{LH}), but does not perform well with a logic 0 input (a large t_{HL}). The ratio between the switching time intervals is

$$\frac{t_{HL}}{t_{LH}} \approx \frac{18}{2.94} \approx 6.11\,, \tag{4.50}$$

which is exactly the inverse of the ratio for the nFET. These results once again confirm our observa-

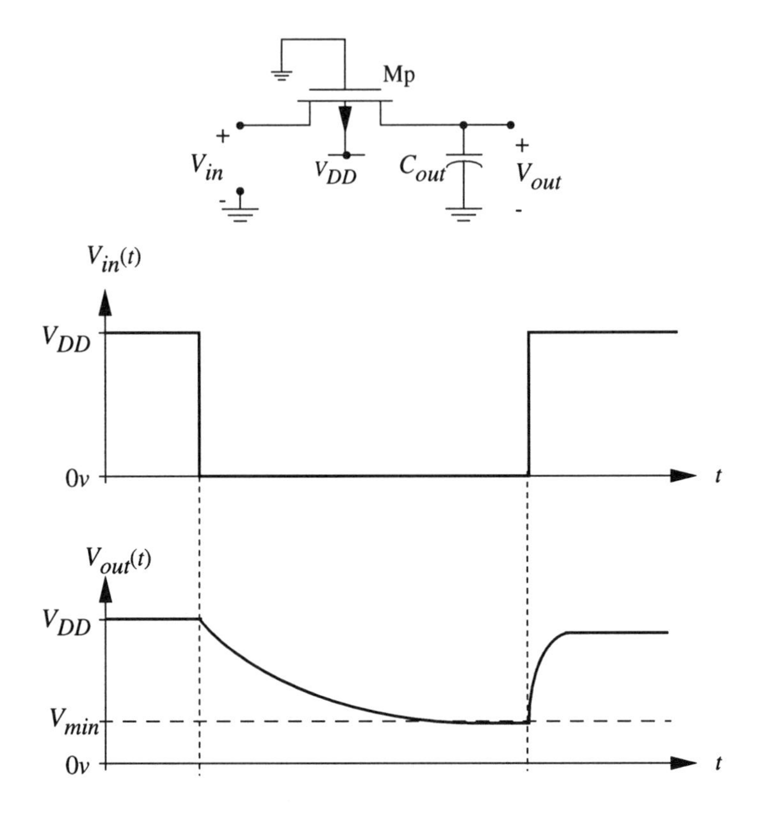

Figure 4.8. Pulse response of a pFET

tion that nFETs and pFETs complement each other.

The plots in Figure 4.8 illustrate the important results from this analysis. Switching the input voltage $V_{in}(t)$ from V_{DD} to 0v results in the slow logic 0 transfer as described above. No such problem occurs when $V_{in}(t)$ is increased from 0v back up to V_{DD}, but we are not interested in the fastest case since it does not limit the performance. As pointed out continuously throughout the book, the circuit designer must concentrate on the worst-case situations to insure that the network operates in all instances.

4.3 The Inverter Revisited

Let us now apply the analysis of nFET and pFET pass transistors to the inverter circuit discussed in the previous chapter. In Figure 4.9(a), the inverter has been redrawn to emphasize that the MOSFETs actually function as pass transistors with fixed inputs. The nFET Mn controls the connection between the ground (0v) and the output *f*, while pFET Mp is responsible for the connection between the power supply V_{DD} and the output *f*. The transistors are thus positioned to allow them to perform what they do the best: the nFET passes a strong logic 0, while the pFET passes a strong logic 1. This connection insures that the output can attain a full-rail output voltage swing.

Note that drawing the circuit in this manner implies that the inverter can be viewed as a 2:1 multiplexer as shown in Figure 4.9(b). The inputs are 0v and V_{DD} which represent logic 0 and 1 values, respectively. The input variable *x* acts as the MUX control signal, and determines which input is sent to the output. When *x*=0, the input labelled "0" (that has V_{DD} applied) is sent to the output; conversely, when *x*=1, the input labelled "1" (that has 0v applied) is sent to the output. The Boolean expression for this network is

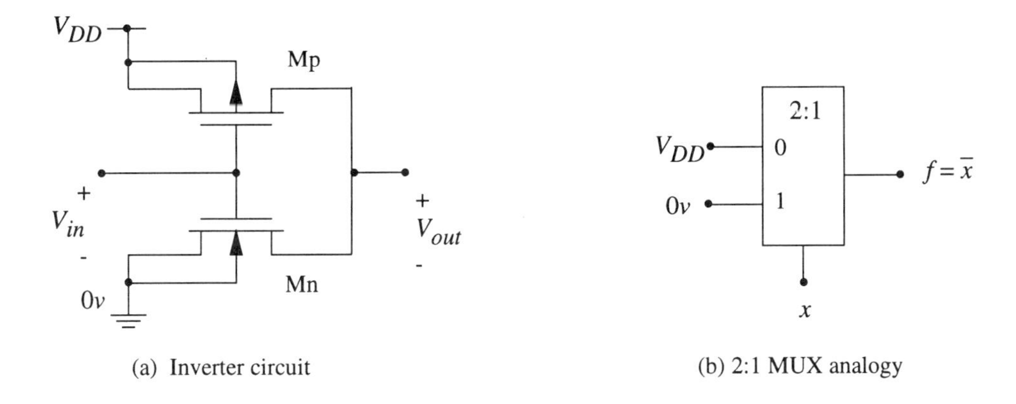

(a) Inverter circuit (b) 2:1 MUX analogy

Figure 4.9. The CMOS inverter using pass FET construction

$$f = \bar{x} \cdot 1 + x \cdot 0 \qquad (4.51)$$

which reduces to

$$f = \bar{x} \qquad (4.52)$$

by noting that ANDing anything with a logic 0 always gives a 0, while a 1 acts as the identity operator. This expression is, of course, the NOT operation as required. As we will see in the next chapter, writing logic expressions using the FET rules is very useful when we construct complex logic gate circuits.

4.4 Series-Connected MOSFETs

The next step in our study of the switching properties of MOSFETs is to create groups of transistors and analyze their behavior. Series-connected arrangements are of particular interest since they present some design limitations. As we will see below, a series chain can be modelled using straightforward techniques.

4.4.1 nFET Chains

Let us start with the series-connected n-channel transistors shown in Figure 4.10. The gate voltages have been set at V_{DD}, giving conduction between the source and drain electrodes. Interest is

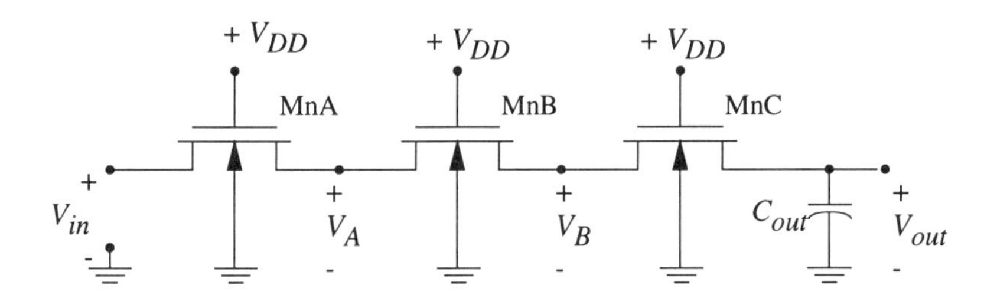

Figure 4.10. Series-connected pass nFETs

directed towards the study of the output voltage V_{out} for different values of the input voltage V_{in}.

Consider first the case where the input voltage is at a value of V_{in} =0v. Since an nFET is capable of passing a strong logic 0 voltage, the 0v level is transferred down the chain:

$$0v \rightarrow V_A \rightarrow V_B \rightarrow V_{out} = 0v \tag{4.53}$$

The output voltage can thus achieve a value of V_{out}=0v.

The situation is quite different if the input voltage is set to $V_{in} = V_{DD}$ due to the problem of threshold voltage loss. Consider the first transistor MnA. Since V_{DD} is applied to the gate, the maximum value of V_A is given by

$$\begin{aligned} \max(V_A) &= V_{max} \\ &= V_{DD} - V_{Tn} \end{aligned} \tag{4.54}$$

Next, consider MOSFET MnB. Since this transistor also has V_{DD} applied to the gate, the maximum value for V_B is also given by

$$\begin{aligned} \max(V_B) &= V_{max} \\ &= V_{DD} - V_{Tn} \end{aligned} \tag{4.55}$$

however, V_A is acting as the input to MnB, and can never exceed V_{max} because of drop induced by MnA. The same argument may be applied to the third FET MnC. This says that only one threshold loss occurs in a series connected chain:

$$V_{DD} \rightarrow V_A = V_{max} \rightarrow V_B = V_{max} \rightarrow V_{out} = V_{max} \tag{4.56}$$

Although the threshold voltage loss reduces the logic 1 noise margin, it may be possible to compensate for the loss by adjusting the electrical design of the following logic gates. This is discussed in more detail in Chapter 7 in the context of "dynamic" CMOS circuits.

4.4.2 pFET Chains

A chain of series-connected p-channel transistors exhibits complementary characteristics. Consider the 3-FET circuit shown in Figure 4.11. All of the gate voltages have been set at 0v to bias the pFETs into the active region of operation. As for the nFET analysis, let us find the output voltage V_{out} for the maximum and minimum values of V_{in}.

A p-channel MOSFET can pass the power supply voltage without reducing the level. If $V_{in} = V_{DD}$ is applied, then it can be transferred down the chain with

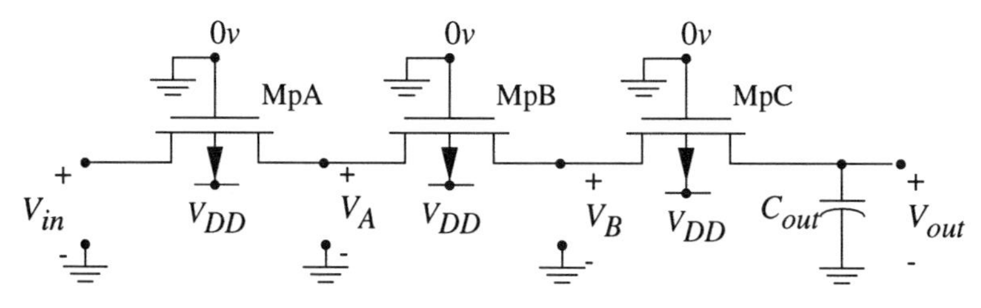

Figure 4.11. Series-connected pFET chain

$$V_{DD} \rightarrow V_A \rightarrow V_B \rightarrow V_{out} = V_{DD} \quad (4.57)$$

As expected, the output voltage can reach the value $V_{out} = V_{in} = V_{DD}$.

If the input voltage is set to $V_{in}=0v$, then having a gate voltage of $0v$ induces the threshold voltage rise. The voltage V_A of MOSFET MpA can only fall to a value of

$$\min(V_A) = V_{min} = |V_{Tp}| \quad (4.58)$$

due to the fact that V_{SGA} must be at least $|V_{Tp}|$ to maintain a conducting channel. Transistor MpB also has this limitation, but due to the threshold limitation on MpA, the second FET MpB receives an input that never drops below $|V_{Tp}|$ anyway. Thus,

$$\min(V_B) = V_{min} = |V_{Tp}| \quad (4.59)$$

with the same conditions on MpC. The result is that only one threshold rise occurs in the series chain:

$$V_{in} = 0 \rightarrow V_A = |V_{Tp}| \rightarrow V_B = |V_{Tp}| \rightarrow V_{out} = |V_{Tp}| \quad . \quad (4.60)$$

This is true regardless of the number of pFETs in the chain.

4.4.3 FETs Driving Other FETs

The above analyses show that we can connect like-polarity MOSFETs in series and cause only a single threshold voltage rise (pFETs) or drop (nFETs) in transferring a voltage through the network. Depending upon the noise margins of the following circuit, this may be an acceptable modification. There are other situations, however, where the threshold drops cannot be tolerated. This occurs in particular when we use the output of a FET as the gate voltage for another transistor.

Figure 4.12 shows the case where the output of nFET Mn1 is used to drive the gate of Mn2. The problem with this connection occurs when we attempt to transfer a voltage V_{DD} through Mn2. Since Mn1 has V_{DD} applied to its gate, the maximum voltage that is applied to the gate of Mn2 is

$$V_{max} = V_{DD} - V_{Tn} \quad . \quad (4.61)$$

Accounting for the threshold loss in Mn2 gives a maximum output voltage of

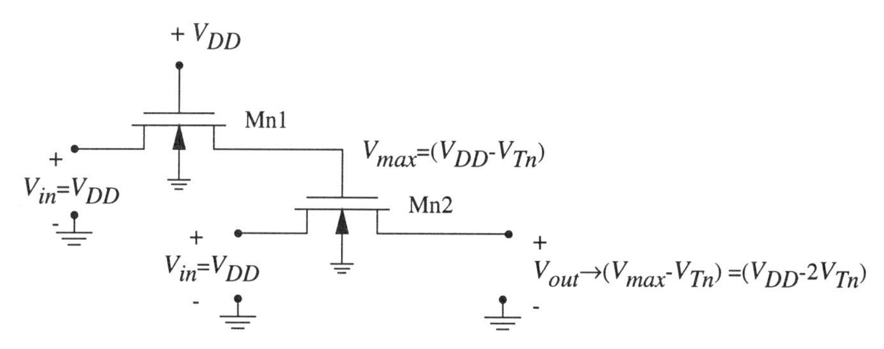

Figure 4.12. Dual threshold-loss pass FET circuit

$$V_{out} = V_{max} - V_{Tn} \\ = (V_{DD} - V_{Tn}) - V_{Tn} \quad (4.62)$$

This shows that connecting nFETs in this manner induces one threshold drop per transistor, resulting in a large reduction in the logic 1 voltage. Including body-bias effects in the calculations shows that the problem gets worse. We thus tend to avoid circuits like this unless we are purposely trying to induce a voltage drop.

An analogous situation occurs when we use the output of a pFET to drive another pFET. As shown in Figure 4.13, the second pFET Mp2 cannot pass a low logic 0 voltage. The reasoning is along the same lines as for the nFET circuit. Applying 0*v* at the gate of Mp1 gives the minimum voltage of

$$V_{min} = |V_{Tp}| \quad (4.63)$$

at the gate of Mp2. Since $\min(V_{SG2}) = |V_{Tp}|$ is needed to keep Mp2 conducting, the output cannot exceed the value

$$V_{out} = V_{min} + |V_{Tp}| \cong 2|V_{Tp}| \quad (4.64)$$

which gives a very large logic 0 voltage. Including body-body in the analysis makes the value of V_{min} even larger than this estimate.

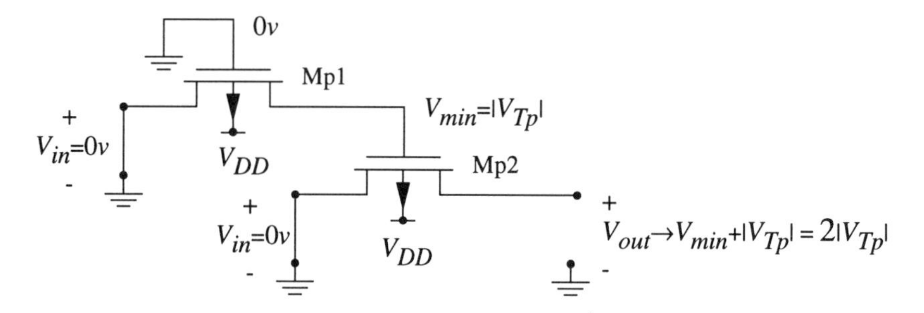

Figure 4.13. pFET dual-threshold problem

Although we generally desire small logic 0 voltages and large logic 1 voltages, threshold voltage drops and rises can be used in certain situations to tailor the operation of a circuit. They do have the advantage that they are relative stable voltage values and are sometimes useful for reference devices.[5]

4.5 Transient Modelling

Now that the DC characteristics of series-connected MOSFETs have been established, let ut turn to the problem of analyzing the transient switching characteristics of these networks. This will prove to be the limiting factor in high performance applications. In practice, there are two levels of modelling that can be used. These are

- Full MOSFET equations with a computer simulation;

---

[5] In that they will track processing variations present in neighboring devices.

and

- Simplified RC ladder networks.

A straightforward engineering methodology is to employ basic RC networks to arrive at the first design choices, simulate the circuit, and then redesign as needed.

4.5.1 The MOSFET RC Model

RC ladder networks provide insight into the performance characteristics and design criteria of multi-transistor networks. It is based on the simple RC model for the MOSFET shown in Figure 4.14 where conduction from the drain to the source (or vice-versa) is represented by the π-network that consists of one resistor and two capacitors. Although there are several approaches that can be used to specify the component values, we shall choose the simplest and examine the consequence and applications later.

A MOSFET is a nonlinear device so that it is not possible to introduce accurate LTI[6] values for either the resistance or the capacitance. However, we can use the concept of an equivalent resistance by defining

$$R = \frac{1}{\beta(V_{DD} - V_T)} \quad \Omega \tag{4. 65}$$

where

$$\beta = k'\left(\frac{W}{L}\right) \tag{4. 66}$$

so long as we are careful about not interpreting it too literally.[7] Note that this gives the dependence

$$R \propto \frac{1}{W} \tag{4. 67}$$

which shows that the resistance decreases with an increasing value of the aspect ratio (*W/L*). This is one of the main reasons that the LTI model is useful: it gives the qualitative result that current flow levels can be increased by using larger aspect ratios.

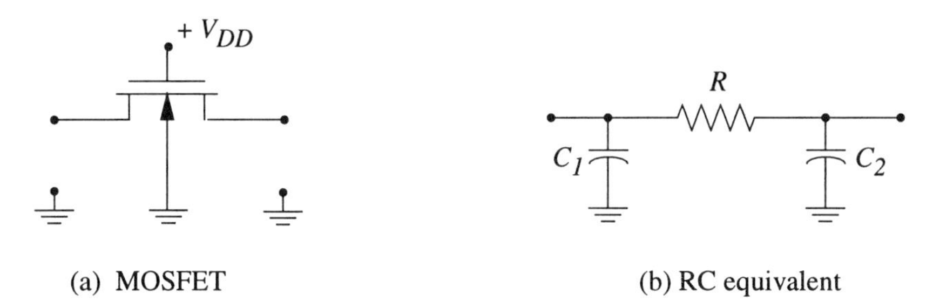

Figure 4.14. nFET RC π-equivalent LTI parasitic model

[6] LTI is an acronym for **linear, time-invariant**, which characterizes the "standard" devices used in basic circuit theory where the *I-V* relationship is linear, and the value of the component does not vary in time.

[7] This expression is widely used in the literature, and is obtained from the transient time constant.

The capacitors C_1 and C_2 represent the device contributions at the drain and source terminals, and vary with the voltage. Rather than be too concerned about the exact values, we will estimate

$$\begin{aligned} C_1 &= \frac{1}{2}C_G + C_{n1} \\ C_2 &= \frac{1}{2}C_G + C_{n2} \end{aligned} \tag{4. 68}$$

where

$$C_G = C_{ox}WL' \tag{4. 69}$$

is the total gate capacitance, with L' the drawn channel length, and C_{n1} and C_{n2} are the junction capacitances (the source-bulk and drain-bulk contributions). In general, we approximate that

$$C_{1,2} \propto W, \tag{4. 70}$$

i.e., the internal capacitances increase with the channel width W. In simple estimates, we use the zero-bias values corresponding to the worst-case (largest) values. Alternately, the voltage-averaged quantites may be employed in an effort to include some of the nonlinear effects.

Now that the basic device model has been established, let us examine the transient response when a voltage of the form

$$V_{in}(t) = V_{DD}u(t) \tag{4. 71}$$

is applied as shown in Figure 4.15. In this expression, $u(t)$ is the unit step function defined by

$$u(t) = \begin{cases} 0 & \text{for } (t < 0) \\ 1 & \text{for } (t \geq 0) \end{cases} \tag{4. 72}$$

The voltage $V_{out}(t)$ across the output capacitor C in the model is simply the exponential response

$$V_{out}(t) = V_{DD}(1 - e^{-t/\tau}) \tag{4. 73}$$

where

$$\tau = RC \tag{4. 74}$$

is the time constant. Rearranging gives

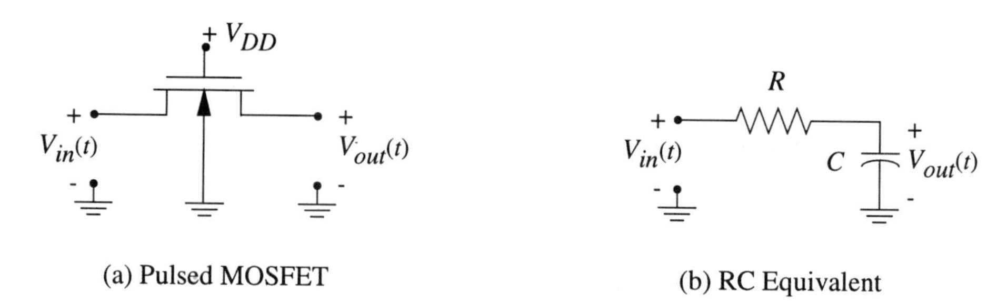

Figure 4.15. Linear step response model for nFET

$$t = \tau \ln \left[\frac{1}{1 - \frac{V_{out}}{V_{DD}}} \right] \tag{4.75}$$

as the time needed for the output to change from 0v to an arbitrary voltage V_{out}. Defining the charge time t_{ch} as the time needed for the output voltage to change from 0v to 90% voltage $0.9V_{DD}$ gives

$$t_{ch} = \tau \ln(10) \approx 2.3\tau \tag{4.76}$$

showing the dependence on the time constant.

An input voltage for a downward transition is obtained by writing

$$V_{in}(t) = V_{DD}[1 - u(t)] \tag{4.77}$$

gives the exponentially decaying response

$$V_{out}(t) = V_{DD}e^{-t/\tau} \tag{4.78}$$

for the capacitor voltage. The discharge time t_{dis} needed for the output voltage to change from V_{DD} to $0.1V_{DD}$ is obtained by solving this equation for time as

$$t = \tau \ln \left[\frac{V_{DD}}{V_{out}} \right] \tag{4.79}$$

so that

$$t_{dis} = \tau \ln[10] \approx 2.3\tau . \tag{4.80}$$

This is identical to t_{ch} due to the symmetry of the RC model.

Although the RC model provides simple results, it ignores the problem of the threshold voltage loss through an nFET. This can be included by substituting $V_{max}=(V_{DD} - V_{Tn})$ for V_{DD} in the above expressions. However, the asymmetry of the MOSFET with regard to logic 0 and logic 1 transfer times is still ignored by the analysis. The bottom line is that one must be very careful when interpreting results that are based on RC equivalent MOSFET modelling. With this in mind, let us proceed to study a very important problem in CMOS circuits.

4.5.2 Voltage Decay On an RC Ladder

The step response of a chain of two nFETs can be approximated by analyzing the RC-ladder network shown in Figure 4.16(a) for the capacitor voltages $V_1(t)$ and $V_2(t)$. For a discharge event, we will assume initial conditions[8] of $V_1(0) = V_{max}$ and $V_2(0) = V_{max}$, and direct our attention to calculating $V_2(t)$ in Figure 4.16(b); note that the input voltage source has been idled and been replaced by a short circuit to ground. Physically, the capacitors will discharge and both $V_1(t)$ and $V_2(t)$ will decay to 0v. It is important to note that C_1 discharges through R_1 only, but that C_2 must discharge through the series combination of R_1 and R_2.

As a simple model, a first guess is that $V_2(t)$ assumes the exponential form

$$V_2(t) \approx V_{max}e^{-t/\tau} \tag{4.81}$$

[8] Recall that $V_{max}= (V_{DD}-V_{Tn})$ is the maximum voltage that can be passed through an nFET with V_{DD} applied to the gate. A pFET chain can be modelled by changing the voltages.

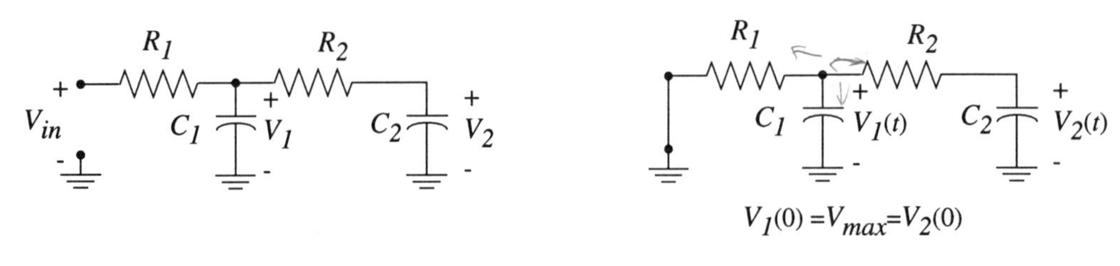

(a) General circuit (b) Discharge network

Figure 4.16. 2-rung RC ladder circuit

where τ is the time constant for the decay. Let us write the time constant in the form

$$\tau \approx R_1 C_1 + (R_1 + R_2) C_2, \tag{4. 82}$$

since this accounts for the RC time constants of both discharge paths: C_1 discharging through R_1, and C_2 discharging through the series chain (R_1+R_2). This form of the time constant is known as the **Elmore formula** for a series RC chain.[9] As shown in the analysis below, this can in fact be justified as a reasonable first approximation to the actual behavior. Moreover, we can use the Elmore formula to find the time constant for arbitrary RC ladder networks.

Analysis

Let us now examine the circuit in more detail to justify the model. As a starting point, let us write the node equations for the circuit as

$$\begin{aligned} -C_1 \frac{dV_1}{dt} &= \frac{V_1}{R_1} - \frac{V_2 - V_1}{R_2} \\ -C_2 \frac{dV_2}{dt} &= \frac{V_2 - V_1}{R_2} \end{aligned} \tag{4. 83}$$

by applying KCL to each node. Laplace transforming from time domain to s-domain, we obtain equations for the transformed voltages $v_1(s)$ and $v_2(s)$

$$\begin{aligned} -sC_1 v_1(s) + C_1 V_1(0) &= \frac{v_1}{R_1} - \frac{v_2 - v_1}{R_2} \\ -sC_2 v_2(s) + C_2 V_2(0) &= \frac{v_2 - v_1}{R_2} \end{aligned} \tag{4. 84}$$

where $V_1(0)$ and $V_2(0)$ are the initial conditions (in time domain).[10] Grouping terms and rearranging gives the matrix equation

[9] W.C> Elmore, "The transient response of damped linear networks with particular regard to wideband amplifiers," *J. Applied Physics*, vol. 19, pp. 55-63, January, 1948.

[10] The analysis can, of course, be performed directly in time domain if desired.

$$\begin{bmatrix} sC_1 + \left(\frac{1}{R_1} + \frac{1}{R_2}\right) & -\frac{1}{R_2} \\ -\frac{1}{R_2} & sC_2 + \frac{1}{R_2} \end{bmatrix} \begin{bmatrix} v_1(s) \\ v_2(s) \end{bmatrix} = \begin{bmatrix} C_1 V_{max} \\ C_2 V_{max} \end{bmatrix} \tag{4. 85}$$

which shows the algebraic system in s-domain. Using Cramer's Rule, the solutions are given by

$$\begin{aligned} v_1(s) &= \frac{1}{\Delta}\det\begin{bmatrix} C_1 V_{max} & -\frac{1}{R_2} \\ C_2 V_{max} & sC_2 + \frac{1}{R_2} \end{bmatrix} \\ v_2(s) &= \frac{1}{\Delta}\det\begin{bmatrix} sC_1 + \left(\frac{1}{R_1} + \frac{1}{R_2}\right) & C_1 V_{max} \\ -\frac{1}{R_2} & C_2 V_{max} \end{bmatrix} \end{aligned} \tag{4. 86}$$

where

$$\Delta = \det\begin{bmatrix} sC_1 + \left(\frac{1}{R_1} + \frac{1}{R_2}\right) & -\frac{1}{R_2} \\ -\frac{1}{R_2} & sC_2 + \frac{1}{R_2} \end{bmatrix} \tag{4. 87}$$

is the determinant of the coefficient matrix. Although the calculations will be straightforward, they do not reveal much about the nature of the solution. It is therefore much more enlightening to study the general characteristics of the solution by first examining the zeroes of the determinant.

To find the roots of the denominator in both expressions, we set the determinant of the coefficient matrix to zero:

$$\det\begin{bmatrix} sC_1 + \left(\frac{1}{R_1} + \frac{1}{R_2}\right) & -\frac{1}{R_2} \\ -\frac{1}{R_2} & sC_2 + \frac{1}{R_2} \end{bmatrix} = 0 \tag{4. 88}$$

Expanding gives

$$\Delta = C_1 C_2 (s^2 + bs + c) = 0 \tag{4. 89}$$

where the coefficients are

$$\begin{aligned} b &= \left[\frac{1}{R_1 C_1} + \frac{1}{R_2 C_2} + \frac{1}{R_2 C_1}\right] \\ c &= \frac{1}{R_1 R_2 C_1 C_2} \end{aligned} \tag{4. 90}$$

In general, the solutions to the quadratic equation

$$s^2 + bs + c = 0 \tag{4.91}$$

give the roots in the form

$$\begin{aligned} s_1 &= -\frac{b}{2} + \frac{1}{2}\sqrt{b^2 - 4c} \\ s_2 &= -\frac{b}{2} - \frac{1}{2}\sqrt{b^2 - 4c} \end{aligned} \tag{4.92}$$

such that

$$\begin{aligned} s_1 + s_2 &= -b \\ s_1 s_2 &= c \end{aligned} \tag{4.93}$$

expresses the relationship between the two. Substituting, these have the explicit form

$$\begin{aligned} s_1 + s_2 &= -\left[\frac{1}{R_1 C_1} + \frac{1}{R_2 C_2} + \frac{1}{R_2 C_1}\right] \\ s_1 s_2 &= \frac{1}{R_1 R_2 C_1 C_2} \end{aligned} \tag{4.94}$$

which allows us to study the characteristics of the roots for different ranges of resistor and capacitor values.

Next, let us make some general observations about the roots. First, since b is positive, s_2 will in general be more negative than s_1, as justified by comparing the expressions in eqn. (4.94); in terms of magnitudes we may write $|s_1| < |s_2|$. Also, the value of the product $s_1 s_2$ is determined by the values of the elements. Let us assume that $C_1 \approx C_2$ is a reasonable situation. Then, the relative values of R_1 and R_2 determine the nature of the roots. For $R_1 \sim R_2$, the two roots satisfy the general relation $|s_1| < |s_2|$, but the expression are somewhat messy. To illustrate the nature of the roots, however, we can study the case where $R_1 >> R_2$, and the opposite situation where $R_2 >> R_1$ as these are a little simpler.

Consider first the case where $R_1 >> R_2$. To identify the roots, we will assume that $|s_1| < |s_2|$. To approximate s_1, we start with the quadratic equation

$$s_1^2 + bs_1 + c = 0 \tag{4.95}$$

and assume that $s_1 << 1$. Then,

$$bs_1 + c \approx 0 \tag{4.96}$$

so that

$$s_1 \approx -\frac{c}{b} \approx -\frac{1}{R_1(C_1 + C_2)} \tag{4.97}$$

is a reasonable estimate. To obtain the other root s_2, we assume that both the square and the linear terms are large compared to the constant c in the quadratic equation, and then solve

$$s_2^2 + bs_2 \approx 0\ . \tag{4.98}$$

Ignoring the trivial solution, this reduces to $s_2 \approx -b$ so that for $R_1 >> R_2$,

$$s_2 \approx -\left(\frac{1}{R_2C_2} + \frac{1}{R_2C_1}\right) \tag{4. 99}$$

Comparing these two roots shows that s_2 is more negative than s_1 as we stated earlier using general arguments.

Now let us examine the opposite case where $R_2 >> R_1$. Following the same procedure as above, The first root s_1 is given by

$$s_1 \approx -\frac{c}{b} \approx -\frac{1}{R_2C_2}. \tag{4. 100}$$

while the second root s_2 is found to be

$$s_2 \approx -\frac{1}{R_1C_1}. \tag{4. 101}$$

This shows that $s_2 < s_1$ also holds for this case, making our general argument about the relative sizes of the roots more plausible without going into any further mathematical analysis.

The significance of the roots can be seen by noting that $v_2(s)$ can be solved as

$$v_2(s) = \frac{(1/C_1C_2)}{(s-s_1)(s-s_2)}\left[sC_1C_2V_{max} + C_2V_{max}\left(\frac{1}{R_1} + \frac{1}{R_2}\right) + \frac{C_1V_{max}}{R_2}\right] \tag{4. 102}$$

This shows the existence of two s-domain poles at s_1 and s_2. Performing a partial fraction expansion gives an expression of the general form

$$v_2(s) = \frac{K_1}{(s-s_1)} + \frac{K_2}{(s-s_2)} \tag{4. 103}$$

so that the time domain equation obtained by inverse transforming $v_2(s)$ is

$$V_2(t) = \left[K_1e^{s_1t} + K_2e^{s_2t}\right]u(t) \tag{4. 104}$$

where $u(t)$ is the unit step function. Since we know that the roots s_1 and s_2 are negative, we may write that, for times $t \geq 0$,

$$V_2(t) = K_1e^{-|s_1|t} + K_2e^{-|s_2|t} \tag{4. 105}$$

Finally, noting that we have chosen the roots such that $|s_1| < |s_2|$, the dominant root that determines the transient response is s_1 since it gives the slowest decaying solution. Thus, we approximate

$$V_2(t) \approx K_1e^{-|s_1|t} \tag{4. 106}$$

as the important term and calculate the coefficient K_1 as

$$K_1 = \frac{V_{max}}{(s_1-s_2)}s_1 + \frac{V_{max}}{(s_1-s_2)}\frac{1}{C_1}\left(\frac{1}{R_1} + \frac{1}{R_2}\right) + \frac{V_{max}}{(s_1-s_2)}\frac{1}{R_2C_2} \tag{4. 107}$$

by standard techniques. To understand the structure of this expression, let us regroup terms so that K_1 reads as

$$
\begin{aligned}
K_1 &= \frac{V_{max}}{(s_1 - s_2)}\left[s_1 + \left(\frac{1}{R_1C_1} + \frac{1}{R_2C_1} + \frac{1}{R_2C_2}\right)\right] \\
&= \frac{V_{max}}{(s_1 - s_2)}[s_1 + b] \\
&= \frac{V_{max}}{(s_1 - s_2)}[s_1 - (s_1 + s_2)] \\
&= V_{max}\left(\frac{-s_2}{s_1 - s_2}\right)
\end{aligned}
\tag{4. 108}
$$

where we have used the definition of the coefficient b that appeared in the quadratic equation. This gives us the final form

$$V_2(t) \approx V_{max}\left(\frac{-s_2}{s_1 - s_2}\right)e^{s_1 t} \tag{4. 109}$$

for the capacitor voltage.

Now let us compare this to an ideal exponential decay where

$$V_{cap}(t) \approx V_{max}e^{-(t/\tau)} \tag{4. 110}$$

where τ is the time constant. When $t = \tau$, $V_{cap}(t) = V_{max}\,e^{-1}$, so let us set the voltage $V_2(\tau')$ to the same value in an effort to find an equivalent time constant τ' for the 2-rung ladder:

$$V_2(t)\,|_{t=\tau'} \approx \frac{-s_2 V_{max}}{(s_1 - s_2)}e^{s_1\tau'} \equiv V_{max}e^{-1}\,. \tag{4. 111}$$

This requires that

$$\frac{-s_2}{(s_1 - s_2)}e^{s_1\tau'} = e^{-1}\,. \tag{4. 112}$$

To solve for the value of the time constant τ, we write this is the form

$$e^{s_1\tau' + \ln[-s_2/(s_1 - s_2)]} = e^{-1} \tag{4. 113}$$

where we note that the argument $[-s_1/(s_1 - s_2)]$ in the logarithm is a positive number. The time constant may thus be estimated by solving

$$s_1\tau' + \ln\left[\frac{-s_2}{(s_1 - s_2)}\right] = -1\,. \tag{4. 114}$$

Using the expansion[11]

$$\ln(x) \approx \frac{x-1}{x} \tag{4. 115}$$

---

[11] The general expansion is valid for $x > (1/2)$ which is satisfied by the roots, and it is assumed that only the first term is significant.

we have

$$-\tau' \approx \left(\frac{1}{s_1}\right)\left[\frac{\frac{-s_2}{(s_2-s_1)}-1}{\frac{-s_2}{(s_2-s_1)}}\right]+\frac{1}{s_1}=\frac{(s_1+s_2)}{s_1 s_2}. \quad (4.116)$$

Finally, substituting the expressions from eqn. (4.94) gives

$$\begin{aligned} -\tau' &= R_1R_2C_1C_2\left(\frac{1}{R_1C_1}+\frac{1}{R_2C_2}+\frac{1}{R_2C_1}\right) \\ &= R_2C_2+R_1C_1+R_1C_2 \\ &= R_1C_1+C_2(R_1+R_2) \end{aligned} \quad (4.117)$$

where the minus sign appears because we used negative roots, i.e., $s_1 = (-1/\tau')$. Since

$$V_2(t) \approx K_1 e^{s_1 t} = K_1 e^{-t/\tau}, \quad (4.118)$$

we see by inspection that τ=-τ', giving the final result

$$\tau \approx R_1C_1+C_2(R_1+R_2) \quad (4.119)$$

for the time constant of the 2-rung ladder, as previously stated. Let us reiterate that this looks like the summation of two time constants, one for each capacitor, with each term representing the discharge through the total resistance in the path. This provides a handy tool for estimating the transient response of RC chains and will be used in many applications later in the book.

Consider the charging problem illustrated in Figure 4.17. With the initial conditions $V_1(0)$=0v and $V_2(0)$=0v the analysis gives the result

$$V_2(t) \approx V_{max}[1-e^{-(t/\tau)}] \quad (4.120)$$

with the same time constant. The result can be derived by simply replacing the driving terms in the original s-domain equation set. This technique provides a useful approximation for characterizing the transient response of both nFET and pFET chains.

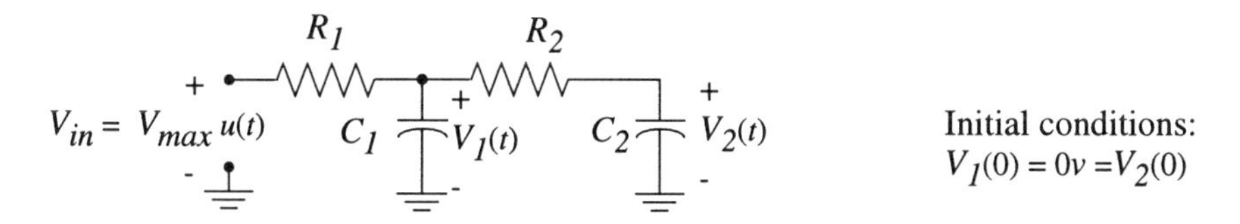

Figure 4.17. Charging of a 2-rung RC ladder network

Extension to Longer RC Chains

The results above may be extended to approximate the charging or discharging through an arbitrary *RC* ladder network. Consider the N-rung ladder shown in Figure 4.18. The charging of the last capacitor C_N from 0v to V_{max} is assumed to be in the exponential form

$$V_N(t) \approx V_{max}[1 - e^{-t/\tau}] \qquad (4.121)$$

where the time constant is computed using the general Elmore formula expresed by the double summation

$$\tau = \sum_{k=1}^{N} C_k \left(\sum_{m=1}^{k} R_m \right) . \qquad (4.122)$$

In words, this says that the time constant is calculated by adding up individual terms where each is a capacitance C multiplied by the sum of the resistances seen to the end of the chain. Note that the estimate assumes that all capacitors are initially uncharged at time t=0. Similarly, the discharge of C_N from V_{max} to $0v$ is approximated by

$$V_N(t) \approx V_{max} e^{-t/\tau} \qquad (4.123)$$

with the same time constant. This result assumes that all capacitors in the chain are initially charged to a voltage of V_{max}.

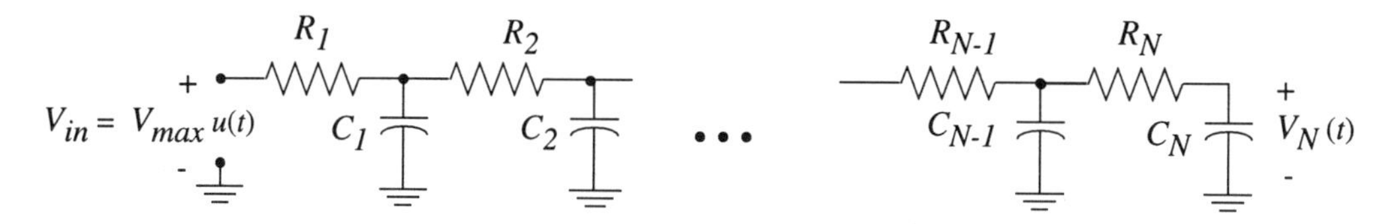

Figure 4.18. Application to a longer RC ladder

Example 4-2

Suppose that we have the 4-rung ladder shown in Figure 4.19. For this case, the time constant is approximated by

$$\tau = C_4(R_1 + R_2 + R_3 + R_4) + C_3(R_1 + R_2 + R_3) + C_2(R_1 + R_2) + C_1 R_1 \qquad (4.124)$$

by directly applying the formula. The meaning of the time constant is as follows. If we apply a step input voltage $V_{in}(t) = V_{max}u(t)$ to the input at R_1, then the voltage $V_4(t)$ across C_4 will charge according to

$$V_4(t) \approx V_{max}[1 - e^{-t/\tau}] \qquad (4.125)$$

where it is assumed that all capacitors in the chain are initially uncharged. Conversely, if we apply an input voltage to the chain of $V_{in}(t) = V_{max}[1 - u(t)]$, then C_4 will discharge with a voltage approximated by

$$V_4(t) \approx V_{max} e^{-t/\tau} \qquad (4.126)$$

if all of the capacitors are initially charged to V_{max}. Note that, if the capacitors are of the same order of magnitude, the first term is the largest, the second term is the second largest, and so on.

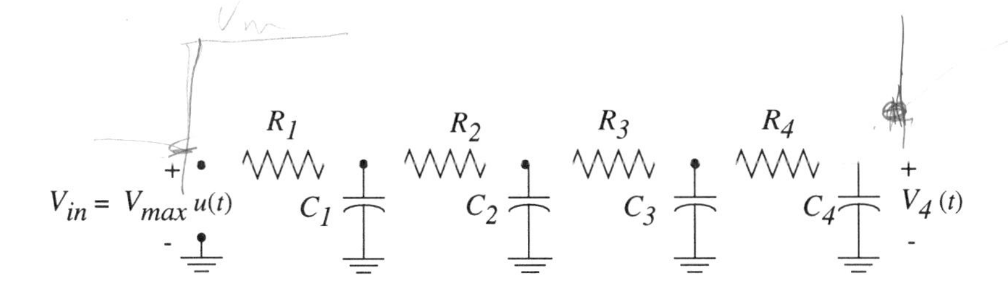

Figure 4.19. Example of a 4-rung RC ladder

Application to FET Chains

The *RC* model provides a simple technique for approximating the behavior of a chain of series connected MOSFETs. However, it is important to remember that the model has several intrinsic limitations such as

- It does not include the threshold voltage modification through the chain;
- The linearized resistance of a MOSFET is at best a crude approximation; and
- The capacitances are assume to all have the same initial condition as the final capacitor at the end of the chain.

The threshold modification can be accounted for by adjusting the input voltage, so this is not a major obstacle.

Perhaps the biggest problem with the accuracy is that modelling the voltage using an exponential time function greatly oversimplifies the operation of the circuit. Consider an nFET being used as a pass transistor. If we apply a logic 0 input with $V_{in} = 0v$, then the discharging of the capacitor can be reasonably approximated as an exponential with

$$V_{1\to 0} \approx V_1 e^{-t/\tau_n} \tag{4.127}$$

where V_1 is the initial high voltage. However, a logic 1 input with $V_{in} = V_{DD}$ induces a charging event with a voltage that depends on time approximated by modifying the time constant and using

$$V_{0\to 1} \approx V_1\left[\frac{t/2\tau_n}{1 + t/2\tau_n}\right]. \tag{4.128}$$

This is much slower (and quite different in form) than the simple exponential time dependence, and should be used for this case if better accuracy is desired.

The model has similar limitations when applied to a pFET. Since a pFET can pass a strong logic 1 voltage, the exponential form

$$V_{0\to 1} \approx V_1\left[1 - e^{-t/\tau_p}\right] \tag{4.129}$$

with the appropriate time constant can be used as a reasonable approximation. For the opposite case of passing a logic 0 voltage, the discharge described by

$$V_{1\to 0} = |V_{Tp}| + \frac{(V_1 - |V_{Tp}|)}{1 + t/2\tau_p} \qquad (4.130)$$

is more accurate than the exponential, and should be used when possible. Finally, note that we have ignored body-bias effects throughout the section.

Although these problems cannot be eliminated from the modelling approach, it remains a very useful basis for design so long as we exercise caution in interpreting the results. This observation is due to the fact that the channel width W is the primary design variable and we have the general dependences

$$R \propto \frac{1}{W} , \qquad C \propto W \qquad (4.131)$$

for the MOSFET resistance and capacitance values. The effect of varying the aspect ratios of the transistors will be seen directly in the RC ladder model as changing the time constant.

An example of the calculations involved in this type of analysis starts with a layout such as that shown in Figure 4.20(a) for a 3-transistor nFET chain. Figure 4.20(b) shows the equivalent RC ladder circuit that we use to model the transient electrical characteristics. To determine the elements, we first note from the layout that all three MOSFETs have the same aspect ratio. Thus,

$$R = \frac{1}{k'_n\left(\frac{W}{L}\right)(V_{DD} - V_{Tn})} \qquad (4.132)$$

is valid for all three transistors; the gate overlap L_o is needed to find the electrical channel length L since the layout only shows the drawn channel length $L' = L + 2L_o$. We may also calculate the total gate capacitance

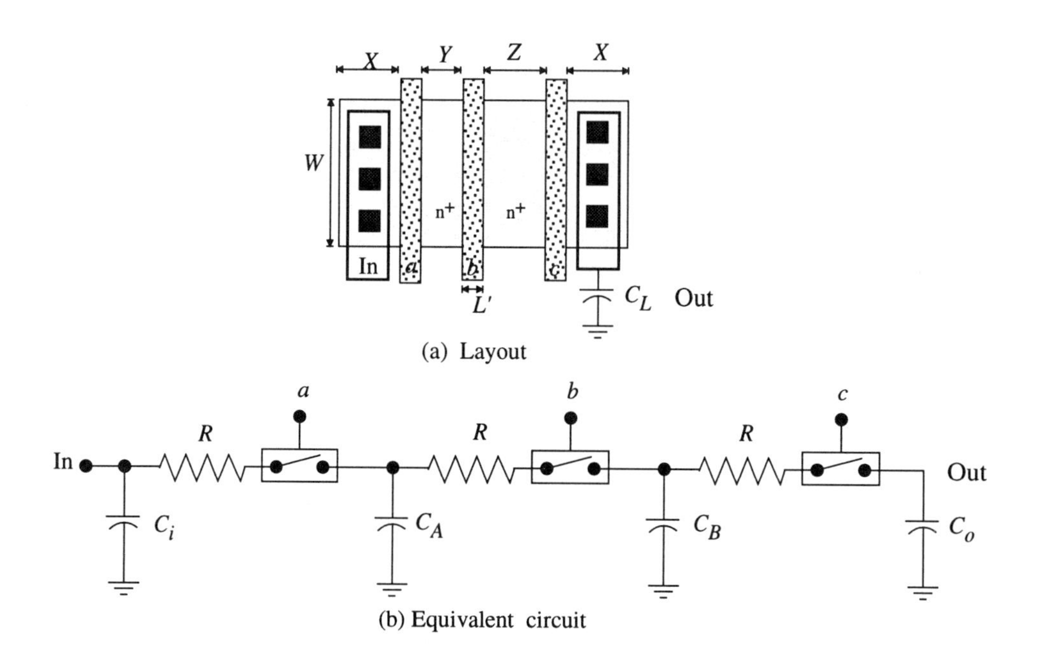

Figure 4.20. RC modelling of a series FET chain

$$C_G = C_{ox}W\dot{L} \qquad (4.133)$$

which is the same for all three transistors, and therefore estimate

$$C_{GS} \approx \left(\frac{1}{2}C_{ox}W\dot{L}\right) \approx C_{GD} \qquad (4.134)$$

as the gate-source and gate-drain contributions.

The junction capacitances depend the dimensions. Since every n^+ has a width W, the differences arise in the horizontal distances that are labelled by X, Y, Z. The LTI values for each region are given by

$$\begin{aligned} C_X &= K_m(0, V_{max}) C_{j0} W(X + L_o) + K_{mj}(0, V_{max}) 2C_{jsw}(W + X + L_o) \\ C_Y &= K_m(0, V_{max}) C_{j0} W(Y + 2L_o) + K_{mj}(0, V_{max}) 2C_{jsw}(W + Y + 2L_o) \\ C_Z &= K_m(0, V_{max}) C_{j0} W(Z + 2L_o) + K_{mj}(0, V_{max}) 2C_{jsw}(W + Z + 2L_o) \end{aligned} \qquad (4.135)$$

where we have included the gate overlap distance L_o in the dimensions. Note that L_o is important every location where the gate borders an n^+ region. This is why the C_X terms only have one factor of L_o while C_Y and C_Z both of which border two FET gates, have a factor of $2L_o$ in the area and perimeter terms.

Now we may use the capacitance contributions to determine the equivalent circuit values. The input capacitance can be estimated as

$$C_{in} = C_X + C_{GD} \qquad (4.136)$$

with both contributions originating from the FET farthest to the left. Similarly, the output capacitance is

$$C_o = C_X + C_{GD} + C_L \qquad (4.137)$$

because the right n^+ region has the same dimensions as the left side, but it also drives the external load capacitance C_L. The internal node capacitances are given by

$$\begin{aligned} C_A &= C_Y + C_{GD} + C_{GS} = C_X + C_G \\ C_B &= C_Z + C_{GD} + C_{GS} = C_Z + C_G \end{aligned} \qquad (4.138)$$

where the terms $(C_{GD}+C_{GS})=C_G$ arise because both regions border two FET gates.

The general philosophy is illustrated by the diagram in Figure 4.21. As we will see in the following chapters, CMOS logic circuits consist only of MOSFETs that are wired together in a specified manner; we will loosely refer to the wiring scheme as the **circuit topology**. For a given logic function, we will be able to construct the required CMOS circuit topology directly using straightforward circuit design procedures as indicated in the first block. This will yield the proper logic behavior, but does not provide information as to the switching performance. The speed depends upon the aspect ratios and layout parameters. The next step illustrated in the flow is to choose aspect ratios for every transistors and then create the RC-equivalent network; the simplified circuit is then used to estimate the transient response. Once the desired switching times are achieved, the CMOS realization is simulated and checked. Any redesigns may be performed on the RC circuit as implied by the loop in the design flow.

The characteristics of the RC-equivalent analysis can be seen using the example of a 2-FET chain shown in Figure 4.22. In (a), the transistor aspect ratios $(W/L)_A$ and $(W/L)_B$ are used to create an RC ladder with elements R_A, R_B, C_1, and C_2; this may be used to estimate the response $V_{out}(t)$. If

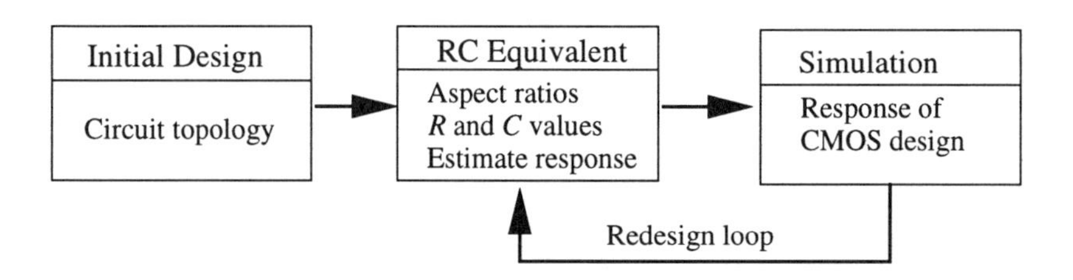

Figure 4.21. Use of the RC equivalent network in the circuit design process

we change the aspect ratios to new values $(W/L)'_A$ and $(W/L)'_B$ as in (b), then the circuit elements are changed to R'_A, R'_B, C'_1, and C'_2 and the output voltage is $V'_{out}(t)$. In both cases, simplified equations can be used to estimate the delay times. The interesting conclusion is that the percentage difference between $V_{out}(t)$ and $V'_{out}(t)$ in the RC networks will be about the same as that in the FET circuits. In other words, the simplified equivalent circuit tracks changes in the FET design, so that it may be used for quick estimates. In a practical engineering environment, this allows one to estimate the device sizes needed to meet the timing specifications early in the design cycle. However, the most important aspect of this approach is the fact that the RC circuit provides a basis for understanding the factors that determine the transient response of the circuit. It is therefore widely used in the literature in the context of high-performance design work.

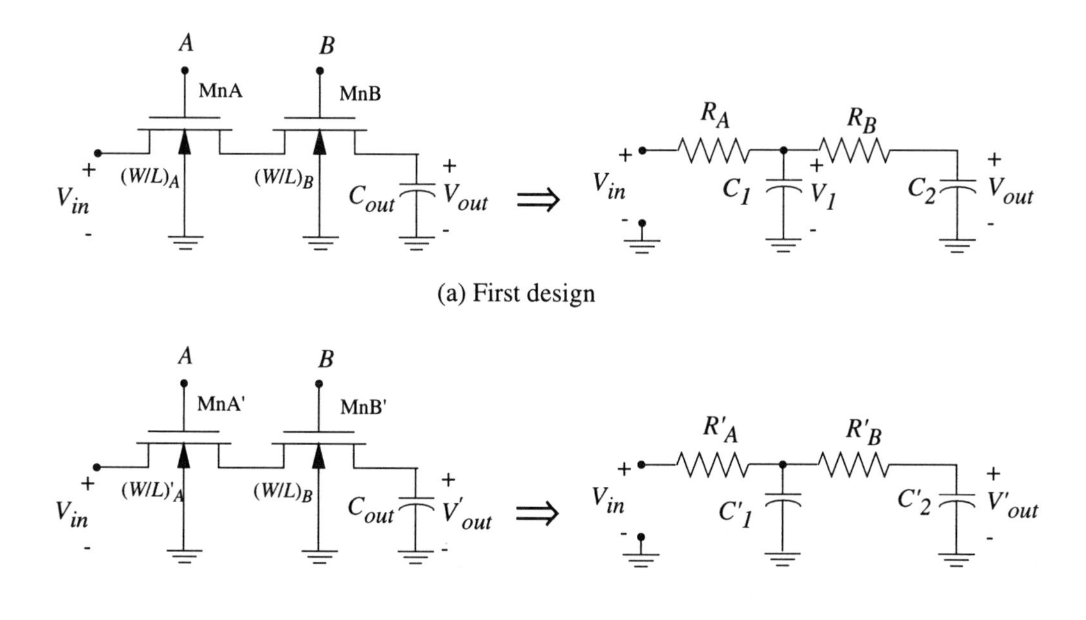

Figure 4.22. Application RC modelling to a 2-FET chain

4.6 MOSFET Switch Logic

It is possible to use the results of the chapter as the foundation for developing a logic design formalism for MOSFET arrays. Let us start with an nFET switch as shown in Figure 4.23(a). Note that we have use simplified nFET circuit symbols in which the bulk electrode is ignored. The state of conduction is controlled by the gate voltage, which is now represented by the Boolean variable G; A is viewed as the input, and B is the output. If $G = 0$ as in (b), then the FET is in cutoff, and no relationship exists between the left and right values A and B. However, placing a value of $G = 1$ on the gate [drawing (c)] turns on the transistor, and the value of A is transferred to B. A simple way to express this is to write

$$G = 1: \ A \rightarrow B \tag{4. 139}$$

where we use the syntax

Condition: *Action*

to express the event. From the design viewpoint, it is more useful to write the output B as

$$B = A \cdot G \tag{4. 140}$$

which is valid iff G=1. Note that this says to take the input A ANDed with the variable G applied to the gate.

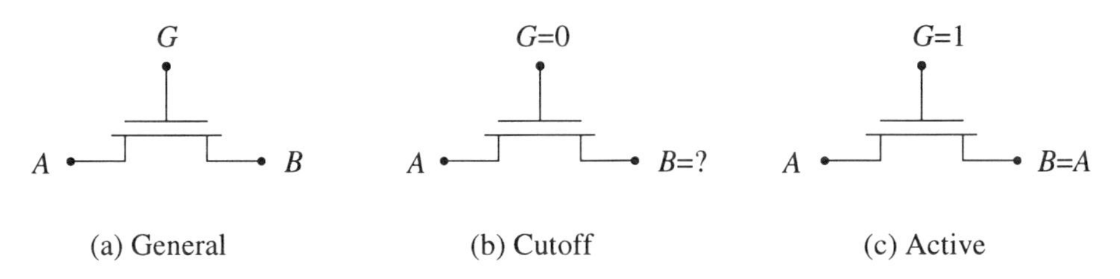

Figure 4.23. Logical behavior of an nFET

A pFET, as shown in Figure 4.24(a) using the simplified symbol (where the bulk electrode is not drawn, and a bubble is added to the gate) has characteristics that are exactly opposite to those of an nFET. For the pFET, applying $G = 1$ places the pFET in cutoff and there is no relationship between the input A and the output B; this is shown in Figure 4.24(b). If, on the other hand, we have $G = 0$ as in drawing (c), then the transistor is active and allows us to state that

$$G = 0: \ A \rightarrow B \tag{4. 141}$$

This may be written in alternate form as

$$B = A \cdot \overline{G} \tag{4. 142}$$

but is only valid for the case where $G = 0$. For the case of the pFET, this equation tells us that the output is obtained by ANDing the input A with the complement $\overline{G}$ of the gate variable. Although these relations are very simple, they are very powerful when applied to FET logic design.

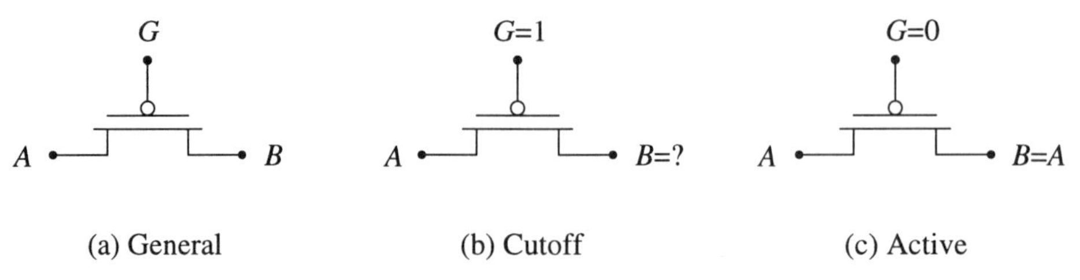

Figure 4.24. Logical behavior of a pFET

4.6.1 Multiplexor Networks

In general, a MUX allows us to choose one of n inputs for transmission to the output using $m=2^n$ control bits. As discussed in Section 4.3 above, the inverter may be viewed as a simple multiplexor where the inputs are tied to V_{DD} and ground. Arrays of pass transistors may be used to create simple switching networks that implement various MUX arrangements.

2:1 Networks

Let us examine how nFETs can be used to construct the simple 2:1 multiplexor network shown in Figure 4.25. Operationally, the input data paths are denoted as P_0 and P_1, while the select bit S determines which input will be directed to the output. This is expressed by writing

$$f = P_0 \cdot \overline{S} + P_1 \cdot S \tag{4. 143}$$

i.e., S=0 gives and output of f=P_0, while S=1 gives an output of f=P_1.

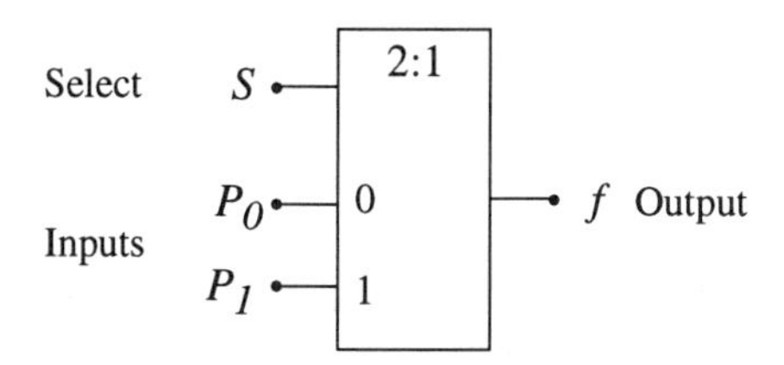

Figure 4.25. General 2:1 multiplexor

Figure 4.26 shows the nFET-based design of a 2:1 multiplexor. By inverting the control signal S and applying it to MOSFET Mn1, we create the term $P_0\overline{S}$, Similarly, transistor Mn2 yields the term P_1S by applying the nFET logic expression. The OR operation (+) exists because either the top line is connected to the output OR the bottom line is connected to the output. Note that this neatly sidesteps the problem of not having a logic condition for an open switch by insuring that there is always a connection to the output.

The drawback of this circuit is the reduced voltage logic swing. The minimum voltage at the output is 0v, but the threshold voltage loss restricts the maximum voltage to

$$V_{max} = V_{DD} - V_{Tn} . \tag{4. 144}$$

In addition, charging the output to a logic 1 voltage is very slow compared to the transition down-

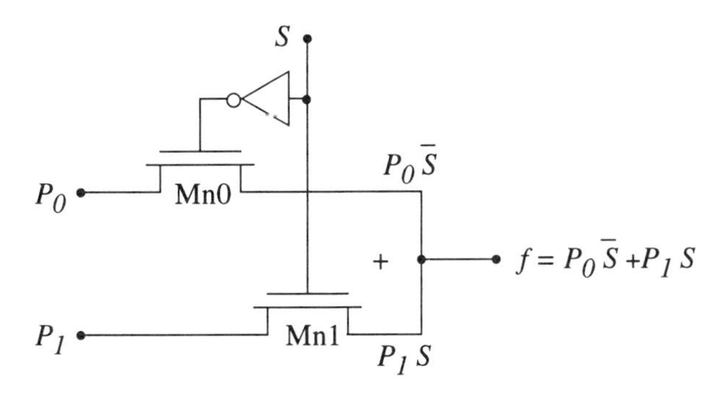

Figure 4.26. nFET-base 2:1 MUX

ward to a logic 0.

A pFET design for a 2:1 network is shown in Figure 4.27. This has the same logical output, but now we apply S to the upper FET Mp1 to switch it ON when S=0. When S=1, the lower FET Mp2 conducts. The output voltage range for this network is between $|V_{Tp}|$ and V_{DD}, with the transition to a logic 0 state being the slowest. It would be rare to find this circuit used in practice because we usually try to avoid pFETs in the signal path. However, if the switching speed is not critical, then it could be used to fill up otherwise vacant real estate on the chip. This layout strategy saves the nFETs for the high-speed circuits.

4:1 MUX Networks

A 4:1 multiplexor uses a two bit word S_1S_0 to switch one of the four inputs paths P_0, P_1, P_2, P_3 t o the output. This is expressed by

$$f = (\overline{S_1}\overline{S_0}) \cdot P_0 + (\overline{S_1}S_0) \cdot P_1 + (S_1\overline{S_0}) \cdot P_2 + (S_1S_0) \cdot P_3 \tag{4. 145}$$

so that the decimal equivalent of S_1S_0 determines the selected path.

Figure 4.28 shows an nFET circuit for the 4:1 MUX network. The output function is obtained by using the nFET switching equation to write each possible output condition. Consider, for example, path P_0. The first MOSFET has an input of P_0 and is active when S_1 =0, giving a factor $\overline{S}_1P_0$; the second nFET uses this as an input, and conducts when S_0=0, so the top line provides the first term $(\overline{S}_1\overline{S}_0)P_0$ in the output function. The remaining terms are obtained in the same manner, and the

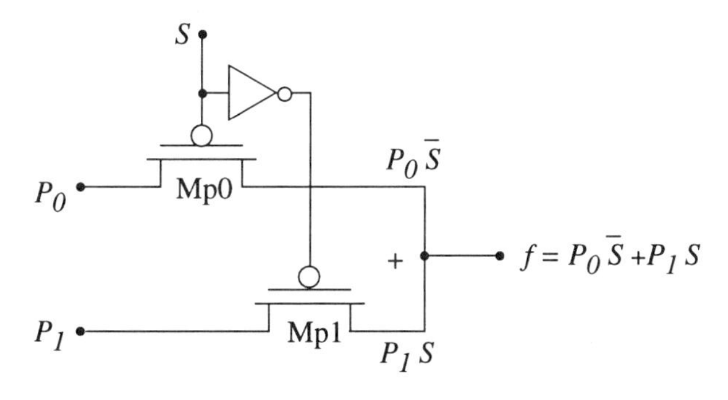

Figure 4.27. pFET-based 2:1 MUX circuit

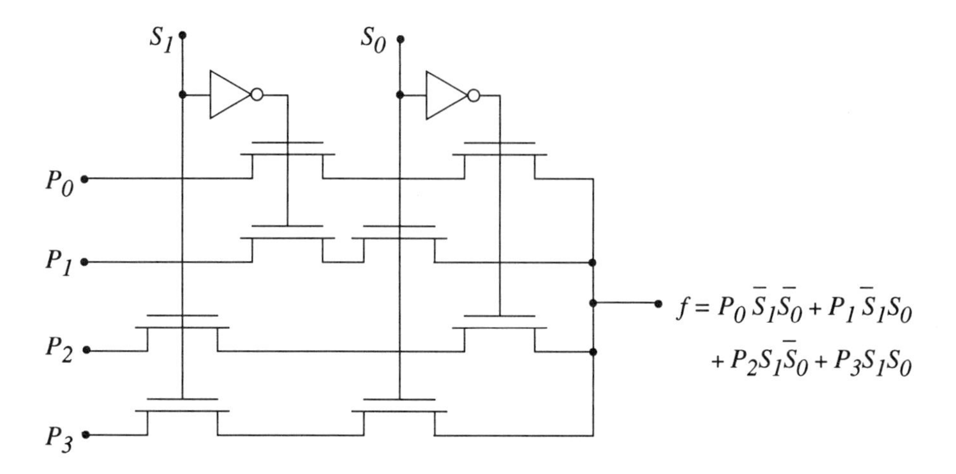

Figure 4.28. nFET-based 4:1 multiplexor circuit

OR function is due to the fact that only one input is connected to the output at a time. The main limitation of this network is the fact that the threshold voltage loss restricts the output voltage to the range [0*v*, (V_{DD}-V_{Tn})].

Now consider the pFET realization shown in Figure 4.29. This provides the same logic switching function, but restricts the transmitted voltage to the range [$|V_{Tp}|$,V_{DD}] due to the poor logic 0 problem. In addition, the pFETs exhibit slower response when compared to identical size nFETs, and must be made using larger aspect ratios to achieve the same switching speeds.

Merging the nFET and pFET arrays into a single network yields the **split array** in Figure 4.30. This network gives output voltages in the full range [0*v*, V_{DD}] by using the complementary properties of the transistors. Consider the input P_0. If P_0=0, then the nFET path allows the output to reach a value of V_{out} = 0*v*; conversely, an value of P_0=1 is transmitted through the pFETs, and $V_{out} = V_{DD}$. The drawbacks of this approach are that

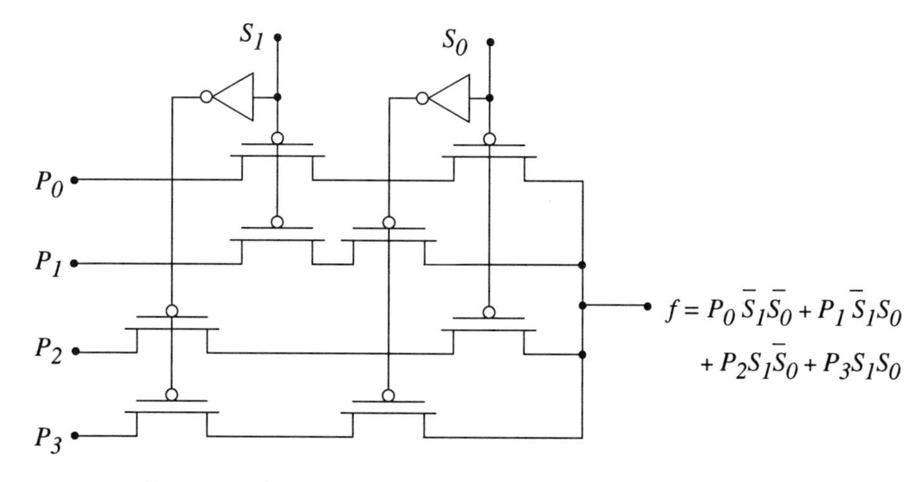

Figure 4.29. pFET-based 4:1 multiplexor circuit

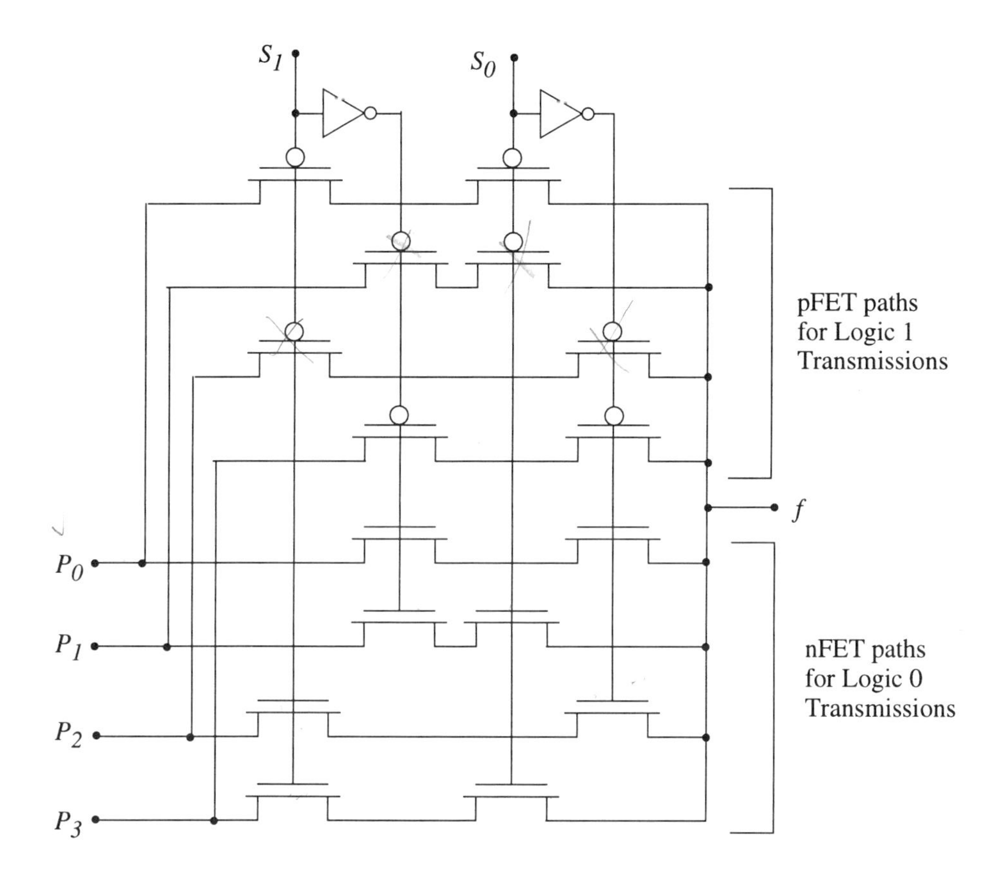

Figure 4.30. Split array 4:1 multiplexor circuit that provides full rail voltage transmission

- Dual arrays of nFETs and pFETs must be used, resulting in a larger area requirement, and,
- The capacitance C_{out} at the output node will be larger than that of a single array, which increases the switching time constants.

Note, however, that even though C_{out} is increased, the output voltage is exponential in time for either low or high input voltages. This is due to the fact that the split array eliminates the poor logic 1 nFET transition and logic 0 pFET transition by routing through the opposite polarity device.

4.7 Problems

[4.1] Consider an nFET used as a pass transistor as shown in Figure 4.1. Assume that V_{DD}=3.3v, V_{T0n}=+0.75v, and γ=0.08$v^{1/2}$, 2$|\phi_F|$=0.58v. Use the iterative approach described in the text to perform the following tasks.

(a) Calculate the value of V_{max} if body-bias is ignored.

(b) Now include body-bias in your calculation of V_{max} and then find the percentage error if body-bias is ignored.

[4.2] Perform the integral and associated algebra as described in the text to verify that eqn. (4.4) that describes $V_{out}(t)$ for a logic 1 transfer through an nFET is correct.

[4.3] Consider the nFET described in Problem 1. As discussed in the text, the threshold voltage drop is due to the fact that the gate voltage V_G is set at the power supply V_{DD}, which is assumed to be the highest voltage in the circuit. Suppose instead that we are allowed to increase V_G to any value.

(a) Calculate the minimum value of the applied gate voltage V_G that would overcome the threshold voltage loss, i.e., that would result in an output voltage of $V_{out} = V_{DD}$ when $V_{in} = V_{DD}$.

(b) What are the effects on the transmission of $V_{in} = V_{DD}$ through an nFET if the applied gate voltage is two times larger than the value found in part (a).

(c) Discuss the disadvantages of using this approach to eliminate the threshold voltage loss.

Use the values V_{DD}=3.3v, V_{T0n}=+0.8v, and γ=0.071$v^{1/2}$, and 2$|\phi_F|$=0.58v in your calculations.

[4.4] Consider the circuit shown in Figure 4.12 where the output of pass nFET Mn1 is used to drive the gate of another pass FET Mn2. Assume parameters V_{DD}=3v, V_{T0n} = +0.70v, 2$|\phi_F|$=0.58v, and γ =0.071$v^{1/2}$. Use the iterative approach described in the text to calculate the maximum voltage transmitted through Mn2 if body bias effects are included in both transistors.

[4.5] Evaluate the integral in eqn. (4.14) and show that the result quoted in eqn. (4.15) for $V_{out}(t)$ is correct.

[4.6] Consider a pFET that is used as a pass transistors as shown in Figure 4.6. Assume that V_{DD}=3.3v, V_{T0p}=-0.85v, 2ϕ_F=0.70v and γ=0.18$v^{1/2}$.

(a) Use numerical iterations to calculate the value of V_{min} that can be transferred through the pFET.

(b) What would be the percentage error if body-bias effects were ignored?

[4.7] Consider the nFET layout in Figure P4.1 where all dimensions are in units of microns. The process parameters are given as

$$k'_n=140\mu A/v^2,\ V_{T0n}=0.70v,\ \mu_n=560\ cm^2/V\text{-}sec,\ C_j=0.3fF/\mu m^2,$$
$$C_{jsw}=0.25fF/\mu m,\ m_j=0.5,\ m_{jsw}=1/3,\ N_a=10^{15}cm^{-3}.$$

(a) Find the LTI-equivalent resistance of the nFET.

(b) Find the voltage-averaged capacitances for the switching model.

(c) Calculate the time constant of the device.

(d) Perform a transient simulation on the circuit using SPICE with a load capacitor value of C_L=10C_G.

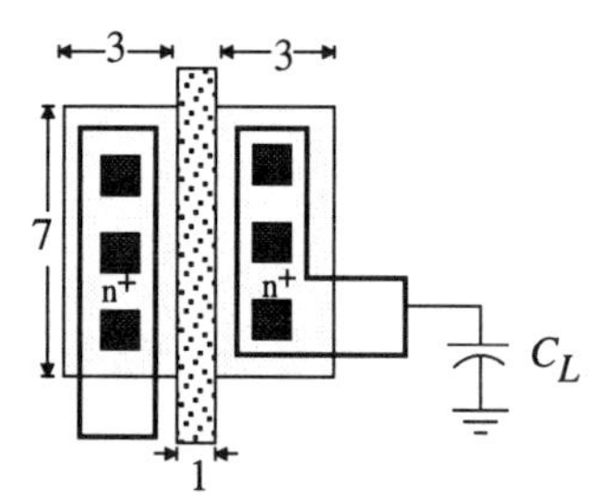

Figure P4.1

[4.8] Consider the circuit shown in Figure P4.2. This is simply an inverter where we have interchanged the nFET and pFET.

(a) What is logic function performed by this circuit?

(b) What is V_{OH} for this circuit?

(c) What is V_{OL} for this circuit?

Assume V_{DD}=3.3v, V_{TOn} = +0.75v, V_{TOn} = -0.90v, $2|\phi_F|$=0.58v and γ=0.071$v^{1/2}$ for the nFET, and $2\phi_F$=0.60v and γ=0.135$v^{1/2}$ for the pFET.

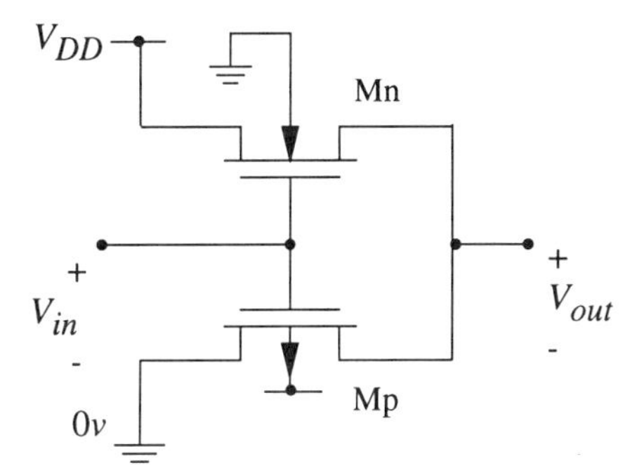

Figure P4.2

[4.9] A 2-FET chain is shown in Figure P4.3 below.

(a) Create the equivalent RC circuit that can be used to estimate the discharge.

(b) Calculate the time constant using the Elmore formula. The use your result to estimate the discharge time if the load capacitor is initially at V_{max}.

(c) Simulate the circuit using SPICE and compare the results with the hand estimates.

Assume parameters of V_{TOn}=0.75v, μ_n=560cm^2/V-sec, C_j=0.42$fF/\mu m^2$, C_{jsw}=0.38$fF/\mu m$, N_a=$10^{15}cm^{-3}$.

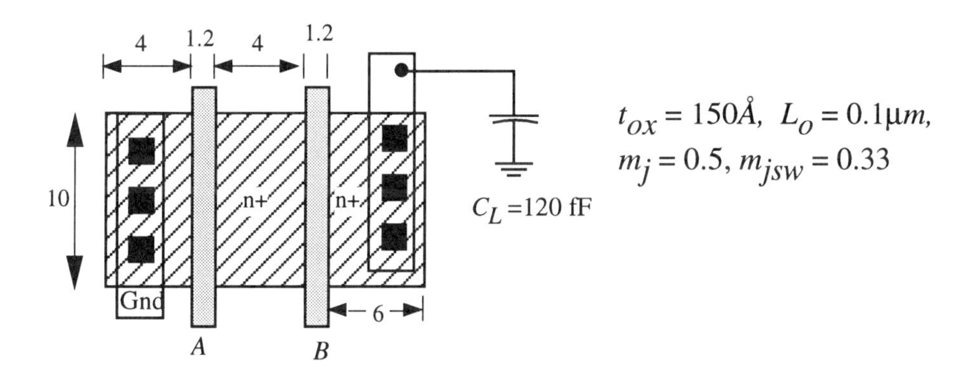

Figure P4.3

[4.10] Suppose that the layout in Figure P4.3 is modified such that the channel widths are increased to a value of W=20μm, but everything else remains the same.

(a) Create the equivalent RC circuit that can be used to estimate the discharge.

(b) Calculate the time constant using the Elmore formula. The use your result to estimate the discharge time if the load capacitor is initially at V_{max}.

(c) If you have solved Problem [4.9], then compare the changes in the circuit performance with the increased width. Assume parameters of k'_n=150$\mu A/v^2$, V_{TOn}=0.75v, μ_n=560 cm^2/V-sec, C_j=0.42$fF/\mu m^2$, C_{jsw}=0.38$fF/\mu m$, N_a=$10^{15}cm^{-3}$.

[4.11] Consider the 3-FET chain in Figure 4.10. The transistor parameters are k'_n=150$\mu A/v^2$, V_{TOn}=+0.75v, μ_n=560 $cm^2/V\text{-}sec$, and $2|\phi_F|$=0.58v. The nFETs are all identical and with a common aspect ratio of (W/L) = 4. When one of these transistors is modelled as an individual device as in Figure 4.14, the capacitances are given as C_I=34fF =C_2. The external load on the right-hand side of the chain has a value of C_l=100fF.

(a) Construct the RC equivalent network for the FET chain

(b) Calculate the time constant for the series network

(c) Suppose that the aspect ratios of all three FETs are increased to (W/L) = 8. Construct the new RC equivalent circuit and find the time constant.

(d) Simulate both circuits on SPICE using lumped-element capacitances (only) and compare the transient response times.

Chapter 5

Static Logic Gates

One extremely powerful aspect of CMOS is the ability to create single gate circuits that can implement functions consisting of several basic logic operations. This makes digital CMOS design quite different from classical logic design techniques, since now the logic expressions and the corresponding circuits become very closely related. A **static logic gate** is one that has a well defined output once the inputs are stabilized and the switching transients have decayed away. Static CMOS logic gates are relatively easy to design and use. This chapter deals with the static logic gates, from simple NAND and NOR operations to complex functions that are quite large and powerful.

5.1 Complex Logic Functions

Complex logic gates provide functions that consist of several primitive NOT, AND, or OR operations. Consider a three-variable function with input variables *a*, *b*, and *c*. Suppose that we construct the function

$$f(a, b, c) = \overline{a \cdot \bar{b} \cdot c + \bar{a} \cdot b \cdot c + a \cdot b \cdot c} \tag{5. 1}$$

this is an example of a canonical AOI (and-or-invert) equation[1] which falls into the category of a complex logic operation. If we reverse the ordering of the logic operations, we obtain an OAI (or-and-invert) expression. For example,

$$g(a, b, c) = \overline{(a + \bar{b} + c) \cdot (\bar{a} + b + c) \cdot (a + b + c)} \tag{5. 2}$$

exhibits this characteristic, and is a complex logic form. It is worthwhile to note that *f* and *g* are

[1] Recall that the "AND" operation has precedence over the "OR" operation.

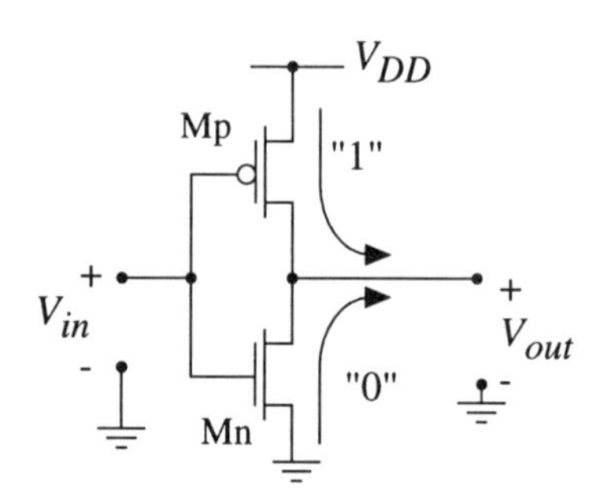

Figure 5.1 Operation of the CMOS inverter circuit

duals of each other, i.e., interchanging the AND and OR operations in one gives the other.

Complex logic gates are constructed using the CMOS inverter as a basis. As discussed in the previous chapter, the nFET and pFET are placed to act as pass transistors. With regards to the circuit shown in Figure 5.1, the input voltage V_{in} controls the conduction modes of both transistors. When $V_{in} = 0v$, the pFET conducts the power supply voltage V_{DD} to the output. When $V_{in}=V_{DD}$, the nFET is ON, and transmits the ground ($0v$) to the output. This provides the foundation for creating logic gates using arrays of FETs. Inputs are used to control whether V_{DD} or ground is connected to the output. As with the inverter, only one conduction path can exist at a time. This specifically eliminates the possibility of both the power supply and ground being simultaneously connected to output. The opposite case where neither is connected to the output is not desirable in a logic gate, but is useful for isolating the circuit; this will lead to the concept of a "tri-state" logic gate later in the chapter.

In order to construct a complex logic gate, let us replace the single inverter nFET by an array of nFETs that are connected to operate as a large switch. Similarly, we will substitute an array of pFETs for the single pFET used in the inverter, and view the pFET array as a "giant" switch. In order to insure proper electrical operation, however, we must exercise care so that operation of the nFET array complements the operation of the pFET array. This means that if one array is a closed switch, the other must be open.

The general structure of a complex logic gate can be created by the following steps.

- Provide a complementary pair (an nFET and a pFET with a common gate) for each input;
- Replace the single nFET with an **array** of nFETs that connects the output to ground;
- Replace the single pFET with an **array** of pFETs that connects the output to V_{DD};
- Design the nFET and pFET switching network so that only one network acts as a closed switch for any given input combination.

This results in the general network shown in Figure 5.2 for the case of three inputs *A*, *B*, and *C*.

With $m > 1$ inputs, the nFET and pFET arrays are viewed as large "composite" switches, with each array containing *m* MOSFETs. For a given input combination, only one composite switch can be closed. If the pFET switching array is closed, then the output voltage is $V_{out}=V_{DD}$, giving a logic 1 results. Conversely, if the nFET array is closed, then the output is a logic 0 with $V_{out}=0v$. Both cases are shown in Figure 5.3. Note that, for proper operation, the arrays must be designed so that the two cases where either (i) both switching arrays are closed, or, (ii) both switching arrays are open, cannot occur, since both situations give an undefined output.

In this chapter, we will first examine the NAND and NOR gates to see how the complementary structures work. Then, we will progress to more complex logic gates, where the integration capabilities of CMOS will become clear. As will be stressed in the treatment, CMOS provides an extremely powerful approach to building complicated digital logic networks in a very efficient

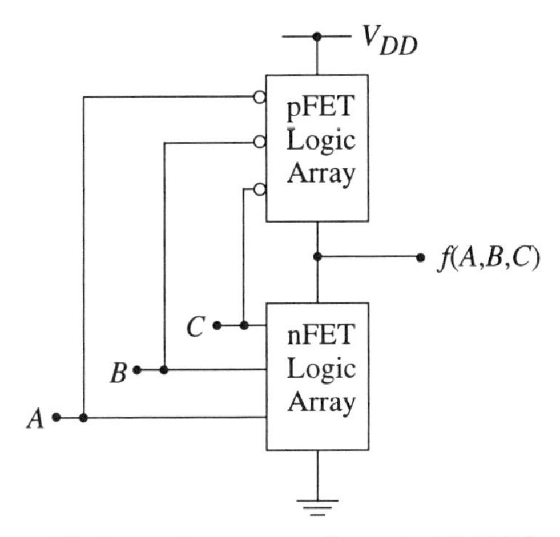

Figure 5.2 General structure of a static CMOS logic gate

manner. It allows for an amazingly large collection of circuit design styles to be developed, which provides the circuit designer with many options to chose from.

5.2 CMOS NAND Gate

The first CMOS logic gate that we will study is the 2-input NAND (**NAND2**). Figure 5.4 shows the logic symbol and the truth table where the inputs are denoted by Boolean variables A and B. The NAND2 operation is described by

$$\begin{aligned} f &= \overline{A \cdot B} \\ &= \text{NOT}(A \cdot B) \\ &= \bar{A} + \bar{B} \end{aligned} \tag{5. 3}$$

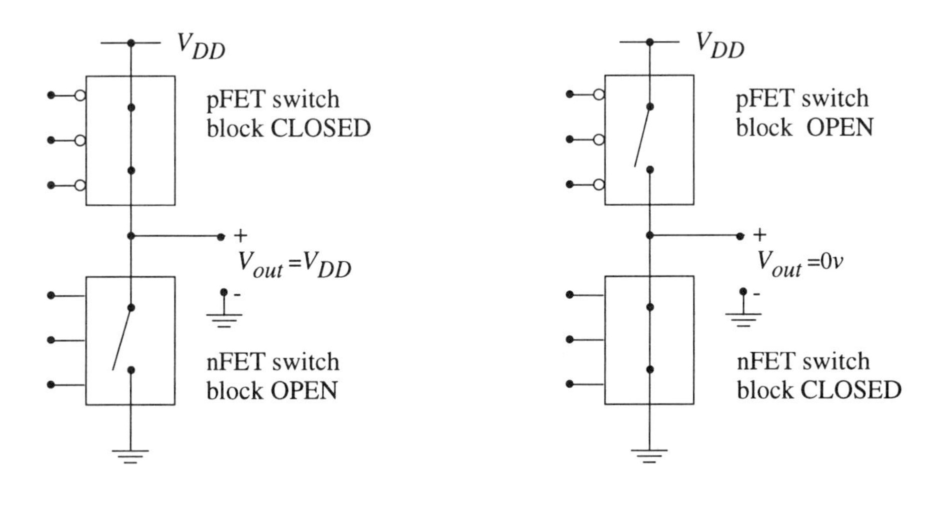

(a) Logic 1 output (b) Logic 0 output

Figure 5.3 General operation using switch logic

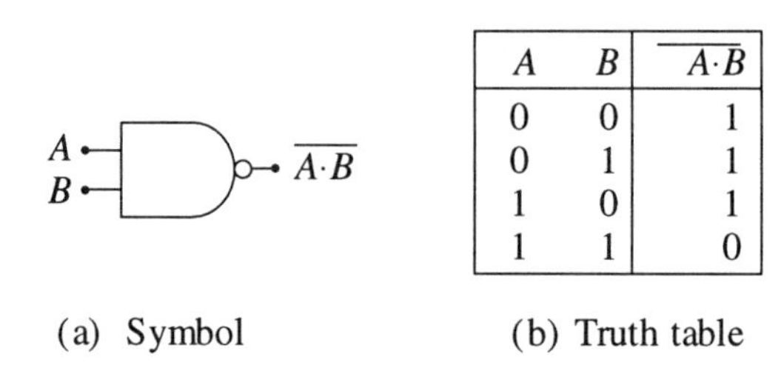

A	B	$\overline{A \cdot B}$
0	0	1
0	1	1
1	0	1
1	1	0

(a) Symbol (b) Truth table

Figure 5.4 NAND2 symbol and truth table

where the last step follows from the DeMorgan rules. This provides us with the information needed to create a CMOS NAND2 gate.

To construct a CMOS circuit that provides this function we will use two complementary pairs, one for each of the inputs A and B, and create the nFET and pFET arrays according to the needed outputs. First, note that there is only a single case where the output is a 0. This occurs when both inputs are at logic 1 values. Translating this observation into voltages then says that the output voltage $V_{out} = 0v$ if and only if both of the two input voltages high, i.e., $V_{in,A} = V_{DD} = V_{in,b}$. Since the nFETs connect the output node to ground, this requires that the two nFETs be connected in series. If either input voltage is low, then $V_{out} = V_{DD}$, indicating that the output node must be connected to the power supply. To accommodate these cases, we will wire the two pFETs in parallel. Combining the requirements for the FETs results in the circuit shown in Figure 5.5.

The logical operation of the circuit can be verified by working in reverse. Consider the series-connected nFETs MnA and MnB. If both $V_{in,A}$ and $V_{in,B}$ are high, then these transistors are active and conduct current while both pFETs are in cutoff. This provides a strong conduction path to ground and gives an output voltage of $V_{OL} = 0v$. However, if either A or B is low (either individually or at the same time) then there is no path to ground; in this case, at least one p-channel device is conducting to the power supply, giving a value of $V_{OH} = V_{DD}$.

Another approach to deriving the logic function is to use the FET switching formalism introduced at the end of Chapter 4. To apply this technique, we associate the power supply voltage V_{DD}

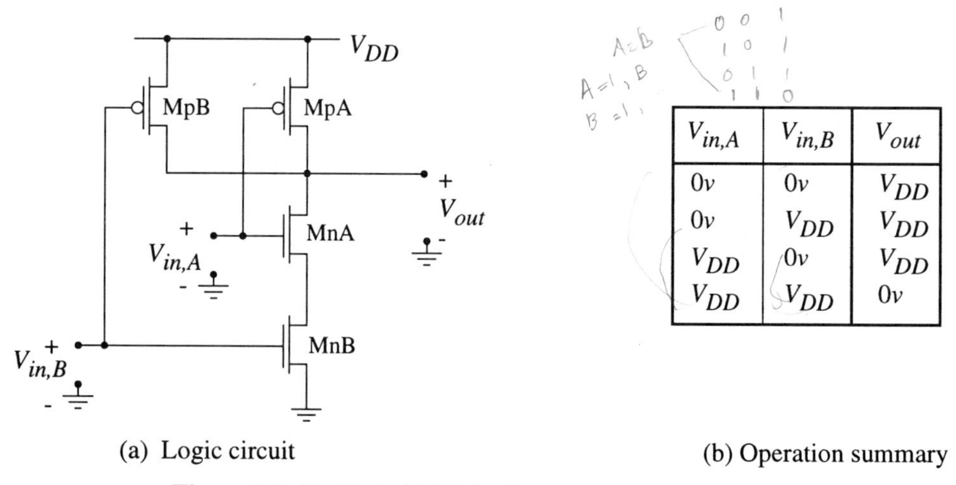

$V_{in,A}$	$V_{in,B}$	V_{out}
0v	0v	V_{DD}
0v	V_{DD}	V_{DD}
V_{DD}	0v	V_{DD}
V_{DD}	V_{DD}	0v

(a) Logic circuit (b) Operation summary

Figure 5.5 CMOS NAND2 logic gate

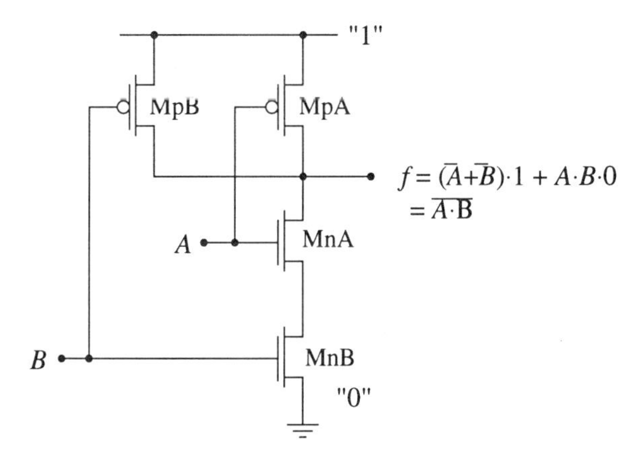

Figure 5.6 Logical operation of the NAND2 gate

with a logic 1, and the ground connection with a logic 0. Denoting the inputs by simply A and B results in the circuit shown in Figure 5.6. The output is viewed as the OR operation between the pFET switches and the nFET switches such that

$$f = \left(\overline{A} + \overline{B}\right) \cdot 1 + A \cdot B \cdot 0 \tag{5. 4}$$

The first terms are obtained by ANDing the complement of the pFET gate variable with the input ("1"), while the second term represents the series nFETs that transmit the "0" to the output. Since the nFET terms are logically 0, we may apply the DeMorgan theorem to the pFET terms to arrive at

$$f = \overline{A \cdot B} \tag{5. 5}$$

which is the NAND operation. The important point to be made here is that the **topology** of the circuit, i.e., the placing and wiring of the transistors, determines the logic function in its entirety.

5.2.1 DC Characteristics

The DC transfer characteristics depend on the input combinations. Figure 5.7 illustrates the VTC for a CMOS NAND gate. There are three input combinations that result in the output voltage changing from a high state to a low state. The possibilities are:

$V_{in,A}$=$V_{in,B}$ are simultaneously switched from 0v towards V_{DD}

$V_{in,A}$= V_{DD} while $V_{in,B}$ is switched from 0v towards V_{DD}

$V_{in,B}$= V_{DD} while $V_{in,A}$ is switched from 0v towards V_{DD}

The differences in the curves are due to the electrical structuring of the gate circuit. Since the nMOS transistors are in series, an additional node X between the nFETs enters into the problem as illustrated in Figure 5.8. Input voltages $V_{in,A}$ and $V_{in,B}$ are referenced to ground so that the gate-source voltages are

$$\begin{aligned} V_{GSA} &= V_{in,A} - V_{DSB} \\ V_{GSB} &= V_{in,B} \end{aligned} \tag{5. 6}$$

To establish conduction through the chain, both MnA and MnB must have a voltage $V_{GS} > V_{Tn}$. Assuming that the p-bulk is grounded,

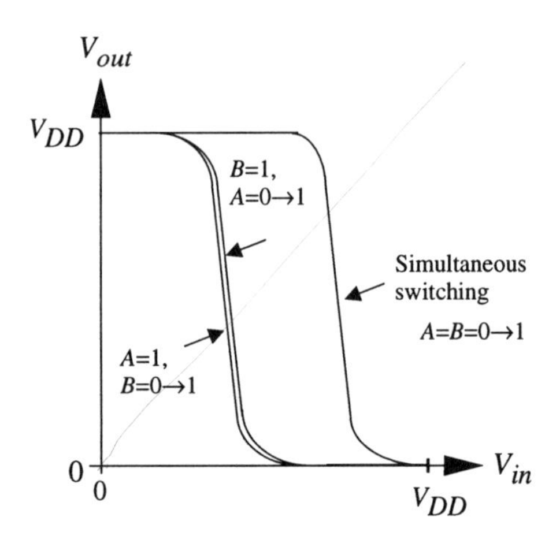

Figure 5.7 Voltage transfer curve for the NAND gate

$$\begin{aligned} V_{SBA} &= V_{DSB} \\ V_{SBB} &= 0v \end{aligned} \tag{5.7}$$

indicates the presence of body bias on MnA. The transistor threshold voltages are thus given by

$$\begin{aligned} V_{TnA} &= V_{T0n} + \gamma\left(\sqrt{2|\phi_F| + V_{DSB}} - \sqrt{2|\phi_F|}\right) \\ V_{TnB} &= V_{T0n} \end{aligned} \tag{5.8}$$

which illustrates that MnA is more difficult to turn on than MnB. Combining this observation with the complementary nMOS-pMOS transistor placement accounts for the distinct voltage-transfer characteristics for the three input switching combinations.

Let us calculate the value of the gate threshold voltage V_I for the case of simultaneous switching

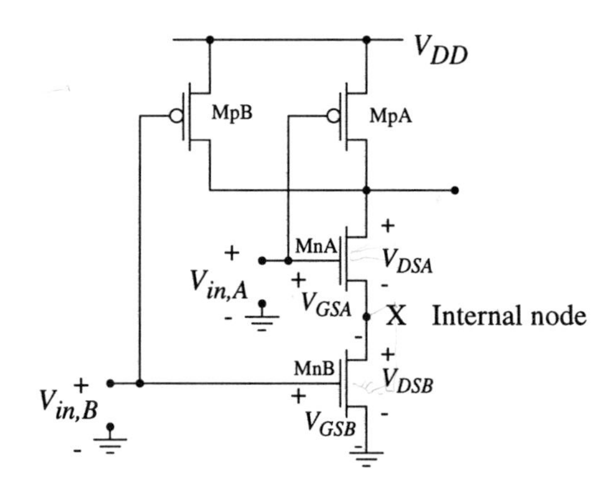

Figure 5.8 Voltages in the NAND2 circuit

by analyzing the circuit shown in Figure 5.9. Calling this voltage value V_I is appropriate since the circuit is in fact acting as an inverter when the inputs are tied together. The input and output voltages have been placed at

$$V_{in,B} = V_{in,A} = V_I = V_{out} \tag{5.9}$$

With these established, the source-gate voltage on both pFETs is given by

$$V_{SGp} = V_{DD} - V_I = V_{SDp} \tag{5.10}$$

so that both are saturated with

$$I_{DpA} = I_{DpB} = \frac{\beta_p}{2}(V_{DD} - V_I - |V_{Tp}|)^2 \tag{5.11}$$

where we will assume that both MpA and MpB have the same value of β_p.

The nFETs are more complicated. We will ignore body bias for simplicity, so that $V_{TnA} \approx V_{TnB}$. Also, we will assume that both devices have the same aspect ratio and device transconductance β_n. First, we need to determine the state of conduction (saturated or non-saturated) of each nFET. Let us write the gate source voltages as

$$\begin{aligned} V_{GSA} &= V_I - V_{DSB} \\ V_{GSB} &= V_I \end{aligned} \tag{5.12}$$

while KVL may be used to sum the drain-source voltages to read

$$V_{DSA} + V_{DSB} = V_I \tag{5.13}$$

at the output. Consider first the terminal voltages on MnA. Solving for V_{DSB} from the output equation and substituting into the gate-source expression gives

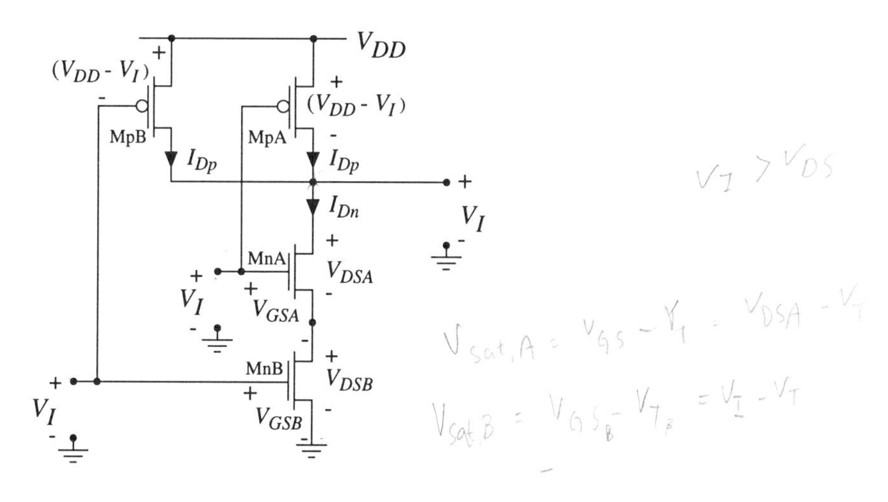

Figure 5.9 NAND2 voltages for the case of simultaneous switching

$$\begin{aligned} V_{GSA} &= V_I - V_{DSB} \\ &= V_I - (V_I - V_{DSA}) \\ &= V_{DSA} \end{aligned} \qquad (5.14)$$

Since the saturation voltage is $V_{sat,A} = (V_{GSA} - V_{Tn})$, we see that $V_{DSA} > V_{sat}$ is automatically satisfied. Thus, we may conclude that MnA is conducting in the saturation mode with

$$I_{DnA} = \frac{\beta_n}{2}(V_I - V_{DSB} - V_{Tn})^2 \qquad (5.15)$$

as its current. The other nFET MnB has a gate-source voltage of $V_{GSA} = V_I$, which is larger than that on MnA. This makes the saturation voltage $V_{sat,B} = (V_I - V_{Tn})$ greater than $V_{sat,A}$. Now then, since MnA and MnB are in series, they must have the same current: $I_{DA} = I_{DB}$. For transistors that have the same value of β_n, this says that MnB is conducting in the non-saturation region with

$$I_{DnB} = \frac{\beta_n}{2}\left[2(V_I - V_{Tn})V_{DSB} - V_{DSB}^2\right] \qquad (5.16)$$

describing the current.[2]

To calculate the gate threshold voltage V_I, we apply KCL to write

$$\begin{aligned} I_{Dn} &= I_{DpA} + I_{DpB} \\ &= \beta_p(V_{DD} - V_I - |V_{Tp}|)^2 \end{aligned} \qquad (5.17)$$

where $I_{Dn} = I_{DnA} = I_{DnB}$ is the current through the nFET chain. Using the equation for I_{DnA} gives

$$V_{DSB} = V_I - V_{Tn} - \sqrt{\frac{2I_{Dn}}{\beta_n}} \qquad (5.18)$$

substituting this into the expression for I_{DnB} and simplifying allows us to eliminate V_{DSB} and results in

$$I_{Dn} = \frac{\beta_n}{2^2}(V_I - V_{Tn})^2 \qquad (5.19)$$

Finally, this may be substituted into the KCL expression to arrive at

$$\beta_p(V_{DD} - V_I - |V_{Tp}|)^2 = \frac{\beta_n}{2^2}(V_I - V_{Tn})^2 \qquad (5.20)$$

Solving gives

$$V_I = \frac{(V_{DD} - |V_{Tp}|) + \frac{1}{2}\sqrt{\frac{\beta_n}{\beta_p}}V_{Tn}}{1 + \frac{1}{2}\sqrt{\frac{\beta_n}{\beta_p}}} \qquad (5.21)$$

[2] This argument assumes that $\lambda=0$, i.e., it ignores all channel-length modulation effects.

for the gate threshold voltage in this case. This shows that the value of V_I is determined by the ratio (β_n/β_p).

The significance of this result becomes clear by recalling the expression for $V_{I,NOT}$ for an inverter as derived in Chapter 3. The analysis gave

$$V_{I,NOT} = \frac{(V_{DD} - |V_{Tp}|) + \sqrt{\frac{\beta_n}{\beta_p}} V_{Tn}}{1 + \sqrt{\frac{\beta_n}{\beta_p}}} \qquad (5.22)$$

Comparing this with the NAND2 result above, we see that the only difference is the factor of (1/2) multiplying the square root term. If we construct an inverter and a NAND2 gate using the same values of (β_n/β_p) for both, then these equations show that $V_{I,NAND2}$ will be larger than $V_{I,NOT}$. This is due to the series-connected nFETs in the NAND2 gate that combine to increase the resistance from the output to ground. Although the numerical difference between the two values depends upon the specific numbers, the comparison in Figure 5.10 illustrates the important idea.

The remaining two cases where (ii) A=1 and B is switched, and, (iii) A is switched while B=1, can be analyzed using the same basic circuit techniques. Although it is possible to use the same analytic techniques above to analyze the circuits, both calculations are somewhat tedious and will not be reproduced here. The most important result of the single-input switching cases is that both are shifted to the left of the simultaneous switching case as shown in Figure 5.7. The separation between cases (ii) and (iii) arises from the stacking order of nFETs in the series connection. In particular, MnB is easier to turn on than MnA due to body-bias effects and the difference between the applied voltage and V_{GSA}.

5.2.2 Transient Characteristics

Switching times can be estimated using the capacitances shown in Figure 5.11. The output capacitance is taken to be the LTI value

$$C_{out} = C_{FET} + C_L \qquad (5.23)$$

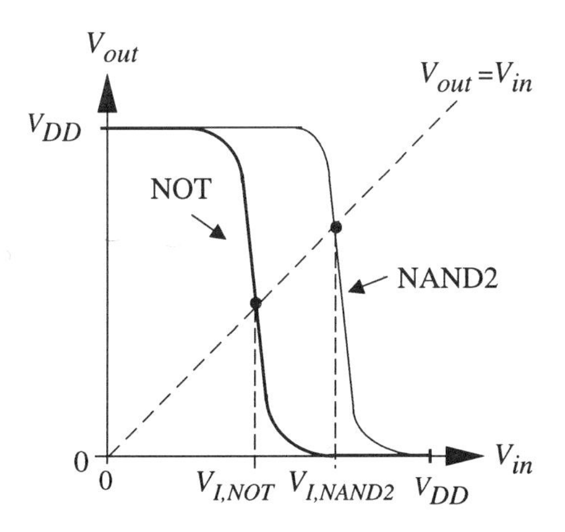

Figure 5.10 Comparison of NOT and NAND2 VTCs

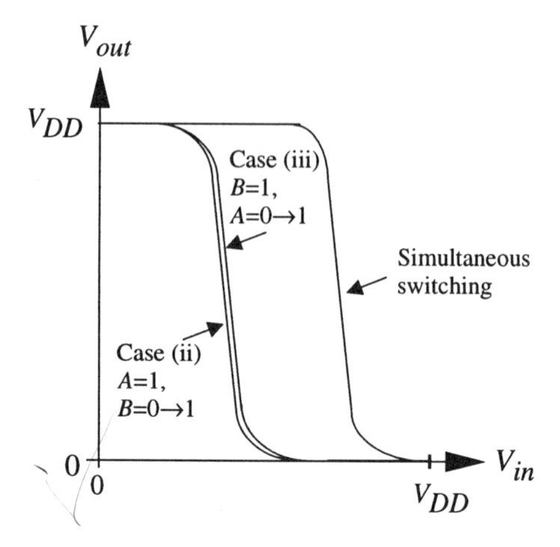

Figure 5.11 NAND2 VTC indicating the different possibilities

where

$$C_{FET} = C_{GDnA} + C_{GDpA} + C_{GDpB} + C_{DBnA} + C_{DBpA} + C_{DBpB} \tag{5. 24}$$

accounts for the MOSFET parasitics by assuming that all are in parallel, and

$$C_L = C_{line} + C_{FO} \tag{5. 25}$$

is due to the external load. As with the inverter, analytic estimates can be made using average LTI values for the depletion capacitances. The internal node capacitance C_X is given by

$$C_X = C_{GSnA} + C_{GSnB} + C_{n+} \tag{5. 26}$$

where C_{n+} is the total capacitance of the $n+$ drain/source region between the series-connected nFETs.

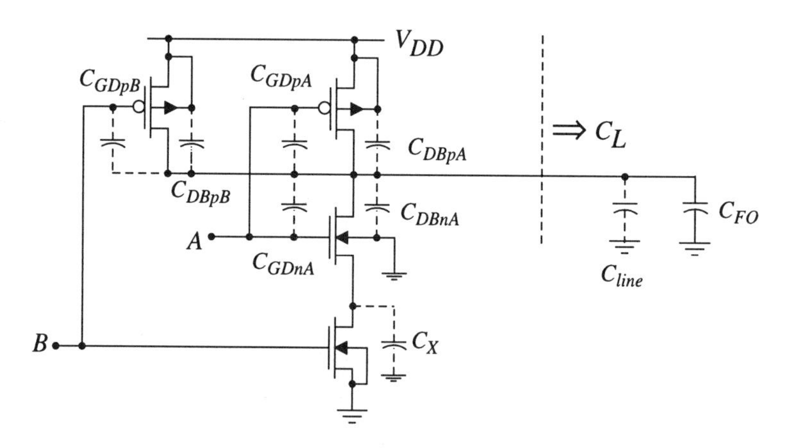

Figure 5.12 Capacitance contributions in the NAND2 gate

Output Charge Time

Consider first the low-to-high time t_{LH}. The worst-case initial condition on the output voltage is that $V_{out} = 0v$ which implies that both $V_{in,A}$ and $V_{in,B}$ are initially at V_{DD}. If either A or B (or both) switch to logic 0 values, C_{out} charges through the appropriate pFET transistors. Figure 5.13(a) shows the case where a single pFET MpA is switched into conduction. In this case, the MOSFET current I_{Dp} charges the output capacitance as seen using the equivalent RC network shown in Figure 5.13(b). The inverter analysis of Section 3.2 may thus be applied to estimate

$$t_{LH} = s_p \tau_p \tag{5. 27}$$

where

$$\tau_p = R_{pA} C_{out} \tag{5. 28}$$

is the time constant with R_{pA} as the equivalent pFET resistance. In the opposite case where pFET MpB is switched into conduction (implying that the nFET MnB is in cutoff), the parasitic capacitance C_X between the nFETs will also charge. This increases the value of t_{LH} since charge is diverted away from the output node.

The best case (shortest) charge time occurs if both pFETs are conducting. If both transistor have the same aspect ratio, then the total charging current is double that passing through a single transistor. This is equivalent to having one-half of a single FET resistance so that the effective pFET resistance is given by

$$R_{pA} \rightarrow \frac{R_{pA}}{2} \tag{5. 29}$$

in the time constant. In practice, however, we usually concentrate on the longest time intervals, since these are the limiting factors in the performance.

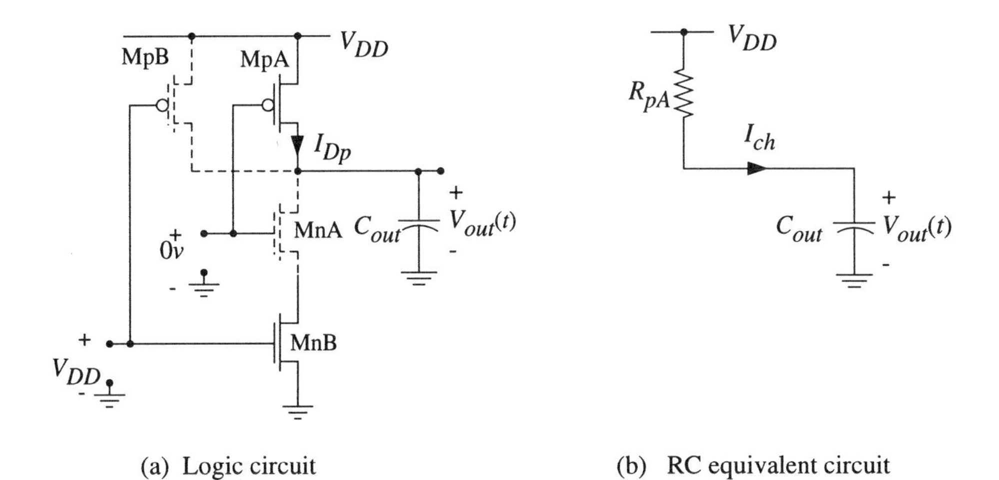

Figure 5.13 Subcircuit for t_{LH} calculation

Output Discharge Delay Time

To calculate the discharge delay time we assume an initial output voltage of $V_{out}(0) = V_{DD}$; this implies that at least one of the inputs is at a logic 0 value. Discharging occurs when both A and B increase to logic 1 voltages. MOSFETs MnA and MnB are both active and provide a conducting path between C_{out} and ground as shown in Figure 5.14(a). The situation can be modeled using the equivalent circuit in Figure 5.14(b). Applying the results of Section 4.5, the time constant is modified to include C_x by writing the Elmore form

$$\tau_n = (R_{nA} + R_{nB}) C_{out} + R_{nA} C_X \tag{5. 30}$$

where R_{nA} and R_{nB} are the equivalent MOSFET resistances, and C_X represents the internal node capacitance between the two transistors. This can be used to estimate the high-to-low time by noting that the use of the Elmore time constant implies that we are modelling the output voltage as an exponential of the form

$$V_{out}(t) = V_{DD} e^{-t/\tau_n} \tag{5. 31}$$

The time required to achieve a particular value of V_{out} is computed from

$$t \approx \tau_n \ln\left[\frac{V_{DD}}{V_{out}(t)}\right] \tag{5. 32}$$

Since t_{HL} is defined as the time required for the voltage to fall from the 90% voltage to the 10% voltage, we have that

$$t_{HL} \approx \ln(9)\,\tau_n \approx 2.2\tau_n, \tag{5. 33}$$

as a reasonable hand approximation. The results must be verified by computer simulation if accurate values are needed. Qualitatively, we see that the output capacitance C_{out} must discharge through the series-connected nFETs, which implies that the value of t_{HL} can be large.

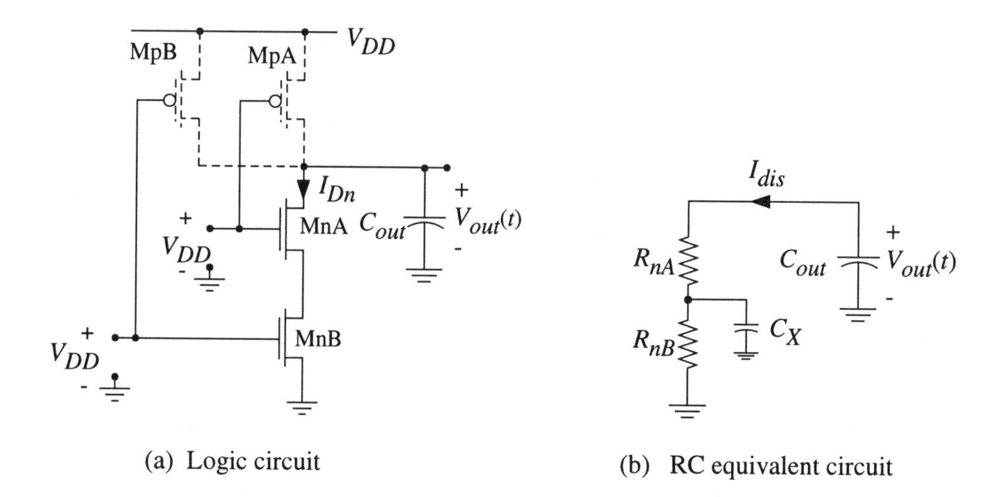

Figure 5.14 Subcircuit for t_{HL} calculation

5.2.3 Design

Designing a NAND gate is straightforward. Since the logic function is a consequence of the circuit structure, the specific choices for MOSFET aspect ratios do not change the logic operation. Instead, the device sizes establish the DC critical voltages such as switching voltage V_I, and directly determine the transient time intervals.

Some circuits require that the DC switching point be in a specific range. In this case, we first adjust the ratio of (β_n/β_p), and then choose the aspect ratios according to the technology limits and the desired transient response. If timing is critical, we may choose the aspect ratios needed to satisfy the switching response and only calculate the DC characteristics as an after thought.

The low-to-high time t_{LH} is controlled by the pFET aspect ratios $(W/L)_{pA}$ and $(W/L)_{pB}$. Since MpA and MpB are in parallel, the worst-case situation occurs when only a single device is conducting. Thus, we can design both pFETs to be the same size $(W/L)_p$ such that either transistor can individually meet the rise time specification. In terms of the charging time constant we have

$$\left(\frac{W}{L}\right)_p = \frac{C_{out}}{\tau_p k'_p (V_{DD} - |V_{Tp}|)} \tag{5. 34}$$

which provides a relationship between the aspect ratio and the value of C_{out}.

The series-connected nFETs limit the discharging response. The simplest design uses identical values of $(W/L)_n$ for both. In this case, the resistances are equal, so that the time constant is

$$\tau_n = R_n (2C_{out} + C_X) \tag{5. 35}$$

The aspect ratio is then given by

$$\left(\frac{W}{L}\right)_n = \frac{(2C_{out} + C_X)}{\tau_n k'_n (V_{DD} - V_{Tn})} \tag{5. 36}$$

where both C_{out} and C_X depend on the value of $(W/L)_n$.

The complicating factor in both procedures is that C_{out} depends upon the choice of (W/L) for every transistor that is connected to the output node. As with the case of the inverter, the design process starts by first estimating the capacitance values and designing the network around these choices; a computer simulation is then used to check the actual behavior with a specified external load. Another approach is to simply choose reasonable values for the device sizes, simulate the circuit, and then use the equations to adjust the aspect ratios as needed for the transient response.

5.2.4 N-Input NAND

The NAND2 structure can be extended to an N-input NAND gate by using N complementary pairs where the nFETs are in series and the pFETs are in parallel. For the case of simultaneous switching, the analysis gives

$$V_I = \frac{(V_{DD} - |V_{Tp}|) + \frac{1}{N}\sqrt{\frac{\beta_n}{\beta_p}}V_{Tn}}{1 + \frac{1}{N}\sqrt{\frac{\beta_n}{\beta_p}}} \tag{5. 37}$$

for the point where the VTC crosses the unity gain line. For the case $N=1$, this is identical to the inverter results, as should be the case. Multiple-input NAND gates are easily implemented in circuit and logic designs. However, the output capacitance C_{out} increases with N due to the pMOS parasitics, slowing down the overall response. The worst-case problem is that the discharge time is

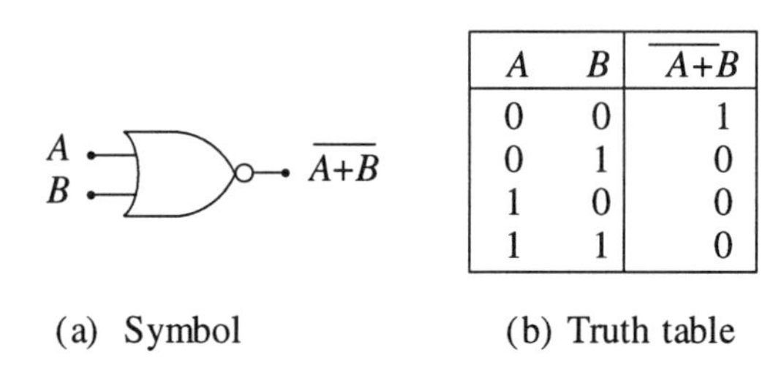

A	B	$\overline{A+B}$
0	0	1
0	1	0
1	0	0
1	1	0

(a) Symbol (b) Truth table

Figure 5.15 Symbol and truth table for the NOR2 gate

limited by the discharge through the nFET chain that consists of series-connected transistors. These considerations generally limit N to a maximum of 3 or 4 inputs in realistic designs.

5.3 CMOS NOR Gate

Now let us examine a 2-input CMOS NOR gate with inputs A and B. Figure 5.15 shows the logic symbol and truth table; we will refer to this as a NOR2 gate. The operation is characterized by the by the fact that a logic 1 at either input (or both inputs) causes a logic 0 output. Logically, the NOR2 operation is denoted by

$$\begin{aligned} f &= \overline{A+B} \\ &= \overline{A} \cdot \overline{B} \end{aligned} \tag{5. 38}$$

where $(A+B)$ denotes the OR operation between A and B; the second step follows from applying the DeMorgan rule.

A CMOS NOR2 gate can be built by using two complementary pairs as shown in Figure 5.16(a). Input A is connected to MnA and MpA, while B controls MnB and MpB. Note that the nFETs are connected in parallel, while the pFETs form a series chain. To understand the operation of the gate, we examine the conduction states of the transistors for different input voltages $V_{in,A}$ and $V_{in,B}$. If $V_{in,A}=V_{DD}$, then MnA is ON and MpA is OFF; since MnA provides a conducting path from the ground to the output, V_{out} = 0v. Setting $V_{in,B}=V_{DD}$ turns MnB ON and MpB OFF and also

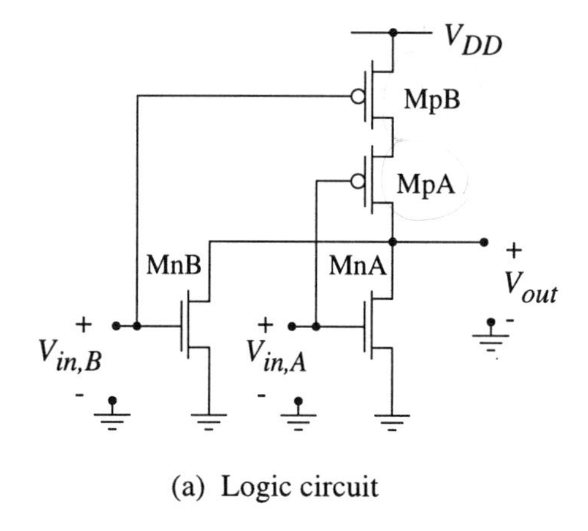

$V_{in,A}$	$V_{in,B}$	V_{out}
0v	0v	V_{DD}
0v	V_{DD}	0v
V_{DD}	0v	0v
V_{DD}	V_{DD}	0v

(a) Logic circuit (b) Operation summary

Figure 5.16 CMOS NOR2 circuit

results in $V_{out} = 0v$. And, if both $V_{in,A}$ and $V_{in,B}$ are high, then both nFETs are ON and the output voltage is $V_{out} = 0v$. The only input combination that results in $V_{out} = V_{DD}$ is when $V_{in,A}=0v=V_{in,B}$, since both pFETs are ON while both nFETs are OFF. As verified by the truth table in Figure 5.16(b), this gives exactly the NOR2 operation.

We may also verify the logic function by viewing the circuit as a simple multiplexor between the power supply V_{DD} ("1") and ground ("0") as shown in Figure 5.17. Using the logic equations for MOSFETs gives the output as

$$g = \overline{A} \cdot \overline{B} \cdot 1 + (A + B) \cdot 0 \tag{5.39}$$

As in the case of the NAND gate, the nFET terms logically evaluate to 0 which leaves

$$\begin{aligned} g &= \overline{A} \cdot \overline{B} \\ &= \overline{(A + B)} \end{aligned} \tag{5.40}$$

using the DeMorgan theorem. This verifies our previous statement that the logic function is determined entirely by the topology of the circuit.

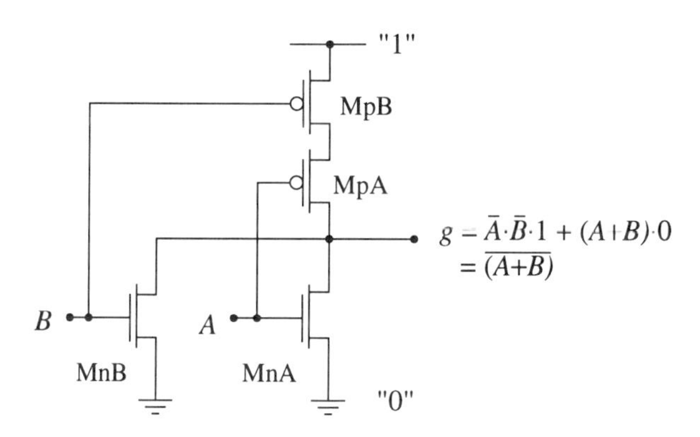

Figure 5.17 Logic operation of the CMOS NOR2 gate

5.3.1 DC Transfer Characteristic

The DC voltage switching characteristics of a NOR gate depends on how the inputs are changed. Three different curves are shown in the VTC of Figure 5.18, each representing a different switching combination. These are

$V_{in,A}=V_{in,B}$ are simultaneously switched from $0v$ towards V_{DD};

$V_{in,A}=0$ while $V_{in,B}$ is switched from $0v$ towards V_{DD};

$V_{in,B}=0$ while $V_{in,A}$ is switched from $0v$ towards V_{DD}.

As can be seen from the drawing, each case is characterized by a different switching point on the VTC. This is similar to the situation encountered with the NAND2 gate, except that the individual switching events are on the right side of the VTC, which is opposite to behavior of the NAND2 gate.

Let us examine the case of simultaneous switching in detail. The starting point for the analysis is to set $V_{in,A}=V_{in,B}=0v$, and then increase both inputs simultaneously. When $V_{in,A}=V_{in,B}=V_I$ the out-

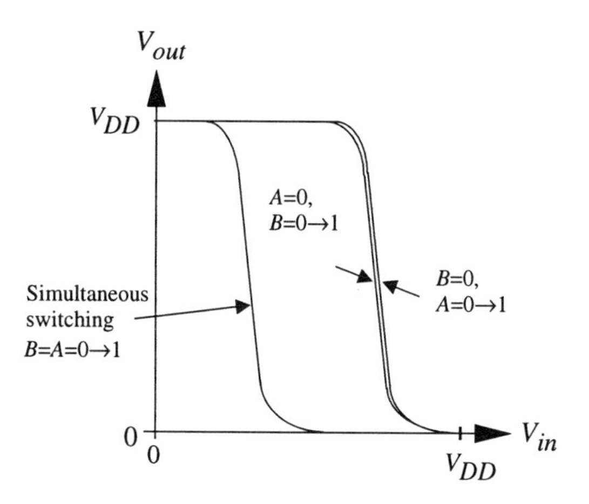

Figure 5.18 Voltage transfer characteristics for the CMOS NOR2 circuit

put voltage is also at the same value with $V_{out} = V_I$. The value of V_I is found by determining the conduction modes of all transistors and then applying the Kirchhoff Laws to obtain the desired equation set. Figure 5.19 shows the circuit with these voltages applied.

Consider first the nFETs. Since the device voltages are defined by

$$\begin{aligned} V_{GSA} &= V_I = V_{DSA} \\ V_{GSB} &= V_I = V_{DSB} \end{aligned} \tag{5.41}$$

both MnA and MnB are saturated with

$$I_{DAn} = \frac{\beta_n}{2}(V_I - V_{Tn})^2 = I_{DnB} \tag{5.42}$$

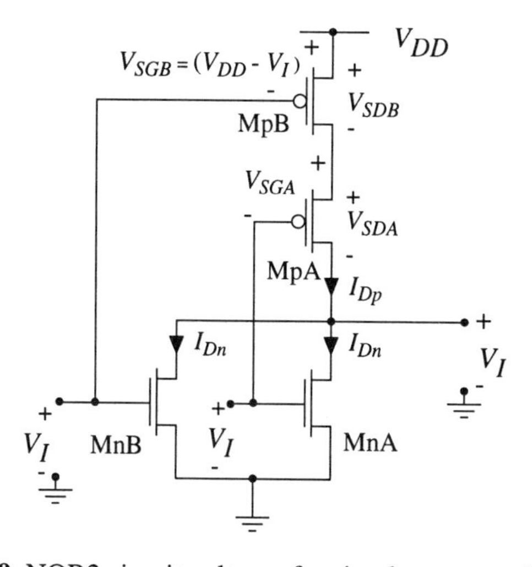

Figure 5.19 NOR2 circuit voltages for simultaneous switching case

as the current through the MOSFETs. We will assume that both nFETs have the same aspect ratio β_n for simplicity.

The pFETs can be analyzed in a similar manner. MpB (which is connected to the power supply) is described by the terminal voltage equation set

$$V_{SGB} = V_{DD} - V_I \qquad V_{SDB} = V_{DD} - V_{SDA} - V_I \tag{5. 43}$$

while MpA (which is closest to the output) has terminal voltages of

$$V_{SGA} = V_{DD} - V_{SDB} - V_I \qquad V_{SDA} = V_{DD} - V_{SDB} - V_I \tag{5. 44}$$

Since $V_{SGA} = V_{SDA}$, MpA is saturated and has a current of

$$I_{DpA} = \frac{\beta_p}{2}(V_{DD} - V_I - |V_{Tp}| - V_{SDB})^2 \tag{5. 45}$$

where we will ignore body bias effects for simplicity. The other pFET MpB is non-saturated at this point and is described by

$$I_{DpB} = \frac{\beta_p}{2}\left[2(V_{DD} - V_I - |V_{Tp}|)V_{SDB} - V_{SDB}^2\right] . \tag{5. 46}$$

Since the pFETs are in series,

$$I_{DpA} = I_{Dp} = I_{DpB} \tag{5. 47}$$

such that applying KCL at the output node gives

$$I_{Dn} = I_{DpA} + I_{DpB} = \beta_n (V_I - V_{Tn})^2 \tag{5. 48}$$

as the primary relationship among the currents.

This expression may be combined with the pFET constraints using the same approach as for the NOR2 gate in the previous section. Analyzing the circuit for this case of simultaneous input switching gives a gate threshold voltage of

$$V_I = \frac{(V_{DD} - |V_{Tp}|) + 2\sqrt{\frac{\beta_n}{\beta_p}}V_{Tn}}{1 + 2\sqrt{\frac{\beta_n}{\beta_p}}} \tag{5. 49}$$

Comparing this with the results for both the inverter and the NAND2 gate shows that the only difference is the factor of 2 multiply the square root factor. This is due to the fact that a NOR gate with its inputs tied together is an inverter from both the logic and circuit viewpoints. Figure 5.20 illustrates the difference between an inverter (NOT) gate and a NOR2 gate designed with the same (β_n/β_p) ratio. The amount of shift depends upon the specific numbers, but we always have $V_{I,NOT} > V_{I,NOR2}$ as shown.

The remaining two cases where (ii) A=0 and B is switched, and, (iii) A is switched while B=0, can be analyzed using the same basic circuit techniques. The resulting VTC comparison is shown in

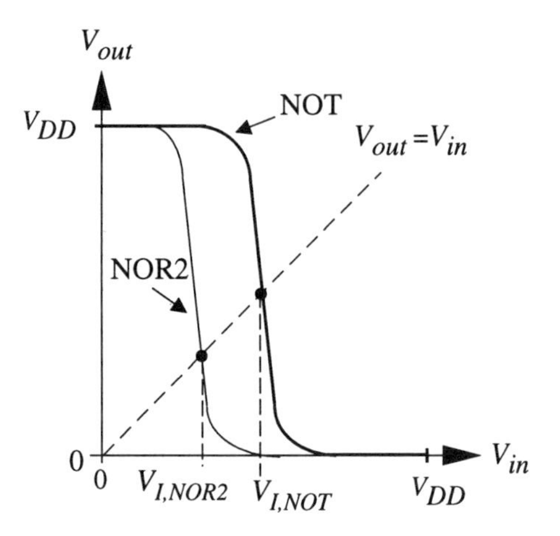

Figure 5.20 Comparison of NOT and NOR2 VTC

Figure 5.21. The differences among the VTC switching points is due to the series connection of MpA and MpB.

5.3.2 Transient Times

To estimate the switching performance of the NOR2 gate, we first examine Figure 5.22 to identify the most important lumped-element contributions to C_{out}. This quantity is estimated by

$$C_{out} = C_{FET} + C_L \tag{5. 50}$$

where

$$C_{FET} = C_{GDnA} + C_{GDnB} + C_{GDpA} + C_{DBnA} + C_{DBnB} + C_{DBpA} \tag{5. 51}$$

are the internal FET contributions, and

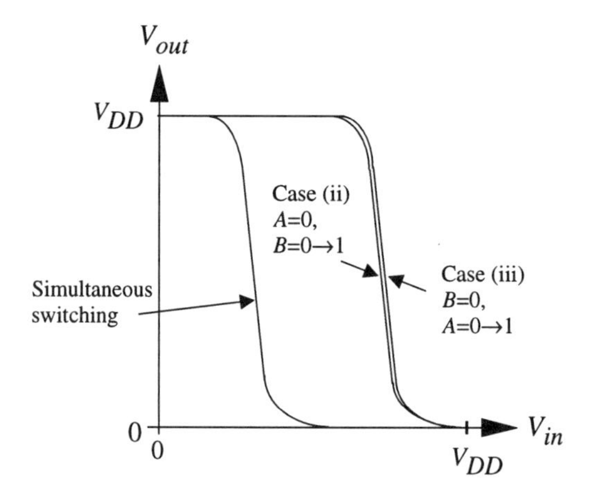

Figure 5.21 NOR2 VTC with the 3 cases identified

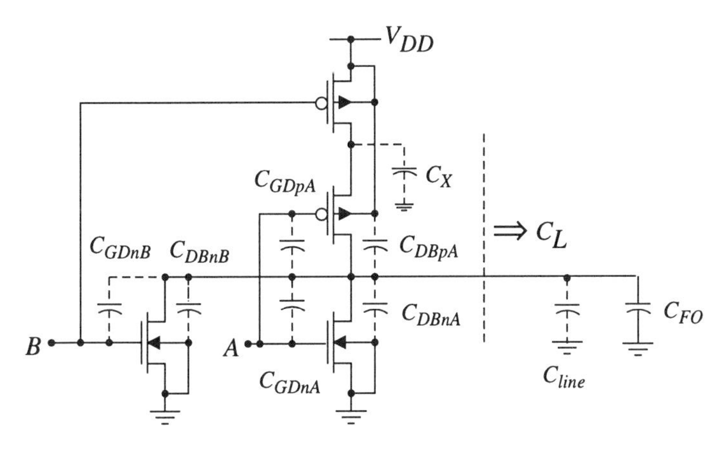

Figure 5.22 Contribution to the NOR2 output capacitance

$$C_L = C_{line} + C_{FO} \tag{5.52}$$

represents the total external load capacitance. As before, hand calculations generally employ voltage-averaged LTI quantities for the best approximations.

Output Discharge Time

The high-to-low time t_{HL} is computed by noting that C_{out} discharges through the nFET MnA or MnB. Since these are in parallel, the worst-case discharge time occurs when only a single nFET is conducting as shown in Figure 5.23. This situation is equivalent to the discharge in a simple inverter circuit (with a larger C_{out}) so that

$$t_{HL} = s_n \tau_n \tag{5.53}$$

where

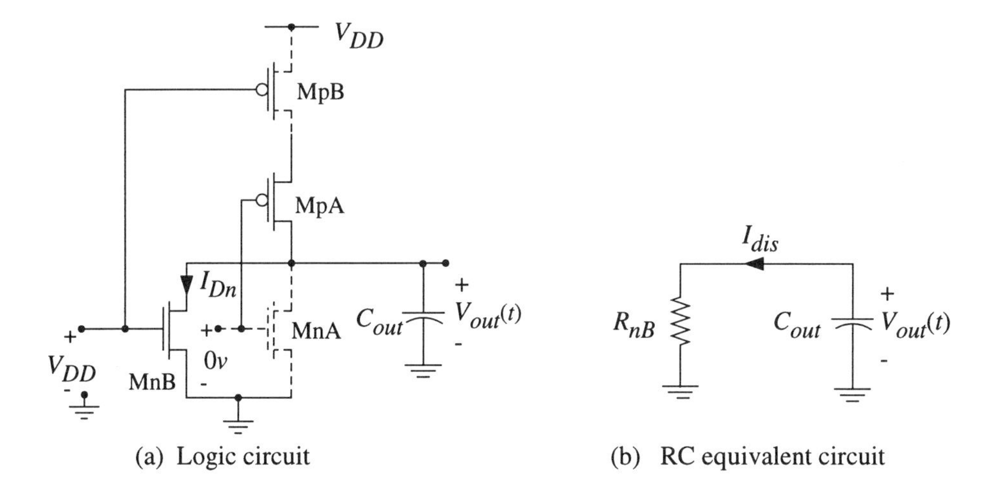

Figure 5.23 Subcircuit for t_{HL} estimate

$$\tau_n = \frac{C_{out}}{\beta_p (V_{DD} - V_{Tn})} \tag{5. 54}$$

is the time constant. If both nFETs are active, then the discharge time is essentially cut in half since the current is doubled. Note that this approximation ignores the possibility of charge on the node between the two pFETs. If this situation occurs, then the value of t_{HL} will be increased.

Output Charge Delay Time

The charging delay time can be estimated using the results obtained in Section 4.5. With both A and B at logic 0 levels, pFETs MpA and MpB provide a conducting path between the output and the power supply as in Figure 5.24(a). Constructing the RC model in Figure 5.24(b) gives a time constant for the discharge event as

$$\tau_p = (R_{pA} + R_{pB})\, C_{out} + R_{pB} C_X \tag{5. 55}$$

where R_{pA} and R_{pA} are the equivalent pFET resistances, and C_X is the capacitance between the two series-connected p-channel transistors such that

$$C_X = C_{GSpA} + C_{GSpB} + C_{p+} \tag{5. 56}$$

with C_{p+} the total depletion capacitance between the two. To estimate the low-to-high time t_{LH} we use the exponential approximation

$$V_{out}(t) \approx V_{DD}\left[1 - e^{-t/\tau_n}\right] \tag{5. 57}$$

which gives

$$t_{LH} \approx \ln(9)\, \tau_p \approx 2.2\tau_p \tag{5. 58}$$

This has the same form as the discharge time in a NAND gate. As always, the equations only provide first order estimates and critical dependences; a computer simulation must be performed to obtain accurate values.

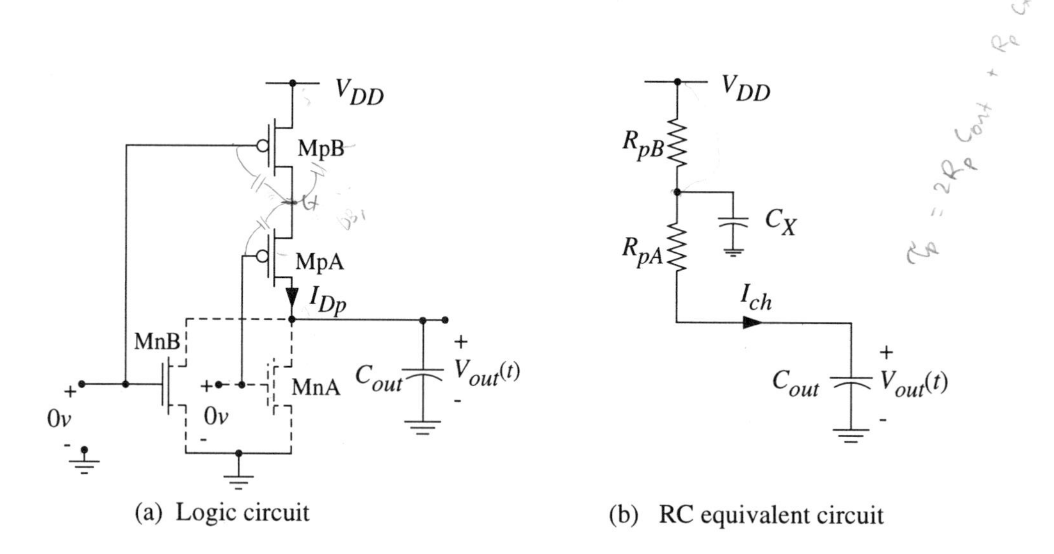

(a) Logic circuit

(b) RC equivalent circuit

Figure 5.24 Subcircuit for t_{LH} estimate

5.3.3 Design

Functional logic design for a NOR gate is automatic with the placement of the transistors in the circuit: nMOSFETs are in parallel and pMOSFETs are in series. The choice of aspect ratios affects the gate threshold voltage V_I and the transient characteristics, as is the case for all static CMOS gates.

NOR gate design is similar to that discussed for NAND circuits. To set the value of V_I, we adjust the ratio (β_n/β_p); the transient specifications dictate the specific values of the aspect ratios. The high-to-low time t_{HL} is set by the values of $(W/L)_{nA}$ and $(W/L)_{nB}$. Since the two are in parallel, the simplest choice is to calculate the aspect ratio

$$\left(\frac{W}{L}\right)_n = \frac{C_{out}}{\tau_n k'_n (V_{DD} - V_{Tn})} \tag{5.59}$$

for both transistors by assuming a reasonable value of C_{out}.

The charging event is controlled by the series-connected pFETs. For identical transistors, the time constant reduces to

$$\tau_p = R_p (2C_{out} + C_X) \tag{5.60}$$

so that the basic design equation is of the form

$$\left(\frac{W}{L}\right)_p = \frac{(2C_{out} + C_X)}{\tau_p k'_p (V_{DD} - |V_{Tp}|)} \tag{5.61}$$

This allows us to choose the transistor size for the first design.

5.3.4 N-Input NOR

An N-input NOR gate can be constructed by using complementary structuring with N nFETs in parallel, and N pFETs in series. For the case of simultaneous switching, the VTC crosses the unity gain line at a voltage given by

$$V_I = \frac{(V_{DD} - |V_{Tp}|) + N\sqrt{\frac{\beta_n}{\beta_p}}V_{Tn}}{1 + N\sqrt{\frac{\beta_n}{\beta_p}}} \tag{5.62}$$

note that the case where N=1 is identical to the inverter. NOR gate logic is very straightforward to implement, and is also very popular. However, it also has an output capacitance problem due to the contributions from the parallel nMOS devices, and requires charging through a pMOS chain which limits the transient response. In practice, the latter consideration usually limits N to a maximum of 2 or 3, unless speed is not an issue.

5.3.5 Comparison of NAND and NOR Gates

Both NAND and NOR gates are easy to implement in CMOS logic. However, for equal numbers of inputs and device sizes, NAND gates have better transient response than NOR gates, making them more popular in high-performance design. The reasoning behind this statement can be understood by recalling the delay times. The series-connected transistors are the limiting factor. In a NAND gate, the discharge delay time t_{HL} is determined by a chain of n-channel MOSFETs. A NOR gate, on the other hand, has a charging time t_{LH} which is due to charging through a chain of p-channel transistors. Since, in general, the resistance of a MOSFET has the functional dependence

$$R = \frac{1}{k'\left(\frac{W}{L}\right)(V_{DD} - V_T)} \tag{5.63}$$

and $(k'_n / k'_p) > 1$, equal-size nFETs and pFETs have resistance values that satisfy

$$R_p > R_n \tag{5.64}$$

Applying this to the case of series-connected MOSFETs shows that the nFET chain always discharges faster than the pFET chain can charge. Thus, for equal area designs, NAND gates are preferable to NOR gates. In practice, however, both NAND and NOR gates are widely used due to the fact that the **system speed** is generally limited by a large complex unit, not the response of a single gate.

5.3.6 Layout

Example layouts for NAND and NOR gates are shown in Figure 5.25. A moment of study will verify that both are based on (i) equal size nFETs and pFETs, and (ii) identical transistor placements. In fact, the only difference between the two gates are the metal connections. While these are not minimum-area layouts, they illustrate the important features quite well.

Consider the NAND2 layout in Figure 5.25(a). The nFETs are wired in series by using the common n^+ region between them as the electrical connector. The spacing between the two gates would be reduced in a minimum-area layout. Parallel pFETs are obtained by using the p^+ region between the two gates as the common output while the left and right sides are connected to the power supply voltage V_{DD}. The NOR2 gate in Figure 5.25(b) is constructed in a similar manner. In fact, if one takes the metal lines on the NAND2 gate and "flips" them around an imaginary horizontal line drawn in the center, the NOR2 gate results. This is an interesting observation and is related to the

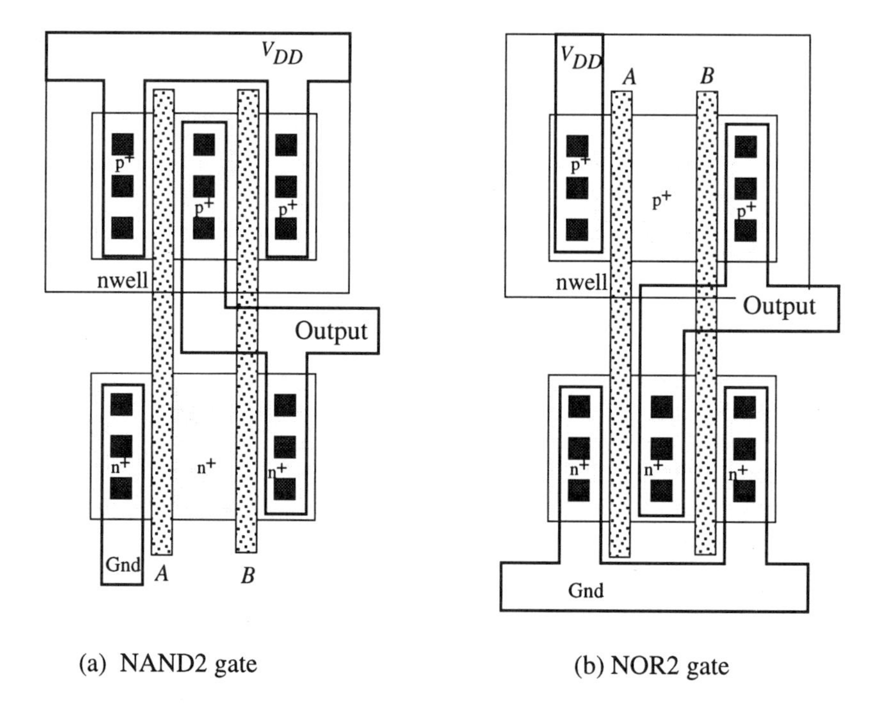

(a) NAND2 gate (b) NOR2 gate

Figure 5.25 Layout examples for NAND2 and NOR2 gates

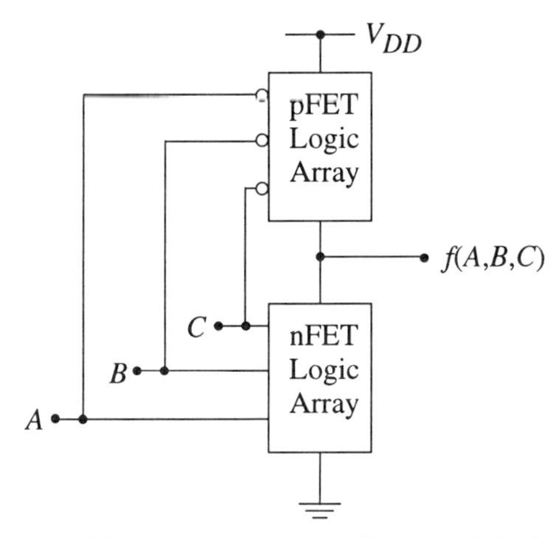

Figure 5.26 General structure of a CMOS logic gate

mathematical fact that the NAND and NOR gates use series-parallel logic with opposite wiring for nFETs and pFETs.

5.4 Complex Logic Gates

Consider the general complex logic gate circuit reproduced for reference in Figure 5.26. As discussed in Section 5.1, the use of complementary pairs requires that every input drives both an nFET and a pFET. Complex logic functions can be implemented by designing the nFET and pFET switching arrays such that only one composite switch is closed for a given set of inputs. The switch equivalents are illustrated in Figure 5.27 to aid in visualizing the operation. If the nFET switch is closed while the pFET switch is open, then the output is a logic 0. Conversely, a closed pFET switch and an open nFET switch results in a logic 1 output. The arrays must be designed to avoid two situations: (1) both switches are open at the same time, since this gives an undetermined value;

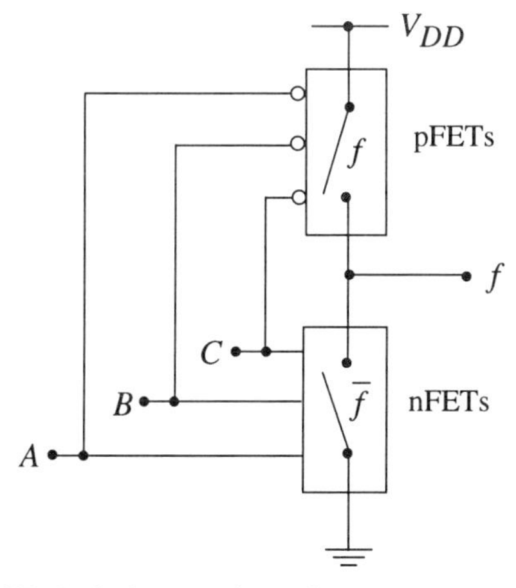

Figure 5.27 Switch operation of a general CMOS logic gate

and, (2) both are closed at the same time, since the voltage will not be a well-defined logic voltage. Note that we have labelled the nFET switching block by $\bar{f}$, the pFET block by f, and the output is also given as f. This notation is defined to mean that if the output f is a logical 1, the nFET block is OPEN and the pFET block is CLOSED. Conversely, if the nFET block is CLOSED while the pFET block is OPEN, the output is at a value of f=0 (i.e., $\bar{f}$ = 1). With regards to the individual blocks themselves, the outcome ($\bar{f}$ for the nFETs and f for the pFETs) is TRUE (a logic 1) if the switch is closed.

A useful set of general logic formation rules can be obtained by extending the lessons learned from the study of the CMOS NAND2 and NOR2 circuits. Figure 5.28 summarizes how nFETs behave in the static gate circuit. When interpreting these sub-circuits, it is important to remember that (1) the source of the "lowest" nFET is connected to ground, and (2) the drain of the "highest" nFET is the output. Now then, recall that connecting nFETs in series gives the AND-NOT function, while wiring nFETs in parallel results in the OR-NOT operation; these are summarized in Figure 5.28(a) and (b). A set of general rules may be obtained by extending these results to groups of nFETs. In other words, the logic formation rules also apply if we replace the individual series- or parallel-connected transistors by blocks of transistors such as the groups illustrated in Figure 5.28(c) and (d) which respectively form the AND-OR-Invert (AOI) function

$$\overline{A \cdot B + C \cdot D} \tag{5. 65}$$

and the OR-AND-Invert (OAI) expression

$$\overline{(A + B) \cdot (C + D)} \tag{5. 66}$$

using the rules. These may be generalized to arbitrary nFET logic blocks.

Now let us examine the pFET rules. Figure 5.29(a) shows that two series-connected pFETs gives the OR-NOT operation, while the two parallel-connected pFETs in (b) results in the AND-

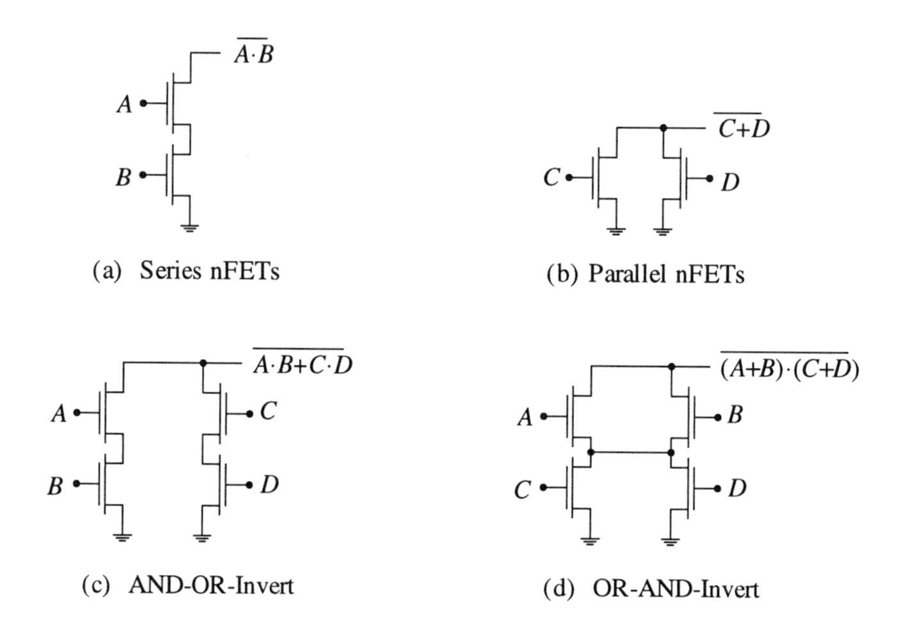

Figure 5.28 Logic formation using nFETs in the CMOS logic gate

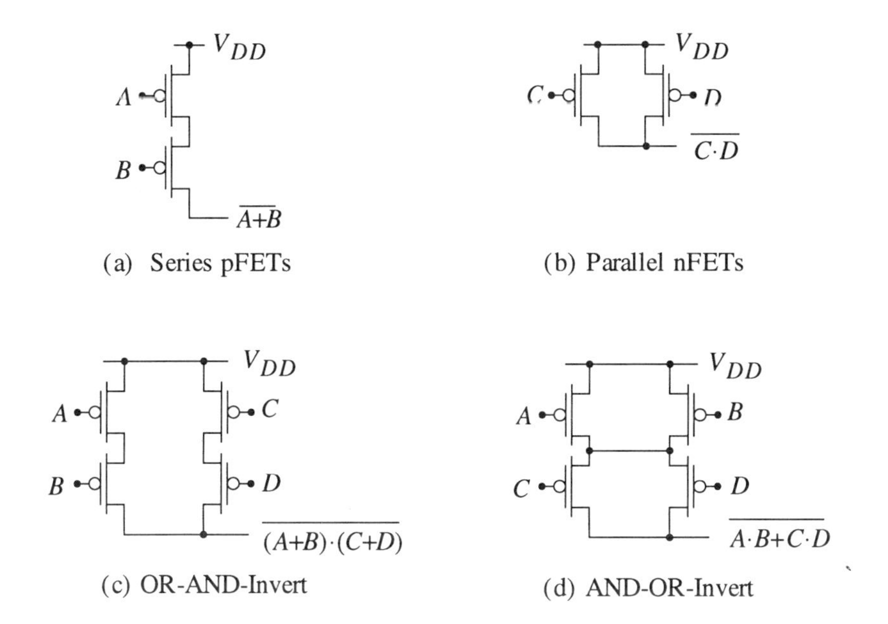

Figure 5.29 Logic formation using pFETs in the CMOS logic gate

NOT operation. This is exactly opposite to the manner in which nFETs behave. When the individual transistors are replaced by arrays of pFETs, we can create more complex functions. For example, in Figure 5.29(c), the series-connected pFETs implement the OR-AND-Invert operation

$$\overline{(A+B) \cdot (C+D)} \tag{5.67}$$

which is the dual of the equivalent nFET arrangement. Similarly, the sub-network in FIgure 5.29(d) gives the AND-OR-Invert term

$$\overline{A \cdot B + C \cdot D} \tag{5.68}$$

A CMOS gate that is created with nFET/pFET groups using this type of structuring implements what is called **series-parallel logic**. Complex logic gates can be created by applying these rules to the complementary pairs; the resulting functions can be verified using switch logic formalism. When the nFET and pFET arrays are created in this manner, the resulting logic gates have a unique function assignment with well-defined outputs. This means that the DC output voltage is either V_{DD} or $0v$; it is never left floating or simultaneously connected to both the power supply and ground.

5.4.1 Examples of Complex Logic Gates

Consider the circuit shown in Figure 5.30. This logic gate implements the function

$$f = \overline{A \cdot B + C} \tag{5.69}$$

since either $A \cdot B$=1 OR C=1 will connect the output to ground by turning on an nFET conduction path. This is verified by the function table shown with the gate. If C=0 AND (A=0 OR B=0), then the pFETs provide a conduction path between the power supply and the output, giving a logic 1 output voltage of V_{DD}. Note that the function has the AOI structuring discussed above. In particu-

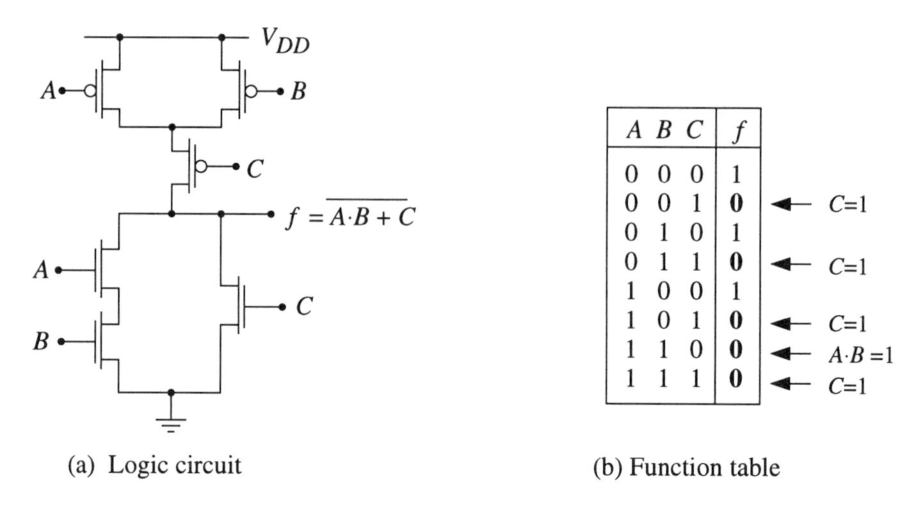

A	B	C	f	
0	0	0	1	
0	0	1	**0**	← C=1
0	1	0	1	
0	1	1	**0**	← C=1
1	0	0	1	
1	0	1	**0**	← C=1
1	1	0	**0**	← $A \cdot B = 1$
1	1	1	**0**	← C=1

(a) Logic circuit (b) Function table

Figure 5.30 Example of a CMOS complex logic gate

lar, the nFETs with input variables A and B are in **series**, so the that pFETs with inputs A and B must be in **parallel**. Working outward in the nFET array we see that the nFET with input C is **parallel** with "*A-B*" group of nFETs, so that the C-pFET must be in **series** with the "*A-B*" group of pFETs. It is seen that this of structuring is very useful for design purposes. A logic gate can be designed by building the nFET logic array to meet the functional specifications. Once the nFET logic wiring is completed, the pFET array can be constructed by applying series-parallel arguments using the nFET connections. In general, the series-parallel design process must be initiated at the simplest sub-block level, and expanded outward until the entire array is in this form. For the above example, this meant that we first looked at the two series "*A-B*" nFETs, and then moved to the next level where the C-input nFET was in parallel with the "*A-B*" group.

The circuit shown in Figure 5.31 provides another example of a complex logic gate. This implements the function

$$g = \overline{A \cdot (B + C)} \tag{5.70}$$

as can be verified by the straightforward reasoning. Consider the viewpoint where we trace the output rail down to the ground connection. An input A=1 is needed to get through the "first level" of nFETs. To complete the path to ground, then either B=1 OR C=1 must be true. This says that the output node will be connected to ground if $A \cdot (B+C)=1$; however, this results in a logic 0 output (since the output node will be connected to ground), so that we must take the complement to arrive at the function g.

Another example of how a complex logic gate can be built is shown in Figure 5.32. Consider the nFET logic array. This group of transistors connects the output node to ground if $A \cdot (B+C)=1$, OR, if $D \cdot E=1$. Since either case gives an output of f=0, we may write by inspection that

$$f = \overline{A \cdot (B + C) + D \cdot E} \tag{5.71}$$

is the Boolean logic expression for the circuit. The series-parallel relationship between the nFET and the pFET arrays can be verified by starting with the smallest nFET subcircuit consisting of the "B+C" parallel transistors, and then working outward with larger and larger groupings. Note that the function may be expanded to read as

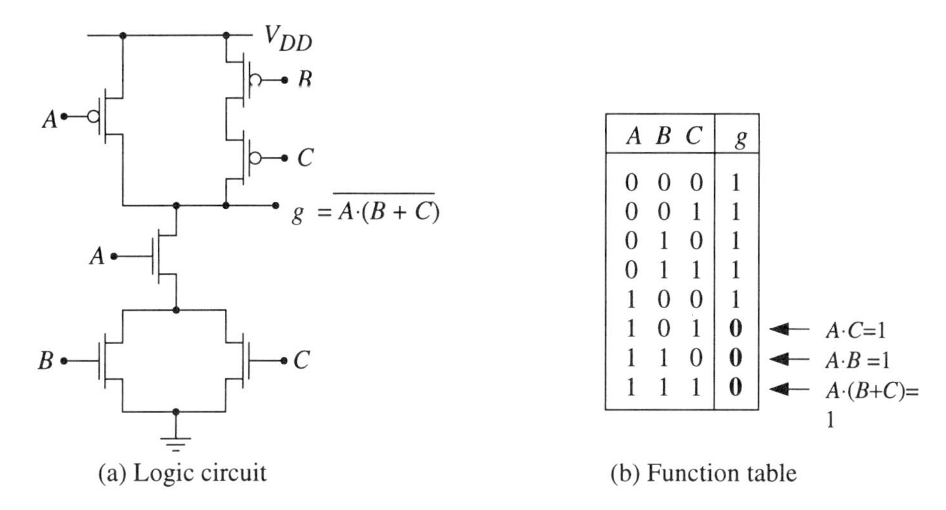

A	B	C	g
0	0	0	1
0	0	1	1
0	1	0	1
0	1	1	1
1	0	0	1
1	0	1	**0**
1	1	0	**0**
1	1	1	**0**

(a) Logic circuit (b) Function table

Figure 5.31 A complex logic CMOS gate circuit

$$f = \overline{A \cdot B + A \cdot C + D \cdot E} \tag{5. 72}$$

which is true AOI form.

These examples should be studied in detail until the nature of the series-parallel nFET-pFET relationship becomes clear. Just remember to start with the largest FET groups, and then work into smaller and smaller groups until the gate is complete.

5.4.2 Logic Design Techniques

The examples above illustrate the technique for designing complex logic gates using function tables and/or occurrences of 0's at the output. Other approaches are easy to formulate. As an example, let us look at the problem of creating a complex logic gate from a logic diagram. Consider the

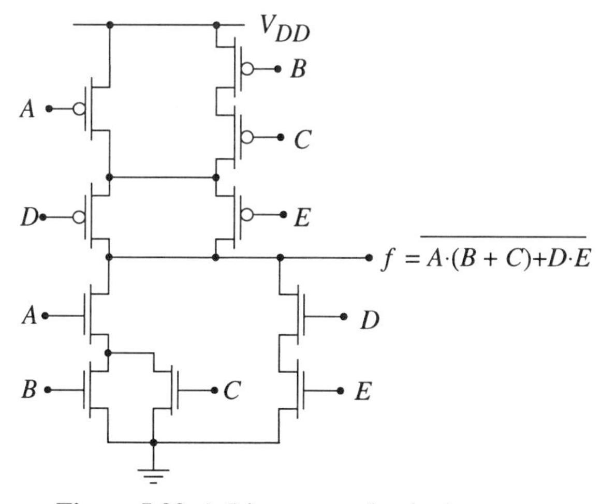

Figure 5.32 A 5-input complex logic gate

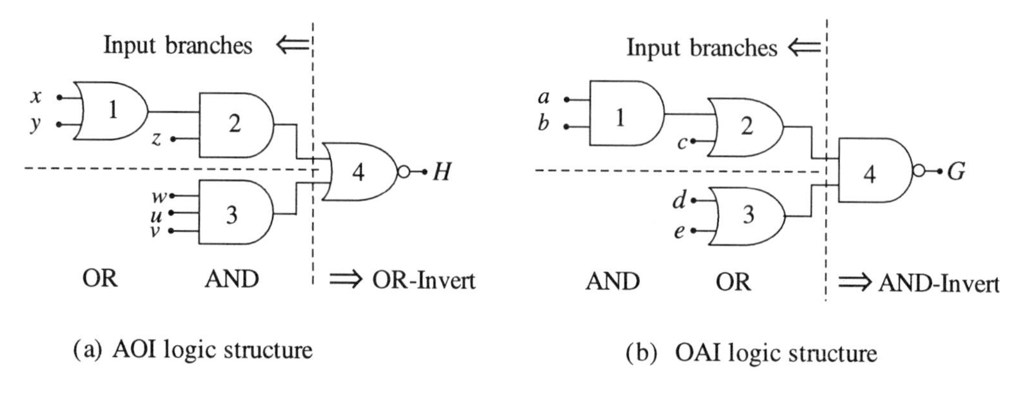

(a) AOI logic structure

(b) OAI logic structure

Figure 5.33 Basic logic cascades

logic diagrams shown in Figure 5.33. The AOI network (a) has been structured to show the logic patterning OR-AND-OR-Invert, while the OAI network (b) implements a flow of AND-OR-AND-Invert. The *Input Branches* in both drawings consist of the gates that have logic variables applied as inputs. These provide the basis for creating the CMOS logic circuit.

Consider first the logic characteristics of nFETs as summarized in Figure 5.34. In general, series-connected nFETs yield the AND operation among the inputs, while parallel-connected nFETs are used for ORing the input variables. This provides us with a one-to-one correspondence between the logic gate diagram and the structure of nFET logic array.

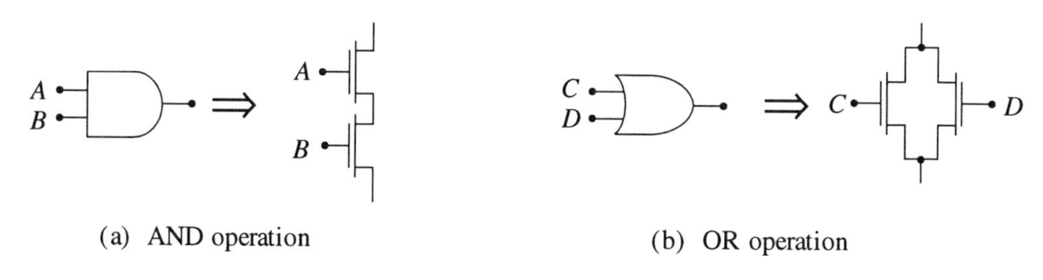

(a) AND operation

(b) OR operation

Figure 5.34 nFET-to- logic gate equivalence

The rules for pFETs are summarized in Figure 5.35, and requires a more detailed explanation. We will adopt the viewpoint that series-connected pFETs still yield the AND operation, but with **assert-low** (bubbled) inputs. For the pFETs shown in Figure 5.35(a), the series grouping corresponds to the AND operation

$$\overline{A} \cdot \overline{B} = \overline{A + B} \tag{5. 73}$$

where the DeMorgan rule has been used to obtain the NOR equivalent. Applying the same reasoning to parallel-connected pFETs as shown in Figure 5.35(b) gives the OR operation with assert-low inputs. This is the same as the NAND operation since

$$\overline{A} + \overline{B} = \overline{A \cdot B} \tag{5. 74}$$

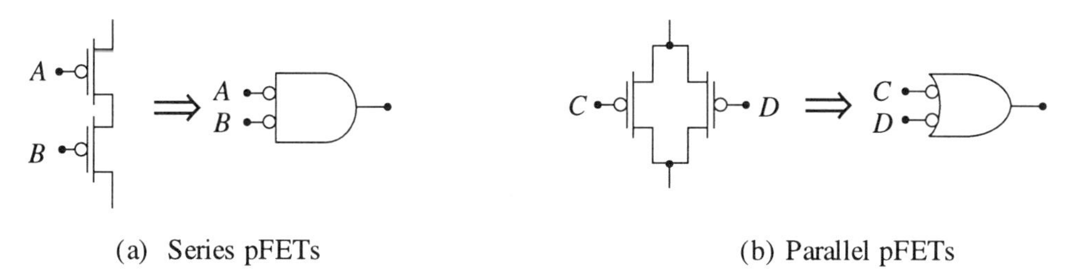

(a) Series pFETs (b) Parallel pFETs

Figure 5.35 pFET-logic gate equivalence

by the basic DeMorgan identity. When the output of a gate is used as an input to another logic gate, it acts in the same manner. The AND operation implies that series groups of FETs are to be used, while OR is accomplished by means of parallel groups of FETs.

To understand how these structuring rules allow us to construct the required nFET and pFET logic arrays, let us examine the logic diagrams that were introduced in Figure 5.33. Since these have assert-high (normal or unbubbled inputs), we may use them directly to create the nFET logic arrays. The pFET circuit is based on assert-low inputs, so that it is helpful to modify the logic diagrams by "pushing the bubble from the output towards the input," i.e., applying the DeMorgan theorems. This results in the logic diagrams shown in Figure 5.36. Since these have assert-low inputs, the pFET rules may be directly applied to arrive at the pFET logic array. This procedure results in the complex logic gates show in Figure 5.37. It is a straightforward matter to verify the correspondence between the primitive logic operations and the resulting FET arrangements.

5.4.3 FET Sizing and Transient Design

Series-parallel logic gates have the characteristic that the output logic function is determined entirely by the circuit topology. Wiring the transistors together in the correct manner is sufficient to set the DC characteristics of

$$\begin{aligned} V_{OL} &= 0v \\ V_{OH} &= V_{DD} \end{aligned} \qquad (5.75)$$

The sizes of the transistors do determine the values of the DC switching voltages such as V_I for the various input combinations. However, most of the design and sizing problems center around the transient switching times because the use of series-connected FETs introduce delays that are intrin-

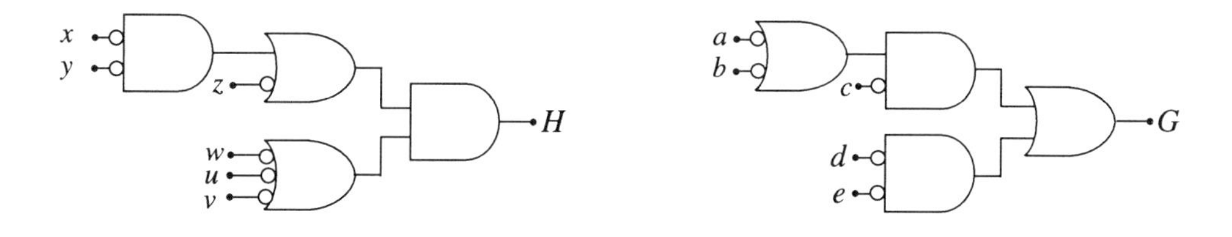

(a) AOI logic structure (b) OAI logic structure

Figure 5.36 Logic diagrams for construction of the pFET arrays

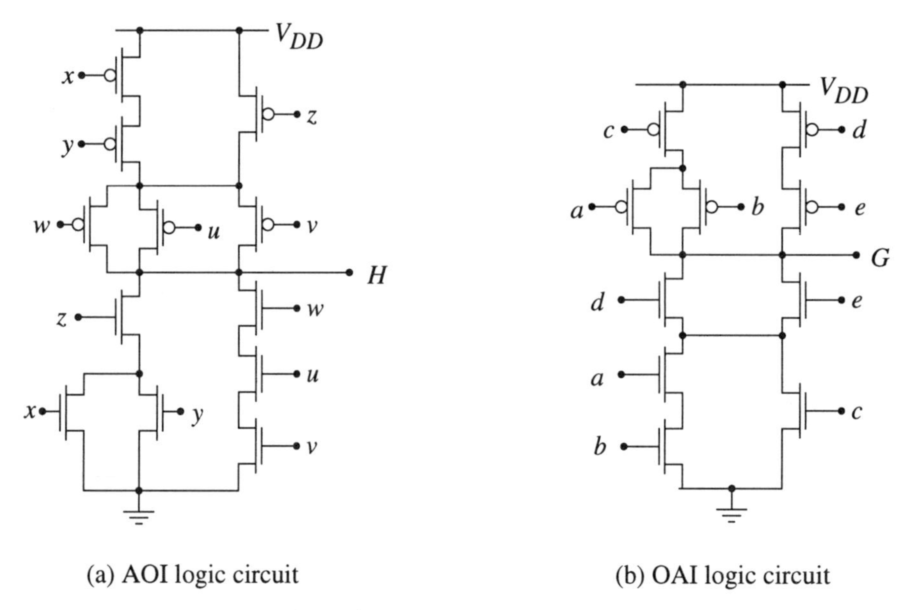

(a) AOI logic circuit (b) OAI logic circuit

Figure 5.37 Final gate designs

sic to the design style.When analyzing the output transients, we note that

- The pFET logic array sets the value of t_{LH}, and,
- The nFET logic array sets the value of t_{HL}.

In general, we are only interested in estimating the worst-case values. This in turn says that we should concentrate our efforts on the longest chain of series-connected MOSFETs for each case, since these result in the largest time constants.

The procedure for estimating the switching times for a static logic gate can be summarized as follows.

1. Estimate the total output capacitance C_{out} for the gate as the sum of all capacitors connected to the output node.

2. To calculate the charge time t_{LH}, find the longest chain of series-connected pFETs between the power supply and the output node. Construct the equivalent RC ladder network, find the time constant τ_p using the Elmore formula, and then estimate

$$t_{LH} \approx 2.2\Upsilon\tau_p . \quad (5.76)$$

3. Use the same procedure to find t_{HL}. Identify the longest chain of series-connected nFETs between the output and ground, construct the equivalent RC ladder network, apply the Elmore formula to find τ_n, and then use

$$t_{HL} \approx 2.2\Upsilon\tau_n . \quad (5.77)$$

This provides a straightforward manner to estimate the important switching times. Of course, a computer simulation is required for an accurate analysis.

An example of this technique is summarized in Figure 5.38 for a gate that gives an output of $\overline{x \cdot (y+z)}$. The structure of the equation implies that the values of both t_{LH} and t_{HL} are determined by series chains consisting of 2 FETs; this is verified by the circuit schematic. Consider first the low-to-high time. Using the equivalent RC network, we see that

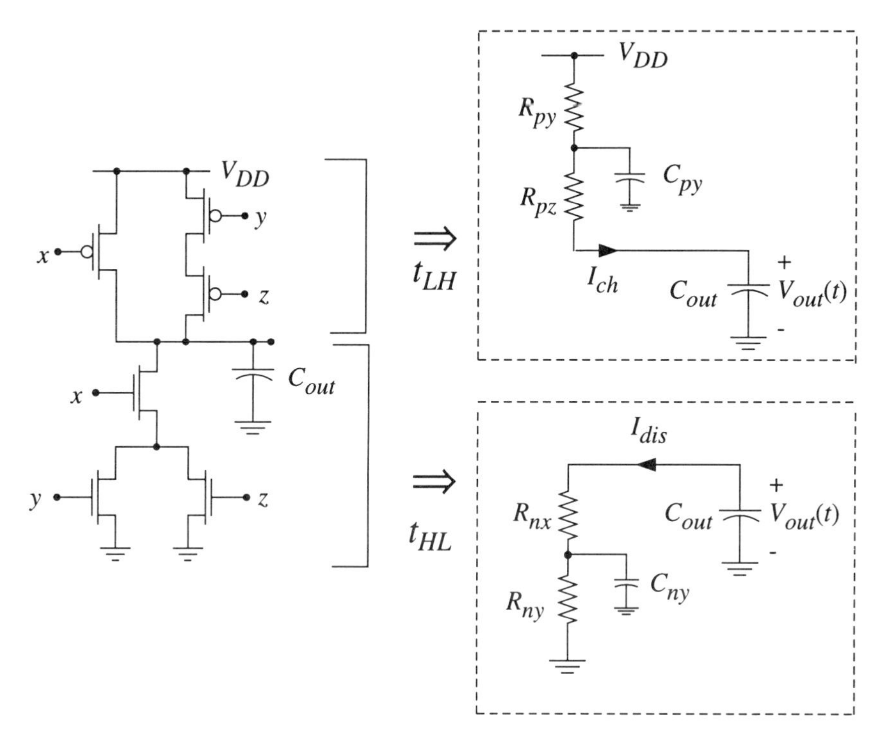

Figure 5.38 Example of switching model for complex gates

$$t_{LH} = 2.2\,[\,(R_{py} + R_{pz})\,C_{out} + R_{py}C_{py}] \tag{5.78}$$

is the desired time interval. Similarly, the high-to-low time is estimated from

$$t_{HL} = 2.2\,[\,(R_{nx} + R_{nz})\,C_{out} + R_{ny}C_{ny}] \tag{5.79}$$

It is worthwhile to mention that these formulas oversimplify the actual problem since they ignore the possibility of influence from the opposite-polarity FETs. For example, it is possible to have an certain input switching combination where t_{LH} depends upon C_{ny}. They are, however, reasonable estimates so long as the results are not over-interpreted.

The simplest connection to the device sizes is through the resistance formula

$$R = \frac{1}{k'\left(\frac{W}{L}\right)(V_{DD} - V_T)} \tag{5.80}$$

A first-cut design procedure can be based upon this observation. First, we design an *inverter* that has the desired values of

$$\begin{aligned} t_{LH} &= s_p\tau_p = s_pR_pC_{out} \\ t_{HL} &= s_n\tau_n = s_nR_nC_{out} \end{aligned} \tag{5.81}$$

The values of R_p and R_n are then used as a basis for calculating the sizes of the transistors in the complex logic gates. For *m* series-connected pFETs, we choose

$$R_{p,m} = \frac{R_p}{m} \tag{5.82}$$

for the individual resistances. This approach thus equates the end-to-end resistance of the series chain to that of the single pFET used in the inverter. To achieve this goal requires that the aspect ratios $(W/L)_{p,m}$ in the series connected chain transistors satisfy

$$\left(\frac{W}{L}\right)_{p,m} = m\left(\frac{W}{L}\right)_p \tag{5.83}$$

where $(W/L)_p$ is the size of the inverter pFET. The longer the chain, the larger the factor m, which can result in large device sizes. Similarly, for r series-connected nFETs, the aspect ratios are chosen as

$$\left(\frac{W}{L}\right)_{n,r} = r\left(\frac{W}{L}\right)_n \tag{5.84}$$

where $(W/L)_n$ is the size of the inverter nFET. This gives

$$R_{n,r} = \frac{R_n}{r} \tag{5.85}$$

so that the end-to-end resistance of the nFET chain equals R_n. Applying this approach to the example circuit of Figure 5.38 above, the device sizes would be

$$\left(\frac{W}{L}\right)_{py} = 2\left(\frac{W}{L}\right)_p = \left(\frac{W}{L}\right)_{pz} \tag{5.86}$$

for the pFETs, and

$$\left(\frac{W}{L}\right)_{nx} = \left(\frac{W}{L}\right)_{ny} = \left(\frac{W}{L}\right)_{nz} = 2\left(\frac{W}{L}\right)_n \tag{5.87}$$

for the nFETs. Note that the size of the pFET with input x is not specified directly by this procedure. It can be the same size as the inverter pFET, or may be chosen to be the same as the other two pFETs to simplify layout.

While this procedure gives a reasonable first estimate of the device sizes, it ignores several important aspects of the transient design problem. The most obvious among these is the dependence of capacitor values on the layout dimensions, and the fact that it ignores the inter-transistor capacitance within a series connected chain. Regardless of these two items, it does serve the purpose of giving us a simple procedure for first design estimates, i.e., a starting point. Once the circuit is designed with this approach, it can be simulated and redesigned as needed.

5.5 Exclusive OR and Equivalence Gates

The Exclusive OR (XOR) operation is not a primitive logic function, but it is used so often that it is deserves a special logic symbol of its own. Figure 5.39(a) shows the logic symbol for a 2-input XOR gate. By definition, the output is a logic 1 when either input is a logic 1, but is 0 for the case where both inputs are a logic 1 states simultaneously. The XOR operation is denoted by[3]

[3] The symbol "⊕" is read as "O-plus".

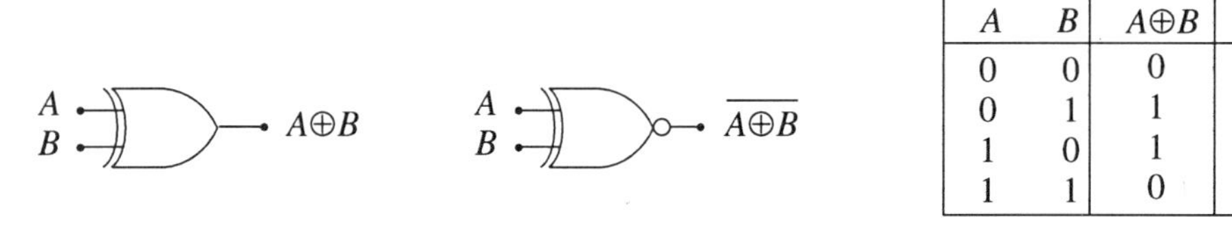

A	B	$A\oplus B$	$\overline{A\oplus B}$
0	0	0	1
0	1	1	0
1	0	1	0
1	1	0	1

(a) XOR symbol (b) XNOR symbol (b) Function table

Figure 5.39 Exclusive-OR and exclusive-NOR gates

$$f_{XOR} = A \oplus B = \overline{A} \cdot B + A \cdot \overline{B} \tag{5.88}$$

Taking the complement of the XOR gives the Exclusive-NOR (XNOR) operation

$$f_{XNOR} = \overline{A \oplus B} = A \cdot B + \overline{A} \cdot \overline{B} \tag{5.89}$$

The XNOR gate symbol is illustrated in Figure 5.39(b). Since f_{XNOR}=1 if the inputs are equal, it is often called the **equivalence function**. The XOR and XNOR operations provide the basis for important system units such as adders and parity networks.

The general logic structuring discussed above may be used to create the CMOS gates shown in Figure 5.40. Both circuits use the input pairs $(A, \overline{A})$ and $(B, \overline{B})$, so that inverters are needed to generate $\overline{A}$ and $\overline{B}$ from the basic inputs A and B. Since these gates create the XOR and XNOR functions using AOI structuring, the logic may not be apparent at first sight. It is therefore useful to work through the algebra to verify the results.

Consider the XOR gate in Figure 5.40(a). Using the logic formation rules gives the output as

$$f = \overline{A \cdot B + \overline{A} \cdot \overline{B}} \tag{5.90}$$

This may be reduced by successive application of the DeMorgan rules and the distribution operation as shown by the following steps:

$$\begin{aligned} f &= \overline{(A \cdot B)} \cdot \overline{(\overline{A} \cdot \overline{B})} \\ &= \left(\overline{A} + \overline{B}\right) \cdot (A + B) \\ &= \overline{A} \cdot B + A \cdot \overline{B} \end{aligned} \tag{5.91}$$

The last reduction was based on the fact that $A \cdot \overline{A} = 0 = B \cdot \overline{B}$. Similarly, the circuit in Figure 5.40(b) gives

$$\begin{aligned} g &= \overline{\overline{A} \cdot B + A \cdot \overline{B}} \\ &= A \cdot B + \overline{A} \cdot \overline{B} \\ &= \overline{(A \oplus B)} \end{aligned} \tag{5.92}$$

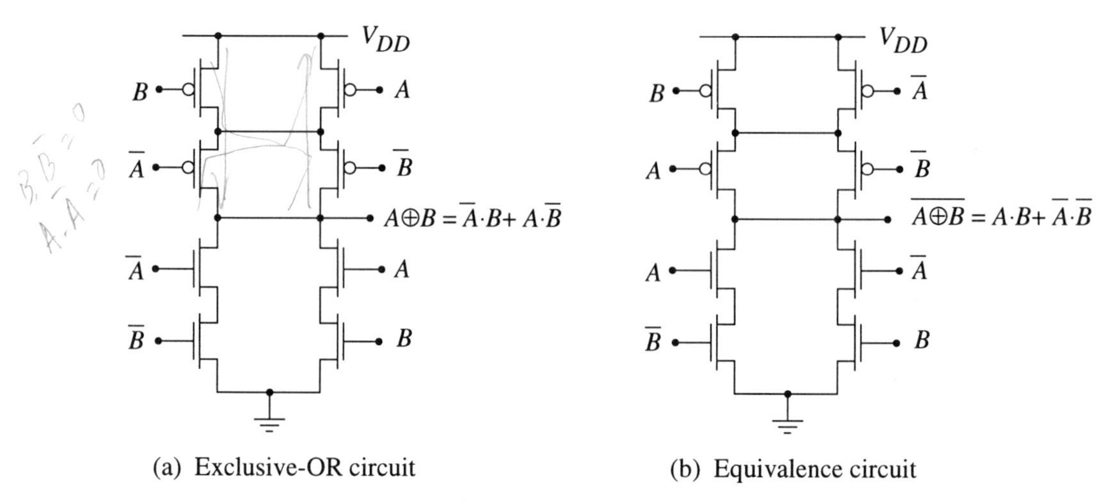

(a) Exclusive-OR circuit (b) Equivalence circuit

Figure 5.40 XOR and XNOR gates with complementary structuring

by applying the DeMorgan rules, expanding, and reducing the expression. These calculations verify our earlier statement that the XOR and XNOR functions can be implemented using the idea of complementary structuring.

5.5.1 Mirror Circuits

The exclusive-OR operation is used extensively in several types of logic networks including adder circuits and parity checkers. Because of the quest for faster and/or more compact circuit structures, alternate CMOS styles have appeared in literature. One of these is the use of **mirror circuits** in which the nFET and pFET arrays have the exactly the same structure. These do not use series-parallel logic formation, but have many of the same characteristics.

The origin of the mirror circuits can be seen by the XOR and XNOR truth tables. With 2 inputs A and B, each given gate has 2 outputs that are high (V_{DD}) and two outputs that are low ($0v$). This implies that we can provide 2 paths from the output to the power supply, and 2 paths from the output to ground with each path consisting of two series-connected FETs. To understand the philosophy, consider a 2 variable gate. The possible input combinations are

$$A \cdot B,\ \ \overline{A} \cdot B,\ \ A \cdot \overline{B},\ \ \overline{A} \cdot \overline{B} \tag{5.93}$$

Since the exclusive-OR function has the form

$$A \oplus B = \overline{A} \cdot B + A \cdot \overline{B} \tag{5.94}$$

this means that the combinations $\overline{A} \cdot B$ and $A \cdot \overline{B}$ should provide connections from the output to the power supply, while $A \cdot B$ and $\overline{A} \cdot \overline{B}$ should connect the output to ground. A circuit constructed with these characteristics is shown in Figure 5.41(a), and constitutes an XOR gate. Similarly, since the equivalence function is given by

$$\overline{A \oplus B} = A \cdot B + \overline{A} \cdot \overline{B} \tag{5.95}$$

the circuit shown in Figure 5.41(b) acts as an XNOR gate. These are called "mirror circuits" since

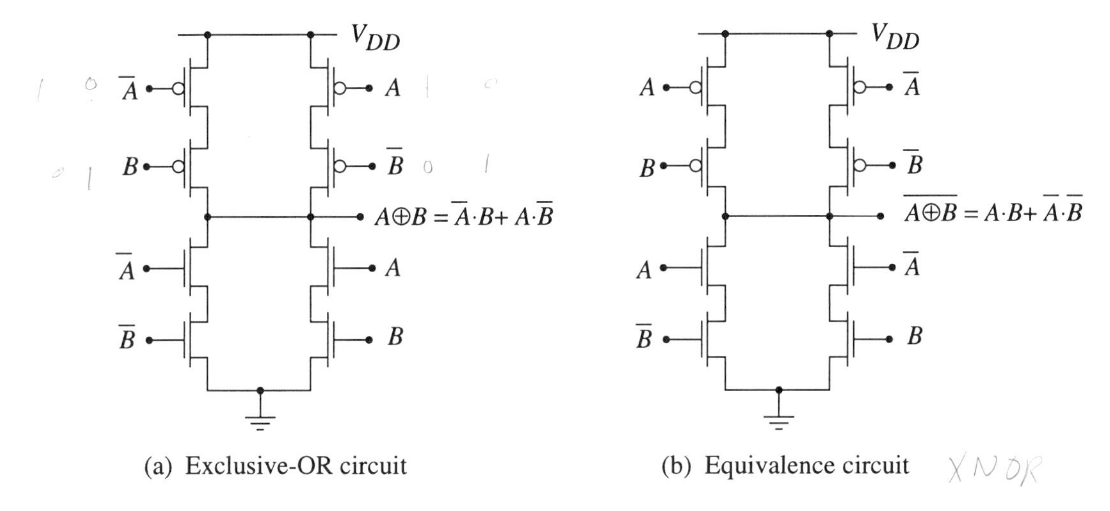

(a) Exclusive-OR circuit (b) Equivalence circuit

Figure 5.41 XOR and XNOR functions using mirror circuits

due to the symmetry above and below the output, i.e., if you place a mirror along the output node, then you will see the other half of the gate as the reflection. Obviously these do not have a series-parallel structuring.

Aside from their departure away from the more general approach to logic formation, these circuits are of interest because they may have shorter switching times. This may be illustrated by using the circuits shown in Figure 5.42. The series-parallel circuit in (a) has a charging time constant of

$$\tau_{p,1} = R_p C_1 + 2R_p C_{out} \tag{5.96}$$

where the pFETs have been assumed to be of equal size with resistance R_p, and C_1 represents the capacitance between the upper and lower pFET groups. In Figure 5.42(b), the time constant for the

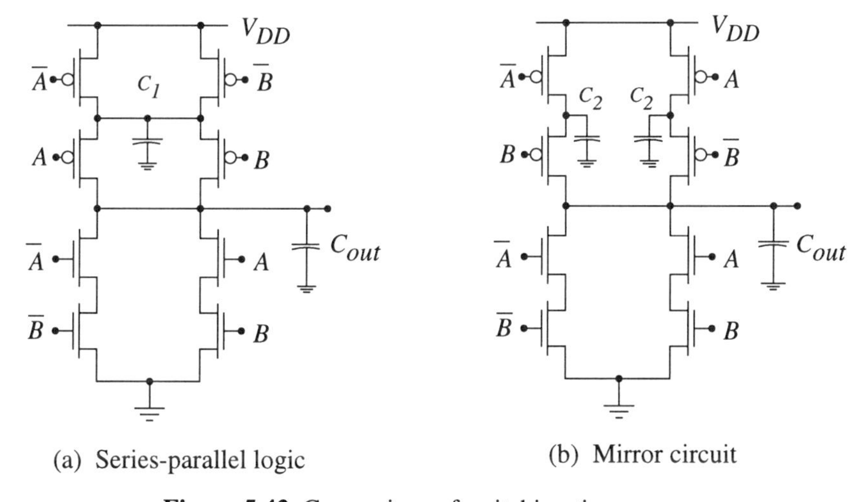

(a) Series-parallel logic (b) Mirror circuit

Figure 5.42 Comparison of switching times

low-to-high transition is given by

$$\tau_{p,2} = R_p C_2 + 2R_p C_{out} \tag{5.97}$$

with C_2 defined as in the drawing. The important difference between these two values is that $C_1 > C_2$ since C_1 has contributions from 4 pFETs, while the mirror circuit splits the left and right sides and leaves C_2 with only 2 pFETs to contribute. Thus, $\tau_{p,2} < \tau_{p,1}$ within the limits of this analysis. In addition, the layout is simpler because of the symmetry of the four branches. Although these arguments may sound convincing, it is important to keep in mind that the use of simple time constants from *RC* ladders is based on step-transitions, and ignores the complexities involved in the true timing of the input signals.

As another example of a mirror circuit, consider the multiple XOR function

$$f = a \oplus b \oplus c \oplus d \tag{5.98}$$

which is used in applications like parity generation. In more general terms, this is the odd function such that an odd number of input 1's gives an output of $f = 1$, with $f = 0$ otherwise. A mirror circuit for this function is shown in Figure 5.43. Note that identical FET arrays are used for each of the 4 connections between the output and the power supply or ground. Although the circuit may appear somewhat complex at first sight, the individual functions can be traced through each branch. For

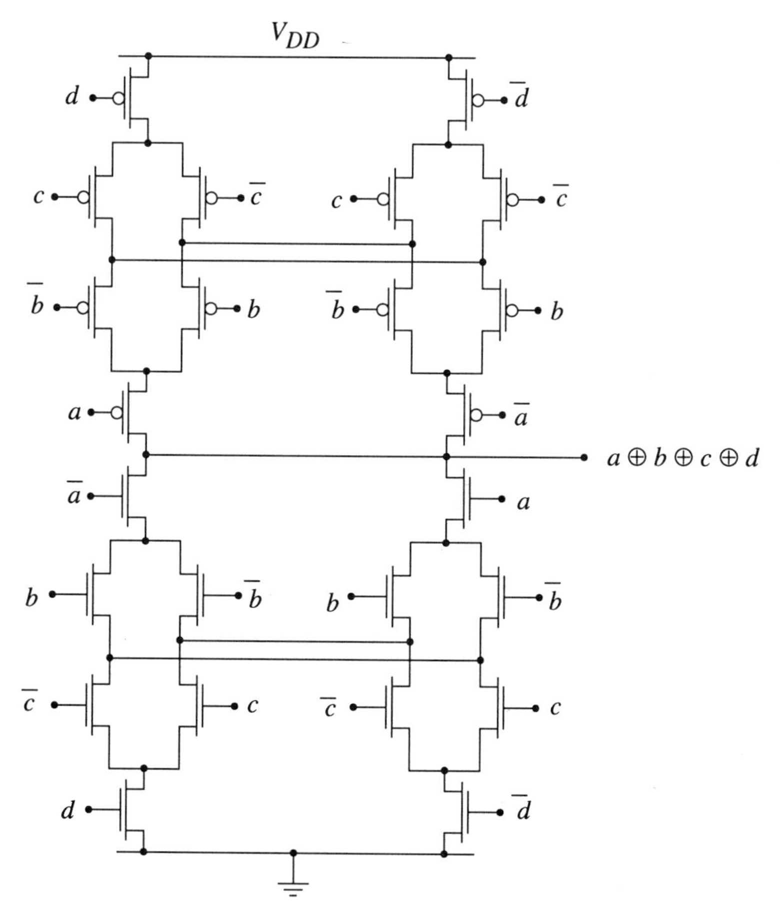

Figure 5.43 4-input odd function using mirror structuring

example, the right-nFET array directly implements the complement of the terms

$$a \cdot b \cdot \bar{c} \cdot \bar{d} + a \cdot \bar{b} \cdot c \cdot \bar{d} = a \cdot (b \cdot \bar{c} + \bar{b} \cdot c) \cdot \bar{d} \\ = a \cdot (b \oplus c) \cdot \bar{d} \tag{5. 99}$$

while the left array gives the complemented form of

$$\bar{a} \cdot b \cdot \bar{c} \cdot d + \bar{a} \cdot \bar{b} \cdot c \cdot d = \bar{a} \cdot (b \oplus c) \cdot d \tag{5. 100}$$

The cross-connections generate the other terms in the same manner.

Although the construction of logic function via FET placement is straightforward, the transient switching times are more complicated to deal with. Figure 5.44 provides the simplest charge and discharge models for the circuit that are created by only looking at the longest *RC* ladder groups. The times constant associated with the charging circuit in (a) is

$$\tau_p = C_{out}(R_{pa} + R_{pb} + R_{pc} + R_{pd}) + C_3(R_{pb} + R_{pc} + R_{pd}) \\ + C_2(R_{pc} + R_{pd}) + C_1 R_{pd} \tag{5. 101}$$

Similarly, the discharge path shown in Figure 5.44(b) has the time constant

$$\tau_n = C_{out}(R_{na} + R_{nb} + R_{nc} + R_{nd}) + C_4(R_{nb} + R_{nc} + R_{nd}) \\ + C_5(R_{nc} + R_{nd}) + C_6 R_{nd} \tag{5. 102}$$

Although these equations ignore capacitors that are associated with opposite polarity FETs that may affect the overall switching times, they are still useful in obtaining simple estimates for the switching delays. In this circuit, the question will revolve around the response time of the single gate versus a conventional cascaded arrangement.

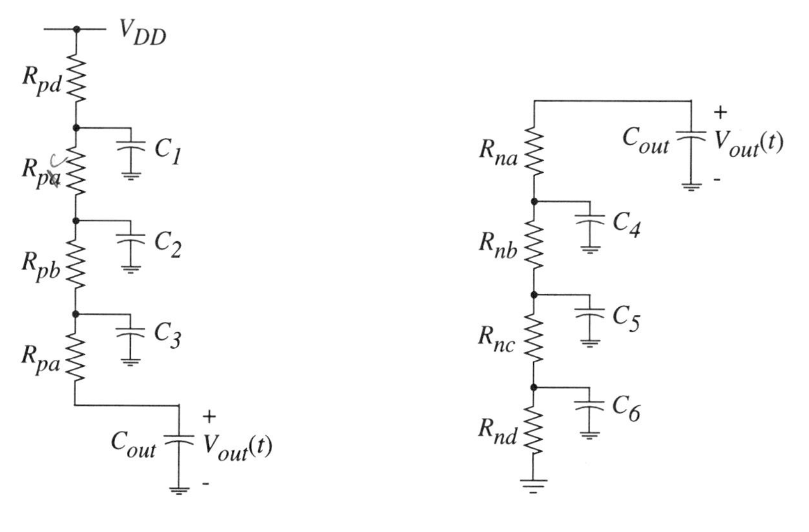

(a) Charge circuit (b) Discharge circuit

Figure 5.44 RC chains for transient modelling

5.6 Adder Circuits

Binary adders provide the basic connection between Boolean operations and arithmetic. They are commonly used for comparing different technologies or design styles since they are of reasonable importance and complexity.

Figure 5.45 shows the basic symbol and function table for a full-adder circuit that uses the inputs a_n and b_n and a carry-in bit c_n to produce the sum bit s_n and the carry-out bit c_{n+1}. The most common SOP expressions obtained directly from the table entries are given by

$$\begin{aligned} s_n &= a_n \oplus b_n \oplus c_n \\ c_{n+1} &= a_n b_n + c_n (a_n \oplus b_n) \end{aligned} \tag{5. 103}$$

as is easily verified by examining the entries that yield a result of 1 for each function. Although it is possible to create the adder circuits from these equations, they are not in AOI form and so do not admit directly to series-parallel logic design. To construct a circuit with this form, we first modify the expressions so that both s_n and c_{n+1} have AOI form, and then use the structured logic design of the previous section.

Let us rewrite the expression for the carry-out bit as

$$c_{n+1} = a_n \cdot b_n + c_n (a_n + b_n) \tag{5. 104}$$

where we have changed the XOR function to an OR function; this still yields the same result due to the fact that the first term is $a_n \cdot b_n$. The equation for the sum bit can be expanded and rearranged to

$$s_n = a_n \cdot b_n \cdot c_n + (a_n + b_n + c_n) \cdot \overline{c_{n+1}} \tag{5. 105}$$

so that it uses the carry-out bit c_{n+1} as an input. The AOI logic diagrams for both the sum and the carry-out are shown in Figure 5.46. Note that the upper (carry-out) and lower (sum) networks are very similar in that they both have OAOI structuring. This allows us to create the series-parallel CMOS circuits shown in Figure 5.47 for both sections of the full adder; FET placement has been accomplished by using the standard rules. This circuit is straightforward. Note, however, that both gates have somewhat long pFET chains, which may result in a slow response.

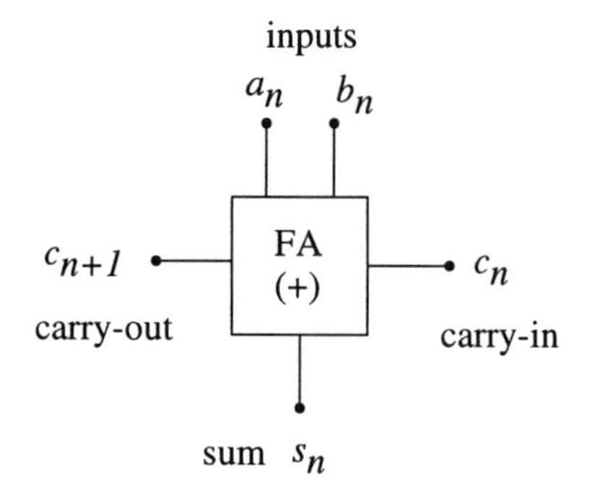

a_n	b_n	c_n	s_n	c_{n+1}
0	0	0	0	0
0	1	0	1	0
1	0	0	1	0
1	1	0	0	1
0	0	1	1	0
0	1	1	0	1
1	0	1	0	1
1	1	1	1	1

Figure 5.45 Full adder operation

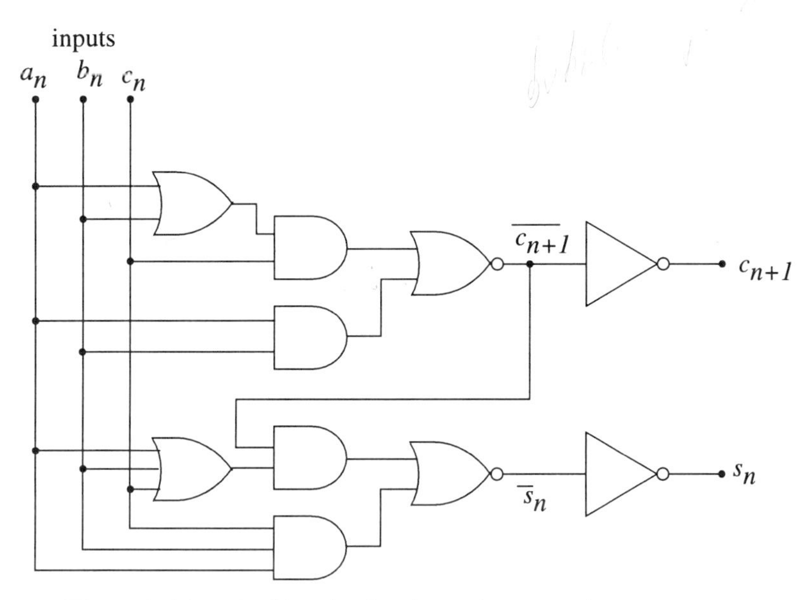

Figure 5.46 Full adder circuit using AOI structuring

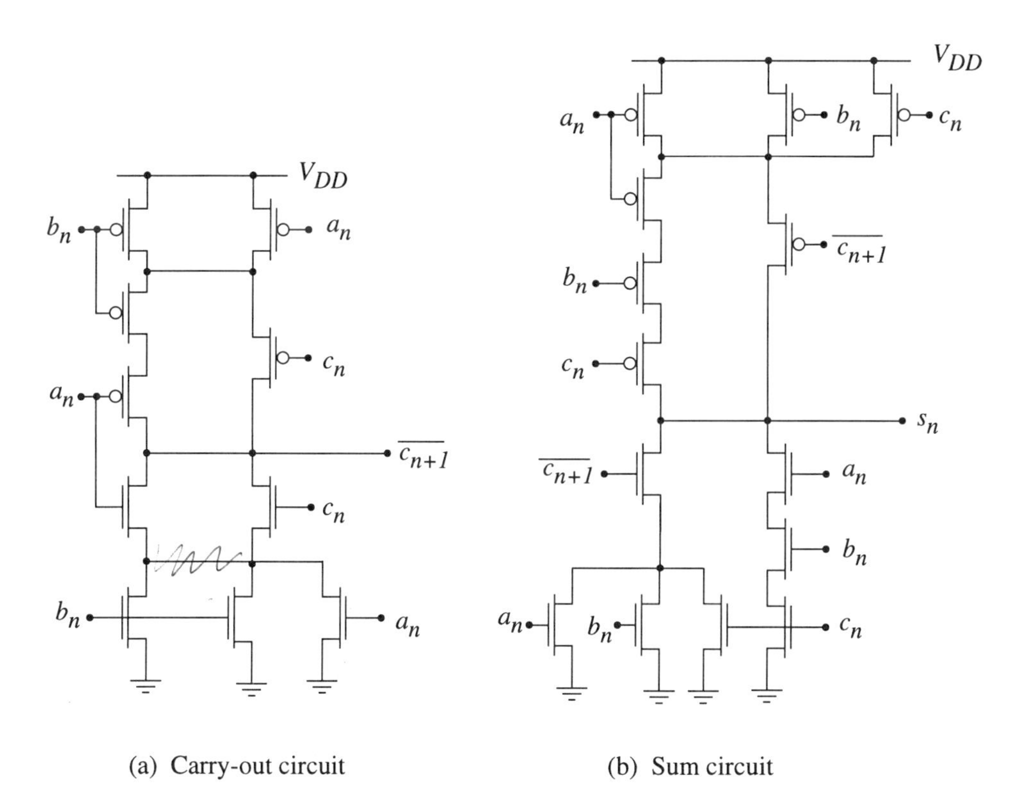

(a) Carry-out circuit (b) Sum circuit

Figure 5.47 CMOS circuits using AOI structuring

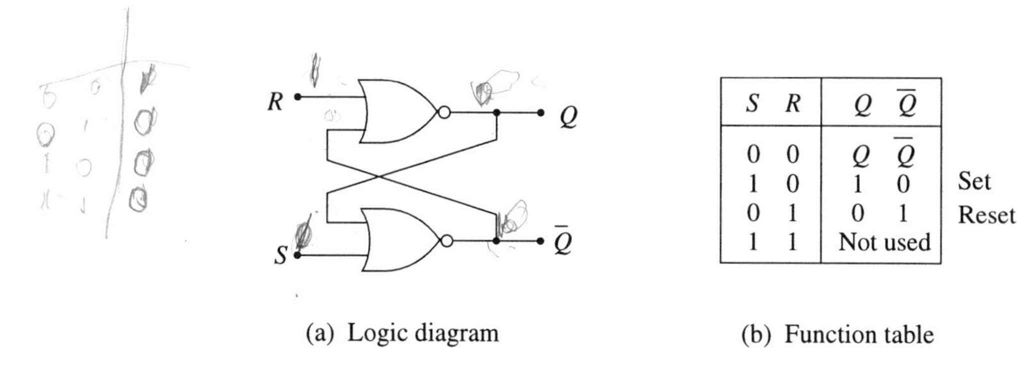

S	R	Q $\overline{Q}$	
0	0	Q $\overline{Q}$	
1	0	1 0	Set
0	1	0 1	Reset
1	1	Not used	

(a) Logic diagram (b) Function table

Figure 5.48 NOR-based SR latch operation

5.7 SR and D-type Latch

A latch is a circuit that can detect and hold a logic 0 or a logic 1 condition that is applied to the input. The **set-reset** (SR) latch is classified as a **bistable** network in that it has two stable states associated with it. The inputs to the SR latch are denoted by S (set) and R (reset), and the output by Q. In general, input values of S=1 and R=0 are used to set the output state to Q=1, while values of S=0 and R=1 resets the circuit and gives an output of Q=0.

Figure 5.48(a) shows an SR latch constructed using two NOR gates. The feedback action provided by the cross-coupled connections provides the necessary latching action The operation summary in (b) defines the set and reset conditions, and also shows the hold state when S=0=R. When both S and R are raised to logic 1 values, the outputs are not complements of each other, so that this input combination is not used.

A CMOS version of the SR latch is straightforward to build by simply wiring two static NOR gates together as in the circuit of Figure 5.49. The sub-circuit within the dashed-line box can be viewed as the portion of the network that provides the holding action. Setting and resetting voltages are denoted by V_S and V_R, respectively. These directly control the switching FETs (MnS, MpS, MnR, MpR) in a manner that allows the hold circuitry to be forced into the desired state.

Figure 5.50 shows the circuit operation for a set operation. With $V_S = V_{DD}$, MnS is biased on and pulls the node $\overline{Q}$ down to 0v; MpS is driven into cutoff. Note that V_R = 0v, so that MnR is cutoff. With $\overline{Q}$ at 0v, both pFETs on the right NOR gate are conducting, which pulls Q up to V_{DD}. This corresponds to a logic output value of Q=1 as desired. The reset operation is identical with only the

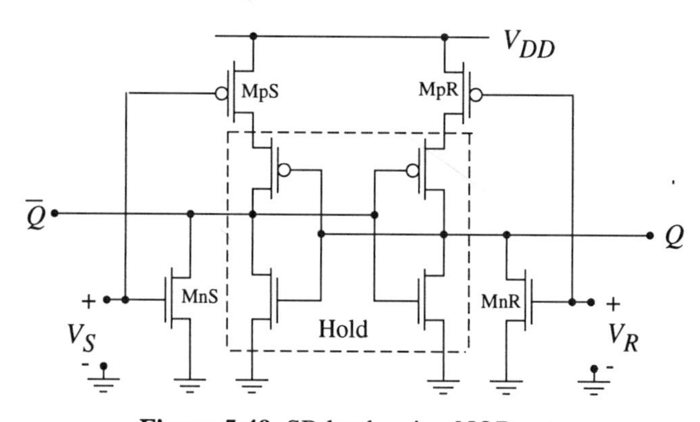

Figure 5.49 SR latch using NOR gates

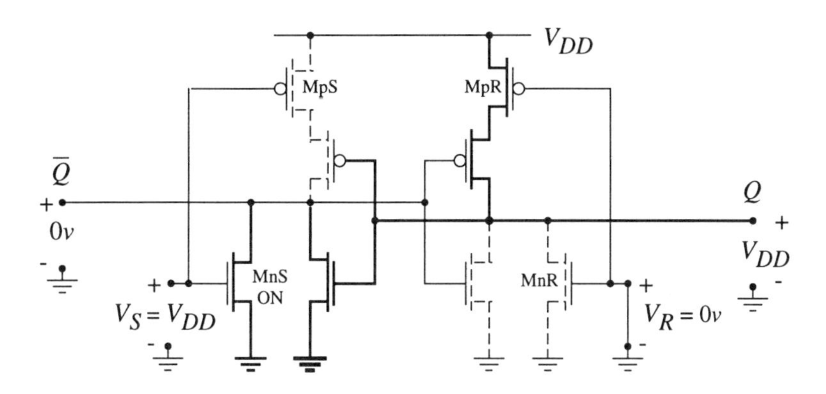

Figure 5.50 Voltages for Set operation

nodes reversed.

The circuit above can be simplified to that shown in Figure 5.51. The input voltages V_S and V_R are now applied only to pull-down nFETs, not to a complementary nFET/pFET pair.The operation is almost the same in that if either MnS or MnR is biased active with a high input voltage, the appropriate drain node ($\overline{Q}$ or Q) is pulled to 0*v*, initiating the latching action of the central hold circuit. Note that the latch has been reduced to a pair of cross-coupled inverters. The switching may take longer because the pFETs have been eliminated; this is due to the fact that in the original circuit, the pFETs were used to disconnect the power supply from the node that was pulled low. This remains a useful circuit regardless.

Another useful bistable element is the D-type latch. This has a single data input *D* that can be held and transferred to the output. Figure 5.52 shows how an SR latch can be used to create a D latch by adding an inverter. This automatically insures that the *S* and *R* inputs are never equal, eliminating the unused input combination. As shown by function table in Figure 5.52(b), the output *Q* follows the input value after an implied circuit-induced delay time. A brute-force CMOS realization of the D latch is drawn in Figure 5.52. This is simply the SR latch with a static inverter added as input logic. Although this is straightforward to construct, other types of D-latches are more common in CMOS.

Various kinds of latches and flip-flops can be built in CMOS. The SR latch illustrated here is based on the "classical" design and provides a useful circuit for many applications. CMOS, however, provides for many alternate circuit design styles so that one is not constrained to use any par-

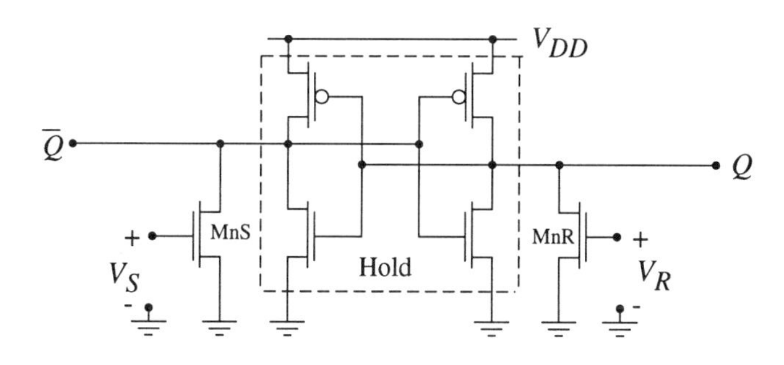

Figure 5.51 Simplified circuit for SR latch

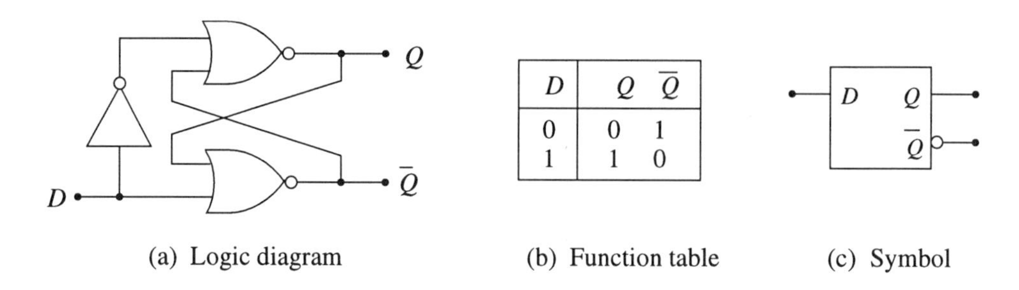

(a) Logic diagram (b) Function table (c) Symbol

Figure 5.52 D-latch logic diagram

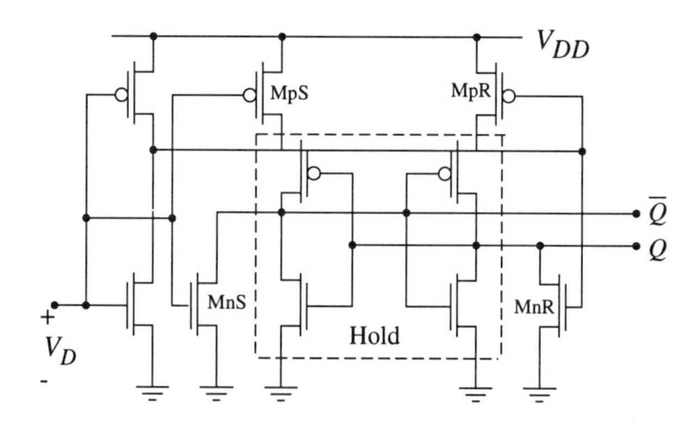

Figure 5.53 CMOS D-latch circuit

ticular circuit or design philosophy. As we will see in our development of CMOS circuit design styles, it is possible to create some very unique circuits that function as well, or better than, the classical SR latch. We note in passing that some of the classical logic components such as JK flip-flops can be constructed in CMOS, but are not commonly found in VLSI circuits because they are relatively cumbersome designs that consume large areas.

5.8 The CMOS SRAM Cell

A random-access memory (RAM) cell is a circuit that has three main operations.

- **Write** - A data bit is stored in the circuit
- **Hold** - The value of the data bit is maintained in the cell
- **Read** - The value of the data bit is transferred to an external circuit

A static RAM (SRAM) cell is capable of holding a data bit so long as the power is applied to the circuit.

Figure 5.54 shows the basic 6 transistor (6T) CMOS SRAM cell. It consists of a central storage cell made up of two cross coupled inverters (Mn1, Mp1 and Mn2, Mp2), and two access transistors MA1 and MA2 that provide for the read and write operations. The conducting state of the access

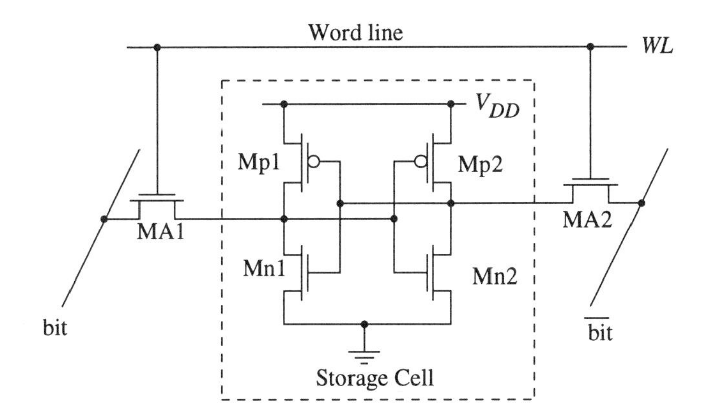

Figure 5.54 CMOS 6T SRAM circuit

transistors is controlled by the signal *WL* on the Word line. When *WL*=1, both MA1 and MA2 conduct and provide the ability to enter or read a data bit. A value of *WL*=0 gives a hold state where both MA1 and MA2 are driven into cutoff, isolating the storage cell. The access FETs connect the storage cell input/output nodes to the data lines denoted as bit and $\overline{\text{bit}}$, which are complements of each other. The operation of the circuit can be summarized as follows.

Write

To write a data bit to the cell, we bring *WL* to a high voltage and placed the voltages on the bit and $\overline{\text{bit}}$ lines. For example, to write a 1 to the cell, $V_{bit} = V_{DD}$ and $V_{\overline{bit}} = 0v$ would be applied as illustrated in Figure 5.55. This drives the internal voltages in a manner that V_1 goes high and $V_2 \rightarrow 0v$. Because of the bistable hold characteristic discussed below, changing the voltages on the left and right sides in opposite directions helps the cell latch onto the state.

Hold

The hold state of the cell is achieved by bringing the word line signal to *WL* =0. This places both access FETs MA1 and MA2 in cutoff, and isolates the storage cell from the bit and $\overline{\text{bit}}$ lines. The basic feature of the storage cell is that it is able to maintain the internal voltages V_1 and V_2 at com-

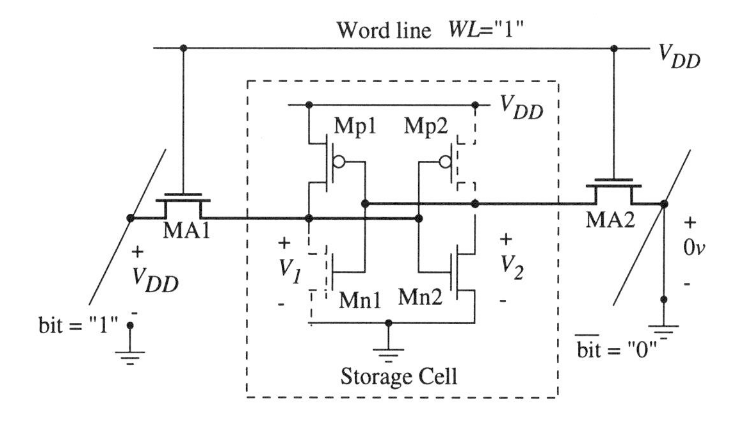

Figure 5.55 Write operation in an SRAM

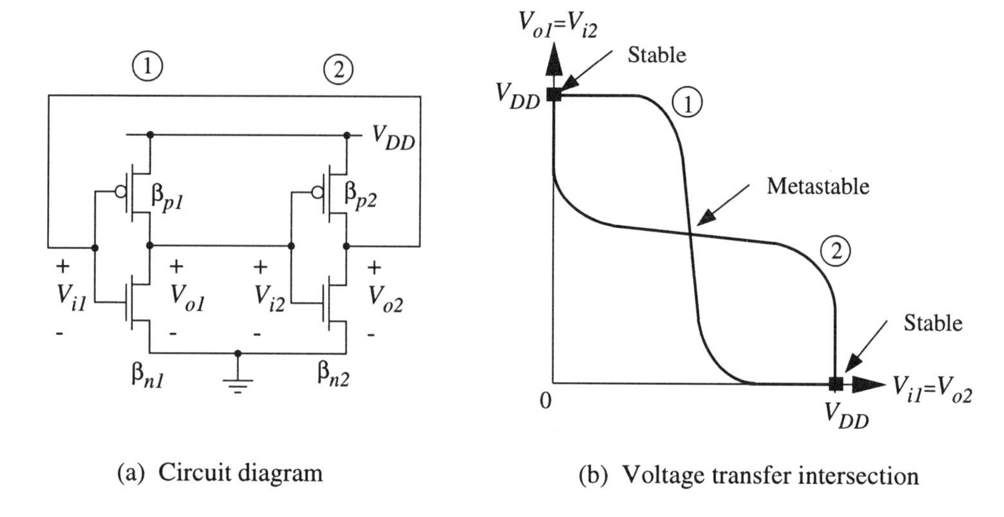

(a) Circuit diagram (b) Voltage transfer intersection

Figure 5.56 Voltage characteristics for an SRAM cell

plementary values (i.e., one high and the other low).

The latching action of the cross-coupled inverters can be studied using the circuit in Figure 5.55(a) where we have cascaded two inverters (1 and 2) and closed the loop to provide feedback. As seen in the drawing, the input voltages (V_{i1} and V_{i2}) and output voltages (V_{o1} and V_{o2}) are related to each other by

$$V_{o1} = V_{i2}$$
$$V_{o2} = V_{i1} \tag{5. 106}$$

This allows us to superpose the voltage transfer curve of the two inverters and arrive at the plot shown in Figure 5.56(b); we have assumed identical inverters, i.e., that β_n/β_p is the same for both. There are three intersection points where the VTCs satisfy the voltage relations. The two stable states occur at coordinates $(0,V_{DD})$ and $(V_{DD},0)$ and either can be used for data storage. The third intersection point occurs along the unity gain line, and is labelled as a "metastable" point in the drawing. In practice, the circuit cannot maintain equilibrium at this point even though it is a solution to the voltage requirements due to the dynamic gain $a_v = (dV_o / dV_i)$ associated with the inverter.

The diagram also illustrates the voltages needed to trigger the circuit during the write operation. To induce a fall in an inverter VTC, the input voltage should be above V_{IH}. Thus we could argue that the voltage received from the access device must exceed this value to insure the writing of a logic 1. Although highly simplified, this is a reasonable statement of the necessary condition; the switching is aided by the fact that the opposite side of the cell will be at 0*v*.

Read

The read operation is used to transfer the contents of the storage cell to the bit and $\overline{\text{bit}}$ lines. Bringing the Word line high with *WL*=1 activates the access transistors, and allows the voltage transfer to take place.

Figure 5.57 shows the read operation for the case where a logic 1 is stored in the cell. This gives internal cell voltages of V_{DD} and 0*v* on the left and right sides, respectively. Since the access FETs act like pass transistors, the line voltages as driven by the cell are

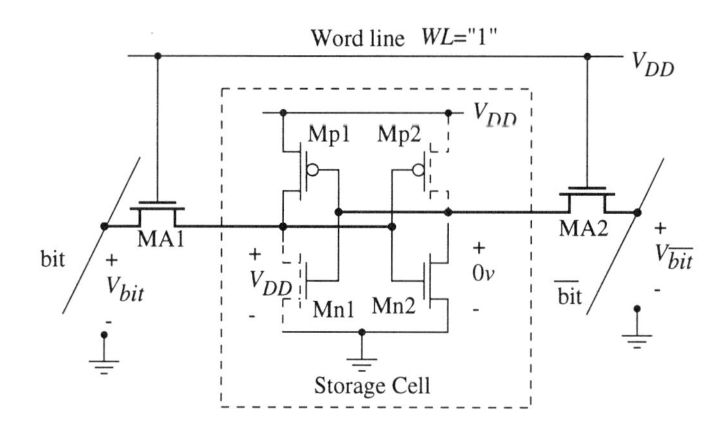

Figure 5.57 Read operation in an SRAM

$$V_{bit} \to V_{max} = (V_{DD} - V_{Tn})$$
$$V_{\overline{bit}} \to 0v \tag{5. 107}$$

In particular, the nFET induces a threshold voltage loss that prevents V_{bit} from reaching V_{DD}. To overcome this problem (and speed up the detection process), V_{bit} and $V_{\overline{bit}}$ are used as inputs to a sensitive differential sense amplifier that can detect the state.[4]

5.8.1 Receiver Latch

A simple extension of SRAM-type latching is shown in the receiver circuit of Figure 5.58. This uses a pair of cross-coupled inverters as a latch to guard against input noise fluctuations.

Consider the input voltage V_{in} shown in the circuit. In a stable state, the ideal values are either V_{in} =0v or V_{in} =V_{DD}. With a noise fluctuation of (ΔV), the worst case 0 and 1 logic voltages would be

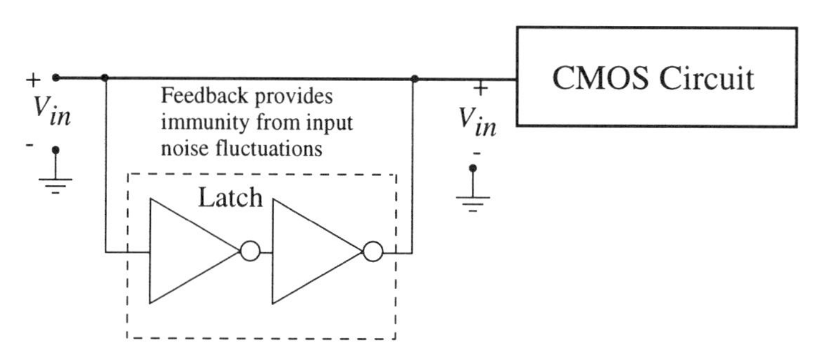

Figure 5.58 Receiver latching circuit

[4] Differential amplifiers are discussed in Chapter 9

$$V_{in,0} = (\Delta V)$$
$$V_{in,1} = V_{DD} - (\Delta V) \tag{5. 108}$$

Using the VTC for the cross-coupled circuit shown previously in Figure 5.56, we see that the state will be maintained so long as (ΔV) is reasonably small with a maximum value of about $(\Delta V) \approx V_{IL}$ or V_{IL} of the inverter. This circuit technique is quite useful in many situations owing to its simplicity and good performance characteristics.

5.9 Schmitt Trigger Circuits

A Schmitt trigger is a circuit whose voltage transfer characteristic exhibits **hysteresis** in which the forward characteristics are different from the reverse behavior. The symbol for a Schmitt trigger and its VTC are illustrated in Figure 5.59(a) and (b), respectively. When the input voltage is increased from 0*v* towards V_{DD}, the transition takes places at the **forward switching voltage** V^+. However, if the input voltage is initially at V_{DD} and is decreased towards 0*v*, the transition is defined by the **reverse switching voltage** V^-. The hysteresis voltage

$$V_H = V^+ - V^- \tag{5. 109}$$

gives the separation between the two switching points.

Figure 5.60 shows an example of how a Schmitt trigger can be used. Although the input voltage $V_{in}(t)$ exhibits small variations in its slope, the circuit resists switching output levels. This makes it useful for rejecting noise variations, or start-up transients when the power is applied. Another common application is in a receiver circuit where the signal enters from a transmission line.

A symmetrical CMOS Schmitt trigger circuit is shown in Figure 5.61; note that the pFET circuit can be viewed as a mirror-image of the nFET circuit when reflected about a horizontal line that passes through the output. The forward switching is controlled by the nFETs (M1, M2, and M3), while the reverse voltage V^- is determined by the pFETs (M4, M5, and M6). As will be seen in the analysis, the numerical values of the V^+ and V^- are set by relative sizes of certain transistors.

The forward switching voltage V^+ can be calculated by analyzing the sub circuit of Figure 5.62. Initially, V_{in} is at 0*v* and $V_{out} = V_{DD}$. Transistor M3 is biased active; it acts as a pass transistor and sets the voltage V_X at node X to an initial value of

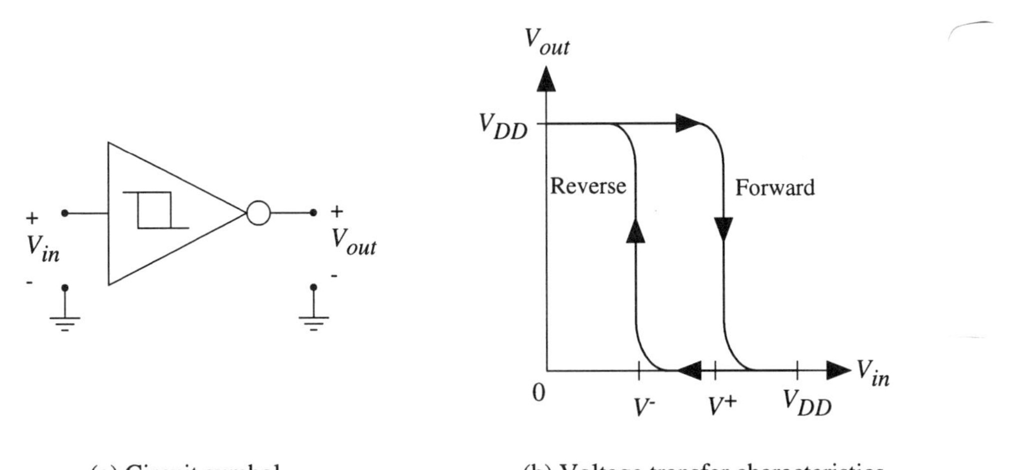

(a) Circuit symbol

(b) Voltage transfer characteristics

Figure 5.59 Characteristics of a Schmitt trigger

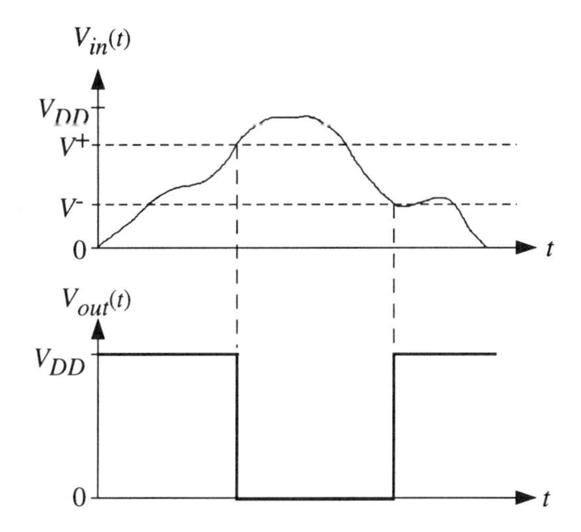

Figure 5.60 Waveform response of a Schmitt trigger

$$V_X = V_{DD} - V_{T3} \tag{5. 110}$$

where V_{T3} is the threshold voltage of M3. Now suppose that V_{in} is increased. When

$$V_{in} = V_{GS1} = V_{T1} \tag{5. 111}$$

M1 turns on. However, the gate-source voltage across M2 is given by

$$V_{GS2} = V_{in} - V_X \tag{5. 112}$$

so it is still in cutoff at this point.

The value of V^+ is based upon the following sequence. When $V_{in}=V^+$, M2 turns on and V_{in} will

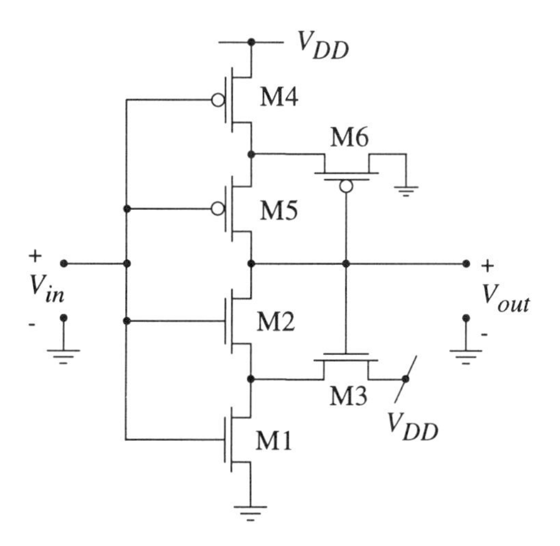

Figure 5.61 A CMOS Schmitt trigger circuit

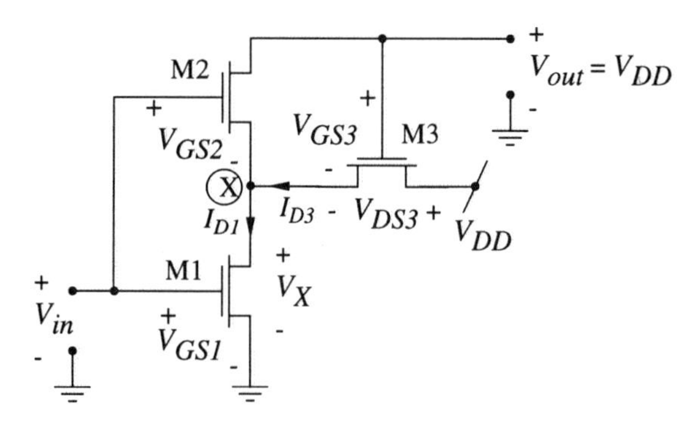

Figure 5.62 Subcircuit for V^+ calculation

fall to 0v since M1 and M2 provide a good conducting path to ground. However, when V_{in} is in the range

$$V_{T1} < V_{in} < V^+ \tag{5. 113}$$

M1 is on but M2 is off. Within this range, transistor M3 is used in the circuit as a controlled feedback device between the power supply rail V_{DD} and the node X. MOSFET M1 acts to lower the value of V_X as V_{in} is increased by providing a conducting path to ground, i.e., it is a pull-down device. In order to find an expression for V^+, we must first analyze the pull down mechanism in more detail. This will also give the design equations for the circuit.

The circuit section consisting of M1 and M3 is redrawn in Figure 5.63. The key to understanding the pull down effect is to find the functional dependence of V_X (V_{in}) as V_{in} is increased above the threshold voltage V_{T1} of M1. First, note that the voltages on M3 are defined by

$$\begin{aligned} V_{GS3} &= V_{DD} - V_X \\ &= V_{DS3} \end{aligned} \tag{5. 114}$$

Since V_{GS3}=V_{DS3}, the voltages on M3 automatically satisfy the condition

$$V_{DS3} > V_{sat,3} \equiv (V_{GS3} - V_{T3}) \tag{5. 115}$$

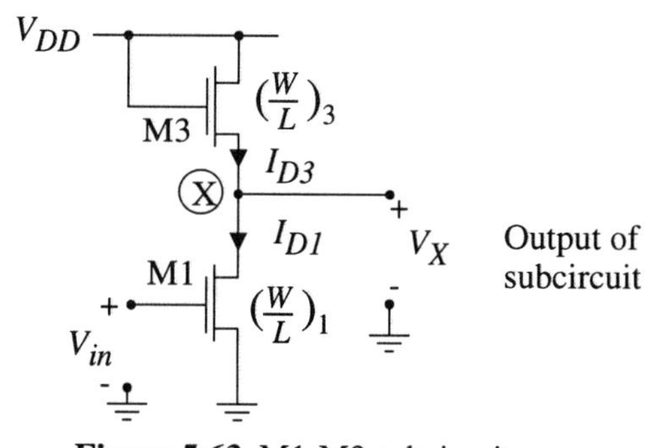

Figure 5.63 M1-M3 subcircuit

indicating that M3 is biased to remain in saturation. Also, for V_{in} slightly above V_{T1}, V_{out} will be relatively high, so that M1 will also be saturated. Equating the drain currents of M1 and M3 gives

$$\frac{\beta_{n1}}{2}(V_{in}-V_{T1})^2 = \frac{\beta_{n3}}{2}(V_{DD}-V_X-V_{T3})^2 \tag{5. 116}$$

where β_{n1} and β_{n3} are the respective transconductance values for M1 and M3. Solving for V_X yields

$$V_X = (V_{DD}-V_{T3}) - \sqrt{\frac{\beta_{n1}}{\beta_{n3}}}(V_{in}-V_{T1}) \,. \tag{5. 117}$$

If body-bias effects in V_{T1} and V_{T3} are ignored, then this shows that V_X decreases linearly with V_{in} so long as M1 is saturated. The rate of decrease is determined by the device ratio (β_{n1}/β_{n3}). If V_X falls to a value

$$V_X \le (V_{in}-V_{T1}) \,, \tag{5. 118}$$

then M1 enters the non-saturated region and the current flow equation is changed to

$$\frac{\beta_{n1}}{2}\left[2(V_{in}-V_{T1})V_X - V_X^2\right] = \frac{\beta_{n3}}{2}(V_{DD}-V_X-V_{T3})^2 \tag{5. 119}$$

which is a quadratic in V_X. The general dependence of V_X as a function of V_{in} is shown in Figure 5.64. This is important to the Schmitt trigger as it is a plot of the feedback voltage as the input is increased towards the forward switching point.

Let us now return to the original problem of calculating the value of V^+. Increasing V_{in} above V_{T1} causes V_X to fall. The series transistor M2 will turn on when its gate-source voltage reaches the value

$$\begin{aligned} V_{GS2} &= V_{T2} \\ &= (V_{in}-V_X) \\ &= \left(V^+ - V_X\right) \end{aligned} \tag{5. 120}$$

Rearranging gives the forward switching voltage as

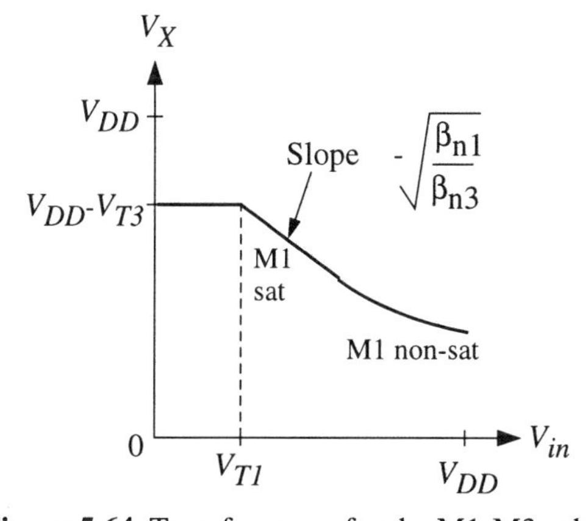

Figure 5.64 Transfer curve for the M1-M3 subcircuit

$$V^{+} = V_{T2} + V_X \tag{5. 121}$$

A closed form expression for V^{+} can be obtained by ignoring body-bias effects and assuming all threshold voltages are the same: $V_{T1}=V_{T2}=V_{T3}=V_{Tn}$. Assuming that M1 is saturated gives the current flow equation as

$$\begin{aligned}\frac{\beta_{n1}}{2}\left(V^{+}-V_{Tn}\right)^2 &= \frac{\beta_{n3}}{2}[V_{DD}-V_X-V_{Tn}]^2 \\ &= \frac{\beta_{n3}}{2}\left[V_{DD}-\left(V^{+}-V_{Tn}\right)-V_{Tn}\right]^2 \\ &= \frac{\beta_{n3}}{2}\left(V_{DD}-V^{+}\right)^2\end{aligned} \tag{5. 122}$$

so that, within the limits of the approximations,

$$V^{+} = \frac{V_{DD}+\sqrt{\frac{\beta_{n1}}{\beta_{n3}}}V_{Tn}}{1+\sqrt{\frac{\beta_{n1}}{\beta_{n3}}}} \tag{5. 123}$$

is our estimate for the forward switching voltage. As mentioned above, the value of V^{+} is determined by the relative sizes of M1 and M3 in the factor (β_{n1}/β_{n3}). Note that only the threshold voltage of M2 is important to the calculation; this is because M2 is the device that blocks the current flow between the output and ground. This equation is a reasonable estimate of the forward switching voltage even though it ignores body-bias effects. Body-bias can be added to arrive at a set of self-iterating equations; this is left as an exercise for the interested reader.

The reverse trigger voltage V^{-} is due to the action of M4, M5, and M6, with the transistors serving rolls analogous to that in the nFET circuit. In particular, M6 acts as the feedback device in conjunction with M4. Analyzing the circuit gives

$$V^{-} = \frac{\sqrt{\frac{\beta_{p4}}{\beta_{p6}}}(V_{DD}-|V_{Tp}|)}{1+\sqrt{\frac{\beta_{p4}}{\beta_{p6}}}} \tag{5. 124}$$

where body-bias effects have been ignored, and V_{Tp} is taken as a common pFET threshold voltage. It is seen that the reverse switching voltage is set by the ratio (β_{p4}/β_{p6}), which is analogous to the characteristics found for the nFET circuits.

A symmetrical Schmitt trigger can be defined by trigger voltages

$$\begin{aligned}V^{+} &= \left(\frac{1}{2}\right)V_{DD}+\Delta V \\ V^{-} &= \left(\frac{1}{2}\right)V_{DD}-\Delta V\end{aligned} \tag{5. 125}$$

This places V^{+} and V^{-} equidistant from the middle of the power supply voltage ($V_{DD}/2$), and corresponds to a hysteresis voltage of

$$V_H = 2(\Delta V) \tag{5. 126}$$

In a symmetrical design, we may set

$$\frac{\beta_{n1}}{\beta_{n3}} = \frac{\beta_{p4}}{\beta_{p6}} = \beta_R \tag{5. 127}$$

as the transconductance ratio for both the nFETs and the pFETs. Assuming that $V_{Tn}=|V_{Tp}| = V_T$ allows us to derive the voltage spacing ΔV from the above equations as

$$\Delta V = \frac{V(1-\sqrt{\beta_R}) + 2\sqrt{\beta_R}V_T}{2(1+\sqrt{\beta_R})} \tag{5. 128}$$

Alternately, we may rearrange this expression to the form

$$\beta_R = \left[\frac{V_{DD} - 2(\Delta V)}{V_{DD} + 2(\Delta V) - 2V_T}\right]^2 \tag{5. 129}$$

which can be used as the design equation.

The most important observation in designing this circuit is that the switching voltages are set by the device ratios (β_{n1}/β_{n3}) and (β_{p4}/β_{p6}); both are less than 1. A large value of (ΔV) means that we need weak feedback, which decreases the required value for both quantities. This in turn implies that M3 and M6 can get quite large.

Example 5-1

Suppose that $V_{Tn}=|V_{Tp}|=0.7v$ and $V_{DD} = 5v$. To design a symmetric inverter with $\Delta V=1v$ we need device ratios of

$$\beta_R = \left[\frac{5-1.4}{5+2-1.4}\right]^2 \approx 0.22$$

This implies that the feedback devices are about 5 times larger than the switching transistors. The small value is used to maintain a weak dependence of V_X on V_{in}. One problem with the circuit is that the series nFETs M1 and M2 need to be large to give a fast discharge, and the feedback nFET M3 must be even bigger, thus consuming a relatively large area. The same comment holds for the pFETs where the problem is even worse due to the smaller process transconductance value.

5.10 Tri-State Output Circuits

Static logic gates provide logic 0 and logic 1 output values by connecting the output node to either ground or to the power supply. A **tri-state** output circuit is designed to give these two logic states, but also provides for a third high-impedance (**Hi-Z**) state in which the output node is floating.

Figure 5.65(a) illustrates a tri-state circuit that uses the enable signal *En* to switch between normal and Hi-Z operation. The operation can be understood by noting that the gate of each FET is controlled by a signal from a logic gate. The pFET is controlled by the function

$$\begin{aligned} f_p &= \overline{D \cdot En} \\ &= \overline{D} + \overline{En} \end{aligned} \tag{5. 130}$$

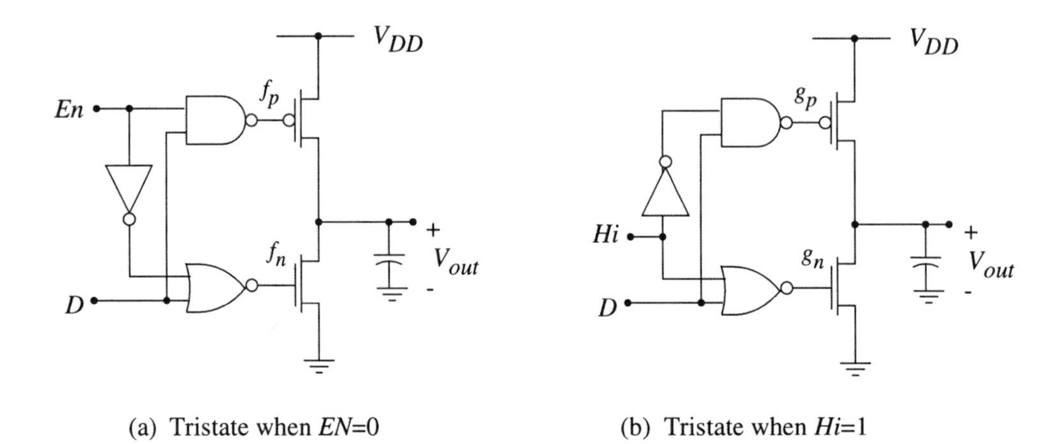

(a) Tristate when *EN*=0 (b) Tristate when *Hi*=1

Figure 5.65 Tri-state output circuits

while the nFET gate has

$$\begin{aligned} f_n &= \overline{D + \overline{En}} \\ &= \overline{D} \cdot En \end{aligned} \tag{5. 131}$$

controlling it. If *En*=0, then

$$f_p = 1 \qquad f_n = 0 \tag{5. 132}$$

which drives both FETs into cutoff, producing the Hi-Z state. On the other hand, a value of *En*=1 yields

$$f_p = \overline{D} \qquad f_n = \overline{D} \tag{5. 133}$$

which allows the input *D* to control the transistors. The circuit is Figure 5.65(b) reverses the roll of the tri-state control by moving the location of the inverter. The FET inputs are controlled by

$$\begin{aligned} g_p &= \overline{D} + Hi \\ g_n &= \overline{D} \cdot \overline{Hi} \end{aligned} \tag{5. 134}$$

which are easy to verify. This results in the circuit that gives a Hi-Z output state when the control bit *Hi* is 1 and normal operation with *Hi* =0.

Another tri-state gate is drawn in Figure 5.66. In this circuit, the Hi-Z control variable *X* is applied directly to the tri-state pFET MpX while $\overline{X}$ is applied to MnX. If *X*=0, then both FETs are active and the gate produces and output of $\overline{D}$. A Hi-Z state is achieve with *X*=1, since this turns both tri-state FETs OFF. This circuit can be used as the basis for creating a CMOS transmission gate, which is the subject of Chapter 6.

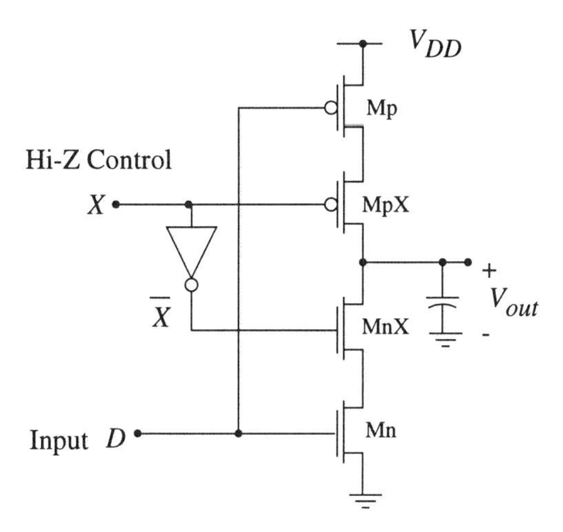

Figure 5.66 A tri-state output circuit

5.11 Pseudo-nMOS Logic Gates

An nMOS logic family is one that uses only nFETs for the circuits. Historically, nMOS preceded CMOS as the dominant technology,[5] but it is now obsolete. Pseudo-nMOS logic is a CMOS technique where the circuits resemble the older nFET-only networks. In order to place pseudo-nMOS into proper perspective, let us first examine the features of ordinary nMOS circuits to understand their characteristics.

An example of a basic nMOS inverter is shown in Figure 5.67. This uses a single nFET MD as a **driver** device that controls the circuit. The output node is connected to the power supply through a **load** resistor R_L that acts as a **pull-up** device, i.e., it always tries to pull the output voltage up to a value of V_{DD}. In more advanced nMOS designs, special transistors were used as load devices. The inverter operation can be understood by varying the input voltage. For $V_{in} < V_{Tn}$, the driver MD is in cutoff giving $I_D \approx 0$. Since the load current I_L is equal to the driver current, the voltage across the load resistor is $V_L = I_L R_L = 0v$. The output voltage is thus given by

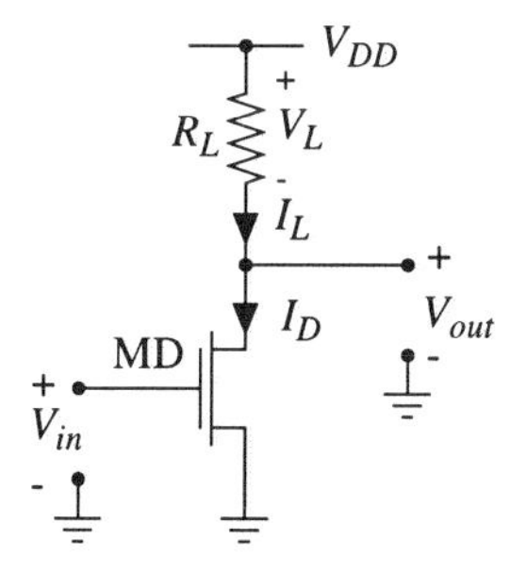

Figure 5.67 A resistive-load nMOS inverter

[5] For example, the Intel 8028/3866 microprocessors were based on advanced nMOS technology.

$$\begin{aligned} V_{OH} &= V_{DD} - V_L \\ &= V_{DD} \end{aligned} \tag{5. 135}$$

which is the output-high voltage for the circuit. When a high input voltage $V_{in} = V_{DD}$ is applied, MD conducts but the resistor still tries to pull up the output voltage. This keeps V_{out} from ever reaching $0v$ so that the output low voltage V_{OL} is always greater than zero: $V_{OL} > 0$. This is a characteristic of the circuit and cannot be overcome. Also note that a high input voltage creates a DC current flow path between the power supply and ground, contributing to power dissipation on the chip.

Pseudo-nMOS logic gates replace the resistor with a biased-on pFET as shown in Figure 5.68(a). Logic formation is achieved using only an nFET array that provides pull-down towards ground; the concept is shown in Figure 5.68(b). The primary advantage to this type of circuit is simplified interconnect wiring due to the absence of a pFET logic array. However, this comes at the cost of more complicated electrical behavior because the exact value of V_{OL} is set by the relative size of the FETs. As we will see below, to attain a small value of V_{OL}, we must use a large nFET driver in the circuit.

To illustrate the operation of a pseudo-nMOS gate, let us analyze the DC characteristics of the inverter shown in Figure 5.69(a). First note that the pFET voltages are given by

$$\begin{aligned} V_{SDp} &= V_{DD} - V_{out} \\ V_{SGp} &= V_{DD} \end{aligned} \tag{5. 136}$$

The value of the source-gate voltage indicates that the pFET is always biased into the active region and cannot be turned off; however, the actual value of the current is controlled by the nFET driver transistor. For $V_{in} < V_{Tn}$, the logic nFET Mn is in cutoff, and the output voltage is $V_{out} = V_{DD} = V_{OH}$ since Mp provides a strong conduction path to the power supply. Increasing V_{in} to a value above V_{Tn} drives Mn into conduction. The output low voltage is computed by setting $V_{in} = V_{OH} = V_{DD}$ and analyzing the circuit. Let us assume that V_{OL} is small, so that Mn is non-saturated; if $V_{OL} < |V_{Tp}|$ also holds, then Mp will be saturated. Equating the currents through the transistors gives

$$\frac{\beta_n}{2}\left[2\left(V_{DD} - V_{Tn}\right) V_{OL} - V_{OL}^2\right] = \frac{\beta_p}{2}\left(V_{DD} - |V_{Tp}|\right)^2. \tag{5. 137}$$

This is a quadratic equation with the solution

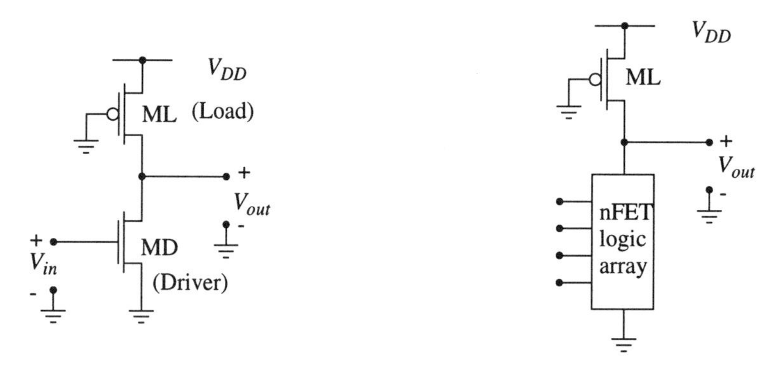

(a) Inverter circuit (b) Complex logic gate

Figure 5.68 Pseudo-nMOS logic gates

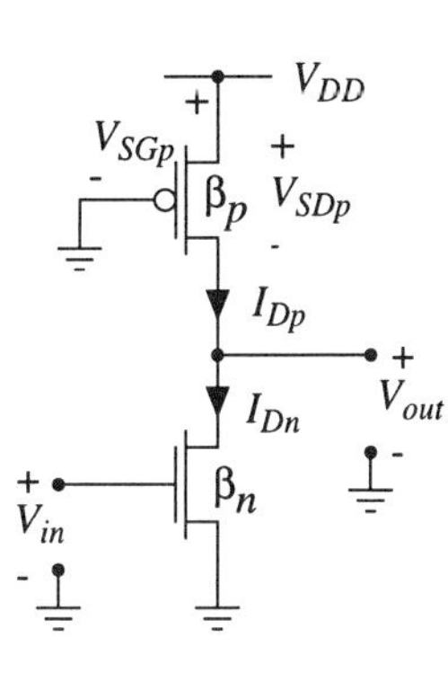

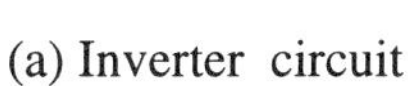

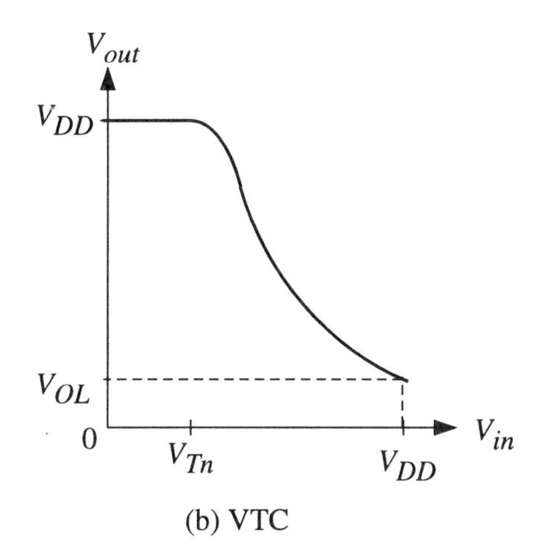

(b) VTC

Figure 5.69 Pseudo-nMOS voltage transfer characteristics

$$V_{OL} = (V_{DD} - V_{Tn}) - \sqrt{(V_{DD} - V_{Tn})^2 - \frac{\beta_p}{\beta_n}(V_{DD} - |V_{Tp}|)^2} \tag{5. 138}$$

for the value of the output-low voltage. Figure 5.69(b) shows the general shape of the voltage transfer curve for this circuit. The analysis above illustrates a few important points about the pseudo-nMOS inverter. First, it is not possible to achieve a value of $V_{OL} = 0v$ since the square root term can never equal $(V_{DD}\text{-}V_{Tn})$. Second, the value of V_{OL} depends upon the **driver-to-load** ratio

$$\frac{\beta_n}{\beta_p} = \frac{k'_n\left(\frac{W}{L}\right)_n}{k'_p\left(\frac{W}{L}\right)_p} \tag{5. 139}$$

such that a small V_{OL} requires a large driver-to-load ratio. Mathematically, this can be seen by noting that increasing the driver-to-load ratio moves the square root term closer to $(V_{DD}\text{-}V_{Tn})$. This corresponds to the physical viewpoint that we must make the nFET more conductive to pull it closer to ground voltage. Finally, note that with V_{in} at a high value, both transistors are conducting, establishing a DC current path between V_{DD} and ground; this implies that DC power dissipation occurs when the input is at a stable logic 1 level.

The transient switching times for the circuit are found by analyzing the circuits in Figure 5.70. Consider the case where the output capacitor initially has a voltage $V_{out}(0\ ) = V_{OL}$ and V_{in} is switched from V_{DD} towards ground. Since the driver Mn is in cutoff, C_{out} charges according to

$$I_{Dp} = C_{out}\frac{dV_{out}}{dt} \tag{5. 140}$$

as shown in Figure 5.69(a). The low-to-high time can be calculated using the integral

$$t_{LH} = C_{out}\int_{V_{OL}}^{V_{DD}} \frac{dV_{out}}{I_{Dp}(V_{out})}\,. \tag{5. 141}$$

This is identical in form to the inverter calculation, but has a different lower limit since $V_{OL} \neq 0$ in the pseudo-nMOS gate. Regardless, the low-to-high time given by

$$t_{LH} \approx s_p \tau_p \tag{5. 142}$$

still provides a reasonable approximation for hand estimates.

The discharge event shown in Figure 5.70(b) is complicated by the fact that the pFET is always biased into conduction and cannot be turned off. Applying KCL to the output node gives

$$I_{Dn} - I_{Dp} = -C_{out}\frac{dV_{out}}{dt} \tag{5. 143}$$

which shows that the presence of the pFET slows down the discharge since it adds charge to the capacitor. There is, however, one factor from the DC analysis that enters the problem. In a functional circuit, the driver-to-load ratio must be greater than 1, so that $\beta_n > \beta_p$. This condition implies that $I_{Dn} > I_{Dp}$, so that the pFET current is smaller than the nFET current. Owing to this observation, we often ignore I_{Dp} in hand estimates and write

$$t_{HL} \approx s_n \tau_n \tag{5. 144}$$

for a first-order approximation, and use a computer simulation if better accuracy is required.

5.11.1 Complex Logic in Pseudo-nMOS

Since the logic function of a pseudo-nMOS gate is determined by the structure of the nFET array, complex logic gates can be built by following the rules established for the complementary circuits. All pseudo-nMOS gates have the characteristic that $V_{OH} = V_{DD}$ but the value of V_{OL} is determined by a driver-to-load ratio.

Let us first examine the structure of a NOR gate. An m-input gate can be obtained by simply replacing the driver nFET by m parallel-connected driver transistors. Figure 5.71 shows a NOR3 pseudo-nMOS circuit constructed in this manner. The operation is straightforward to understand. If any of the three input voltages V_a, V_b, or V_c are high, then a conduction path is established between the output node and ground, resulting in $V_{out} = V_{OL}$. The worst-case value of V_{OL} occurs when only

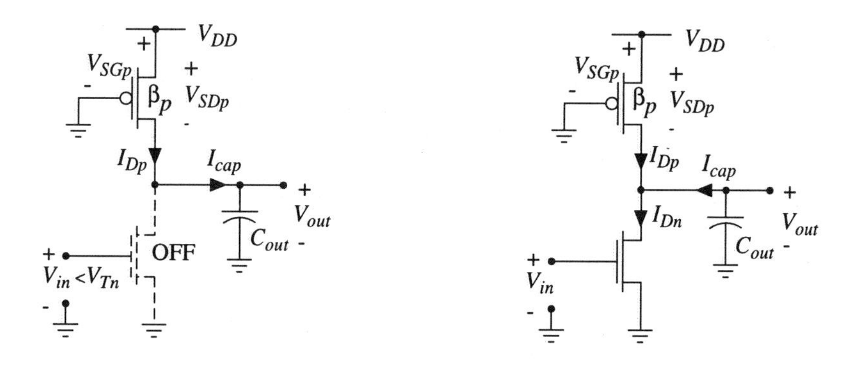

Figure 5.70 Subcircuits for transient response calculations

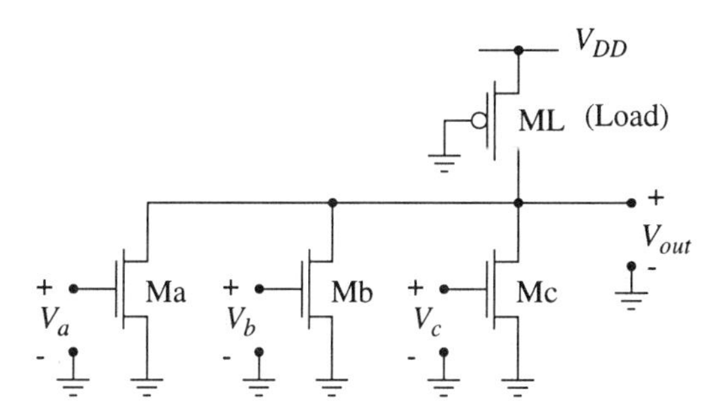

Figure 5.71 NOR3 gate in pseudo-nMOS

a single FET is biased on. This is identical to the inverter, and gives

$$V_{OL} = (V_{DD} - V_{Tn}) - \sqrt{(V_{DD} - V_{Tn})^2 - \frac{\beta_p}{\beta_n}(V_{DD} - |V_{Tp}|)^2} \tag{5.145}$$

where β_n is the device transconductance of the active nFET. The best-case (i.e., lowest) value of V_{OL} is when all three nFETs are biased on. Assuming identical drivers, the analysis gives

$$min(V_{OL}) = (V_{DD} - V_{Tn}) - \sqrt{(V_{DD} - V_{Tn})^2 - \frac{\beta_p}{3\beta_n}(V_{DD} - |V_{Tp}|)^2} \tag{5.146}$$

This value is of academic interest only, as one should always design around the worst-case situation. Both analyses show that the aspect ratios from the design of an inverter can be directly translated to a NOR3 gate.

A NAND gate requires series-connected nFETs, which complicates the circuit design somewhat. Figure 5.72(a) shows a NAND2 gate that is constructed by replacing the driver FET in the inverter with two nFETs. The logic function itself is automatically created by the topology: the output is connected to V_{DD} if either input (or both) are low. When both V_A and V_B are high, both nFETs are biased active and pull the output node to a voltage $V_{out} = V_{OL}$.

To calculate the value of V_{OL} for the NAND2, let us use the circuit voltages portrayed in Figure 5.72(b). First note that the KVL gives the output voltage as

$$V_{OL} = V_{DSA} + V_{DSB} \tag{5.147}$$

Assuming that V_{OL} is small (i.e., that the gate is properly designed), we may conclude that both MA and MB are non-saturated. Since both transistors have the same current, we can write two expressions for I_{Dn}:

$$\begin{aligned} I_{Dn} &= \frac{\beta_{nA}}{2}\left[2(V_{DD} - V_{Tn})V_{DSA} - V_{DSA}^2\right] && \text{(for MA)} \\ &= \frac{\beta_{nB}}{2}\left[2(V_{DD} - V_{Tn})V_{DSB} - V_{DSB}^2\right] && \text{(for MB)} \end{aligned} \tag{5.148}$$

where we will ignore body-bias effects in MA for simplicity. Since KCL gives

$$I_{Dn} = I_{Dp} = \frac{\beta_p}{2}(V_{DD} - |V_{Tp}|)^2 \tag{5.149}$$

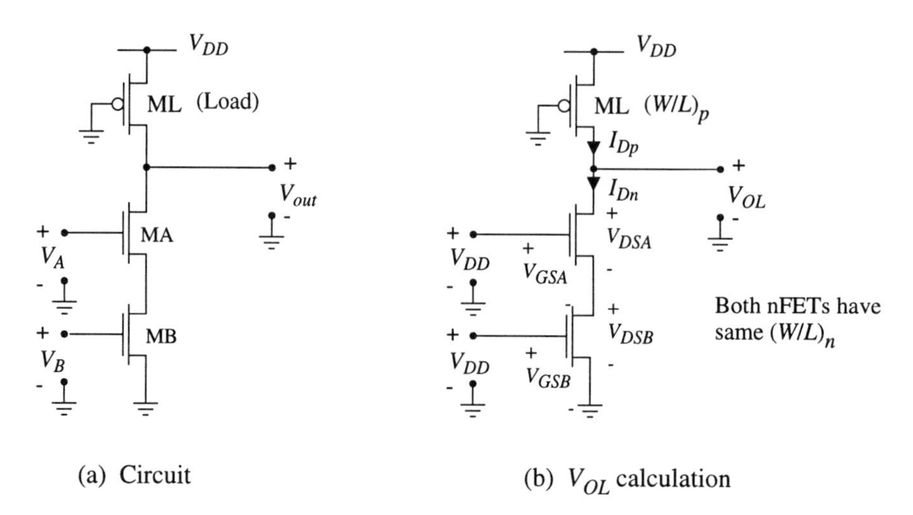

Figure 5.72 Pseudo-nMOS NAND2 gate

we can approximate the nFET relation as providing two independent quadratic equations for the drain-source voltages with solutions

$$V_{DSA} = (V_{DD} - V_{Tn}) - \sqrt{(V_{DD} - V_{Tn})^2 - \frac{\beta_p}{\beta_{nA}}(V_{DD} - |V_{Tp}|)^2}$$
$$V_{DSB} = (V_{DD} - V_{Tn}) - \sqrt{(V_{DD} - V_{Tn})^2 - \frac{\beta_p}{\beta_{nB}}(V_{DD} - |V_{Tp}|)^2} \tag{5. 150}$$

Assuming identical nFETs with $\beta_{nA}=\beta_{nB}$ allows us to sum the two and arrive at the output low voltage expression

$$V_{OL} = V_{DSA} + V_{DSB}$$
$$= 2\left[(V_{DD} - V_{Tn}) - \sqrt{(V_{DD} - V_{Tn})^2 - \frac{\beta_p}{\beta_{nA}}(V_{DD} - |V_{Tp}|)^2}\right] \tag{5. 151}$$

where we see that the additional nFET adds the factor of 2 multiplying the inverter value of V_{OL} (in the square brackets) for the same aspect ratios. To reduce V_{OL} this says that we must increase the sizes of the nFETs.

This result is a consequence of the ratioed nature of the V_{OL} circuit and can be understood by using LTI resistor equivalent-networks. Consider first the inverter equivalent circuit shown in Figure 5.73(a). The value of V_{OL} is give by the voltage divider relationship

$$V_{OL} = \left(\frac{R_D}{R_D + R_L}\right)V_{DD} \tag{5. 152}$$

so that a small value V_{OL} of requires that $R_D << R_L$. This is equivalent to the requirement that $\beta_D = \beta_n >> \beta_p$ in the more rigorous (correct) current flow analysis. If we construct a resistor-equivalent circuit for the NAND2 gate as in Figure 5.73(b), the output voltage becomes

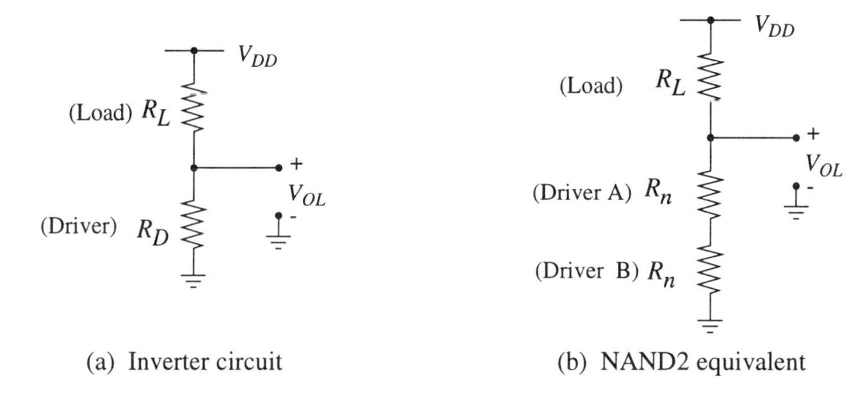

Figure 5.73 Resistive analogy to illustrate the behavior of ratioed circuits

$$V_{OL} = \left(\frac{2R_n}{2R_n + R_L}\right)V_{DD} \tag{5. 153}$$

since the two driver resistors are in series. Now we see that the important quantity is value of $2R_n$ as compared to the load resistor $R_L = R_p$. Although resistor-equivalent circuits cannot be used for accurate calculations, this analysis does imply that if we design the NAND2 circuit using nFETs with aspect ratios $\beta_{nA}=\beta_{nB}$ that are related to inverter value $\beta_D=\beta_n$ by

$$\beta_{nA} = 2\beta_n \tag{5. 154}$$

then the value of V_{OL} for the two would be approximately the same. This means that the aspect ratios should satisfy

$$\left(\frac{W}{L}\right)_{n,\ NAND2} = 2\left(\frac{W}{L}\right)_{n,\ INV} \tag{5. 155}$$

which would give

$$V_{OL} = 2\left[(V_{DD} - V_{Tn}) - \sqrt{(V_{DD} - V_{Tn})^2 - \frac{\beta_p}{2\beta_n}|V_{Tp}|^2}\right] \tag{5. 156}$$

for the circuit. Although this is not exactly the same as that for an inverter, it still represents a reasonable design. This procedure can be extended to an m-input NAND by using series-connected driver FETs that satisfy

$$\left(\frac{W}{L}\right)_{n,\ NANDm} = m\left(\frac{W}{L}\right)_{n,\ INV} \tag{5. 157}$$

This illustrates the fact that NAND gates are not area-efficient in pseudo-nMOS circuits. In addition, the value of t_{HL} will be relatively large due to the series resistance and inter-FET capacitance.

5.11.2 Simplified XNOR Gate

A particularly interesting pseudo-nMOS circuit is the XNOR gate shown in Figure 5.74. This achieves the equivalence function using only two logic transistors Mn1 and Mn2 and a single load pFET Mp. The operation of the circuit hinges upon the fact that the input voltages V_A and V_B establish the gate-source voltages on the logic transistors as

$$\begin{aligned} V_{GS1} &= V_B - V_A \\ V_{GS2} &= V_A - V_B \end{aligned} \quad (5.\ 158)$$

If $V_A = V_B$, then both nFETs are in cutoff since the gate-source voltages are given by $V_{GS1} = V_{GS2} = 0v$; the output voltage for this case is $V_{out} = V_{DD} = V_{OH}$.

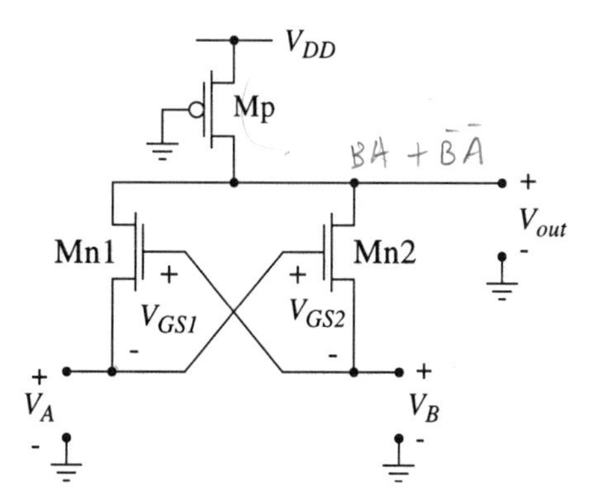

Figure 5.74 Compact XNOR circuit

If one of the input voltages is high while the other is low, then one of the nFETs is active and provides a pull-down path through transistors in the driving circuits (which are not part of the XNOR gate itself). Consider the case where the circuit is fed by inverters as shown in Figure 5.75. If we arbitrarily choose the case where V_A is small and $V_B = V_{DD}$, then Mn1 is active while Mn2 is in cutoff. Tracing the currents shows that

$$I_{Dp} = I_{D1} = I_{Dn} \quad (5.\ 159)$$

where I_{Dn} is the current through the driver FET on the left side. To analyze the circuit, we will make the reasonable assumption that V_{OL} is small enough to give a saturated pFET while leaving

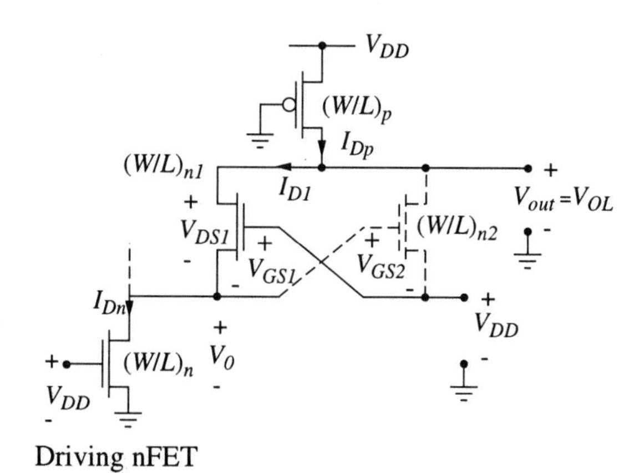

Figure 5.75 Subcircuit for V_{OL} calculation

both nFETs in the non-saturated mode. We may then write

$$\begin{aligned}\frac{\beta_p}{2}(V_{DD}-|V_{Tp}|)^2 &= \frac{\beta_{n1}}{2}\left[2(V_{GS1}-V_{Tn})V_{DS1}-V_{DS1}^2\right] \\ &= \frac{\beta_n}{2}\left[2(V_{DD}-V_{Tn})V_0-V_0^2\right]\end{aligned} \quad (5.160)$$

where the first line is due to M1 and the second expression arises from the driver FET. The output voltage is

$$V_{OL} = V_{DS1}+V_0 \quad (5.161)$$

This can be completed by solving the equations for each term. The second line gives us

$$V_0 = (V_{DD}-V_{Tn}) - \sqrt{(V_{DD}-V_{Tn})^2-\frac{\beta_p}{\beta_n}(V_{DD}-|V_{Tp}|)^2} \quad (5.162)$$

by directly solving the quadratic equation. To obtain V_{DS1}, we note that

$$V_{GS1} = V_{DD}-V_0 \quad (5.163)$$

so that the M1 quadratic has the solution

$$V_{DS1} = (V_{DD}-V_0-V_{Tn}) - \sqrt{(V_{DD}-V_0-V_{Tn})^2-\frac{\beta_p}{\beta_{n1}}(V_{DD}-|V_{Tp}|)^2} \quad (5.164)$$

Since we already know V_0 at this point, we can calculate V_{DS1}. Adding the two values thus gives the desired result for V_{OL}. Note that in this circuit, the ratioed property arises twice through the factors

$$\frac{\beta_p}{\beta_{n1}} \quad \text{and} \quad \frac{\beta_p}{\beta_n} \quad (5.165)$$

in the equations. In both cases, the nFET must be larger than the pFET to achieve small values.

A similar XNOR circuit can be constructed using complementary pairs as shown in Figure 5.76. In this configuration, the single pFET load is replaced by two series-connected pFETs that are switched by the input signals. If V_A and V_B are both low, then both nFETs are OFF while both pFETs are active. This allows a strong pull-up of the output to $V_{out} = V_{DD}$. On the other hand, if V_A and V_B are both high, then nFET Mn2 is responsible for transmitting the high voltage to the output. This limits the value to

$$V_{OH} = V_{max} = V_{DD}-V_{Tn} \quad (5.166)$$

due to the threshold loss through the n-channel device.

5.12 Compact XOR and Equivalence Gates

Several types of alternate XOR and XNOR circuits can be constructed in CMOS. This arises in part due to the balanced nature of the functions, and has been spurred on by extensive use of these gates in circuits such as adders and error correction/detection systems. The circuits in this section are classified as static logic gates, but are not based on the approach to logic that we have seen thus far. They have been developed in an effort to provide more compact circuits.

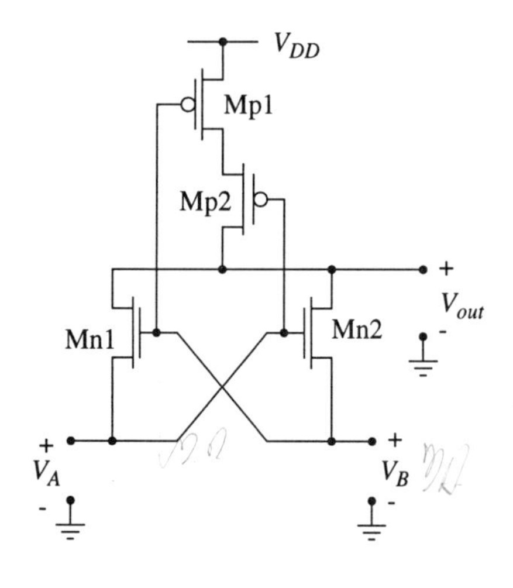

V_A	V_B	Mn1	Mp1	Mn2	Mp2	V_{out}
0	0	Off	On	Off	On	V_{DD}
0	V_{DD}	On	Off	Off	On	0v
V_{DD}	0	Off	On	On	Off	0v
V_{DD}	V_{DD}	Off	Off	On	Off	V_{max}

(a) Circuit (b) Operating summary

Figure 5.76 Operation of the complementary XNOR circuit

The first pair of gates is shown in Figure 5.77. Consider the XOR circuit shown in (a). The A input is applied to an inverter, and also to the top of pFET Mp2; the complement $\overline{A}$ from the inverter is applied to the bottom of nFET Mn2. The other input B is used to switch the FET pairs Mp2 and Mn2. With B=0, Mp2 conducts and A is transferred to the output; this produces the term $A \cdot \overline{B}$. If B=1, Mn2 is active and connects $\overline{A}$ to the output, giving $\overline{A} \cdot B$. ORing the two gives the function

$$A \oplus B = A \cdot \overline{B} + \overline{A} \cdot B \tag{5. 167}$$

as stated. The second circuit in Figure 5.77(b) reverses the rolls of A and $\overline{A}$, giving the XNOR function

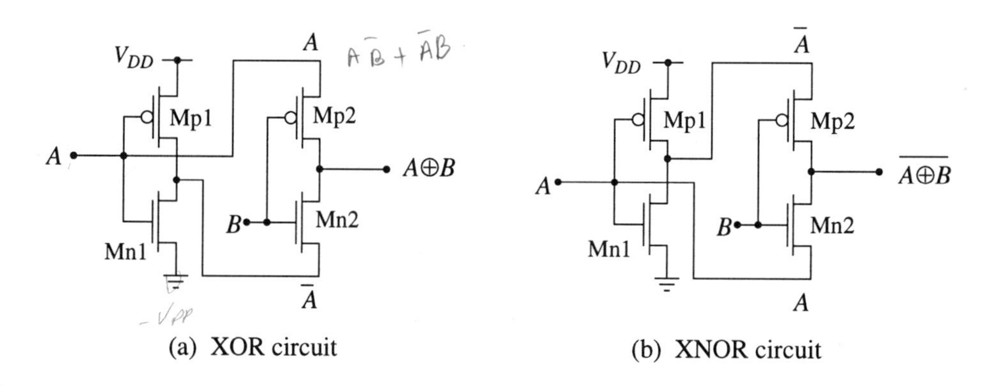

(a) XOR circuit (b) XNOR circuit

Figure 5.77 Alternate XOR and XNOR circuits

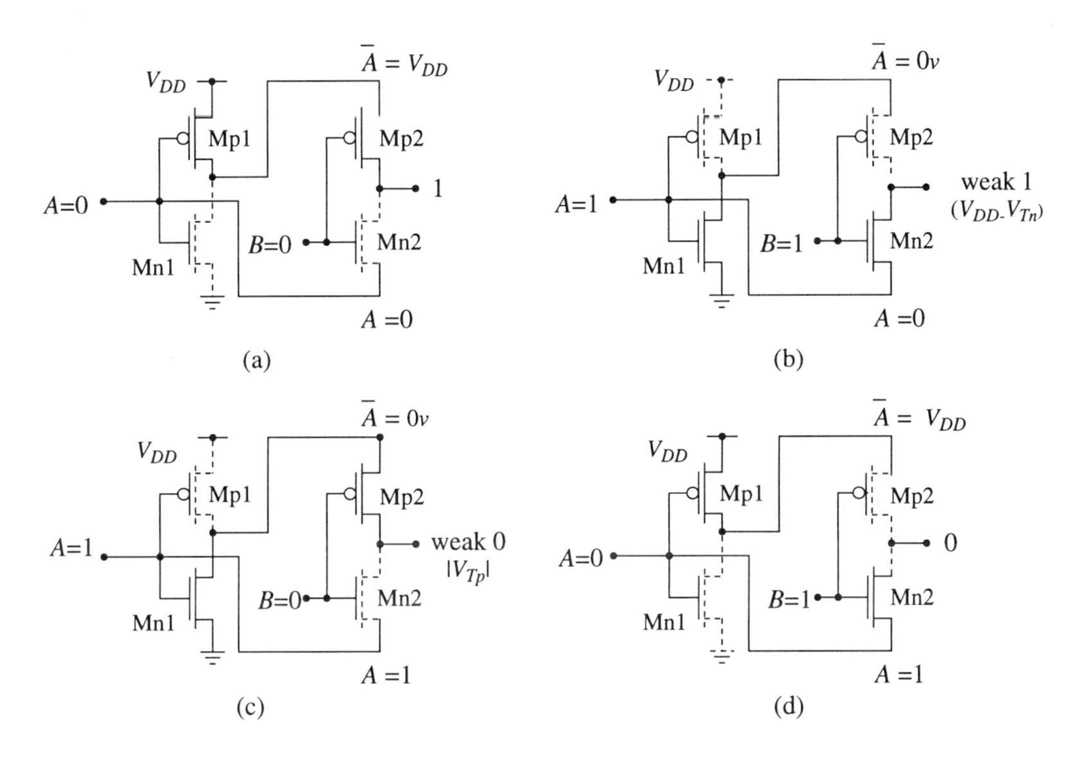

Figure 5.78 Summary of operational characteristics

$$\overline{A \oplus B} = A \cdot B + \overline{A} \cdot \overline{B} \tag{5. 168}$$

by direct calculation

Although the logic formation is straight forward, the electrical characteristics require a deeper study. As with the case of the circuits in the previous section, these do not exhibit full-rail output swings. Figure 5.78 portrays the XNOR circuit for each of the 4 possible input combinations. In (a), the input AB=00 yields a logic 1 at the output with a voltage value of V_{DD}. This is because the pFET Mp2 is used to pass the value of $\overline{A}$=1. However, with AB=11 as shown in (b), the high output is from the input voltage $A = V_{DD}$ which is being passed through nFET Mn2. This induces a threshold voltage loss such that the output voltage is reduced to

$$V_{max} = V_{DD} - V_{Tn} \tag{5. 169}$$

i.e., a weak logic 1. Figures 5.78(c) and (d) show the circuit when the inputs are not equal. In (c), the input combination AB=10 has a logic 0 output voltage that is created by connecting the drain of pFET Mp2 to ground. This restricts the output to discharging to a minimum level of

$$V_{min} = |V_{Tp}| \tag{5. 170}$$

which is a weak 0 voltage. If the inputs are reversed to AB=01, then the output can attain an ideal logic 0 level of 0v. This simple analysis demonstrates that the output voltage swing can be the limiting factor in using these circuits. In practice, a static inverter may be added to the output to rebuild the voltage levels. Designing an inverter threshold voltage of $V_I = (V_{DD}/2)$ by choosing $\beta_n = \beta_p$ would satisfy the necessary requirements.

A similar set of XOR/XNOR gates is shown in Figure 5.79. These use FETs as pass transistors and only have a direct connection to one of the power supply voltages V_{DD} or ground. The other level is obtained from the signal A. In Figure 5.79(a), the maximum output voltage is V_{max} since A must pass through Mp2. In (b), the logic 1 output with AB=10 is provided by passing $A = V_{DD}$ through the nFET Mn2. This results in a threshold loss and gives an output of V_{max}. The other combinations yield strong logic voltages. As before, the outputs can be reshaped to full rail values by using a standard CMOS inverter.

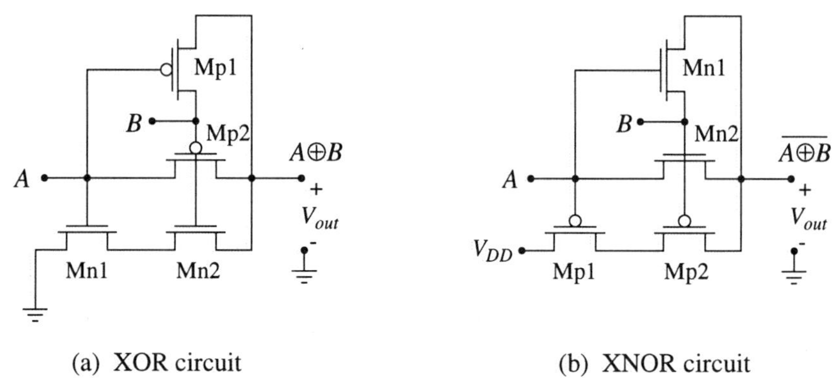

(a) XOR circuit (b) XNOR circuit

Figure 5.79 Alternate approach to XOR and XNOR circuits

The passFET switch logic formalism may be used to verify the operation of these somewhat unique circuit arrangements. For example, the circuit in Figure 5.79(a) gives an output of

$$A \cdot B \cdot 0 + \overline{B} \cdot A + \overline{A} \cdot B = A \oplus B \tag{5. 171}$$

which is the desired XOR function. In the same manner, the circuit in (b) evaluates to

$$\overline{A} \cdot \overline{B} \cdot 1 + A \cdot B + A \cdot B = \overline{A \oplus B} \tag{5. 172}$$

as required. The same technique can be applied to any other network to verify proper logical behavior. Note, however, that this technique cannot provide much information as to the electrical performance.

5.13 Problems

Use the following set of basic CMOS parameters for the problems unless other data is provided.

nFETs: $k_n' = 135\ \mu A/V^2$, $V_{T0n} = +0.7v$, $\gamma = 0.080 V^{1/2}$, $N_a = 10^{15} cm^{-3}$, $C_j = 0.4 fF/\mu m^2$, $C_{jsw} = 0.36 fF/\mu m$, $m_j = 0.5$, $m_{jsw} = 0.33$

pFETs: $k_p' = 65\ \mu A/V^2$, $V_{T0p} = -0.8v$, $\gamma = 0.051 V^{1/2}$, $N_d = 10^{16} cm^{-3}$, $C_j = 0.5\ fF/\mu m^2$, $C_{jsw} = 0.42 fF/\mu m$, $m_j = 0.5$, $m_{jsw} = 0.33$

Common parameters for the process are $x_{ox} = 150\ Å$, $V_{DD} = +3.3v$.

[5-1] A CMOS NAND2 gate is built using $(W/L)_n$=14 and $(W/L)_p$=12.

(a) Calculate the value of V_I for the case of simultaneous switching.

(b) Calculate the value of V_I for an inverter created using the same size FETs.

(c) What would be the nFET aspect ratio needed to force the NAND2 V_I found in (a) to be the same as the inverter value in (b)?

(d) The output capacitance is estimated to be $C_{out}=120fF$. The capacitance between the nFETs is estimated to be $C_X=32fF$. Find t_{HL} and t_{LH} for the gate.

[5-2] A CMOS NOR2 gate is built using $(W/L)_n=10$ and $(W/L)_p=20$.

(a) Calculatc thc valuc of V_I for the case of simultaneous switching.

(b) Calculate the value of V_I for an inverter created using the same size FETs.?

(c) The output capacitance is estimated to be $C_{out}=120fF$. The capacitance between the pFETs is estimated to be $C_X=40fF$. Find t_{HL} and t_{LH} for the gate.

[5-3] Design the circuitry for a CMOS logic gate that gives the function

$$f = \overline{A \cdot (B+C) + B \cdot D}. \tag{5. 173}$$

[5-4] Design a CMOS logic gate for the following functions

(a) $g = \overline{(A+B) \cdot (C+D) \cdot X}$

(b) $h = \overline{[x + (y \cdot z)] \cdot w}$

(c) $u = \overline{x + y + z \cdot w}$

[5-5] Use the function table in Figure P5.1 to find *f*. Then design a CMOS gate for this function.

A	B	C	f
0	0	0	0
0	0	1	1
0	1	0	1
0	1	1	0
1	0	0	0
1	0	1	0
1	1	0	1
1	1	1	1

Figure P5.1

x	y	z	g
0	0	0	0
0	0	1	0
0	1	0	1
0	1	1	1
1	0	0	0
1	0	1	0
1	1	0	0
1	1	1	0

Figure P5.2

u	v	w	T
0	0	0	1
0	0	1	0
0	1	0	0
0	1	1	0
1	0	0	1
1	0	1	0
1	1	0	1
1	1	1	0

Figure P5.3

[5-6] Use the function table in Figure P5.2 to find *g*. Then design a CMOS gate for this function.

[5-7] Use the function table in Figure P5.3 to find *T*. Then design a CMOS gate for this function.

[5-8] Construct a CMOS half-adder that uses inputs a_n and b_n to produce the sum s_n and carry-out c_{n+2}. Use a CMOS mirror circuit.

[5-9] Construct a full-adder using mirror circuits and compare the structure to the AOI approach. What is the difference in FET count?

[5-10] Consider the NAND2 gate shown in Figure P5.4.

(a) Find the value of V_M for the case of simultaneous switching.

(b) Find t_{HL} for the circuit

(c) Find t_{LH} for the circuit.

[5-11] Perform a SPICE simulation on the circuit shown in Figure P5.4. Include both .DC and .TRAN simulations.

[5-12] Consider the NOR2 gate shown in Figure P5.5.

(a) Find the value of V_M for the case of simultaneous switching.

(b) Find t_{HL} for the circuit

(c) Find t_{LH} for the circuit.

[5-13] Perform a SPICE simulation on the circuit shown in Figure P5.5. Include both .DC and .TRAN simulations.

[5-14] Design a Schmitt trigger circuit that has $V^+ = 1.8v$ and $V^- = 1.1v$.

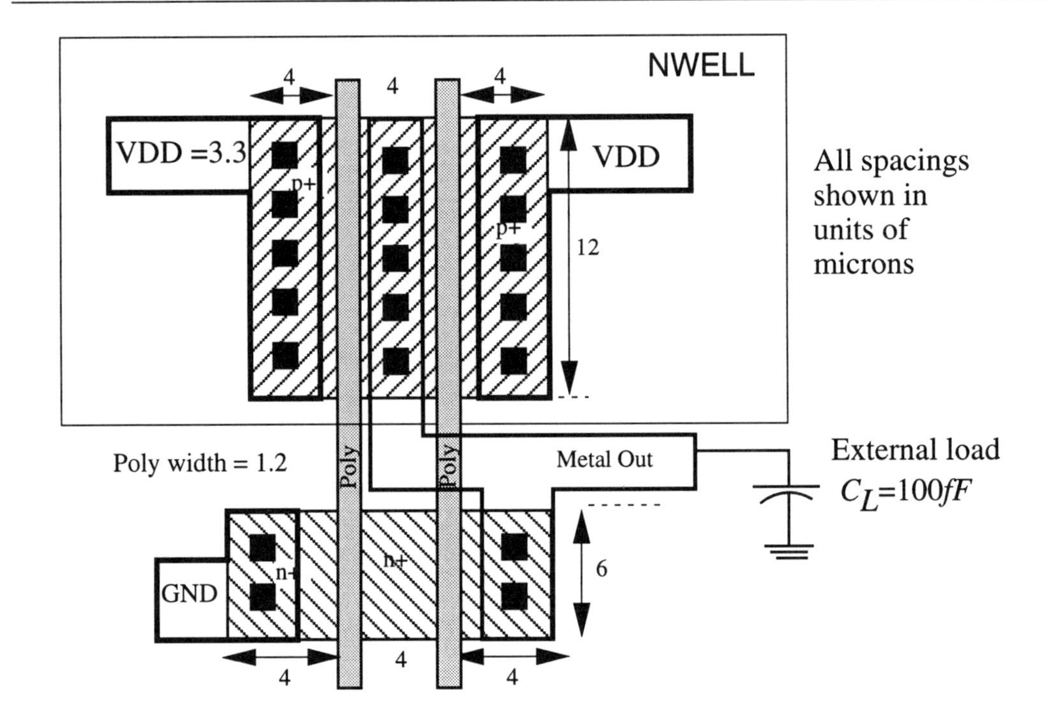

Figure P5.4

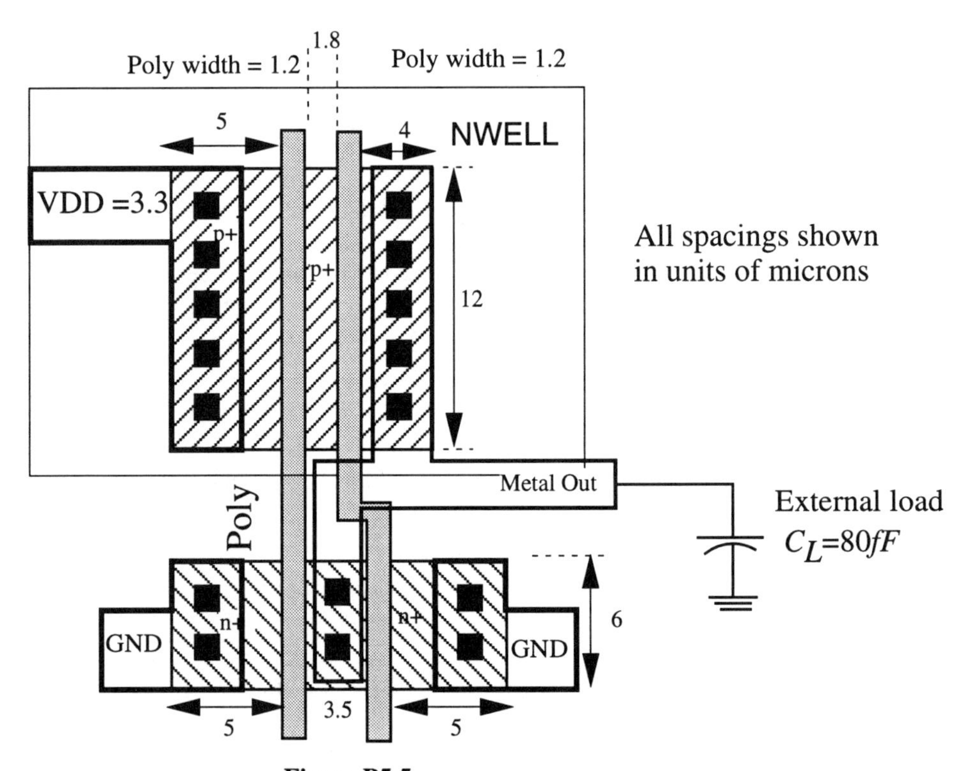

Figure P5.5

Chapter 6

Transmission Gate Logic Circuits

Conventional static logic gates provide the foundation for many system designs. CMOS, however, allows for the designer to choose from several different types of logic circuits. Some provide greater flexibility while others may give superior performance. In this chapter we will examine a class of CMOS logic circuits that are based on the concept of an ideal switch using a pair of MOSFETs wired in parallel to form a **transmission gate** (TG). This provides us with an alternate approach to building static logic circuits.

6.1 Basic Structure

The structure of a CMOS transmission gate is shown in Figure 6.1. It consists of an nFET Mn in parallel with a pFET Mp such that the gates are controlled by the complementary voltages

V_G applied to the nFET, and

$(V_{DD} - V_G)$ applied to the pFET.

The TG is designed to act as a voltage-controlled switch. When V_G is high, both Mn and Mp are biased into conduction and the switch is closed; this gives an electrical conduction path between the left and right sides. If V_G is low, then both MOSFETs are in cutoff and the switch is open; in this case, there is no direct relationship between the voltages V_A and V_B.

The philosophy for using a parallel combination of an nFET and a pFET is straightforward to understand. Recall from Chapter 4 that an nFET cannot pass a strong logic 1 voltage, while a pFET cannot pass a strong logic 0 voltage. By paralleling the two devices, the full voltage range from 0*v* to V_{DD} can be transmitted. Setting $V_G = V_{DD}$ gives the ideal situation where V_B will reach the same value as V_A.

Figure 6.2 shows the transmission gate logic symbol that will be used in this book. It is created using a pair of oppositely directly arrowheads to stress the fact that the device is bidirectional. From the circuit viewpoint, this means that current flow can be established in either direction. Conduction from one side to the other is controlled by the complementary switching signals X and $\overline{X}$

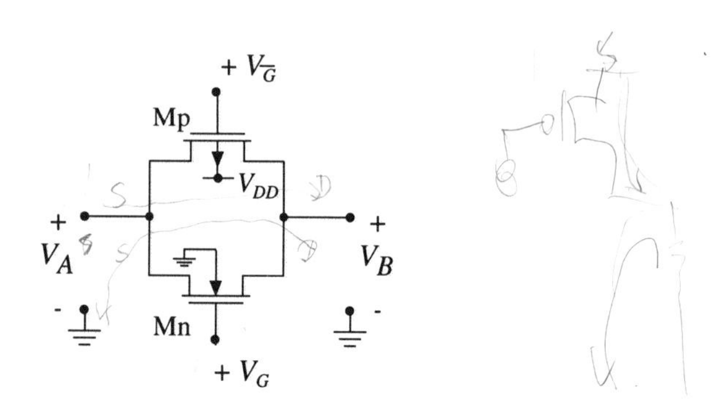

Figure 6.1 CMOS transmission gate (TG)

that are applied to the gates of the nFET and pFET, respectively; in the symbol, the bubble indicates the gate of the pFET. By definition, the TG acts as a closed switch when X=1, while it acts as an open switch when X=0. This can be expressed by writing

$$X = 1:\ A \rightarrow B \tag{6. 1}$$

which gives the condition $X = 1$ (on the left side of the colon) needed to obtain the bit transfer $A \rightarrow B$ action (on the right side of the colon). An alternate logic expression that is more useful for describing and analyzing TG-based logic networks is given by

$$B = A \cdot X \tag{6. 2}$$

which is valid iff X=1. Note that since we have referenced the switch control variable X to the nFET gate, the logic expressions are identical to those developed in Chapter 4 to describe nFET pass transistors. Logically, the two are interchangeable, but the electrical characteristics are distinct.

There are two areas of interest when studying TG-based CMOS networks. First, it is important to understand the circuit aspects of the parallel nFET-pFET pair, since this establishes the DC, transient, and layout characteristics of the switch. The second aspect deals with the use of TGs in constructing various logic gates and networks. As we will see later in the chapter, transmission gate logic provides a unique approach to building many useful logic functions and switching operations.

6.1.1 The TG as a Tri-State Controller

Before progressing deeper into the analysis, it is useful to examine the transmission gate in the context of static logic circuits. Figure 6.3 shows a tri-state static inverter circuit that is controlled by the

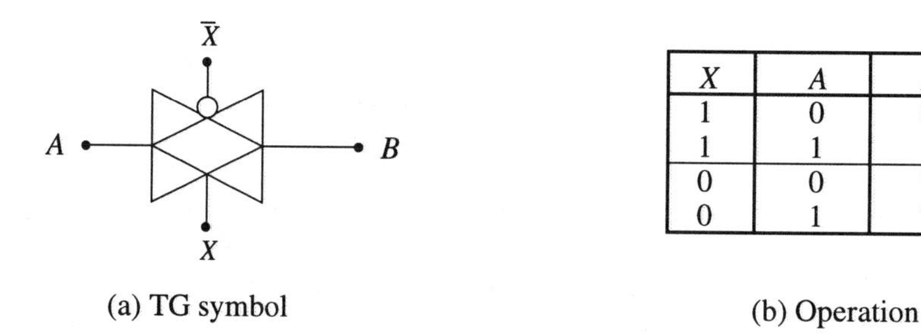

X	A	B
1	0	0
1	1	1
0	0	?
0	1	?

(a) TG symbol

(b) Operation

Figure 6.2 Transmission gate symbol

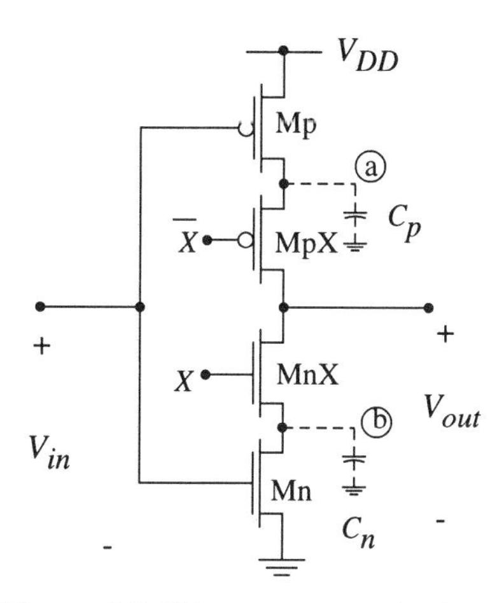

Figure 6.3 Tri-state output circuit

variable X. This circuit operates as an inverter when X=1, but produces the Hi-Z state when X=0.

The operational modes of the circuit are summarized in Figure 6.4. When X=1 as in (a), both of the central FETs MnX and MpX are active, and the output voltage is set by the value of V_{in}. In this case, the circuit is an inverter with additional parasitics due to MnX and MpX as shown in Figure 6.4(b).

If, on the other hand, if X=0 then both MnX and MpX are in cutoff, isolating the output node. This is shown in Figure 6.5(a). The relationship to the TG circuit is shown in Figure 6.5(b) where we have deformed the circuit and connected nodes a and b together to yield an inverter with its output directed through a transmission gate. The TG may thus be used as a switch to control the data flow through a static logic network. However, it is possible to use the transmission gate as a general logic-controlled switch to synthesize complex logic functions.

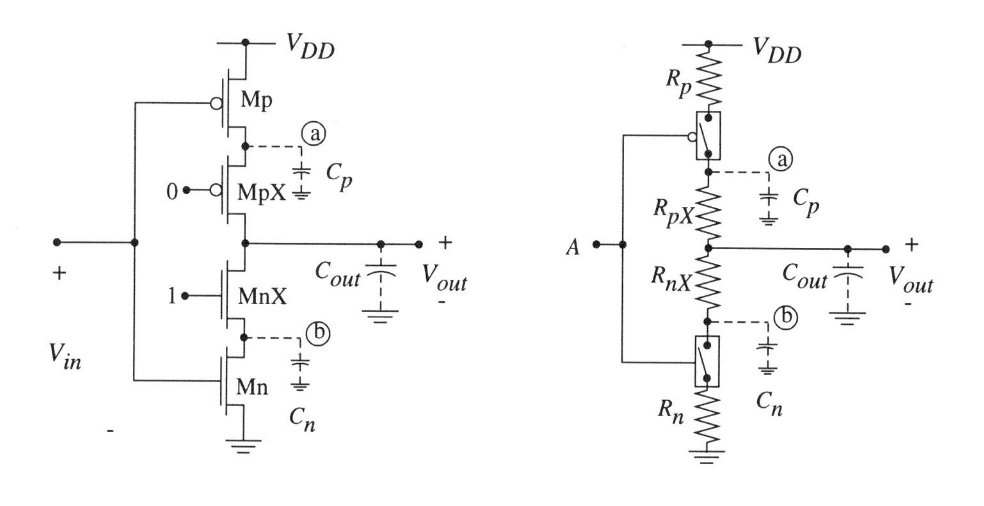

(a) Normal operation

(b) RC equivalent circuit

Figure 6.4 Normal output from a tri-state circuit

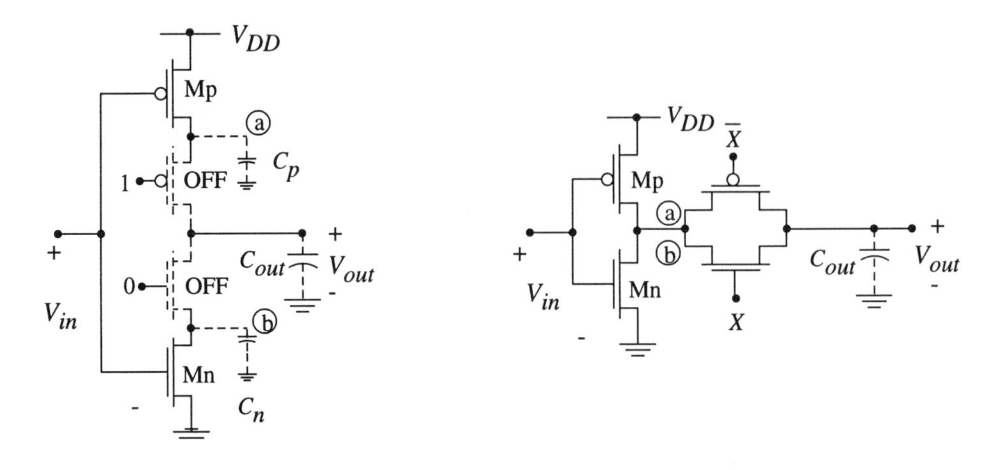

(a) Isolated output node

(b) Relationship to TG output

Figure 6.5 High impedance state and TG analogy

6.2 Electrical Analysis

Let us examine the problem of transmitting logic 1 and logic 0 voltages through a transmission gate that has a capacitive load as shown in the circuit of Figure 6.6. The input voltage is assumed to be V_{in}, while the output voltage V_{out} is taken across the capacitor C_{out}. It is assumed that both transistors are biased into conduction with V_{DD} applied to the gate of the nFET, and $0v$ applied to the gate of the pFET. Using KCL at the output node with the transistor currents as shown gives

$$I_{cap} = C_{out}\frac{dV_{out}}{dt} = I_{Dn} + I_{Dp} \tag{6. 3}$$

for the output voltage $V_{out}(t)$. The complicating factor in solving this equation is that both I_{Dn} and I_{Dp} depend upon V_{in} and V_{out}, and that the conduction states of MOSFETs change as the output capacitance C_{out} is charged or discharged. To understand the behavior of the transmission gate, we will separately study the cases for $V_{in} = V_{DD}$ and $V_{in} = 0v$ which correspond respectively to transferring a logic 1 through the TG, followed by a logic 0 transfer.

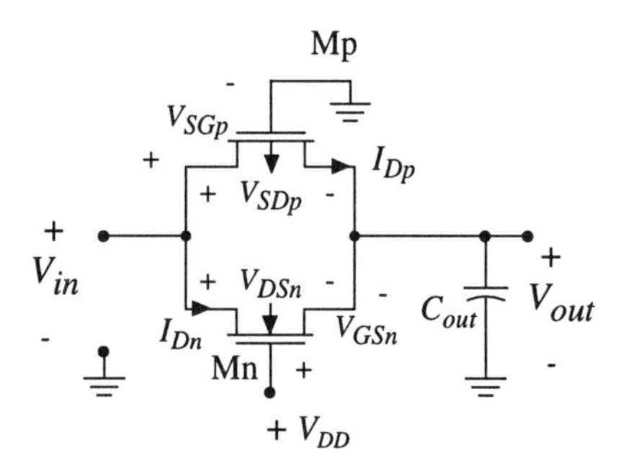

Figure 6.6 General voltage and current definitions

6.2.1 Logic 1 Transfer

Transferring a logic 1 through the TG is defined by setting $V_{in}=V_{DD}$ at the input. This forces current through the TG to charge C_{out}. We will assume the initial condition $V_{out}(0)=0v$, which defines the respective drain and source sides for both MOSFETs as shown in Figure 6.7.

Let us examine the states of conduction for the logic 1 transfer. Consider first the pFET Mp, which is connected to pass a high voltage V_{DD}. Since

$$V_{SGp} = V_{DD} \\ V_{SDp} = V_{DD} - V_{out}(t) \tag{6.4}$$

Mp starts conducting in the saturation region of operation with

$$I_{Dp} = \frac{\beta_p}{2}(V_{DD} - |V_{Tp}|)^2 \tag{6.5}$$

However, V_{out} increases with time, which decreases V_{SDp}. Since the saturation voltage is given by $V_{sat} = (V_{DD} - |V_{Tn}|)$, the pFET changes from saturation to non-saturation when $V_{out} = |V_{Tp}|$. After this occurs,

$$I_{Dp} = \frac{\beta_p}{2}[2(V_{DD} - |V_{Tp}|)(V_{DD} - V_{out}) - (V_{DD} - V_{out})^2] \tag{6.6}$$

until V_{out} reaches V_{DD}.

The nFET operates in a different manner, as it cannot pass the high voltage V_{DD}. In general, the Mn is characterized by terminal voltages of

$$V_{GSn} = V_{DD} - V_{out}(t) \\ V_{DSn} = V_{in} - V_{out}(t) \\ = V_{DD} - V_{out}(t) \tag{6.7}$$

so that for arbitrary times t,

$$V_{GSn} = V_{DSn} \tag{6.8}$$

Since $V_{sat} = (V_{DSn} - V_{Tn})$ and V_{Tn} is a positive number, $V_{DSn} > V_{sat}$ is satisfied so long as the transistor is biased active. Mn thus remains saturated with

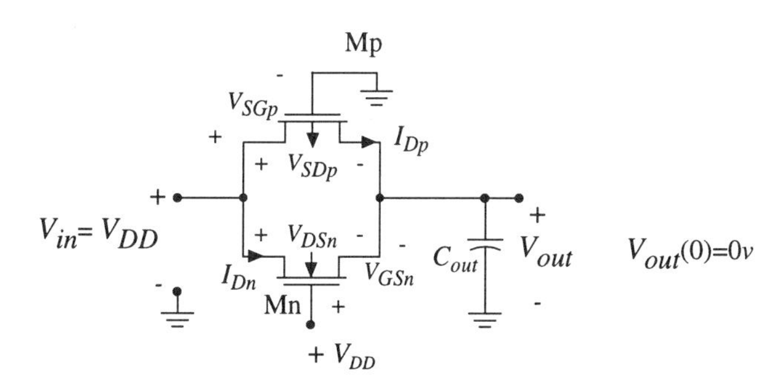

Figure 6.7 Logic 1 transfer

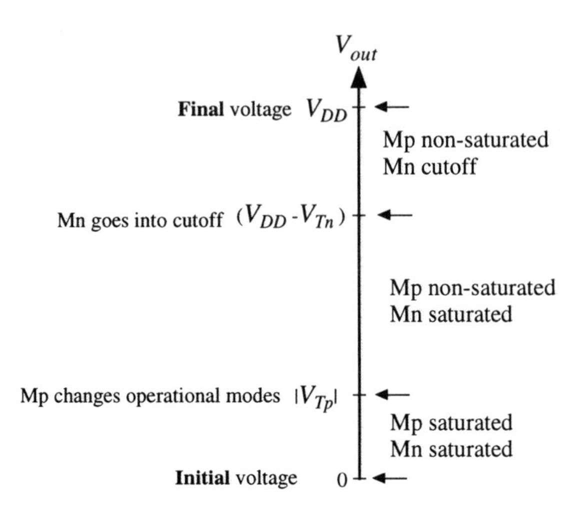

Figure 6.8 FET regions of operation for a logic 1 transfer

$$I_{Dn} = \frac{\beta_n}{2}(V_{DD} - V_{out} - V_{Tn})^2 \tag{6. 9}$$

so long as with $V_{GSn} > V_{Tn}$. However, when $V_{out} \geq (V_{DD} - V_{Tn})$, the gate-source voltage falls to a level $V_{GSn} < V_{Tn}$ which forces Mn into cutoff. During this time, C_{out} is charged entirely by the pFET Mp. The FET regions of operation are summarized by the plot in Figure 6.8.

6.2.2 Logic 0 Transfer

Now let us examine the opposite case where we apply an input 0 voltage of $V_{in} = 0v$ (i.e., it is grounded) with the output capacitor initially at a logic 1 voltage value of $V_{out}(0) = V_{DD}$. The circuit for this case is shown in Figure 6.9. By inspection, we see that the nFET voltages are now given by

$$\begin{aligned} V_{GSn} &= V_{DD} \\ V_{DSn} &= V_{out}(t) \end{aligned} \tag{6. 10}$$

so that Mn initially conducts in saturation with

$$I_{Dn} = \frac{\beta_n}{2}(V_{out} - V_{Tn})^2 \tag{6. 11}$$

Since the output capacitor C_{out} is discharging, Mn changes to non-saturation when V_{out} falls to a value of $(V_{DD} - V_{Tn})$. For the remainder of the discharge event,

$$I_{Dn} = \frac{\beta_n}{2}\left[2(V_{DD} - V_{Tn})V_{out} - V_{out}^2\right] \tag{6. 12}$$

describes the nFET current.

The pFET transistor Mp has terminal voltages of

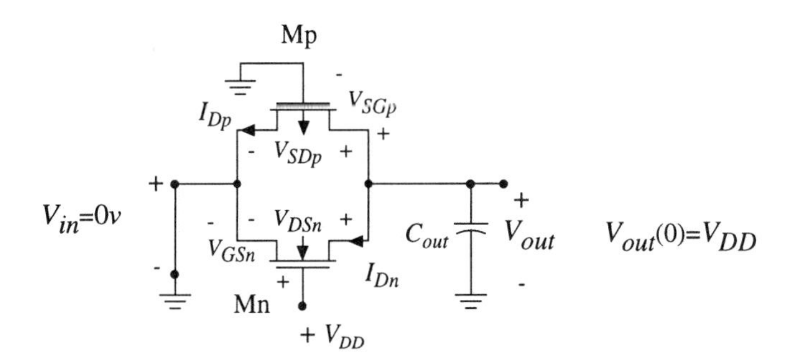

Figure 6.9 Device voltages for a logic 0 transfer

$$V_{SGp} = V_{out}(t)$$
$$V_{SDp} = V_{out}(t) \quad (6.13)$$

so that it is initially saturated with

$$I_{Dp} = \frac{\beta_p}{2}(V_{out} - |V_{Tp}|)^2 \quad (6.14)$$

which is valid so long as $V_{SGp} = V_{out} \geq |V_{Tp}|$. Since the nFET allows for a complete discharge of C_{out}, Mp will go into cutoff when V_{out} falls below $|V_{Tp}|$, and $I_{Dp}=0$. The operational modes of the transistors for this case are summarized in Figure 6.10.

We can, in principle, use the above equations for the transistor currents and solve

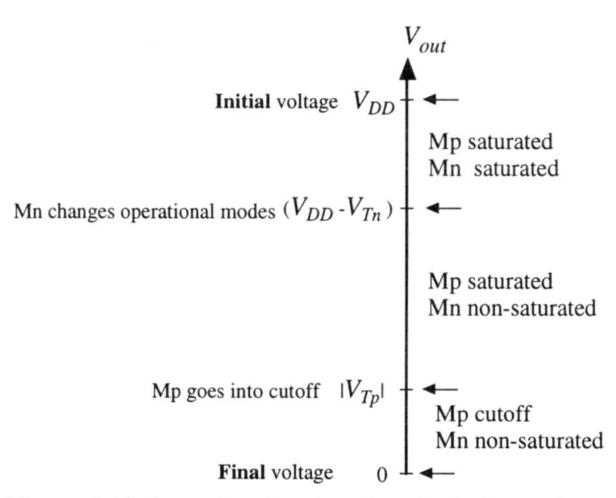

Figure 6.10 Operational regions for a logic 0 transfer

$$I_{Dn} + I_{Dp} = C_{out}\frac{dV_{out}}{dt} \quad (6.15)$$

in each voltage range. The results may then be pieced together to model the entire process. However, it is much more efficient to use either a simplified RC model for design estimates, or a full SPICE simulation for accurate analysis as the situation requires. Since SPICE is straightforward to apply, we will concentrate on modelling the circuit here.

6.3 RC Modelling

The simplest model for a TG is the resistor-switch combination illustrated in Figure 6.11. Logic transfer is controlled by $(X, \overline{X})$ such that $(X, \overline{X}) = (1, 0)$ gives a closed circuit and data transmission, while $(X, \overline{X}) = (0, 1)$ blocks the data path. Let us denote the equivalent transmission gate resistance by R_{TG} and assume that it can be represented by the linear, time-invariant resistor shown. The capacitances that are denoted by C_A and C_B originate from the MOSFETs in this model, and consist of both MOS and depletion capacitance contributions.

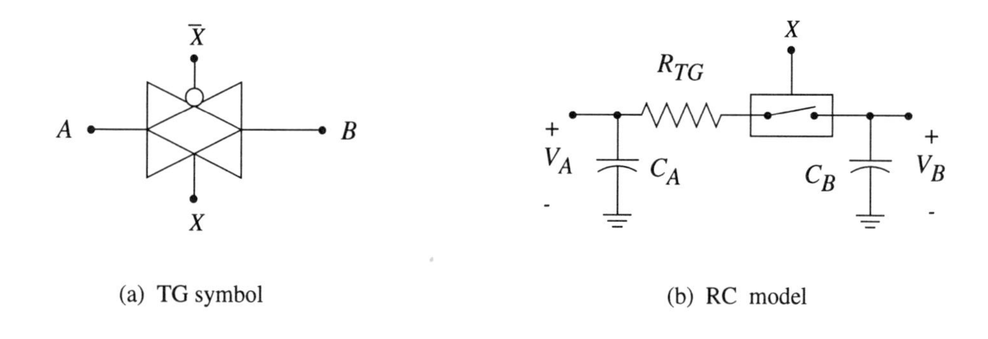

Figure 6.11 Basic RC model for the transmission gate

6.3.1 TG Resistance Estimate

Consider the general application of the RC model shown in Figure 6.12. In this circuit, the input voltage is taken to have step-like characteristics of the form

$$V_{in}(t) = V_{DD}u(t) \quad (6.16)$$

for a 0-to-1 transition, where $u(t)$ is the unit step function. Alternately, we may write

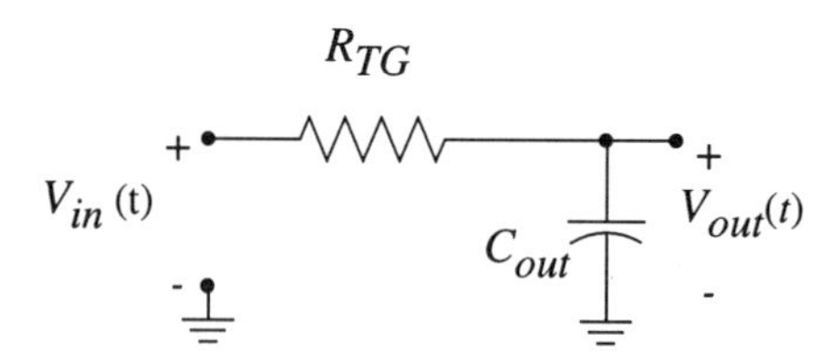

Figure 6.12 RC transmission gate model

$$V_{in}(t) = V_{DD}[1-u(t)] \tag{6.17}$$

for a 1-to-0 input. Transfer of a logic 1 state through the TG is equivalent to charging the capacitor C_{out} through the resistor R_{TG}, so that the voltage may be described by the simple RC exponential solution

$$V_{out}(t) = V_{DD}\left[1-e^{-t/\tau_{TG}}\right] \tag{6.18}$$

In this equation,

$$\tau_{TG} = R_{TG}C_{out} \tag{6.19}$$

is the transmission gate time constant for the RC model. In the same manner, a logic 0 transfer corresponds to discharging the capacitor and the voltage is given by

$$V_{out}(t) = V_{DD}e^{-t/\tau_{TG}} \tag{6.20}$$

The input and output voltage plots for this model are shown in Figure 6.13. The transfer time is limited by the time constant, which is in turn set by the geometry and layout. Note that both transitions are characterized by the same time constant within this model.

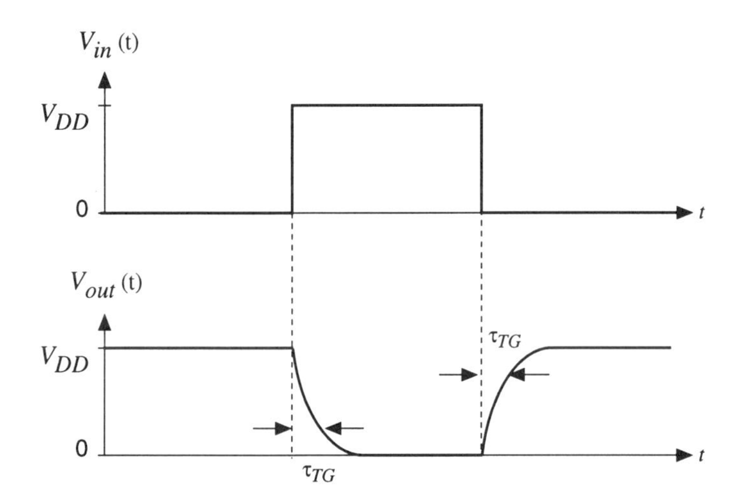

Figure 6.13 Switching waveforms for the simple RC model

6.3.2 Equivalent Resistance

The equivalent transmission gate resistance R_{TG} can be defined by

$$R_{TG} = \frac{V_{TG}}{I_{Dn}+I_{Dp}} \tag{6.21}$$

where V_{TG} is the voltage across the TG. The MOSFETs are nonlinear, so that R_{TG} is itself a nonlinear function of the voltages. Although the functional dependence can be determined, it is easier to approximate R_{TG} as a constant. This is sufficient for initial design or analysis estimates; computer simulations may be used for greater accuracy.

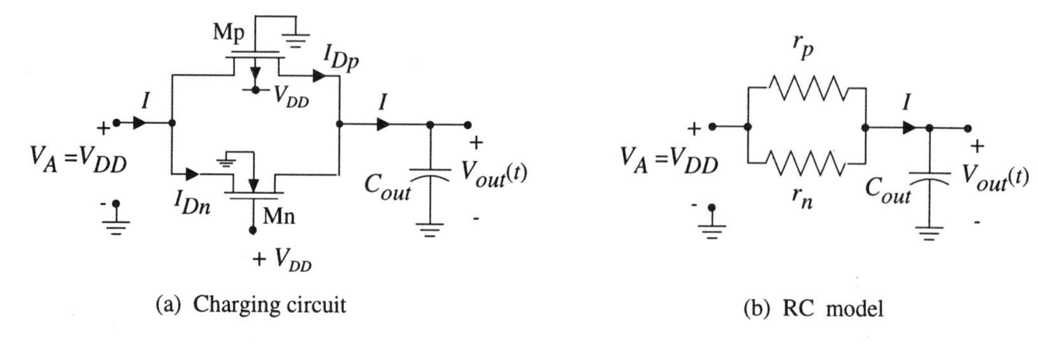

Figure 6.14 Resistances in a logic 1 transfer

To obtain a value of R_{TG} we will analyze the current flow through the transistors and attempt to determine a reasonable formulation of the resistance. Since the FETs are nonlinear, let us introduce the nonlinear (voltage-dependent) resistances r_n and r_p for the nFET and pFET, respectively, and analyze how they behave in the logic 1 transfer event shown in Figure 6.14. By definition,

$$r_n = \frac{V_{DSn}}{I_{Dn}}, \qquad r_p = \frac{V_{SDp}}{I_{Dp}} \tag{6.22}$$

and these equations allow us to express the values in terms of the device parameters. In this approach, the transmission gate resistance is obtained by paralleling these two resistances; it is important to remember, however, that we have defined R_{TG} to be a constant.

Consider first the nFET. In this situation, it is always conducting in the saturated mode so that

$$r_n = \frac{2(V_{DD} - V_{out})}{\beta_n (V_{DD} - V_{out} - V_{Tn})^2} \tag{6.23}$$

which is valid in the voltage range $V_{out} \leq (V_{DD} - V_{Tn})$. In general, r_n is inversely proportional to V_{out}, so that it increases as C_{out} charges. For $V_{out} \geq (V_{DD} - V_{Tn})$, the nFET is in cutoff, forcing $I_{Dn}=0$. The resistance is then infinite: $r_n \to \infty$. The pFET resistance can be written down using the same arguments. For $V_{out} \leq |V_{Tp}|$, Mp is saturated and

$$r_p = \frac{2(V_{DD} - V_{out})}{\beta_p (V_{DD} - |V_{Tp}|)^2} \tag{6.24}$$

gives the proper dependence. In this region, r_p decreases linearly as V_{out} increases. When V_{out} rises above $|V_{Tp}|$, the pFET goes non-saturated, and the resistance is given by

$$r_p = \frac{2(V_{DD} - V_{out})}{\beta_p \left[2(V_{DD} - |V_{Tp}|)(V_{DD} - V_{out}) - (V_{DD} - V_{out})^2\right]} \tag{6.25}$$

Dividing by $(V_{DD}-V_{out})$ gives

$$r_p = \frac{2}{\beta_p [2(V_{DD} - |V_{Tp}|) - V_{DD} + V_{out}]} \tag{6.26}$$

which shows that the pFET resistance decreases as C_{out} charges.

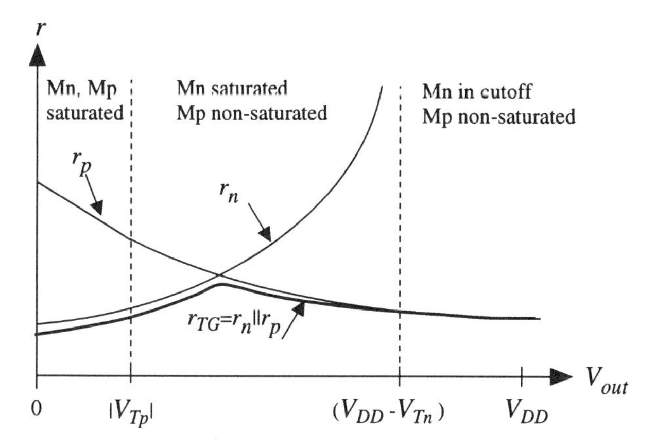

Figure 6.15 TG resistances for a logic 1 transfer

Figure 6.15 illustrates the behavior of r_n, r_p, and the value

$$r_{TG} = r_n||r_p = \frac{r_n r_p}{r_n + r_p} \tag{6.27}$$

as functions of the output voltage V_{out} during a logic 1 transfer event. Although r_{TG} does not remain constant over the range of voltages, it does have the characteristic that it is always less than the smaller of two components. This helps to minimize variations and allows us to use an average or a maximum value of the LTI element R_{TG} in initial circuit modeling.

Different approaches may be used to estimate R_{TG}. A simple approximation is to choose V_{out}=0*v* so that

$$r_{n0} = \frac{2V_{DD}}{\beta_n (V_{DD} - V_{Tn})^2} \qquad r_{p0} = \frac{2V_{DD}}{\beta_p (V_{DD} - |V_{Tp}|)^2} \tag{6.28}$$

and then take

$$R_{TG} \approx r_{n0}||r_{p0} \tag{6.29}$$

as the TG resistance. In addition to being simple, it contains the proper dependences on the FET aspect rations $(W/L)_n$ and $(W/L)_p$. This illustrates the important fact that the transmission gate resistance decreases as the aspect ratios (or, equivalently, the channel widths) are increased. Another simple technique is to choose the output voltage to be

$$V_{out} = (V_{DD} - V_{Tn}) \tag{6.30}$$

corresponding to the point where Mn has entered cutoff. Since the nFET resistance is infinite,

$$R_{TG} = r_p$$

$$= \frac{2}{\beta_p [2(V_{DD} - |V_{Tp}|) + V_{Tn}]} \tag{6.31}$$

The drawback to this formulation is that only the pFET size appears. This is acceptable if the nFET is large enough so that the pFET resistance dominates over most of the voltage range.

A curve fitting approach can also be used to estimate R_{TG}. First, SPICE is used to simulate the transient characteristics and obtain a plot of V_{out}. The response curve is then used to estimate the time constant τ_{TG}, which can in turn be used to find an LTI TG resistance if the output capacitance is known. Estimating C_{out} is discussed below.

6.3.3 TG Capacitances

The values of the transmission gate capacitances C_A and C_B are sensitive to the device sizes and the layout geometry. Figure 6.16 shows the general contributions from both FETs where we have arbitrarily chosen the drain and source sides. For analytic estimates, we construct the values

$$\begin{aligned} C_A &= C_{GDn} + C_{GSp} + K_n(0, V_{DD}) C_{DBn} + K_p(0, V_{DD}) C_{SBp} \\ C_B &= C_{GSn} + C_{GDp} + K_n(0, V_{DD}) C_{SBn} + K_p(0, V_{DD}) C_{DBp} \end{aligned} \tag{6.32}$$

by simply paralleling all of the contributions (using LTI values) that are connected to the node of interest. The values of the MOS and the depletion capacitances depend upon the channel widths, the extent of the doped regions away from the gate, and the process parameters, and may be extracted directly from the layout. As always, these capacitors increase with increasing channel width.

The value of C_{out} introduced above in Figure 6.14 can be estimated by adding the contributions of the external load capacitance C_L. This gives

$$C_{out} = C_B + C_L \tag{6.33}$$

which can be used in the time constant expression $\tau_{TG}=R_{TG}C_{out}$. This allows us to write the time constant as the sum of internal and external contributions via

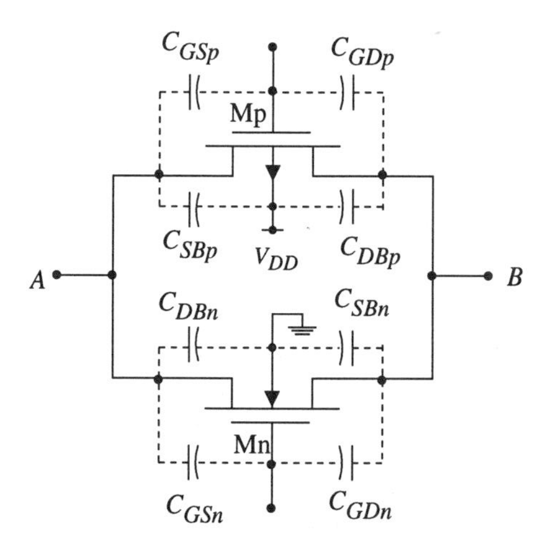

Figure 6.16 Device capacitance contributions in a transmission gate

$$\begin{aligned}\tau_{TG} &= R_{TG}(C_B + C_L) \\ &= \tau_{TG,\,int} + \tau_{TG,\,L}\end{aligned} \tag{6.34}$$

The internal time constant $\tau_{TG,int}$ is relatively insensitive to changes in the device sizes, while the value of $\tau_{TG,L}$ decreases as the aspect ratios of the FETs are increased. Switching performance is limited by the value of C_L in the external load.

6.3.4 Layout Considerations

Design trade-offs are not apparent until the layout is considered. Figure 6.17 shows a basic transmission gate with the important dimensions labeled. To see the overall problem, recall that large values of (W/L) reduce the resistance R_{TG}. However, this implies that W_n and W_p are large. Applying the basic MOSFET capacitance model shows that the MOSFET parasitic capacitances are approximately proportional to W. Increasing either W_n or W_p decreases the resistance of the transistor, but increases the capacitance.

The transient performance is strongly dependent on the value of the external load capacitance term C_L. Since the transmission gate does not have direct support from the power supply or ground, the stage driving the TG circuit must be large enough to drive the following stage. Although we can equalize β-values by adjusting the device sizes, the formula for R_{TG} shows that neither device dominates the circuit. Consequently, there is no overriding reason to use unequal aspect ratios to achieve (β_n/β_p) as in the inverter circuit. Equal size transistors are thus common in TG layouts. The most important aspect to remember is that the TG acts as a parasitic RC addition to the signal path, so that large transistors may be well worth the cost of the real estate.

6.4 TG-Based Switch Logic Gates

Transmission gate logic is based on controlling the flow of data by using the TG as a voltage-controlled switch. To construct logic gates, we use the logical expression

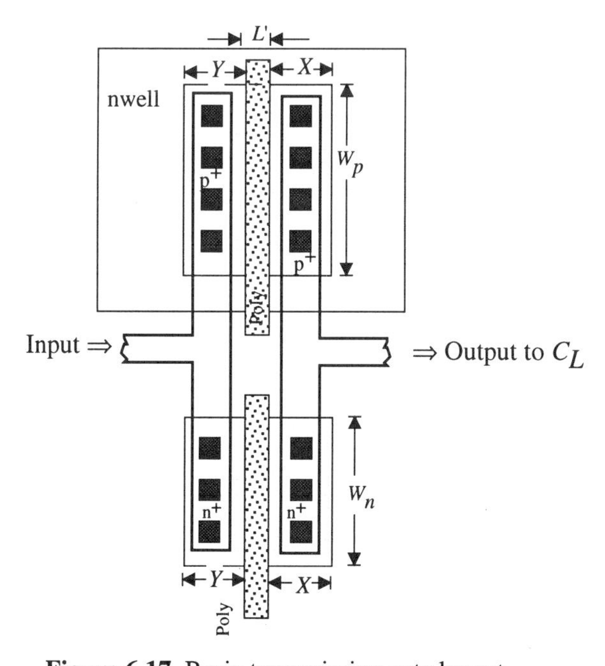

Figure 6.17 Basic transmission gate layout

$$B = A \cdot X \tag{6.35}$$

which is valid iff X=1; the output B is not defined if X=0. This is not really a problem if we design our circuits to avoid the situation of creating an **isolated node** so we do not have to worry about the case where X=0. Violating this rule and allowing nodes to float can lead to significant circuit-level errors in the logic performance.

Static TG-based logic networks are created around switched arrays that steer the data. This idea can be used to implement logic functions with a structured approach. Gate control can be obtained using any set of logical variables available in the system. We will present a few of the more common TG-based logic gates here.

6.4.1 Basic Multiplexors

A 2-input path selector can be created using the circuit shown in Figure 6.18. In this scheme, the input data lines P_0 and P_1 are controlled by the select variable S such that

$$f = P_0 \cdot \overline{S} + P_1 \cdot S \tag{6.36}$$

When $S = 0$, the output is $f = P_0$, while a value of $S = 1$ gives $f = P_1$. From the logic viewpoint, this network provides the same function as the single-polarity 2:1 MUX circuits discussed in Chapter 4. However, the use of transmission gates insures that the output is capable of a full rail output voltage swing from 0v to V_{DD}.

The concept can be extended to create the 4:1 MUX shown in Figure 6.19. In this case, the decimal value of the binary control word ($S_1 S_0$) selects the input using the logic expression

$$f = P_0 \cdot \left(\overline{S_1}\,\overline{S_0}\right) + P_1 \cdot \left(\overline{S_1}\, S_0\right) + P_2 \cdot \left(S_1 \overline{S_0}\right) + P_3 \cdot \left(S_1 S_0\right) \tag{6.37}$$

This is equivalent to the split-array MUX shown in Figure 4.30 of Chapter 4. In principle, we can use TG chains to create multiplexors of arbitrary size. In general, the structure will have n inputs ports controlled by m select lines such that n=2^m. For example, an 8:1 MUX with 8 inputs P_0 through P_7 would require a 3-bit control word S_2 S_1 S_0 such that the output function g is described by

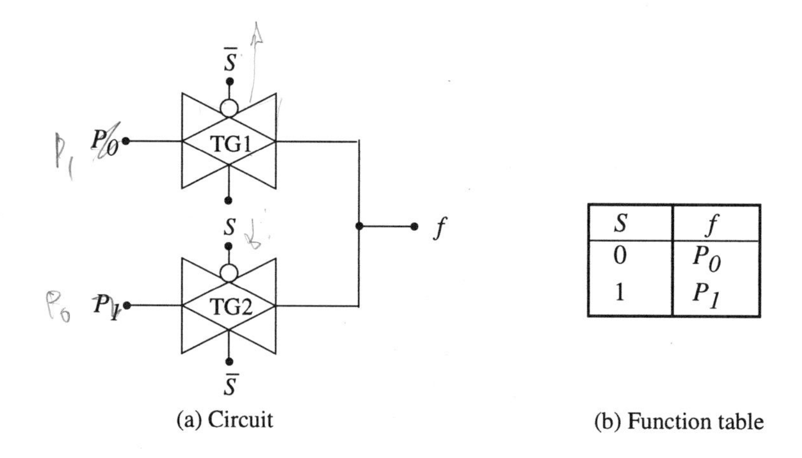

S	f
0	P_0
1	P_1

Figure 6.18 A 2:1 MUX implemented using transmission gates

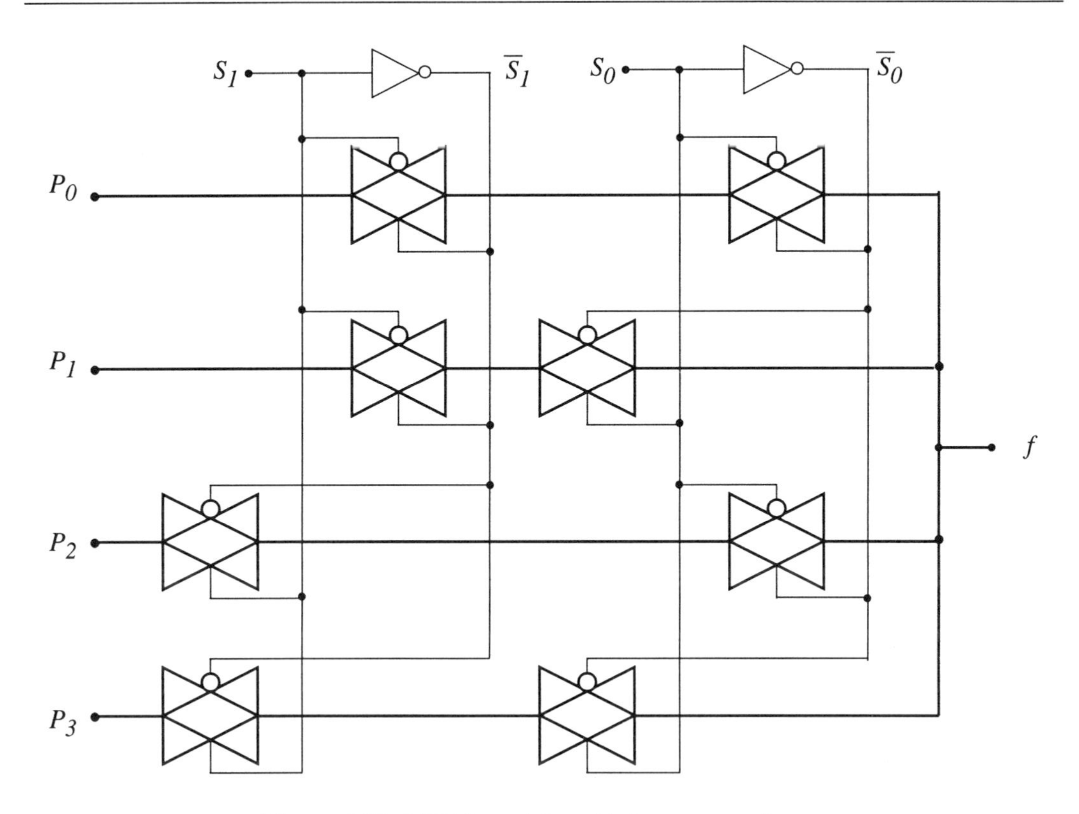

Figure 6.19 A TG-base 4:1 multiplexor network

$$\begin{aligned} g &= P_0 \cdot \left(\overline{S_2}\,\overline{S_1}\,\overline{S_0}\right) + P_1 \cdot \left(\overline{S_2}\,\overline{S_1}\,S_0\right) + P_2 \cdot \left(\overline{S_2}\,S_1\,\overline{S}_0\right) + P_3 \cdot \left(\overline{S_2}\,S_1\,S_0\right) \\ &+ P_4 \cdot \left(S_2\,\overline{S_1}\,\overline{S_0}\right) + P_5 \cdot \left(S_2\,\overline{S_1}\,S_0\right) + P_6 \cdot \left(S_2 S_1 \overline{S_0}\right) + P_7 \cdot \left(S_2 S_1 S_0\right) \end{aligned} \tag{6. 38}$$

To design the TG network, each path from an input P_i to the output must have 3 series-connected transmission gates with each controlled by one of the select bits. While this type of structuring is possible, the transmission delay through the chain is usually quite long, limiting its application.

6.4.2 OR Gate

Transmission gate logic directly yields the logical OR function using the circuit shown in Figure 6.20. The input variable A and its complement $\overline{A}$ are used to control both the pMOS pass transistor Mp and the transmission gate. The upper branch transmits when A=1, while the lower TG circuit propagates B to the output when A=0. Since the pMOS transistor only passes a high voltage corresponding a logic value of A=1, the weak logic 0 characteristics are not important.

The OR function itself can be formally derived by viewing the circuit as a 2:1 MUX and applying the basic logic expressions. The upper branch gives a net output of A, while the lower branch gives an output of $\overline{A}B$. Combining

$$\begin{aligned} g &= A + \overline{A} \cdot B \\ &= A + B \end{aligned} \tag{6. 39}$$

by applying the absorption theorem. Even though this circuit uses a pFET as a pass transistor, it

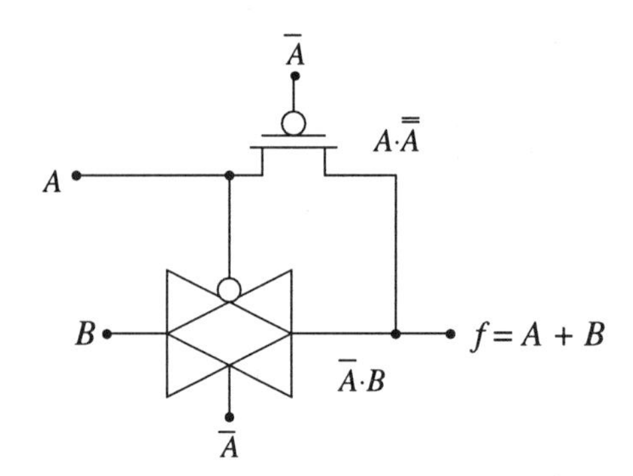

Figure 6.20 A TG-based OR gate circuit

only transmits a high voltage; thus, the output ranges from $0v$ to V_{DD}. Note that the use of the two transmission paths (one through the pFET, the other through the nFET) insures that the output voltage is always well defined.

6.4.3 XOR and Equivalence

The Exclusive-OR (XOR) and Equivalence functions can be implemented by using an input variable to control the transmission gates as shown in Figure 6.21. Recall that the exclusive-OR function is given by

$$A \oplus B = A \cdot \overline{B} + \overline{A} \cdot B \tag{6.40}$$

The network in Figure 6.21 is simply a 2:1 MUX networks with inputs of A and $\overline{A}$, while B and $\overline{B}$ are used to control the transmission gates. Since the Equivalence function is given by

$$\overline{A \oplus B} = A \cdot B + \overline{A} \cdot \overline{B} \tag{6.41}$$

the network in Figure 6.21 can be used to create the XNOR operation by simply reversing B and $\overline{B}$ in the XOR gate; this results in the circuit drawn in Figure 6.22.

Alternate XOR and XNOR circuits are shown in Figure 6.23(a) and (b), respectively. Both net-

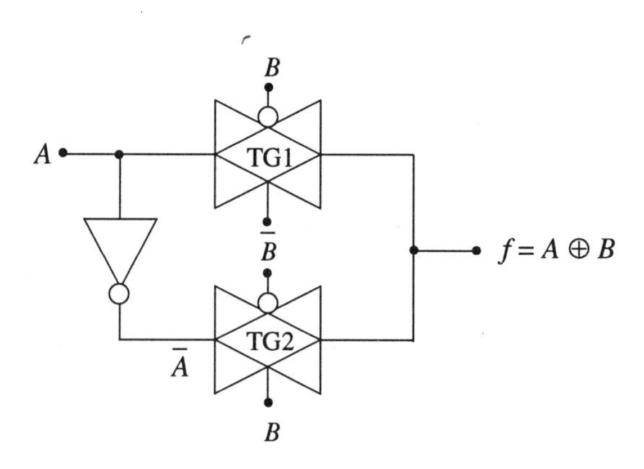

Figure 6.21 The XOR operation based on a 2:1 TG MUX

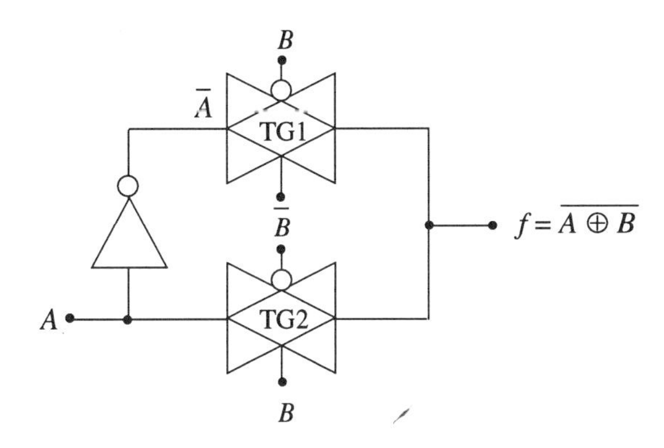

Figure 6.22 A TG-base equivalence (XNOR) network

works use a single transmission gate and a complementary MOSFET pair (Mp,Mn) at the output. Note, however, that the output node does not receive voltage support from the power supply. Rather, the input variable B (and its complement) must supply all necessary current to drive the output capacitance. To understand the operation of these circuits, let us examine the XOR gate in more detail by choosing the value of B and investigating the consequences.

Figure 6.23(a) shows the XOR circuit when B=0. The transmission gate is switched into conduction, so that the output is given by $A \cdot \overline{B}$ by using our TG logic expression. If B=1, then the TG is OFF as illustrated in Figure 6.23(b). In this case, B=1 is applied to the top of Mp, and $\overline{B}$=0 is connected to the bottom on Mn. Since these correspond to respect voltage levels of V_{DD} and ground, B acts as a power supply voltage for Mn and Mp, which form an inverter circuit with A as the input. The output for this case is given by $\overline{A} \cdot B$, so that combing both possibilities gives

$$f = A \oplus B \tag{6.42}$$

as advertised. The operation of the XNOR circuit is identical since only B and $\overline{B}$ are reversed.

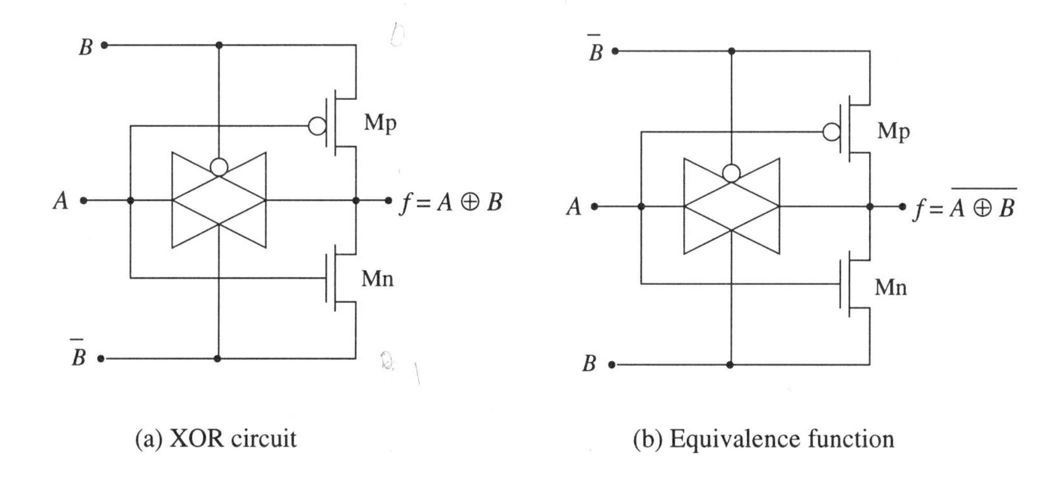

Figure 6.23 TG-based XOR and XNOR circuits

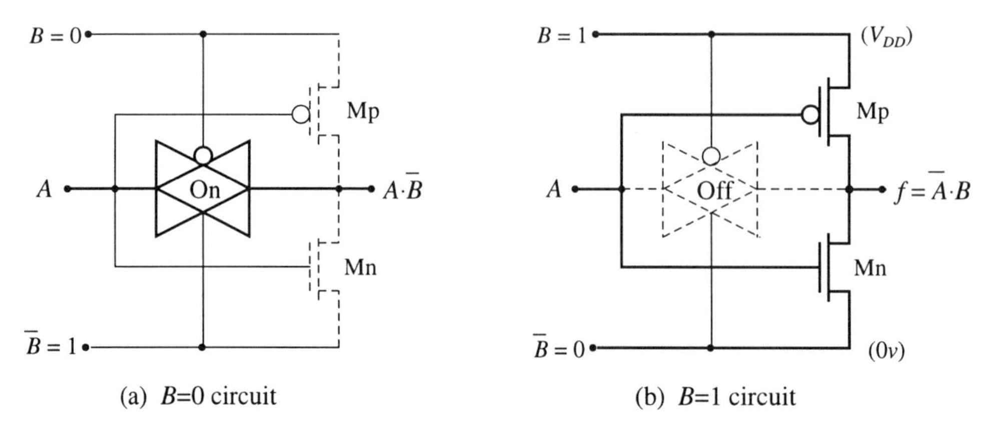

(a) B=0 circuit (b) B=1 circuit

Figure 6.24 Operation of the TG XOR circuit

6.4.4 Transmission-gate Adders

Consider two logic bits a_n and b_n, and a carry-in bit c_n. Recall that the full-adder produces the sum s_n and carry-out c_{n+1} by means of

$$\begin{aligned} s_n &= a_n \oplus b_n \oplus c_n \\ c_{n+1} &= a_n b_n + c_n (a_n \oplus b_n) \end{aligned} \tag{6. 43}$$

Since transmission gate logic allows for direct implementation of the XOR function, various full-adder networks can be constructed using the TG circuits discussed above.

Figure 6.24 shows a full adder that creates the sum bit s_n directly using two XOR operation expressed in

$$\begin{aligned} s_n &= (a_n \oplus b_n) \oplus c_n \\ &= (a_n \oplus b_n) \cdot \overline{c_n} + \overline{(a_n \oplus b_n)} \cdot c_n \end{aligned} \tag{6. 44}$$

where $(a_n \oplus b_n)$ and $\overline{(a_n \oplus b_n)}$ are generated in the left circuits, and the last XOR is obtained using the two TGs in the upper right portion of the circuit. The carry out bit c_{n+1} is obtained by rewriting the expression in the form

$$c_{n+1} = (a_n \oplus b_n) \cdot c_n + \overline{(a_n \oplus b_n)} \cdot a_n \tag{6. 45}$$

One unique aspect of this adder is that the outputs s_n and c_{n+1} are available at approximately the same time. This is due to the fact that the propagation delays from the inputs to the outputs are almost equal because of the symmetry used to construct the circuits.

6.5 TG Registers

Transmission gates can be used as simple switches to create circuits that have at least two distinct operational modes:

- **Load** - The value of a bit D is used as an input to the circuit, and,
- **Hold** - The input line is disconnected from the circuit, and the value is held.

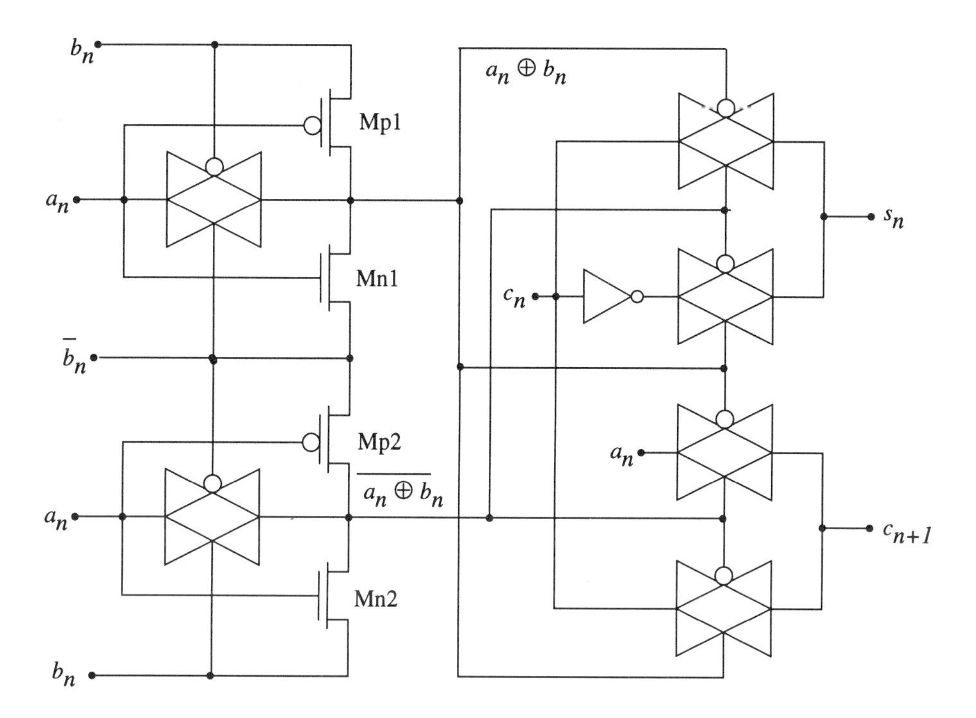

Figure 6.25 Transmission gate full-adder circuit

This allows us to create memory circuits that can be used for latches, registers, and other state elements. Figure 6.26 shows the basic circuit for a level-sensitive D latch. The operation is controlled by the load signal *LD*. A value of $LD = 1$ accepts the input bit *D* into the network. Switching the control to a value of $LD = 0$ allows the circuit to hold the value.

The operation of the latch is detailed in Figure 6.27. The load operation is shown in Figure 6.27(a). When $LD = 1$, the input transmission gate TG1 acts as a closed switch, while TG2 is open. The value of the input data bit *D* enters the circuit, and is available at the outputs *Q* and $\overline{Q}$. Since the outputs change in response to a change in *D*, the latch is said to be **transparent**. The hold operation

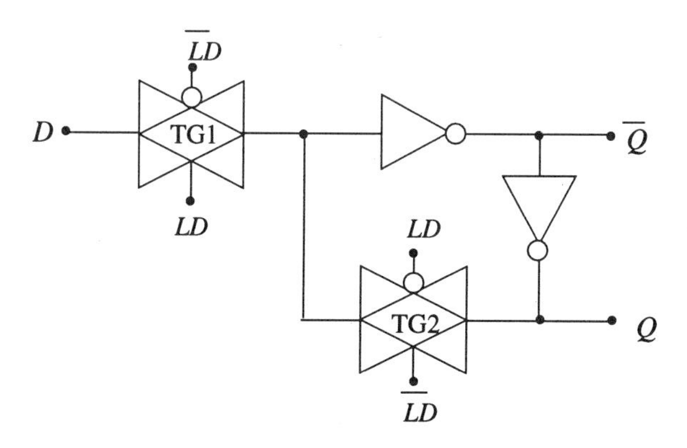

Figure 6.26 Basic TG latch circuit

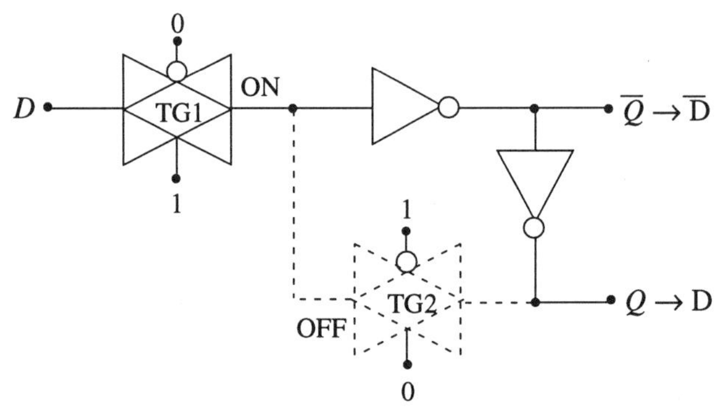

(a) Load operation with *LD*=1

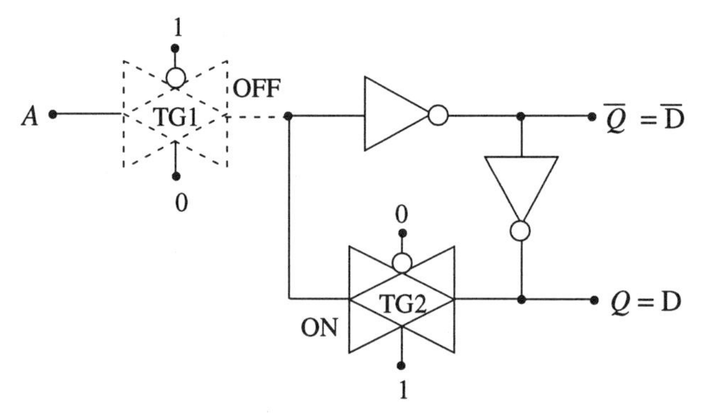

(b) Hold operation with *LD*=0

Figure 6.27 Operation of the TG D-type latch

occurs when *LD*=0, giving the circuit in Figure 6.27(b). In this case, TG1 is open, disconnecting the input line from the circuit; the new input bit *A* cannot enter the circuit. Transmission gate TG2 is closed, and completes the feedback loop between the inverters, changing it into a bistable circuit that can store either a 0 or 1 state. The actual value in the circuit is to *D* that was acquired when the *LD* signal changed from a 1 to a 0.

6.6 The D-type Flip-Flop

Many CMOS circuits use a **clock signal** $\phi(t)$ as shown in Figure 6.28 for their operation. The clock provides a simple way to synchronize operations in a digital network relative to an absolute time base. The period of the clock is T seconds corresponding to a frequency f=(1/T) *Hz*. The complement $\overline{\phi}(t)$ is also shown in the drawing. Since transmission gates can be switched ON or OFF with a complementary pair, the clock signal provides a method for synchronizing data flow. This is a common technique in classical CMOS design, and is a straightforward application that demonstrates the utility of a TG in timing circuits.

We will create a D flip-flop (DFF) that triggers on the positive edge by (a) cascading two oppositely phased D-latches as shown in Figure 6.29. This is a standard master-slave arrangement that allows us to create a non-transparent latching circuit. The master and slave circuits are identical, but are out of phase due to the clock signal distribution.

The operation of the circuit can be understood using the circuits shown in Figure 6.30. When

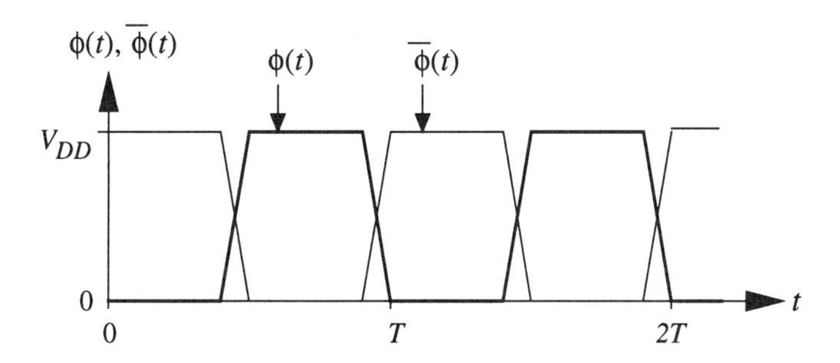

Figure 6.28 TG clocking signals

the clock is at a value of ϕ=0 ($\overline{\phi}$=1), TG1 is in a conducting mode and passes the data input bit D to the first (master) latch as shown in Figure 6.30(a). Since TG2 and TG3 are both OPEN during this time, no further data transfer takes place. When the clock makes a transition to ϕ=1 ($\overline{\phi}$=0) as in drawing (b), TG1 acts like an open circuit and block changes in the data. During this time, TG2 closes and completes the feedback latching circuit, while TG3 is CLOSED allowing the data voltage to be transmitted into the second (slave) latch. The output Q at this time is the value of D that was present when the clock made a transition from ϕ=0 to ϕ=1, making this a **positive edge triggered** storage element.[1] Figure 6.31 shows the symbol and clock diagram for the DFF. A negative-edge triggered DFF is easily constructed by just reversing ϕ and $\overline{\phi}$ everywhere in the circuit.

Logically, we describe the operation of the DFF using the expression

$$D(t) = Q(t+T) \qquad (6.46)$$

This states that the present value of D becomes the next value of Q at the output; this implies that the time t when the clock changes. While this is useful for latching data for the next clock cycle, it does not provide a HOLD state where the value may be stored in the circuit. This is easily remedied using the circuit in Figure 6.32. Everything is identical except that the control signals applied to TG1 have been modified to the composite signals

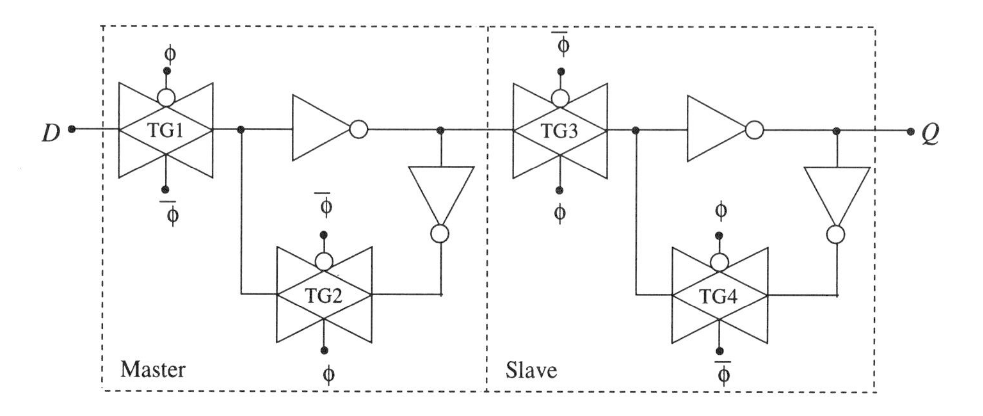

Figure 6.29 TG-based master-slave flip-flop

[1] Strictly speaking, the master-slave flip-flop is not the same as an edge-triggered flip-flop, but the terminology will be used interchangeably here.

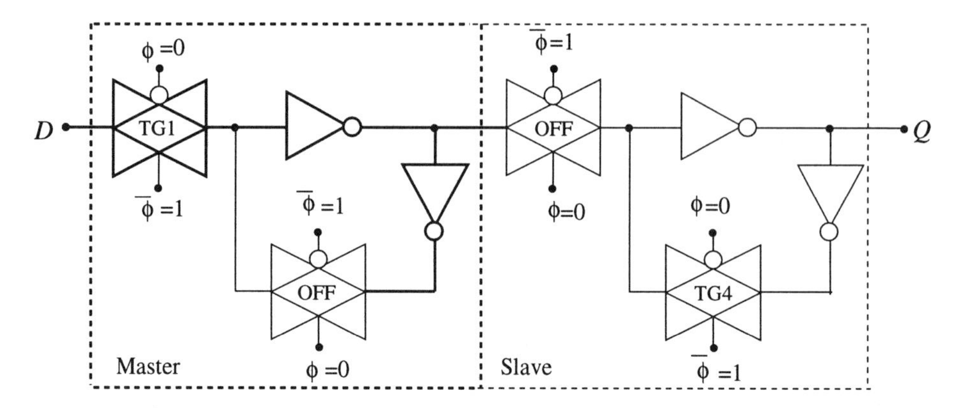

(a) Load master when ϕ=0

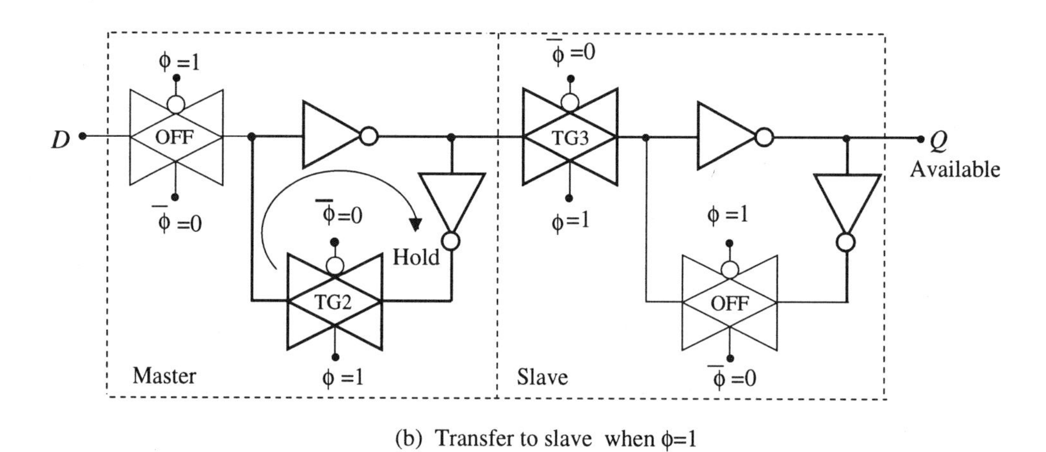

(b) Transfer to slave when ϕ=1

Figure 6.30 Operation of the TG flip-flop

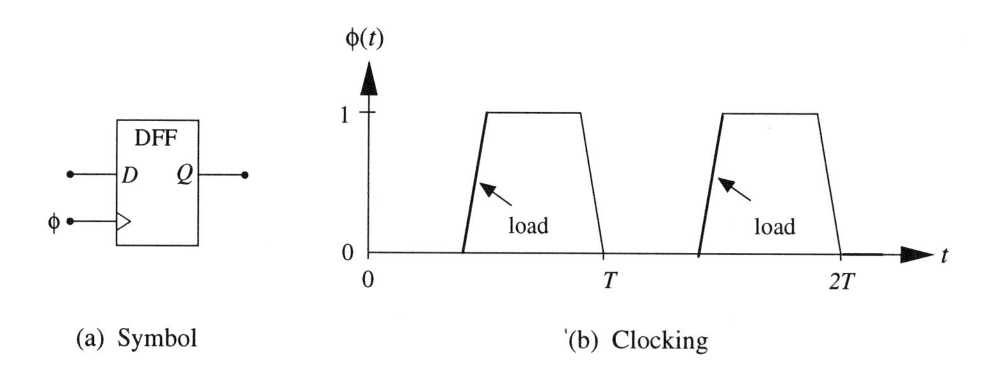

(a) Symbol

(b) Clocking

Figure 6.31 D-Flip-flop timing

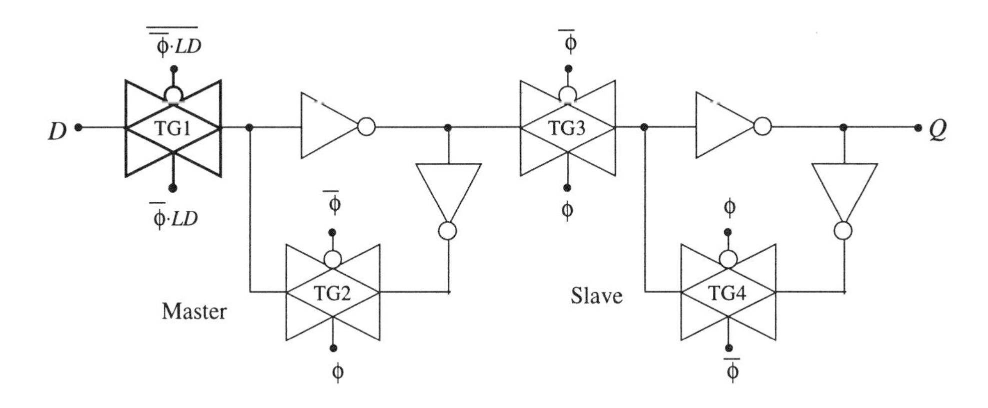

Figure 6.32 DFF with Load control

$$\bar{\phi} \cdot LD \quad \text{and} \quad \overline{\bar{\phi} \cdot LD} \tag{6. 47}$$

where LD is the LOAD control. When $LD = 1$, the circuit operates as a normal DFF. However, if LD=0, then TG1 is kept OPEN which prohibits the entry of a new data bit into the master circuit.

6.7 nFET-Based Storage Circuits

The TG-based storage circuits above can also be constructed using only nFETs instead of transmission gates. A basic latch is shown in Figure 6.33. The input transistor M1 is controlled by the load signal LD, while M2 is controlled by $\overline{LD}$ and is used for the feedback loop in the inverter latch. The operation of the circuit is identical to the TG equivalent, and is summarized in Figure 6.34. When LD=1, the input data bit D is allowed into the circuit [see Figure 6.34(a)]. A value of LD=0 blocks the input path and simultaneously closes the feedback loop. This gives the hold condition shown in Figure 6.34(b).

Although this is simpler than the equivalent TG implementation in that it uses two less transistors (and also eliminates the additional wiring), we must be more careful at the circuit design level since the nFET only passes a limited range of voltages. In particular, we recall that applying V_{DD} to the gate allows the nFET to pass voltages in the range [0, V_{max}] where

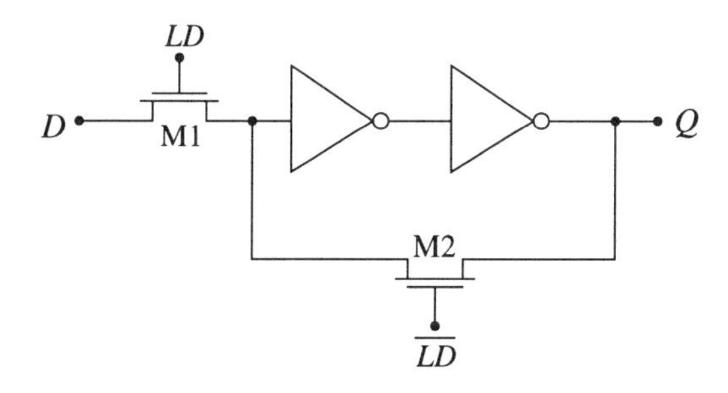

Figure 6.33 nFET-based latching circuit

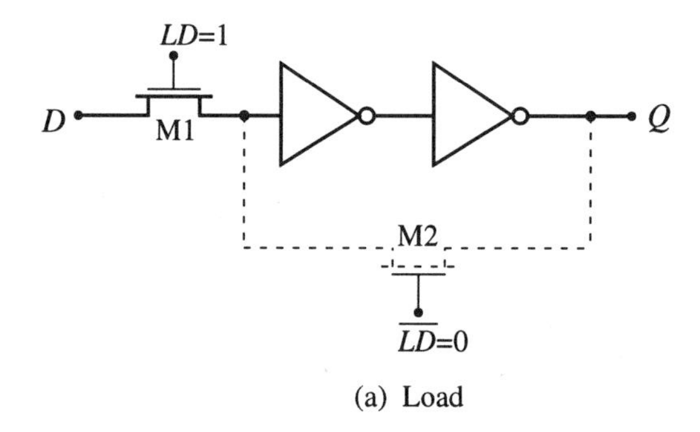

(a) Load

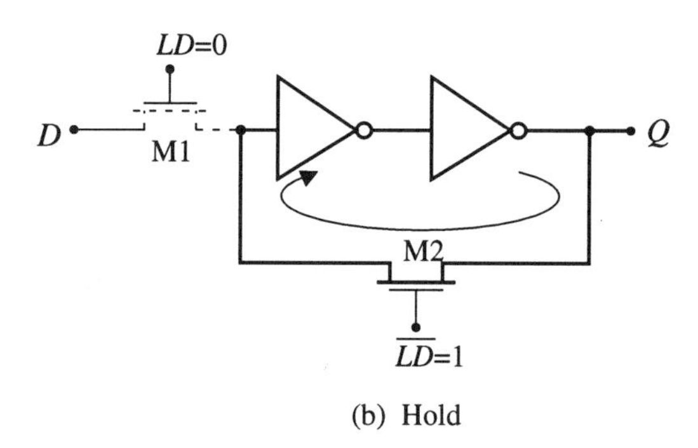

(b) Hold

Figure 6.34 Operation of the nFET-controlled latch

$$\begin{aligned} V_{max} &= V_{DD} - V_{Tn} \\ &= (V_{DD} - V_{T0n}) - \gamma\left(\sqrt{2|\phi_F| + V_{max}} - \sqrt{2|\phi_F|}\right) \end{aligned} \tag{6.48}$$

This threshold loss can affect the operation of the circuit. Consider the situation in Figure 6.35(a) where LD=1 and we are attempting to transmit the data voltage V_D through M1. If $V_D = 0v$, then there is no problem. However, a high voltage of $V_D = V_{DD}$ only transmits as V_{max}, which must be interpreted as a logic 1 voltage by Inverter 1. In terms of the inverter DC VTC, we see that

$$V_{max} > V_{IH} \tag{6.49}$$

where V_{IH} is the input-high voltage. To insure that the circuit operates properly, we must design Inverter 1 by choosing the proper ratio of $(\beta_n/\beta_p) > 1$ to have a VTC such as that shown in Figure 6.35(b) where $V_I < (V_{DD}/2)$. An easy way to accomplish this is use identical-sized transistors with $(W/L)_n = (W/L)_p$. Another problem that must be accounted for is the fact that nFETs are intrinsically slow in passing logic 1 voltages. In particular, we recall from Chapter 4 that

$$V_{in}(t) = V_{max}\left[\frac{t/2\tau_n}{1 + t/2\tau_n}\right] \tag{6.50}$$

describes the time dependence. This transition is slow even if the time constant τ_n can be made small. The same philosophy may be applied to creating the nFET-based DFF shown in Figure 6.34.

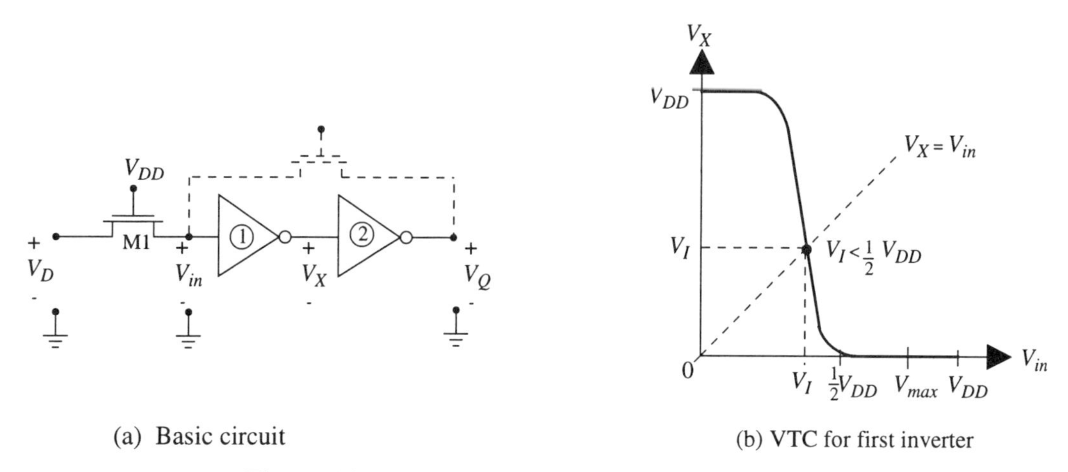

(a) Basic circuit

(b) VTC for first inverter

Figure 6.35 Circuit design considerations

6.8 Transmission Gates in Modern Design

Transmission gates were originally introduced for use as ideal switches that are capable of passing the entire range of voltages from $0v$ to V_{DD}. Early CMOS designs used TG networks extensively for many basic functions. If one studies the data books of CMOS integrated SSI and MSI chips, many examples of TG circuits can be found. Moreover, ASIC libraries often provide several of the TG-based functions presented here. This is because TGs provide a reliable switch that passes the full range of voltages and are conceptually easy to understand.

Modern high-density, high-performance chip designs, however, are severely limited by interconnect. This constraint has led designers to question the need for using the nFET/pFET pair required in the TG. The FET itself is not a problem because of its small size. The wiring, on the other hand, can be significant, especially when TGs are distributed throughout a complex system layout. Owing to this consideration, many modern designs tend to shy away from using TGs, opting instead for single nFETs in their place. In principle, any TG-based network can be converted to using nFETs only so long as we modify the electrical characteristics where needed. Thus, all of the circuits in this chapter are still considered valid.

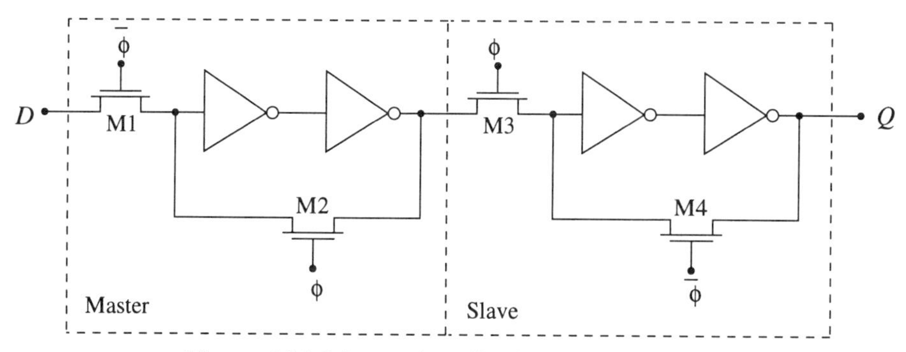

Figure 6.36 Master-slave flip-flop circuit

This short observation brings up a point concerning the evolution of CMOS. Up to this point in the book, we have generally attempted to create logic circuits using "true" complementary networks based on nFET/pFET pairs. Replacing a TG by a single nFET appears to go against this idea, reverting us back to nMOS-only style logic design. This, in fact, is the manner in which CMOS has developed in recent years. The superior performance of nFETs due to their higher mobility has led designers to turn to single-polarity circuits or sub-circuits in critical data paths. In fact, we shall see later that most modern CMOS design styles tend to avoid pFETs as much as possible. We still use a few pFETs where they help us control power dissipation or charge transfer. And, they are mandatory if we wish to charge a node up to the power supply voltage V_{DD}, so that we still maintain the acronym "CMOS" to describe the technology. However, many classical ideas about symmetry and complementary structuring have been replaced by more pragmatic considerations such as layout area and speed.

This chapter thus marks the end of our treatment of basic CMOS. In Chapters 7 through 9, we will investigate how the circuitry has evolved to produce the high-performance, high-density logic networks that are common in modern chip design.

6.9 Problems

[6-1] A transmission gate is made in a process characterized by the basic parameters

$$k'_n=100\times10^{-6}\ \mu A/V^2,\ k'_p=42\times10^{-6}\ \mu A/V^2,\ V_{TOn}=0.7v,\ V_{TOp}=-0.8v,\ V_{DD}=5v,$$

The FET aspect ratios are $(W/L)_n$=8 and $(W/L)_n$=10.

(a) Calculate the LTI resistance values R_n and R_p.

(b) Suppose that you wanted to model the TG using only the LTI resistors? What value would you choose for R_{TG}? Explain your choice.

[6-2] Consider a TG made with the basic process parameters described Problem 6-1. The aspect ratios are given by $(W/L)_n$=6 and $(W/L)_n$=8.

(a) Calculate the value of the pFET resistance R_p at the point where the nFET goes into cutoff as shown in the plot of Figure 6.15. Assume a gate oxide thickness of 200Å and a substrate acceptor doping of $10^{15}\ cm^{-3}$ for the calculation of the body-bias coefficient.

(b) Compare the value for R_p in (a) with that you would obtain by taking $R_{TG}=R_n \| R_p$ at the 0 voltage point in the plot of Figure 6.15 by calculating the percentage difference between the two [based on the value found in (a)].

[6-3] A transmission gate layout drawing is illustrated in Figure P6.1; all values are in units of microns. For the process, the important parameters are given as

$$k'_n = 200\times10^{-6}\ \mu A/V^2,\ k'_p = 90\times10^{-6}\ \mu A/V^2,\ V_{TOn}= 0.7v,\ V_{TOp}= -0.8v,\ V_{DD} = 3.3v,$$

$$C_{jn} = 4\ fF/\mu m^2,\ C_{jswn} = 0.6fF/\mu m,\ C_{jp} = 5\ fF/\mu m^2,\ C_{jswp} = 0.8fF/\mu m,\ t_{ox} = 100Å,\ L_o = 0.1\mu m$$

Step-profile junctions are assumed for simplicity, and the built-in potentials are taken as $\phi_o = 0.9v$ for the nFET junctions and $\phi_o = 1v$ for the pFET junctions.

(a) Calculate the gate capacitances C_{Gn} and C_{Gp} for the two FETs.

(b) Calculate the voltage-averaged p+ junction capacitance for one side of the pFET.

(c) Calculate the voltage-averaged n+ junction capacitance for one side of the nFET.

(d) Combine your results above to obtain the values of C_A and C_B in Figure 6.12.

[6-4] Simulate the pulse characteristics of the TG described in Problem 6-3 driving a 200*fF* load using SPICE. Specify a LEVEL 2 model and let SPICE calculate all of the parasitics from the .MODEL information (be careful to place all capacitances in strict units of farads, and all lengths must be in meters). Can you use the pulse response plot to estimate a linear time constant for the TG with this load?

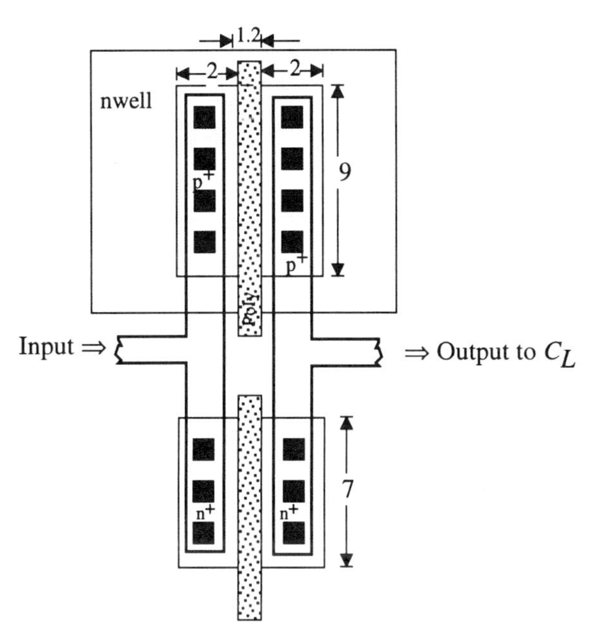

Figure P6.1

[6-5] The layout for a TG cascaded into an inverter is shown in Figure P6.2 below. The process parameters are identical to those given in Problem [6-3].

(a) Calculate the midpoint voltage V_M of the inverter.

(b) Find the input capacitance C_{in} seen looking into the inverter.

(c) A load of $C_L = 4C_{in}$ is attached to the output of the inverter. A unit step voltage of $V(t)=V_{DD}u(t)$ is applied to the input of the TG. Estimate the values of t_{HL} and t_{LH} at the output of the inverter. [There are different approaches than can be used to solve this problem.]

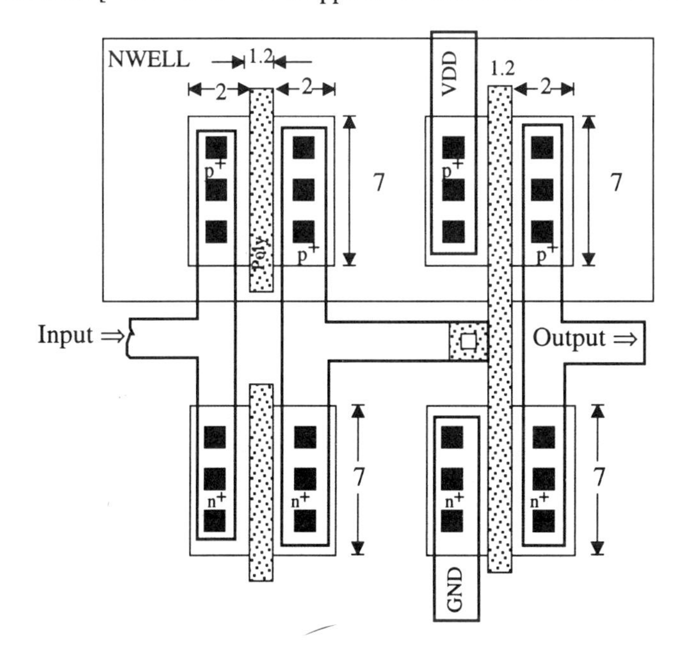

Figure P6.2

[6-6] Simulate the circuit described in Problem 6-6] on SPICE. Perform a transient analysis with the step input voltage and plot the voltages at the input and the output of the inverter.

[6-7] Design a TG-based D-type flip-flop that has a clear control input *CLR* that clears the contents to 0 when *CLR*=1.

[6-8] Design a TG-based D-type flip-flop that has a set input *SET* that sets the contents to 1when *SET*=1.

Chapter 7

Dynamic Logic Circuit Concepts

In the static logic circuits discussed up to this point, the output is valid so long as the inputs are well defined. A **dynamic** logic circuit, on the other hand, gives a result at the output that is only valid for a short period of time. If the result is not used immediately, the voltage may change in time and give an incorrect output value. A dynamic logic circuit uses capacitive nodes to store electrical charge. The charge transfer and retention characteristics of these nodes are critical in the design and operation of advanced logic families. The discussion in this chapter provides a detailed analysis of charge storage nodes and their applications. We also introduce the concept of clocked logic families, and examine some important ideas in data transfer and movement. While the material is important in its own right, it also provides the foundation for the advanced dynamic logic circuits presented in the next chapter.

7.1 Charge Leakage

Up to this point, we have modelled the MOSFET as a voltage-controlled switch that is capable of providing an OPEN circuit by driving the transistor into cutoff. Cutoff is achieved by requiring that $V_{GS} < V_{Tn}$, with the best case situation being $V_{GS} = 0v$. In reality, MOSFETs cannot block the current flow completely due to leakage paths that exist in the device. Although the leakage currents levels are usually very small, they may be critical in certain types of CMOS circuits. In this section, we will analyze the problem in detail.

Consider the simple circuit shown in Figure 7.1 where a pass transistor is connected to the input of an inverter. When the MOSFET is in cutoff, the inverter input node is isolated from both the power supply and ground.[1] Since the node is capacitive, we model it as a **storage capacitor** C_s that can be used to hold a charge

[1] in that there are no low-resistance connections that act as a sink or source for charge.

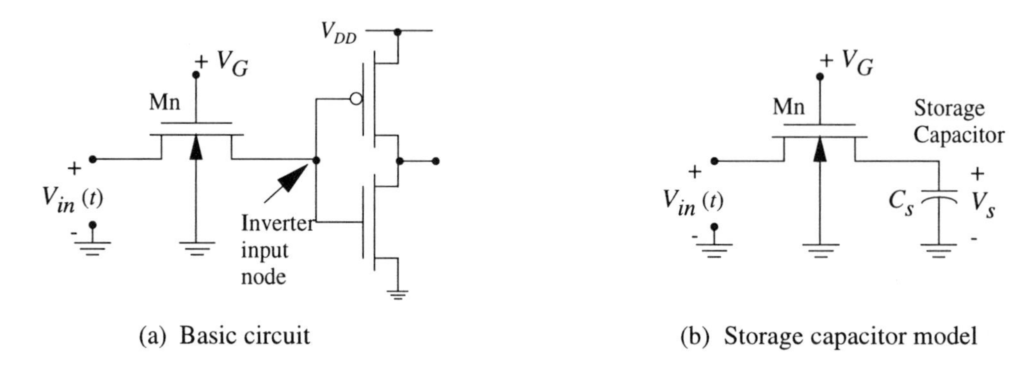

Figure 7.1 Storage node at the input of an inverter

$$Q_s = C_s V_s \tag{7. 1}$$

with V_s the voltage across C_s. This is identical to the circuit shown in Figure 4.1, so that the analyses of Section 4.1 are valid. Let us consider the case where the input is a transition to a logic 1 voltage given by the form $V_{in}(t) = V_{DD}u(t)$. The voltage across the capacitor increases according to

$$V_s(t) = V_{max}\left[\frac{t/2\tau_n}{1 + t/2\tau_n}\right] \tag{7. 2}$$

where

$$V_{max} = (V_{DD} - V_{Tn}) \tag{7. 3}$$

is the final voltage across the capacitor, and τ_n is the time constant. This corresponds to a total stored charge of

$$Q_s = C_s V_{max} \tag{7. 4}$$

on the capacitor.

Charge leakage occurs if we turn off the MOSFET by reducing the gate-source voltage to a value $V_{GSn} < V_{Tn}$, and then attempt to hold this charge on the storage capacitor. As illustrated by the general situation in Figure 7.2(a), the presence of any leakage current I_{leak} removes charge from C_s, which in turn reduces the voltage as described by the *I-V* equation

$$I_{leak} = -\frac{dQ_s}{dt} = -C_s\frac{dV_s}{dt}\ . \tag{7. 5}$$

The problem with this situation is that we use voltages to define Boolean logic values. Initially, V_s was at V_{max}, indicating that a logic 1 was stored on the capacitor. However, since leakage currents remove charge, V_{max} cannot be held and V_s will decrease in time. Eventually, V_s will fall to a level where it will be incorrectly interpreted as a logic 0 value.

To understand the problem in more detail, let us integrate the equation under the assumption that both I_{leak} and C_s are constants; it is important to note that in reality, both are functions of V_s itself, so that this is only a rough approximation. Using the initial condition $V_s(0) = V_{max}$ gives

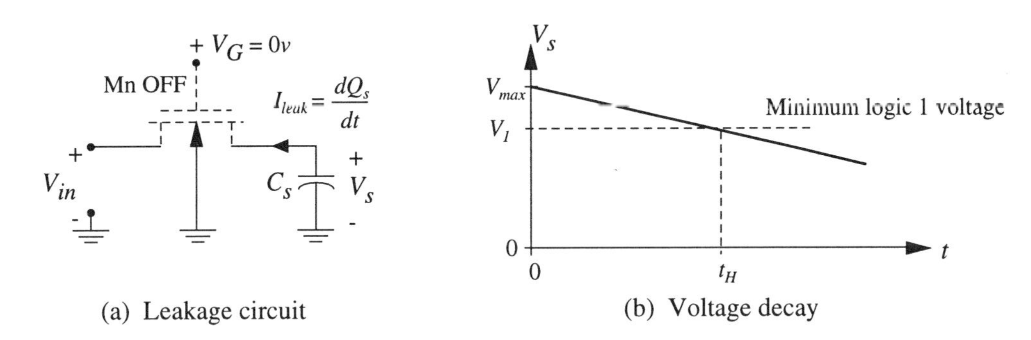

(a) Leakage circuit (b) Voltage decay

Figure 7.2 Charge leakage through an nFET

$$V_s(t) \approx V_{max} - \frac{I_{leak}}{C_s}t \tag{7. 6}$$

indicating a linear decay in the voltage as plotted in Figure 7.2(b). Denoting the smallest logic 1 voltage as V_1 we see that the maximum logic 1 **hold time** t_H is given by

$$t_H \approx \frac{C_s}{I_{leak}}(V_{max} - V_1) \ . \tag{7. 7}$$

Since normal CMOS capacitance values are in the range of 10-200 *fF*, even small leakage currents on the order of tenths of picoamperes (10^{-13} *A*) indicate maximum hold times on the order of a few a few hundred milliseconds (10^{-1} *s*). To get around this problem, a circuit must either (a) avoid the use of **isolated** capacitors at the output of a MOSFET, or (b) not attempt to hold the stored charge for any extended period of time. Static logic gates avoid isolated capacitors and always provide a good connection to either V_{DD} or ground. In the more advanced **dynamic logic** circuits examined in Chapter 8, charge leakage effects may be controlled to extend the hold time by using various types of special circuits and precise clocking.

The existence of a leakage current is due to the internal physics and construction of a MOSFET and I_{leak} itself consists of several terms depending upon the technology. A more accurate analysis of the charge leakage problem requires that we examine these in more detail.

7.1.1 Junction Reverse Leakage Currents

Cutoff in an nFET is achieved by setting $V_{GSn} < V_{Tn}$ which eliminates the inversion electron charge layer between the source and drain. Although the direct conduction path no longer exists, a reverse-biased pn junction is now responsible for holding the charge on C_s as illustrated in Figure 7.3.

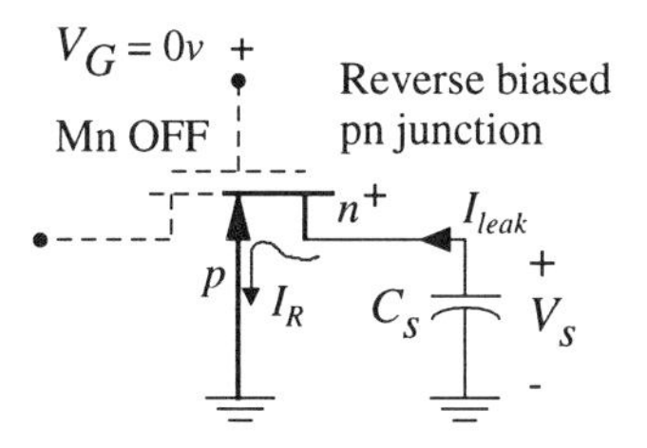

Figure 7.3 Junction leakage in a cutoff nFET

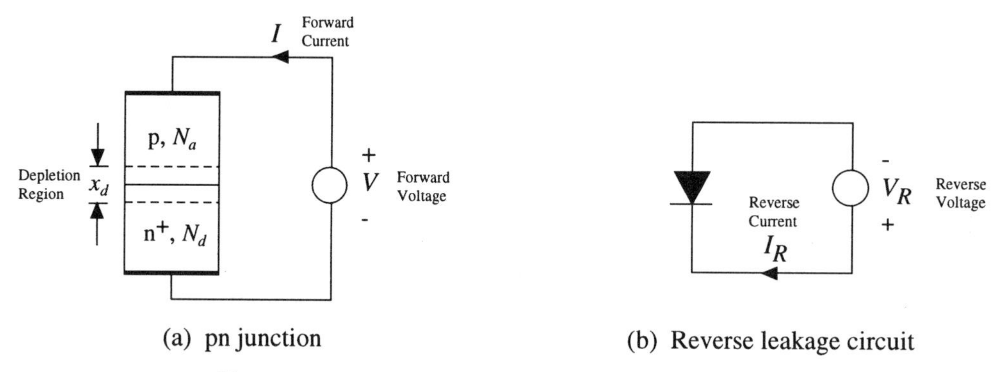

(a) pn junction (b) Reverse leakage circuit

Figure 7.4 Definition of pn junction parameters

While a reverse-biased diode can block most of the current, a small reverse leakage current I_R always exists, leading to unavoidable leakage from the capacitor. In order to analyze the charge leakage problem, let us first review the current-flow characteristics of a pn junction as discussed in Chapter 1, and restated here for convenience.

Reverse Diode Current

Consider the step profile pn junction diode shown in Figure 7.4(a). With a forward voltage V applied, the current through the device is given by

$$I = I_o\left(1 - e^{V/V_{th}}\right) + I_{dep} \tag{7. 8}$$

where I_o is the reverse saturation current,

$$V_{th} = \frac{kT}{q} \tag{7. 9}$$

is the thermal voltage, and I_{dep} is the current due to phenomena in the depletion region. Applying a reverse-bias voltage $V = -V_R$ to the device gives the reverse current of $I = -I_R$ where

$$I_R = I_o + I_{gen} \tag{7. 10}$$

We have introduced I_{gen} as the generation current (originating in the depletion region) that can be approximated by

$$I_{gen} \approx I_{go}\left(\sqrt{1 + \frac{V_R}{\phi_o}} - 1\right) \tag{7. 11}$$

in a step profile junction. In this expression,

$$\phi_o = \left(\frac{kT}{q}\right)\ln\left[\frac{N_d N_a}{n_i^2}\right] \tag{7. 12}$$

is the built-in voltage, and I_{go} is given by

$$I_{go} = \frac{qAn_i}{2\tau_o}x_{do} \tag{7. 13}$$

Also, A is the cross-sectional area of the junction, τ_0 ισ the carrier lifetime in seconds, and x_{do} is the zero-bias depletion width

$$x_{do} = \sqrt{\frac{2\varepsilon_{Si}\phi_o}{q}\left(\frac{1}{N_a} + \frac{1}{N_d}\right)} \tag{7. 14}$$

In a silicon diode at room temperature, the saturation current is very small such that[2]

$$I_{gen} \gg I_o \tag{7. 15}$$

Owing to this fact, we will approximate

$$I_R \approx I_{gen}(V_R) \tag{7. 16}$$

A plot of I_{gen} as a function of V_R is shown in Figure 7.5. Note that the reverse leakage current increases with increasing reverse bias voltage.

7.1.2 Charge Leakage Analysis

Consider now the charge leakage problem illustrated in Figure 7.6. The physical structure in Figure 7.6(a) is used to create the circuit model constructed in Figure 7.6(b). The pn junction of interest is that created by the drain-bulk boundary. With the contributions shown in the drawing, the storage capacitance can be written as

$$C_s = C_n + C_{MOS} + C_L \tag{7. 17}$$

where C_L is the external load capacitance, while C_n and C_{MOS} are the device parasitic contributions; note in particular that C_n is always a function of V_s. Using this expression, the charge leakage equation is now given by

$$I_R(V_s) = -[C_n(V_s) + C_{MOS} + C_L]\frac{dV_s}{dt}, \tag{7. 18}$$

or,

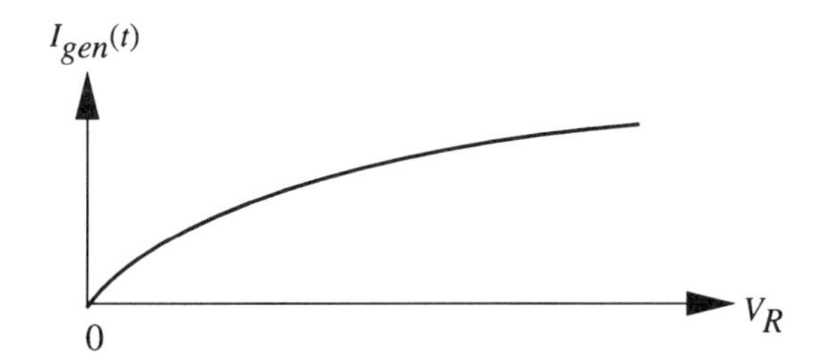

Figure 7.5 Generation current in a reverse-biased junction

[2] A well-known consequence of this equation is that I_o cannot be measured by reverse biasing a diode.

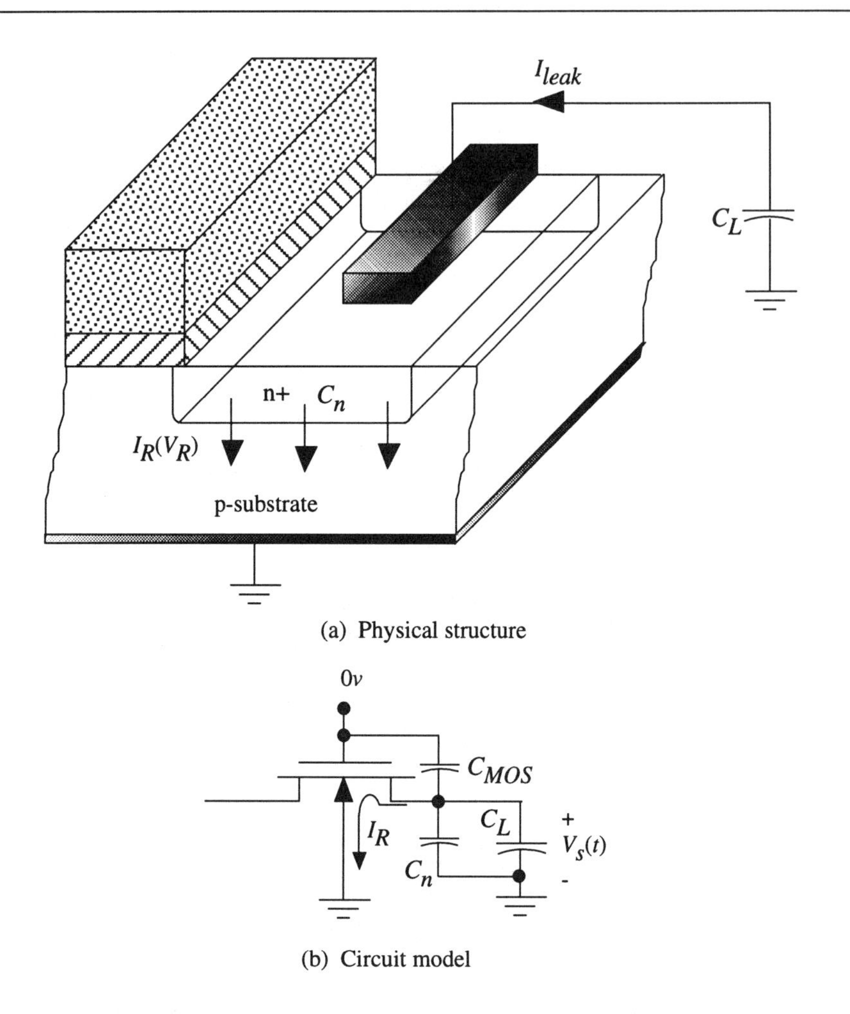

Figure 7.6 Structure of the junction leakage problem

$$I_{go}\left(\sqrt{1+\frac{V_s}{\phi_o}}-1\right) \approx -\left[\frac{C_{j0}}{\sqrt{1+\frac{V_s}{\phi_o}}}+C_{MOS}+C_L\right]\frac{dV_s}{dt} \tag{7. 19}$$

where we have included the voltage-dependence of both the I_{gen} and C_n terms.[3] Although this looks somewhat complicated, the physics of the leakage is relatively straighforward to decipher. The key lies in remembering the initial condition $V_s(t=0) = V_{max}$. This indicates that I_{gen} starts to flow immediately, reducing V_s in the process. However, as V_s falls, so does the magnitude of I_{gen}, so that the decay rate decreases in time.

Generation Current Effects

One interesting aspect of the problem is the dependence of the reverse current I_R on the reverse

[3] The depletion capacitance term has been simplified for this analysis by ignoring sidewall contributions.

voltage. Depletion capacitance *decreases* with increasing V_R through the general relation

$$C_d = \frac{C_{d0}}{\left(1 + \frac{V_R}{\phi_o}\right)^{m_j}} \qquad (7.20)$$

where $m_j = (1/2)$ in the present discussion. The leakage current I_R, on the other hand, is dominated by the generation term that *increases* with the reverse voltage. If the storage capacitor initially has a voltage of $V_s = V_{max}$ on it, the junction capacitance is at its minimum value while the generation current is maximized. As V_s decays because of charge leakage, the junction capacitance increases while the reverse leakage decreases. This means that the decay will be slower than predicted by an analysis that uses constant maximum values. Also, it is important note that the generation current equation

$$I_{gen} \approx I_{go}\left(\sqrt{1 + \frac{V_s}{\phi_o}} - 1\right) \qquad (7.21)$$

will flow so long as $V_s > 0$, so that the leakage to $V_s = 0v$ is unavoidable.

To see the consequences of the voltage dependence of the current, let us use the average capacitance

$$C_s = K(V_1, V_2)\, C_{n0} + C_{MOS} + C_L \qquad (7.22)$$

for C_s in our equation. Our program will then be to solve the simplified differential equation

$$-C_s \frac{dV_s}{dt} \approx I_{go}\left(\sqrt{1 + \frac{V_s}{\phi_o}} - 1\right) \qquad (7.23)$$

Rearranging and integrating the right side gives the form

$$\int_{V_{max}}^{V_s(t)} \frac{dV_s}{\left(\sqrt{1 + \frac{V_s}{\phi_o}} - 1\right)} \approx -\int_0^t \frac{I_{go}}{C_s} dt = -\frac{I_{go}}{C_s} t \qquad (7.24)$$

To integrate the left side without expending a disproportionate amout of effort, let us ignore the constant "-1" term since this reduces our expression to the simple integral

$$\int_{V_{max}}^{V_s(t)} \frac{dV_s}{\left(1 + \frac{V_s}{\phi_o}\right)^{1/2}} = 2\phi_o\left[\left(1 + \frac{V_s}{\phi_o}\right)^{1/2} - \left(1 + \frac{V_{max}}{\phi_o}\right)^{1/2}\right] \approx -\frac{I_{go}}{C_s} t \qquad (7.25)$$

Solving for $V_s(t)$ the yields the approximation

$$V_s(t) \approx \phi_o\left[\frac{V_{max}}{\phi_o} - \left(1 + \frac{V_{max}}{\phi_o}\right)^{1/2} \frac{I_{go}}{C_s} t + \frac{I^2_{go}}{4\phi^2_o C^2_s} t^2\right] \qquad (7.26)$$

where we note that $V_s(0) = V_{max}$ as required. This shows a parabolic dependence due to the t^2 terms, and the leakage is no longer linear. This storage voltage is illustrated in Figure 7.7. The most important result of this estimate is that the hold time will be longer than that obtained from a simple

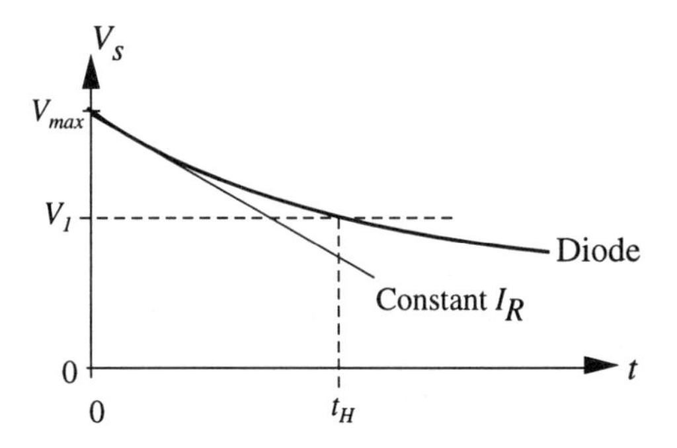

Figure 7.7 Increase in hold time due to a parabolic response

linear decay using the maximum reverse current.

Average Leakage Current

Another approach is to calculate a weighted average value for the reverse current using

$$I_{R,av} = \frac{1}{(V_2 - V_1)} \int_{V_1}^{V_2} I_{gen}(V_R)\, dV_R \tag{7.27}$$

as was done for the depletion capacitance in MOSFET. Substituting for $I_R(V_R)$ gives

$$\begin{aligned} I_{R,av} &= \frac{I_{go}}{(V_2 - V_1)} \int_{V_1}^{V_2} \left[\left(1 + \frac{V_s}{\phi_o}\right)^{1/2} - 1 \right] dV_R \\ &= \frac{I_{go}}{(V_2 - V_1)} \frac{2}{3} \left[\left(1 + \frac{V_2}{\phi_o}\right)^{3/2} - \left(1 + \frac{V_1}{\phi_o}\right)^{3/2} \right] - I_{go} \end{aligned} \tag{7.28}$$

or,

$$I_{R,av} = H_{1/2}(V_1, V_2)\, I_{go} \tag{7.29}$$

where

$$H_{1/2}(V_1, V_2) = \frac{1}{(V_2 - V_1)} \frac{2}{3} \left[\left(1 + \frac{V_2}{\phi_o}\right)^{3/2} - \left(1 + \frac{V_1}{\phi_o}\right)^{3/2} \right] - 1 \tag{7.30}$$

is the weighting factor. Using the LTI equivalent for the storage capacitance then gives the leakage equation as

$$I_{R,av} = -C_s \frac{dV_s}{dt} \tag{7.31}$$

which again gives the linear decay with time. However, the magnitude of the slope will be slightly less than using maximum values of C_s and I_R because of the use of average values.

7.1.3 Subthreshold Leakage

Subthreshold leakage current I_{sub} is drain-to-source current that flows when $V_{GS} < V_{Tn}$ is applied to the gate (hence its name). It is increasingly important as the channel length is decreased, and becomes a critical factor in short channel MOSFETs with $L < 1\mu m$. Let us review the characteristics of subthreshold current to apply it to the problem at hand. In the analysis below, we will make the simplifying assumption that the leakage is due entirely to subthreshold effects. While this cannot be realized in practice (since other terms cannot be eliminated), it does allow us a glimpse of subthreshold leakage behavior.

Consider the circuit shown in Figure 7.8 with $V_{GS} < V_{Tn}$. The subthreshold current can be approximated by

$$I_{sub} = I_x\left(1 - e^{-V_{DS}/V_{th}}\right)e^{(V_{GS} - V_{Tn})/S} \tag{7. 32}$$

which shows the dependence on the voltages. In this expression, the current I_x is proportional to the device width, and S is the subthreshold slope.[4] To understand the effect of subthreshold on the charge leakage, let us assume that I_{sub} dominates the other contributions so that

$$I_{leak} \approx I_{sub} \tag{7. 33}$$

Substituting into the charge leakage equation gives

$$I_x\left(1 - e^{-V_S/V_{th}}\right)e^{(V_{GS} - V_{Tn})/S} \approx -C_s\frac{dV_s}{dt} \tag{7. 34}$$

since the drain-source voltage is the same as the voltage V_s across the storage capacitor. Rearranging,

$$\int_{V_{max}}^{V_s(t)} \frac{dV_s}{\left[1 - e^{-V_s/V_{th}}\right]} \approx -\int_0^t \frac{\tilde{I}_x}{C_s}dt \tag{7. 35}$$

where we have use the initial condition that

$$V_s(0) = V_{max} \tag{7. 36}$$

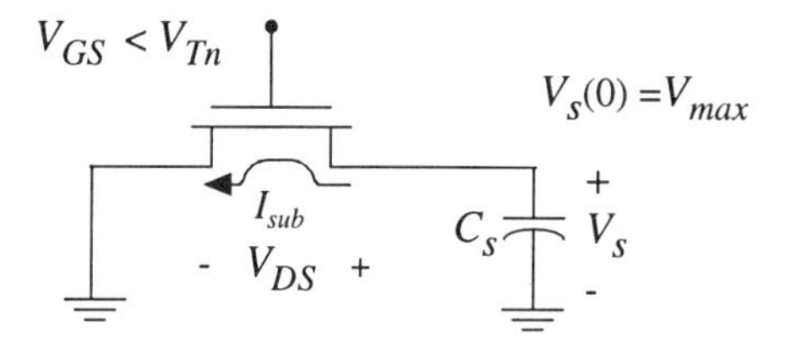

Figure 7.8 Subthreshold leakage circuit

[4] S is the amount of change in the gate-source voltage needed to reduce the current by one decade.

for the lower limit on the voltage integral, and we have defined

$$\tilde{I}_x = I_x e^{(V_{GS} - V_{Tn})/S} \tag{7. 37}$$

Using the general integral

$$\int \frac{dx}{1 - e^x} = \ln\left(\frac{e^x}{1 - e^x}\right) \tag{7. 38}$$

gives

$$\ln\left[\left(\frac{e^{-V_s/V_{th}}}{1 - e^{-V_s/V_{th}}}\right)\left(\frac{1 - e^{-V_{max}/V_{th}}}{e^{-V_{max}/V_{th}}}\right)\right] \approx \frac{\tilde{I}_x}{V_{th} C_s} t \tag{7. 39}$$

Exponentiating both sides yields

$$\left(\frac{e^{-V_s/V_{th}}}{1 - e^{-V_s/V_{th}}}\right) \approx \left(\frac{e^{-V_{max}/V_{th}}}{1 - e^{-V_{max}/V_{th}}}\right) e^{(\tilde{I}_x/V_{th} C_s)\, t} \tag{7. 40}$$

Finally, a straightforward rearrangement gives the storage capacitor voltage as

$$V_s(t) \approx V_{th} \ln\left[\frac{1}{1 - \left[1 - e^{-V_{max}/V_{th}}\right] e^{-\left(\tilde{I}_x/V_{th} C_s\right) t}}\right] \tag{7. 41}$$

Note that

$$V_s(0) = V_{max} \tag{7. 42}$$

as required. Although the form of the decay is somewhat complicated, we can extract some of the time dependence by noting that denominator increases with increasing time *t*. This indicates that the subthreshold leakage is initially large, but decreases as V_s decreases. Since $V_{DS} = V_s$, this behavior is consistent with the physics.

Summary

Charge leakage from a storage node is a critical aspect of designing dynamic logic circuits. The discussion in this section illustrates how to estimate holding times, but the full-blown analysis is much too complicated to analyze by hand. Owing to this situation, we usually rely on data provided from wafer probe testing to extract leakage current densities. The bottom line is that an nFET cannot be used to maintain a high voltage on an isolated node, but the hold time varies with the process.

7.1.4 pFET Leakage Characteristics

Now let us examine the charge retention characteristics of a pFET by analyzing the circuit drawn in Figure 7.9. In the pFET, the n-well constitutes the bulk terminal of the transistor, giving a reverse biased pn-junction diode as shown. As in the case of the nFET, the leakage through the diode cannot be eliminated. However, since the n-well is biased by the power supply V_{DD}, the effects are opposite to those found in the nFET analysis.

Consider the case where initially the voltage across C_s is small with a value $V_s(0) = V_{min}$. Using the pn junction shown in Figure 7.9(a), the basic leakage equation is

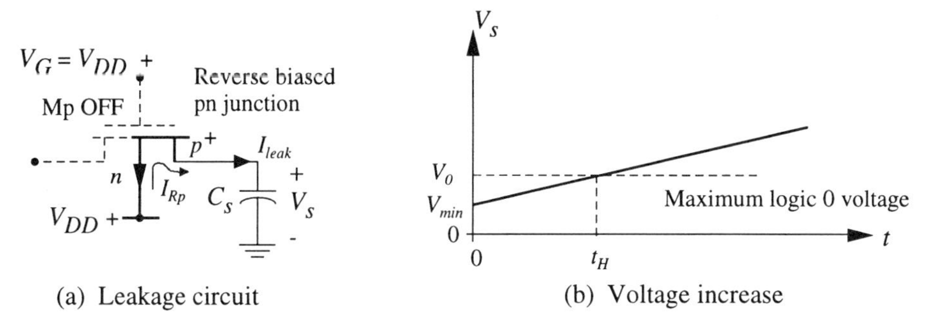

(a) Leakage circuit (b) Voltage increase

Figure 7.9 Charge leakage in a pFET

$$I_{Rp}(V_s) = C_s(V_s)\frac{dV_s}{dt} \tag{7. 43}$$

where I_{Rp} is the reverse leakage current through the pFET. Since

$$\frac{dV_s}{dt} = \frac{C_s(V_s)}{I_{Rp}(V_s)} > 0 \tag{7. 44}$$

shows that the derivative is a positive number, V_s increases with time t; physically, this means that I_{Rp} **adds** charge to C_s. The simplest approximation is obtained by assuming that both I_{Rp} and C_s are constants. Integrating gives

$$V_s(t) \approx V_{min} + \frac{I_{Rp}}{C_s}t \tag{7. 45}$$

which says that the capacitor voltage increases linearly with time as plotted in Figure 7.9(b). We may conclude that a pFET cannot hold a low voltage on the capacitor, but has no problem maintaining a high voltage level since there is a leakage path from the power supply. Let us denote the maximum allowed logic 0 voltage as V_0. The maximum hold time t_H for a logic 0 state is then estimated by

$$t_H \approx \frac{C_s}{I_{Rp}}(V_0 - V_{min}) \tag{7. 46}$$

which is on the order of tens of milliseconds with typical CMOS values. It is important to note that the pFET holding characteristics are exactly opposite to those found for an nFET.

The analysis techniques introduced to model leakage in an nFET can also be applied to a pFET. At the circuit design level, however, we usually use parameters that are extracted from laboratory measurements using wafer probes. This sidesteps the problem of having to perform an accurate analysis to that of using the data in the design process.

7.1.5 Junction Leakage in TGs

Since a transmission gate consists of a parallel nFET-pFET combination, reverse junction charge leakage will occur whenever a TG is used to hold charge on an isolated node. Figure 7.10 illustrates the problem. When the TG is OFF as in Figure 7.10(a), both transistors are in cutoff. The MOSFET

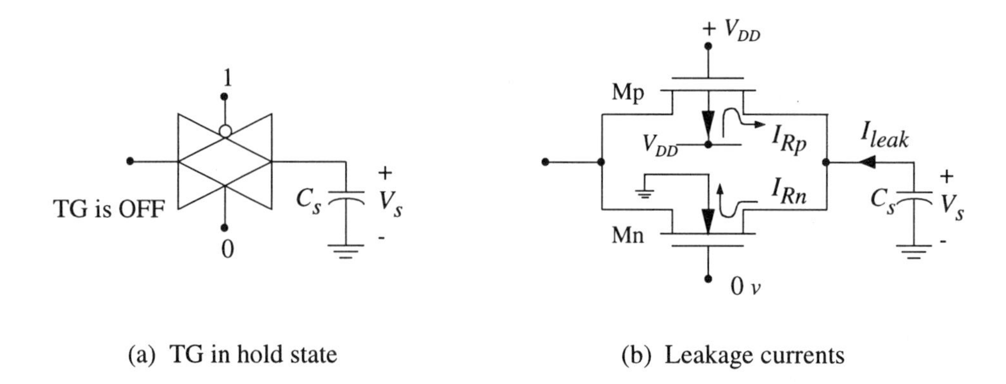

(a) TG in hold state (b) Leakage currents

Figure 7.10 Leakage currents in a transmission gate

circuit shown in Figure 7.10(b) illustrates the existence of two leakage paths exist, one through each device. The leakage current I_{Rp} through the pFET adds charge to the node, while the nFET current I_{Rn} removes charge from the node. Assuming that I_{leak} is a a net flow off of the node with the direction shown, applying KCL gives

$$-C_s\frac{dV_s}{dt} = (I_{Rn} - I_{Rp}) \tag{7.47}$$

with I_{Rn} and I_{Rp} being the nFET and pFET reverse currents, respectively. The sign of the derivative depends upon which current is dominant, so that V_s can increase or decrease, depending upon the situation.

RC Equivalent Circuit

It is instructive to analyze the TG leakage problem using the concept of an RC equivalent network. Figure 7.11(a) provides the starting point for the analysis and shows the leakage paths through reverse biased pn junction diodes. In Figure 7.11(b), each pn junction has been replaced by simple diode model consisting of a junction capacitor C_j in parallel with a linear resistor that has a very small conductance $G = (1/R)$. The load capacitor C_L represents the external load such that the total storage capacitor C_s is given by

$$C_s = C_{jn} + C_L \tag{7.48}$$

The voltage V_s is measured across this parallel combination. The initial voltage is denoted as $V_s(0)$, with specific values substituted after the general analysis is completed.

The reverse voltage V_{Rn} across the nFET diode is given by

$$V_{Rn} = V_s \tag{7.49}$$

so that the current through G_n is

$$I_{n1} = G_n V_s \tag{7.50}$$

This represents the leakage current through the pn junction in the RC equivalent model. Similarly, the reverse bias voltage across the pFET junction is

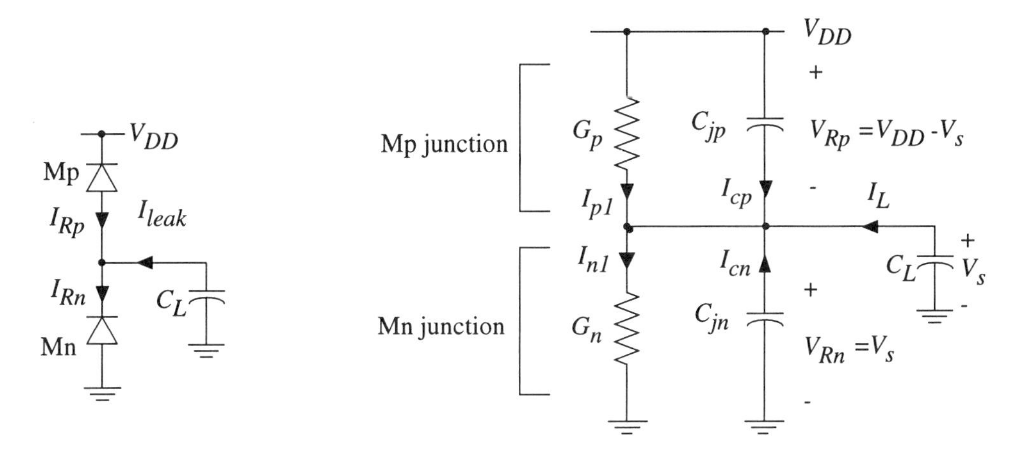

(a) pn junction circuit (b) Equivalent RC network

Figure 7.11 Linear RC model for the TG leakage problem

$$V_{Rp} = V_{DD} - V_s \tag{7.51}$$

so that the leakage current through Mp is approximated as

$$I_{p1} = G_p(V_{DD} - V_s) \tag{7.52}$$

Applying KCL gives the balance equation

$$I_{n1} - I_{p1} = I_{cn} + I_{cp} + I_L \tag{7.53}$$

so noting that

$$I_{cn} = -C_{jn}\frac{dV_s}{dt}$$
$$I_{cp} = +C_{jp}\frac{d}{dt}(V_{DD} - V_s) \tag{7.54}$$
$$I_L = -C_L\frac{dV_s}{dt}$$

gives

$$G_nV_s - G_p(V_{DD} - V_s) = -[C_{jn} + C_L]\frac{dV_s}{dt} + C_{jp}\frac{d}{dt}(V_{DD} - V_s) \tag{7.55}$$

by direct substitution. Rearranging gives the differential equation for $V_s(t)$ of the form

$$C\frac{dV_s}{dt} + GV_s = G_pV_{DD} \tag{7.56}$$

where we have defined

$$C = C_{jn} + C_{jp} + C_L$$
$$G = G_n + G_p \tag{7. 57}$$

as the total capacitance and total conductance, respectively,

Let us solve the equation subject to the initial condition $V_s(0) = V_{DD}$ which corresponds to the storage of a logic 1 voltage. Laplace transforming to s-domain gives

$$sCv(s) - CV_s(0) + Gv(s) = \frac{G_p V_{DD}}{s} \tag{7. 58}$$

where $v(s)$ is the transformed voltage. Solving,

$$\begin{aligned} v(s) &= \frac{G_p V_{DD}}{s(sC+G)} + \frac{CV_{DD}}{(sC+G)} \\ &= \frac{\left(\frac{G_p}{C}\right)V_{DD}}{s\left(s+\frac{G}{C}\right)} + \frac{V_{DD}}{\left(s+\frac{G}{C}\right)} \end{aligned} \tag{7. 59}$$

so expanding into partial fractions and inverse transforming back into time domain yields the storage capacitor voltage

$$V_s(t) = \frac{V_{DD}}{G}\left[G_n e^{-t/\tau} + G_p\right] \tag{7. 60}$$

where the time constant for the circuit has been defined as

$$\tau = \frac{C}{G} = \frac{C_{jn} + C_{jp} + C_L}{G_n + G_p} \tag{7. 61}$$

As a check, we note that $V_s(0) = V_{DD}$ as required. The voltage across the pFET junction can be obtained using

$$\begin{aligned} V_{Rp}(t) &= V_{DD} - V_s(t) \\ &= \frac{G_n}{G}V_{DD}[1 - e^{-t/\tau}] \end{aligned} \tag{7. 62}$$

We note that, as $t \to \infty$,

$$V_s \to \frac{G_p}{G}V_{DD} \tag{7. 63}$$

as the final value. Similarly,

$$V_{Rp} \to \frac{G_n}{G}V_{DD} \tag{7. 64}$$

such that

$$V_s(t) + V_{Rp}(t) = V_{DD} \tag{7. 65}$$

is valid for all times t. This illustrates the fact that the final equilibrium voltage depends upon the relative values of G_n and G_p. In general, the final storage voltage would satisfy

$$0 < V_s < V_{DD} \tag{7.66}$$

indicating some reduction as determined by the conductance values. These are shown in Figure 7.12 for $G_p > G_n$ and $G_n > G_p$. In the extreme case $G_p >> G_n$ corresponding to a large leakage current through the pFET junction that dominates the current through the nFET, then

$$V_s \to V_{DD} , \qquad V_{Rp} \to 0v \tag{7.67}$$

indicating that the logic 1 voltage can be maintained. conversely, if $G_n >> G_p$ we have

$$V_s \to 0v , \qquad V_{Rp} \to V_{DD} \tag{7.68}$$

and the logic 1 voltage is lost.

We may also examine the storage of a logic 0 on C_s by changing the initial condition to be $V_s(0)=0v$. In this case, the expression for the s-domain voltage $v(s)$ is

$$sCv(s) + Gv(s) = \frac{G_p V_{DD}}{s} \tag{7.69}$$

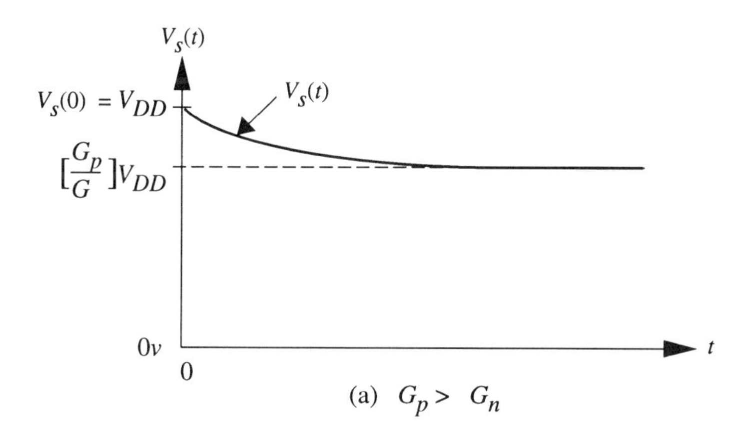

(a) $G_p > G_n$

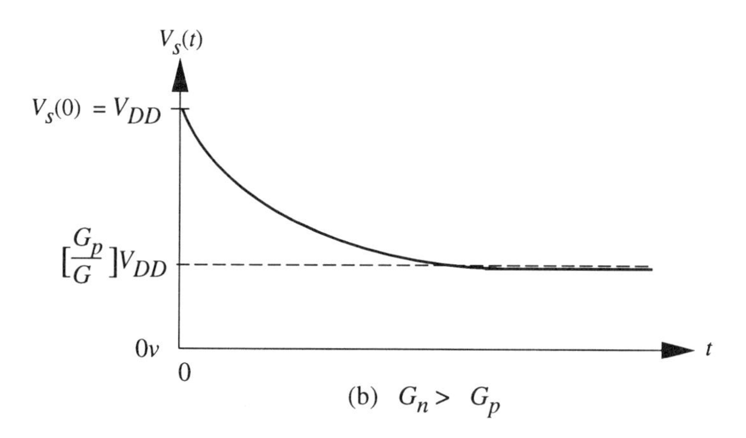

(b) $G_n > G_p$

Figure 7.12 Equilibrium voltage due to leakage effects in a TG

which gives

$$v(s) = \frac{\left(\frac{G_p}{C}\right)V_{DD}}{s\left(s+\frac{1}{\tau}\right)} = \left(\frac{G_p}{G}\right)V_{DD}\left[\frac{1}{s} - \frac{1}{\left(s+\frac{1}{\tau}\right)}\right] \tag{7. 70}$$

Preforming the inverse transform back into the time domain gives the expression

$$V_s(t) = \left(\frac{G_p}{G}\right)V_{DD}[1-e^{-t/\tau}] \tag{7. 71}$$

In this case, the voltage increases with an asymptotic limit of

$$V_s \to \left(\frac{G_p}{G}\right)V_{DD} \tag{7. 72}$$

from its original value of $V_s(0)=0v$. In other words, the TG cannot hold a logic 0 voltage either due to the pFET leakage path. Although the RC-equivalent is highly simplified, it does give a reasonable overview of the charge storage capabilities.

Diode Analysis

A more detailed understanding of the behavior of the TG holding circuit can be understood by looking at the voltage dependence of the leakage currents. Consider the equivalent circuit shown in Figure 7.13. At this level of analysis, a pn junction is represented by a nonlinear junction capacitor C_j in parallel with a voltage-controlled current source with a value of I_R. Although the nonlinearity of the depletion capacitance affects the final value of the capacitor voltage, the interesting physics is controlled by the current sources.

As in the RC analysis, the reverse voltage V_{Rn} across the nFET pn junction is

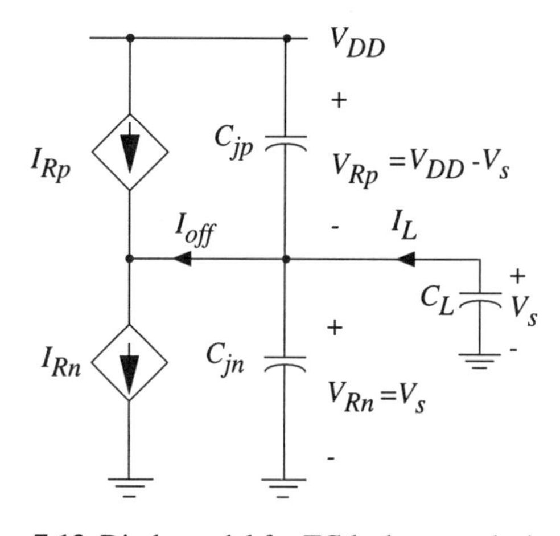

Figure 7.13 Diode model for TG leakage analysis

$$V_{Rn} = V_s \tag{7.73}$$

and the reverse bias across the pFET junction is

$$V_{Rp} = V_{DD} - V_s \tag{7.74}$$

The net current I_{off} that flows off of the junction is thus given by

$$(I_{Rn} - I_{Rp}) = I_{gon}\left(\sqrt{1 + \frac{V_s}{\phi_{on}}} - 1\right) - I_{gop}\left(\sqrt{1 + \frac{(V_{DD} - V_s)}{\phi_{op}}} - 1\right) \tag{7.75}$$

where the additional n and p subscripts in the generation currents are used to denote values for the respective junctions. This expression shows that the direction of the current flow depends upon the voltage V_s. Let us examine the two cases for logic 1 and logic 0 charge states stored on the node.

First, suppose that a logic 1 voltage $V_s(0)=V_{DD}$ is placed on the storage capacitor C_s. When the TG is opened at t=0, the net current flow off of the node is

$$(I_{Rn} - I_{Rp})\Big|_{t=0} = I_{gon}\left(\sqrt{1 + \frac{V_{DD}}{\phi_{on}}} - 1\right) \tag{7.76}$$

showing that V_s will fall. However, as V_s decreases, the reverse current I_{Rp} from V_{DD} to C_s through the pFET increases. Eventually, equilibrium will be reached when $I_{Rn}=I_{Rp}$. At this point, the capacitor voltage has a value $V_s=V_{eq}$ that can be found by solving

$$I_{gon}\left(\sqrt{1 + \frac{V_{eq}}{\phi_{on}}} - 1\right) = I_{gop}\left(\sqrt{1 + \frac{(V_{DD} - V_{eq})}{\phi_{op}}} - 1\right) \tag{7.77}$$

Qualitatively we see that $V_{eq} < V_{DD}$, so that the TG cannot hold a logic 1 voltage without decaying.

The opposite case is where a logic 0 is stored on C_s with $V_s(0)=0v$. The initial leakage current off of the node is now given by

$$(I_{Rn} - I_{Rp})\Big|_{t=0} = -I_{gop}\left(\sqrt{1 + \frac{V_{DD}}{\phi_{op}}} - 1\right) \quad , \tag{7.78}$$

i.e., there is a net charge flow into C_s (since the current is negative), increasing the voltage $V_s(t)$ with time. This verifies our earlier observation that the TG-isolated node cannot hold a logic 0 voltage due to leakage through the p-channel transistor.

Summary

The above analyses demonstrates that a capacitive node that is isolated using a transmission gate cannot hold either logic 0 or logic 1 charge states. This limits the use of the nodes to situations where the voltage state does not have to be held longer than a few hundred milliseconds.

7.2 Charge Sharing

Another important problem that occurs in dynamic circuit is that of **charge sharing**. This occurs when the charge on an isolated capacitive node is used to drive another isolated capacitive node. Consider the situation shown in Figure 7.14. The two capacitors C_1 and C_2 represent parasitic contributions due to the physical structure of the transistors. Suppose that initially nFET M1 is ON while M2 is in cutoff as in Figure 7.14(a). Capacitor C_1 charges to a voltage $V_1 = V_{max}$ giving it a total charge of

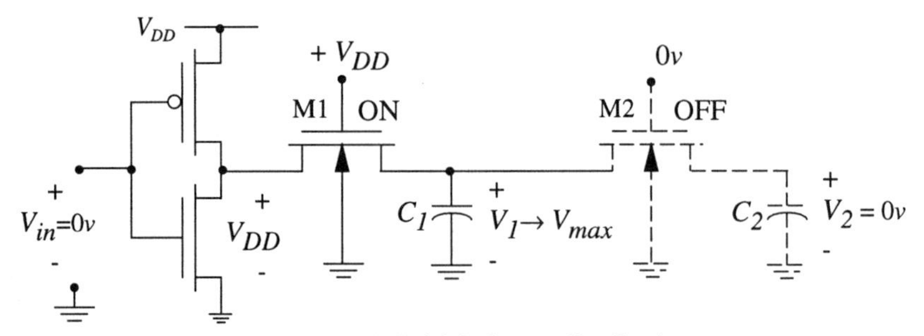

(a) Initial charge distribution

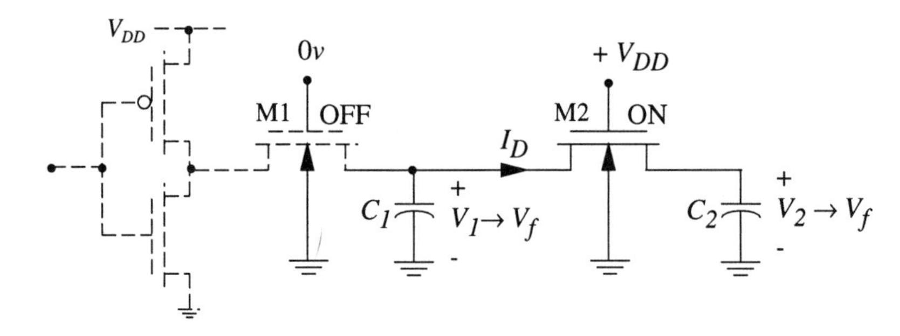

(b) Charge sharing

Figure 7.14 Charge sharing in a FET chain

$$Q_T = C_1 V_{max}. \tag{7.79}$$

Since C_2 is isolated during this event, it is assumed that $V_2 = 0v$.

Now then, suppose that the FETs are switched at time t=0 so that M1 is OFF and M2 is ON; this is shown in Figure 7.14(b). Since there is a difference in voltage between the left and right sides of the transistor, drain current I_D flows in the direction shown. This removes charge from C_1 and adds it to C_2, so that the charge is *shared* between the two capacitors. Since the individual charges are given by

$$Q_1 = C_1 V_1 \quad \text{and} \quad Q_2 = C_2 V_2 \tag{7.80}$$

we see that the current off of C_1 causes V_1 to decrease while V_2 increases. Eventually, equilibrium is reached where the two capacitors have the same final voltage

$$V_f = V_1 = V_2 \tag{7.81}$$

and the current falls to zero. The total charge in the network is now distributed according to

$$\begin{aligned} Q_T &= Q_1 + Q_2 \\ &= C_1 V_f + C_2 V_f \end{aligned} \tag{7.82}$$

However, by conservation of charge, the total charge remains constant, so that this must be equal to the charge originally placed on C_1. Equating the two expressions for Q_T and rearranging gives the final voltage across the two capacitors as

$$V_f = \left(\frac{C_1}{C_1 + C_2}\right) V_{max} \tag{7. 83}$$

By inspection it is seen that the factor in parentheses must be less than one. Thus, the final voltage is characterized by

$$V_f < V_{max} \tag{7. 84}$$

i.e., it is less than the original voltage across C_1. This may reduce the voltage to the point where it is no longer large enough to be a valid logic 1 value. To insure that the voltage remains high, the capacitor values should be such that

$$C_1 \gg C_2 \tag{7. 85}$$

is satisfied. Since the parasitic capacitances of a MOSFET are determined by the layout geometry, satisfying this condition often requires very careful layout planning.

Figure 7.15 provides a convenient way in which to view the changes in the voltages $V_1(t)$ and $V_2(t)$. The initial conditions $V_1(0) = V_{max}$ and $V_2(0) = 0v$ are shown in (a). As charge transfer takes place, $V_1(t)$ decreases since it loses charge, while $V_2(t)$ increases as it receives charge from C_1. Equilibrium is portrayed in Figure 7.15(b), where the two voltages have reached the same final voltage $V_1 = V_2 = V_f$. Since the voltage difference between the two sides is zero, the current flow also goes to zero, indicating that the charge sharing processes has been completed.

7.2.1 RC Equivalent

It is useful to examine the charge sharing problem by using an RC model for the MOSFET. The circuit is shown in Figure 7.16(a). For times $t < 0$, the switch is open with a gate signal $G = 0$ and the voltages are given by $V_1 = V_{max}$ and $V_2 = 0v$. Closing the switch at $t = 0$ [Figure 7.16(b)] gives the voltage across the resistor as

$$V_R = V_1(t) - V_2(t) \tag{7. 86}$$

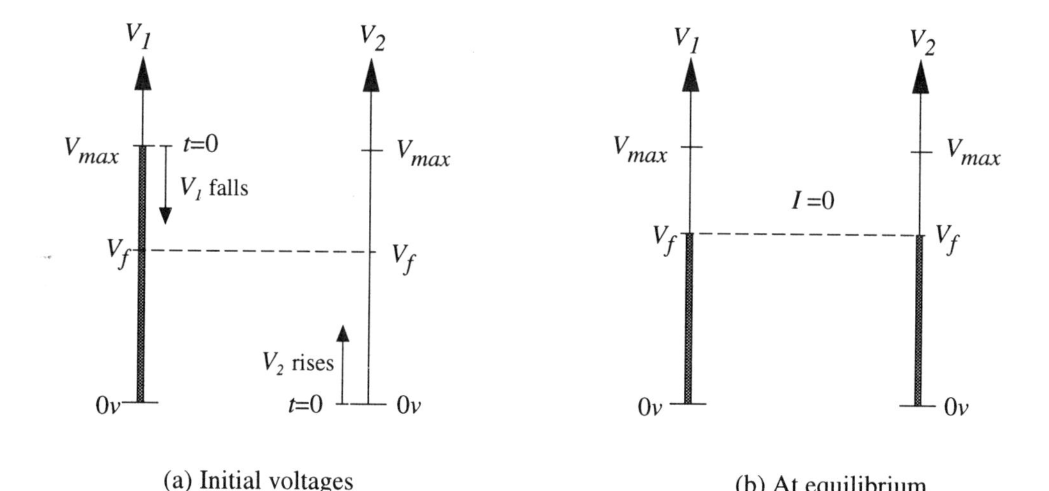

Figure 7.15 Evolution of voltages in the charge sharing problem

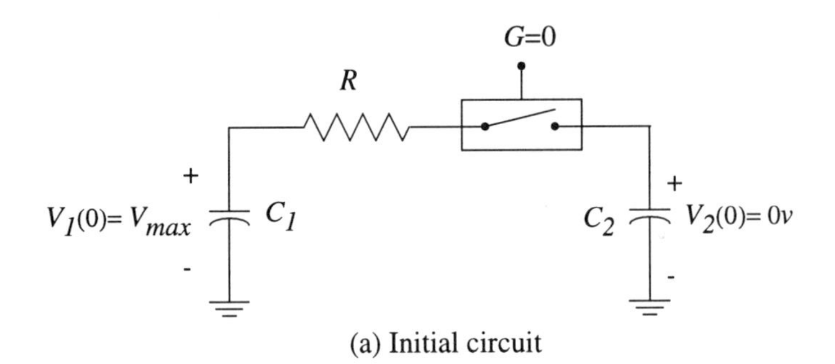

(a) Initial circuit

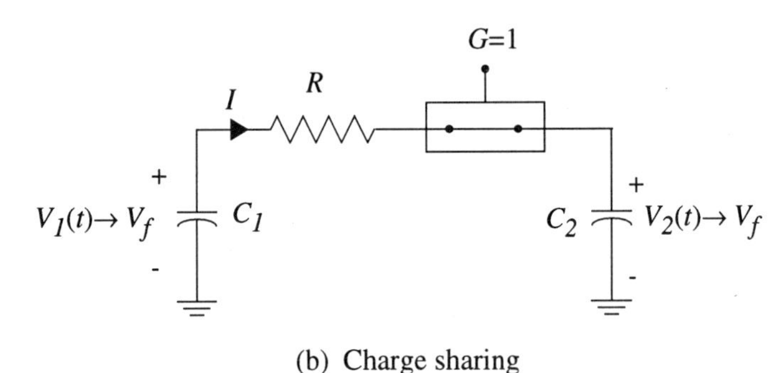

(b) Charge sharing

Figure 7.16 RC modelling of the charge sharing problem

such that the current I through the resistor is

$$I = \frac{V_1 - V_2}{R} = -C_1 \frac{dV_1}{dt} = C_2 \frac{dV_2}{dt} \tag{7. 87}$$

The circuit can be analyzed in either time domain or s-domain; we will choose the former here since the mathematics it straightforward. First, let us write the difference in time derivatives as

$$\begin{aligned} \frac{dV_1}{dt} - \frac{dV_2}{dt} &= -\left(\frac{I}{C_1} + \frac{I}{C_2}\right) \\ &= -\frac{I}{C_{eq}} \end{aligned} \tag{7. 88}$$

where

$$C_{eq} = \frac{C_1 C_2}{C_1 + C_2} \tag{7. 89}$$

shows that the two capacitors are in series for the charge sharing event. Using the equation for I gives the differential equation

$$\frac{d(V_1 - V_2)}{dt} = -\frac{(V_1 - V_2)}{\tau} \tag{7. 90}$$

where

$$\tau = RC_{eq} \tag{7.91}$$

is the time constant. Solving and applying the initial conditions gives

$$V_1(t) - V_2(t) = V_{max}e^{-t/\tau} \tag{7.92}$$

which shows that in the limit as $t \to \infty$, $V_1 \to V_2$, i.e., the voltages have the same final value.

To examine the behavior of the individual voltages, we first note that

$$\frac{dV_1}{dt} = -\left(\frac{C_2}{C_1}\right)\frac{dV_2}{dt} \tag{7.93}$$

follows from the basic I-V relations. The derivative of the difference voltage $(V_1 - V_2)$ may then be written as

$$\frac{d}{dt}(V_1 - V_2) = -\left(1 + \frac{C_2}{C_1}\right)\frac{dV_2}{dt}. \tag{7.94}$$

so that using the solution for $(V_1 - V_2)$ and integrating gives

$$V_2(t) = \left(\frac{C_1}{C_1 + C_2}\right)V_{max}(1 - e^{-t/\tau}). \tag{7.95}$$

The voltage across C_1 is then easily obtained as

$$V_1(t) = \frac{V_{max}}{(C_1 + C_2)}\left(C_1 + C_2 e^{-t/\tau}\right) \tag{7.96}$$

using simple algebra. To check these results, note that at $t = 0$, $V_1(0) = V_{DD}$ and $V_2(0) = 0v$ as required. In the limit where $t \to \infty$,

$$\lim_{t \to \infty} V_1(t) = \lim_{t \to \infty} V_2(t) = \left(\frac{C_1}{C_1 + C_2}\right)V_{max} = V_f \tag{7.97}$$

which is the same result that was obtained using charge conservation.

The dynamics of the charge sharing problem are illustrated by the plots in Figure 7.17. The charge redistribution time is characterized by the value of $\tau = RC_{eq}$. Note that τ is less than both of the individual times constants RC_1 and RC_2, so that the sharing event is relatively fast. Figure 7.17(a) shows the case where $C_1 > C_2$; this gives

$$V_f = \left(\frac{C_1}{C_1 + C_2}\right)V_{max} > \frac{1}{2}V_{max} \tag{7.98}$$

as the final voltage. This will result in a "high" voltage as desired. In Figure 7.17(b), on the other hand, we consider the case where $C_1 < C_2$ so that

$$V_f = \left(\frac{C_1}{C_1 + C_2}\right)V_{max} < \frac{1}{2}V_{max} \tag{7.99}$$

This may be incorrectly be interpreted as a logic 0 level because charge sharing has decreased the voltage to a low value.

Charge sharing can result in incorrect voltage levels being transmitted through a logic chain. As

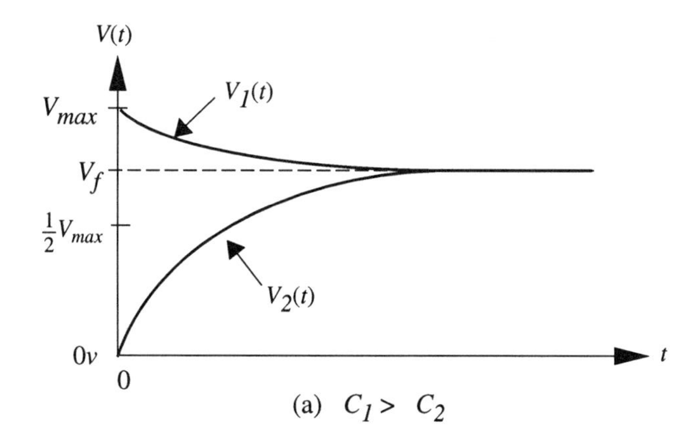

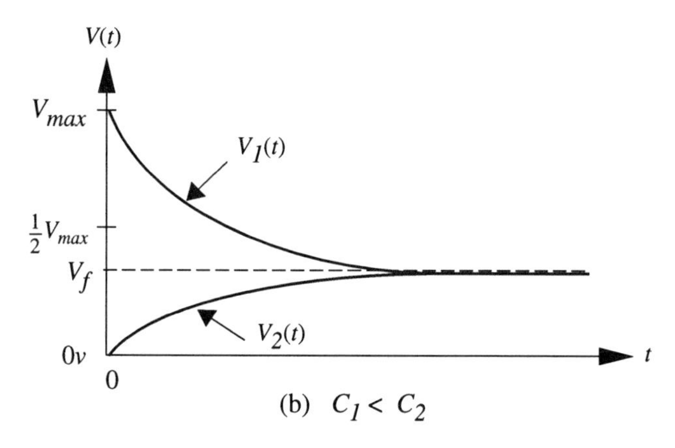

Figure 7.17 Evolution of the final voltage due to charge sharing

an example, let us consider the situation in Figure 7.18 where we have chosen numerical values of $C_1 = 60fF$ and $C_2 = 15fF$. With the assumed initial voltage of $V_1(0) = V_{max} = 4v$, then the final voltage is easily calculated to be

$$V_f = \left(\frac{60}{15+60}\right)4 = 3.2v \tag{7. 100}$$

As indicated in the drawings, this is a high voltage and corresponds to transmitting the logic "1" on C_1 to the right node defined by C_2. Now consider the situation shown in Figure 7.19 where we have switched the values of the capacitors to $C_1 = 15fF$ and $C_2 = 60fF$. Again assuming $V_1(0) = V_{max} = 4v$, the final voltage on both capacitors is now given by

$$V_f = \left(\frac{15}{15+60}\right)4 = 0.8v \tag{7. 101}$$

which will probably be interpreted as a logic "0". The problem is now clear. Initially, we had a logic "1" voltage on the left and a logic "0" on the right. After charge sharing, both nodes are at a logic "0" voltage. This makes it look like the "0" originally on the right node has "propagated backwards" to the left node! Obviously, this situation must be avoided at all costs as it will lead to errors and glitches.

The charge sharing analysis can be extended to the case of driving multiple capacitors such as

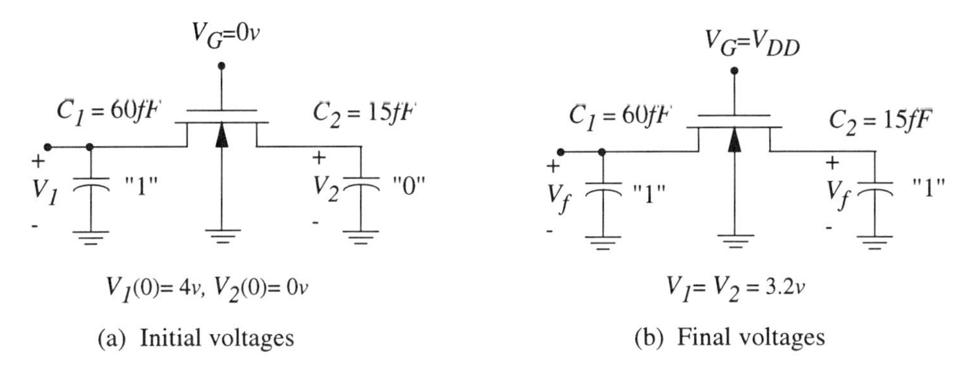

Figure 7.18 Numerical example of a charge sharing calculation

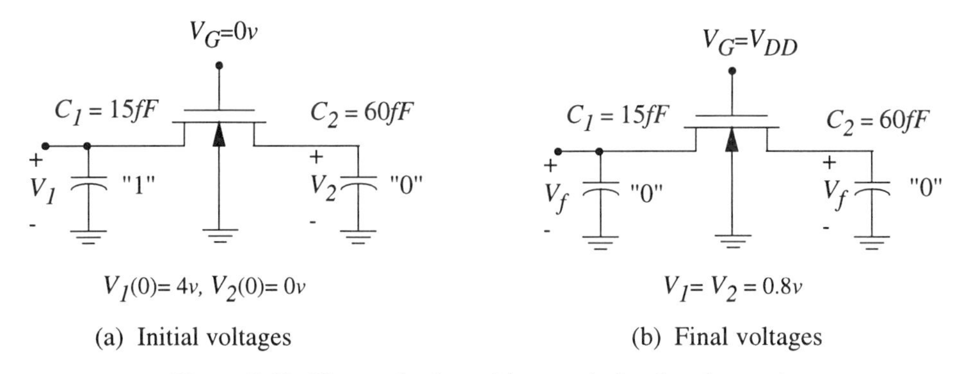

Figure 7.19 Charge sharing with oppositely-placed capacitors

that shown in Figure 7.20. In this circuit, the charge on C_1 is shared with three other capacitors C_2, C_3, and C_4. The initial conditions are shown in drawing (a) as

$$V_1(0) = V_{max}\,, \quad V_2(0) = V_3(0) = V_4(0) = 0v \tag{7. 102}$$

so that the total charge in the system is

$$Q_T = C_1 V_{max} \tag{7. 103}$$

After charge sharing takes place, all of the capacitors will have the same voltage V_f. The charge is distributed according to

$$Q_T = (C_1 + C_2 + C_3 + C_4)\, V_f \tag{7. 104}$$

so that equating the two expressions gives

$$V_f = \left(\frac{C_1}{C_1 + C_2 + C_3 + C_4}\right) V_{max} \tag{7. 105}$$

as the final voltage. To keep V_f high now requires that

$$C_1 \gg (C_2 + C_3 + C_4) \tag{7. 106}$$

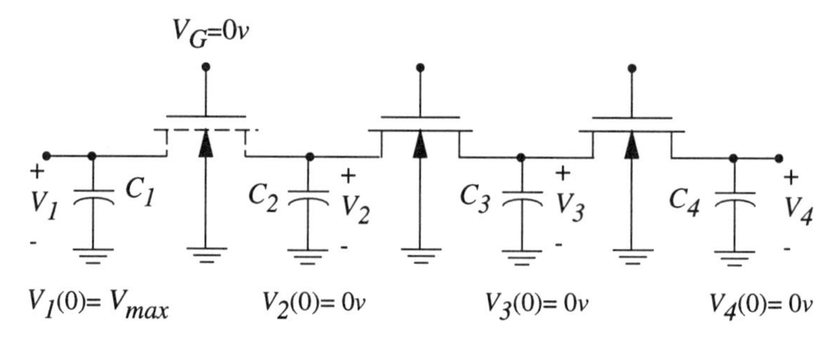

(a) Initial voltages

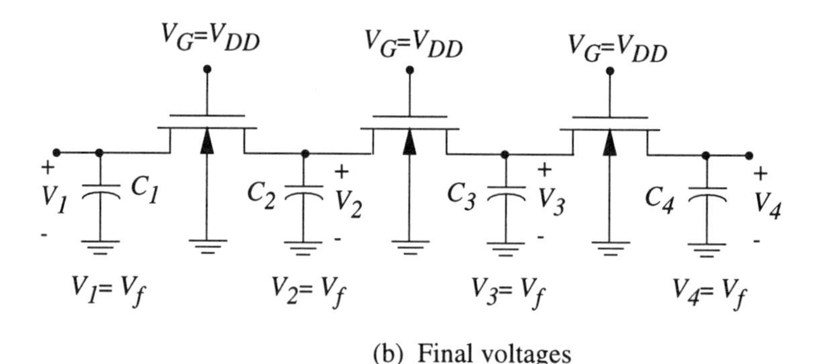

(b) Final voltages

Figure 7.20 Charge sharing among several nodes

i.e., capacitor C_1 must be much larger than the *sum* of the other capacitors. This shows that charge sharing becomes more severe as additional charge sharing nodes are added.

One way to battle charge sharing voltage reductions is to provide more initial charge to the network by charging more than one capacitor. In general, this technique is called **precharging**, and used extensively in dynamic circuit arrangements. Figure 7.21 shows an example of this is shown for the four capacitor chain. In this case, C_1 is charged to V_{max} while C_4 has an initial voltage V_y as in Figure 7.21(a). The total charge is now increased to

$$Q_T = C_1 V_{max} + C_4 V_y \tag{7. 107}$$

After charge sharing takes place, all capacitors have the same voltage so that

$$Q_T = (C_1 + C_2 + C_3 + C_4) V_f \tag{7. 108}$$

is still valid. Equating then gives

$$V_f = \frac{C_1 V_{max} + C_4 V_y}{(C_1 + C_2 + C_3 + C_4)} \tag{7. 109}$$

as the final value. It is clear that this helps maintain the voltage at a logic 1 level. Note, however, that if we wish to discharge the capacitor(s) to 0v, it will take longer because of the additional charge introduced by the precharge operation.

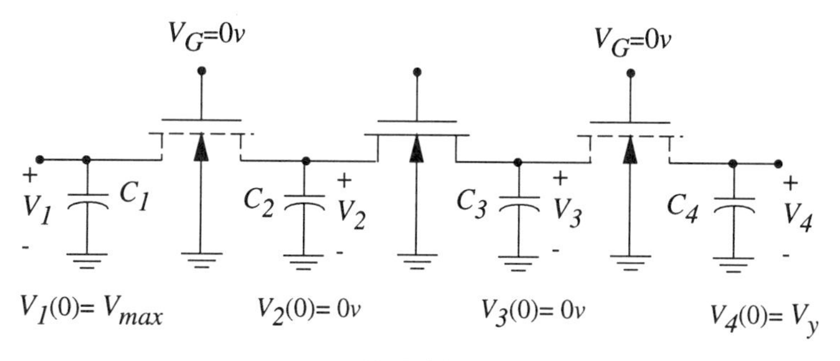

(a) Initial voltages

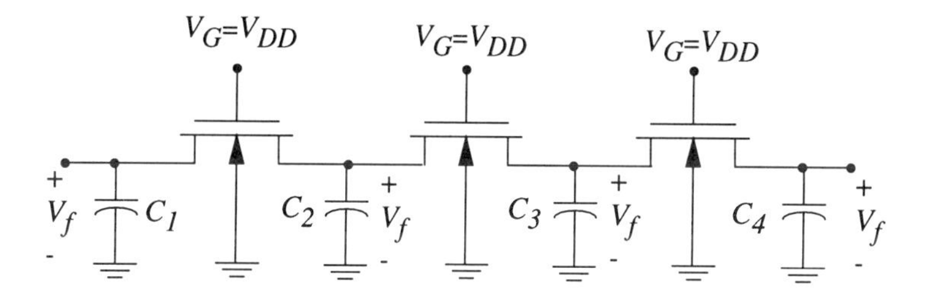

(b) Final voltages

Figure 7.21 Charge sharing with two precharged capacitors

7.3 The Dynamic RAM Cell

A dynamic random-access memory (DRAM) cell is a storage circuit that consists of an access transistor MA and a storage capacitor C_s as shown in Figure 7.22. The access FET is controlled by the Word line signal *WL*, and the bit line is the input/output path. The simplicity of the circuit makes it very attractive for high-density storage. As with any RAM, there are three distinct operations for a cell:

- **Write** - A data bit is stored in the circuit;
- **Hold** - The value of the data bit is maintained in the cell;

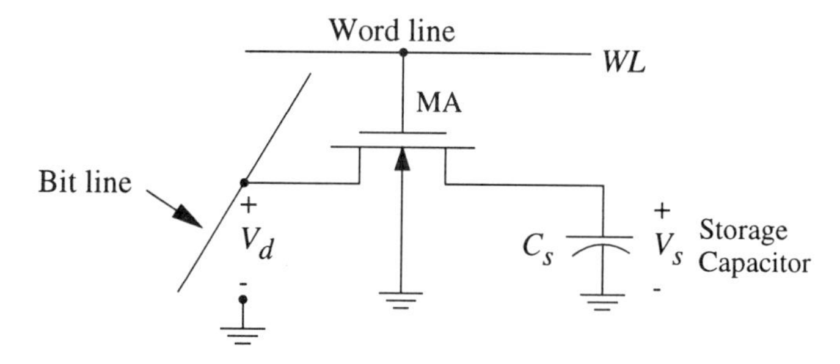

Figure 7.22 Basic DRAM cell

and

- **Read** - The value of the data bit is transferred to an external circuit.

Access to the capacitor is controlled by the word line signal *WL* that is connected to the gate of the access transistor. Aside from this change in nomenclature, the cell itself is identical to the circuit in Figure 7.1, so that the operation is easily understood by applying the results of this chapter to each of the three operational modes.

Write

The write operation is illustrated in Figure 7.23. The word line voltage is elevated to a value of V_{DD} (corresponding to a logic a level $WL = 1$) to turn on the access FET; the input data voltage V_{in} on the bit line then establishes the required charge Q_s on the capacitor. To store a logic 1 in the cell, V_{in} is set to the value of V_{DD} so that the storage cell voltage V_s increases according to

$$V_s(t) = (V_{DD} - V_{Tn}) \left[\frac{t/2\tau_s}{1 + t/2\tau_s}\right] \tag{7. 110}$$

where the time constant is given by

$$\tau_s = \frac{C_s}{\beta_n (V_{DD} - V_{Tn})}. \tag{7. 111}$$

Storage of a logic 0 is accomplished by using an input voltage of $V_{in} = 0v$, so that the capacitor is discharged as described by

$$V_s(t) = V_s(0) \left[\frac{2e^{-t/\tau_s}}{1 + e^{-t/\tau_s}}\right] \tag{7. 112}$$

where we have designated the voltage at the beginning of the write operation as $V_s(0)$. From our earlier discussions, we may conclude that writing a logic 1 value requires more time than writing a logic 0 state.

Hold

A DRAM cell holds the charge on the capacitor by turning off the access transistor using $WL = 0$ as shown in Figure 7.24. This creates an isolated node, and charge leakage occurs if a logic 1 high voltage is stored on C_s. The maximum hold time t_H for a logic 1 bit can be estimated by

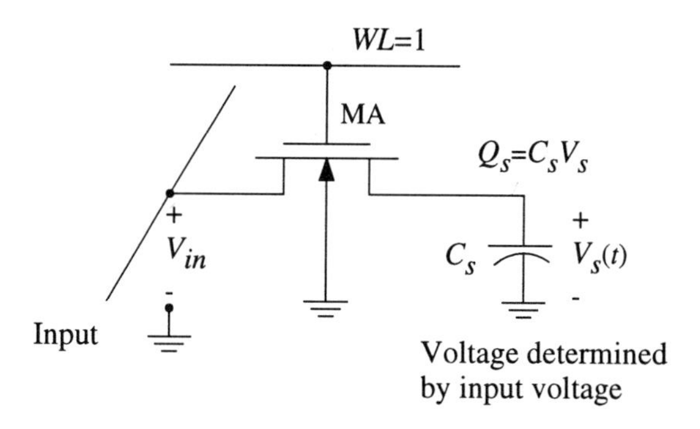

Figure 7.23 Write operation in a DRAM

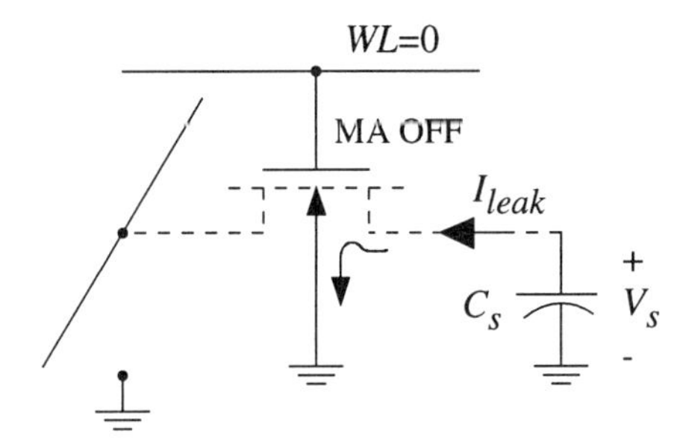

Figure 7.24 Hold operation in a DRAM cell

$$t_H \approx \frac{C_s}{I_{leak}}(V_{max} - V_1) \tag{7. 113}$$

and is usually limited to a duration on the order of 100 milliseconds. In physical DRAM cell designs, several contributions to the leakage current are found, and much time is dedicated to the problem of increasing the charge retention time. This change in the stored charge (and hence, the logic level) in time is the origin of the name **dynamic**.[5] Special **refresh** circuits that periodically read the bit, amplify the voltage, and rewrite the data to the cell must be added so that the that the circuit can be used to store data for longer periods of time.

Read

When a read operation is performed, the data bit line is connected to the input of a high-gain **sense amplifier** that is designed to provide amplification of the voltage level. During this operation, the line is connected to FET gates terminals so that the line itself is capacitive. This is included in Figure 7.25 by including a line capacitance C_{line}. During a read operation, charge sharing will take place between the storage cell and the output line, resulting in a final voltage data voltage V_d of

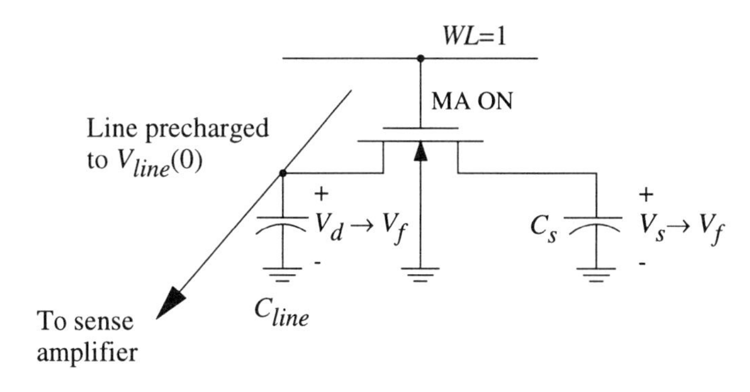

Figure 7.25 Read operation in a DRAM cell

[5] Which means *varying in time.*

$$V_d = \left(\frac{C_s}{C_s + C_{line}}\right)V_s \tag{7.114}$$

A major difficulty in designing high-density DRAM chips is that the cell storage capacitance C_s is relatively small compared to the parasitic line capacitance C_{line}. For example, a ratio with a value $(C_{line}/C_s)=8$ would be considered reasonable in modern high-density design. This implies that the difference between a logic 0 and a logic 1 data voltage is very small. One way to help this situation is to precharge the line to an initial voltage $V_{line}(0)$ before the read operation. The final voltage after charge sharing takes place is then given by

$$V_d = \left(\frac{C_s}{C_s + C_{line}}\right)V_s + \left(\frac{C_{line}}{C_s + C_{line}}\right)V_{line}(0) \tag{7.115}$$

For example, suppose that we choose $V_{line}(0) = (1/2)V_{DD}$. Assuming that the line capacitance is large compared to the storage capacitance, the final voltage on the line will be

$$V_d \approx \Delta V_s + \left(\frac{1}{2}\right)V_{DD} \tag{7.116}$$

where ΔV_s is the change due to the stored charge. This gives an output voltage that changes around the reference value of $(1/2)V_{DD}$; this change can be detected using a comparator as a sense amplifier.

7.3.1 Cell Design and Array Architecture

High-density DRAMs are often viewed as the **technology drivers** for the discipline of silicon integrated circuits. The conflicting requirements of increasingly larger cell arrays and smaller die size means that the most sophisticated chip designs are created using the most advanced fabrication technology available.

Let us examine the problem of creating a high-capacity storage cell as needed to combat both charge leakage and charge sharing problems in the DRAM circuit. From the chip viewpoint, only a finite area A_{chip} is available to implement the design, and this should be as small as possible to ensure a reasonable yield[6] in the fabrication line. A generalized DRAM layout is portrayed in Figure 7.26. Storage cells are arranged in convenient groups, with the largest group called a **block**. A block consists of several **sub-blocks**, and so on. The storage density of the DRAM chip is ultimately determined by the surface area A_{cell} of a single cell. The cell area A_{cell} is called the **footprint** and (from the design perspective) represents the largest surface area that is allocated for one storage circuit. Obviously, we want the footprint to be as small as possible, less than $1\mu m^2$ in a modern design. Since the storage capacitance is typically 50-60*fF*, this presents a unique challenge to the chip layout designer.

To see how the limit on A_{cell} affects the chip design, consider the simple parallel-plate capacitor shown in Figure 7.27. The capacitance is estimated by

$$C = \frac{\varepsilon A_s}{d} \tag{7.117}$$

which ignores fringing fields, trapped charge, and non-uniformities in the electric field. Based on this equation, we see that the capacitance can be increased by three obvious techniques:

[6] The die yield Y was discussed in Chapter 2.

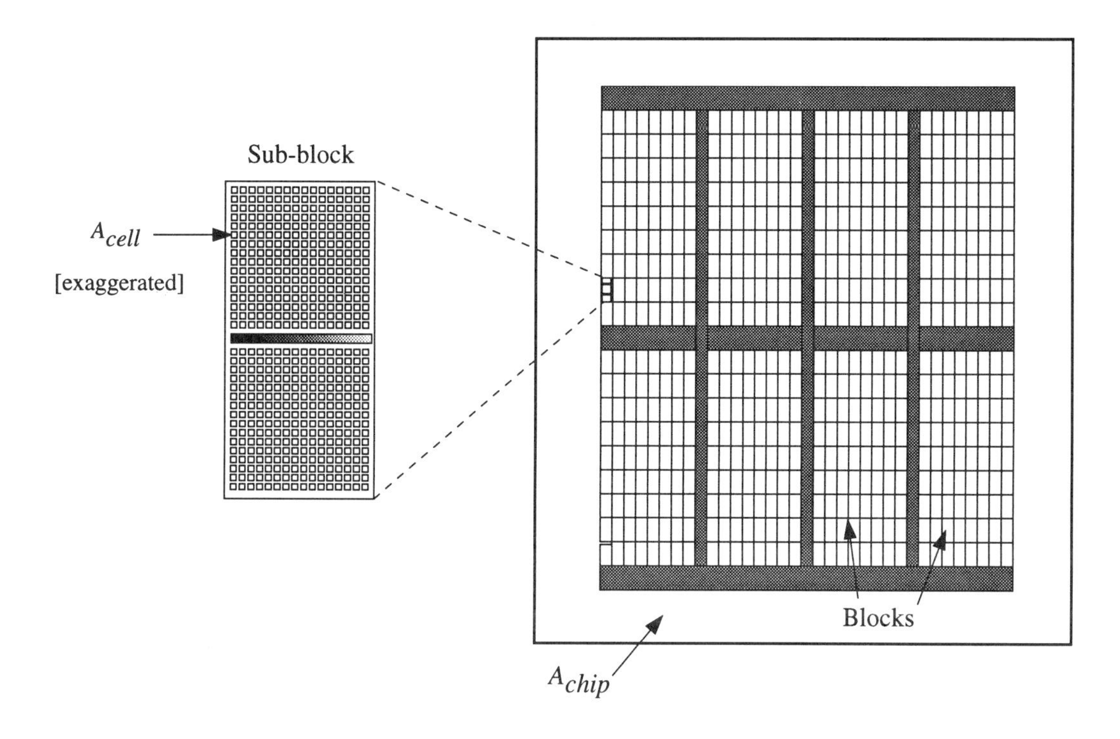

Figure 7.26 Chip layout for a DRAM array

- increase the permittivity ε of the dielectric;
- decrease the distance of separation d between the plates; or ,
- increase the surface area A_s of the plates.

Modern DRAM cell design employs all three techniques along with some additional considerations to increase C_s to a reasonably large value.

Composite Insulators

Classically, silicon dioxide (generically called **oxide** here) has served as the insulating dielectric in MOS capacitors. This is due to many reasons, including ease of growth, excellent insulating characteristics, and uniform coverage. However, the permittivity of silicon dioxide is relatively small

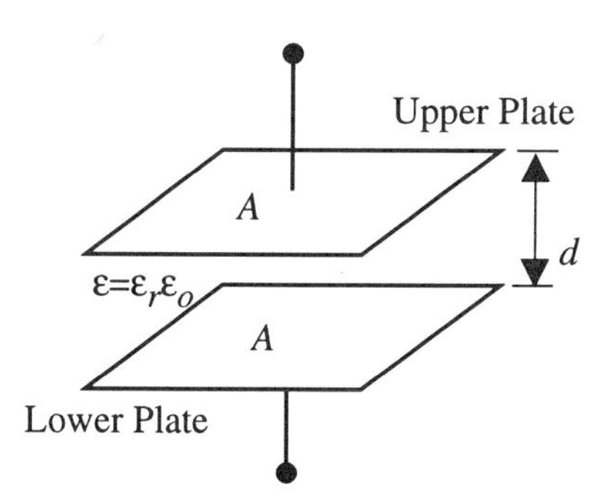

Figure 7.27 Parallel-plate capacitor

($\varepsilon_{ox} \approx 3.9\varepsilon_o$), and extremely thin oxides are subject to electrical breakdown[7] and tunnelling problems, so that there has always been interest in finding an insulator that has superior characteristics. One popular material is silicon nitride (Si_3N_4 or simply **nitride**) which has a permittivity of $\varepsilon_N \approx 7.8\varepsilon_o$. Nitride has been well studied because of its extensive use in device isolation techniques, and as a passivation layer that covers and protects the finished die. If we replace an oxide insulator with a nitride layer, then the capacitance approximately doubles. There are, however, overriding technology problems that arise.

One approach to using this technology is the ON (oxy-nitride) structure shown in Figure 7.28. The plates of DRAM storage capacitors are made from polysilicon, which is easy to oxidize using thermal growth techniques. Insulating nitride layers are more difficult to create, but can be grown using rapid thermal nitridation (RTN) where a nitrogen gas flow is established over the poly surface and heat is used as a catalyst for the reaction. Oxide is then deposited over the nitride layer which helps to "seal" the layer by filling pinholes. The top polysilicon plate competes the structure. The main idea is to create a composite insulator with a thickness

$$d_{ON} = d_{ox} + d_{Nitride} \tag{7. 118}$$

that has an effective permittivity

$$\varepsilon_{ON} = x\,\varepsilon_{ox} + y\,\varepsilon_{Nitride} \tag{7. 119}$$

where x and y are multipliers that are determined by the relative thicknesses. The actual value of the permittivity will be in the range

$$\varepsilon_{ox} < \varepsilon_{ON} < \varepsilon_{Nitride} \tag{7. 120}$$

so that the structure will exhibit a larger capacitance than if only a simple oxide insulator of the same thickness were used. ONO and other structures have also been published in the literature. While these techniques help increase the value of C_s, they are not sufficient by themselves to boost the capacity to the necessary values.

Storage Capacitor Design

Modern storage cell design techniques center around increasing the capacitor plate area by creating 3-dimensional capacitor structures. This leads to larger values of C_s without increasing the foot-

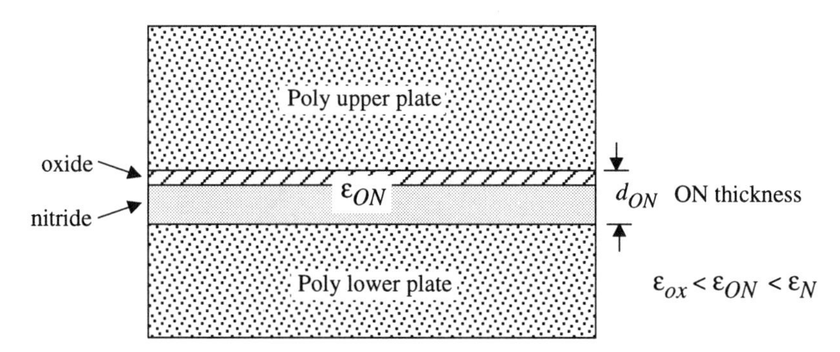

Figure 7.28 Structure of an ON-insulator capacitor

[7] Oxide breakdown is discussed in Chapter 10 in the context of input protection networks.

print. There are two main approaches. One is to create a **trench capacitor** by first etching away a portion of the of the silicon substrate, and then constructing the capacitor using the walls of the trench. The other is to build the capacitor above the substrate plane, creating what is called a **stacked capacitor**. Although both are found in practice, stacked capacitors are more common than trench structures.

An integrated pair of DRAM cells that use trench capacitors is illustrated in Figure 7.29. Let us examine Cell 1 on the left side in more detail to understand the structure. The lower plate of the storage capacitor is the extended n^+ region implanted in the silicon substrate; this becomes one side of the access transistor MA1, which is connected to the bit line on the other side. The storage cell itself is created by etching a trench into the silicon substrate and then oxidizing the surface to give the insulating oxide layer. Doped polysilicon is deposited as a filler, and also acts as the upper capacitor plate. A metal layer then provides electrical contact to the top plate of the cell.

The philosophy behind the trench capacitor structure is easily understood by referring to the simplified geometry illustrated in Figure 7.30. It is seen that the plate area of the capacitor is given by

$$A_s = A_{bot} + A_{side} \tag{7. 121}$$

This shows explicitly that the plate area is larger than the footprint area

$$A_{footprint} = A_{bot} \tag{7. 122}$$

because of the sidewall area A_{side}. With the dimensions shown in the drawing, $A_{bot}=XY$ while

$$A_{side} = 2XD + 2YD \tag{7. 123}$$

where D is the depth of the trench. Deep trenches are useful for increasing the capacitance, but are more difficult to fabricate.

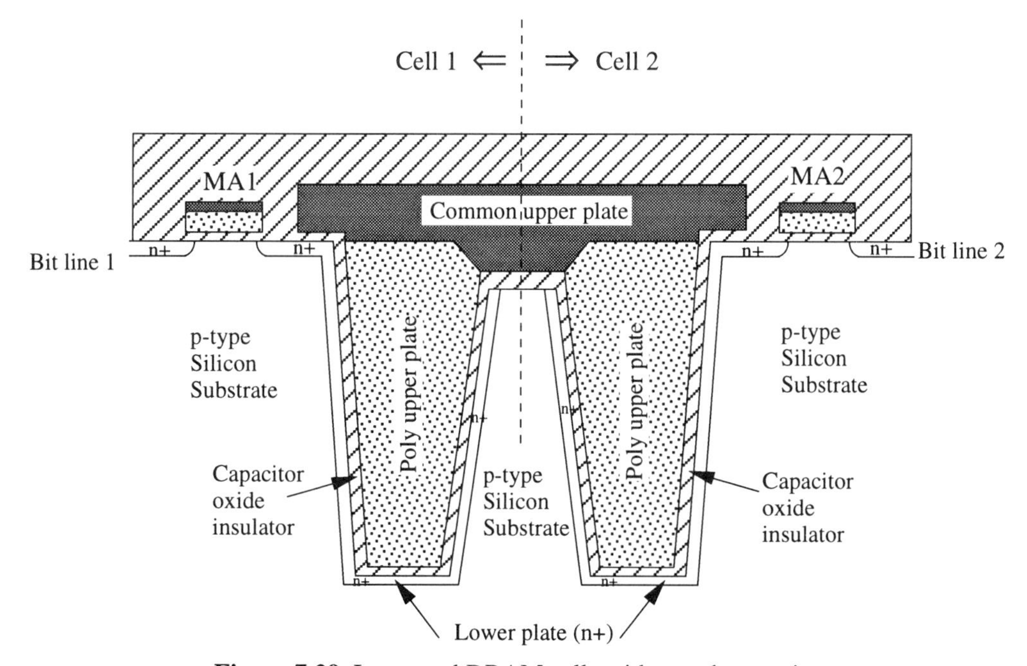

Figure 7.29 Integrated DRAM cells with trench capacitors

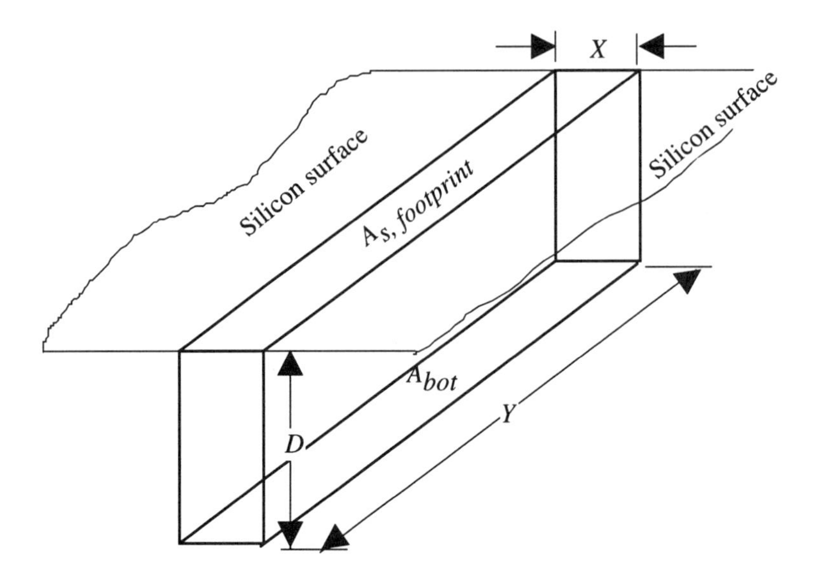

Figure 7.30 Geometry of a trench capacitor

Stacked-capacitor cells provide an alternate to using trench capacitors. These preserve the small footprint area by building a capacitor above the silicon surface and stacking it on top of the FET. The general idea is shown in Figure 7.31. Many variations in stacked capacitor structures have been published in the literature, and large numbers of patents have been issued for specific structures. All have the same idea: increase the surface area of the storage capacitor C_s without violating the footprint budget. To achieve this, various 3-dimensional polysilicon structures have been built. Also, corrugated and "bumpy" surfaces have been created as a means to increase the capacitor plate area, and to introduce non-uniform electric field densities, as both increase the device capacitance. Micro-structure engineers have produced many ingenious ideas in this area.

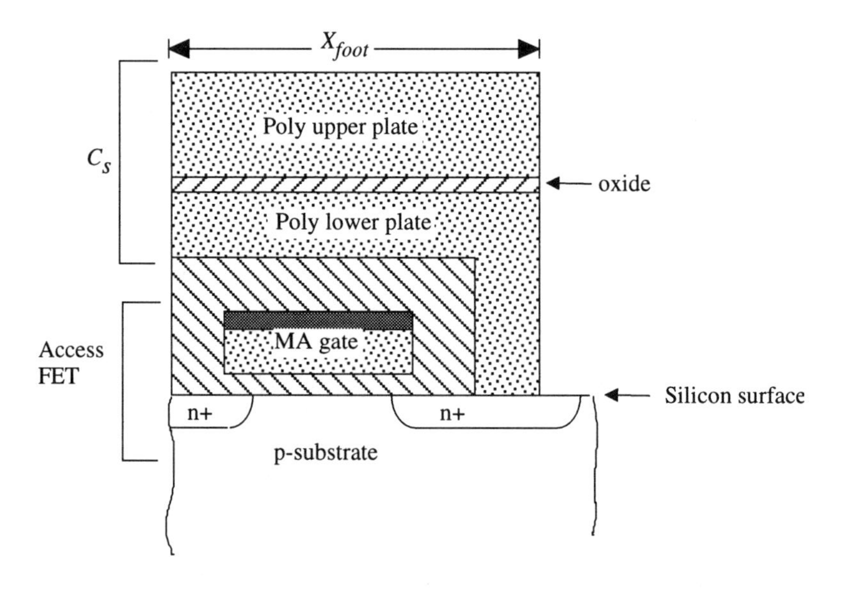

Figure 7.31 Cross-sectional view of a stacked capacitor DRAM cell

With modern DRAM designs exceeding 1Gbit per chip, this has become a highly specialized field for circuit designers and micro-structure fabrication engineers. Most chip designs that have embedded DRAM use library entries rather than resorting to a full-custom design. This is, in fact, the recommended route for most chip design engineers.[8]

7.3.2 DRAM Overhead Circuits

The refresh operation is worth mentioning because it counts as overhead in the DRAM array itself. Because of the charge leakage problem, we must periodically read the contents of a cell, amplify it, and rewrite it back to the same cell. This circuitry is provided on-chip to make the use of the DRAM integrated circuit straightforward to the board designer. At the chip level, however, one must always remember that the refresh circuitry must be accounted for in the planning stages.

Another type of overhead circuits that are commonly found in DRAM are for implementing error-detection codes (EDC) and error-correction codes (ECC). The former are designed to detect the presence of an error in the output, while the latter are algorithms that correct the errors and provide the correct output word.

7.4 Bootstrapping and Charge Pumps

The term **bootstrapping** usually refers to "pulling up" the value of a physical parameter.[9] In the present context, we will analyze a situation where the voltage on an isolated node is boosted to a value well above the power supply value of V_{DD} by means of dynamic switching.

The bootstrapping mechanism can be illustrated by analyzing the circuit shown in Figure 7.32(a). The input voltage $V_{in}(t)$ is taken to be a square wave that switches between 0*v* to V_{DD}, although a full-rail swing is not necessary for the operation. Since the switching transistor M1 acts as a logic device, $V_{out}(t)$ will range from V_{OL} to V_{OH} as set by the DC characteristics of the circuit. What makes this circuit distinct from others we have studied is that V_{OH} will be a function of time *t* due to the bootstrapping of the internal node voltage V_X.

To understand the operation, let us first extract the DC characteristics. With V_{in} = 0*v*, M1 is in cutoff and transistor MX acts like a pass transistor between the power supply and node X. This

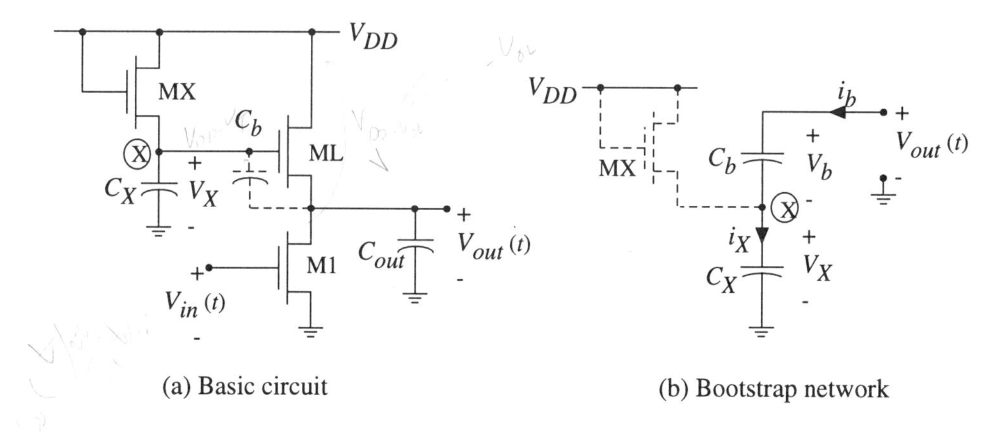

Figure 7.32 Basic bootstrapping circuit

[8] See B. Prince, **Semiconductor Memories**, 2nd ed., John Wiley & Sons, 1991, for a good overview.

[9] It originates from the idea of "pulling up one's boots using the straps" that are supplied on the sides.

gives a value for voltage V_X across C_X as

$$V_X = V_{DD} - V_{TX} \tag{7.124}$$

where V_{TX} is the threshold voltage of MX; note that C_X represents the parasitic capacitance at the node X. The output voltage is then given by

$$V_{out} = V_X - V_{TL} \tag{7.125}$$

with V_{TL} being the threshold voltage of ML. Ignoring body-bias effects allows us to write the output-high voltage as

$$\begin{aligned} V_{OH} &= V_{DD} - V_{TX} - V_{TL} \\ &= V_{DD} - 2V_T \end{aligned} \tag{7.126}$$

where V_T is the threshold voltage.

The output-low voltage is found by setting $V_{in} = V_{DD}$. This turns on M1 and pulls the output voltage V_{out} towards ground; when $V_{out} = V_{OL}$, M1 is non-saturated. Since the voltage V_X is given by (V_{DD} -V_{TX}), it is reasonable to assume that ML is saturated. Equating currents gives

$$\frac{\beta_{n1}}{2}\left[2(V_{DD} - V_{T1})V_{OL} - V_{OL}^2\right] = \frac{\beta_{nL}}{2}(V_X - V_{TL})^2 \tag{7.127}$$

which is a quadratic that can be solved for V_{OL}. Note that the actual value of V_{OL} depends upon the aspect ratios, and that V_{OL} is always greater than zero.

Bootstrapping is a dynamic event that occurs because of coupling between V_{out} and the node X by means of a bootstrap capacitor C_b. Figure 7.32(b) shows the capacitors C_X and C_b that form the basic network of interest. As mentioned above, C_X represents the capacitance at node X, while the bootstrap capacitance C_b in this circuit is due to the parasitic gate-source contributions of ML. Dynamic switching enters the problem by noting that V_{out} will periodically rise and fall in response to the changing input voltage $V_{in}(t)$.

The main features can be extracted by first analyzing the currents in the simpler circuit shown in Figure 7.32(b). The current flowing into C_X is

$$i_X = C_X\frac{dV_X}{dt} \tag{7.128}$$

while the current flow into the bootstrap capacitor is

$$i_b = C_b\frac{d(V_{out} - V_X)}{dt} \tag{7.129}$$

Noting that $i_b \approx i_X$ gives

$$C_X\frac{dV_X}{dt} \approx C_b\frac{d(V_{out} - V_X)}{dt} \tag{7.130}$$

which may be rearranged to give

$$\frac{dV_X}{dt} = \left(\frac{C_b}{C_b + C_X}\right)\frac{dV_{out}}{dt} \tag{7.131}$$

This shows that the internal voltage $V_X(t)$ is dynamically coupled to $V_{out}(t)$.

The bootstrap dynamics can be studied by assuming that initially $V_{in} = V_{DD}$ so that $V_{out}(0)=V_{OL}$. If the input voltage is switched to $V_{in} = 0v$, then V_{out} increases in time because ML is charging C_{out}. This in turns implies that $(dV_{out}/dt) > 0$, so that (dV_X/dt) is also greater than 0, i.e., $V_X(t)$ increases with time. Multiplying by dt and integrating both sides gives.

$$\int_{(V_{DD}-V_{TX})}^{V_X(t)} dV_X = \left(\frac{C_b}{C_b + C_X}\right)\int_0^t \left(\frac{dV_{out}}{dt}\right)dt \tag{7. 132}$$

where we have retained the derivative on the right-hand side to illustrate the important dependence on the charging rate. Completing the integral on the left side and rearranging gives

$$V_X(t) = (V_{DD} - V_{TX}) + r\int_0^t \left(\frac{dV_{out}}{dt}\right)dt \tag{7. 133}$$

which allows us to calculate $V_X(t)$ from $V_{out}(t)$. We have introduced the capacitance ratio

$$r = \left(\frac{C_b}{C_b + C_X}\right) \tag{7. 134}$$

as a parameter that indicates the strength of the coupling. As we will see below, the best coupling occurs when $C_b >> C_X$, and $r \approx 1$. Now that the general dependence has been portrayed, let us cancel the dt terms on the right side and integrate to arrive at

$$V_X(t) = (V_{DD} - V_{TX}) + r[V_{out}(t) - V_{OL}] \tag{7. 135}$$

where we have used the initial condition $V_{out}(0) = V_{OL}$. Given $V_{out}(t)$, this allows us to find $V_X(t)$ explicitly.

Since MOSFET ML acts as a pass transistor during the charging event $V_{out}(t)$ can be calculated by using the device equations to solve for the transient response. The output circuit in Figure 7.32 shows that

$$\begin{aligned} V_{GSL} &= V_X - V_{out} \\ V_{DSL} &= V_{DD} - V_{out} \end{aligned} \tag{7. 136}$$

Since

$$V_X = V_{DD} - V_{TX} \tag{7. 137}$$

the saturation voltage can be written as

$$V_{sat} = V_{DS} - V_{TX} - V_{TL} \tag{7. 138}$$

This shows that $V_{DSL} > V_{sat}$ is always true, so that ML charges C_{out} while conducting in the saturation region. Equating currents gives

$$\frac{\beta_{nL}}{2}(V_{DD} - V_{TX} - V_{TL} - V_{out})^2 = C_{out}\frac{dV_{out}}{dt} \tag{7. 139}$$

so that $V_{out}(t)$ can be found using the integral

$$\int_{V_{OL}}^{V_{out}(t)} \frac{dV_{out}}{(V_{DD}-V_{TX}-V_{TL}-V_{out})^2} = \int_0^t \frac{\beta_{nL}}{2C_{out}} dt \tag{7.140}$$

Integrating and rearranging gives

$$V_{out}(t) \approx V_m \left[\frac{t/2\tau_L}{1+t/2\tau_L}\right] + \frac{V_{OL}}{1+t/2\tau_L} \tag{7.141}$$

where V_m is the maximum output voltage given by

$$\begin{aligned} V_m(t) &= V_X(t) - V_{TL} \\ &= V_{DD} - V_{TX} - V_{TL} \end{aligned} \tag{7.142}$$

and

$$\tau_L = \frac{C_{out}}{\beta_{nL}(V_{DD} - V_{TX} - V_{TL} - V_{OL})} \tag{7.143}$$

is the time constant for the circuit.

Now let us use this expression for $V_{out}(t)$ to find $V_X(t)$. The complicating factor that should be considered is that the amplitude V_m varies with time as V_X changes. To account for this variation, let define the value at the beginning of the charge cycle at time t=0 and write

$$V_{m,0} = V_{DD} - V_{TX} - V_{TL} \tag{7.144}$$

for the first waveform cycle. Substituting then gives

$$V_X(t) \approx (V_{DD} - V_{TX}) + r(V_{m,0} - V_{OL})\left(\frac{t/2\tau_L}{1+t/2\tau_L}\right) \tag{7.145}$$

which shows the increasing voltage as a function of time. At the end of the charging cycle when t=t_1, V_X has a value

$$V_{X,1} \approx (V_{DD} - V_{TX}) + r\left[(V_{DD} - V_{TX} - V_{TL} - V_{OL})\left(\frac{t_1/2\tau_L}{1+t_1/2\tau_L}\right)\right] \tag{7.146}$$

Since r >1, it is obvious that

$$V_{X,1} > (V_{DD} - V_{TX}) \tag{7.147}$$

i.e., the voltage has increased over its value at the beginning of the cycle. This is the bootstrapping effect. In general, the value of $V_{X,1}$ is given by

$$V_{X,1} \approx (1+r)V_{DD} - (1+r)V_{TX} - r(V_{TL} + V_{OL}) \tag{7.148}$$

note in particular the factor of $(1+r)V_{DD}$ that has appeared in the analysis. The best case value is

$$V_{X,1} \approx 2V_{DD} - (2V_{TX} + V_{TL} + V_{OL}) \tag{7.149}$$

which occurs for $t_1 >> 2\tau_L$ and $r \approx 1$.

When the input voltage V_{in} goes to a high value, logic FET M1 turns on and provides a discharge path for C_{out} as shown in Figure 7.33. This action gives $V_{out} = V_{OL}$ as final value. Applying

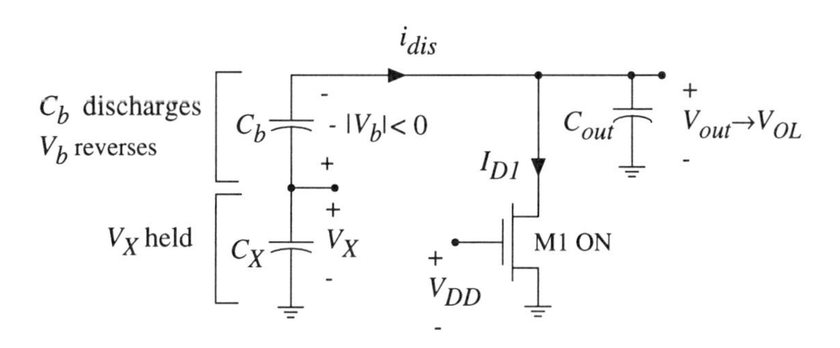

Figure 7.33 Discharge circuit at the output

KVL to the left loop gives the final value

$$V_X + V_b = V_{OL} \tag{7. 150}$$

with V_{OL} assumed to be a small (logic 0) voltage. As C_{out} discharges, this can be satisfied by V_b changing polarity (to that shown in the drawing) because C_b has a discharge path to ground as evidenced by i_{dis} flowing through M1. Our equation then becomes

$$V_X - |V_b| = V_{OL} \tag{7. 151}$$

where we have used the absolute value of V_b. This in turn allows C_X to remain charged at a voltage V_X which will be the initial condition for the next bootstrapping event.

During the next charging cycle, the bootstrapped voltage increases further. This is seen by noting that the output voltage at the beginning of the cycle is again of the form

$$V_{out}(t) \approx V_{m,1}\left[\frac{t/2\tau_L}{1 + t/2\tau_L}\right] + \frac{V_{OL}}{1 + t/2\tau_L} \tag{7. 152}$$

with appropriate time shifting. The main difference is that the amplitude is given by the increased value

$$V_{m,1} = V_{X,1} - V_{TL} \tag{7. 153}$$

since $V_{X,1}$ is still on C_X (assuming negligible charge leakage). At the end of the second cycle at time $t = t_2$, the voltage is

$$V_{X,2} \approx V_{X,1} + r\left[(V_{X,1} - V_{TL} - V_{OL})\left(\frac{t_2/2\tau_L}{1 + t_2/2\tau_L}\right)\right] > V_{X,1} \tag{7. 154}$$

showing a further increase. The maximum value of V_X is restricted by the output circuitry to

$$max(V_{out}) = V_{DD} \tag{7. 155}$$

as seen by a simple application of KVL. Using this value gives a maximum final voltage of

$$V_{X,max} = (V_{DD} - V_{TX}) + r(V_{DD} - V_{OL}) \tag{7. 156}$$

If $r = 1$, then we obtain

$$V_{X,\,max} = 2V_{DD} - V_{TX} - V_{OL} \tag{7. 157}$$

as the highest value. It is important to remember that it may take several cycles to achieve this voltage. Figure 7.34 shows the ideal plots of $V_{in}(t)$, $V_X(t)$, and $V_{out}(t)$ for this circuit. The important parameter to track is the voltage on the bootstrapped node $V_X(t)$. This clearly illustrates the bootstrapping effect where the voltage of an isolated node may be increased to a value well above the DC limits by means of transient capacitive coupling.

7.4.1 Physics of Bootstrapping

Now that we have completed the basic mathematics of the bootstrapping process, let us pause for a moment and examine the physics to obtain a qualitative understanding of the results. Consider the two capacitor circuit shown in Figure 7.35(a) which represents the basic circuit that gives rise to bootstrapping. From the external viewpoint, the two capacitors can be combined to give the single equivalent capacitor drawn in Figure 7.35(b) which has a value

$$C = \frac{C_1 C_2}{C_1 + C_2}; \tag{7. 158}$$

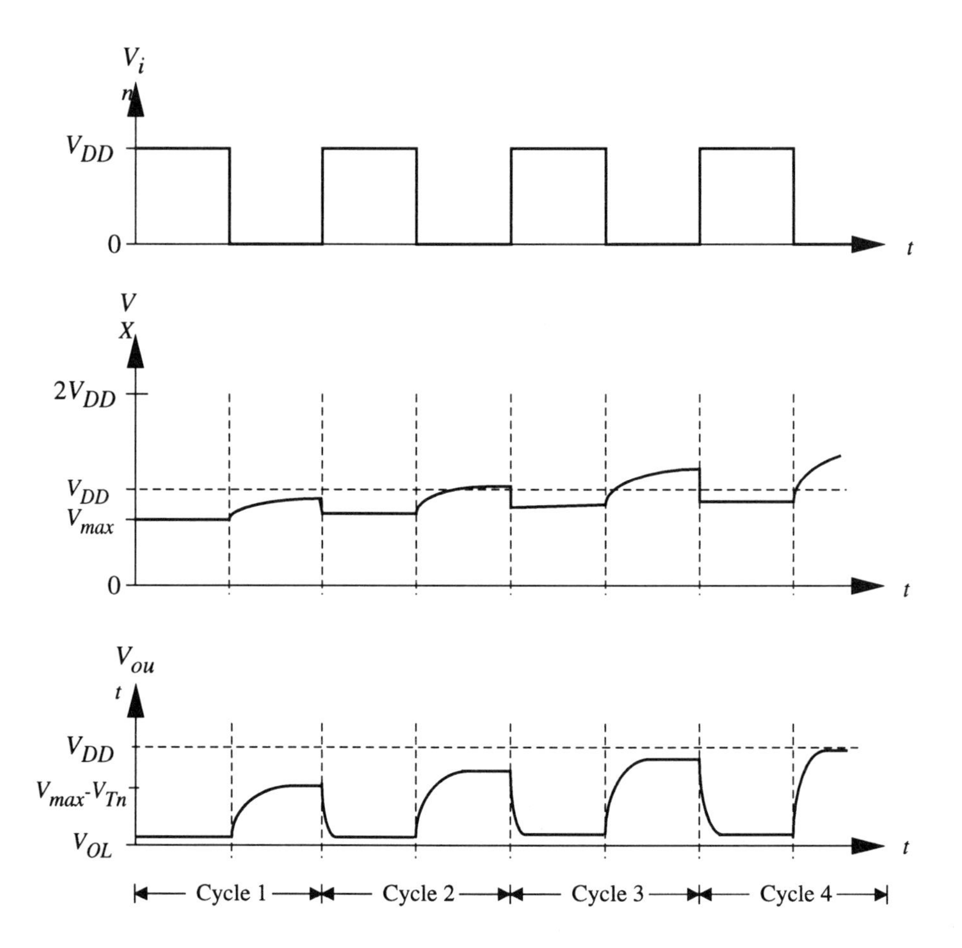

Figure 7.34 Buildup of the internal node voltage by bootstrapping

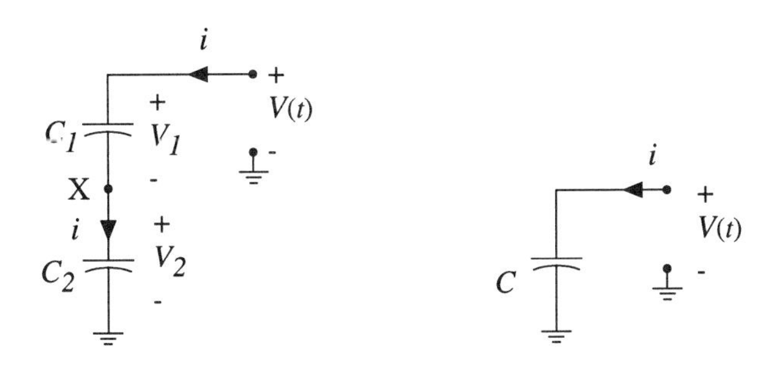

(a) Original problem (b) Reduction loses physics

Figure 7.35 Physics of the bootstrap circuit

but this masks the physics of the circuit. The important difference between the original and the equivalent circuit is the isolated node X that exists between the two elements.

When the voltage V is varied in time, the current i into the equivalent capacitor C is

$$i = C\frac{dV}{dt} = C\left(\frac{dV_1}{dt} + \frac{dV_2}{dt}\right) \tag{7. 159}$$

where we have used the obvious relation

$$V = V_1 + V_2 \tag{7. 160}$$

Now note that the individual currents are given by

$$i_1 = C_1\frac{dV_1}{dt}, \qquad i_2 = C_2\frac{dV_2}{dt}, \tag{7. 161}$$

such that

$$i = i_1 = i_2 \tag{7. 162}$$

This shows that the rate of charge flow is the same into both capacitors. Substituting these into the general equation (7.160) yields the two voltage derivatives as

$$\begin{aligned}\frac{dV_1}{dt} &= \frac{C_2}{C_1 + C_2}\left(\frac{dV}{dt}\right)\\ \frac{dV_2}{dt} &= \frac{C_1}{C_1 + C_2}\left(\frac{dV}{dt}\right)\end{aligned} \tag{7. 163}$$

These can be interpreted as demonstrating that the rate of change of voltage across a particular capacitor is proportional to the value of the *other* capacitor. If $C_1 > C_2$ as required for an effective bootstrapping network, then

$$\frac{dV_2}{dt} = \left(\frac{C_1}{C_2}\right)\frac{dV_1}{dt} \tag{7. 164}$$

by simply equating the currents. With this choice, the voltage V_2 on C_2 increases at a faster rate than the voltage on C_1. The amount of charge Q_2 on C_2 increases according to

$$\frac{dQ_2}{dt} = C_2\left(\frac{dV_2}{dt}\right) = i \qquad (7.165)$$

and can be held even after the transient event is over since the node X is isolated from ground. Each successive pulse in $V(t)$ provides additional charge to the node, adding to the voltage across the capacitor. Since CMOS integrated circuits are inherently capacitive, this type of bootstrapping situation can occur unintentionally and induce some odd-looking transient effects. Many circuits of this type are called **charge pumps** because they move charge to isolated nodes for storage.

7.4.2 Bootstrapped AND Circuit

Another application of dynamic bootstrapping is the AND circuit shown in Figure 7.36. In this circuit, the clock $\phi(t)$ is used to power a CMOS inverter circuit, while a logic signal A is input from the left. If $A = 0$, then nFET M1 is on and the output is 0. If $A = 1$, M1 is off, and capacitor C_X charges to a voltage of

$$V_X = V_{DD} - V_{Tn} \qquad (7.166)$$

through MA. The dynamic clocking signal $\phi(t)$ bootstraps this node to a higher voltage through C_b which turn drives ML into conduction. The output $A \cdot \phi$ is then high, giving the ANDing characteristics.

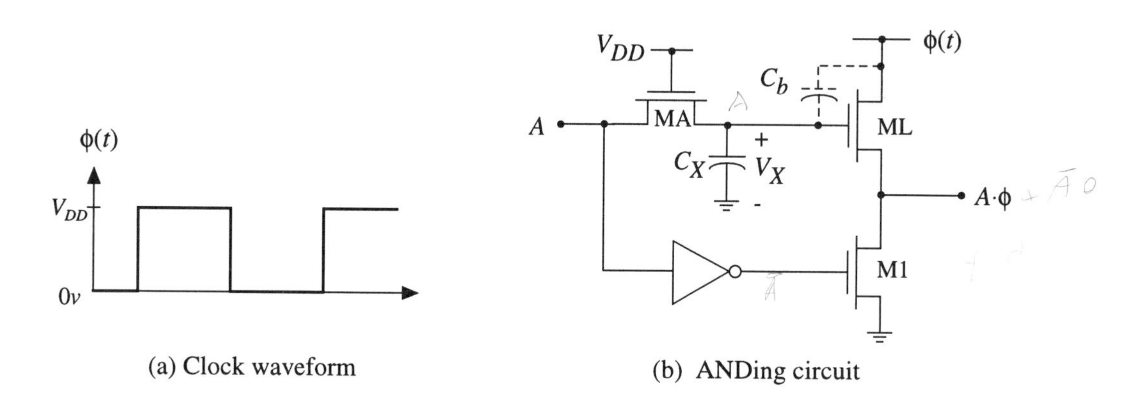

(a) Clock waveform

(b) ANDing circuit

Figure 7.36 Bootstrapped AND gate

7.5 Clocks and Synchronization

Data flow through a complex logic network is usually controlled by a clock signal $\phi(t)$. Figure 7.37 shows a clock voltage that has a **period** T in seconds that defines the time for waveform to repeat itself. The **frequency** f of the clock is related to the period by

$$f = \frac{1}{T} \qquad (7.167)$$

and has units of **Hertz** (Hz), where $1Hz$ means that the clock undergoes one cycle in one second. We will assume that the clock voltage swings the entire range from $0v$ to V_{DD}. CMOS circuits often require that the complement $\overline{\phi}(t)$ also be available for use in the timing. In terms of voltages, this means that the voltage $V_{\overline{\phi}}(t)$ is given by

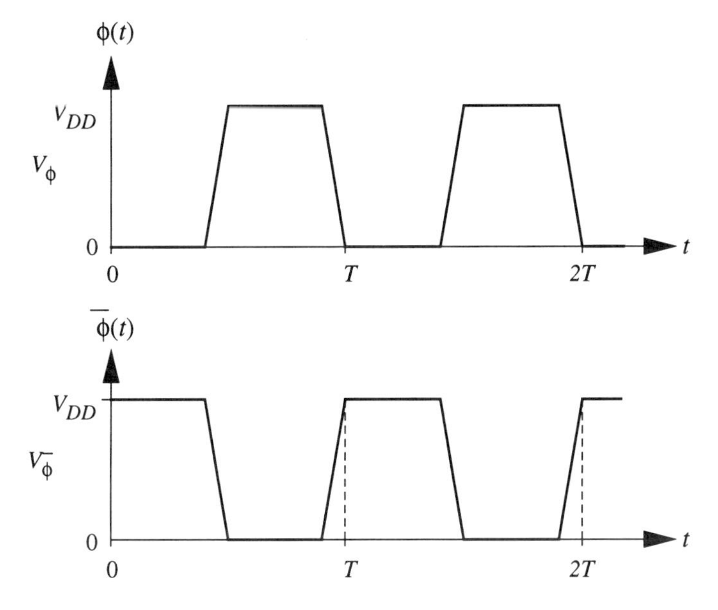

Figure 7.37 Clock voltages

$$V_{\bar{\phi}}(t) = V_{DD} - V_{\phi}(t) \qquad (7.168)$$

where $V_{\phi}(t)$ is the voltage associated with $\phi(t)$. Ideally,

$$\phi \cdot \bar{\phi} = 0 \qquad (7.169)$$

for all times; however, this cannot be achieved in practice since the two will overlap during the transition times.

Data synchronization in a CMOS circuit can be achieved in several ways. A straightforward approach is to use clocked transistors to control the transfer of data from one gate to the next. This is achieved by alternating ϕ and $\bar{\phi}$ on a set of clocked FETs such the transistors alternately pass the signal from one stage to the next.

7.5.1 Shift Register

A 4-stage shift register circuit is shown in Figure 7.38. This network is designed to move a data bit one position to the right during each half-cycle of the clock. A data bit is admitted to the first stage when ϕ=1, and is transferred to stage 2 when ϕ goes to 0. Each successive bit entered into the system follows the previous bit, resulting in the movement from left to right. It is clear from the operation of the circuit that the electronic characteristics of the circuit will place some limitation on the clock frequency f. These constraints can be obtained by examining each clock state separately.

Let us first analyze a data input event as shown in Figure 7.39(a). During this time, ϕ=1 and $V_{\phi}=V_{DD}$, which turns on M1 and allows C_1 to charge to the appropriate value. Since M1 is acting as a pass transistors, the worst-case situation will be a logic 1 input voltage as described by

$$V_1(t) = V_{max}\left[\frac{t/2\tau_n}{1+t/2\tau_n}\right] \qquad (7.170)$$

where

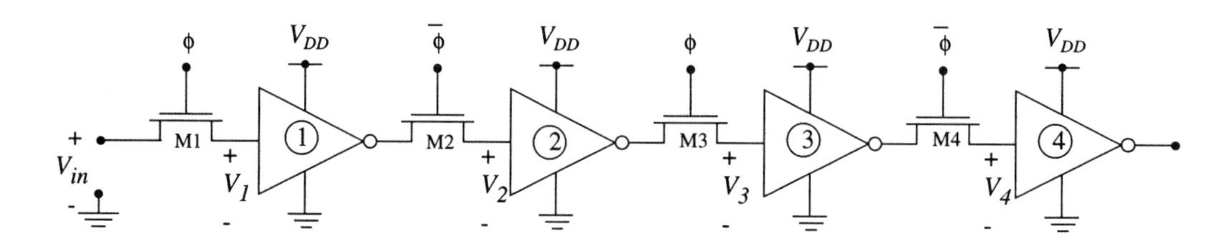

Figure 7.38 Shift register circuit

$$V_{max} = V_{DD} - V_{Tn}. \tag{7.171}$$

The charge time for this event may be estimated by

$$t_{ch} = 18\tau_n \tag{7.172}$$

as found in Chapter 4. There is also a delay time t_{HL} associated with the inverter nFET discharging the output capacitor $C_{0,1}$. Combining these two contributions gives

$$t_{in} = t_{ch} + t_{HL} \tag{7.173}$$

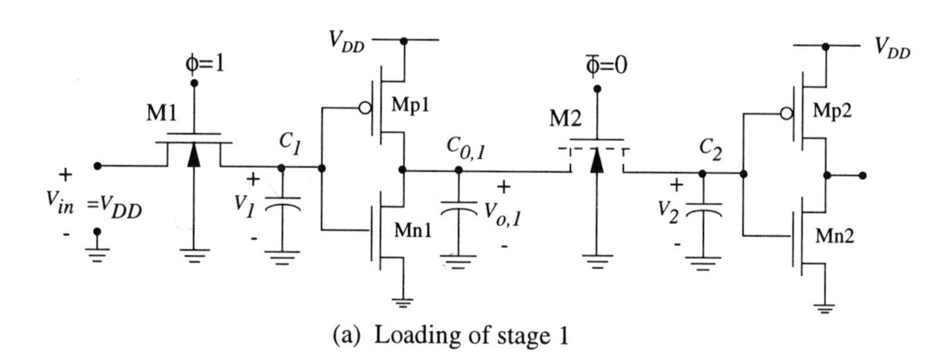

(a) Loading of stage 1

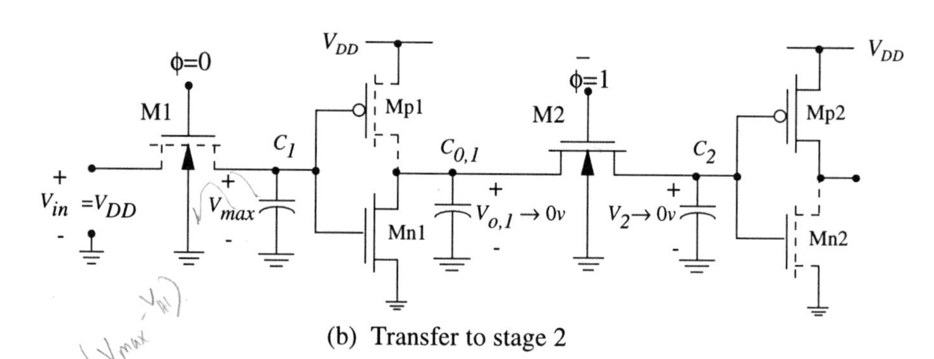

(b) Transfer to stage 2

Figure 7.39 Shift register circuit voltages

as the time needed to allow the output to react to a change in the input state.

During the next portion of the clock cycle when $\phi = 0$ and $V_{\phi} = 0v$, the pass transistor M1 is in cutoff as illustrated in Figure 7.39(b). During this time, charge leakage will occur and the voltage V_1 across C_1 will decay from its original value of V_{max}. The minimum value at the inverter input that will still be interpreted as a logic 1 value is V_{IH}, so that the maximum hold time is estimated by

$$t_H \approx \frac{C_1}{I_{leak}}(V_{max} - V_{IH}) \tag{7. 174}$$

where we have used the simplest charge leakage analysis. So long as $V_1 > V_{IH}$ the output voltage $V_{o,1}$ is held $0v$, which is transferred to the input of Stage 2 through pass FET M2. This results in the voltage $V_2 = 0v$.

Let us now examine the consequences of this analysis with respect to the clock frequency f. Consider the clocking waveform in Figure 7.40 where we assumed a 50% **duty cycle**; this means that the clock has a high value for 50% of the period. As applied to the circuit in Figure 7.39(a), we must have

$$min\left(\frac{T}{2}\right) = t_{in} \tag{7. 175}$$

in order to insure that the voltage has sufficient time to be transferred in and passed through the inverter. This sets the **maximum clock frequency** as

$$f_{max} = \frac{1}{T_{min}} \approx \frac{1}{2(t_{ch} + t_{HL})}, \tag{7. 176}$$

which acts as the upper limit for the system data transfer rate. The charge leakage problem of Figure 7.39(b) acts in exactly opposite manner. In this case, the maximum time that the clock can be at ϕ=0 is given by

$$max\left(\frac{T}{2}\right) = t_H \tag{7. 177}$$

since the node cannot hold the logic 1 state any longer. This then sets the **minimum clock frequency** as

$$f_{min} = \frac{1}{T_{max}} \approx \frac{1}{2t_H} \tag{7. 178}$$

in order to avoid charge leakage problems. Note that the time intervals are reversed when applied to the second stage, but otherwise have the same limitations. In general, this analysis shows that the clock frequency must be chosen in the range

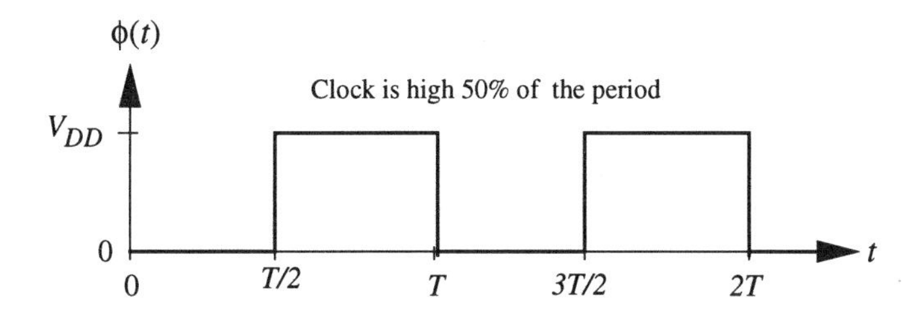

Figure 7.40 Clock with 50% duty cycle

$$f_{min} < f < f_{max} \tag{7. 179}$$

for proper operation. Although high-performance design usually dictates that we employ the highest clock frequency possible, other considerations such as chip testing can be easier to handle at low clocking rates.

7.5.2 TGs as Control Elements

Transmission gates may also be used to control data flow as illustrated in Figure 7.41. These have the advantage of passing the entire range of voltages $[0,V_{DD}]$ at the expense of additional wiring and layout complexity due to the use of complementary FETS and control signals. The limitations on the clock frequency are similar to those found for the nFETs circuits; only the numerical values need to be changed to account for the differences in the pass characteristics of the TGs.

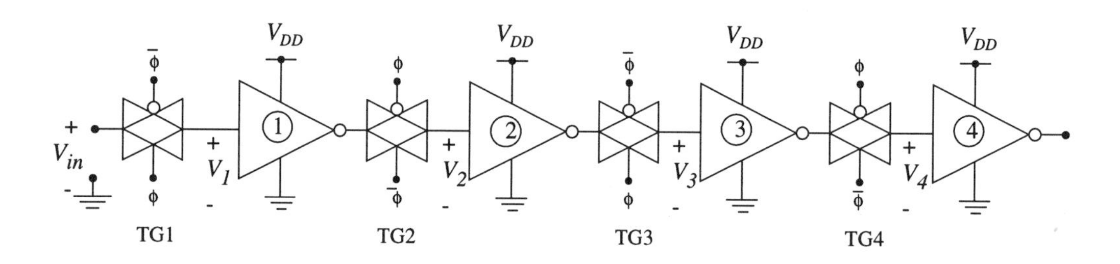

Figure 7.41 TG-based shift register

7.5.3 Extension to General Clocked Systems

The data transfer scheme used in the shift register can be extended more complex systems as illustrated by the example in Figure 7.42. In network (a), the inputs and outputs of a static full-adder circuit are controlled by FETs with opposite clock phases. The transistors thus act to control the data flow through the circuit when it is placed in a larger system design. The maximum frequency is now determined by the longest delay time t_{delay} through the circuit such that

$$f_{max} = \frac{1}{2t_{delay}}. \tag{7. 180}$$

The minimum frequency f_{min} is still set by charge leakage at the isolated nodes at the outputs of the pass transistors. An equivalent TG-controlled circuit is shown in Figure 7.42(b); it has similar clocking limitations.

A more generalized extension is the multi-stage cascade illustrated in Figure 7.43 where the clocked nFETs control the data flow through the system. In this case, the value of f_{max} is set by the slowest circuit in the chain, which is an important characteristic of clocked systems: the system throughput is limited by the slowest data path.

Pipelining can accomplished by providing a set of input registers to each stage. Figure 7.44 illustrates a simple approach for a pipeline stage design. The inputs are fed into transparent clock-controlled D-latches. A clock level of ϕ=1 allows the data to be accepted, while ϕ=0 defines a hold state by closing the feedback loop. One problem with this arrangement is that the use of transparent latches may cause signal race problems through a stage. This can be avoided by replacing the latches with master-slave flip-flops, resulting in the pipelined input circuits drawn in Figure 7.45. This employs the standard circuit using oppositely phased clocks for input and hold conditions. The

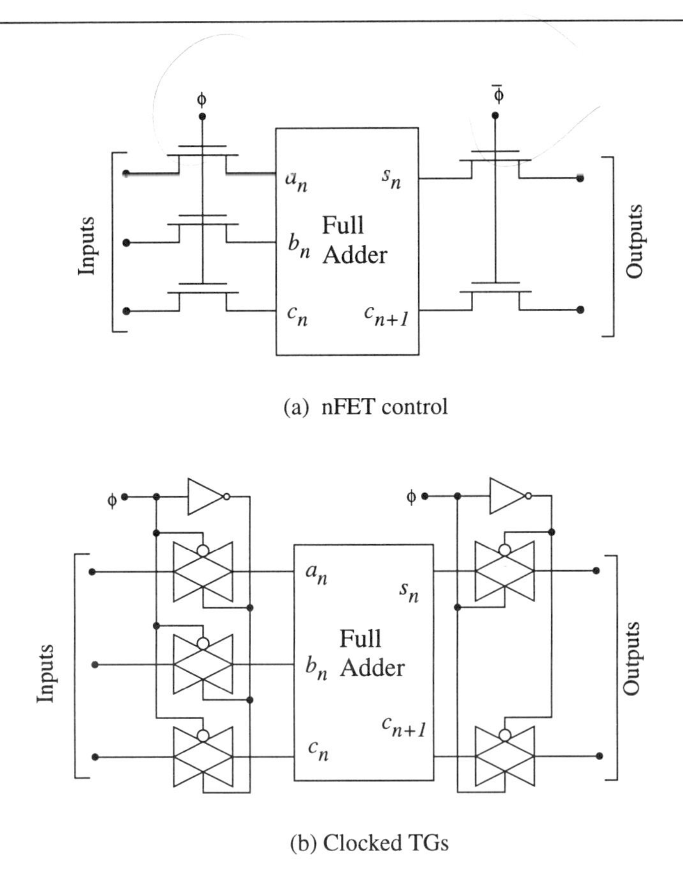

(a) nFET control

(b) Clocked TGs

Figure 7.42 Data flow control using TGs

overall structure of the pipeline path is obtained by cascading oppositely phased pipeline stages as shown in Figure 7.46. The clocking notation in this drawing is meant to imply that when one stage is admitting new data values, the following stage is in a hold condition.

7.6 Clocked-CMOS

Clocked-CMOS (C^2MOS) is a logic family that combines static logic design with the synchronization achieved by using clock signals. In the early days of CMOS, many SSI and MSI chips were based on C^2MOS. In modern design, the technique is still useful in certain applications, such as

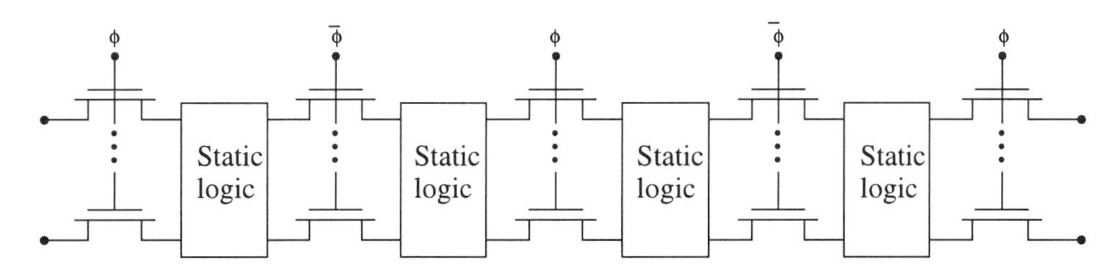

Figure 7.43 Clock synchronization using nFETs

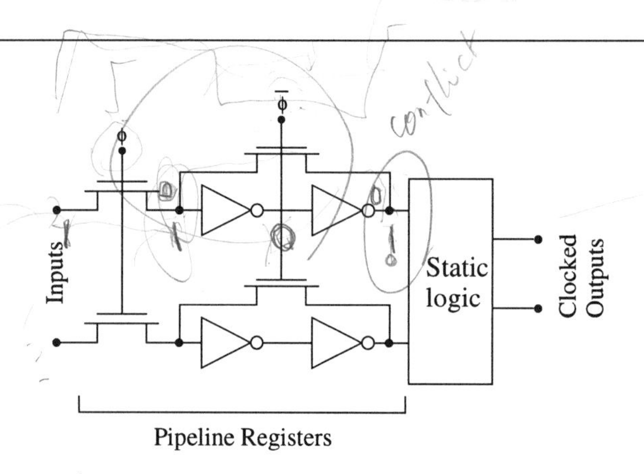

Figure 7.44 Pipeline register example

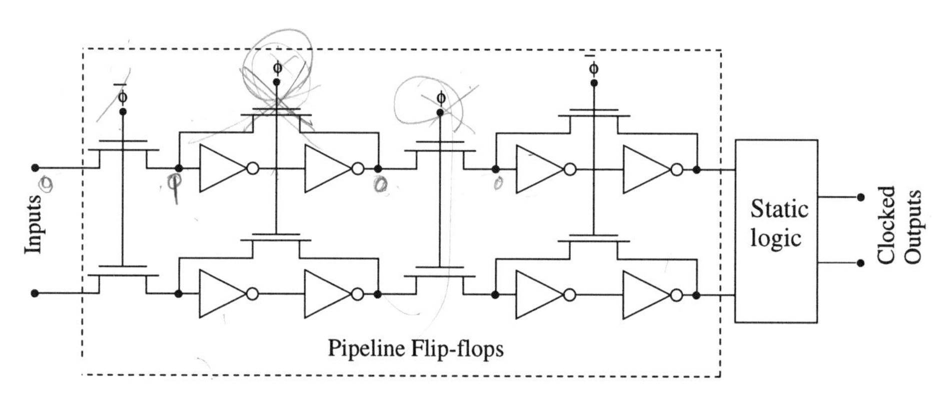

Figure 7.45 Pipeline registers implemented with flip-flops

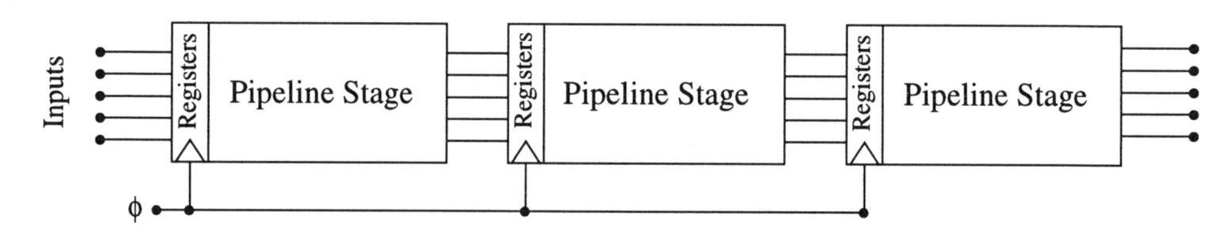

Figure 7.46 General pipeline structure

dynamic "NORA" circuits discussed in Chapter 8.

Figure 7.47 show the general structure of a C^2MOS logic gate. The inputs A, B, and C are connected to complementary nFET/pFET pairs as in ordinary static design where they act like open or closed switches. The only modification is the insertion of two clocked FETs between the logic arrays and the output. Mp is controlled by $\overline{\phi}$ and separates the pFET logic block and $C_{out,}$ while Mn is controlled by ϕ and serves the same function for the nFET logic block. The operation of the gate

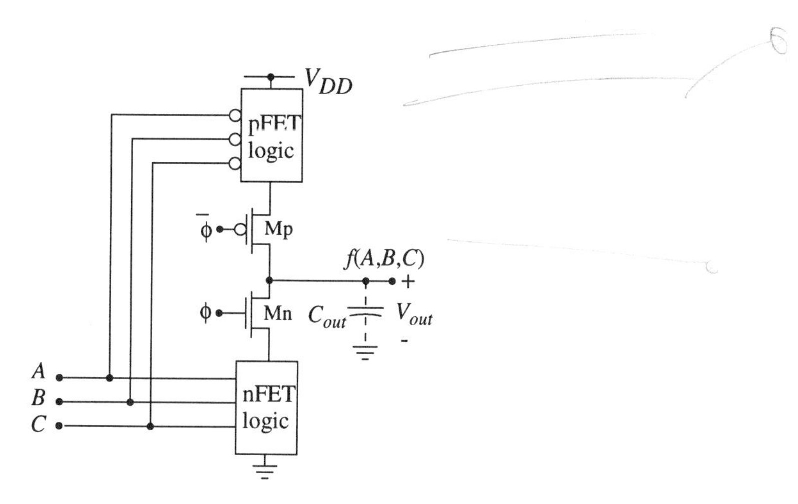

Figure 7.47 Clocked CMOS gate circuit

can be understood by studying the effects of the clock $\phi(t)$ as summarized in Figure 7.48. When the clock is at a level of ϕ=1 as in Figure 7.48(a), both Mn and Mp are biased active. This connects both logic arrays to the output node, and the gate degenerates to its static equivalent circuit; the main difference are longer switching times due to the additional parasitics. After the transients have decayed, the output capacitor C_{out} will be charged to a voltage V_{out} = 0v or $V_{out} = V_{DD}$. Figure 7.48(b) shows the circuit when ϕ = 0, and both Mn and Mp are in cutoff. This isolates the output node from both logic arrays, and the value of $V_{out} = V_{Result}$ is held on C_{out}. However, a moment's reflection will verify that this is identical to the problem of maintaining charge on a capacitive node using an OFF transmission gate, so that the value of V_{out} will change in time. The result is only valid for the hold time t_H, which is an important characteristic of this type of circuit.

A basic C^2MOS inverter is shown in Figure 7.49. This is constructed by using a single complementary pair Mp1 and Mn1 in the usual manner. We have show the internal node capacitances C_p and C_n explicitly in the drawing. The operation of the clocked-CMOS inverter is summarized in

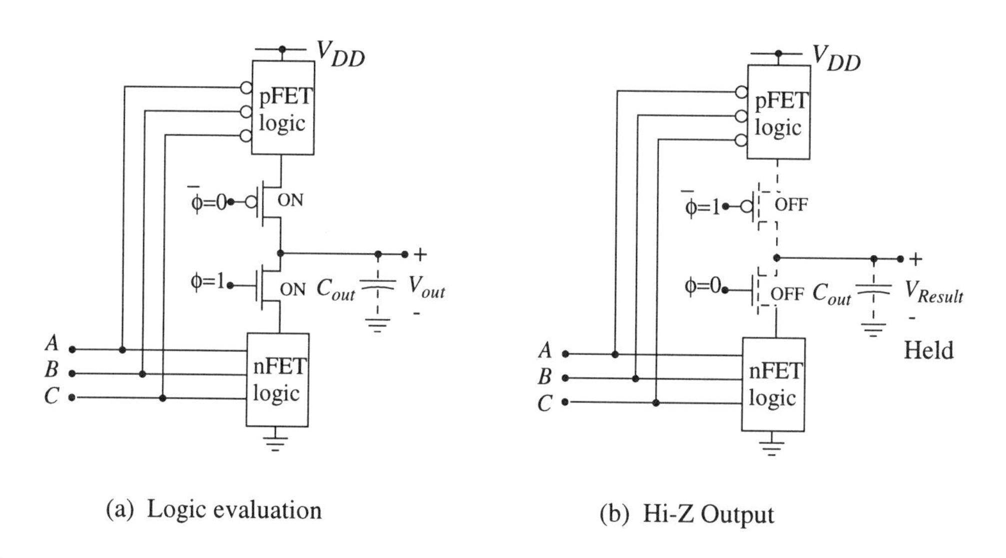

Figure 7.48 Operation of a clocked CMOS gate

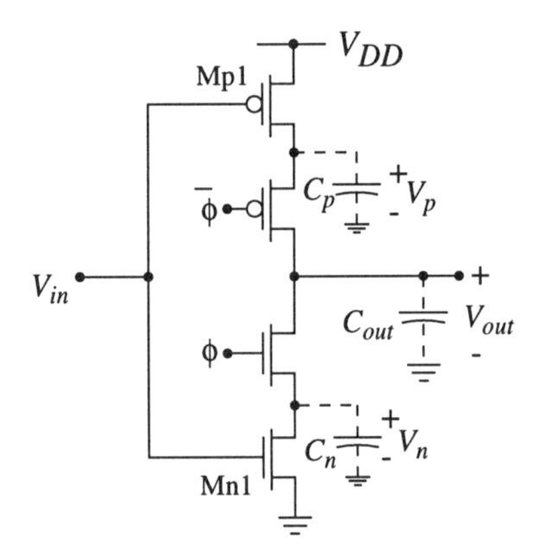

Figure 7.49 Clocked CMOS inverter

Figure 7.50. When $\phi = 1$, both clocked FETs are active and the input controls the logic FETs Mp1 and Mn1. If $V_{in} = 0v$, then $V_{out} \rightarrow V_{DD}$, while an input of $V_{in} = V_{DD}$ results in $V_{out} \rightarrow 0v$; the general values are shown in Figure 7.50(a). Owing to the presence of C_p and C_n, both t_{LH} and t_{HL} are larger than for a static inverter. For example, the charging time constant is given by

$$\tau_p = (R_{p1} + R_p) C_{out} + R_{p1} C_p \tag{7. 181}$$

where R_p and R_{p1} are the pFET resistances. The output is driven into a Hi-Z state when the clock makes a transition to ϕ=0. As shown in Figure 7.50(b), the output voltage $V_{out} = V_X$ is maintained by charge storage on C_{out}. The voltages V_p and V_n across the capacitors C_p and C_n, respectively, can be changed by the input, but the output remains isolated. The operation is identical to what can be achieved using a clocked transmission gate at the output of a static logic gate. Combing both aspects into a single gate allows the designer to automatically synchronize the data flow.

Cascading two oppositely-phased inverters results in the clocked latch shown in Figure 7.51. The timing is chosen to insure that when one inverter is accepting inputs, the other is in a Hi-Z hold state. The operation is similar to a master-slave DFF and is summarized in Figure 7.52. A clock

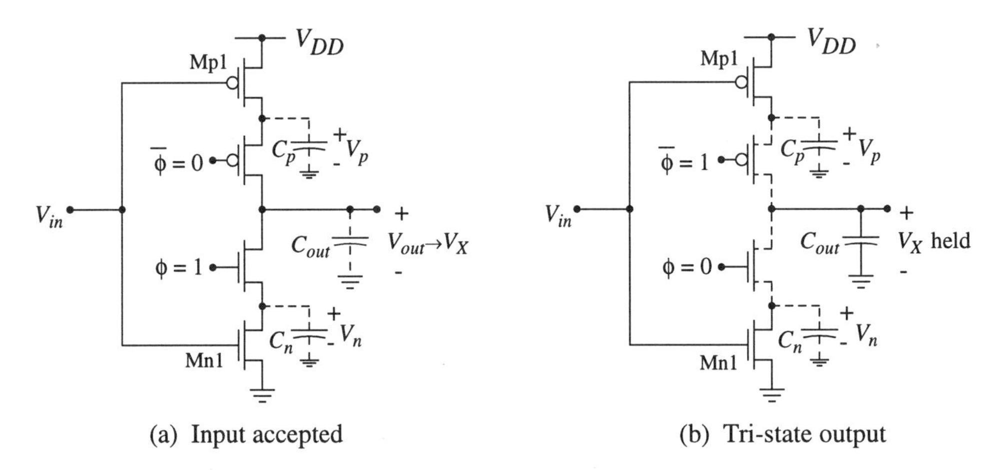

(a) Input accepted (b) Tri-state output

Figure 7.50 Clocked CMOS timing behavior

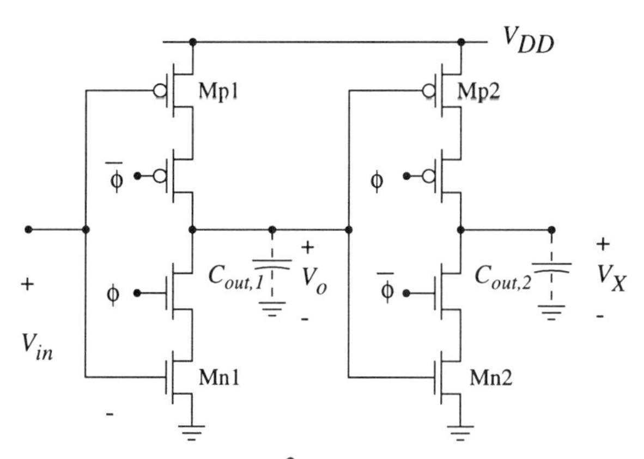

Figure 7.51 A C^2MOS latch circuit

signal of $\phi = 1$ allows the first inverter to accept the input data voltage V_D which results in a value of $V_{\overline{D}}$ on $C_{out,1}$ as shown in drawing (a) of Figure 7.52. This is true for both logic 0 and logic 1 inputs. When the clock changes to a value of ϕ=0, the first stage is driven into a Hi-Z state which the second stage accepts the value of $V_{\overline{D}}$ as an input. The voltage on $C_{out,2}$ then corresponds to the original value of V_D; this is illustrated in Figure 7.52(b). The output voltage can be maintained so long as the voltage $V_{\overline{D}}$ across $C_{out,1}$ remains within the appropriate voltage range, i.e.,

$$0 < V_{\overline{D}} < V_{IL} \qquad (7.\ 182)$$

for a logic 0, and

$$V_{IH} < V_{\overline{D}} < V_{DD} \qquad (7.\ 183)$$

for a logic 1. Both V_{IL} and V_{IH} are determined by the VTC that is calculated when the circuit is operating like a static inverter. Charge leakage limits the minimum clock frequency f_{min}, while the charging and discharging delay times combine to specify the maximum clock frequency f_{max}.

Since C^2MOS is based on static logic, it is a simple matter to design an entire family of gates with the same characteristics. Examples of these are the NAND2 gate in Figure 7.53(a) and the NOR2 gate in Figure 7.53(b). In principle, any AOI or OAI logic circuit may be created using the formalism developed in Chapter 5. However, since the additional delay introduced by the clocking FETs cannot be eliminated, the logic family is automatically limited to slower systems.

A variation of the C^2MOS latch is shown in Figure 7.54. This uses a static inverter between two clocked circuits as the second stage to produce the output, which is not a tri-state node. This allows for the output to be taken at any time. The third stage circuit (which is the second C^2MOS inverter in the chain) is now being used to provide clocked controlled feedback.

C^2MOS logic provides a straightforward approach to synchronizing data flow while maintaining static logic ideas. Moreover, the Hi-Z characteristic of the output makes it useful for controlling signal races in the data path. As we shall see in the next chapter, the NORA (No Race) dynamic logic family employs C^2MOS latches for just that purpose.

7.7 Clock Generation Circuits

Clocking signals are used to synchronize data flow and, in the case of logic cascades such as C^2MOS, also serve to control the internal operation of the gates. In general, a single external clock *CLK* is applied to a pin on the IC package. All timing signals must then be generated by on-chip cir-

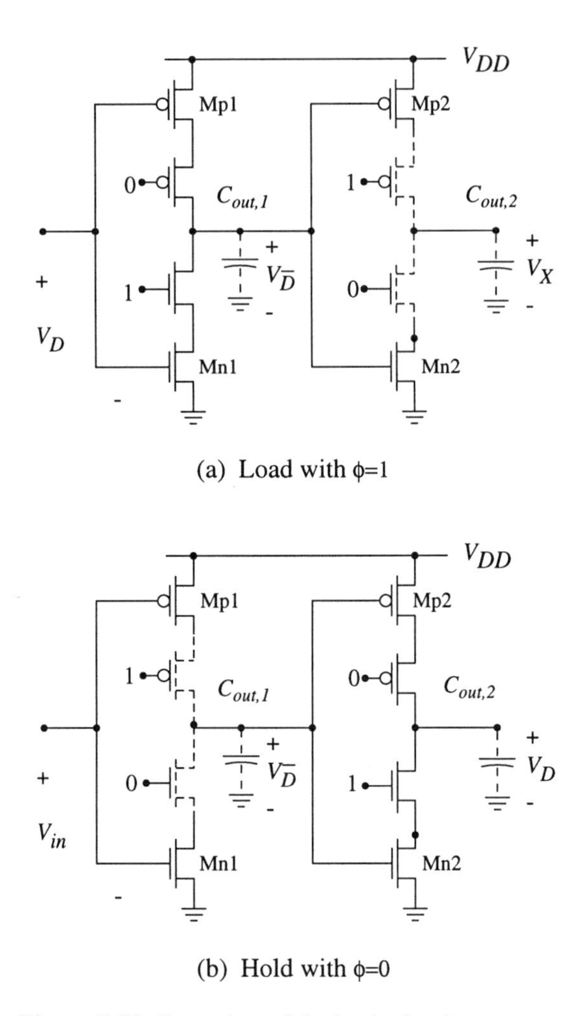

Figure 7.52 Operation of the latch circuit

cuitry. This consideration brings into play an important problem in chip engineering: how does one generate the required clocking signals and then distribute them over the area of the chip? In this section, we will examine techniques for deriving ϕ and $\overline{\phi}$ from the input *CLK*. Clock distribution is more of a global problem and is discussed in Chapter 10.

To understand the details of the clock generation problem, consider the block diagram shown in Figure 7.55. At first sight, this seems quite straighforward. For example, the input clock *CLK*. can be used directly as $\phi(t)$, while $\overline{\phi}(t)$ can be obtained using an inverter. The resulting circuit is shown in Figure 7.56(a) Logically, this is a nice, simple solution. However, the circuit designer will recognize that since an inverter is used to generate $\overline{\phi}$ from ϕ, the propagation delay t_p delays the phase of $\overline{\phi}(t)$ as shown in Figure 7.56(b). The delay is known as **clock skew**. For large periods $T >> t_p$, the difference will be small and the clock skew will not cause any major problems. At high frequencies, on the other hand, the delay may affect the data flow or circuit operation, and must be dealt with. Controlling the clock skew is critically important for high-speed circuits.

One approach to solving this problem is shown in Figure 7.57. This circuit uses Inverter 1 to drive two branches. The upper branch through Inverters 2 and 3 gives the clock $\phi(t)$. The lower branch is used to generate the complement $\overline{\phi}(t)$ that has the proper phase relationship to $\phi(t)$. This is

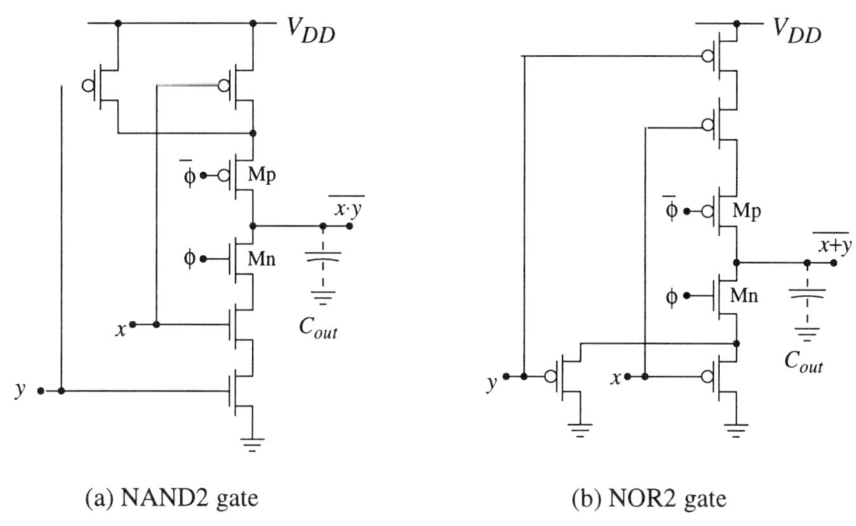

(a) NAND2 gate (b) NOR2 gate

Figure 7.53 C^2MOS logic gate examples

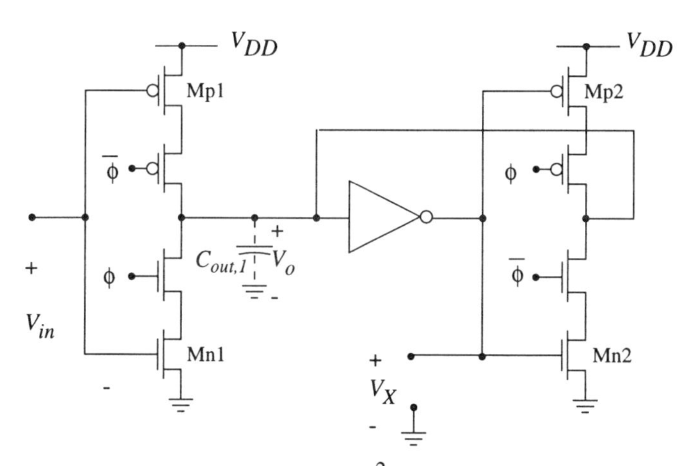

Figure 7.54 Variation on the C^2MOS latch

achieved by cascading the signal through a transmission gate that is biased active before it is transmitted through Inverter 4. The TG is used as a "delay element" that introduces a time delay t_d to automatically compensate for the propagation delay t_p through Inverter 3 in the upper branch. If the delay satisfies $t_p = t_d$, then the outputs of Inverter 3 and the TG (indicated by the dashed vertical line in the drawing) will have the proper phase relation.

The delay mechanism can be understood by referring to the equivalent circuit in Figure 7.58(a) where we have replaced the TG with its RC-equivalent network consisting of resistor R_{TG} and capacitors C_A and C_B. The additional RC parasitics introduced by the TG are used to delay the signal the desired amount. Note that we have indicated other elements in the circuit. C_{out} is the output capacitance of Inverter 1 due to the FETS Mn1 and Mp2, and capacitors C_2, C_3, and C_4 represent

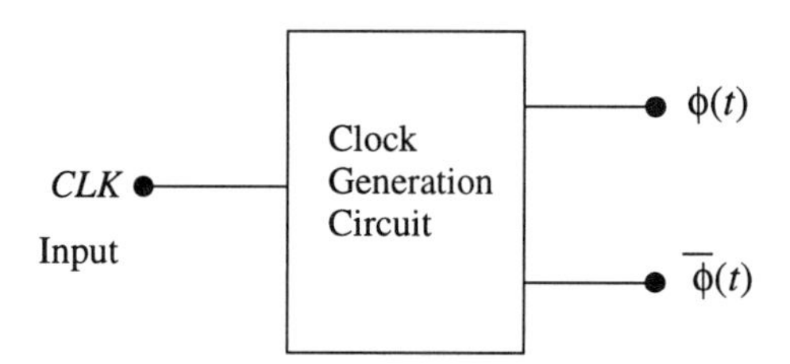

Figure 7.55 Clock generation problem

the total capacitance seen at the input nodes to Inverters 2, 3, and 4, respectively. The circuit is then designed such that the voltages $V_3(t)$ and $V_4(t)$ are exactly $(T/2)$ out of phase. This is accomplished by selecting the transistor sizes to insure that the delay time t_d from node X to C_4 is the same as t_p through Inverter 2.

Let us concentrate on the effects of the transmission gate by further reducing the circuit to that shown in Figure 7.58(b). To arrive at this simplified network we have combined capacitances together such that

$$C_X = C_A + C_2 + C_{out} \tag{7.184}$$

and

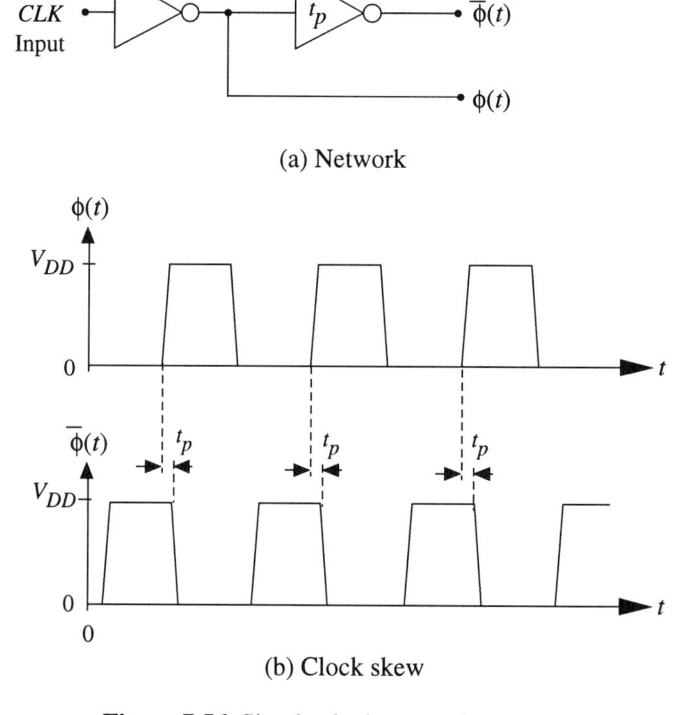

Figure 7.56 Simple clock generation network

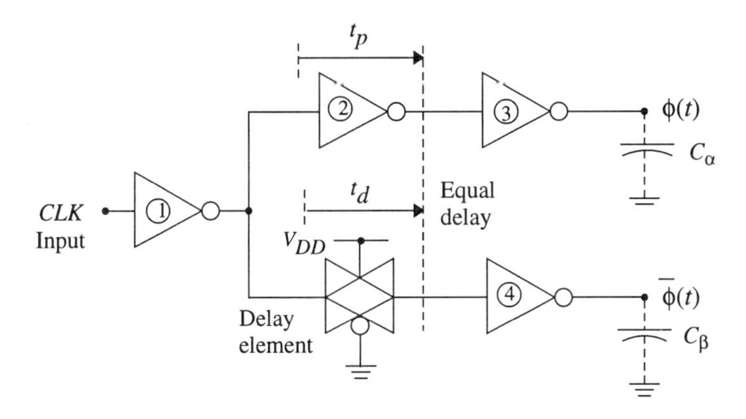

Figure 7.57 TG as a delay element to control clock skew

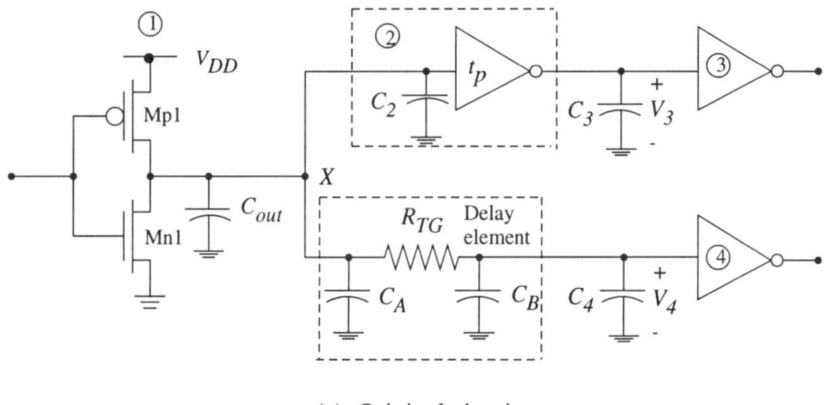

(a) Original circuit

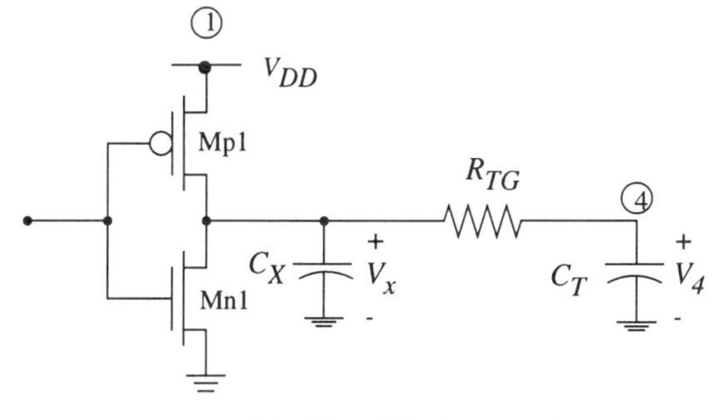

(b) Simplified network

Figure 7.58 Analysis of the TG delay circuit

$$C_T = C_B + C_4 \quad (7.185)$$

is the total value at node 4. We thus direct our interest towards the voltage $V_4(t)$. Since this is at the end of an RC chain, we will assume that $V_4(t)$ can be modelled as an exponential waveform and then calculate the time constants. A direct application of the Elmore formula for the charging time constant gives

$$\tau_{ch} = (R_{p1} + R_{TG})\, C_T + R_{p1} C_X \quad (7.186)$$

where

$$R_{p1} = \frac{1}{k'_p \left(\frac{W}{L}\right)_{p1} (V_{DD} - |V_{Tp}|)} \quad (7.187)$$

is the resistance of Mp1. Similarly, the discharge time constant is

$$\tau_{dis} = (R_{n1} + R_{TG})\, C_T + R_{n1} C_X \quad (7.188)$$

with

$$R_{n1} = \frac{1}{k'_n \left(\frac{W}{L}\right)_{n1} (V_{DD} - V_{Tn})} \quad (7.189)$$

representing the resistance of Mn1. These equations demonstrate that the delay depends upon the aspect ratios $(W/L)_{n1}$ and $(W/L)_{p1}$ of Inverter 1, in addition to the aspect ratios of the TG FETs through R_{TG} and the parasitic capacitance contributions. Moreover, rewriting the time constants in the forms

$$\tau_{ch} = R_{p1}\left[\left(1 + \frac{R_{TG}}{R_{p1}}\right) C_T + C_X\right]$$
$$\tau_{dis} = R_{n1}\left[\left(1 + \frac{R_{TG}}{R_{n1}}\right) C_T + C_X\right] \quad (7.190)$$

shows that it is the *ratios* of R_{TG} to R_{p1} and R_{n1} that are important in designing the circuit delays. These equations may be used as a basis to analyze the problem. However, it is instructive to simplify the circuit even farther in a manner that clearly illustrates the interplay among the transistors.

Consider the driven RC circuit in Figure 7.59. At this level of modelling, the output of Inverter 1 is replaced by an independent voltage source $V_X(t)$ that is modelled as an exponential with simple time constants τ_n and τ_p defined by

$$\tau_p = R_{p1} C_X \quad \text{and} \quad \tau_n = R_{n1} C_X \quad (7.191)$$

that depend only on Mn1 and Mp1. The passive portion of the circuit, on the other hand, is characterized by the time constant

$$\tau_{TG} = R_{TG} C_T \quad (7.192)$$

which is determined by the transmission gate parameters. The forced response of the circuit will then show the relationship among the three time constants.

To analyze the circuit, we write the current through the resistor as

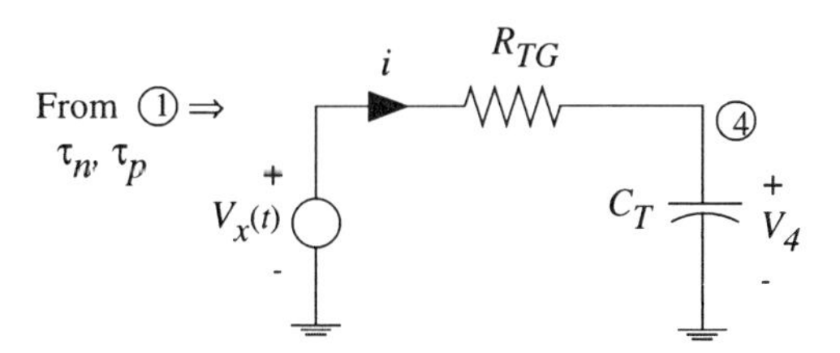

Figure 7.59 Simplified RC model

$$i = \frac{V_X(t) - V_4}{R_{TG}} = C_T \frac{dV_4}{dt} \tag{7. 193}$$

Rearranging gives the differential equation

$$\frac{dV_4}{dt} + \frac{1}{\tau_{TG}} V_4(t) = \frac{V_X(t)}{\tau_{TG}} \tag{7. 194}$$

for $V_4(t)$. The forcing function $V_X(t)$ has two possible forms. For a logic 1 transfer (a charging event), we write

$$V_X(t) = V_{DD}\left[1 - e^{-t/\tau_p}\right] \tag{7. 195}$$

with the initial condition $V_4(0) = 0v$. A logic 0 discharge, on the other hand, is described by

$$V_X(t) = V_{DD} e^{-t/\tau_n} \tag{7. 196}$$

with $V_4(0) = V_{DD}$. The important point is that the forcing function depend upon the aspect ratios of the transistors used in Inverter 1 in this level of approximation.

To solve, we will transform to s-domain and work with the transformed voltage $v_4(s)$. A logic 0 transfer is described by the time-domain equation

$$\frac{dV_4}{dt} + \frac{1}{\tau_{TG}} V_4(t) = \frac{V_{DD} e^{-t/\tau_n}}{\tau_{TG}} \tag{7. 197}$$

Transforming into s-domain gives

$$s v_4(s) - V_{DD} + \frac{1}{\tau_{TG}} v_4(s) = \frac{V_{DD}}{\tau_{TG}\left(s + \frac{1}{\tau_n}\right)} \tag{7. 198}$$

where we see that the forcing term mixes the time constants τ_n and τ_{TG}. Solving for $v_4(s)$ results in

$$v_4(s) = \frac{(V_{DD}/\tau_{TG})}{\left(s + \frac{1}{\tau_n}\right)\left(s + \frac{1}{\tau_{TG}}\right)} + \frac{V_{DD}}{\left(s + \frac{1}{\tau_{TG}}\right)} \tag{7. 199}$$

so that we may employ a partial fraction expansion of the form

$$v_4(s) = \frac{V_{DD}+K}{\left(s+\frac{1}{\tau_{TG}}\right)} - \frac{K}{\left(s+\frac{1}{\tau_n}\right)} \tag{7.200}$$

It is easily shown that the coefficient K is

$$K = \frac{V_{DD}}{\left(\frac{\tau_{TG}}{\tau_n}-1\right)} \tag{7.201}$$

Inverse transforming back into time domain then gives the capacitor voltage as

$$V_4(t) = V_{DD}\left[1+\frac{1}{\left(\frac{\tau_{TG}}{\tau_n}-1\right)}\right]e^{-t/\tau_{TG}} - \frac{V_{DD}}{\left(\frac{\tau_{TG}}{\tau_n}-1\right)}e^{-t/\tau_n} \tag{7.202}$$

This expresses the voltage as the superposition of two terms, each of which is characterized by a distinct time dependence.

A logic 1 charging event may be analyzed in the same manner. The time-domain equation is

$$\frac{dV_4}{dt} + \frac{1}{\tau_{TG}}V_4(t) = \frac{V_{DD}\left[1-e^{-t/\tau_p}\right]}{\tau_{TG}} \tag{7.203}$$

which transforms to s-domain as

$$sv_4(s) + \frac{1}{\tau_{TG}}v_4(s) = \frac{V_{DD}}{\tau_{TG}}\left(\frac{1}{s} - \frac{1}{\left(s+\frac{1}{\tau_p}\right)}\right) \tag{7.204}$$

Performing the algebra and inverse transforming gives the time-domain result

$$V_4(t) = V_{DD}\left[1-e^{-t/\tau_{TG}}\right] + \frac{V_{DD}}{\left(\frac{\tau_{TG}}{\tau_p}-1\right)}\left[e^{-t/\tau_p} - e^{-t/\tau_{TG}}\right] \tag{7.205}$$

This is also a superposition of terms with different time constants.

The meaning of these results can be seen in the plots of Figure 7.60 where the TG-induced delays are shown as the relative values of (τ_{TG}/τ_n) [in plot (a)] and (τ_{TG}/τ_n) [in plot (b)] are varied. In both cases, increasing τ_{TG} relative to the nFET or pFET time constant of the forcing function slows the change of the signal in time. Since our objective is to obtain a TG-delay that is equal to that through an inverter, the analysis tells us that the design should initially center on decreasing the time constants τ_n and τ_p in the inverter design. The expression

$$\begin{aligned}\tau_{TG} &= R_{TG}C_T \\ &= R_{TG}(C_B + C_4)\end{aligned} \tag{7.206}$$

for the TG time constant depends upon the sizes of the FETs. Combining this with the inverter time constant equations

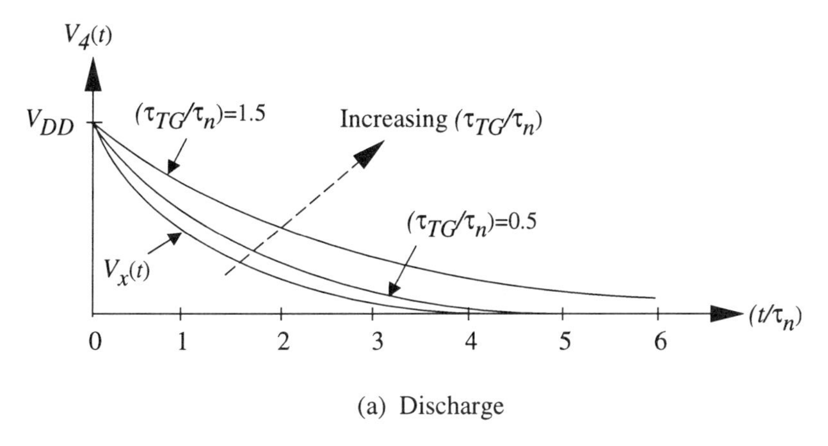

(a) Discharge

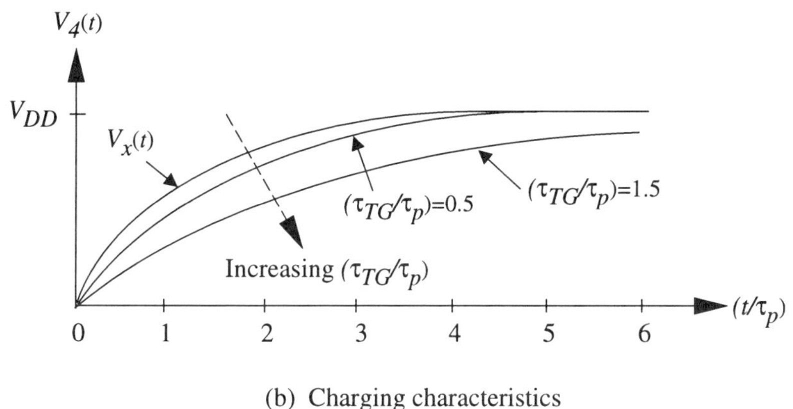

(b) Charging characteristics

Figure 7.60 TG-induced delay plots

$$\begin{aligned} \tau_p &= R_{p1}(C_A + C_2 + C_{out}) \\ \tau_n &= R_{n1}(C_A + C_2 + C_{out}) \end{aligned} \tag{7.207}$$

shows the interplay among the design variables. Note in particular that C_A depends upon the TG transistors, while C_{out} is the internal capacitance of the driving inverter. This means that both circuits must be adjusted to meet the equal delay specification.

Another clock generation circuit is shown in Figure 7.61.This uses a simple D-latch with the *CLK* input to create both ϕ and $\overline{\phi}$. The outputs are skewed slightly because of the time delay introduced by the inverter, but the scheme itself is easy to use for moderate clock speeds.

The idea may be extended to the network shown in Figure 7.62(a) where we have added a pair of inverters in both branches. These are used to delay the signal through the feedback loop by an amount $\Delta t = 2t_p$ which increases the time between rising edges. In other words, the circuit extends the time where

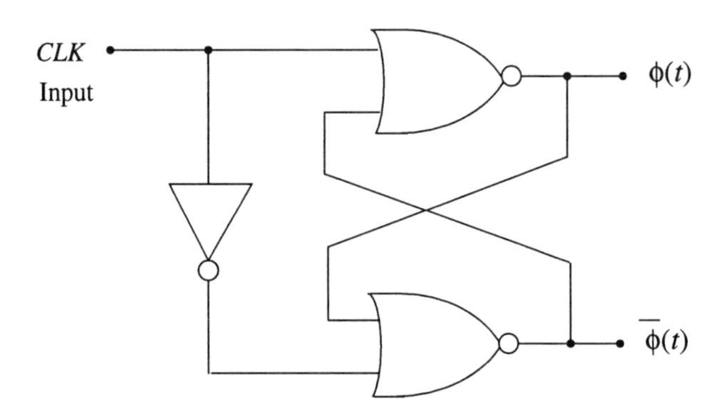

Figure 7.61 D-latch as a clock driver

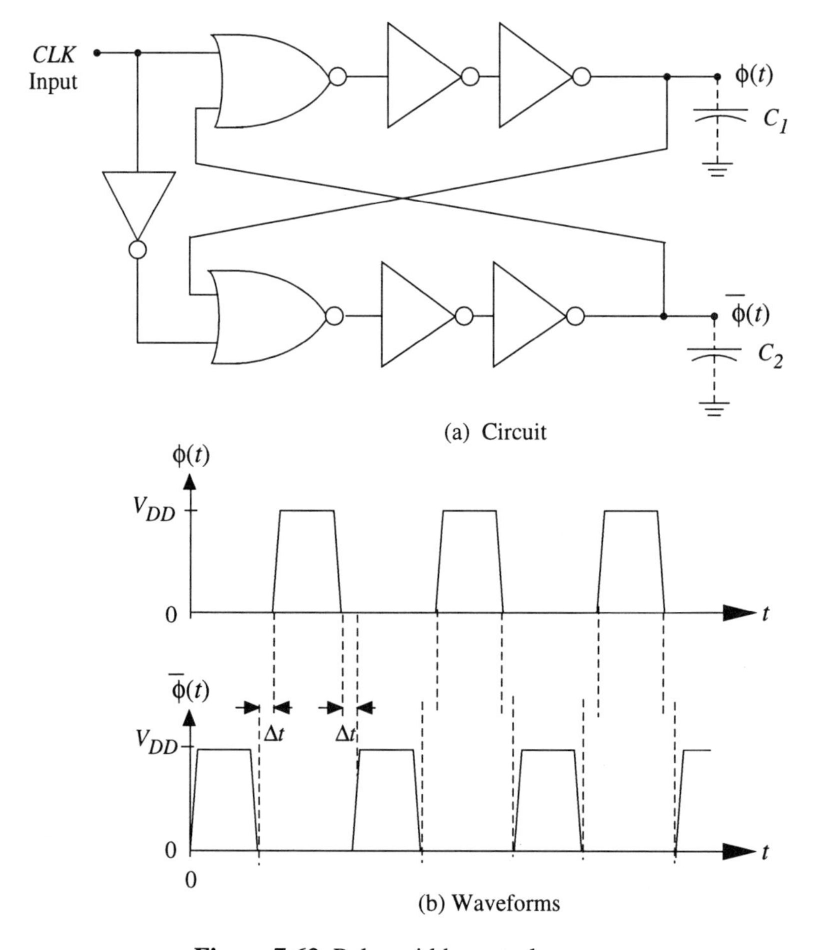

Figure 7.62 Pulse width control

$$\phi(t) \cdot \bar{\phi}(t) = 0 \tag{7.208}$$

to insure that the two are non-overlapping. Additional buffers may be added as required.

Clock generating circuits are very important in high-speed logic, and are still the subject of research articles in the open literature. These are usually combined with the problem of signal delays introduced by the distribution interconnect lines. Providing clock signals to the various logic circuits on a large chip can be difficult due to the parasitic resistance and capacitance of the distribution lines. This problem is discussed in more detail in Chapter 10 where we perform the characterization and modelling of on-chip interconnect structures. In addition, many papers in this area are being published in the current literature because of the increased importance of maintaining accurate control of the data in high-speed digital networks.

7.8 Summary Comments

In this chapter we have examined many of the features of dynamic logic circuits. In general, dynamic effects arise whenever there are isolated capacitive nodes. Since these are easily created in CMOS, dynamic effects may affect various aspects of circuit design.

In the next chapter, we take a different approach to CMOS logic by creating logic design styles that take advantage of dynamic charge storage nodes. Chapter 8 will introduce you to fascinating variations on CMOS logic circuits that illustrate some advanced techniques for high-speed design.

7.9 Problems

[7-1] Consider the charge leakage circuit shown in Figure P7.1. Assume a power supply voltage of V_{DD}=5v.

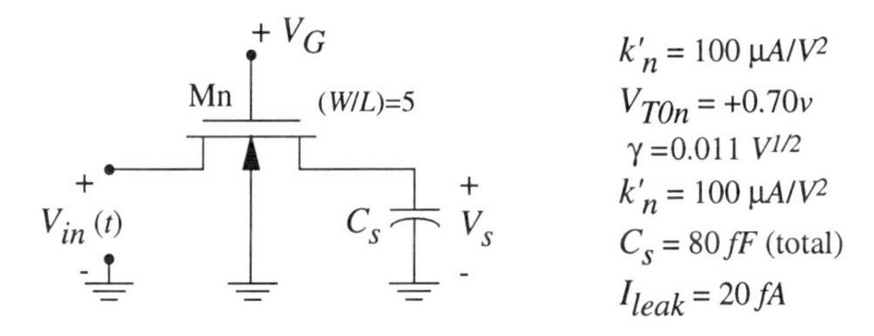

Figure P7.1

(a) Calculate the maximum logic 1 charge that can be stored on C_s in fC.

(b) Find the charging time t_{LH} for the logic 1 transfer.

(c) The minimum logic 1 voltage is assumed to be $0.6V_{DD}$. Find the maximum hold time t_H for this circuit.

(d) Repeat the calculation if the parameters are changed to C_s=50fF and I_{leak} =25fA.

[7-2] Consider an nFET that is described by the following parameters:

$$k'_n = 120\ \mu A/v^2,\ \mu_n = 560\ cm^2/V\text{-}sec,\ 2|\phi_F| = 0.58v,\ V_{TOn} = 0.80v,\ (W/L) = 8.$$

The power supply voltage is chosen to be 3.3v. The nFET is used as pass transistor such that the total storage capacitance is estimated to be C_s=85fF .

(a) Calculate the maximum logic 1 charge Q_s in fC.

(b) Suppose that C_s is charged to V_{max} as computed in (a). The leakage current from the capacitor is estimated by the linear function

$$I_{leak} = I_x\left(\frac{V_s}{V_{max}}\right) \tag{7.209}$$

where $I_x = 0.1pA$. The minimum logic 1 voltage is taken to be $0.6V_{max}$. Find the hold time t_H for the circuit.

[7-3] The storage capacitor C_s of a DRAM cell is initially charged to a value of $V_s = V_{max}$. The leakage current is taken to be of the form

$$I_{leak} = I_x e^{\left(\frac{V_s}{V_{max}}\right)} \tag{7.210}$$

Find the voltage $V_s(t)$ across the capacitor.

[7-4] Consider the TG chain shown below in Figure P7.2. For times $t<0$, the inputs are specified to be $(x,y,z)=(1,0,0)$, which charges C_1 which has a voltage of $V_1 = V_{DD} = 5v$. At time $t = 0$, the signals are changed to $(x,y,z)=(0,1,1)$. Calculate the final voltage V_f on the capacitors after charge sharing has taken place.

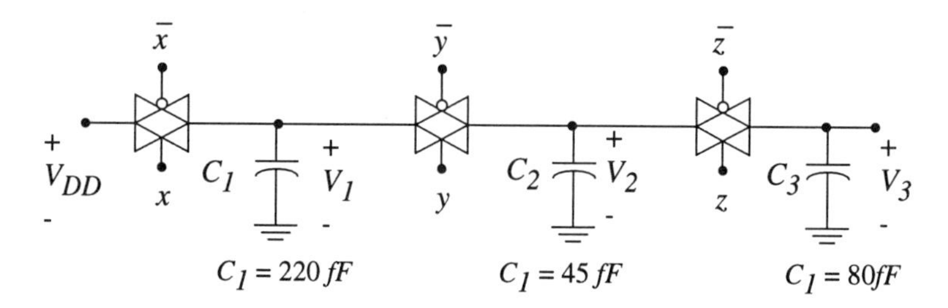

Figure P7.2

[7-5] Repeat the calculation in Problem [7-3] if the initial conditions are changed to $V_1(0)=3.3v$, $V_2(0)=1v$, and $V_3(0)=0.5v$.

[7-6] Use Laplace transforms to perform the s-domain analysis of charge sharing in Section 7.2.1 and verify the time-domain results.

[7-7] Verify the expression for $V_{out}(t)$ given in eqn. (7.141). Ignore body bias effects throughout the entire calculation.

[7-8] Consider the bootstrapping analysis of the circuit shown in Figure 7.32. Suppose that the output voltage is approximated by the simple exponential dependence

$$V_{out}(t) \approx V_m\left[1 - e^{-t/\tau_L}\right] \tag{7.211}$$

instead of the more complicated expression. Discuss how this will change the results of the bootstrapping discussion.

[7-9] A 2-FET chain is shown in Figure P7.3; all dimensions are in microns. The important parameters are given as

$$\mu_n = 580\ cm^2/V\text{-}sec,\ V_{T0n} = 0.7v,\ V_{DD} = 3.3v,\ C_j = 0.4fF/\mu m^2,\ C_{jsw} = 0.32fF/\mu m$$

and body bias can be ignored in the calculations. The built-in voltage is $0.9v$ for both the bottom and the sidewall regions.

(a) Calculate the maximum voltage transmitted to the load capacitance if $V_{in} = V_{DD}$ if both FETs are ON.

(b) Suppose now that $V_{in} = V_{DD}$ is applied when $A = 1$and $B = 0$. The signals are then changed to $A = 0$and $B = 1$. Calculate the voltage across the load capacitor in this case.

Use voltage-averaged LTI capacitors for the calculations.

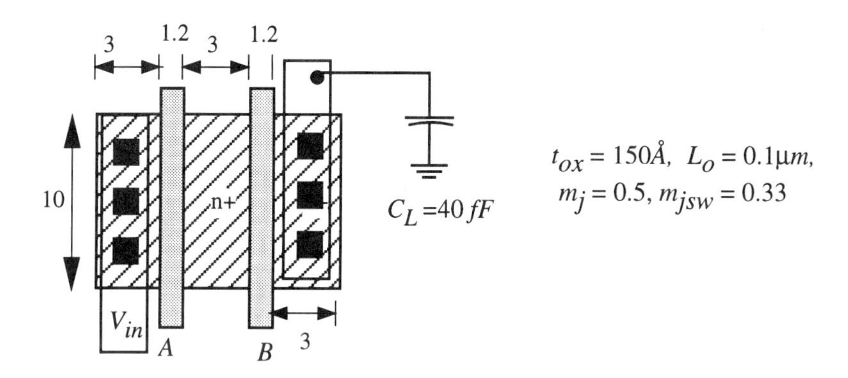

Figure P7.3

[7-10] A basic DRAM cell is shown in Figure P7.4. Use the parameters given in Problem [7.9] for all calculations.

(a) Calculate the total storage capacitance C_s using LTI values.

(b) Calculate the maximum amount of charge that can be stored on C_s.

(c) The sense amplifiers require a minimum voltage of 250 mV to distinguish a logic 0 from a logic 1. The line capacitance is known to be $C_{line} = 200fF$. What is the minimum voltage on the storage capacitor that will still be sensed as a logic 1?

(d) The reverse junction leakage current density is $J_{go} = 25\ fA/\mu m^2$. Use the LTI average value of the reverse current to estimate the logic 1 hold time for the cell.

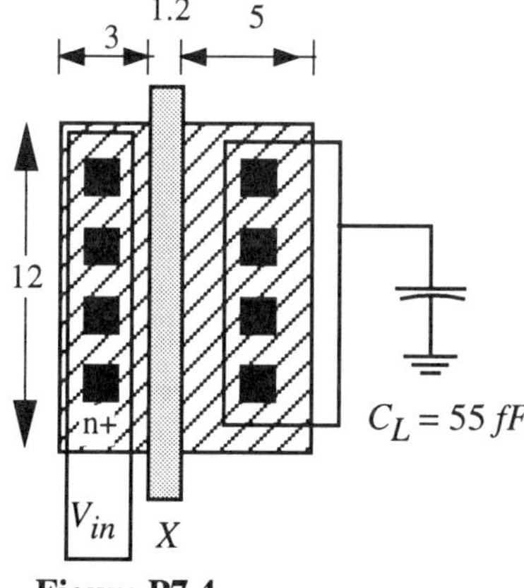

Figure P7.4

[7-11] Perform a literature search for the past 15 months and find two DRAM papers that specify the values of the storage capacitor. Describe the structure of cells.

Chapter 8

CMOS Dynamic Logic Families

A dynamic logic gate is one in which the output is only valid for a short amount of time after the result is produced. Although this sounds quite restrictive, dynamic CMOS networks are useful for high-speed system design. In this chapter, we will analyze several dynamic CMOS logic families. The presentation starts with basic ideas and logically extends them to very advanced circuit design techniques. The important concepts and circuit techniques combine to provide a powerful set of alternatives for modern high-performance system design.

8.1 Basic Philosophy

Designing a high-density, high-speed CMOS logic network requires that we take a critical look at the lessons learned from our study of MOSFETs and static logic gates. Some are obvious. For example,

- Avoid the use of pFETs as they are slow and/or take up too much area.

We do not want to eliminate pFETs completely from our circuits as they are useful for passing the power supply voltage V_{DD} to interior nodes. However, it is plausible to adopt the viewpoint where we concentrate on using nFETs as much as possible in logic blocks and in all critical situations.

Another important observation is that the switching speed of static circuits is limited by two related factors: the current conduction level through a MOSFET and the parasitic capacitances in the network. Since the numerical values of the parasitic elements originate from the technology, one may be led to believe that increased speed can only be attained by moving to a new and improved process. Of course, this is true only if we adhere to the same circuit design styles.

Dynamic circuits take a different viewpoint. Instead of fighting the time constant limits induced by the RC parasitics, we accept the presence of capacitances and use them as integral parts of the circuit. The simple DRAM cell studied in Section 7.3 of the previous chapter illustrates how a capacitor can be used to hold a logic variable. A dynamic logic gate takes this type of operation to the next level of sophistication by using clocks and power supplies to provide charge to a few selected nodes during pre-specified clock times, and then using the stored charge itself to control the movement of other charges through the network. The voltage is still used as the logic variable, but our viewpoint is radically changed because the power supply is only connected to the gate for short time intervals, and is not used to directly drive the outputs.

Dynamic circuits are of interest because they ***may*** provide more compact designs with faster switching speeds and reduced power consumption. On the other hand, it is also possible to design a dynamic logic unit that is smaller than an equivalent static design, but is slower or consumes more power, or both! Good dynamic logic circuits usually require more thought and care to create, and one must be intimately familiar with the characteristics of dynamic charge storage and clocking.

The design techniques introduced in this chapter are properly classified as **system design styles**, and should only be applied in this manner. As will be seen in the discussion, the circuits only work well when the same (or related) approach is used to design an **entire logic cascade**, not just a few gates. Moreover, the characteristics of the circuits can be quite different, so that mixing the styles should only be done using great care. Failure to adhere to these simple guidelines may result in logic glitches and other problems that may render the entire network unusable.

8.2 Precharge/Evaluate Logic

The basis for creating a dynamic logic family can be found in the general circuit shown in Figure 8.1. This consists of a single clock-driven complementary pair made up of FETs Mp and Mn, and an nFET logic network that acts as a composite switch from top to bottom. The load at the output node is modelled as a capacitor C_{out} with the voltage V_{out} across it. As we shall see in the analysis, the output voltage V_{out} is only valid for a short period of time, giving the gate its classification as a dynamic circuit.

The clock signal $\phi(t)$ controls both the internal operation of the logic gate and the data flow through a cascaded chain. This is achieved by the complementary action of Mp and Mn as the clock cycles through each period. When ϕ=0*v*, the pFET Mp is active while Mn is in cutoff; this defines the **precharge** event, where the output capacitor is charged to a voltage V_{DD}. Alternately, with $\phi = V_{DD}$, Mp is in cutoff while Mn is biased into conduction; during this time interval, the gate will **evaluate** the inputs and establish the output voltage. Since the clock alternates between 0 and 1 states, this circuit provides the basis for **Precharge/Evaluate** (or P/E) logic cascades. The two operational regions are labelled in the clock waveform shown in Figure 8.2.

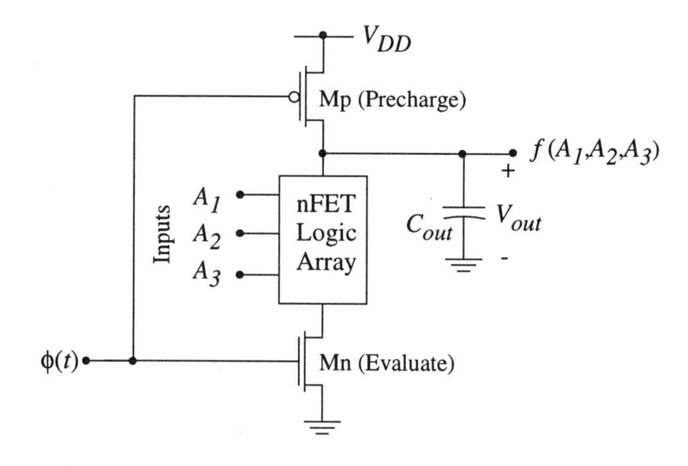

Figure 8.1 General dynamic logic gate

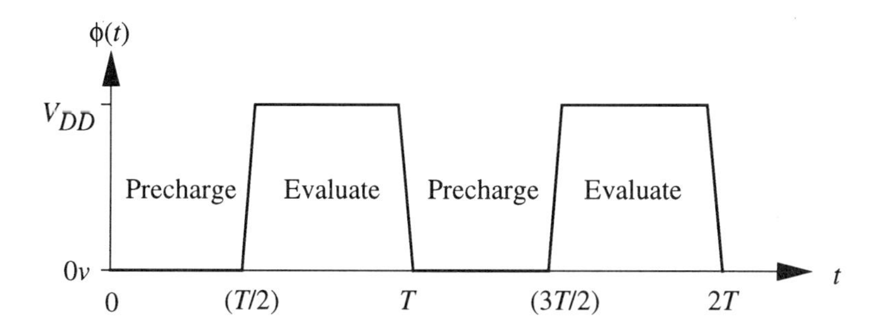

Figure 8.2 Clocking sequence for precharge/evaluate logic

Precharge

Let us examine the precharge event using the drawing in Figure 8.3. With the clock at $\phi=0v$, the output node capacitance C_{out} is charged through Mp (which is called the precharge device) to a final value of $V_{out}=V_{DD}$. During this time, Mn is cutoff, severing the DC path to ground. The output voltage always rises to a final value of V_{DD}, but this does not mean that the output is at a logic 1 level. Rather, this should be viewed as "pre-conditioning the node" for a later sequence of events. Owing to this characteristic, the input signals, are not valid during the precharge event, regardless of their origin. The sole purpose of the precharge interval is to add the charge

$$Q = C_{out}V_{DD} \tag{8. 1}$$

to the output node.

Evaluate

Evaluation of the logic inputs takes place when the clock goes to a value of $\phi=1$ corresponding to a clock voltage of $V_\phi = V_{DD}$. During this time, the precharge transistor Mp is in cutoff, while the evaluate FET Mn is active with $V_{GSn}=V_{DD}$. Since the output capacitor was precharged to a voltage

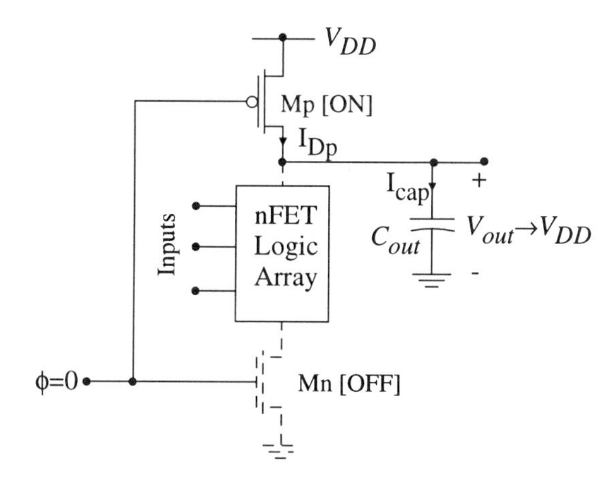

Figure 8.3 Circuit undergoing precharge

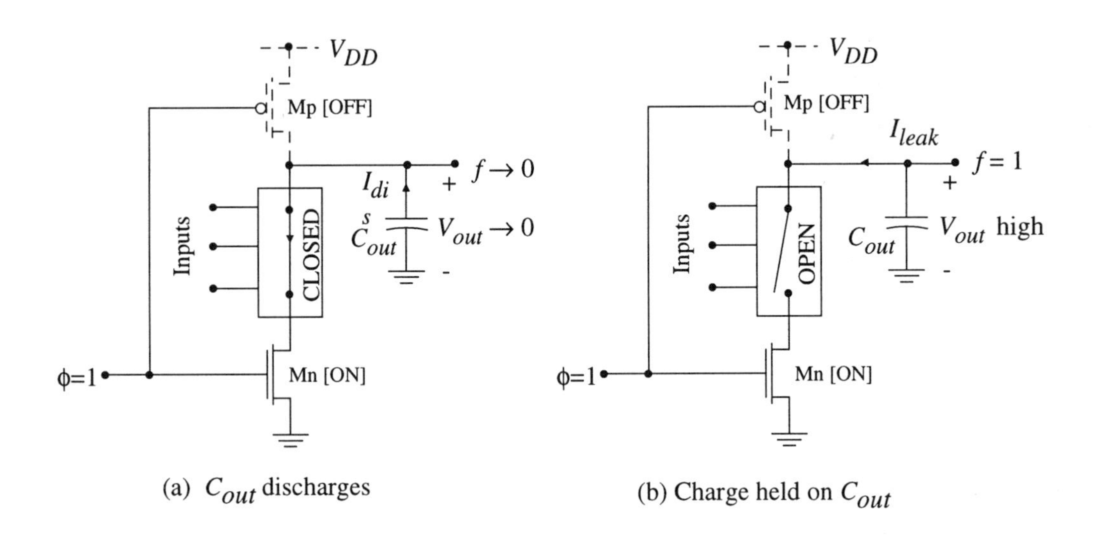

(a) C_{out} discharges

(b) Charge held on C_{out}

Figure 8.4 Evaluation mode of operation

of $V_{out} = V_{DD}$, there are two possible actions that can take place. If the inputs to the nFET logic block result in a closed circuit from the top to the bottom, then C_{out} will discharge through the logic array and result in an eventual logic 0 output voltage of $V_{out} = 0v$. This is shown in Figure 8.4(a), and indicates a logical output of $f = 0$.

If, on the other hand, the inputs combine to give an open circuit between the top and the bottom of the nFET array, then there is no direct path to ground and the charge is held on C_{out}. This results in a high logic 1 voltage at the output, and is illustrated by the circuit in Figure 8.4(b). Directly after the beginning of the evaluation event, the actual value of V_{out} may be smaller than V_{DD} because of charge sharing. Moreover, the charge on C_{out} cannot be held indefinitely due to charge leakage through the transistors. During evaluation, the circuit is described as undergoing a **conditional discharge** event. The characterization of this circuit as being dynamic is due to the fact that the logical value of the output voltage V_{out} is only valid for a short period of time, and must be used before charge leakage corrupts the value. The minimum switching frequency for this type of dynamic logic circuit is limited by charge sharing and leakage during the evaluation portion of the clocking waveform.

8.2.1 NAND3 Analysis

Let us analyze the switching times involved in the dynamic NAND3 gate of Figure 8.5 as an example of a typical P/E logic circuit. The three series nFETs form the logic array using the same rules as in static gate design. Denoting the inputs as A_1, A_2, and A_3, the output is given by the logic expression

$$f = \overline{A_1 \cdot A_2 \cdot A_3} \qquad (8.2)$$

since the output can only discharge if all three inputs are a logic 1 values. This expression is only valid during the evaluation portion of the clock cycle when ϕ=1. A more correct statement that includes this fact would be

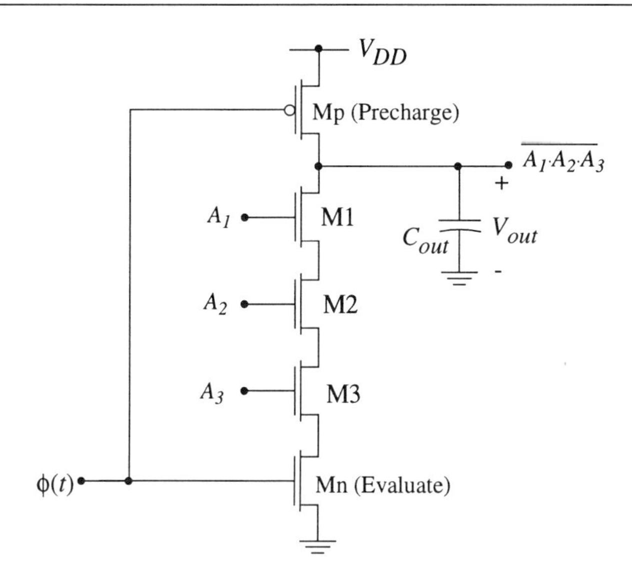

Figure 8.5 A dynamic NAND3 logic gate

$$f = \overline{A_1 \cdot A_2 \cdot A_3} \cdot \phi \tag{8. 3}$$

but we will continue to use the simpler notation where the clock is understood to be at a value of ϕ=1 for the logic to be valid. In our discussion, we will assume for simplicity that the precharge and evaluate FETs (Mp and Mn) have aspect ratios of $(W/L)_p$ and $(W/L)_n$, respectively, and that the three logic transistors M1, M2, and M3 all have the same aspect ratio (W/L). Of course, all device sizes may be adjusted as required by the transient specifications.

Precharge

The precharge interval is designed to charge the output node capacitance. The clock value has a value of ϕ=0*v*, which biases Mp with a source-gate voltage of

$$V_{SGp} = V_{DD} \tag{8. 4}$$

so that it is active and can conduct current. The simplest situation is where A_1=0 as shown in Figure 8.6, since the charging of C_{out} is identical to that found in an inverter. The low-to-high time is thus given by

$$t_{LH} = s_p \tau_p \tag{8. 5}$$

which is a reasonable approximation for the precharge time. The time constant is given by the usual expression

$$\tau_p = \frac{C_{out}}{\beta_p (V_{DD} - |V_{Tp}|)} \tag{8. 6}$$

If instead A_1=1 during this time interval, then the precharge time is increased because some of the current is diverted to charge the internal parasitic device capacitances that exist between the FETs. This is true for any of the input combinations $(A_1 A_2 A_3)$=(100), (101), (110), or (111) since all of these provide some conduction to the internal nodes of the logic chain. Analytic estimates for these

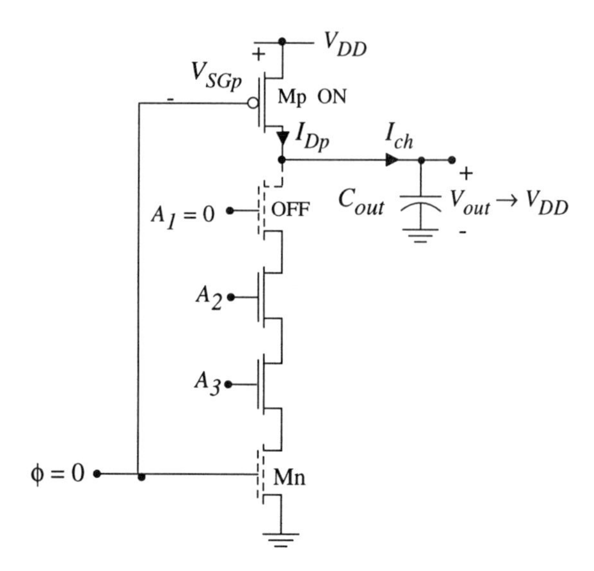

Figure 8.6 Precharge of the NAND3 gate

cases can be derived using the techniques developed in Chapter 4, but are not reproduced here. In practice, it is easier to make reasonable assumptions during the initial design work, and then test the more complicated cases using a computer simulation.

Evaluation

The evaluation interval starts when the clock makes a transition to a value of ϕ=1. The evaluation nFET Mn is now active, while Mp is driven into cutoff, setting the stage for a conditional discharge event to take place. Let us analyze both possibilities to understand the behavior of the output voltage V_{out} during this time.

Suppose that the inputs are given by $(A_1\ A_2\ A_3)$=(111). Since this will turn on all three logic transistors, C_{out} will discharge to ground and V_{out} will fall to 0*v*. The general situation is shown in Figure 8.7(a). Replacing the MOSFETs by their LTI equivalent circuits gives the RC network shown in Figure 8.7(b). The Elmore formula may then be used to approximate the voltage decay by the simple exponential function

$$V_{out}(t) \approx V_{DD} e^{-t/\tau_n} \tag{8.7}$$

where the time constant is given by

$$\tau_n = C_{out}(3R_A + R_n) + C_1(2R_A + R_n) + C_2(R_A + R_n) + C_3 R_n \tag{8.8}$$

In this formula, R_A is the resistance of a single logic transistor, and R_n is the resistance of the evaluate FET Mn. The capacitors C_1, C_2, and C_3 represent the nodal contributions between each pair of transistors, and include both depletion and MOS terms. With the exponential approximation, the time needed for the voltage to fall from V_{DD} to a particular output voltage $V_{DD}(t)$ is given by

$$t \approx \tau_n \ln\left[\frac{V_{DD}}{V_{out}(t)}\right] \tag{8.9}$$

the actual value of a logic 0 voltage (that is equivalent to V_{IL} in a static gate) is determined by the

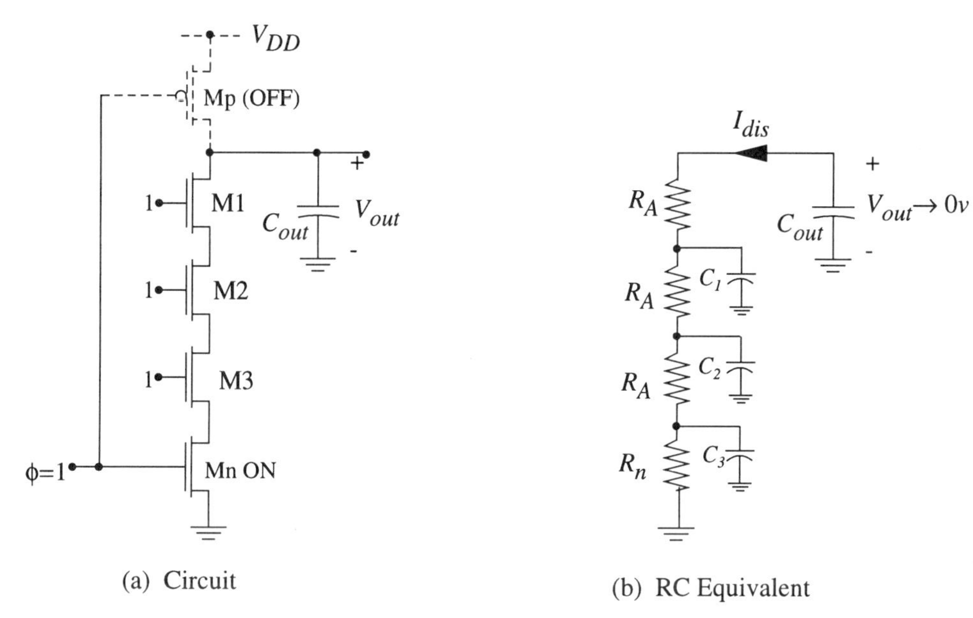

Figure 8.7 Discharge event in the NAND3 gate

characteristics of the next gate in the cascade. If the same type of logic gate is used, then the output will be connected to the gate of an nFET in a similar logic chain. $V_{out} = V_{Tn}$ is thus the lowest possible value for V_{IH}, so that[1]

$$t_{1 \to 0} \approx \tau_n \ln\left[\frac{V_{DD}}{V_{Tn}}\right] \tag{8. 10}$$

provides an estimate of the transition time needed for the output to fall from an ideal logic 1 voltage to a logic 0 input level.

If any of the inputs A_1, A_2, or A_3 are at a logic 0 voltage, then the output voltage V_{out} should remain high corresponding to the NAND operation

$$\overline{A_1 \cdot A_2 \cdot A_3} = 1 \tag{8. 11}$$

Ideally, this would result in the case where V_{out}=V_{DD}. However, both charge sharing and charge leakage will reduce the value of V_{out} below this ideal level. The worst-case situation for charge sharing is that where the largest number of capacitors are switched into the network. For the NAND3 gate, this occurs when $(A_1\,A_2\,A_3)$ = (0xx) during the precharge interval (with "x" a don't care) and then the inputs are switched to $(A_1\,A_2\,A_3)$ = (110) for the evaluation interval; these values also give the worst-case charge leakage situation.

Charge sharing is induced immediately after the clock makes the transition $\phi \to 1$. During the precharge, the worst-case (i.e., minimum) charge stored in the gate is

$$Q = C_{out} V_{DD} \tag{8. 12}$$

[1] Note that body-bias effects will be present for all nFETs in the logic array.

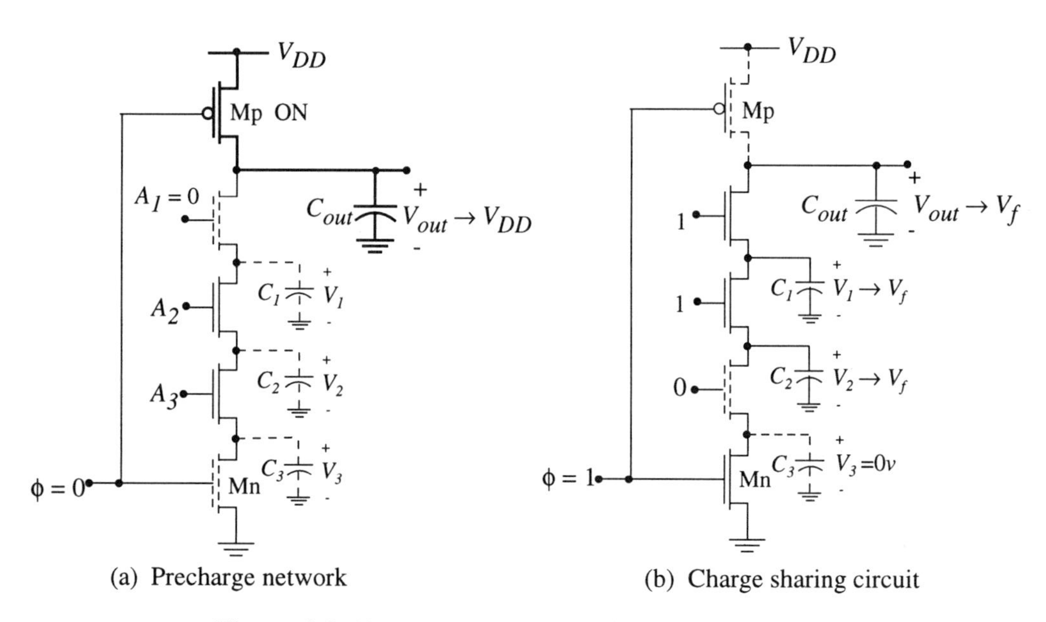

(a) Precharge network (b) Charge sharing circuit

Figure 8.8 Charge sharing in the NAND3 gate

which occurs when A_1=0 and no internal capacitors are charged. Figure 8.8 shows the charge sharing circuit that occurs during evaluation with the inputs at $(A_1\, A_2\, A_3) = (110)$. It is seen that C_{out} must share charge with C_1 and C_2. Let us denote the final voltage across C_{out} by V_f. Two possibilities exist depending upon the level of charge sharing. If $V_f < (V_{DD} - V_{Tn})$, then all of the capacitors will have the same voltage and the final charge distribution may be written as

$$Q = C_{out}V_f + C_1V_f + C_2V_f \tag{8. 13}$$

Equating this to the original charge expression and rearranging gives

$$V_f = \frac{C_{out}}{(C_{out} + C_1 + C_2)} V_{DD} \tag{8. 14}$$

for the final value. In this case, we must have

$$C_{out} \gg C_1 + C_2 \tag{8. 15}$$

to insure that V_f is maintained at a sufficient value to still be interpreted as a logic 1 output. The actual values of the capacitors depend upon the details of the layout geometry, and must be considered when designing the circuit.

The other possibility is that where only a small amount of charge is lost from C_{out}, and the final voltage satisfies $V_f \geq (V_{DD} - V_{Tn})$. Since M2 and M3 are connected as pass transistors, the maximum voltage that can be transmitted to C_1 and C_2 is $V_{max} = (V_{DD} - V_{Tn})$. Thus, the final charge distribution is described by

$$Q = C_{out}V_f + (C_1 + C_2)\,V_{max} \tag{8. 16}$$

which gives the final voltage as

$$V_f = V_{DD} - \left(\frac{C_1 + C_2}{C_{out}}\right)(V_{DD} - V_{Tn}) \tag{8.17}$$

The value of C_{out} relative to the sum (C_1+C_2) determines which situation will take place. For $C_{out} >> C_1+C_2$, the final voltage will be that just obtained as equation (8.17); this can be understood by noting that the relative sizes of the capacitors indicate that only a small amount of charge is needed to come to equilibrium. If C_{out} is the same order of magnitude as C_1+C_2, then eqn. (8.14) will apply. The border between the two is at the value

$$C_{out} = (C_1 + C_2)\left[\frac{V_{DD}}{V_{Tn}} - 1\right] \tag{8.18}$$

For C_{out} less than this value, the charge sharing is severe and $V_f < V_{max}$. The analysis shows that a large C_{out} is needed to maintain the output voltage at high value, as expected on physical grounds.

After charge redistribution takes place, the value of V_{out} continues to fall because of charge leakage. Figure 8.9 shows the worst-case charge leakage paths where it has been recognized that every pn junction admits reverse current flow. The precharge transistor Mp provides charge to C_{out} from the power supply in the form of I_{Rp}, but three of the nFETs act to drain charge to ground. Assuming a leakage current of I_{Rn} at every pn junction gives

$$5I_{Rn} - I_{Rp} = -(C_{out} + C_1 + C_2)\frac{dV_{out}}{dt} \tag{8.19}$$

by summing the currents at the output node; we have assumed that all capacitors are at the same voltage for this calculation. Note that all three capacitors hold the charge, so that the sum of capacitance values is used in the equation. If the nFET and pFET leakage currents are about the same order of magnitude,[2] then this indicates that there is a net flow of charge off of the node. Assuming that I_{Rp} and I_{Rn} are constants gives

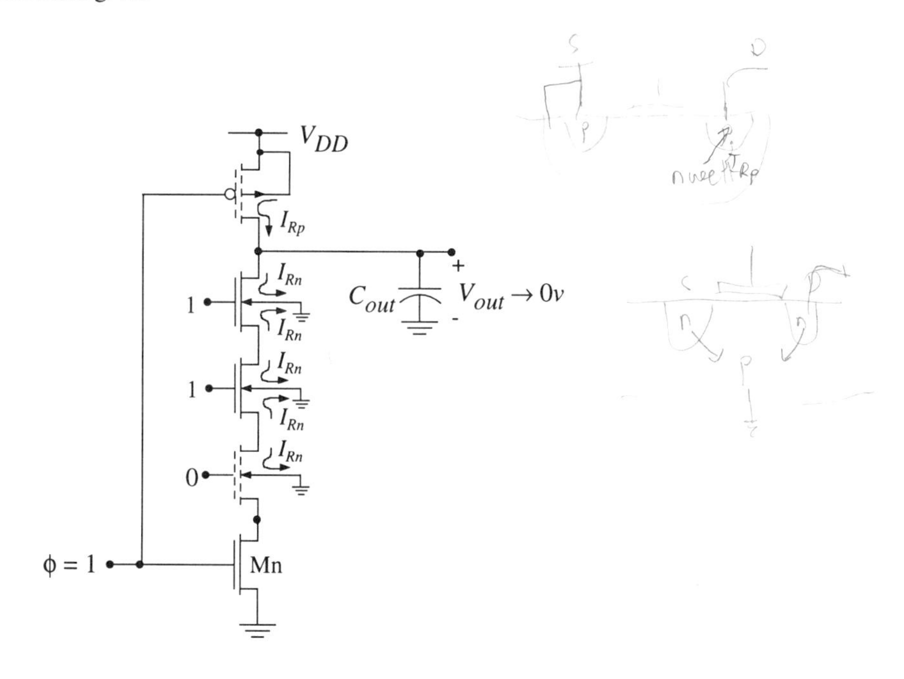

Figure 8.9 Charge leakage paths in the NAND3 hold state

$$V_{out}(t) = V_f - \frac{(5I_{Rn} - I_{Rp})}{(C_{out} + C_1 + C_2)}t \quad (8.20)$$

so that $V_{out} \rightarrow 0v$. This clearly illustrates the dynamic characteristic of the gate. We cannot hold a high value of V_{out} too long as charge leakage always makes the output fall to a logic 0 voltage. The maximum hold time t_H depends upon the value of V_{IH}-equivalent voltage of the next gate in the chain through

$$t_H \approx \frac{(C_{out} + C_1 + C_2)(V_f - V_{IH})}{(5I_{Rn} - I_{Rp})} \quad (8.21)$$

This, in turn, sets the minimum clock frequency.

The most important feature of this type of dynamic logic gate is that the clock controls the precharge and evaluate sequence, and hence synchronizes the data flow through a cascade that consists of similar gates. The problems of discharge time constants, charge sharing, and charge leakage are the most critical aspects of the circuit behavior.

8.2.2 Dynamic nMOS Gate Examples

Since we use nFET arrays to create the logical switching action, all of the logic array design techniques introduced in Chapter 5 for static circuits are valid for dynamic nMOS gates.

Figure 8.10 gives the circuit diagram for a dynamic NOR3 gate. This is constructed by simply paralleling the three logic transistors as shown. The general operation is identical to that discussed above for the NAND3 circuit. A clock signal value of ϕ=0 defines the precharge event during which C_{out} is charged to a voltage of $V_{out} = V_{DD}$. When the clock makes a transition to ϕ=1, the gate evaluates the inputs. If any one (or more) of the input variables A_1, A_2, or A_3 is high (at a logic 1 voltage), then C_{out} discharges and $V_{out} \rightarrow 0v$. The only time that V_{out} is held high is if $A_1 = A_2 = A_3 = 0$. For this gate, there are no charge sharing paths. However, charge leakage will occur with a hold time estimate of

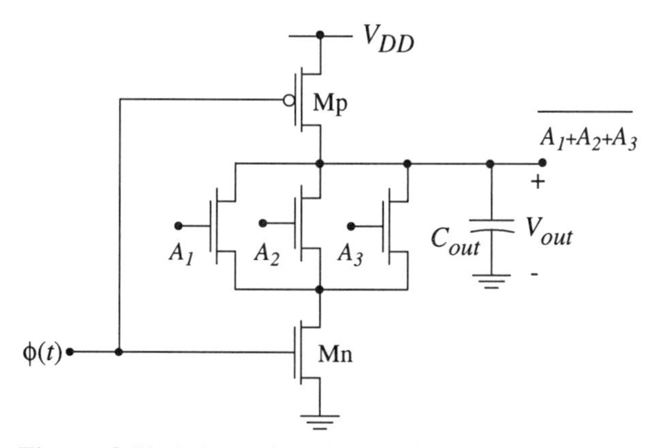

Figure 8.10 A dynamic NOR3 logic gate

[2] This assumption is not always valid. Indeed, it is easy to find processes where one leakage current dominates the other.

$$t_H \approx \frac{C_{out}(V_f - V_{IH})}{(3I_{Rn} - I_{Rp})} \tag{8. 22}$$

due to the three nFET leakage paths. Also note that C_{out} has contributions from Mp and all three logic transistors.

Complex logic gates can be designed in the same manner. Figure 8.11 shows two examples. In Figure 8.11(a), the transistors are arranged to provide the function

$$f = \overline{(x+y) \cdot z} \tag{8. 23}$$

when the clock swings to the evaluation interval with ϕ=1. The circuit in Figure 8.11(b) is the exclusive-OR

$$g = x \oplus y \tag{8. 24}$$

as can be verified by applying the nFET rules. The important point being reiterated here is that the rules remain the same as in static logic design, but only the nFET array is needed. Eliminating the logic pFETs gives much simpler layout wiring and reduced capacitance levels.

8.2.3 nMOS-nMOS Cascades

Let us now examine the concept of cascading P/E gates to form a logic chain. The simplest approach is to cascade dynamic stages as shown in Figure 8.12. This drawing only shows one of the datapaths through the logic chain. This means that every input to Stages 2, 3, and 4 is assumed to originate from a preceding logic gate that is electrically similar to Stage 1.

The clock is distributed to every stage in the chain with the same phase, which gives the operation shown in Figure 8.13. Precharge occurs with ϕ=0 as shown in (a). Since the stages are all synchronized to the clock, every stage undergoes precharge at this time. This causes the output voltages to rise to V_{DD} as shown. A clock transition to ϕ=1 causes the chain to undergo evaluation.

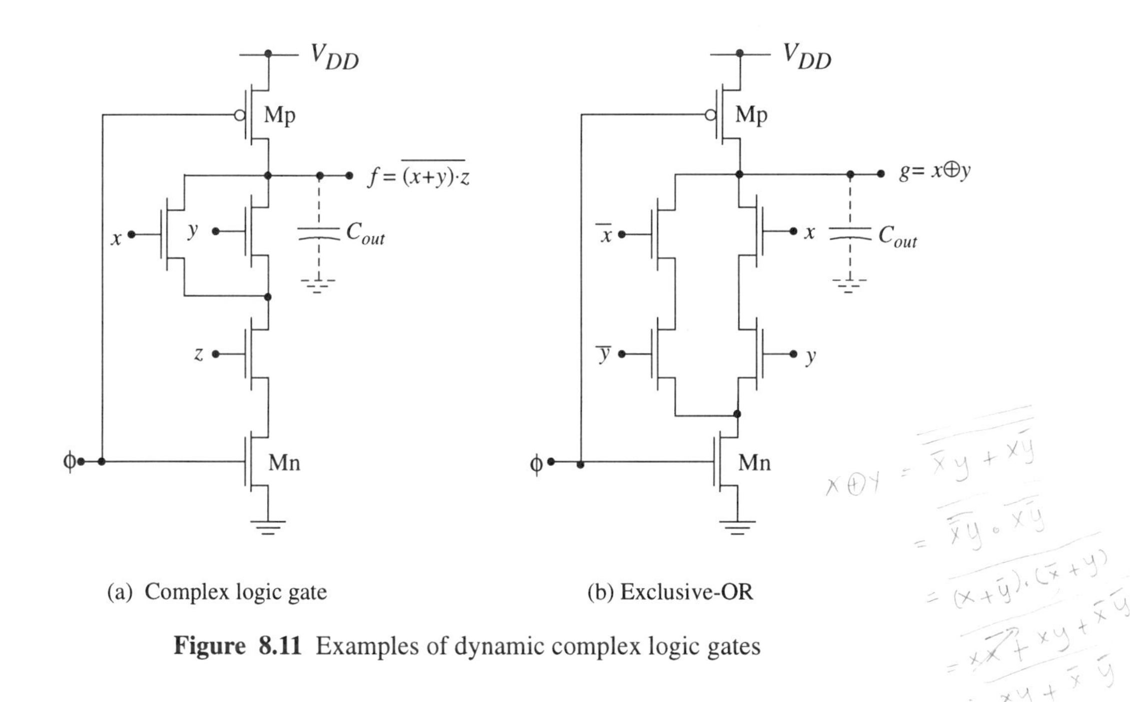

(a) Complex logic gate (b) Exclusive-OR

Figure 8.11 Examples of dynamic complex logic gates

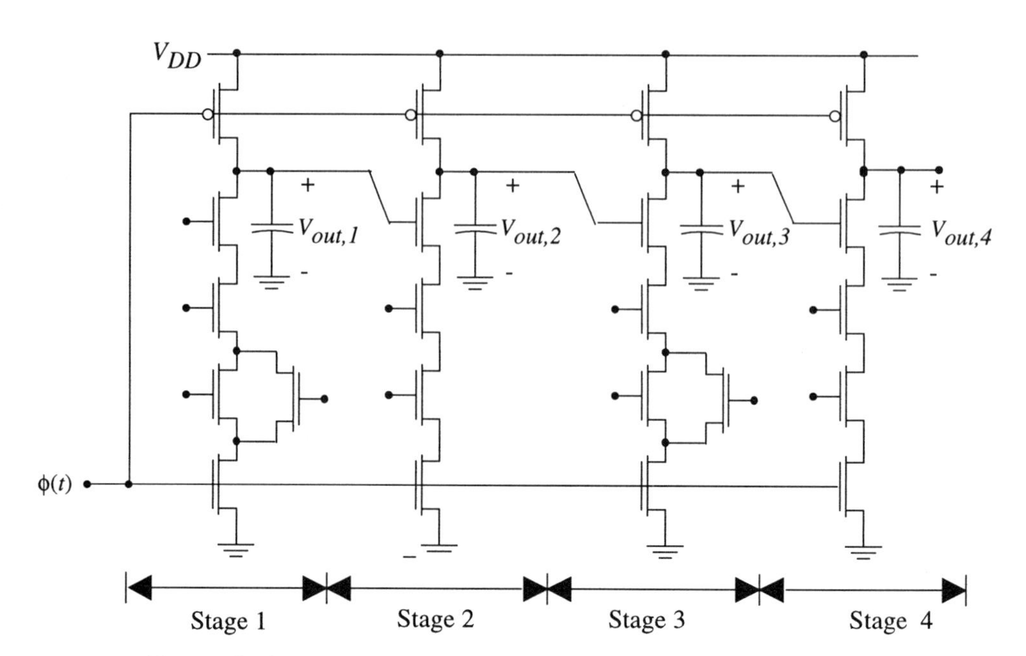

Figure 8.12 Cascade of dynamic logic gates

During this time, the output of each circuit is transferred to the input of the next stage(s), so that the logic "ripples" down the chain from left to right. In Figure 8.13(b), we have indicated the conditional discharge event by showing the currents $I_{dis,m}$ for m = 1, 2, 3, 4. If current flow is permitted through an nFET logic array, then the output voltage of that stage decays to 0*v*. A hold condition in the *j*-h stage is characterized by $I_{dis,j}= 0$. This acts to keep the output voltage $V_{out,j}$ at a high value, indicating a logic 1 result from the output of the gate.

Although this appears to be a straightforward scheme, cascading one nMOS stage into another nMOS logic stage introduces the possibility of a logic **glitch** in which an incorrect logic value is created by a circuit problem. To understand the origin of the problem, let us analyze the 2-stage NAND-NAND cascade shown in Figure 8.14(a). During the precharge event, ϕ=1 drives both Mp1 and Mp2 into conduction, charging both output nodes to V_{DD}. Note in particular that, since

$$V_{out,1} = V_{DD} \tag{8. 25}$$

the logic nFET MX in the second stage is driven into an active conduction state.

The possibility of a logic glitch can be seen by the situation Figure 8.14(b) where all of the inputs have been set at logic 1 levels. Consider the first stage. Since all inputs are high, $C_{out,1}$ will discharge in a characteristic time

$$\tau_{dis,1} = 4R_nC_{out} + 3R_nC_1 + 2R_nC_2 + R_nC_3 \tag{8. 26}$$

where we have assumed that all of the logic transistors have the same aspect ratio for simplicity. The value of $V_{out,1}(t)$ decays according to

$$V_{out,1}(t) = V_{DD}e^{-t/\tau_{dis,1}} \tag{8. 27}$$

However, the logic nFET MX in the second stage conducts current until a time t_1 that is defined by the condition $V_{out,1}(t_1) = V_{Tn}$. Solving,

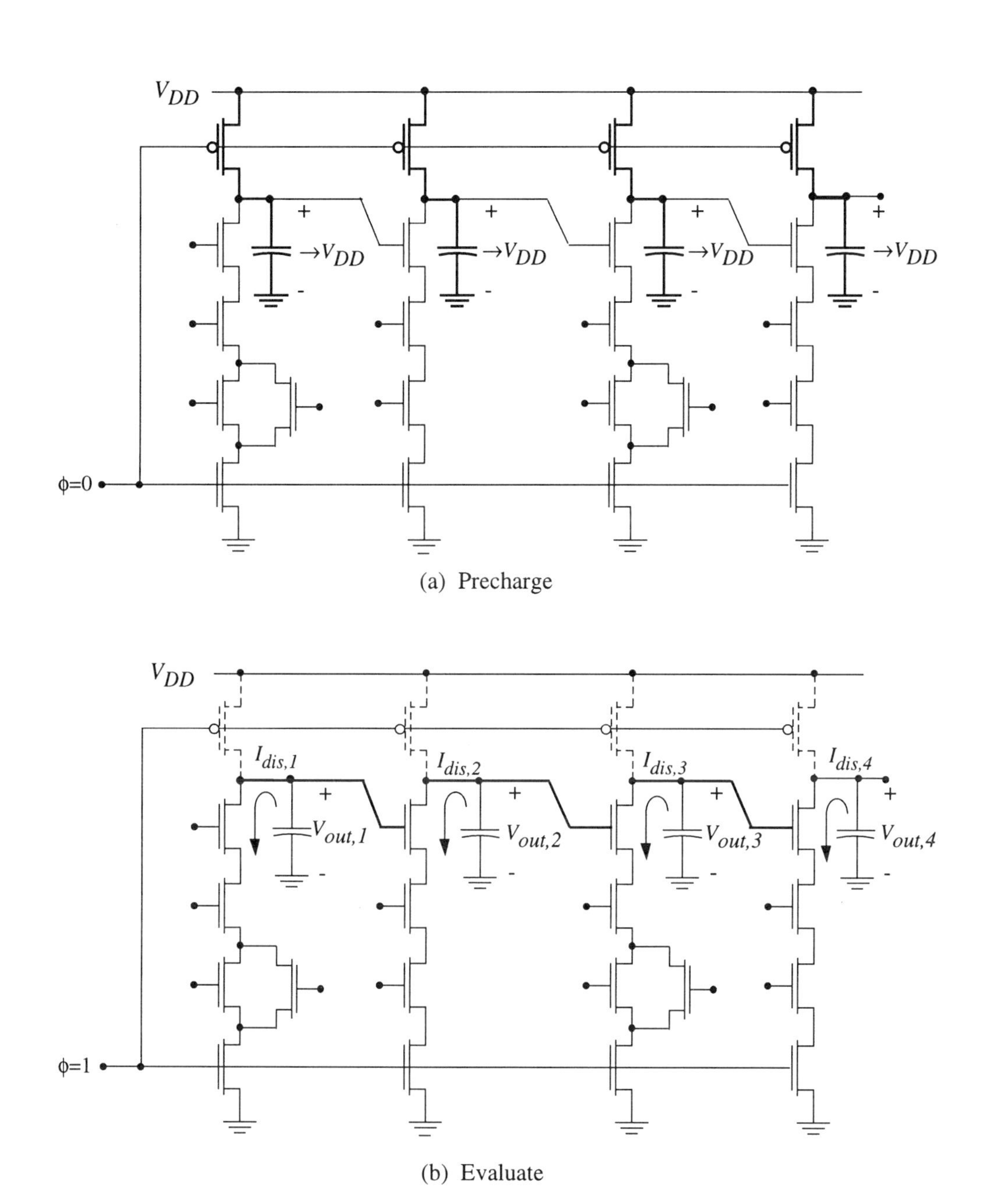

Figure 8.13 Internal operation of the dynamic logic cascade

$$t_1 = \tau_{dis,1} \ln\left[\frac{V_{DD}}{V_{Tn}}\right] \tag{8.28}$$

The potential for a problem can be seen by examining the behavior of the second stage while $C_{out,1}$ in discharging. During the time that $V_{out,1}$ is falling from V_{DD} to V_{Tn}, $C_{out,2}$ is discharging as long as MX is active. A glitch occurs if the final value of $V_{out,2}$ is so low that it is incorrectly interpreted

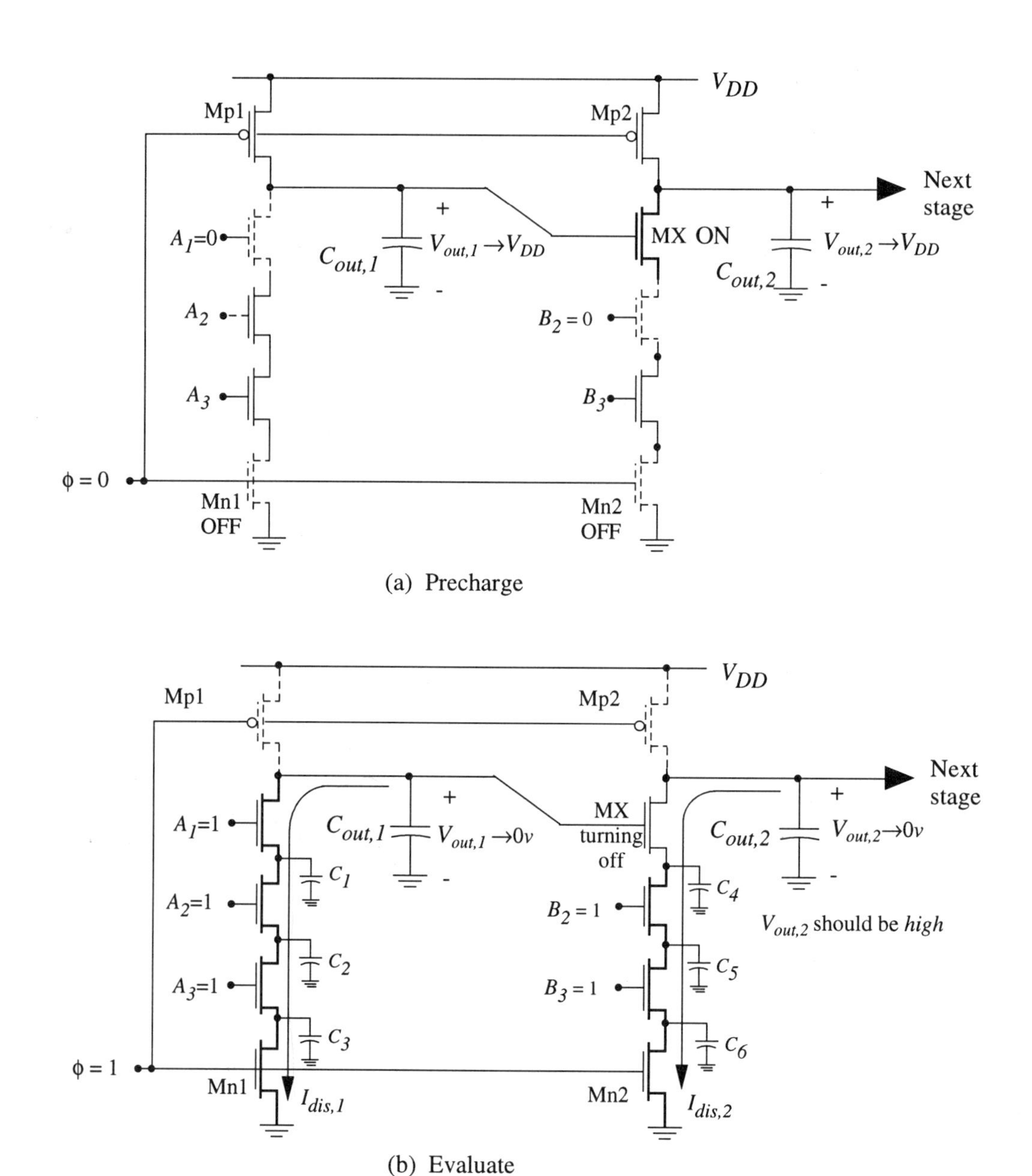

Figure 8.14 The glitch problem in an nMOS-nMOS cascade

as a logic 0 voltage in the next (third) stage of the cascade.

To understand the problem, let us approximate the output voltage by the exponential form

$$V_{out,2}(t) = V_{DD}e^{-t/\tau_{dis,2}} \tag{8.29}$$

where $\tau_{dis,2}$ is the discharge time constant for the second stage characterizing the conducting path with $C_{out,2}$, C_4, C_5, and C_6. A glitch will occur if $V_{out,2}(t_1)$ falls below V_{IH} of the next stage. The critical condition is expressed by

$$V_{out,2}(t_1) = V_{DD}e^{-t_1/\tau_{dis,2}} = V_{IH} \quad (8.30)$$

If $V_{IH} = V_{Tn}$, then avoiding the glitch requires that

$$\tau_{dis,1} < \tau_{dis,2} \quad (8.31)$$

This is intuitively obvious: the first stage should discharge faster than the second stage to insure that MX shuts off before $V_{out,2}$ drops to a logic 0 voltage. If, on the other hand, the next stage is a different type of logic gate (such as a static circuit), then the condition will be defined the specific value of V_{IH}. The problem can be controlled for the 2-stage network by careful design of the discharge networks, but the complexity of the FET array is closely related to the logic equations. In general, the problem gets worse as we increase the number of cascaded stages since the delays are additive from the input to the output. This problem restricts the use of dynamic nMOS-nMOS cascades because they require logic chains that always progress monotonically from simple gates to more complex gates.

8.2.4 Dynamic pMOS Logic

A simple solution to the possible glitch discussed above is obtained by examining the cause of the problem. This is shown in the circuit of Figure 8.15(a). Precharging the output capacitor C_{out} to a value $V_{out} = V_{DD}$ places the nFET M1 in the active region of operation, which in turn can cause a glitch in the next stage. If we replace M1 with a pFET M2, as in Figure 8.15(b), then the precharge voltage is no longer a problem: applying a voltage V_{DD} to a pFET drives it into cutoff. In other words, if the next stage is based on pMOS logic, then the glitch problem is automatically eliminated.

With this observation in hand, let us now construct a dynamic logic gate that uses pFETs by combining a clocked complementary pair to control the precharge and evaluate operations with a pMOS logic array as shown in Figure 8.16. Although this may appear similar to the nMOS gate, there are several differences. First, note that the roles of the clocked transistors are reversed; Mp is now labelled as Evaluate, while Mn is the Precharge device. Second, the clock signal has been chosen to be $\overline{\phi}(t)$, which is not mandatory, but will make it easier to interface to nMOS circuits in a logic cascade. Finally, the output node is taken at the top of Mn, or equivalently, at the bottom of the lowest pFET in the logic chain.

The operation of the pMOS dynamic logic gate is similar to that of an nMOS gate except that everything is reversed. Consider first the precharge event. This takes place when $\overline{\phi}=1$ (i.e., $\phi=0$) as shown in Figure 8.17(a). The clock biases the precharge transistor Mn into conduction, which allows a discharge path for C_{out} to ground and results in the final precharge output of

$$V_{out} \rightarrow 0v \quad (8.32)$$

This is opposite to that which takes place in the nMOS equivalent logic gate during a precharge event. Evaluation of the inputs occurs when $\overline{\phi}=0$ (meaning that $\phi=1$). Since Mp is biased active while Mn is driven into cutoff, the output capacitor C_{out} is subjected to a **conditional charge** event as illustrated in Figure 8.17(b). The pFET logic array acts like a composite switch between Mp and C_{out}. If the switch is closed, then current flows and C_{out} charges to a final voltage of $V_{out} = V_{DD}$, indicating a logic 1 result. In the opposite case where the switch is kept open, the voltage on V_{out} is initially low and interpreted as a logic 0 output. However, leakage currents that flow from the power supply through the pFETs (via the n-well and pn junctions) will act to increase the charge on C_{out}, so that the low voltage can only be maintained for a short period of time.

Dynamic pMOS logic gates can be designed using the pFET rules of static logic that were pre-

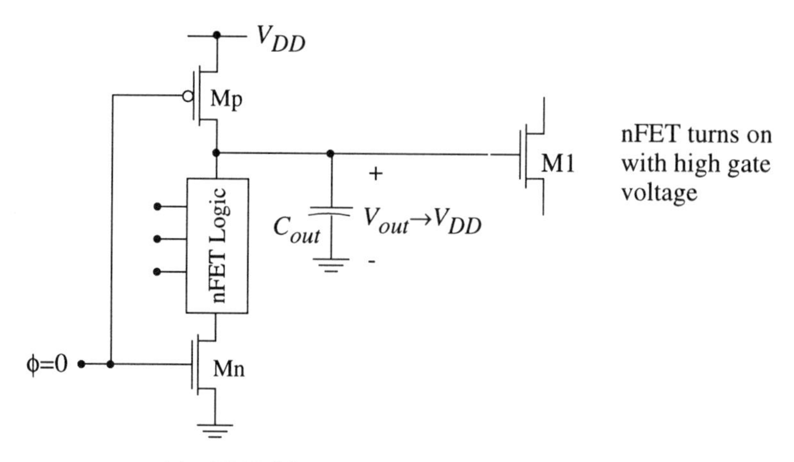

(a) nFET driven active during precharge

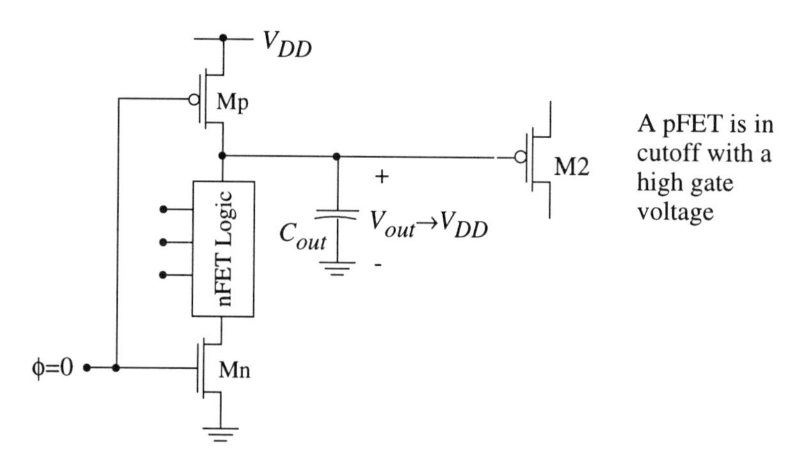

(b) Replacement of logic transistor with a pFET

Figure 8.15 One solution to solving the glitch problem

sented in Chapter 5. Two examples are shown in Figure 8.18. In the gate drawn in (a), the series-connected pFETs yield the NOR operation that is valid when $\overline{\phi}$=0. The logic network used to construct the gate in Figure 8.18(b) provide an output function of

$$g = \overline{a \cdot b + c} \tag{8. 33}$$

in a similar manner. Using this approach, we can design pFET-based dynamic logic gates as required.

The limiting factor in pMOS logic design are the pFETs themselves. Recall that the process transconductance for a pFET is always smaller than that for an nFET: $k'_n > k'_p$. Given equal size nFETs and pFETs, the p-channel transistors have a higher drain-source resistance. When applied to the dynamic pMOS logic circuits above, this implies that the precharge time t_{HL} is small since the time constant

$$\tau_n = R_n C_{out} \tag{8. 34}$$

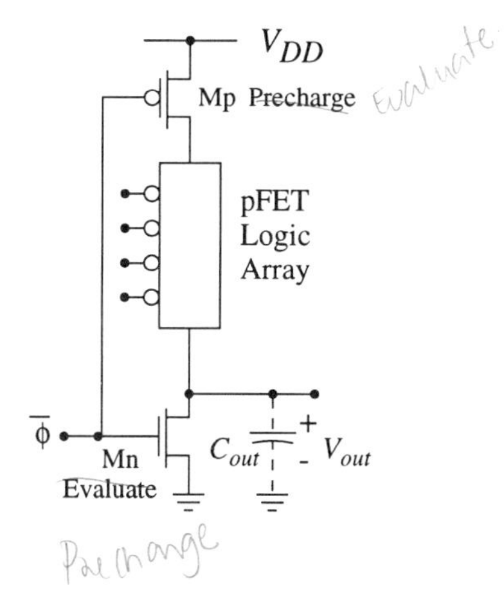

Figure 8.16 A general dynamic pMOS logic gate

is due to the nFET resistance only. However, when the output is a logic 1, the output capacitor C_{out} must charge to a voltage of V_{DD} through the pFET array where the individual devices have a (relatively) high resistance. This reduces the clock frequency that can be used for the circuit. An example of the problem is shown in Figure 8.19 for the NOR2 gate. When both inputs are 0 as in drawing (a), the output capacitor charges through Mp, M1, and M2. Modelling the event using RC-equivalents yields the subcircuit shown in Figure 8.19(b). The charging time constant for this circuit is seen to be given by

$$\tau_p = (R_p + R_{p2} + R_{p1})\, C_{out} + (R_p + R_{p2})\, C_2 + R_p C_p \tag{8. 35}$$

such that the charging time is approximately

$$t_{0 \to 1} \approx 2.2\tau_p \tag{8. 36}$$

Since the clocking transistor Mp always contributes to the charging time, we are faced with the

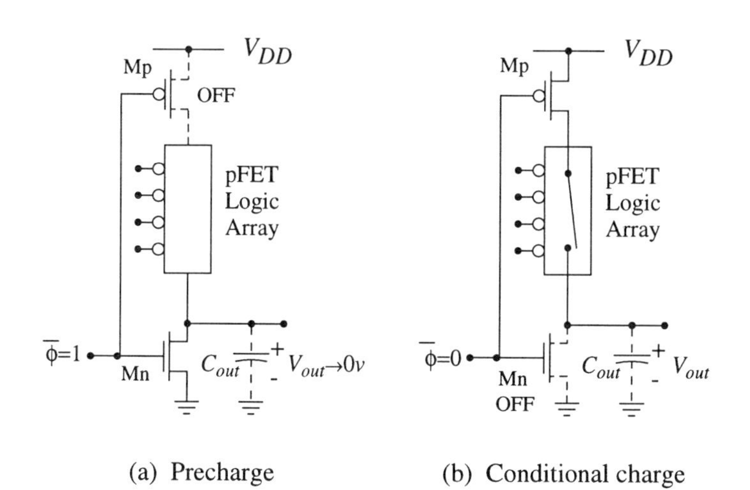

(a) Precharge (b) Conditional charge

Figure 8.17 Operational modes of a dynamic pMOS gate

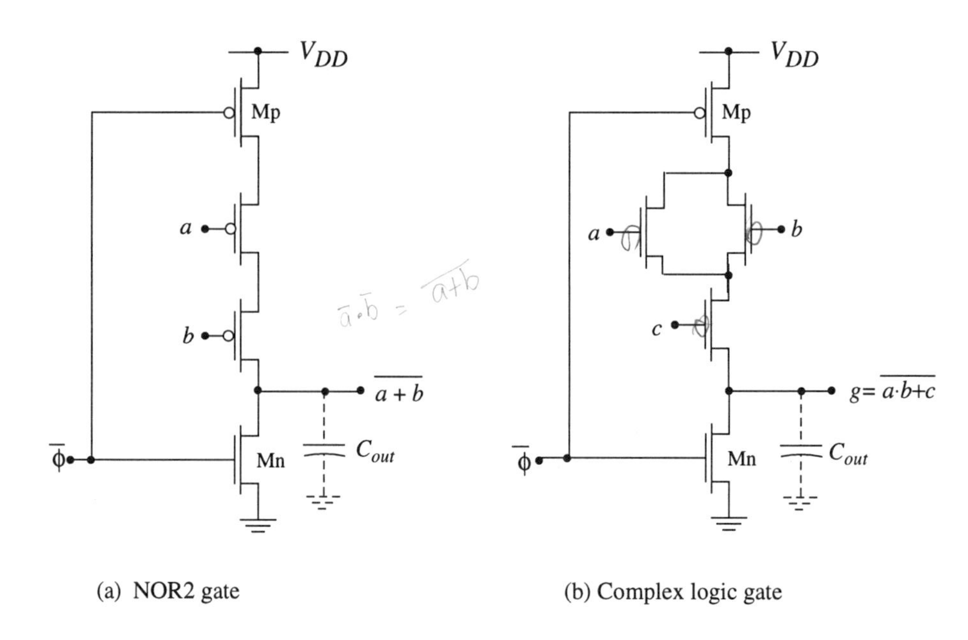

(a) NOR2 gate (b) Complex logic gate

Figure 8.18 Examples of pMOS logic gate circuits

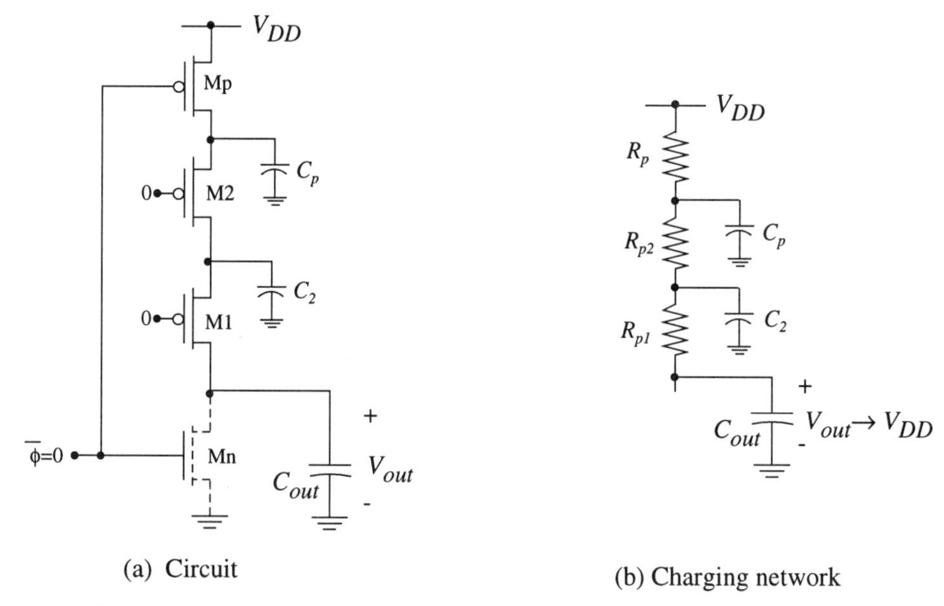

(a) Circuit (b) Charging network

Figure 8.19 Modelling of the conditional charge event

problem of charging through a group of series-connected pFETs, which can lead to large effective resistances and time constants. However, this does not mean that the circuits are never useful. Rather, it tells us that we must be careful to avoid long pFET chains to insure fast operation. Note, for example, that a NAND3 pMOS logic gate only has two series-pFETs in the charging circuit (one is the logic transistor and the other is an evaluation pFET) while an nMOS NAND3 stage has a 4-transistor discharge path.

8.2.5 nMOS-pMOS Alternating Cascades

Recall that our motivation for introducing pMOS logic gates in the first place was to overcome the glitch problem found in nMOS-nMOS cascades. Now that we have characterized a pFET-based dynamic gate, we may create a glitch-free logic chain by alternating the stages using the pattern nMOS-pMOS-nMOS and so on, as shown in figure 8.20. This is a natural extension of the discussion above, and is straightforward to analyze.

First, note that the pMOS stage is controlled by the complementary clock $\bar{\phi}(t)$ that drives a complementary pair consisting of Mp2 and Mn2. This is chosen so that every stage in the chain (whether nMOS or pMOS based) is in precharge (or evaluate) at the same time. The clocking of the chain is described by the plot in Figure 8.21. A value of $\phi=0$ allows every stage to precharge in a simultaneous manner. With regard to the output voltages shown in the schematic of Figure 8.20, this means that

$$\begin{aligned} &V_{out,1} \to V_{DD} \quad \text{(nMOS, Stage 1)} \\ &V_{out,2} \to 0v \quad \text{(pMOS, Stage 2)} \\ &V_{out,3} \to V_{DD} \quad \text{(nMOS, Stage 3)} \end{aligned} \tag{8. 37}$$

When the clock changes to $\phi=1$, every stage is driven into the evaluation mode. However, since the stages are connected as a cascade, the logic "ripples" down the circuit from left to right as indicated

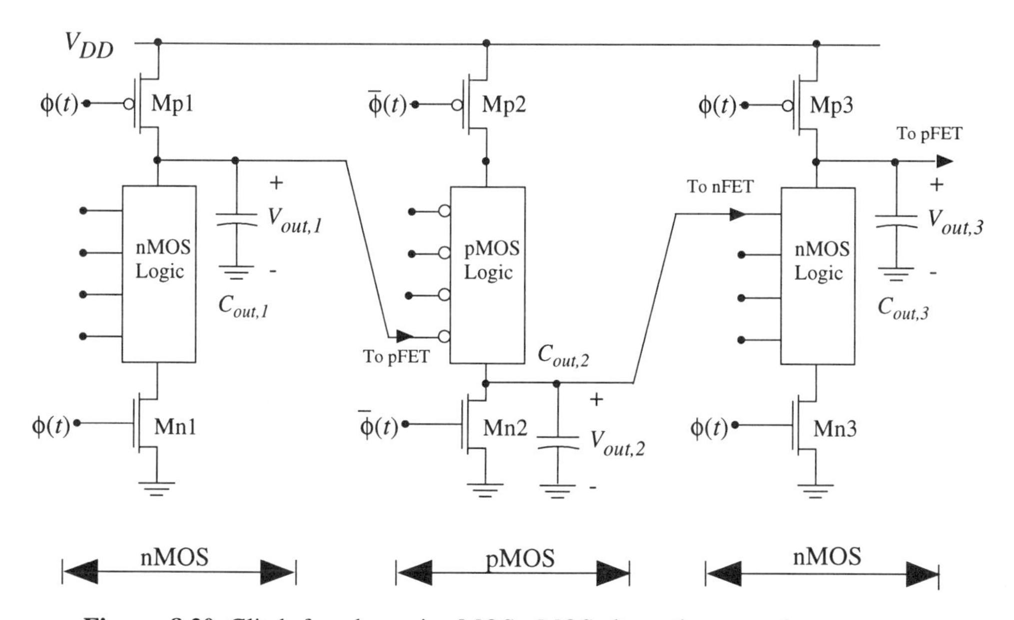

Figure 8.20 Glitch-free dynamic nMOS-pMOS alternating cascade

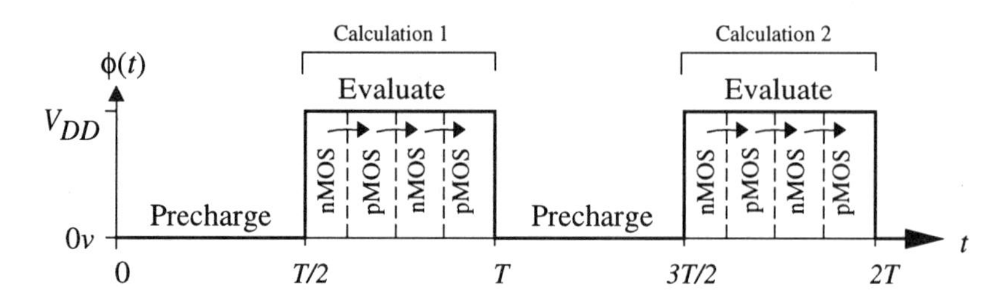

Figure 8.21 General clock timing for an nMOS/pMOS logic chain

in the clocking diagram ("Calculation 1"). This is due to the fact that the output from Stage 1 must be valid before Stage 2 can use it, and the output from Stage 2 must be finalized before Stage 3 can accept it. The operation repeats with the next clock cycle such that it takes one clock period T for the logic to propagate through the logic chain.

This discussion illustrates the fact that the clock frequency f is limited by the dynamic characteristics of the entire cascade, not just a single gate. The constraints are due to the events taking place during the Evaluate portion of the clock cycle. Let us denote the propagation delay for the j-th stage by $t_{p,j}$. For the 3-stage circuit shown above, it takes a total time of

$$t_p = t_{p,1} + t_{p,2} + t_{p,3} \tag{8. 38}$$

for the logic to stabilize. Since we have chosen the evaluation interval to be $(T/2)$ seconds in duration, this gives the minimum clock period as

$$T_{min} \approx 2t_p \tag{8. 39}$$

which corresponds to a maximum clock frequency of

$$f_{max} \approx \frac{1}{2t_p} = \frac{1}{2\,(t_{p,1} + t_{p,2} + t_{p,3})} \tag{8. 40}$$

This verifies that the system clocking limit is determined by the sum of delay times, which are in turn established by the time constants of the FET chains in each stage. Note, however, that the logic chain can perform several operations during one clock cycle. The logic throughput of the cascade thus depends upon both the clock frequency and the number of stages used in the chain of gates.

Charge leakage limits the minimum clock frequency to a value

$$f_{min} \approx \frac{1}{2t_H} \tag{8. 41}$$

where t_H is a hold time. The actual value of t_H is chosen to be long enough to avoid changes in any of the output logic states. In most designs, Stage 1 provides the critical value since it is the first gate to have a stable output that must be held until the next precharge event. However, a complex logic gate with several leakage paths may turn out to be the limiting factor.

The main drawback of nMOS-pMOS cascades is the reliance on pFET logic arrays for every other stage. If the delays or real-estate penalties introduced by the pMOS logic stages are acceptable, then this provides a glitch-free design style that is relatively easy to use. However, in our quest to eliminate pFETs from the logic arrays, let us progress in our studies to another circuit design style that is based on nFET-only logic array.

8.3 Domino Logic

Domino logic is a system design style that eliminates the nMOS-nMOS glitch problem without introducing pMOS-type logic stages. The basis for domino circuits arises from once again studying the origin of the glitch problem in the nMOS-nMOS cascade.

Consider the basic domino logic circuit shown in Figure 8.22. This consists of a dynamic nMOS gate with the output cascaded into a static inverter. The output of the domino gate is taken to be at the output of the inverter, which is directly connected to the input nFET Min of the next stage. Without the inverter, capacitor C_X is precharged to V_{DD}, which would turn on Min . However, with the inverter added as shown, a precharge event ϕ=0 allows the capacitor C_X to precharge to a voltage of

$$V_X \rightarrow V_{DD} \tag{8. 42}$$

which then gives

$$V_{out} \rightarrow 0 \tag{8. 43}$$

at the output of the inverter. This drives the nFET Min into cutoff, eliminating the possibility of a glitch in the next stage. Cascading domino stages thus allows for all nFET glitch-free logic.

Domino logic circuits do have several characteristics that complicates their use. One is due to the fact that the inverter complements the output of the dynamic nMOS gate. Consider the evaluate circuit shown in Figure 8.23 where we assume that the precharge has already established voltages of $V_X = V_{DD}$ and $V_{out} = 0v$ on C_X and C_{out}, respectively. Let us direct our interest to the internal node variable X and examine the circuit when a discharge takes place. To discharge C_X, the nFET logic array must act as a closed switch. This in turn requires that at least some of the inputs x, y, z are at logic 1 values to turn on the required transistors; the actual combination depends upon the logic function. When C_X discharges, $V_X \rightarrow 0v$ corresponding to $X \rightarrow 0$. The output of the dynamic logic stage is X so we see that the interior dynamic nMOS circuit automatically provides the NOT operation: logic 1's at the input give a value of $X = 0$. However, in the domino configuration, the internal node X is *not* the output. Instead, the output of the gate is taken after the inverter, and

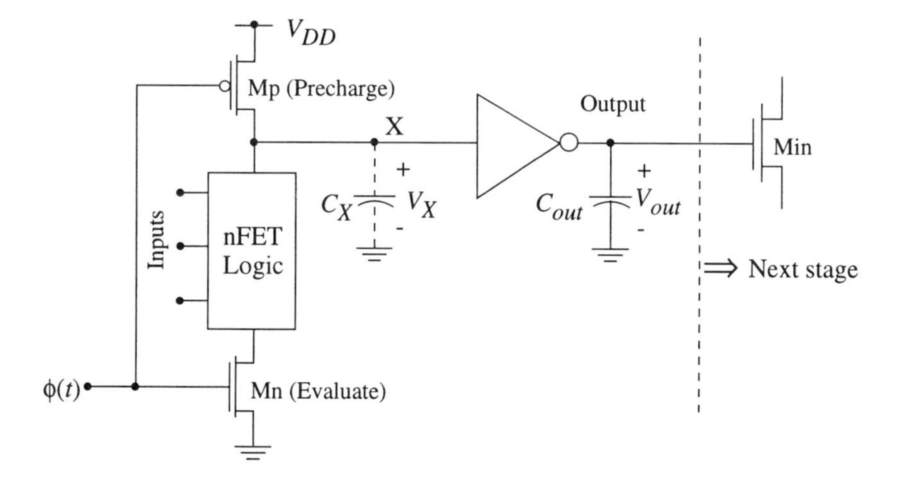

Figure 8.22 Basic domino logic stage as a solution to the glitch problem

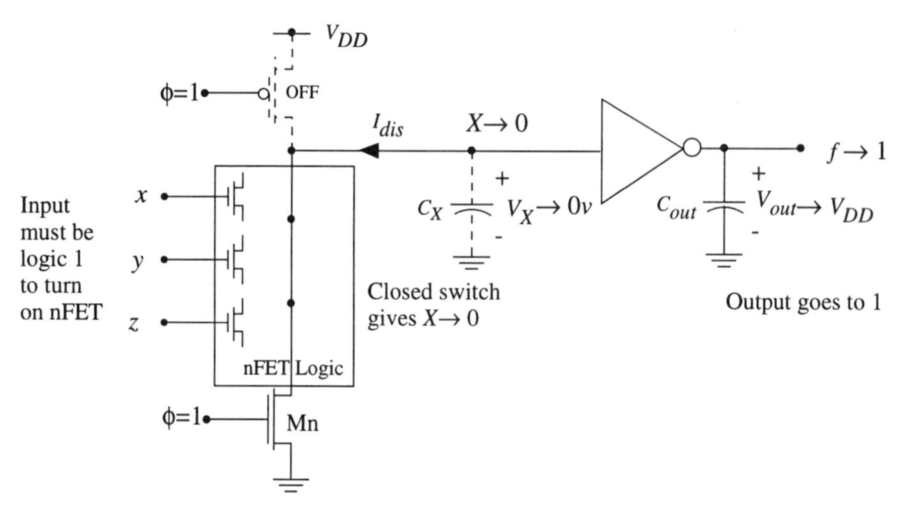

Figure 8.23 Circuit-to-logic operation of a domino logic stage

$$f = \overline{X} \tag{8.44}$$

A discharge of C_X thus results in an output of f=1, illustrating the fact that domino logic is ***non-inverting***. This means, for example, that we can create gates for the basic AND and OR operations, but cannot implement the NOT function. The only way to obtain the NOT is to add another inverter to the output, but this takes us back to the glitch problem. From the viewpoint of logic design, this can be tricky to deal with since the NOT operation is required to form a complete logic set.[3] This characteristic can make domino logic designs somewhat tricky. We must provide the NOT operation somewhere, but it cannot be placed *within* the chain. The solution is to restrict inverters to the beginning or end of a domino chain.

Figure 8.24 shows an OR3 gate as an example of a domino logic gate. Precharge takes place when φ=0 with C_X charging to a voltage of $V_X = V_{DD}$. If any one or more of the inputs x, y, or z is at a logic 1 level when φ=1, C_X discharges and $V_X \rightarrow 0v$. This forces the output voltage to change from the precharge value of $0v$ to a final value of $V_{out} = V_{DD}$ which is interpreted as a logic 1 output. Thus, we see that

$$g = x + y + z \tag{8.45}$$

describes the operation as stated. Another example of a domino gate is drawn in Figure 8.25. Using the same arguments gives

$$f = x \cdot y + z \tag{8.46}$$

which is in AO (AND-OR) form. OA and other functions can be created in the same manner.

---

[3] A complete logic set is a group of operations that can be used to form any logic function. For example, AND and NOT constitute a complete set, as does OR and NOT. The two operations AND and OR by themselves do not form a complete set.

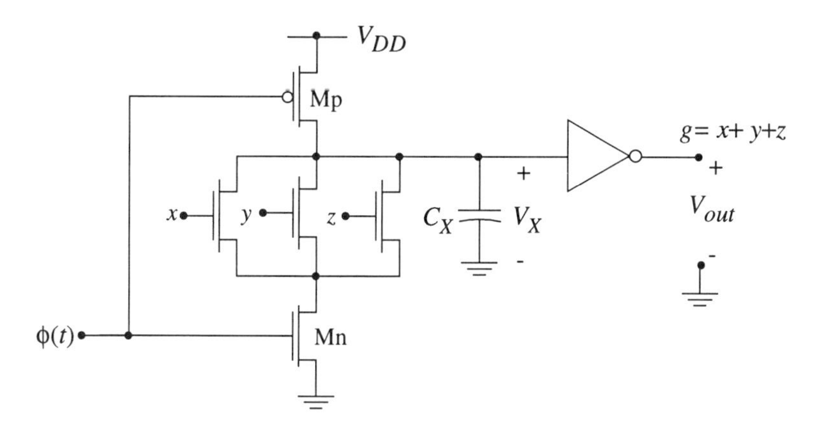

Figure 8.24 Domino OR3 logic gate

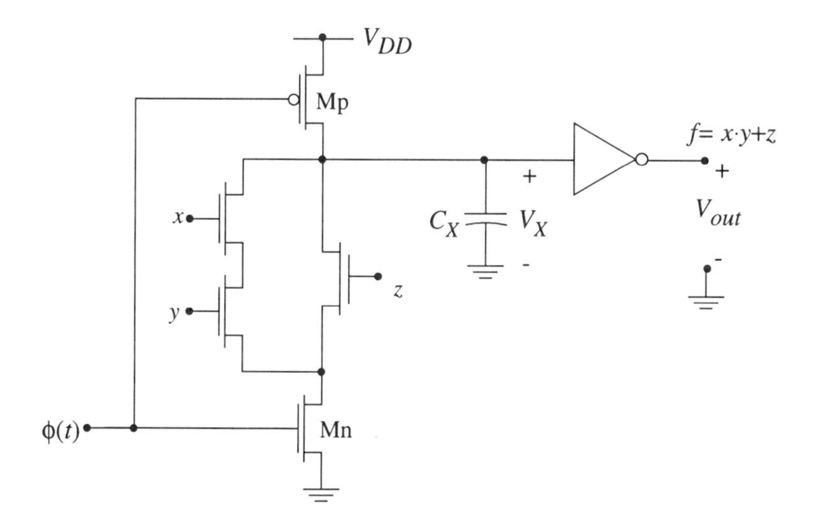

Figure 8.25 AND-OR domino logic gate

8.3.1 Gate Characteristics

To understand the important operating characteristics of a domino logic gate, let use analyze the AND3 circuit shown in Figure 8.26. Since the logic function is determined by the circuit topology, we will direct our attention to the transient calculations.

Precharging of the circuit occurs when ϕ=0 and the important circuit characteristics are shown in Figure 8.27. The precharge pFET Mp is biased into conduction and I_{Dp} flows. Since Mn is biased into cutoff, there is no conducting path to ground. This insures that C_X charges to a final voltage of $V_X = V_{DD}$ as indicated. The actual percentage of I_{Dp} that is initially steered to C_X depends upon the inputs (which in turn depends upon their origin). If $A_1 = 0$, then we may estimate the charge time by

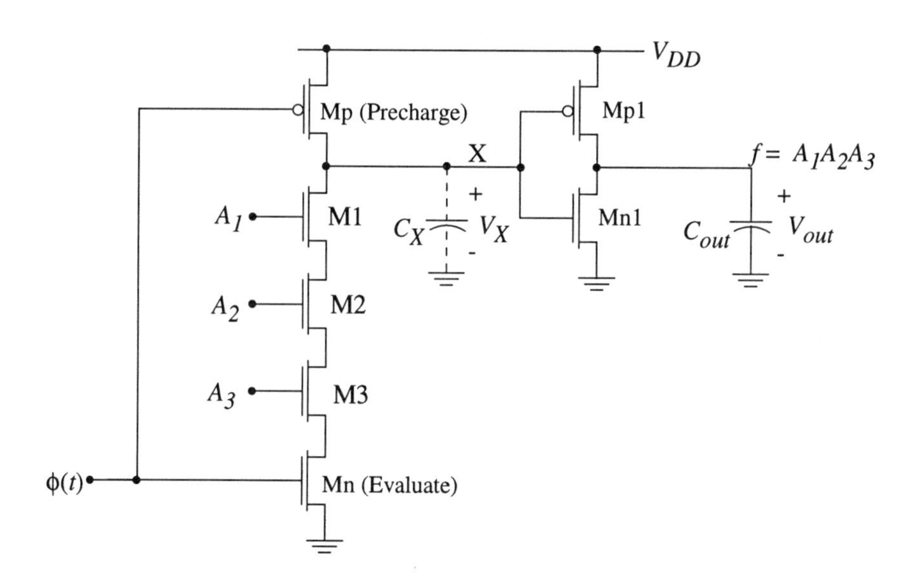

Figure 8.26 Domino logic AND3 gate circuit

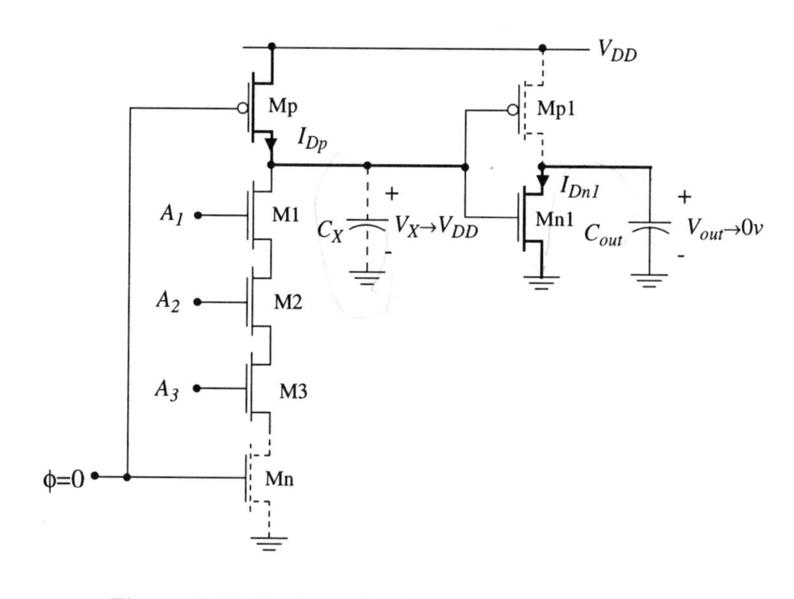

Figure 8.27 Precharge in the AND3 domino gate

$$t_{ch} \approx s_p \tau_p \tag{8.47}$$

where the time constant is

$$\tau_p = R_p C_X \tag{8.48}$$

with C_X the total capacitance of the node; this would be the best-case precharge time. Once V_X increases to V_{IH} of the inverter, Mn1 acts to discharge the output capacitor C_{out} to ground. This yields a final value of $V_{out} = 0v$ after a time t_{HL} (which is the worst-case value). The total precharge time may then be estimated by

$$t_{pre} \approx t_{ch} + t_{HL} \tag{8.49}$$

where we have just added the two individual event times for simplicity.

When ϕ increases to 1, the circuit goes into the evaluation mode. Referring to Figure 8.28, we see that the internal node capacitor C_X is subjected to a conditional discharge that depends upon the values of the input variables A_1, A_2, and A_3. If all three inputs are at logic 1 high voltages, then C_X discharges and V_X drops to 0v. This drives Mp1 into conduction and Mn1 into cutoff, and C_{out} charges to a final voltage of $V_{out} = V_{DD}$ corresponding to a logic 1 result. On the other hand, if any one or more of the inputs is 0, then the charge is held on C_X and V_X is high; the actual value changes due to charge sharing and leakage. However, so long as $V_X > V_{IH}$ of the inverter, the output voltage remains at a logic 0 level of $V_{out} = 0v$.

The behavior of this logic gate can also be explained using the timing diagrams in Figure 8.29. This shows how the voltages $V_X(t)$ and $V_{out}(t)$ react to the clock and to the inputs. During the precharge time intervals, V_X always rises to V_{DD} which causes V_{out} to fall to 0v. The first evaluation period shown in the plot represents the case where C_X discharges and causes C_{out} to charge to V_{DD}, representing a logic 1 output. The second cycle shows the situation where C_X attempts to hold the charge state, but undergoes charge sharing (the first drop) and then charge leakage which reduces the voltage V_X. V_{out} remains at a logic 0 level so long as we can maintain V_X sufficiently high.

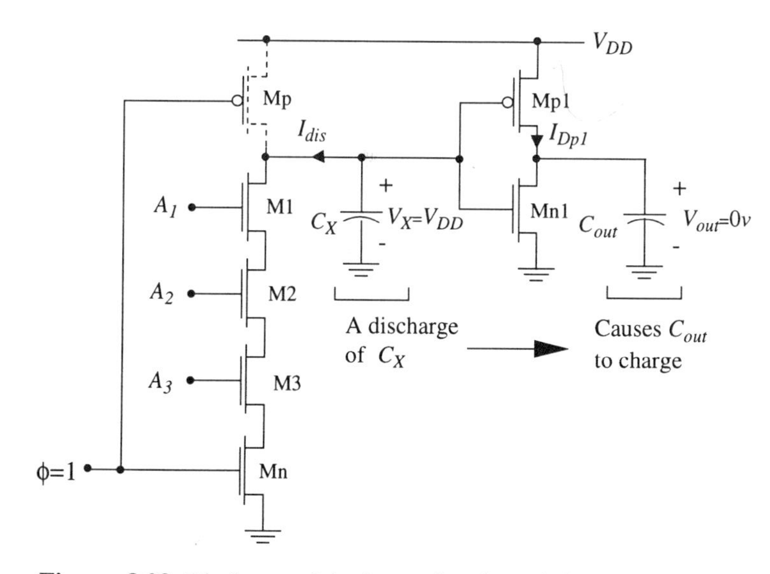

Figure 8.28 Discharge of the internal node and change in output

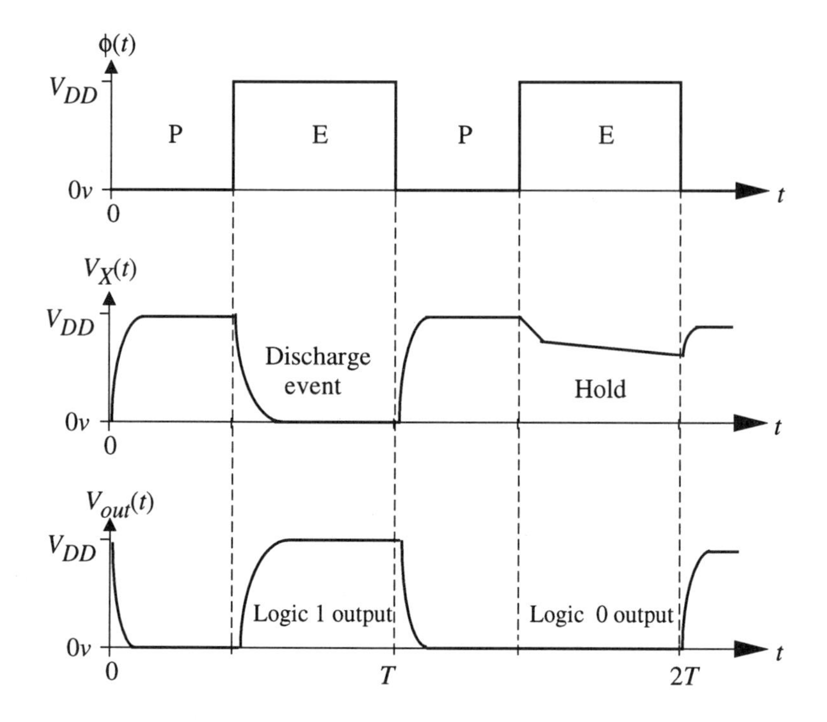

Figure 8.29 Timing diagram for a domino logic gate

8.3.2 Domino Cascades

Now that we understand the operation of a single isolated domino gate, it is a straightforward matter to create a glitch-free cascade as shown in Figure 8.30. In the example, we have created a 5-stage cascade of individual domino AND3 gates for simplicity.[4] The clock φ is wired to each gate in the same way so that every stage is in the same operational mode (precharge or evaluate) at a

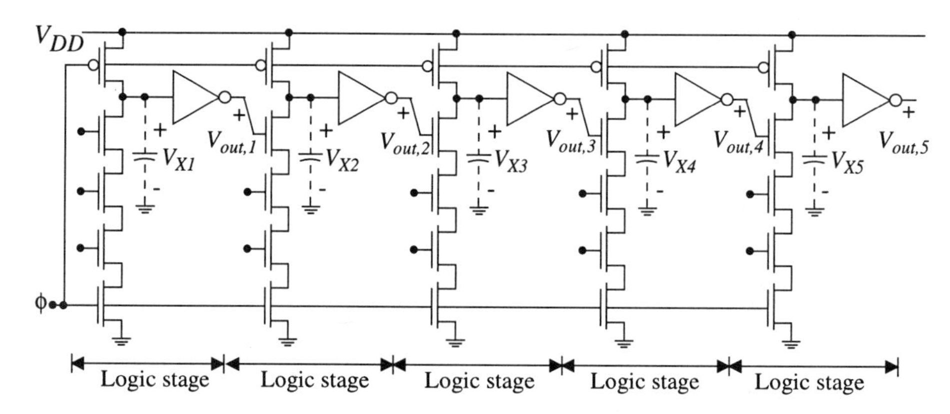

Figure 8.30 Example of a domino logic cascade

[4] Of course, we can easily substitute other gates as desired.

given time. Because of this property, it is important to view the cascade as a single logic chain and characterize its behavior as a unit, not as a set of independent gates. When ϕ=0, every stage in the chain is simultaneously in precharge, so that every output voltage is reset to a value of $V_{out} = 0v$; this insures that the logic nFET in the following stage is in cutoff. Evaluation takes place when the clock makes a transition to ϕ=1. If any output voltage changes from a high voltage to $V_{out} \to 0v$, then this means that every stage preceding it has undergone an internal discharge event and has also undergone the same change in output voltage.

Let us clarify these statements using a few simple equations as they apply to the 4-stage chain portrayed in Figure 8.31. A precharge event with ϕ=0 results in

$$V_{X,i} \to V_{DD}$$
$$V_{out,i} \to 0v \tag{8.50}$$

for i = 1, 2, 3, 4. Now suppose that the first stage undergoes a discharge resulting in

$$V_{X,1} \to 0v$$
$$V_{out,1} \to V_{DD}; \tag{8.51}$$

this drives logic FET M2 into conduction, and sets up the possibility for stage 2 to undergo a discharge event. Note that if $V_{out,1}$ remained at $0v$ then $C_{X,2}$ could not discharge since M2 would remain in cutoff. This line of reasoning can be applied to every successive stage in the chain, and demonstrates the fact that the n-th stage must discharge in order to allow the possibility of a discharge in the $(n+1)$-th stage. When applied to the entire chain, this says that a change of the output voltage $V_{out,4} \to V_{DD}$ can only occur if

$$V_{X,1} \to 0v, \quad V_{X,2} \to 0v, \quad V_{X,3} \to 0v,$$
$$V_{out,1} \to V_{DD}; \quad V_{out,2} \to V_{DD}; \quad V_{out,3} \to V_{DD}; \tag{8.52}$$

have all taken place.

This behavior is the origin for the name "domino logic" in analogy with the game where domi-

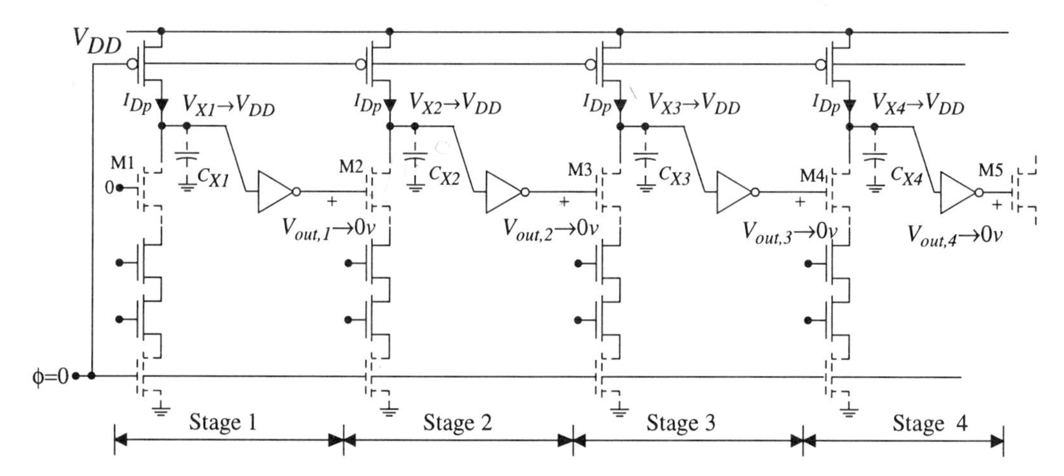

Figure 8.31 Precharge conditions in a 4-stage domino chain

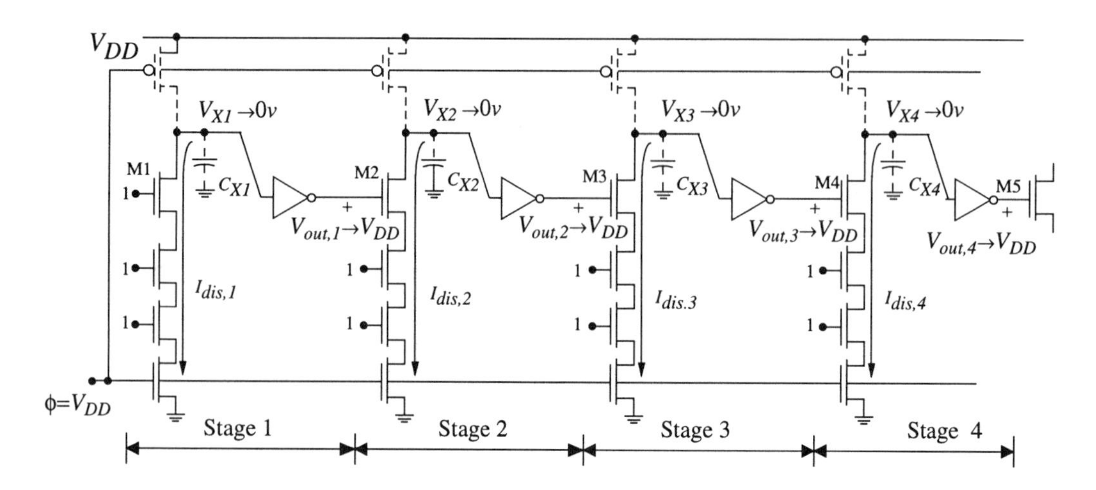

Figure 8.32 Rippling of discharge events down the chain

nos are stood on-end in rows. The initial setup of a "chain" of dominos as illustrated in Figure 8.33(a) corresponds to the precharge event. Evaluation is represented by the action illustrated in Figure 8.33(b). A discharge event corresponds to a falling domino that may cause the next one in line to also fall. If every stage undergoes an internal node discharge, then every domino topples in order from left to right. The significance of the analogy can be seen by the situation shown in Figure 8.34. In this example, the precharge still corresponds to setting up all of the dominos as in Figure 8.34(a). However, in this case, the first two dominos fall (discharge), but the third domino remain upright (no discharge) as in Figure 8.34(b). Since the third domino does not topple, the fourth domino also remains upright, corresponding to an output voltage of $V_{out,4} = 0v$ which is unchanged from the precharge value. In other words, once the toppling is halted anywhere within the chain, all dominos to the right will be unaffected. In terms of the logic circuit, the output will be a 0 unless every stage preceding it undergoes a discharge event.

Now let us examine the timing for the entire 4-stage logic chain using the waveform in Figure 8.35. The precharge interval when ϕ=0 does not create any problems to deal with since every stage precharges at the same time. However, the domino effect during the evaluation portion of the clock requires that we allocate sufficient time for the entire chain to react to the inputs. Labelling the individual propagation times by t_{dj} for $j = 1, 2, 3, 4$ then gives the constraint that the minimum clock period must satisfy

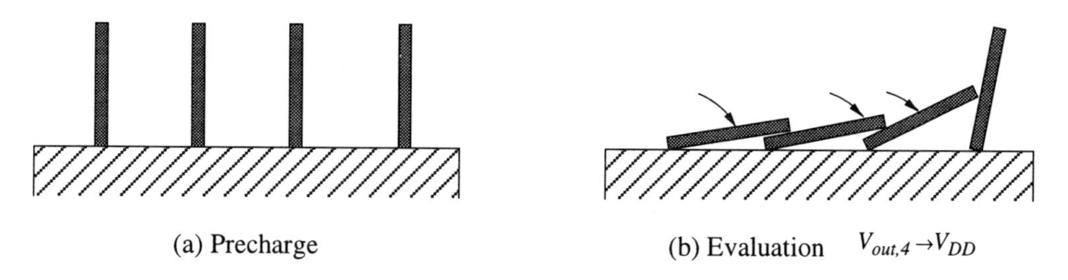

Figure 8.33 Domino analogy for discharge of every stage

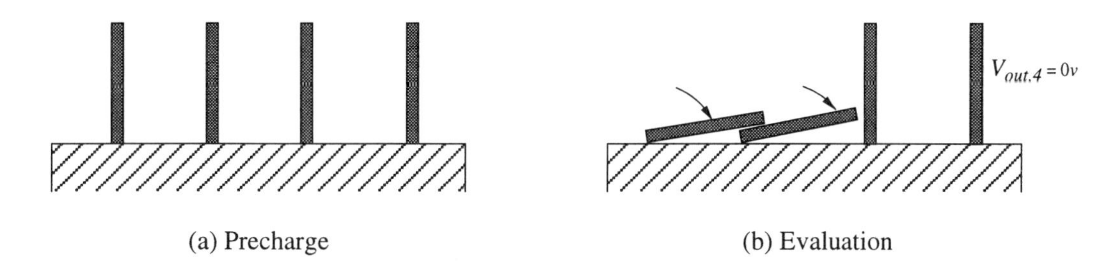

Figure 8.34 Domino analogy for discharge in only the first two stages

$$\frac{T_{min}}{2} \approx t_{d1} + t_{d2} + t_{d3} + t_{d4} \tag{8. 53}$$

which gives a maximum clock frequency of

$$f_{max} \approx \frac{1}{2\,(t_{d1} + t_{d2} + t_{d3} + t_{d4})} \tag{8. 54}$$

This is similar to the constraint discussed for the nMOS-pMOS cascade. The minimum clock frequency is set by the maximum hold time of the internal dynamic nodes (C_{Xj}).

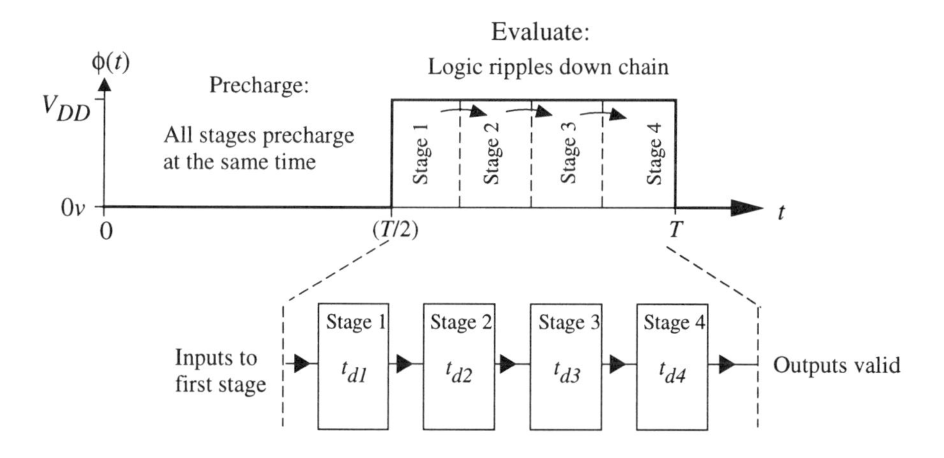

Figure 8.35 Ripple logic and the clock timing problem

8.3.3 Charge Sharing and Charge Leakage Problems

Domino logic chains are susceptible to the usual problems of charge sharing and charge leakage. In fact, the analysis of the domino AND3 gate is identical to the simpler nMOS NAND3 circuit, except that the results apply to the internal node capacitor C_X. Charge sharing problems are approached by adjusting the layout dimensions, and perhaps restructuring the logic functions if necessary. Charge sharing effects combine with charge leakage to set the maximum time allowed for the evaluate mode that is defined by the clock. Since we must allow sufficient time for the entire chain to complete the evaluation of the inputs, the holding period may be quite long, especially if long cascades are used. The ability to maintain the charge on the internal (C_X-type) nodes is crucial for the proper operation of a domino design.

Let us examine one aspect of the problem using the circuit shown in Figure 8.36(a). The critical node is identified as having a voltage V_X. Since this acts as the input voltage to the inverter consisting of Mp1 and Mn1, we may construct the DC voltage transfer curve shown in the upper portion of Figure 8.36(b). The dynamics of the problem are indicated by the plot of $V_X(t)$ in the lower portion of the same drawing. This shows V_X decaying in time due first to charge sharing, and then because of charge leakage. The objective of the circuit design at this point is to keep V_{out} low as long as possible. With regards to the to the VTC, this means that we want to keep $V_X > V_{IH}$ since this defines the maximum hold time identified as t_{max} in the drawing. To achieve this goal, we

- Minimize charge sharing problems as much as possible to maximize V_f
- Minimize charge leakage currents by using small junction areas, and
- Design the inverter to have a relatively low value of V_{IH}

The last objective is met by choosing the aspect ratios $(W/L)_{p1}/(W/L)_{n1}$ of the inverter to move the VTC toward the left, i.e.,

$$\frac{\beta_n}{\beta_p} = \frac{k'_n\left(\frac{W}{L}\right)_{n1}}{k'_p\left(\frac{W}{L}\right)_{p1}} > 1 \tag{8. 55}$$

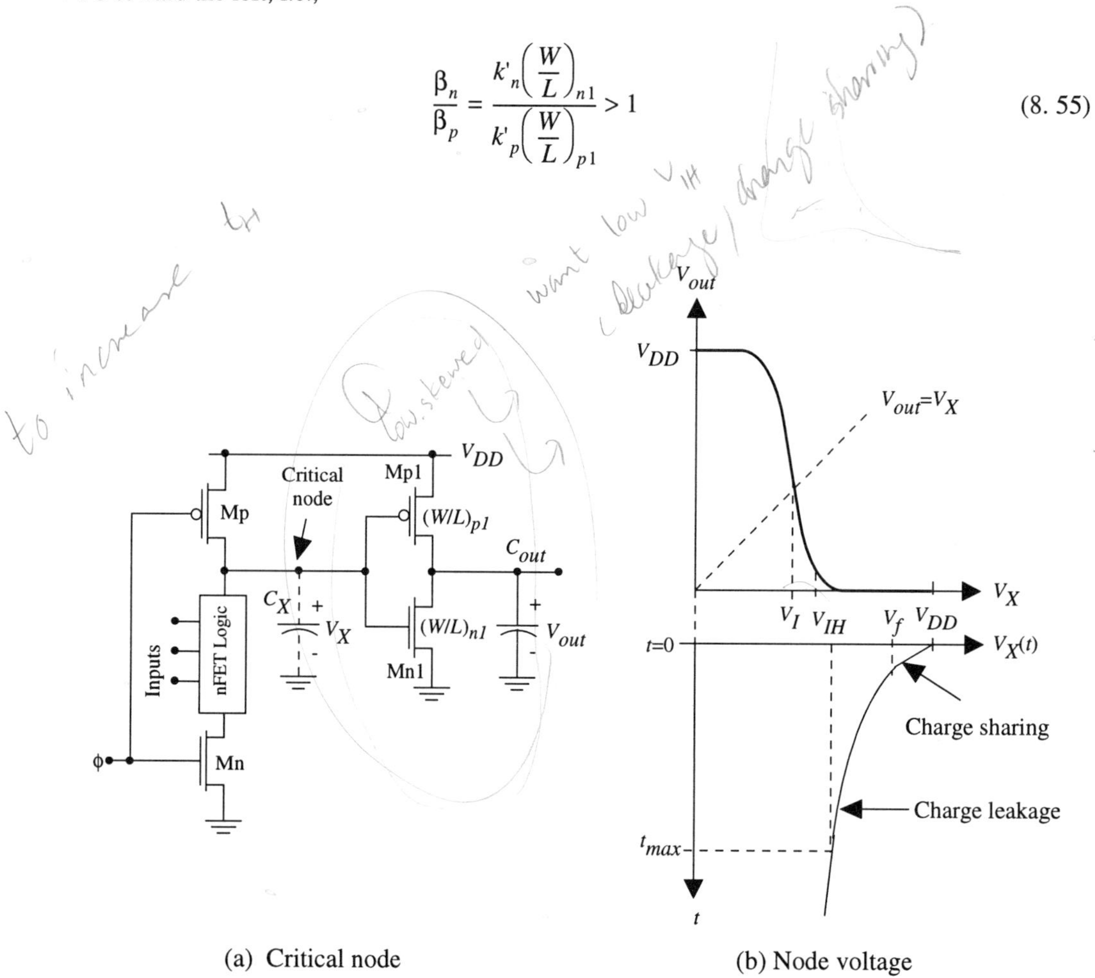

(a) Critical node (b) Node voltage

Figure 8.36 Inverter design considerations for domino logic

This helps lower V_{IH} as can be seen by recalling that the inverter threshold voltage V_I is computed from

$$V_I = \frac{V_I - |V_{Tp}| + \sqrt{\frac{\beta_n}{\beta_p}} V_{Tn}}{1 + \sqrt{\frac{\beta_n}{\beta_p}}} \tag{8.56}$$

such that a ratio $(\beta_n/\beta_p)>1$ yields a value for V_I that is less than $(V_{DD}/2)$.

Since charge leakage problems can lead to errors, it is often worthwhile to provide additional circuitry to combat the problem. The circuit in Figure 8.37 uses a pFET MX to provide charge to C_X to overcome charge leakage effects. MX is biased into conduction by grounding the gate, so that the device voltages are

$$\begin{aligned} V_{SGp} &= V_{DD} \\ V_{SDp} &= V_{DD} - V_X \end{aligned} \tag{8.57}$$

Directly after the precharge interval, $V_X = V_{DD}$. If V_X falls below V_{DD}, then the transistor conducts a current

$$I_{DX} = \frac{k'_p}{2}\left(\frac{W}{L}\right)_{pX}\left[2(V_{DD} - |V_{Tp}|)(V_{DD} - V_X) - (V_{DD} - V_X)^2\right] \tag{8.58}$$

to recharge the node. The main circuit design problem that arises is the fact that the logic may lead to a discharge of C_X such that V_X is supposed to fall to 0*v*. If I_{DX} is too large, it will keep V_X high, and the node will be stuck at a logic 1 voltage. To overcome this, we must choose the pFET to have an aspect ratio $(W/L)_{pX}$ that is small enough to still allow the discharge, but large enough to help maintain the voltage if the charge is held on C_X. This is accomplished by using the value of the leakage current I_{leak} to calculate $(W/L)_{pX}$ for an acceptable minimum value of V_X. We call this

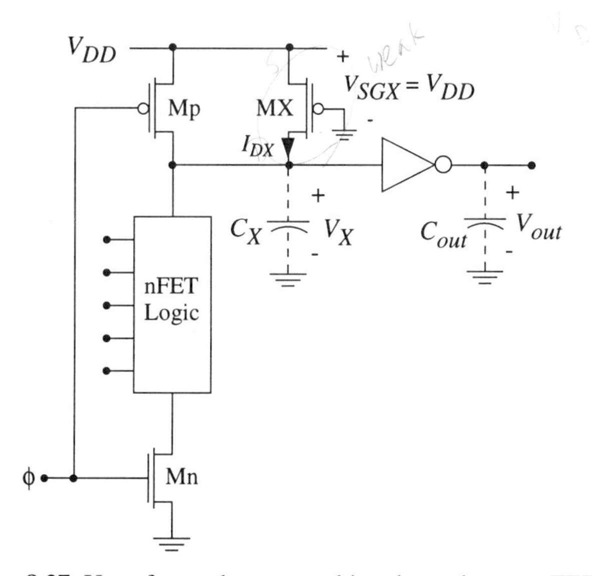

Figure 8.37 Use of a weak constant-bias charge keeper pFET

device a **weak** pFET since it allows only small compensation currents to flow, making it a weak conductor.

An improved design is obtained by using a feedback loop to control the conduction of the pFET. This results in the "charge-keeper" circuit shown in Figure 8.38. In this circuit, MX is biased by the output voltage V_{out}. When V_{out} is low, then the gate of MX is at 0v, and I_{DX} flows if $(V_{DD}-V_X) > 0$ is to help maintain charge on C_X. If a discharge occurs, then V_X falls towards 0v, and the gate voltage on the pFET will eventually change to $V_{out} = V_{DD}$. This drives MX into cutoff and allows the rest of the discharge to proceed without any hinderance.

Another design for the charge keeper is shown in Figure 8.39. Instead of using the output voltage V_{out} to control the gate of MX, we have added an extra inverter to provide the feedback voltage. This helps keep the signal path free from as many parasitic effects as possible; in this case, it frees the output from a slowdown that would be induced by "flipping" the state of the feedback network. This philosophy adds extra circuitry, but allows us to "tweak" every picosecond possible out of the delay. Since the extra inverter is small, it will probably be worth the real estate it consumes.

Regardless of the control circuitry used to switch MX, the important design parameter is the value aspect ratio $(W/L)_{pX}$ of the device itself. This must be chosen to be large enough to compensate for charge sharing and leakage, but not so large that it overwhelms the discharge event. These competing requirements make the problem an interesting exercise in design. Moreover, the value of $(W/L)_{pX}$ needs to be adjusted for each logic gate in a manner that accounts for the number of junctions that actually contribute to the leakage.[5] An empirical technique for selecting this value is to sum all of the aspect ratios in the logic array

$$\left(\frac{W}{L}\right)_T = \sum_{\alpha = 1, 2, \ldots} \left(\frac{W}{L}\right)_a \tag{8.59}$$

where α is a dummy index that includes all of the nFETs in the logic array. Since the value of the leakage current is proportional to the area of a pn junction, this gives us a measure of the total leak-

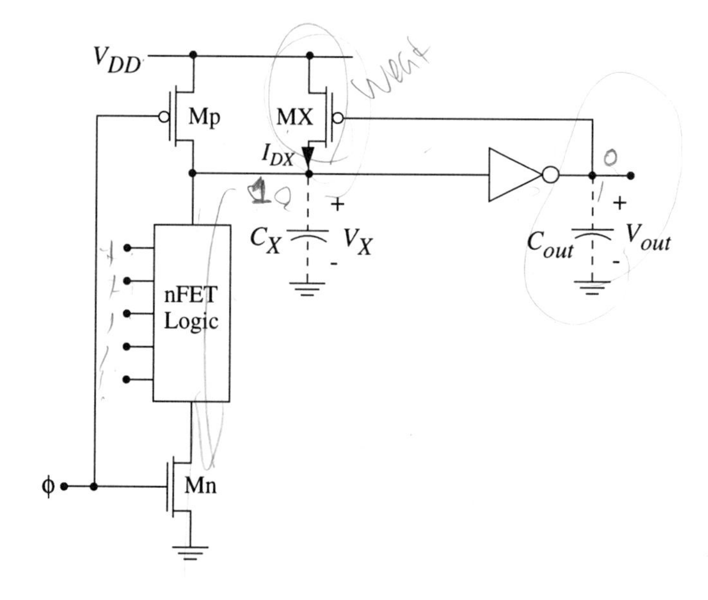

Figure 8.38 Charge keeper circuit with feedback control

[5] Subthreshold currents can also be accounted for using this technique

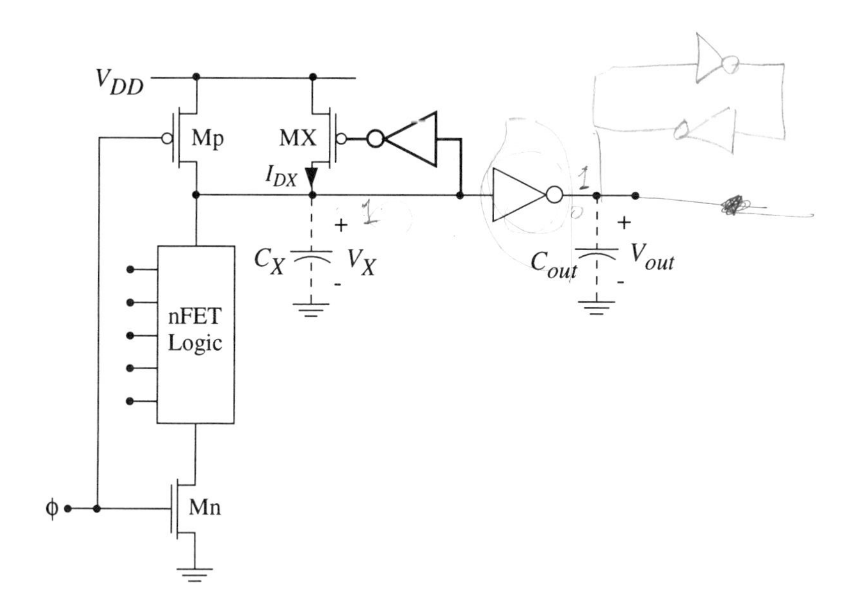

Figure 8.39 Improved charge-keeper feedback circuit

age potential in the logic array. A rule-of-thumb is then to choose the charge keeper size by

$$\left(\frac{W}{L}\right)_{pX} = x\left(\frac{W}{L}\right)_T \tag{8.60}$$

where x is a multiplier with a value less than 1. For example, x =0.15 gives

$$\left(\frac{W}{L}\right)_{pX} = 0.15\left(\frac{W}{L}\right)_T \tag{8.61}$$

which would indicate that the aspect ratio of the pFET charge keeper is 15% of the total value of all logic FETs. Although this may sound quite arbitrary, the actual value of x can be adjusted according to the process parameters. This simplifies the design, and can always be changed if the computer simulations indicate a problem.

8.3.4 Sizing of MOSFET Chains

The worst-case delay through a domino chain depends upon the discharge times in the individual stages. The AND3 gate in Figure 8.40(a) can be used to illustrate the problem. The internal capacitor C_X will discharge when all inputs are at logic 1 values. Using the equivalent circuit in Figure 8.40(b), the time constant is

$$\tau = (R_1 + R_2 + R_3 + R_4)\,C_1 + (R_2 + R_3 + R_4)\,C_2 + (R_3 + R_4)\,C_3 + R_4 C_4 \tag{8.62}$$

where we assume that the voltage fall is described by

$$V_X(t) = V_{DD}e^{-t/\tau}. \tag{8.63}$$

The output V_{out} is initially at 0v and will not begin to switch until V_x falls to around V_{IL}. This requires a delay time

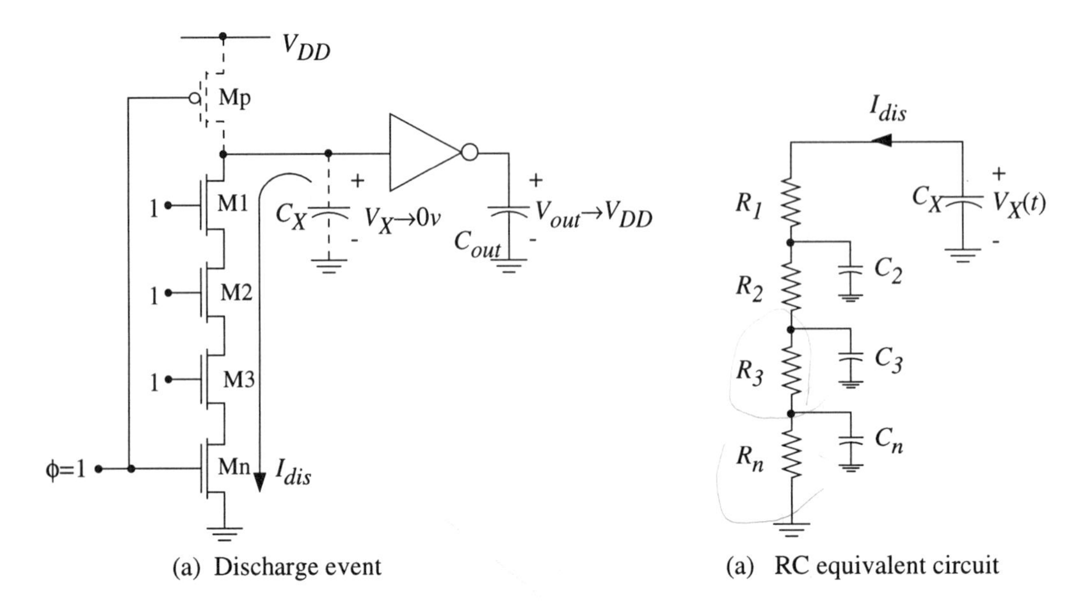

Figure 8.40 Analysis of an nFET discharge network

$$t_d \approx \tau \ln\left(\frac{V_{DD}}{V_{IL}}\right), \tag{8.64}$$

so that the value of the time constant is critical to the system speed.

Let us turn our attention to designing the nFET chain in a manner that reduces the time constant τ of the network. The key to this approach is noting that the parasitic resistance and capacitance of a MOSFET depend upon the channel width *W*. Recall that the drain-source resistance formula

$$R = \frac{1}{k'_n\left(\frac{W}{L}\right)(V_{DD} - V_{Tn})} \tag{8.65}$$

shows that the resistance follows

$$R \propto \frac{1}{W} \tag{8.66}$$

i.e, it is inversely proportional to *W*. The drain/source capacitance is more complicated and consists of both depletion and MOS terms in the form

$$C \approx C_n + C_{MOS} \tag{8.67}$$

The MOS contribution is either gate-source or gate-drain capacitance, and can be estimated by

$$C_{MOS} \approx \frac{1}{2} C_{ox} WL \tag{8.68}$$

while the depletion capacitance is given by

$$C_n = C_j A_n + C_{jsw} P_n \tag{8. 69}$$

where A_n and P_n are the area and perimeter, respectively, of the n^+ region. For a rectangular n^+ region with an area of $A_n = (W \times X)$, the perimeter is given by $P_n = 2(W + X)$, yielding

$$C_n = C_j WX + 2C_{jsw}(W + X) \ . \tag{8. 70}$$

Combining this with the MOS contribution, we see that we can approximate the dependence on channel width W by the proportionality

$$C \propto W \tag{8. 71}$$

even though the sidewall contribution doesn't follow this exactly. The expressions for resistance and capacitance of a FET illustrate that the values of the parasitics in the time constant depend on the individual aspect ratios. Let us therefore use these relations to choose the relative sizes of the MOSFETs in an effort to reduce τ.

The starting point of the analysis will be the FET chain in Figure 8.41(a) that is discharging capacitor C_X. This can be modelled to first order using the RC ladder network shown in Figure 8.41(b). The elements are defined such that the aspect ratio $(W/L)_j$ of transistor Mj is used to calculate the associated resistance R_j and capacitance C_j in the equivalent circuit. Note that the model ignores the fact that the capacitance of a node between adjacent FETs has contributions from both devices. The output capacitance C_X is related to this circuit by

$$C_X = C_1 + C_0 \tag{8. 72}$$

with C_0 representing all other capacitance contributions at the node.

Let us now turn our attention to the time constant for discharging capacitor $C_1 = (C_X - C_0)$ through the chain. This is of interest because the simplified circuit then consists only of transistor parasitics. The time constant is then given by

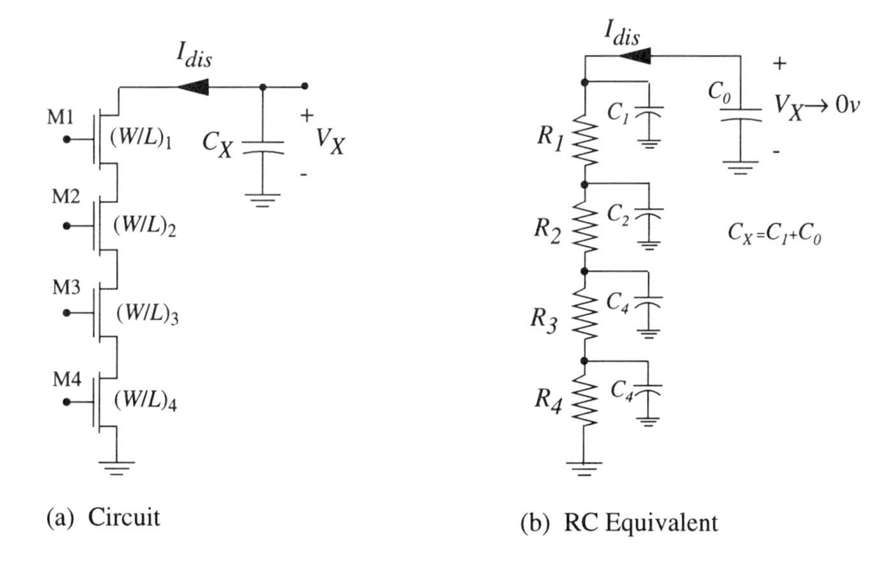

Figure 8.41 Generalized problem for a FET chain response analysis

$$\tau_n = C_1(R_1 + R_2 + R_3 + R_4) + C_2(R_2 + R_3 + R_4) + C_3(R_3 + R_4) + C_4R_4 \qquad (8.73)$$

Now note that the last term R_4C_4 depends only on $(W/L)_4$. Suppose that we increase the size of M4 to a new aspect ratio

$$\left(\frac{\tilde{W}}{L}\right)_4 = \alpha\left(\frac{W}{L}\right)_4 \qquad (8.74)$$

where $\alpha > 1$. The resistance decreases to a value

$$\tilde{R}_4 = \frac{R_4}{\alpha} \qquad (8.75)$$

while the capacitance increases to

$$\tilde{C}_4 = \alpha C_4 \qquad (8.76)$$

such that

$$\tilde{R}_4\tilde{C}_4 = \left(\frac{R_4}{\alpha}\right)(\alpha C_4) = R_4C_4 \qquad (8.77)$$

i.e., the RC product for M4 remains the same. However, since R_4 appears in every term in the time constant expression, it is modified to

$$\tilde{\tau}_n = C_1\left(R_1 + R_2 + R_3 + \frac{R_4}{\alpha}\right) + C_2\left(R_2 + R_3 + \frac{R_4}{\alpha}\right) + C_3\left(R_3 + \frac{R_4}{\alpha}\right) + C_4R_4 < \tau_n \qquad (8.78)$$

Increasing the size of M4 thus yields a smaller time constant, indicating a faster discharge. Note that this is possible because the capacitor C_4 only appears in the last term, so it does not increase the time constant.

This conclusion can be understood by physical arguments using the currents shown in Figure 8.42. In the drawing, the current flowing out of capacitor C_j is denoted by i_{cj}, while I_j represents the

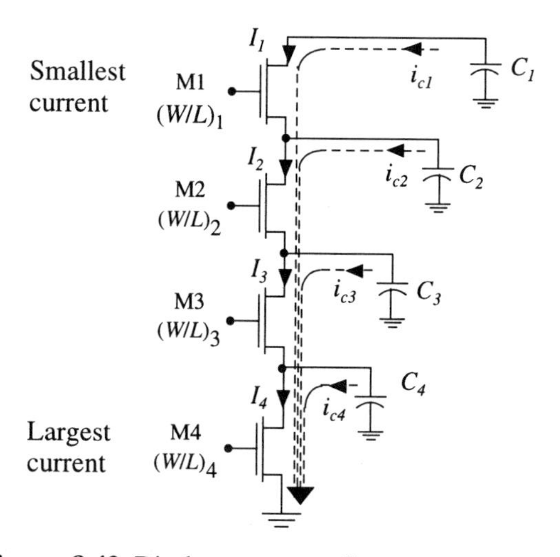

Figure 8.42 Discharge current flow components

total current through FET Mj. Applying KCL to each node gives

$$\begin{aligned} I_1 &= i_{c1} \\ I_2 &= i_{c1} + i_{c2} \\ I_3 &= i_{c1} + i_{c2} + i_{c3} \\ I_4 &= i_{c1} + i_{c2} + i_{c3} + i_{c4} \end{aligned} \quad (8.\ 79)$$

which shows that M4 handles the largest current level, M3 the second largest, and so on. Increasing the size of M4 reduces the discharge time because we are allowing for the higher current flow level. Once we see this behavior, it is natural to apply the scaling to the entire chain.

Consider the scaling shown in Figure 8.43(a); this uses α to scale the size of every transistor in the chain by the same amount relative to its neighbors. In this scheme, MOSFET M1, which has an aspect ratio given by $(W/L)_1$, is chosen as the smallest device since it has the smallest current flow. Note that this also helps to reduce the parasitic capacitance seen at the output (top) node. M2 is scaled to be α-times larger than M1:

$$\left(\frac{W}{L}\right)_2 = \alpha\left(\frac{W}{L}\right)_1 \quad (8.\ 80)$$

Similarly, M3 is larger than both M2 and M1 with the same relative scaling factor:

$$\left(\frac{W}{L}\right)_3 = \alpha\left(\frac{W}{L}\right)_2 = \alpha^2\left(\frac{W}{L}\right)_1 \quad (8.\ 81)$$

The end (bottom) transistor M4 is the largest device in the chain with

$$\left(\frac{W}{L}\right)_4 = \alpha\left(\frac{W}{L}\right)_3 = \alpha^3\left(\frac{W}{L}\right)_1 \quad (8.\ 82)$$

since it carries the largest current. The equivalent RC ladder network in Figure 8.43(b) has a gen-

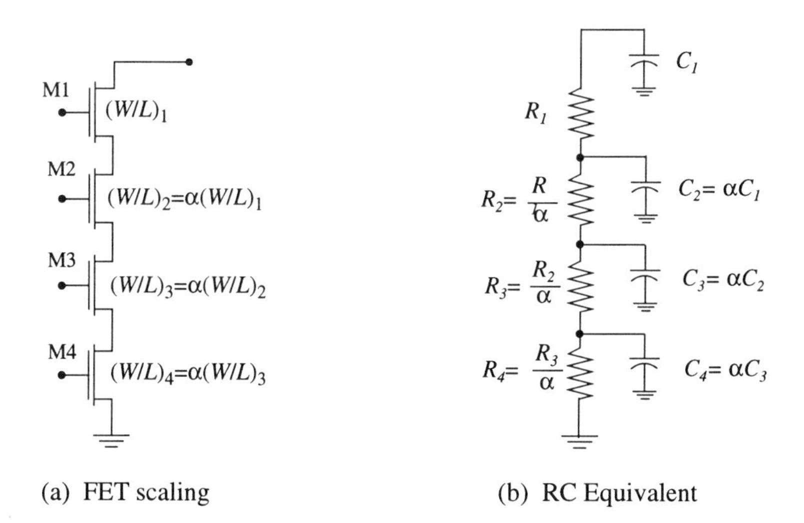

Figure 8.43 Device sizing for improved current flow

eral time constant of

$$\tau = C_1(R_1 + R_2 + R_3 + R_4) + C_2(R_2 + R_3 + R_4) + C_3(R_3 + R_4) + C_4R_4 \tag{8. 83}$$

where the resistance and capacitance values are related by

$$\begin{aligned} R_j &= \frac{R_{j-1}}{\alpha} \\ C_j &= \alpha C_{j-1} \end{aligned} \tag{8. 84}$$

Alternately, we may reference the values to M1 so that

$$\begin{aligned} R_j &= \frac{R_1}{\alpha^{(j-1)}} \\ C_j &= \alpha^{(j-1)} C_1 \end{aligned} \tag{8. 85}$$

for $j = 2, 3$, and 4. Substituting into the time constant equation and reducing gives

$$\begin{aligned} \tau &= C_1\left(R_1 + \frac{R_1}{\alpha} + \frac{R_1}{\alpha^2} + \frac{R_1}{\alpha^3}\right) + \alpha C_1\left(\frac{R_1}{\alpha} + \frac{R_1}{\alpha^2} + \frac{R_1}{\alpha^3}\right) + \alpha^2 C_1\left(\frac{R_1}{\alpha^2} + \frac{R_1}{\alpha^3}\right) + \alpha^3 C_1\left(\frac{R_1}{\alpha^3}\right) \\ &= R_1C_1\left[4 + \frac{3}{\alpha} + \frac{2}{\alpha^2} + \frac{1}{\alpha^3}\right] \end{aligned} \tag{8. 86}$$

Let us rewrite this as

$$\tau = R_1C_1\delta(\alpha) \tag{8. 87}$$

where

$$\delta(\alpha) = 4 + \frac{3}{\alpha} + \frac{2}{\alpha^2} + \frac{1}{\alpha^3} \tag{8. 88}$$

describes the effects of the scaling.

To understand the significance of this expression, suppose that we instead have a chain where all transistors have the same (small) aspect ratio

$$\left(\frac{W}{L}\right)_1 = \left(\frac{W}{L}\right)_2 = \left(\frac{W}{L}\right)_3 = \left(\frac{W}{L}\right)_4 \tag{8. 89}$$

The time constant for this case is

$$\begin{aligned} \hat{\tau} &= C_1(R_1 + R_1 + R_1 + R_1) + C_1(R_1 + R_1 + R_1) + C_1(R_1 + R_1) + C_1R_1 \\ &= 10R_1C_1 \end{aligned} \tag{8. 90}$$

so that we should examine values of $\delta(\alpha)$ for different scaling factors $\alpha > 1$ and compare it to 10. If we set $\alpha = 1.2$ corresponding to 20% increase in size from one transistor to the next, then

$$\delta(\alpha)\big|_{\alpha = 1.2} = 4 + \frac{3}{1.2} + \frac{2}{1.2^2} + \frac{1}{1.2^3} \approx 8.46 \tag{8. 91}$$

so that $\tau = 8.46\,(R_1C_1)$. With $\alpha = 1.3$ we obtain

$$\delta(\alpha)\big|_{\alpha = 1.3} = 4 + \frac{3}{1.3} + \frac{2}{1.3^2} + \frac{1}{1.3^3} \approx 7.95 \quad (8.92)$$

and $\tau = 7.95(R_1 C_1)$. This shows that the scaling can reduce the time constant, thus decreasing the discharge time. A value of $\alpha = 1.3$ indicates a 30% scaling and would be typical in practice.

Although we have chosen uniform scaling where adjacent transistors are related by the same scale factor α throughout the chain, it is possible to use different values for each. For example,

$$\left(\frac{W}{L}\right)_2 = \alpha_2\left(\frac{W}{L}\right)_1 , \qquad \left(\frac{W}{L}\right)_3 = \alpha_3\left(\frac{W}{L}\right)_2 \quad (8.93)$$

with $\alpha_2 \neq \alpha_3$. The analysis follows the same approach, so that the details are left for the interested reader to pursue.

Layout and Practical Limitations

The above analysis appears to provide an approach that can be used to reduce the discharge time in any dynamic circuit by appropriate scaling of the FET sizes in a chain. While transistor scaling is a useful technique in certain applications, the simplicity of the model used to derive the result needs to be examined in more detail in order to illustrate the practical limitations.

If we translate the circuit schematic to a chip layout, we arrive at the patterning illustrated in Figure 8.44(a). Each transistor has an active area ($n+$) that is sized to give the proper aspect ratio through the channel width W. When this is compared to the simple RC modelling used above, we see that we have ignored the fact that there is FET capacitance at both the top and bottom of the gate region. This results in the equivalent circuit shown in Figure 8.44(b) where we have used a complete π-model for the RC contributions of each transistor. The time constant for this circuit is given by the more complicated expression

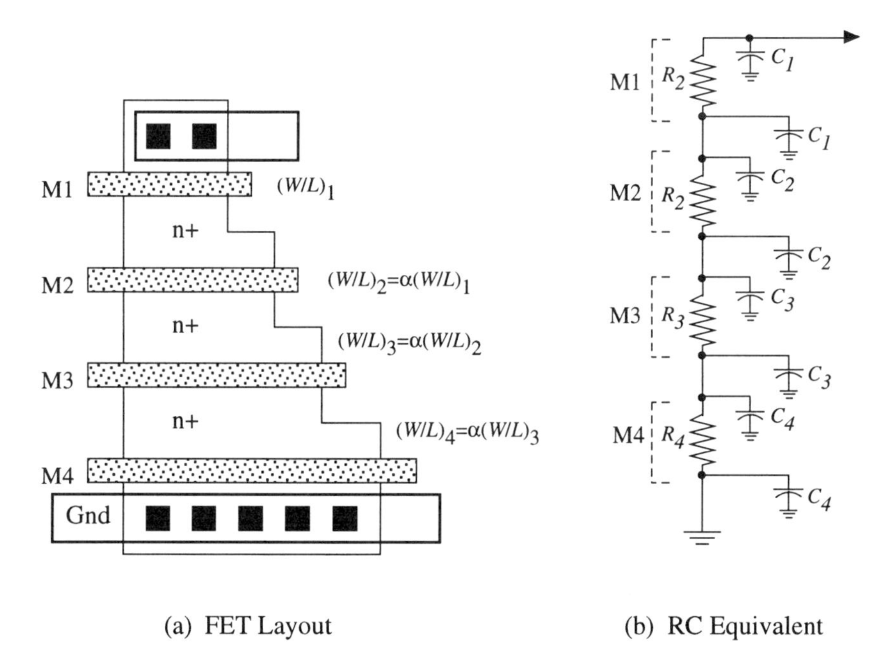

(a) FET Layout (b) RC Equivalent

Figure 8.44 Improved RC modelling of the scaled FET chain

$$\tau = C_1 (R_1 + R_2 + R_3 + R_4) + (C_1 + C_2) (R_2 + R_3 + R_4) + (C_2 + C_3) (R_3 + R_4) + (C_3 + C_4) R_4 \quad (8.94)$$

This can still be scaled by applying the same formulas, but the conclusion are not quite as simple.

A more important question deals with the usefulness of FET scaling in a modern process in which the minimum poly-to-poly gate spacing distance $S_{p\text{-}p}$ is small. This consideration is illustrated in Figure 8.45. The scaled FET layout is illustrated in drawing (a). The design rule that dictates the spacing between the transistor gates is the spacing $S_{p\text{-}a}$ between the edge of the poly and the 90-degree turn in the active area boundary. As discussed in Chapter 2, this is required because the self-aligned transistor process must account for slight mask registration errors when placing the poly. The minimum distance between two neighboring gates is seen to be $2S_{p\text{-}a}$. Figure 8.45(b) shows a layout where all MOSFETs have the same (small) channel width W, but the gates are spaced by the distance $S_{p\text{-}p}$ instead. This comparison brings up the important questions of (i) whether a scaled chain actually improves the performance, or, (ii) if it does help, how much improvement can be achieved. The reasoning for these statements can be understood by noticing that the $n+$ area and perimeter of adjacent FETs in the scaled chain may in fact be larger than the same quantities in the minimum area layout. If $S_{p\text{-}p} < S_{p\text{-}a}$, then the junction capacitance contributions may be smaller in the constant-W design, and therefore give faster switching. Even if the design rule values are about equal, the small decrease in the discharge time may not be worth the time and effort needed to implement the scaling.

This simple analysis demonstrates that CMOS design techniques are very sensitive to evolution of processing technology. As such, the *savvy designer*[6] is always aware of the close interplay among circuits, parasitics, layout, and processing.

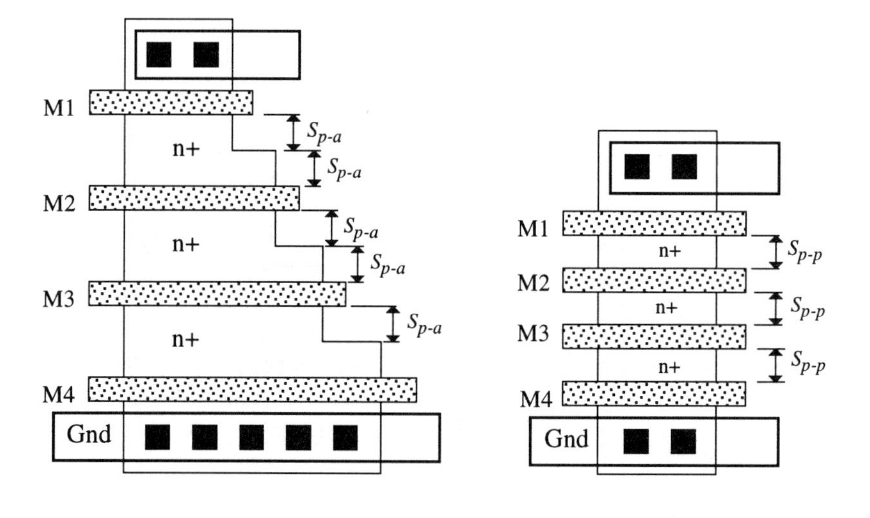

(a) Scaled chain (b) Minimum gate spacing

Figure 8.45 Comparison of scaled layout to minimum area design

[6] Read as "employed chip engineer"

8.3.5 High-Speed Cascades

The above sections illustrate that the gate discharge time is a limiting factor on the speed of a domino logic chain. Owing to this fact, it is useful to re-examine the design of a single gate so as to gain a deeper understanding of the operation.

Consider the circuit shown in Figure 8.46 which shows a domino gate undergoing precharge. MOSFET Mp is important to the operation of the circuit since it charges C_X to the voltage $V_X = V_{DD}$. The evaluation nFET Mn is in cutoff during this time, and is used to block the current flow to ground. This was, in fact, the main reason for using the clocked complementary pair. Now suppose that the clock is switched to $\phi=1$ and we have a discharge event. Mn is biased on, allowing C_X to discharge to ground. However, in this situation it acts solely as a parasitic element that adds both resistance and capacitance to the time constant. This can be seen by the AND2 example in Figure 8.47. With both logic transistors active, I_X flows and the discharge is described by the time constant

$$\tau = (R_2 + R_1 + R_n) C_X + (R_1 + R_n) C_2 + R_n (C_n + C_1) \tag{8.95}$$

where we have split up the capacitance contributions on the last term to clearly indicate their origin. The effect of Mn is obvious: the resistance R_n appears in every term, and the capacitance C_n also contributes to the time constant expression. Thus, in summary, Mn provides a needed function during the precharge, but its very presence slows down the discharge. If we could somehow eliminate the evaluation nFET in a domino stage, then it would increase the switching speed of the chain.

One approach to achieving this goal is shown in Figure 8.48. The first stage is a domino circuit with standard structuring. The second stage, however, has been altered in two ways: the evaluation nFET has been removed, and the clock signal applied to the precharge pFET has been changed from ϕ to a delayed clock signal ϕ'. This type of an arrangement yields a glitch-free cascade with an improved second stage. The key to understanding the operation is to note that a non-inverting buffer (that consists of two cascaded inverters) is used to delay $\phi(t)$ by an amount t_d before it is applied to the second stage. The time shift between the original clock $\phi(t)$ and the new clock $\phi'(t)$ is shown by the waveforms in Figure 8.49. Consider the operation of the first stage. A clock value of $\phi=0$ induces precharging of the internal node which takes a time

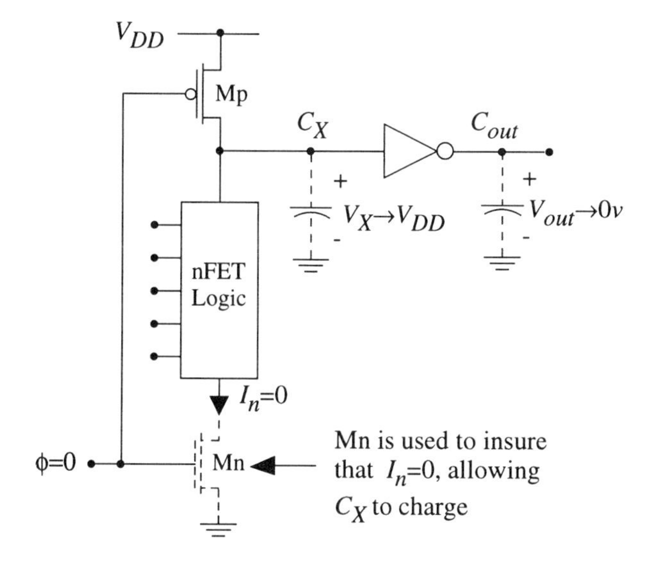

Figure 8.46 The purpose of the evaluate nFET

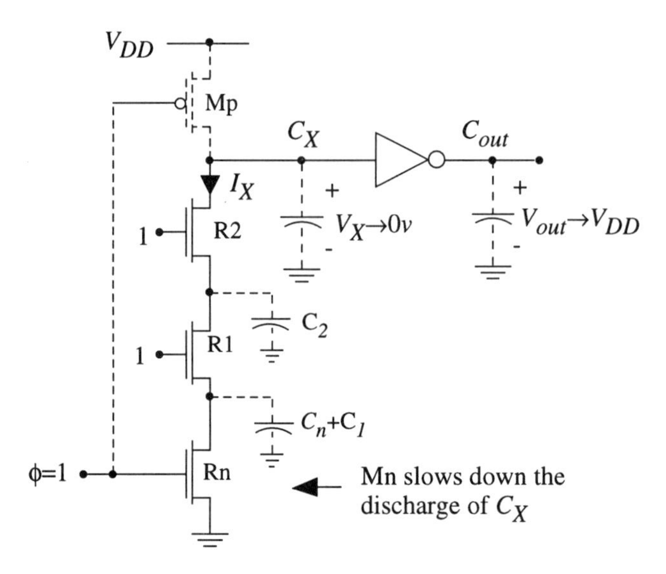

Figure 8.47 Slowdown in discharge due to evaluate nFET

$$t_{ch} \approx s_p \tau_p \tag{8.96}$$

This drives output voltage $V_{out,1}$ to a value of $0v$ after a time t_{LH} of the inverter. The second clock $\phi'(t)$ is delayed by an amount t_d that is long enough to insure that $V_{out,1} = 0v$ has been achieved.

The overall effect of this delay is shown explicitly in Figure 8.50. An output voltage of $V_{out,1} = 0v$ insures that the logic FET M2 of the second stage is in cutoff. This severs the path between the

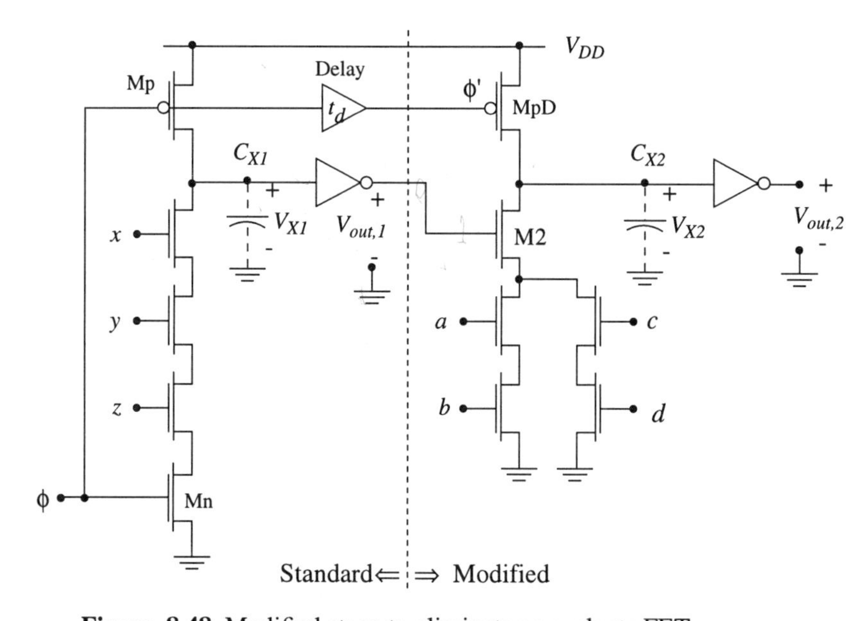

Figure 8.48 Modified stage to eliminate an evaluate FET

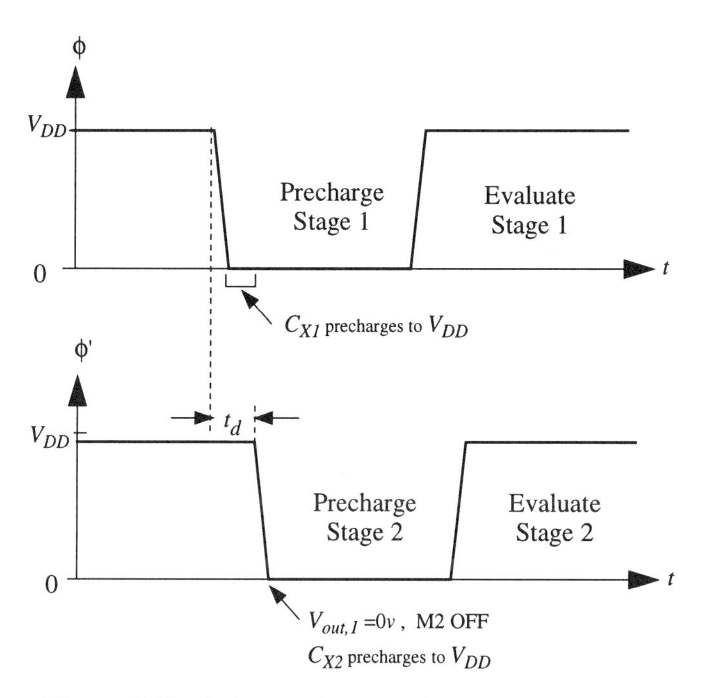

Figure 8.49 Timing requirements for high-speed cascade

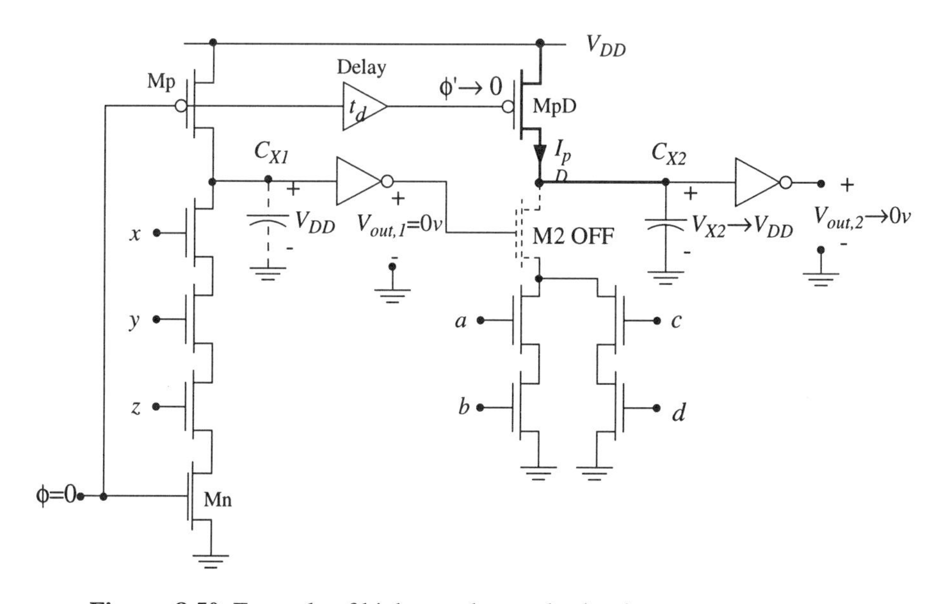

Figure 8.50 Example of high-speed cascade circuitry

internal node capacitor C_{X2} of the delayed stage and ground. By the time the delayed clock makes a transition to $\phi'\rightarrow 0$, M2 is off so that all of the precharge current I_{pD} through the clocking pFET MpD is directed towards C_{X2}. This insures proper operation of the circuit when the (delayed) evaluation takes place.

The design of this type of cascade centers around determining the proper delay t_d between the clocks. The value must be long enough to insure that the output of the first stage has fallen below the threshold voltage of the nFET in the next logic chain. Note that body-bias effects will increase this voltage. The complicating factor is to insure that the process variations do not change the delay to the point where a glitch is introduced or where the timing is off. Moreover, all signal and clock path delays are very sensitive to layout, so that this is really a chip design technique, not just a circuit solution.

8.4 Multiple-Output Domino Logic

All of the logic gates that we have studied thus far are restricted by their very nature to have a single output. **Multiple-output domino logic (MODL)** is a variation of standard domino logic that accepts multiple inputs and can produce two or more distinct output functions.

A generic MODL gate is shown in Figure 8.51. Several changes in the basic domino circuit have been made to arrive at this expanded gate. First, the nFET logic array in this example has been split into three distinct blocks that are labelled by f_1, f_2, and f_3, such that a value of $f_i = 1$ ($i = 1, 2,$ or 3) means that the corresponding nFET block acts as a closed switch from top to bottom. The second modification is that two output inverter circuits have been added, along with precharge pFETs Mp2 and Mp3 that charge capacitors C_{X2} and C_{X3}, respectively.

The operation of the circuit is similar to that of a standard domino gate. When the clock is low with $\phi=0$, all of the precharge FETs are active. The internal capacitors C_{X1}, C_{X2}, and C_{X3} are charged to

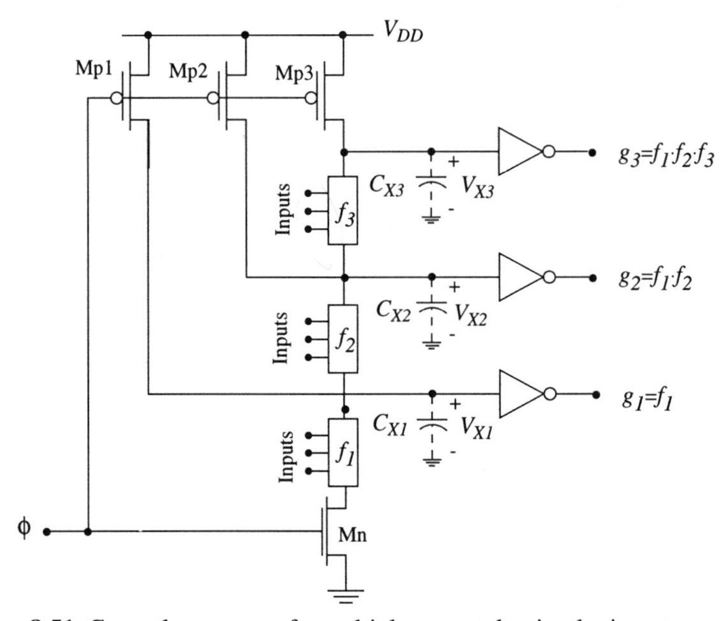

Figure 8.51 General structure of a multiple-output domino logic gate

$$V_{Xi} \rightarrow V_{DD} \tag{8.97}$$

for $i = 1, 2, 3$, so that the output voltages are all at

$$V_{out,\,i} = 0v \tag{8.98}$$

when the circuit goes into the evaluation mode of operation. When the clock changes to a value of $\phi=1$, every capacitor is subject to a conditional discharge event. However, since the logic blocks are in series, f_2 depends upon f_1, and f_3 depends upon both f_1 and f_2. To understand this dependence, let us examine the behavior of the internal nodes.

Suppose that $f_1 = 1$, allowing capacitor C_{X1} to discharge. In this case,

$$V_{X1} \rightarrow 0v \tag{8.99}$$

and

$$V_{out,\,1} \rightarrow V_{DD} \tag{8.100}$$

giving the proper output voltage. Now then, if $f_1 = 1$ AND $f_2 = 1$, then both C_{X1} and C_{X2} discharge giving

$$V_{out,\,1} \rightarrow V_{DD}\text{, and} \qquad V_{out,\,2} \rightarrow V_{DD}\,. \tag{8.101}$$

The third possibility that can occur is if $f_1 = 1$ AND $f_2 = 1$ AND $f_3 = 1$, which allows all three of the internal node capacitors C_{X1}, C_{X2}, and C_{X3} discharge to ground and results in

$$V_{out,\,1} \rightarrow V_{DD}, \qquad V_{out,\,2} \rightarrow V_{DD}\,, \qquad V_{out,\,3\,(1)} \rightarrow V_{DD} \tag{8.102}$$

In terms of the logical expressions for the outputs, we see that

$$\begin{aligned} g_1 &= f_1 \\ g_2 &= f_1 \cdot f_2 \\ g_3 &= f_1 \cdot f_2 \cdot f_3 \end{aligned} \tag{8.103}$$

so that MODL requires that the output functions can be expressed with a nested AND structure; it cannot be applied to an arbitrary set of logic functions.

Figure 8.52 shows the interdependence of the logic blocks described above. The subcircuit in Figure 8.52(a) corresponds to the case where logic block f_1 is closed while the f_2 block is open. This allows C_{X1} to discharge so that $g_1 \rightarrow 1$, but holds the charge on C_{X2} and gives $g_2 = 0$. In case (b), both the f_1 and f_2 blocks are closed, forcing both outputs g_1 and g_2 to logic 1 output values. Note that the evaluation nFET Mn is required to sink all current components, so that its aspect ratio must be chosen accordingly.

An example of a simple 2-output MODL circuit is shown in Figure 8.53. The lower logic block provides the logic function

$$f_1 = A \cdot B + C \tag{8.104}$$

while the upper block uses two parallel nFETs to give

$$f_2 = x + y \tag{8.105}$$

using standard logic formation rules. The two possible outputs are thus seen to be

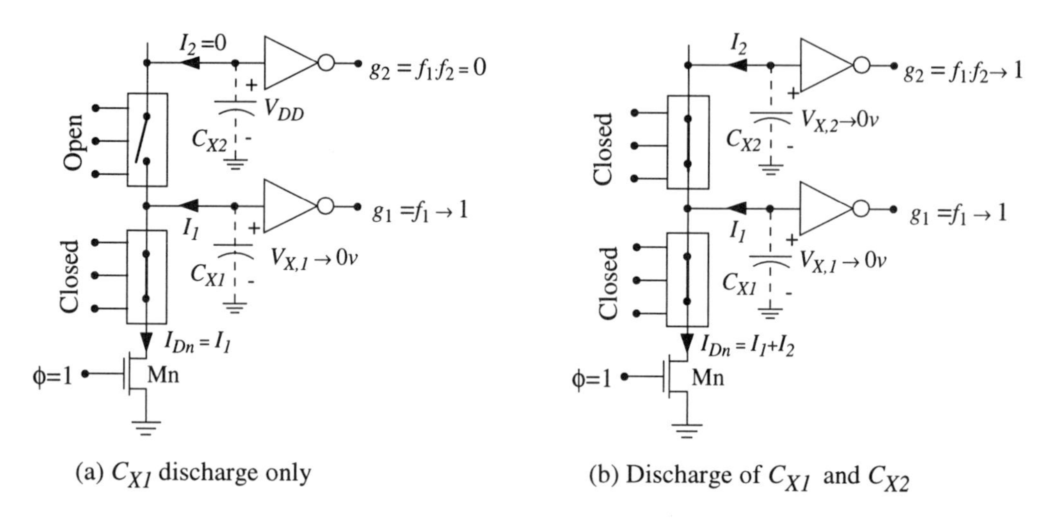

(a) C_{X1} discharge only

(b) Discharge of C_{X1} and C_{X2}

Figure 8.52 Circuit operation for the AND function

$$
\begin{aligned}
g_1 &= f_1 = A \cdot B + C \\
g_2 &= f_1 \cdot f_2 = (A \cdot B + C) \cdot (x + y)
\end{aligned}
\tag{8. 106}
$$

As with standard domino logic, MODL is non-inverting. It can be directly interfaced to a domino chain and driven with the same clocks, making it easy to use when nested AND logic functions are required.

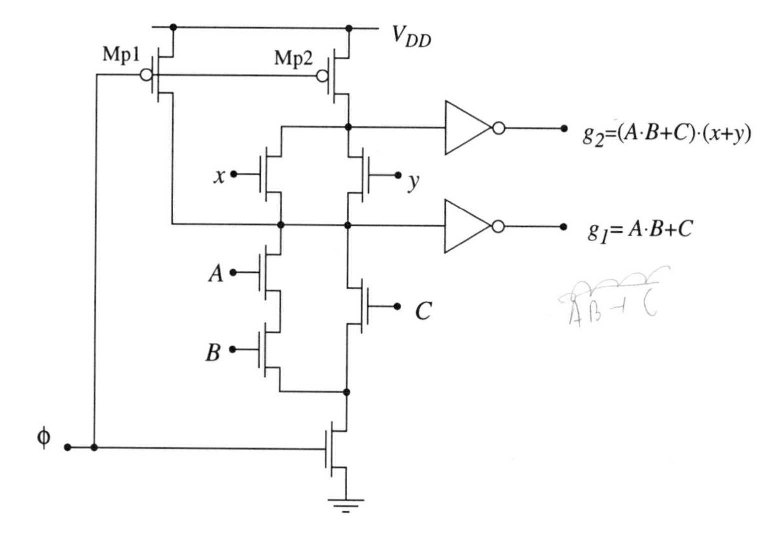

Figure 8.53 Example of an MODL logic gate

8.4.1 Charge Sharing and Charge Leakage

The use of multiple precharge transistors in MODL helps overcome the bad effects of both charge sharing and charge leakage problems by precharging several capacitive nodes at the same time. Consider the MODL stage shown in Figure 8.54 as an example. During precharge, the clock is at a value ϕ=0 so that both Mp1 and Mp2 conduct. This results in both internal nodes charging to give

$$V_{X1} \rightarrow V_{DD}$$
$$V_{X2} \rightarrow V_{DD} \tag{8. 107}$$

so that the total charge on the internal nodes is

$$Q_T = (C_{X1} + C_{X2}) V_{DD} \tag{8. 108}$$

Since the logic inputs are 0 during precharge (assuming that the stage is being driven by another domino circuit), a value of A=0 implies that C_p remains uncharged. Now suppose that $\phi \rightarrow 1$ and the inputs are at x=1, A=1, and B=0. The latter condition blocks any discharge from taking place, and keeps both F_1 and F_2 at 0 values. In this circuit, charge sharing will occur because C_p is connected to the output nodes. However, both C_{X1} and C_{X2} are precharged to V_{DD} so that charge sharing with C_p will not have as much of an effect; a second consideration is that we expect both C_{X1} and C_{X2} to be large compared with C_p owing to the number of device parasitics contributing to each term. Charge leakage occurs at every junction, and is still important. Denoting the leakage through a reverse biased drain or source nFET junction by I_n and that by a reverse biased pFET drain or source junction by I_p, we have a net leakage off of the C_{X1} and C_{X2} nodes of

$$I_{leak} = 9I_n - 2I_p \tag{8. 109}$$

by simply counting drain/source nodes. This indicates that both V_{X1} and V_{X2} will decay in time, limiting the duration of the evaluation interval. The maximum hold time can be determined by using the minimum voltage

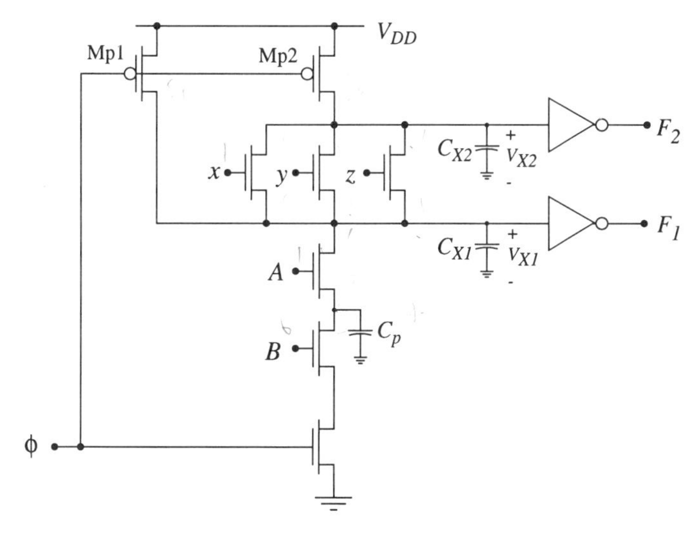

Figure 8.54 MODL gate used to illustrate charge problems

$$\min(V_{X1}, V_{X2}) = V_{IH} \tag{8.110}$$

since this the condition for switching the inverter input from a logic 1 to a logic 0 state. Of course, charge keeper circuits can be added if charge leakage becomes a problem.

Charge sharing may be more of a problem if long series-connected logic arrays are used. Figure 8.55 shows a logic gate with only two precharge nodes (with capacitances C_{X1} and C_{X2}), but where both logic blocks use series FETs to implement AND operations. During the evaluate time (ϕ=1) indicated, the input levels induce severe charge sharing with the inter-FET capacitors C_a, C_b, C_c, C_d, and C_e. The most critical voltage will probably be the value of V_{X2} across C_{X2} since the upper logic block is closed and permits charge sharing with the lower logic block. The problem is compounded by charge leakage. Owing to these considerations and increased discharge times, long chains of logic nFETs are generally avoided when designing the switching arrays.

8.4.2 Carry Look-Ahead (CLA) Adder

The effective use of MODL requires that a logic block employ algorithms that consist of nested functions ANDed together. A particularly useful and interesting application of the technique is for constructing **carry look-ahead** (**CLA**) parallel adders.

First, let us recall the operation of a full adder unit with the symbol shown in Figure 8.56. The binary inputs are denoted by a_n and b_n, and the carry-in bit is c_n. The outputs are the sum bit function

$$s_n = (a_n \oplus b_n) \oplus c_n \tag{8.111}$$

and the carry-out bit that can be computed from

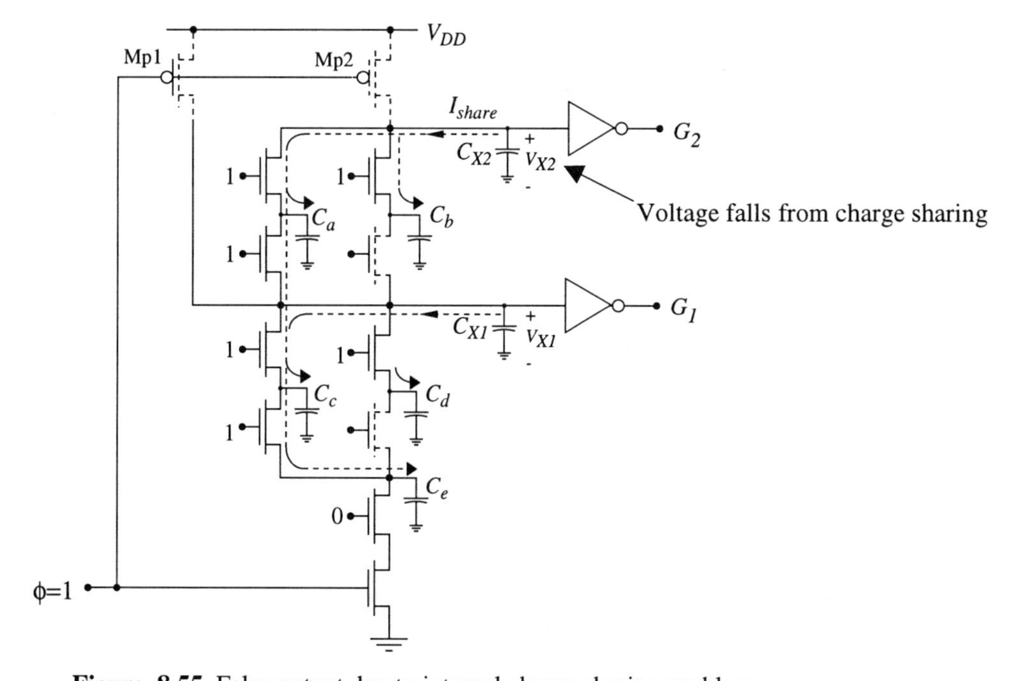

Figure 8.55 False output due to internal charge sharing problems

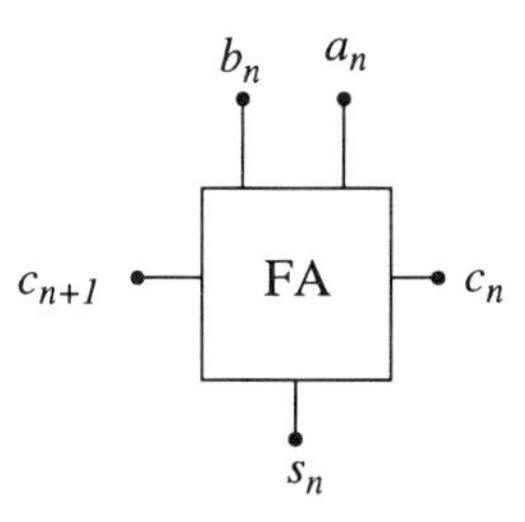

Figure 8.56 Full adder logic symbol

$$c_{n+1} = a_n \cdot b_n + (a_n \oplus b_n) \cdot c_n \tag{8. 112}$$

Suppose that we want to add two 4-bit words as described by

$$\begin{array}{r} a_3\ a_2\ a_1\ a_0 \\ +\ b_3\ b_2\ b_1\ b_0 \\ \hline c_4\ s_3\ s_2\ s_1\ s_0 \end{array} \tag{8. 113}$$

where c_4 is the carry-out bit. Since the n-th sum requires uses the (n-1)-th carry bit, the simplest way to construct a 4-bit parallel adders is using the ripple carry scheme drawn in Figure 8.57 where the full adder cells are connected to directly provide the carry bit to the next FA unit. The latency associated with obtaining the total sum word $s_3\, s_2\, s_1\, s_0$ and the carry-out bit c_4 is due to the fact that all of the carry bits are created in sequence from c_1 to c_4, and the observation that the output sum bit s_n is not valid until the carry-in bit c_n from the (n-1) full adder is valid.

The carry look-ahead algorithm provides an alternate approach to constructing parallel adders by calculating the carry bits using separate circuits, and then feeding them to logic gates that produce the sum bits. The basis for the CLA is obtained by studying the conditions that lead to a value of c_{n+1}=1. We see immediately that the OR operation in

$$c_{n+1} = a_n \cdot b_n + (a_n \oplus b_n) \cdot c_n \tag{8. 114}$$

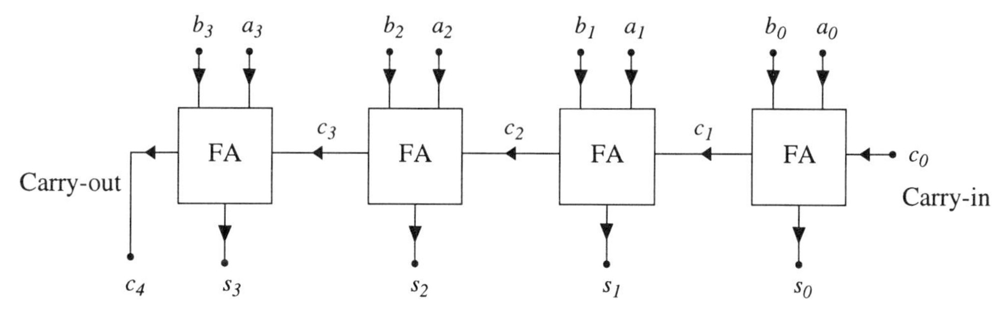

Figure 8.57 4-bit ripple carry adder design

shows that a carry-out bit with a value of 1 can be "created" in two ways. First, $c_{n+1} = 1$ if the inputs satisfy $a_n \cdot b_n = 1$; this is called a **carry generation**, since the carry-out is "generated" by the input bits of the word itself. This is illustrated in Figure 8.58(a). The second case that causes $c_{n+1} = 1$ is where $(a_n \oplus b_n) = 1$ AND the input carry bit has a value of $c_n = 1$ from the previous $(n\text{-}1)$-st unit; this is called a **carry propagation** since we can view the input carry $c_n = 1$ as being propagated through the unit to the output as in Figure 8.58(b).

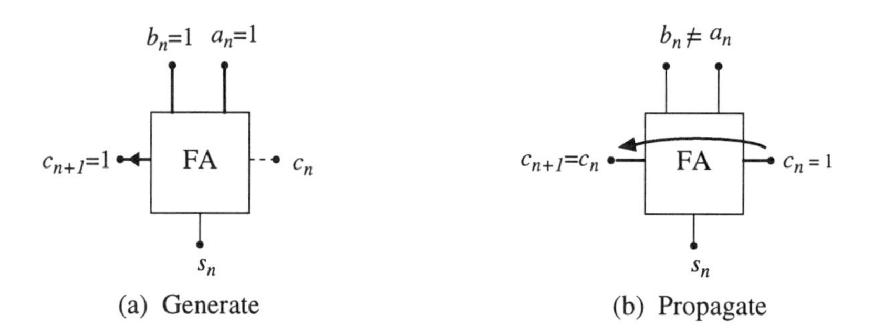

Figure 8.58 Origin of the carry-out bit c_{n+1}

The CLA algorithm is based on these simple observations. To develop the basic equations, we define the **generate bit** g_n by

$$g_n = a_n \cdot b_n \tag{8.115}$$

and the **propagate bit** p_n as

$$p_n = a_n \oplus b_n \tag{8.116}$$

The carry-out bit is then given by

$$c_{n+1} = g_n + p_n \cdot c_n \tag{8.117}$$

and the sum is calculated using

$$s_n = p_n \oplus c_n \tag{8.118}$$

for each bit in the word.

Using the CLA approach leads to the block diagram for a 4-bit adder that is drawn in Figure 8.59. The input words are denoted by $a = (a_3\, a_2\, a_1\, a_0)$ and $b = (b_3\, b_2\, b_1\, b_0)$, and the carry-in bit into the 0-th stage is labelled as c_{in}. The first logic network uses $a_3\, a_2\, a_1\, a_0$ and $b_3\, b_2\, b_1\, b_0$ to calculate the generate and propagate bits, $g_3\, g_2\, g_1\, g_0$ and $p_3\, p_2\, p_1\, p_0$, respectively. These are then fed to a logic network that is dedicated to calculating the carry bits $c_4\, c_3\, c_2\, c_1$, where we note that

$$c_0 = c_{in} \tag{8.119}$$

defines the carry-in bit. The outputs from this unit are then used to find the result $s_3\, s_2\, s_1\, s_0$. The carry-out bit

$$c_4 = c_{out} \tag{8.120}$$

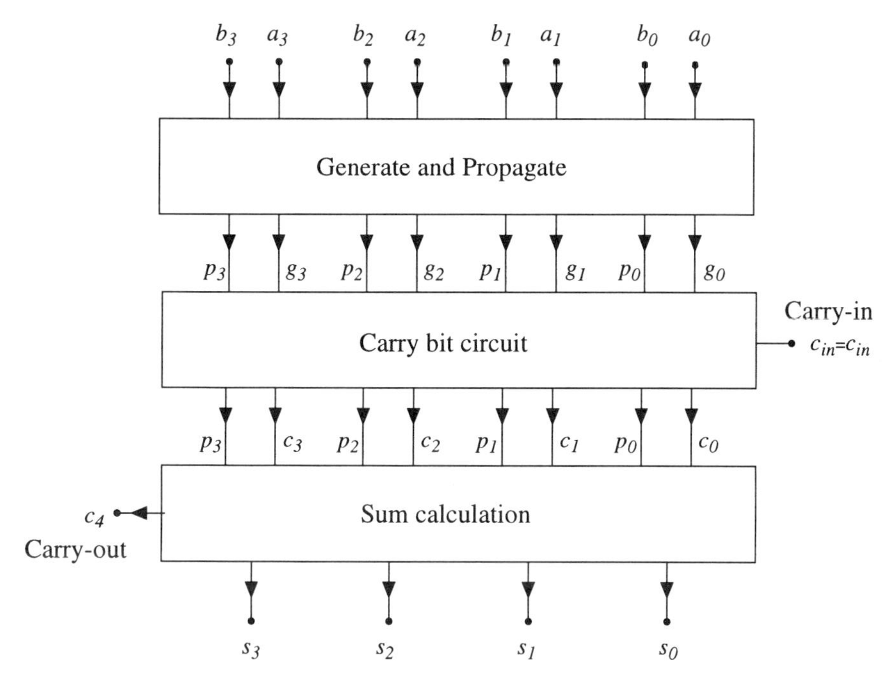

Figure 8.59 Block diagram for a 4-bit carry look-ahead adder

is used to indicate overflow, or as the carry-in bit to the next group of a larger word.

The application of MODL circuits to the CLA network can be seen by examining the equations used to compute the carry bits. The carry-out bit c_1 for the $n = 0$ bit position is given by

$$\begin{aligned} c_1 &= g_0 + p_0 \cdot c_0 \\ &= g_0 + p_0 \cdot c_{in} \end{aligned} \tag{8. 121}$$

and is used as an input by stage 1. Since the carry-out bit c_2 from stage 1 is given be

$$c_2 = g_1 + p_1 \cdot c_1 \tag{8. 122}$$

we may substitute for c_1 to arrive at

$$\begin{aligned} c_2 &= g_1 + p_1 c_1 \\ &= g_1 + p_1 (g_0 + p_0 c_0) \\ &= g_1 + p_1 g_0 + p_1 p_0 c_0 \end{aligned} \tag{8. 123}$$

This shows that c_2 can be computed solely from the generate and propagation bits. In the same manner, c_3 is given by

$$\begin{aligned} c_3 &= g_2 + p_2 c_2 \\ &= g_2 + p_2 (g_1 + p_1 g_0 + p_1 p_0 c_0) \\ &= g_2 + p_2 g_1 + p_2 p_1 g_0 + p_2 p_1 p_0 c_0 \end{aligned} \tag{8. 124}$$

while

$$\begin{aligned} c_4 &= g_3 + p_3 c_3 \\ &= g_3 + p_3 (g_2 + p_2 g_1 + p_2 p_1 g_0 + p_2 p_1 p_0 c_0) \\ &= g_3 + p_3 g_2 + p_3 p_2 g_1 + p_3 p_2 p_1 g_0 + p_3 p_2 p_1 p_0 c_0 \end{aligned} \tag{8. 125}$$

gives the carry-out bit c_{out} from the 4-bit unit. These equations show explicitly that each bit can be calculated by ANDing generate and propagate factors together with the nesting required for an MODL circuit.

Let us first investigate the circuitry required to implement the algorithm using standard domino logic. As usual, every output is generated by a separate gate, which results in the four circuits shown in Figure 8.60. Each gate creates the appropriate expression with AND-OR series-parallel nFET structuring. While this approach is a valid one, the expanded equations show that MODL can be used to derive all four carry bits from a single circuit. The evolution from several single-output gates to one multiple-output circuit is based on the simplified equation set

$$\begin{aligned} c_1 &= g_0 + p_0 \cdot c_0 \\ c_2 &= g_1 + p_1 \cdot c_1 \\ c_3 &= g_2 + p_2 \cdot c_2 \\ c_4 &= g_3 + p_3 \cdot c_3 \end{aligned} \tag{8. 126}$$

that shows how c_n depends on c_{n-1}. Expanding each c_n into generate and propagate terms allows us to combine the standard domino circuits into a single MODL gate that gives the necessary carry bits as shown in Figure 8.61. This circuit provides the same logical outputs as the four individual DL gates, so that the advantage is obvious.

Of course, reducing the number of transistors may not be significant unless it can be achieved without sacrificing performance. Let use examine the transient switching times using the circuit shown in Figure 8.62. The worst-case circuit MODL discharge is shown in Figure 8.62(a) for the case where $c_4 = 1$; the equivalent RC ladder network is shown in Figure 8.62(b). We will approximate the voltage across C_X using the simplified exponential form

$$V_X(t) = V_{DD} e^{-t/\tau} \tag{8. 127}$$

so that the total discharge time is

$$t_{HL} \approx 2.2\tau \tag{8. 128}$$

The time constant τ is given by

$$\begin{aligned} \tau = {} & C_X (R_3 + R_2 + R_1 + R_0 + R + R_n) + C_3 (R_2 + R_1 + R_0 + R + R_n) \\ & + C_2 (R_2 + R_1 + R_0 + R + R_n) + C_2 (R_1 + R_0 + R + R_n) \\ & + C_1 (R_0 + R + R_n) + C_0 (R + R_n) + C_n R_n \end{aligned} \tag{8. 129}$$

If we analyze the c_4 circuit designed in standard domino logic, we will find that the time constant

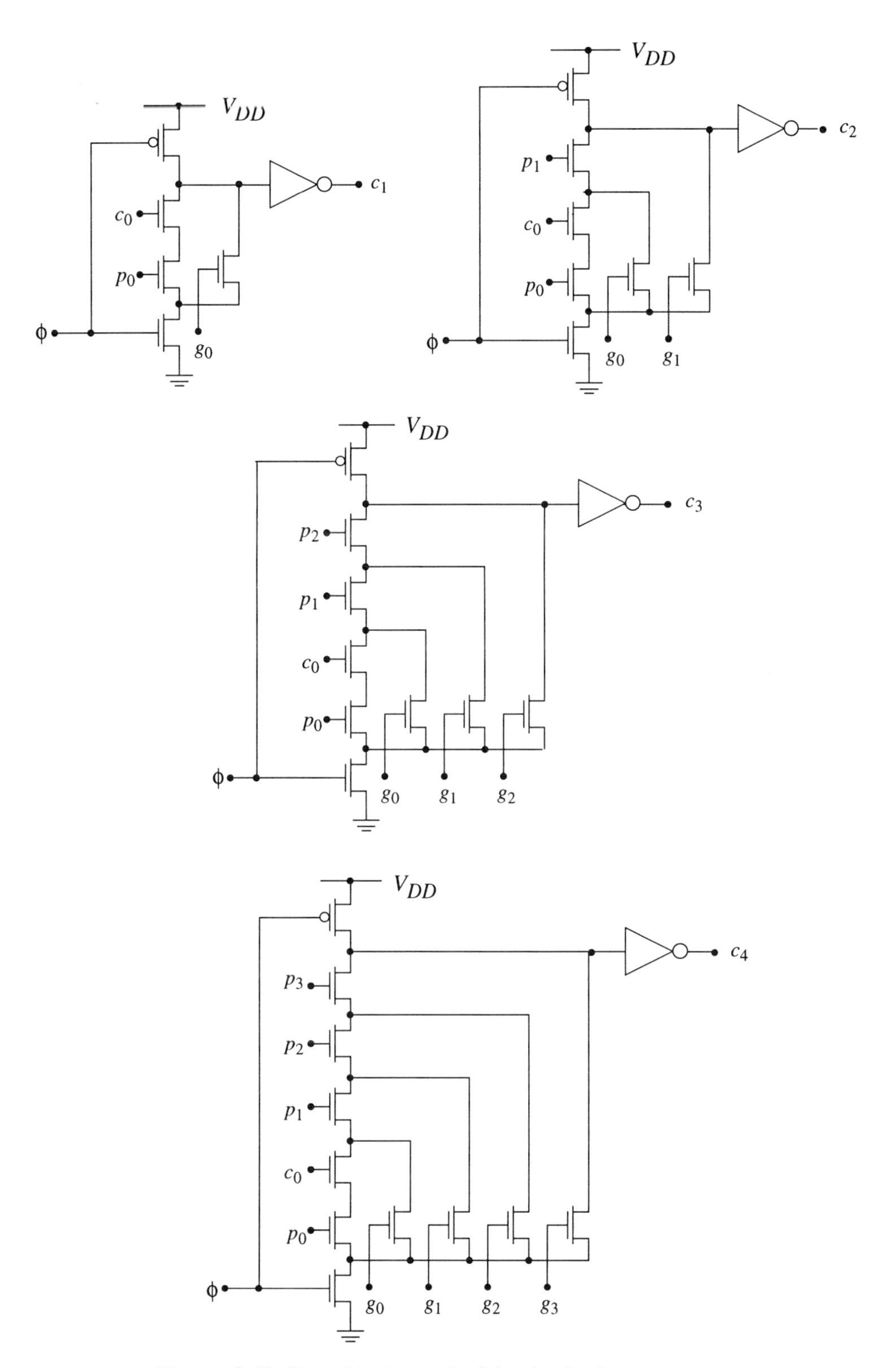

Figure 8.60 Carry-bits in standard domino logic

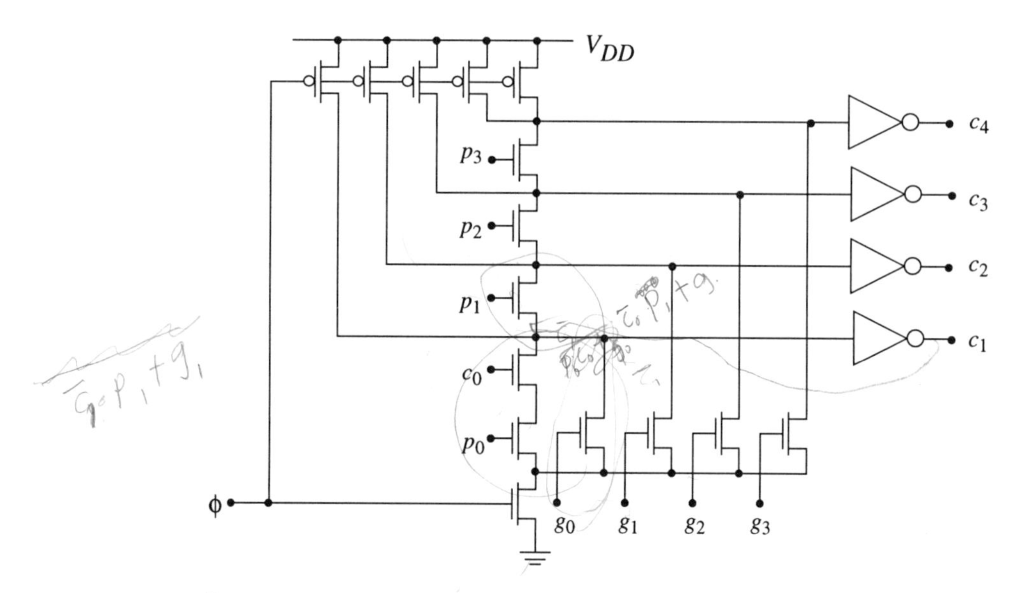

Figure 8.61 Single MODL gate that calculates all 4 carry-out bits

has the same form, but with different capacitance values. Comparing the values in the two circuits shows that

$$C'_{i,\,MODL} > C_{i,\,DL} \tag{8. 130}$$

because there are more device contributions at the nodes in the MODL circuit than in the single-output DL circuit. This illustrates the fact that the MODL circuit will be slightly slower than a sin-

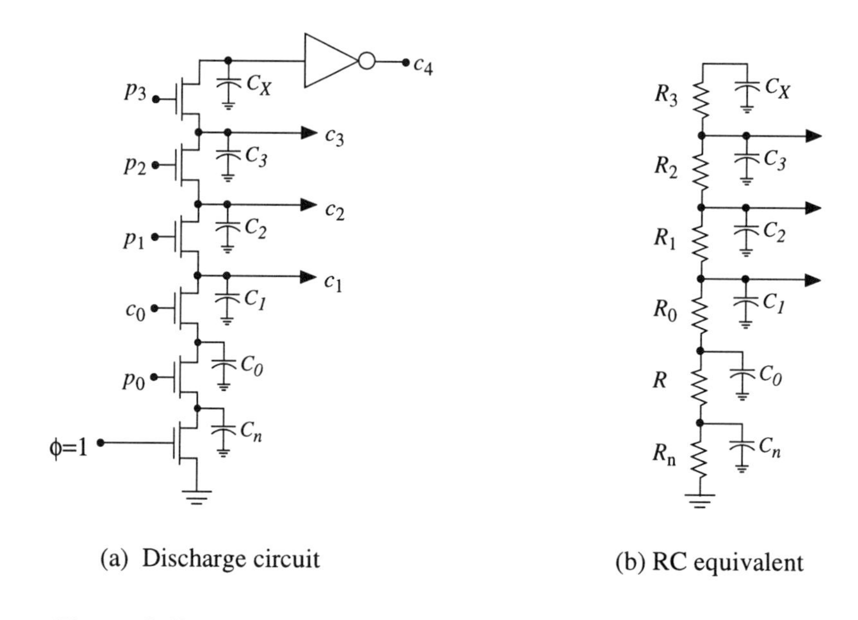

Figure 8.62 RC modelling of the MODL discharge circuit

gle-output circuit that uses the same size FETs. However, standard design techniques may be used to reduce the time constant, and the MODL approach requires much less area, so that it constitutes a viable choice for high-performance designs.

The carry circuit is only one portion of the complete CLA network, so let us examine the remaining circuits for completeness. The generate and propagate circuits are straightforward to design in standard domino logic, and are shown in Figure 8.63. If the input bits a_n and b_n are taken from static logic circuits (or latches) that can also provide the complements $\overline{a}_n$ and $\overline{b}_n$, then there are no major problems associated with either gate. The output sum circuit for s_n is more difficult to include in the domino chain since

$$
\begin{aligned}
s_n &= p_n \oplus c_n \\
&= p_n \cdot \overline{c_n} + \overline{p_n} \cdot c_n
\end{aligned}
\tag{8. 131}
$$

shows that we need to invert both signals in the cascade. Figure 8.64 shows the necessary circuit. Since p_n and c_n are outputs from DL or MODL circuits, using inverters introduces the possibility of a glitch. Once solution to this problem is to terminate the domino chain and use a static logic circuit to calculate the sum bits as in Figure 8.65. This uses a mirror XOR circuit which is relatively fast owing to its simplicity. When the sum bits are calculated, the values of p_n and $\overline{p}_n$ are already established. The carry bit c_n is 0 during the precharge. If c_n remains at 0, then s_n is immediately valid. If c_n makes a transition to a logic 1 value, then the output s_n changes and the circuit dissipates power during the switching event. We note in passing that clocked output latches can be added if needed.

Although the discussion here has centered around a 4-bit adder, MODL circuits have been shown to be useful in constructing high-speed wide adders (e.g., 32-bits or 64-bits or more) in which the carry-out bits of n-bit segments are calculated using multiple-output gates for use by higher position segments. The general problem is shown in Figure 8.66(a) where we attempt to construct an 8-bit adder using two 4-bit circuits. Even if we use 4-bit CLA circuitry, there is still a delay involved in transferring the carry-out bit c_4 left to the next 4-bit segment. The highly simplified block diagram in Figure 8.66(b) provides a qualitative view of how this can be overcome. The inputs a_n and b_n are first used to calculate all of the generate g_n and propagate p_n terms for $n = 0$ to 8. These are then used as inputs to calculate all of the necessary carry bits needed for each of the sum terms s_n. Some realizations can be shown to be identical to the Manchester carry scheme. The

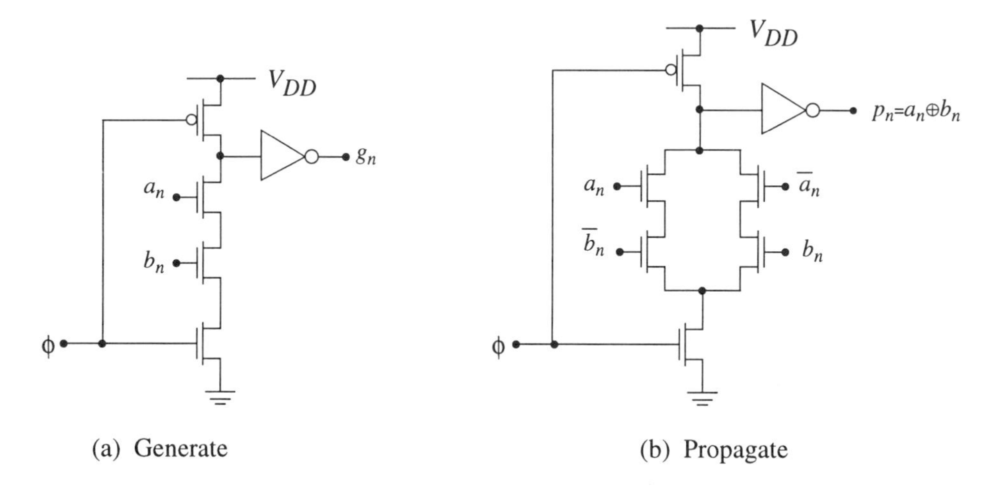

Figure 8.63 Generate and propagate circuits

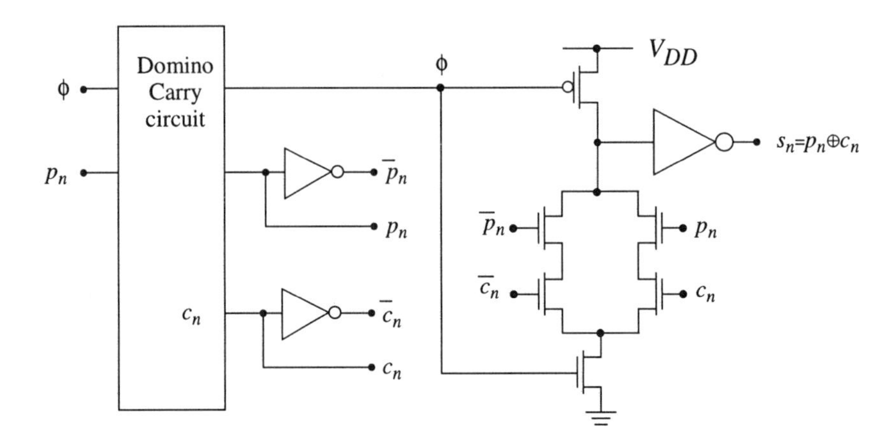

Figure 8.64 Circuit for the providing the sum

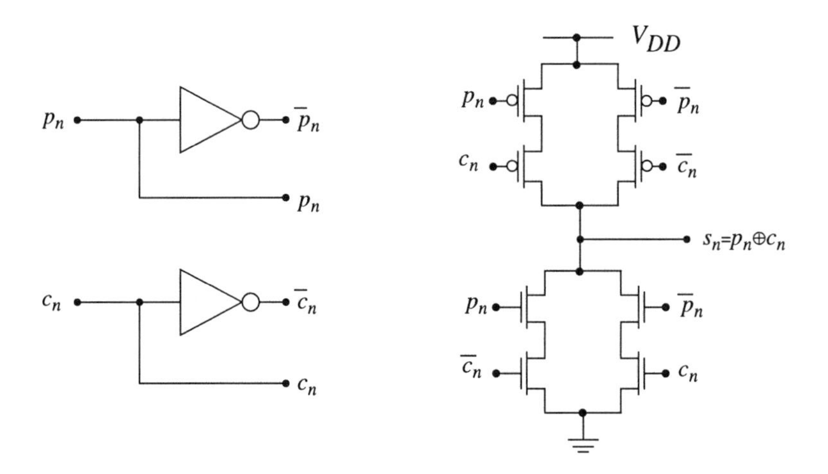

Figure 8.65 Use of a static mirror XOR circuit for the sum

interested reader is directed toward the literature, as a discussion here would be far beyond the intended scope of the book.

8.5 Self-Resetting Logic

Self-resetting logic (SRL) can be classified as a variant of domino logic that allows for asynchronous operation. A basic SRL circuit is shown in Figure 8.67. A careful inspection of the schematic shows that the primary differences between this gate and the standard domino circuit are (a) the addition of the inverter chain that provides feedback from the output voltage $V_{out}(t)$ to the gate of the **reset** pFET MR, and (b) the elimination of the evaluation nFET. Note that an odd number of inverters (3) is used in the feedback. As discussed below, the feedback loop has a significant effect on both the internal operation of the circuit and the characteristics of the output voltage.

Precharging of C_X occurs when the clock is at a value ϕ=0 and the circuit conditions are shown

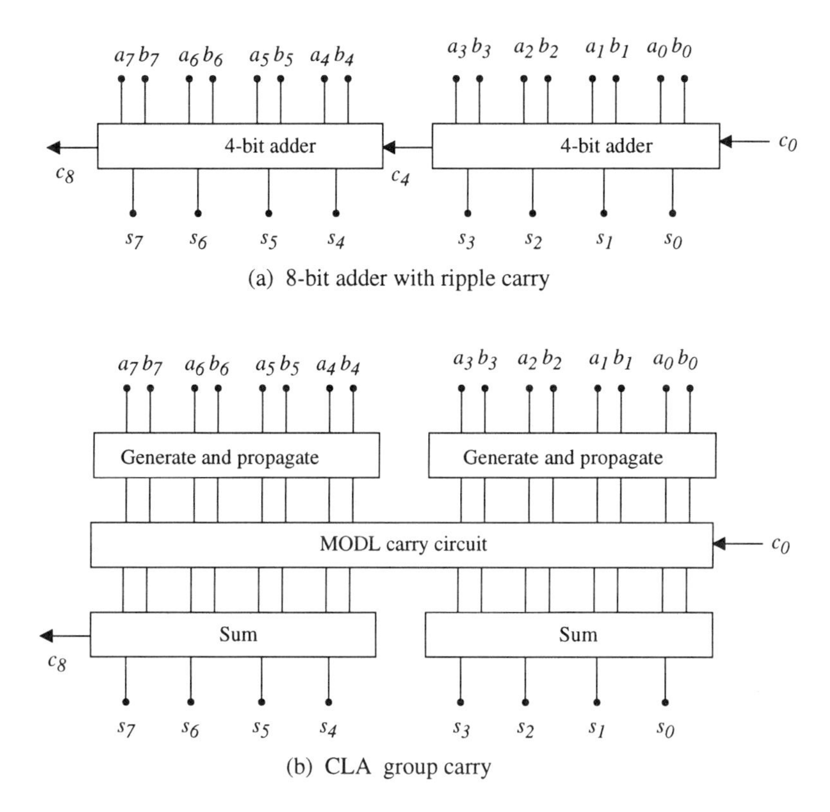

Figure 8.66 Creation of an 8-bit adder network

in Figure 8.68(a). During this time, $V_X \rightarrow V_{DD}$, and $V_{out} \rightarrow 0v$, which is identical to the event in a standard domino circuit. As we will see, the timing of the input signals precludes the possibility of a DC discharge path to ground by insuring that the inputs to all logic nFET are 0 during precharge. The voltage on the gate of MR is at a value of V_{DD}, so that

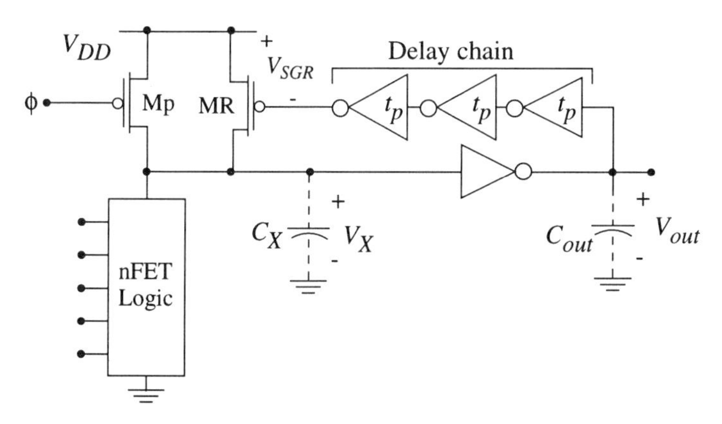

Figure 8.67 Basic structure of a self-resetting logic circuit

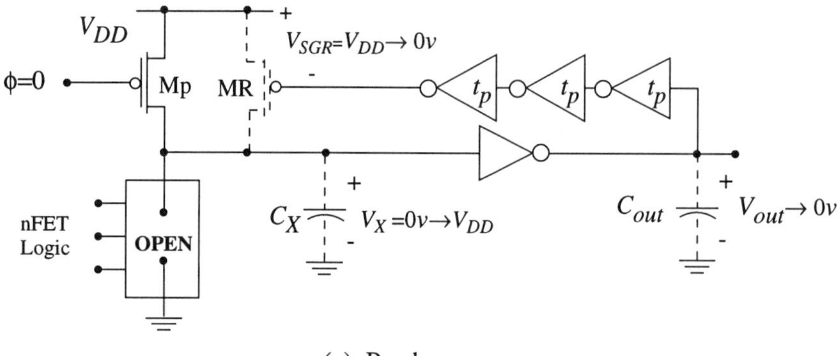

(a) Precharge

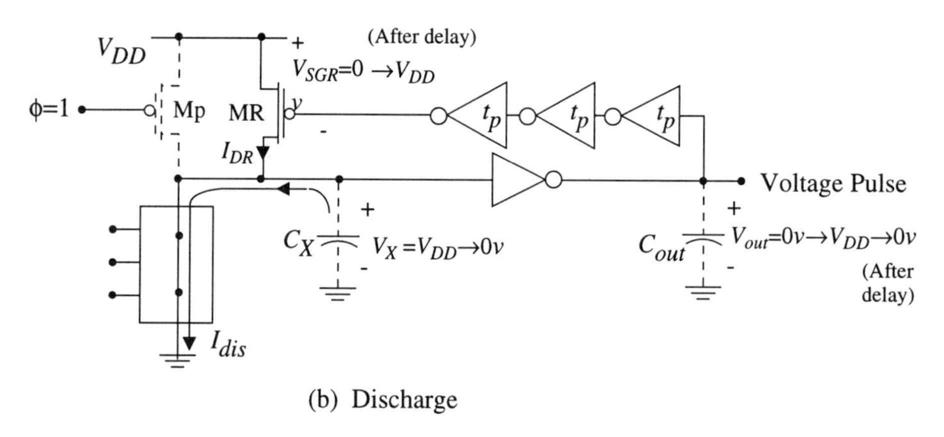

(b) Discharge

Figure 8.68 Operational modes of the self-resetting circuit

$$V_{SGR} = 0v \tag{8. 132}$$

insure that MR is in cutoff during this time.

The distinct features of SRL arise when a discharge occurs. The circuit conditions are shown in Figure 8.68(b). In this case, $V_X \to 0v$ and C_{out} charges to give an output voltage of $V_{out} \to V_{DD}$. This is fed through the triple-inverter chain to drive the gate voltage of MR to 0v after a delay of

$$t_D = 3t_d \tag{8. 133}$$

where t_d is the delay through one inverter. Since now we have that

$$V_{SGR} = V_{DD} \tag{8. 134}$$

MR is active which allows I_{DR} to flow and recharge C_X back up to a voltage of $V_X = V_{DD}$. This action resets the output voltage to its original precharge value of V_{out} = 0v, giving the logic family its name: it automatically resets its output to 0.

To gain a better understanding of the voltage transitions involved in a self-resetting logic gate, let us analyze the circuit shown in Figure 8.69 which implements the AO function

$$f = x \cdot y + z \tag{8. 135}$$

for the case where $x{\cdot}y \rightarrow 1$. The voltage waveforms are shown in Figure 8.70. When ϕ=0, the circuit undergoes a "normal" precharge where

$$V_X \rightarrow V_{DD} \text{ and } V_{out} \rightarrow 0v \tag{8. 136}$$

as shown. During the precharge, the source-gate voltage V_{SGR} applied to MR is driven to V_{DD} which drives MR into cutoff. Now let us examine the effects of a discharge event that is initiated by the input variables during evaluation. Assuming that the inputs x and y are taken from a self-resetting circuit, the condition $x{\cdot}y = 1$ is described by a pulse whose duration is set by the driving circuit. This allows C_X to discharge to ground, yielding the transition $V_X \rightarrow 0v$. Since the output voltage is symbolically given by

$$V_{out} = \text{NOT}\,(V_X) \tag{8. 137}$$

V_{out} rises to a value of V_{DD}. This change is fed through the inverter chain so that the source-gate voltage of MR switches to a value $V_{SGR} \rightarrow V_{DD}$ after a delay of $t_D = 3t_d$. Note that MR is in a feedback loop. The reset transistor charges C_X back up to a voltage $V_X = V_{DD}$, which then forces the transition $V_{out} \rightarrow 0v$, giving a *pulsed output* as shown. The new output voltage is fed through the inverter chain, resulting in $V_{SGR} \rightarrow 0v$, which then shuts MR off. This illustrates that

- MR is placed in a feedback to automatically recharge the internal node capacitor C_X, which resets the output to 0, and,
- The output voltage of a self-resetting gate is a voltage pulse whose width is determined by the circuit delays.

Self-resetting logic can be more difficult to use because of its sensitivity to the delay times. However, it does allow for interesting variations in the design of the circuit. For example, self-resetting circuits can be used for asynchronous logic. Also note that the power supply current used to precharge C_X is spread out in time, and not concentrated during a single portion of the precharge cycle as in standard dynamic logic. This aids "current spike" problems in large chips.

It is possible to make a domino analogy that describes a self-resetting logic chain. This is shown by the upright (precharged) dominos in Figure 8.71(a) where we have added springs between the top of the dominos and a "ceiling" to account for the self-resetting feature. Discharges are still

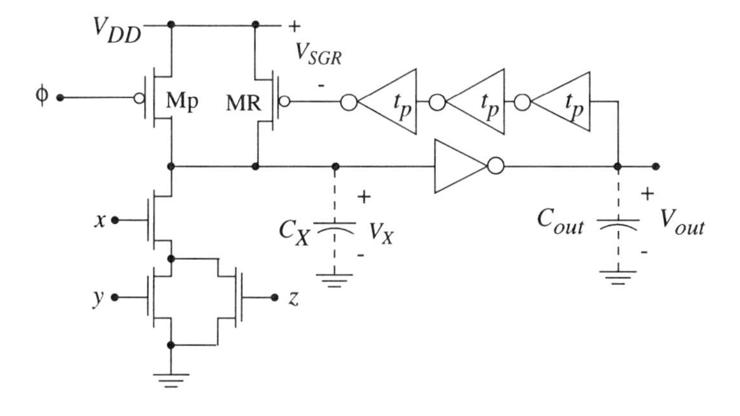

Figure 8.69 Logic design example in SR logic

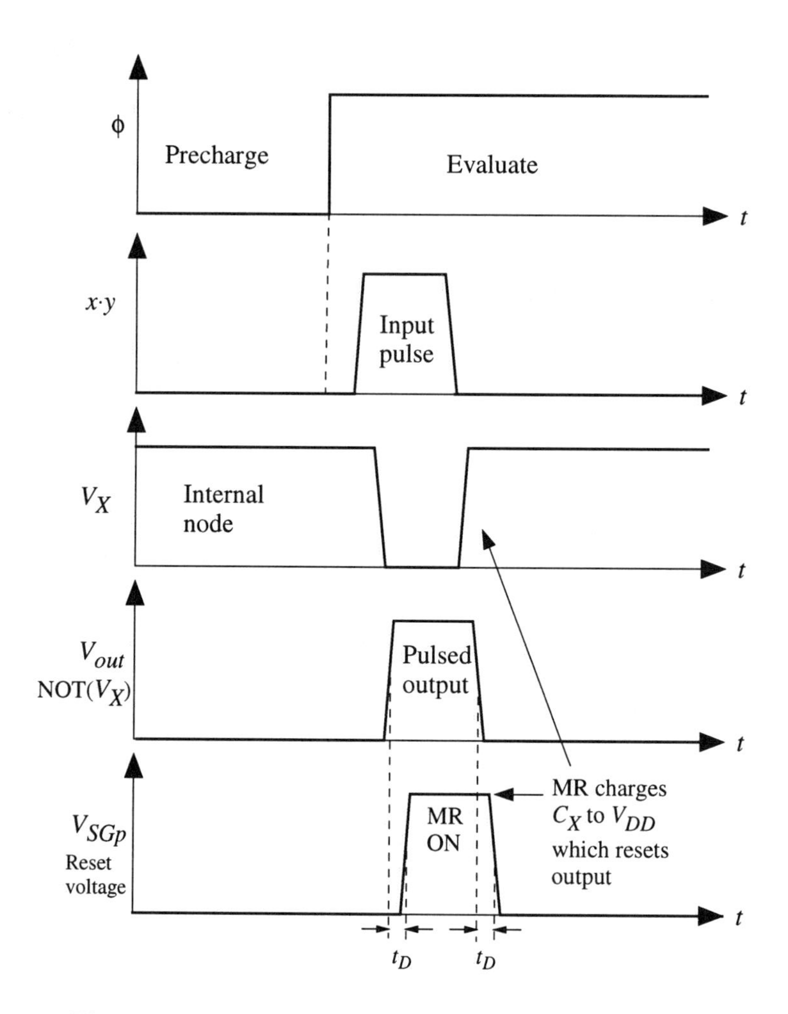

Figure 8.70 Timing diagram for self-resetting logic gate

described by falling dominos as in Figure 8.72(b). However, we can visualize the springs as stretching and then "pulling" the dominos upright again. This is the reset operation, and is portrayed in Figure 8.73(c). Note the analogy is more complicated to interpret since the output of a discharging stage is a pulse rather than a transition to a constant logic 1 voltage.

8.6 NORA Logic

No-Race (NORA) logic is another dynamic design style that was introduced to overcome **signal race** problems associated with using clocked FETs or TGs to control data flow. Consider the circuit shown in Figure 8.72. This uses oppositely clocked nFETs to control the data flow into (with ϕ), and out of (with $\overline{\phi}$), the static logic chain. The problem of signal races arises when we look at the delay times through the logic gates from the input to the output in relationship to the time interval needed to open or close the pass transistors.

Consider the clocking waveforms illustrated in Figure 8.73. The inputs are controlled by ϕ which control passFETs MnA, MnB, and MnC. The logic variables *a*, *b*, and *c* are allowed to enter the logic chain when the nFETs are conducting. In terms of voltages, this requires that the voltage be above the device threshold V_{Tn}. A signal race may occur if one or more of the inputs changes

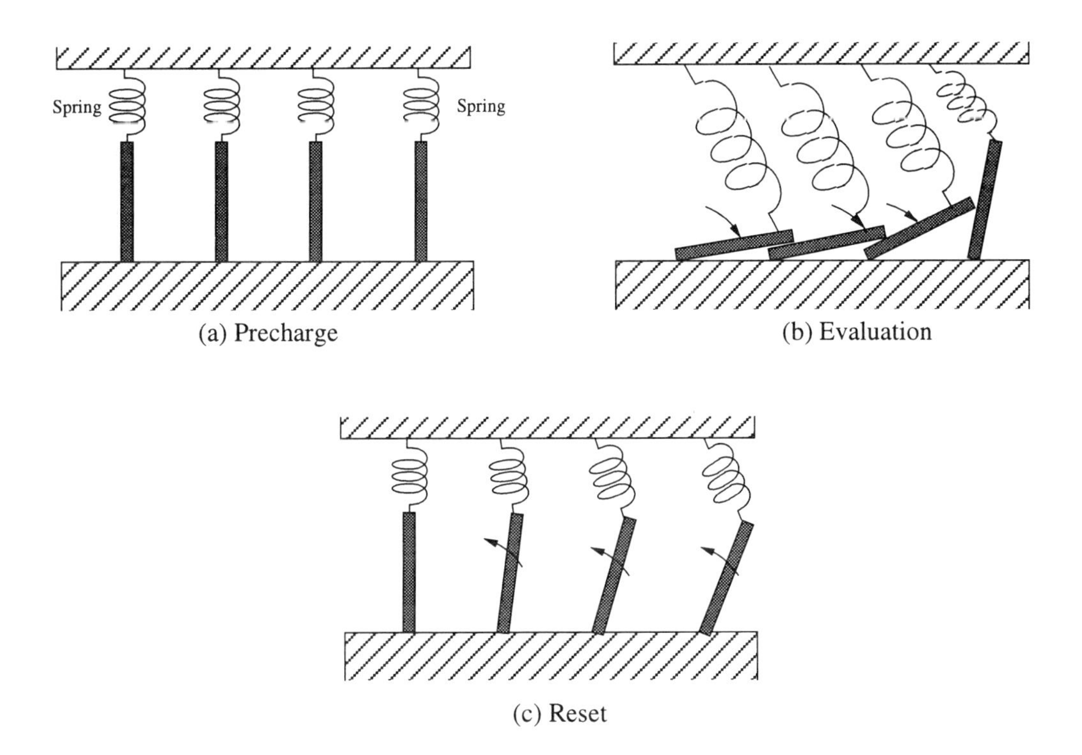

Figure 8.71 Domino analogy for self-reset logic

when the clock ϕ is falling from V_{DD} towards $0v$, starting at the time t_X indicated in the drawing, and extending a time interval t_{in}. If this occurs, then the new signal propagates through the logic chain. If the new result makes it to the output passFET in time, then the new (incorrect) value will be transferred on to the next stage. This is called a signal race as the new value is viewed as "racing" through the logic chain to "beat" the old (correct) result.

To further illustrate the problem, consider the input variable c in the original circuit. Suppose

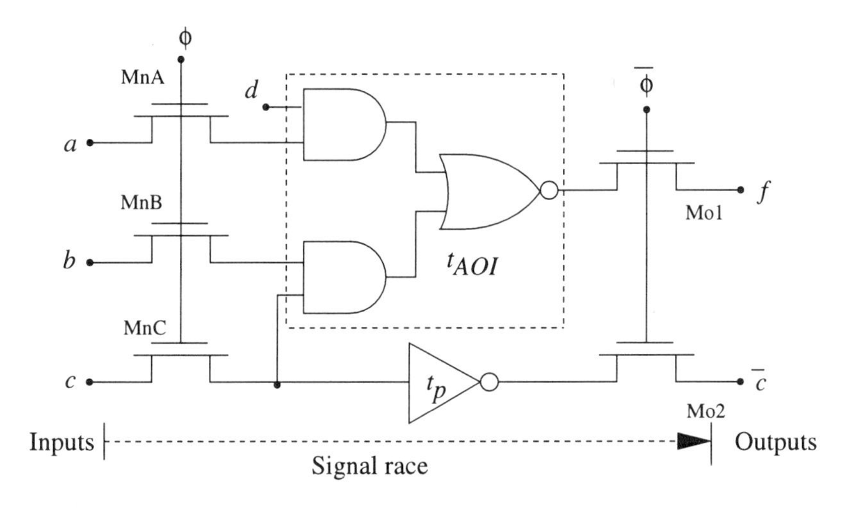

Figure 8.72 Signal flow control using pass FETs and static logic gates

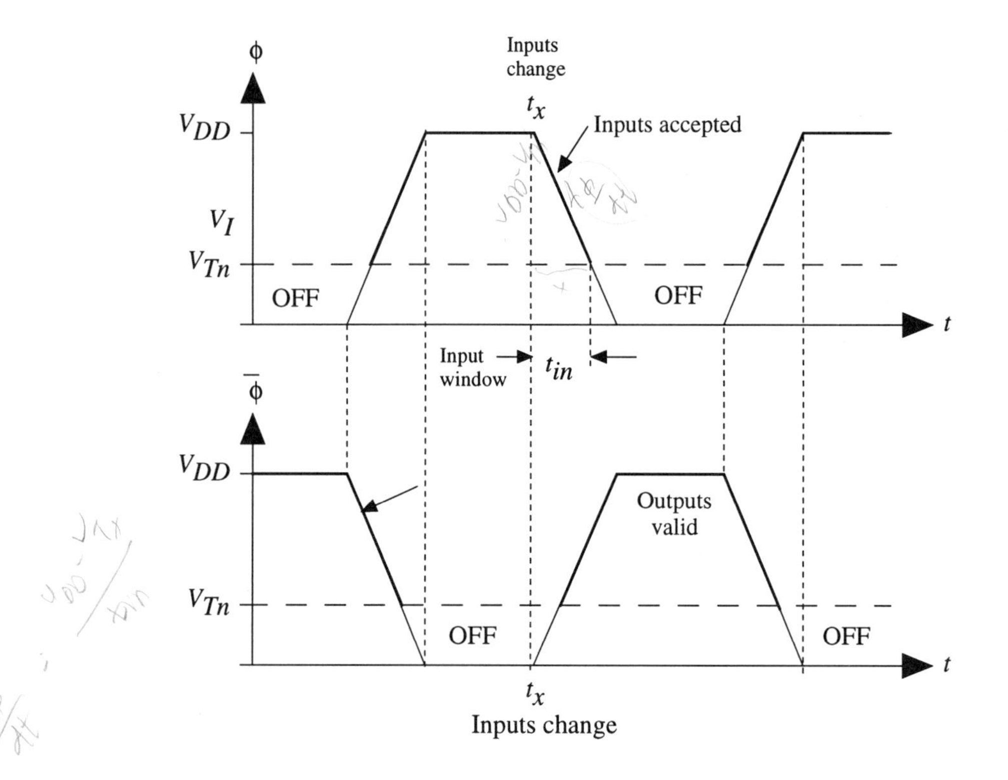

Figure 8.73 Clocking signals applied to pass FETs

that this is initially at a stable value $c = 1$, and then makes a transition $c \to 0$ at time t_X. Since MnC will pass a 0 voltage until ϕ falls to V_{Tn}, the important parameters will be the relative values of the slew rates

$$\frac{d\phi}{dt} \quad \text{and} \quad \frac{dc}{dt} \tag{8. 138}$$

that describe the rates of change of the clock and the input signal, respectively. From time t_x, the clock will turn off MnC in a time

$$t_{in} = \frac{V_{DD} - V_{Tn}}{(d\phi/dt)} \tag{8. 139}$$

If c falls low before the passFET is turned off, then it enters the logic chain as a new value. Winning the signal race then depends on the delay times through the network. These are shown as t_{AOI} through the AOI circuit, and t_p through the inverter. Since $t_p < t_{AOI}$, the NOT output $\bar{c}$ is more susceptible to the problem.

The signal race problem is complicated by the presence of **clock skew** in which the clock signals are slightly displaced from each other as illustrated in Figure 8.74. A large value of the skew time t_{skew} widens the race window, and enhances the probability of a race problem in the system. Skew is particularly troublesome in high-speed circuits as it limits the clock frequency.

NORA logic is easily understood using the background provided by the preceding sections in this chapter. The basic building block of NORA logic is the "**φ-section**" logic network illustrated in Figure 8.75. This consists of a dynamic nMOS logic state that is cascaded into a dynamic pMOS

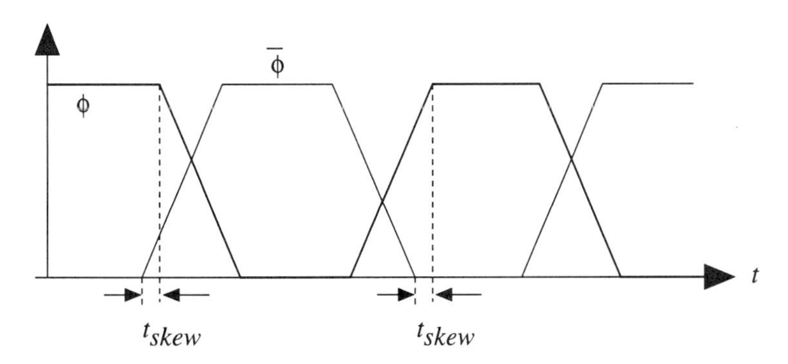

Figure 8.74 Clock skew

logic gate; a C^2MOS inverter is used as an output latch.[7] Note that optional inverters are provided at the outputs of both logic gates if, for example, one wishes to use a glitch-free domino nMOS-nMOS cascade. Moreover, the ordering of the logic gates may be reversed (i.e., pMOS to nMOS) without loss of generality.

The main features of NORA logic arise from the manner in which the clocks are applied to the logic gates and the C^2MOS latch. A clock value of ϕ=0 defines the precharge interval for the ϕ-section; the main features are shown in Figure 8.76(a). The output capacitors of the logic stages are precharged to values of

$$\begin{aligned} C_1&: \ V_1 \rightarrow V_{DD} \\ C_2&: \ V_2 \rightarrow 0v \end{aligned} \tag{8. 140}$$

for the nMOS and pMOS gates, respectively. The most important aspect of the precharge is noting that the output of the C^2MOS latch is in the Hi-Z state at this time. This means that the voltage V_{out}

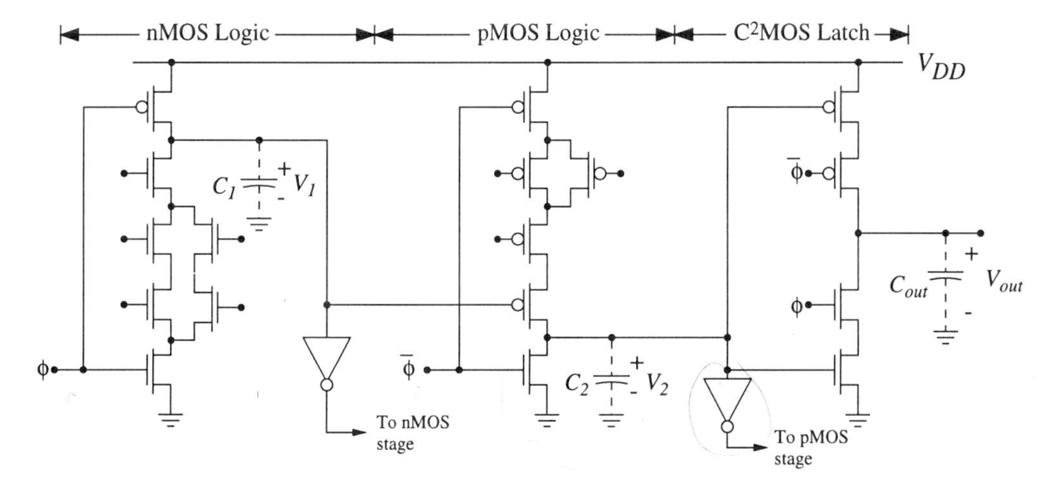

Figure 8.75 General structure of a NORA ϕ-section

[7] Clocked CMOS (C^2MOS) circuits were discussed in Section 7.6 of the previous chapter.

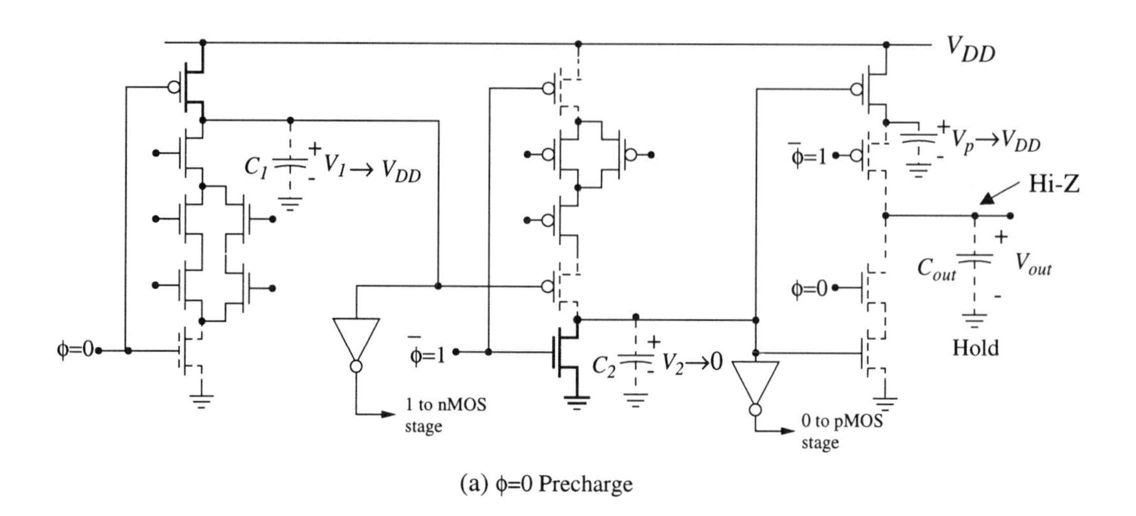

(a) ϕ=0 Precharge

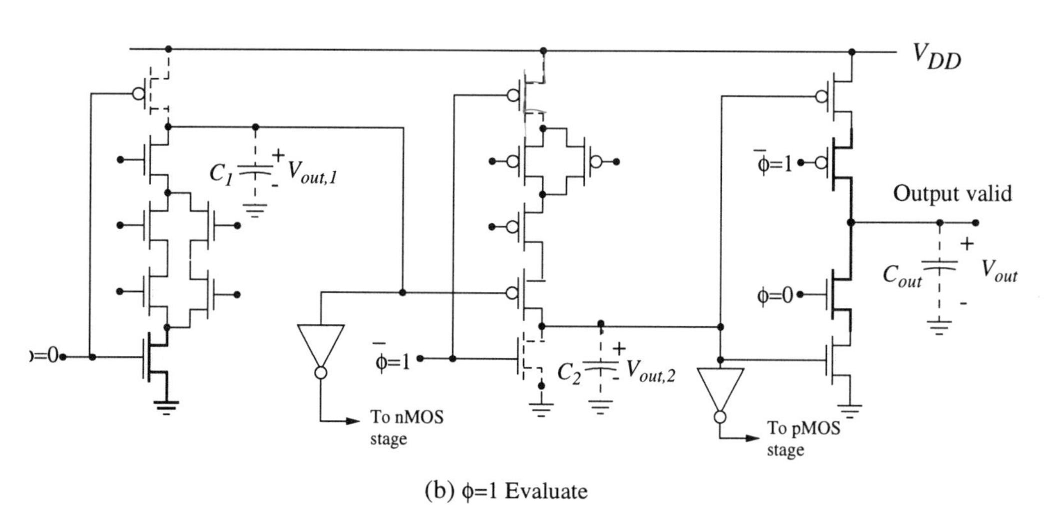

(b) ϕ=1 Evaluate

Figure 8.76 Operational modes for a NORA ϕ-section

on C_{out} is not affected by the precharge states. The actual value of V_{out} is due to charge held on the capacitor. When the clock makes a transition to ϕ=1, the entire section goes into evaluation as illustrated in Figure 8.76(b). During this time, the inputs are valid and the output result from the logic chain is given by the voltage V_{out} on the output capacitor C_{out}. This is held when the clock changes back to ϕ=0 for the next precharge event. The operation of the ϕ-section is summarized by the simplified block diagrams in Figure 8.77.

Next, we construct a NORA **$\overline{\phi}$-section** that has the general features shown in Figure 8.78. We again use a cascade of dynamic logic gates with alternating polarities (nMOS to pMOS, etc.), and provide a tri-state C^2MOS latch at the output. The only difference is that the clock phases ϕ and $\overline{\phi}$ have been reversed everywhere throughout the circuit. This means that a $\overline{\phi}$-section precharges when ϕ=1 and undergoes evaluations when ϕ=0, exactly opposite to the behavior of a ϕ-section.

The no race characteristic of the design style is obtained by creating an alternating cascade of ϕ- and $\overline{\phi}$-sections as in Figure 8.80. The timing of the two section types automatically ensures that signal races cannot occur through either section. To understand this comment, consider the operational drawings in Figure 8.81. When the clock is at a value ϕ=1, the ϕ-sections are in evaluation and the

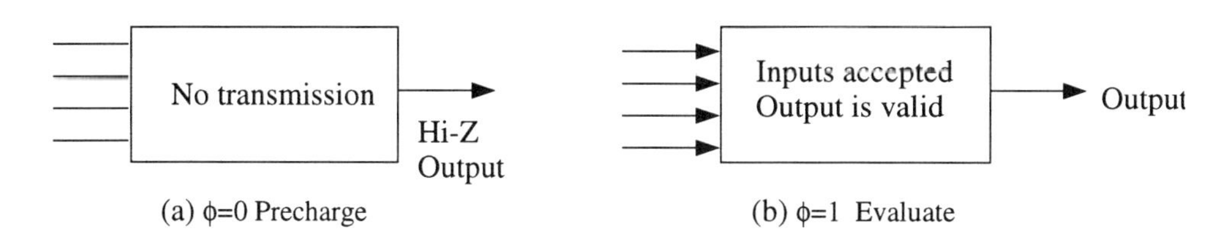

Figure 8.77 Block diagram for NORA ϕ-section behavior

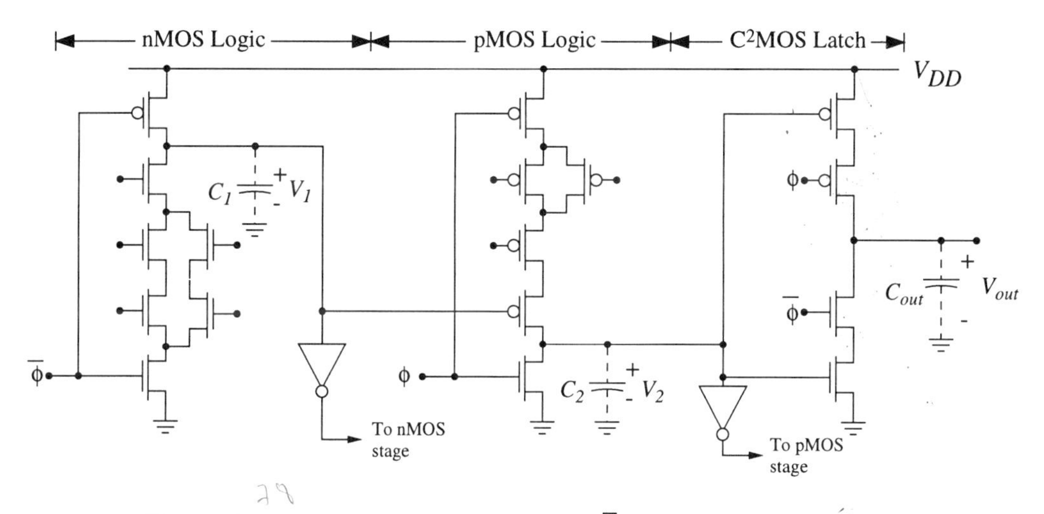

Figure 8.79 General structure of a NORA $\overline{\phi}$-section

$\overline{\phi}$-sections are undergoing precharge. This is shown in Figure 8.81(a). Consider the first logic section in the chain. During this time, the inputs are valid and yield results at the output. However, since the next logic group is a $\overline{\phi}$-section, it is in precharge with ϕ=1 and does not accept input data values. This eliminates race problems through the $\overline{\phi}$-sections. Similarly, when the clock changes to ϕ=0 as in Figure 8.81(b), the ϕ-sections are in precharge and block data transmission while the $\overline{\phi}$-sections undergo evaluation. As the clock oscillates, the sections take turns evaluating the inputs and blocking data transmission. The race-free characteristics remain even in the presence of clock

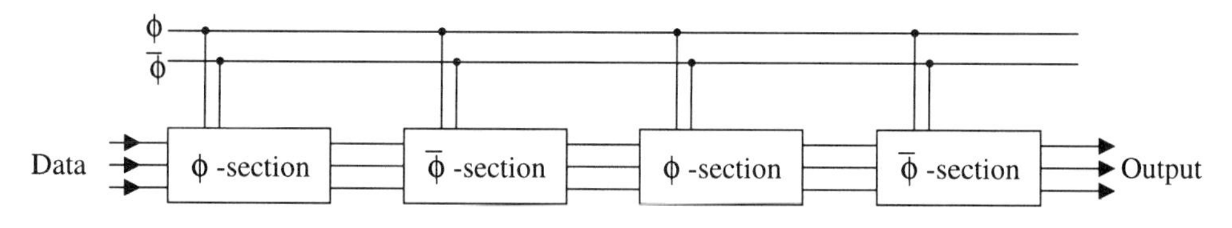

Figure 8.80 A NORA pipeline cascade

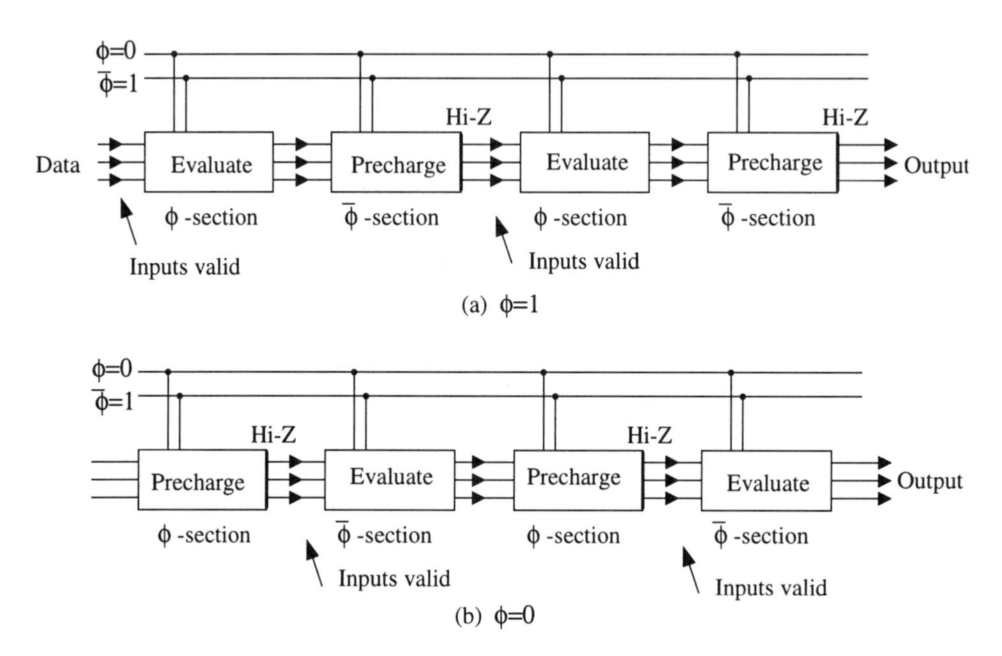

Figure 8.81 Block diagram of NORA pipeline operation

skew, and the structuring of the logic into separate ϕ and $\overline{\phi}$-sections is convenient for designing pipelined systems.

8.6.1 NORA Series-Parallel Multiplier

An interesting example of a NORA circuit is a series-parallel multiplier which accepts one word in parallel format and multiplies it by a second word that is in a serial format. Let us first develop an understanding of the algorithm, and then study the NORA circuit implementation.

Consider two 4-bit words $a = a_3\, a_2\, a_1\, a_0$ and $b = b_3\, b_2\, b_1\, b_0$. The standard binary product $a \times b$ is calculated by

$$\begin{array}{r} a_3a_2a_1a_0 \\ \times\, \underline{b_3b_2b_1b_0} \\ p_7p_6p_5p_4p_3p_2p_1p_0 \end{array} \qquad (8.\ 141)$$

where the individual terms are given by the expressions

$$\begin{aligned} p_0 &= a_0 b_0 \\ p_1 &= a_0 b_1 + a_1 b_0 + c_0 \\ p_2 &= a_0 b_2 + a_1 b_1 + a_2 b_0 + c_1 \\ p_3 &= a_0 b_3 + a_1 b_2 + a_2 b_1 + a_3 b_0 + c_2 \\ p_4 &= a_1 b_3 + a_2 b_2 + a_1 b_3 + c_3 \\ p_5 &= a_2 b_3 + a_3 b_2 + c_4 \\ p_6 &= a_3 b_3 + c_5 \\ p_7 &= c_6 \end{aligned} \tag{8. 142}$$

for each of the product bits p_i ($i = 0 - 7$). In these equations, the "+" signs indicate binary addition, and c_{n-1} is the carry bit from the p_n addition. The multiplication may be summarized by writing

$$p_i = \sum_{i = \alpha + \beta} a_\kappa b_\beta + c_{i-1} \tag{8. 143}$$

with $c_{-1} = 0$ by definition.

Let us now can construct a 4-bit series-parallel multiplier that uses an input word $a = a_3 a_2 a_1 a_0$ as a single 4-bit grouping, and multiplies it by the bits of the second word b, one bit at a time, i.e., first b_0, then b_1, and so on. As each b-bit is provided, the multiplier calculates the product

$$(a_3 a_2 a_1 a_0) \times b_i \tag{8. 144}$$

This yields the individual products

$$\begin{aligned} (a_3 a_2 a_1 a_0) \times b_0 &= \qquad\qquad\qquad (a_3 b_0,\ a_2 b_0, a_1 b_0, a_0 b_0) \\ (a_3 a_2 a_1 a_0) \times b_1 &= \qquad\qquad (a_3 b_1,\ a_2 b_1, a_1 b_1, a_0 b_1) \\ (a_3 a_2 a_1 a_0) \times b_2 &= \qquad (a_3 b_2,\ a_2 b_2, a_1 b_2, a_0 b_2) \\ (a_3 a_2 a_1 a_0) \times b_3 &= (a_3 b_3,\ a_2 b_3, a_1 b_3, a_0 b_3) \end{aligned} \tag{8. 145}$$

which are all the terms required in the product. In fact, the above equation has been written so that adding each column gives the product terms in order from right (p_0) to left (p_7).

To implement the circuitry for this circuit, we first recall that multiplication of two bits $a_n \times b_m$ is equivalent to the AND operation, making it easy to provide the required products. To add the product terms, we will use the bit-serial adder shown in Figure 8.82. This clocked circuit takes the inputs $x_n(t)$ and$y_n(t)$ and calculates outputs of

$$\begin{aligned} s_n(t) &= x_n(t) + y_n(t) \\ c_{out}(t) &= x_n(t) \cdot y_n(t) + c_n(t)\,[x_n(t) \oplus y_n(t)] \end{aligned} \tag{8. 146}$$

where "+" means "plus" in the sum bit s_n. Note that the c_{out} equation uses the carry-in bit $c_n(t)$ that exists at the time t. The carry-out bit $c_{out}(t)$ is sent to a clocked Delay Unit Δ where it is held for one clock period, and output as

$$c_n(t + T) = c_{out}(t) \tag{8. 147}$$

with T is the clock period: This is just a D-type flip-flop. The new carry-in bit $c_n(t + T)$ is used during the next clock cycle to compute

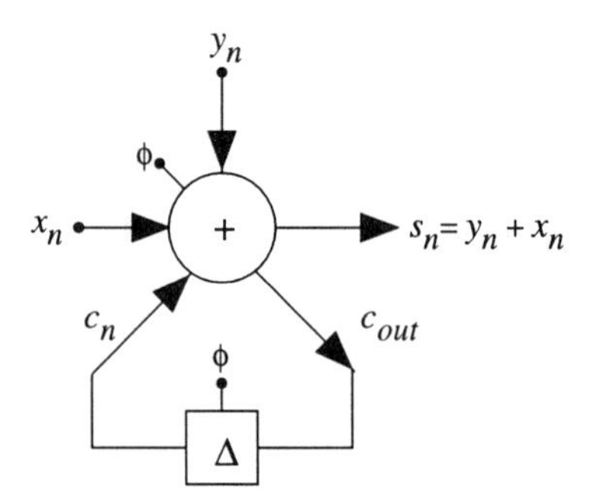

Figure 8.82 Bit serial adder

$$\begin{aligned} s_n(t+T) &= x_n(t+T) + y_n(t+T) \\ c_{out}(t+T) &= x_n(t+T) \cdot y_n(t+T) + c_n(t+T)\,[x_n(t+T) \oplus y_n(t+T)] \end{aligned} \tag{8. 148}$$

and so on. The bit-serial NORA adder circuit is shown in Figure 8.83 using a pMOS-nMOS-C^2MOS cascade and AOI logic structuring. The delay unit is created by cascading two oppositely phased C^2MOS latches that collectively delay the $\bar{c}_{out}$ bit by one clock period. The static inverter is added to provide c_{out} as an input during the next clock cycle. Note that a control signal denoted as "Clear c " allows us to initialize the delay unit at a starting value of $c_{out} = 0$.

This NORA series-parallel multiplier network can be constructed by using the general architecture illustrated in Figure 8.84. In this scheme, the entire word $a = a_3\, a_2\, a_1\, a_0$ is available at the same time, while b is sent to the circuit in a serial manner. AND2 gates are used to calculate the bit products $a_n \times b_m$, which are then sent to the NORA serial adder units. The first serial input bit is b_0. This results in the first output term

$$p_0(t) = a_0 b_0 \tag{8. 149}$$

since that is immediately available at the output of the right-most adder. The other AND2 gates calculate $a_3 \cdot b_0$, $a_2 \cdot b_0$, and $a_1 \cdot b_0$ left to right, respectively. The next input bit is b_1 during the next clock cycle. This gives the second output as

$$p_1(t+T) = a_0 b_1 + a_1 b_0 + c_0 \tag{8. 150}$$

where the first term is given by the right-most AND2 gate, second term was calculated during the first input bit time, and the carry bit c_0 is delayed from the first bit time. This cycle is repeated until all of the product bits are calculated. It is noted in passing that the same NORA circuits can be used to create a pipelined adder.

8.7 Single-Phase Logic

The CMOS families present up to this point are classified as **single clock, dual phase circuits**. This is because a single clock ϕ is the basis, but both ϕ and $\bar{\phi}$ have been used for timing the circuits (hence the name dual-phase). In contrast, **single-phase logic** gates that only require one clock signal ϕ can be created by adding additional circuitry to perform the required functions. These are of interest because the use of a single clock phase simplifies the clock distribution and associated interconnect requirements. Also, the clock generation problem is simplified. These are other features can lead to higher speeds in carefully designed circuits.

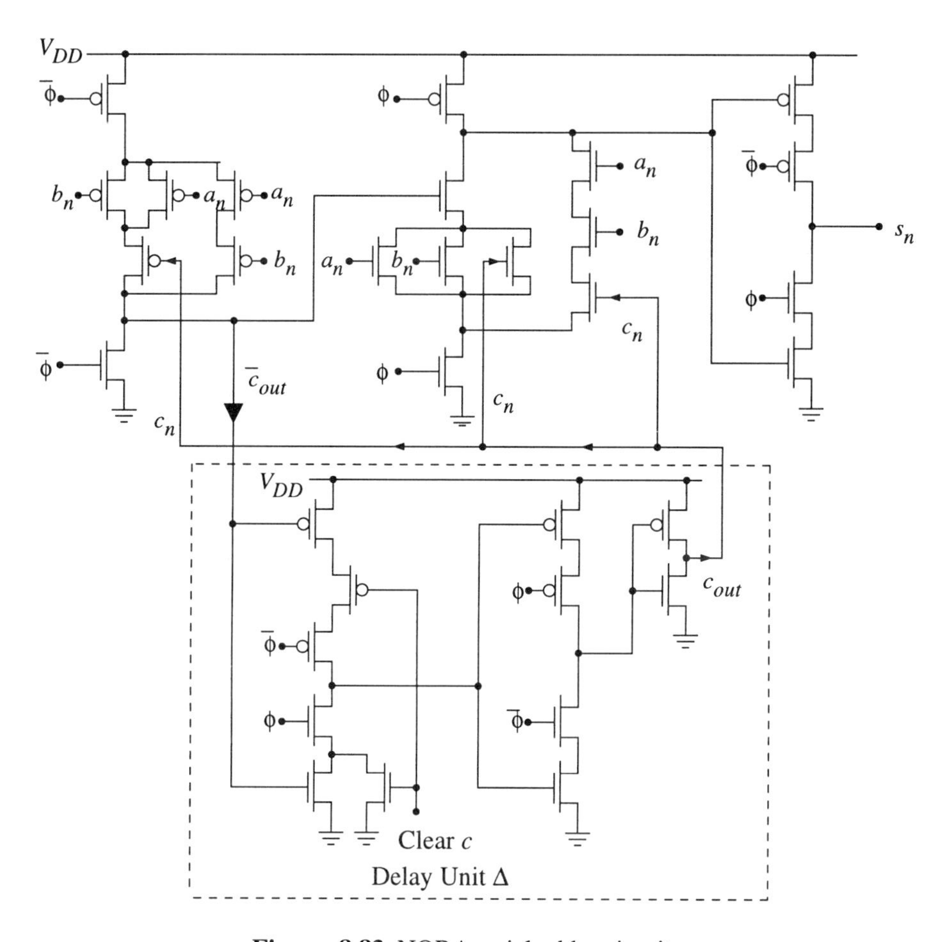

Figure 8.83 NORA serial adder circuit

Single-phase logic networks can be described using the simple block diagrams in Figure 8.85. In (a), two SP latches with opposite characteristics are employed. As the clock oscillates between 0 and 1 values, one latch is transparent while the other is opaque with a Hi-Z (tri-state) output. The optional logic block inserted between the two latches provides the logic operation. In Figure 8.85(b), the concept is extended to single-phase logic gates and cascades. Given a value of the clock signal ϕ, one circuit accepts inputs and provides logical outputs while the other is opaque. In both examples, the opaque condition is most easily achieved by applying the clock ϕ to either a single nFET or a single pFET, not to a complementary pair.

Let us consider the problem of constructing single-phases latches, as these can be used as the basis for general logic circuits. One approach is to use the "doubled-C^2MOS" circuits shown in Figure 8.86 for nMOS (a) and pMOS (b). Although these are similar to the C^2MOS delay latch used in the series-parallel multiplier of the previous section, the entire circuit is controlled by a single clock phase ϕ. The operation is easily understood by analyzing a circuit for the two clock values. Consider the nFET latch in Figure 8.86(a). When ϕ=1, both clock FETs are active, and the circuit is logically identical to two cascaded inverters. The output Q assumes the value of D after the transient delays. When the clock switches to ϕ=0, both clock nFETs are in cutoff, and the latch is opaque. The output Q is dynamically held on the output capacitance, subject, of course, to variations due to charge leakage. The pFET latch is identical in operation except that it is oppositely

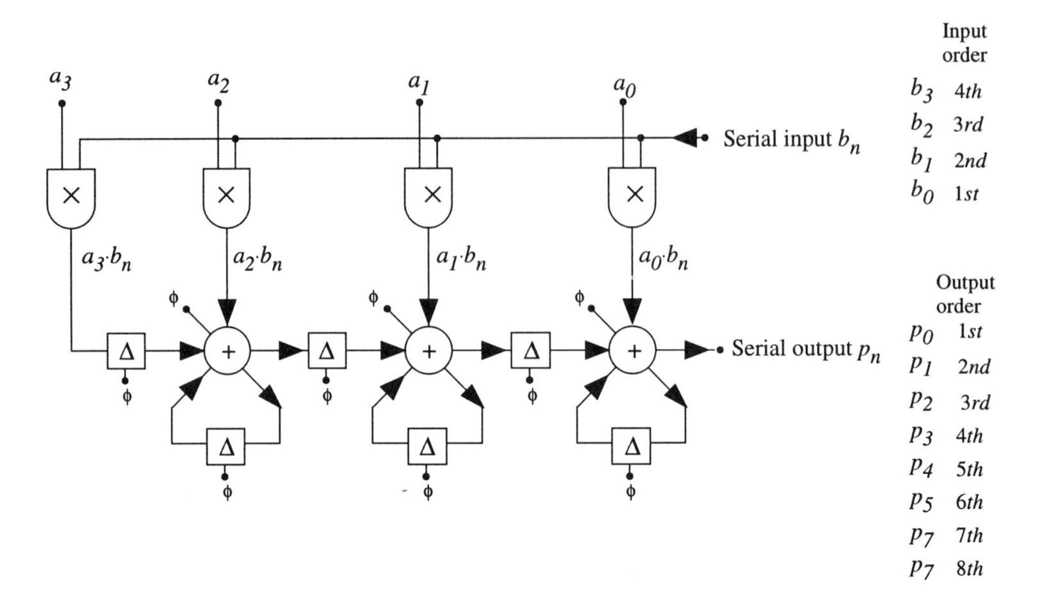

Figure 8.84 A series-parallel multiplier network

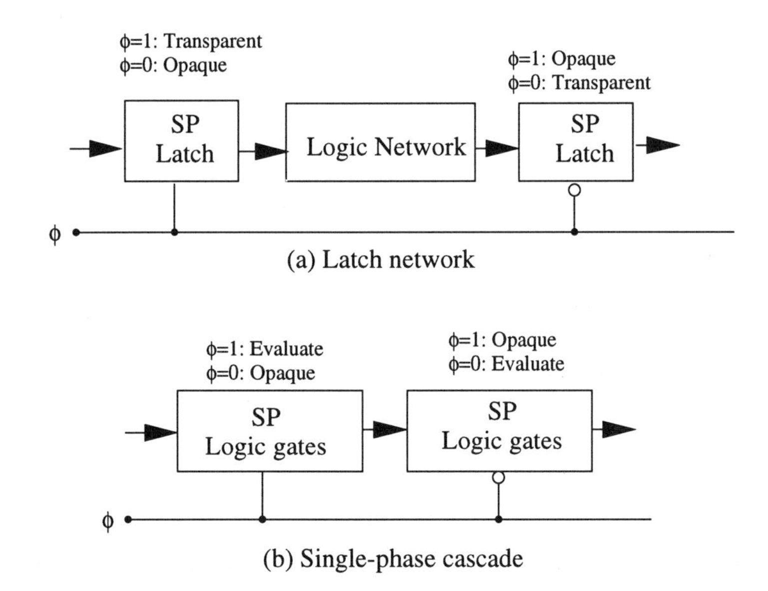

Figure 8.85 General behavior of single-phase circuits

phased. This is accomplished by using ϕ applied to only pFETs, eliminating the need for the opposite phase $\overline{\phi}$. A direct variation of this are the **split-output** latches shown in Figure 8.87. These circuits use separate outputs from the first stage to control the logic devices in the second stage circuit, eliminating the need for a clocked FET there. The design reduces the load on the clock driver circuitry, but has the drawback of passing both 0 and 1 voltages through a single FET to drive the output stage. This results in reduced drive capacity of the output circuit. For example, the output nFET

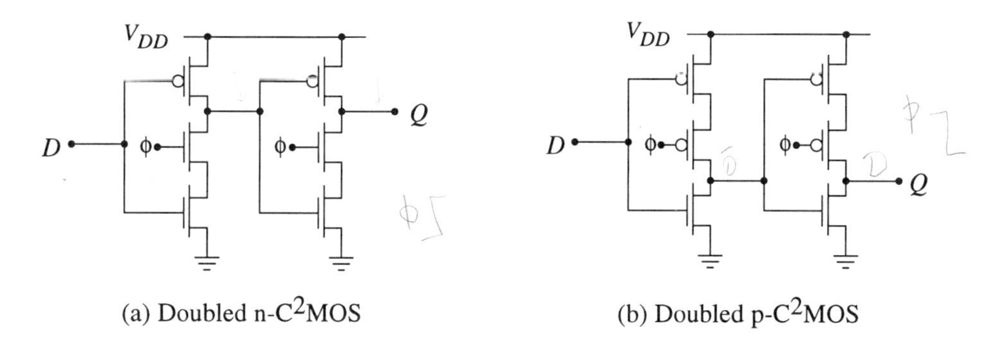

(a) Doubled n-C^2MOS (b) Doubled p-C^2MOS

Figure 8.86 Single-phase "doubled-n and -p" latches

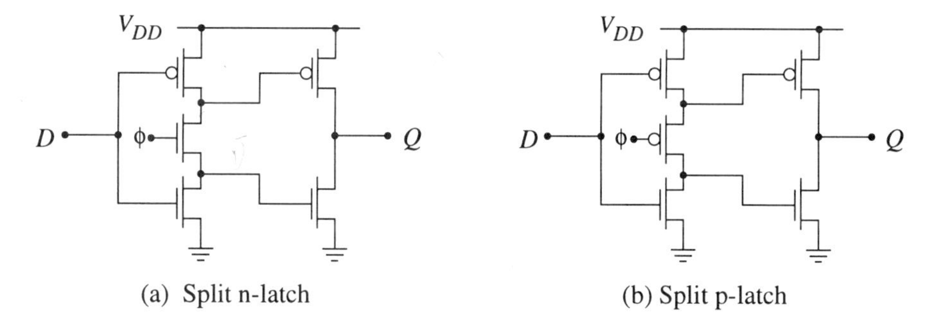

(a) Split n-latch (b) Split p-latch

Figure 8.87 Split single-phase latch circuits

in Figure 8.87(a) has a maximum gate voltage of (V_{DD} - V_{Tn}) due to the presence of the clocking nFET. Similarly, the output pFET in Figure 8.87(b) has a maximum source-gate voltage of (V_{DD} - $|V_{Tp}|$). In both cases, the current flow in the output transistors will be lower than "normal" gate bias ranges would produce.

Figure 8.88 shows a different scheme for creating what is termed **TSPC (True Single-Phase Clock)** logic that uses the concepts of precharge and evaluate intervals. The circuit in Figure

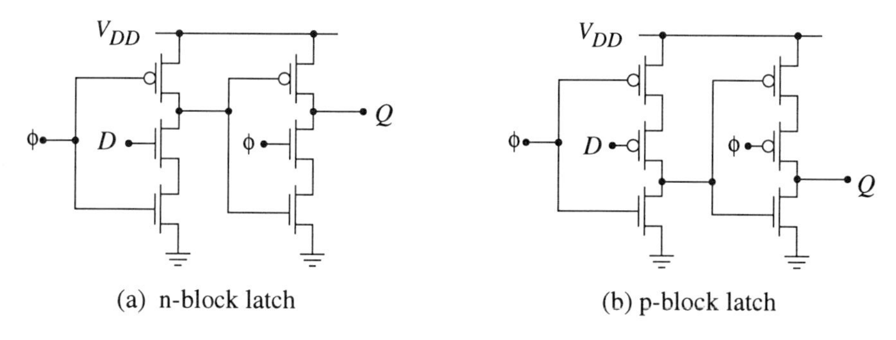

(a) n-block latch (b) p-block latch

Figure 8.88 TSPC latch circuits

8.88(a) accepts the input D using an nFET, and also contains a clock-controlled nFET in the second stage. The p-block circuit shown in Figure 8.88(b) is exactly opposite in that it uses pFETs for both the input and clocking signals.

The operation of the TSPC latches can be understood by examining the state of the circuit for each clock interval. We will choose the n-block latch for this purpose. The opaque state of this circuit occurs when ϕ=0 and the resulting operation can be seen with the aid of Figure 8.89(a). Mp1 is active with the applied clock signal and precharges C_I to a voltage $V_I = V_{DD}$; this drives Mp2 into cutoff and isolates the output node from the power supply. At the same time, clock FET M2 is in cutoff, so that the output is in a Hi-Z state. When the clock changes to ϕ=1, the second stage of circuit is transparent and the input voltage V_{in} applied to M1 determines the operation; this is shown in Figure 8.89(b). If V_{in} =0v, M1 is in cutoff and V_I is dynamically held at V_{DD}. This biases Mn2 active and Mp2 into cutoff. Since M2 is ON, the output capacitor discharges to a voltage of V_{DD} = 0v. Conversely, if V_{in} =V_{DD}, V_I discharges to 0v which turns on Mp2. Since Mn2 is OFF, the output capacitor is charged to a high voltage $V_{out} = V_{DD}$. This shows that the circuit operates as a basic latch that is opaque when ϕ=0 and transparent with ϕ=1. The p-block circuit in Figure 8.88(b) is exactly opposite: it is opaque when ϕ=1 and transparent with ϕ=0.

The TSPC latch can be used as a basis for designing more complex logic gates. This is accomplished by replacing the single input transistor by a FET logic array. Figure 8.90(a) illustrates the general structure of a logic gate using nFETs, while Figure 8.90(b) provides the analogous circuit for pFET-based logic. Both gates behave in the same manner as the latching circuits, alternating between opaque and transparent modes as the clock signal changes.

TSPC logic gates can be used to create NORA-like structured logic cascades that require only a single clock phase. As an example, let us construct a TSPC serial adder circuit by starting with the AOI-structured equations

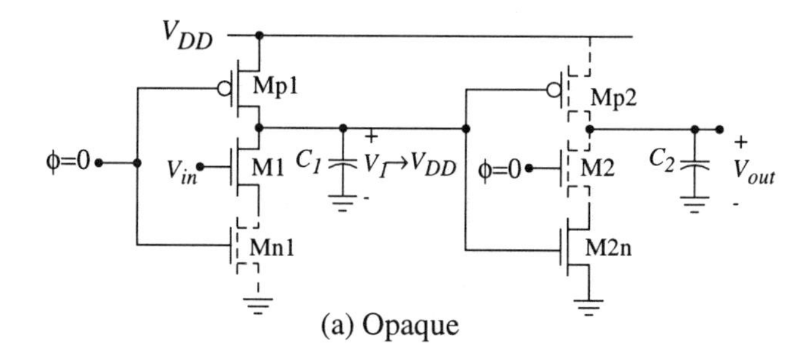

(a) Opaque

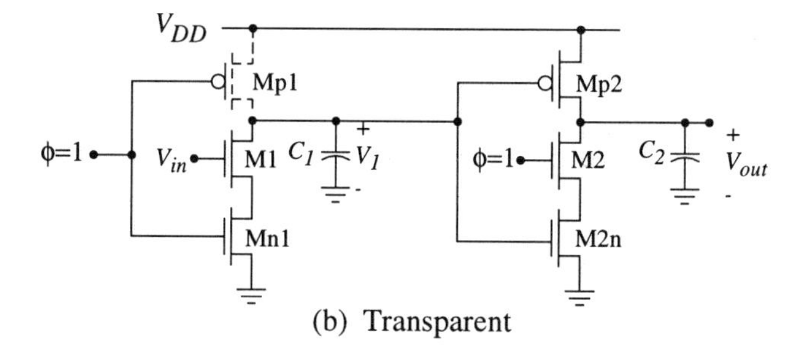

(b) Transparent

Figure 8.89 Operation of a TSPC latch

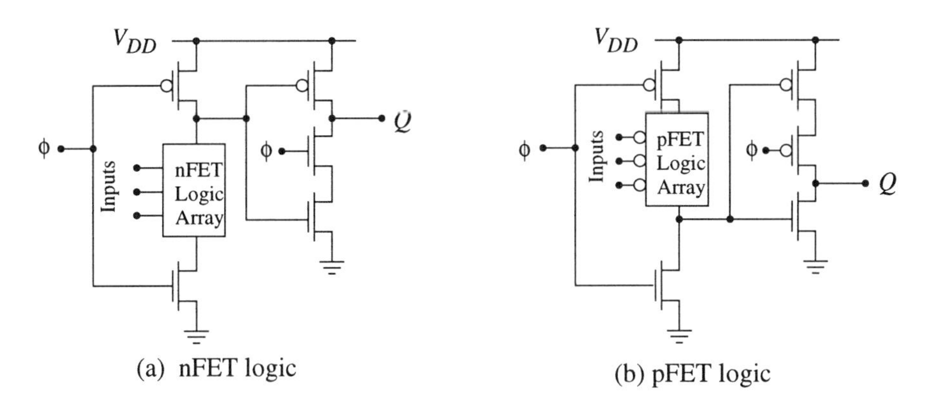

Figure 8.90 TSPC logic gates

$$
\begin{aligned}
c_{n+1} &= a_n \cdot b_n + c_n \cdot (a_n + b_n) \\
s_n &= \overline{c_{n+1}} \cdot (a_n + b_n + c_n) + a_n \cdot b_n \cdot c_n \\
&= \overline{a_n \cdot b_n + c_n \cdot (a_n + b_n)} \cdot (a_n + b_n + c_n) + a_n \cdot b_n \cdot c_n
\end{aligned}
\tag{8.151}
$$

for the sum s_n and carry-out c_{n+1}, respectively. A circuit that directly implements these equations is shown in Figure 8.91. The input stage is a static circuit based on mirrored logic arrays that provide the carry-out bit $\overline{c}_{n+1}$. This is used as input into a TSPC circuit that calculates the sum bit s_n. The

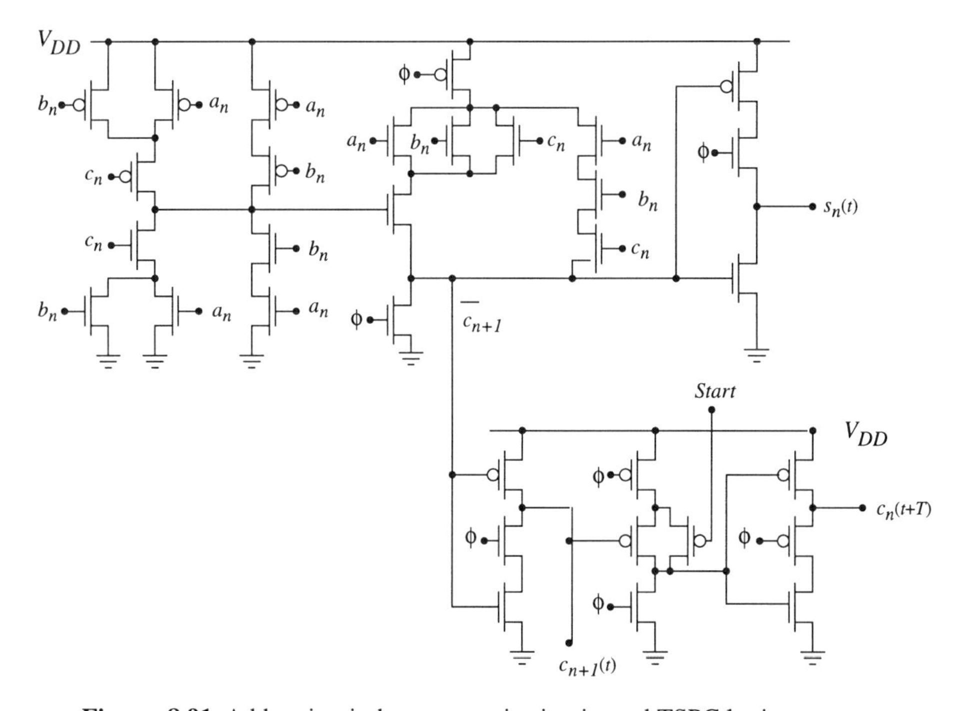

Figure 8.91 Adder circuit that uses static circuits and TSPC logic

lower circuit is used to delay the carry-out bit by one clock cycle; this provides the carry-in bit c_n at time $(t+T)$ for the next calculation in the serial network.

Another approach to TSPC logic is base on the latching circuits shown in Figure 8.92. This variation uses the same number of FETs, but is designed to provide a common clocked FET for both sections of the circuit. As with the other design style, it is possible to use nFET inputs [Figure 8.92(a)] or pFET inputs [Figure 8.92(b)] to create circuits that have opposite clocking characteristics.

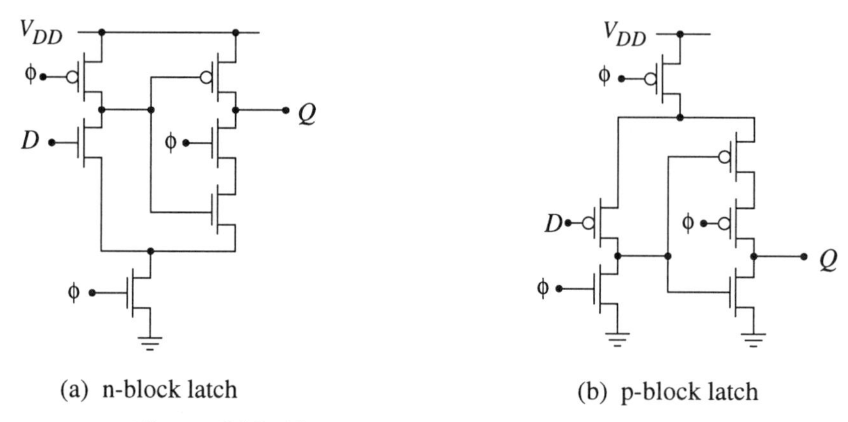

Figure 8.92 Alternate TSPC latch circuits

The operation of the nFET-based circuit is summarized in Figure 8.93. With a clock of $\phi=0$, capacitor C_1 is charged to a voltage $V_1 = V_{DD}$, which places the output in a high-impedance state; this can be verified by the drawing in Figure 8.93(a). The value of V_{out} across C_{out} is held at the previous value. A clock state of $\phi=1$ drives the circuit into logic operation, where it accepts the input voltage. As shown in Figure 8.93(b), the value of D determines the operation of the second stage latch, which in turn establishes the value of V_{out} that will be held when the clock returns to $\phi=0$ during the next cycle. As with most design styles, the single-input circuits provide the basis for more complex logic gates that use nFET and pFET logic arrays. The general circuits for this TSPC style are shown in Figure 8.94. Both arrays employ the logic design rules developed in Chapter 5 for static CMOS gates.

An example of a practical application for this variation is the serial adder drawn in Figure 8.95. This implements the FA algorithm by first defining

$$e_n = a_n \cdot \bar{c}_n + \bar{a}_n \cdot c_n \tag{8. 152}$$

and then using this quantity to compute the sum and carry-out by

$$\begin{aligned} s_n &= e_n \cdot \bar{b}_n + \bar{e}_n \cdot b_n \\ c_{n+1} &= e_n \cdot b_n + \bar{e}_n \cdot \bar{c}_n \end{aligned} \tag{8. 153}$$

This approach interfaces well with the timing since e_n can be computed during the first half-clock cycle, and the sum and carry in the following portion of the clock cycle. A major part of the latency in this circuit occurs from the use of pFET logic arrays. A variation that is based on nFET logic is shown in Figure 8.96. The general features remain the same, but the timing has been changed so that the start-up requires one full clock cycle.

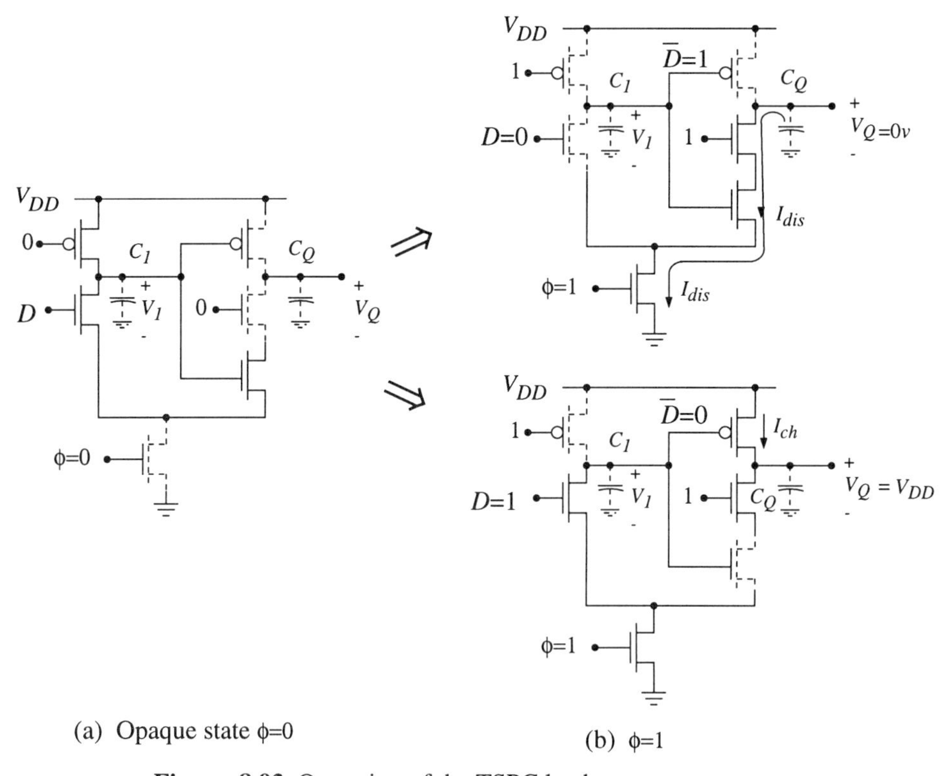

(a) Opaque state φ=0 (b) φ=1

Figure 8.93 Operation of the TSPC latch

The TSPC logic circuits can be used as a basis for other circuit design styles. One such variation is based on the latches shown in Figure 8.97. An **active-high** latch [see (a)] is designed to provide transparent operation when the clock is at a value φ=1; a clock condition φ=0 defines an opaque circuit. Conversely, the **active-low** latch shown in Figure 8.97(b) is transparent when φ=0 and opaque with φ=1. Both circuits are non-inverting as shown, but NOT-latches can be created by simply adding a static inverter at the output; this does not affect the dynamic operation of the circuit. A single-phase pipelined system can designed by using embedding logic gates between alternating latch

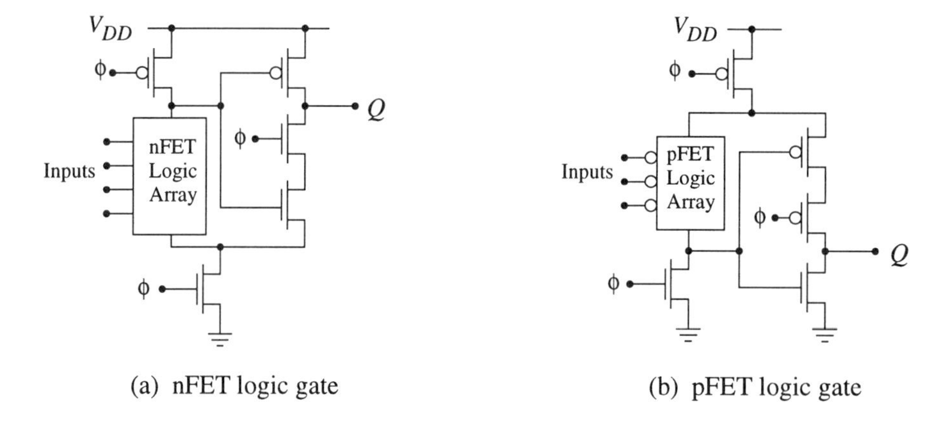

(a) nFET logic gate (b) pFET logic gate

Figure 8.94 Circuits for alternate TSPC latch design

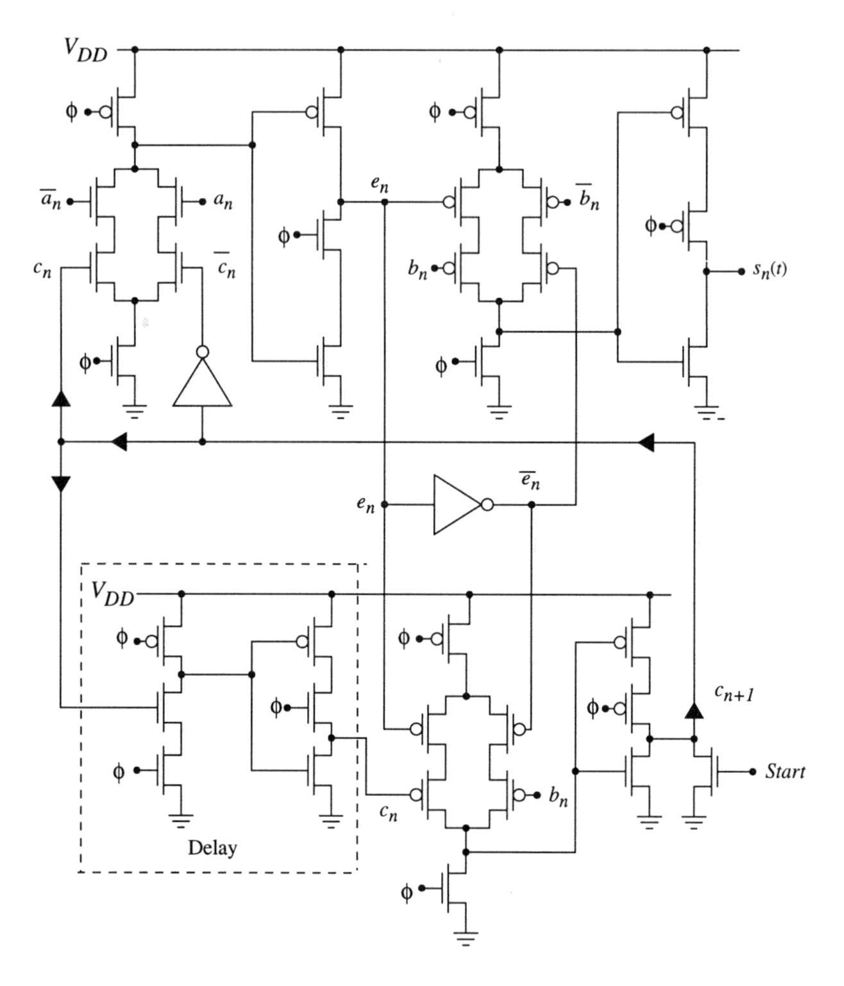

Figure 8.95 Serial adder circuit in alternate TSPC design style

types. Or, as we will see below, it is possible to create clock-sensitive logic gates.

This type of single-phase circuit can be understood by analyzing the circuit for both clock values.The latch is transparent with ϕ=1 and both clock transistors Mc1 and Mc2 are active. The operation of the circuit is shown in Figure 8.98 for $D = 0$ (upper) and $D = 1$ (lower); both result in $Q = D$ at the output. When the clock makes a transition to $\phi \rightarrow 0$, the latch is driven into the opaque state. These are shown on the right side of the drawing for both initial input values. Note that both clock FETs Mc1 and Mc2 are in cutoff. For the first case where D is initially low, the value of V_X starts out at V_{DD}. If D increases, then Mp1 is biased into cutoff, and V_X is held at a high level; the charge keeper transistor MK helps maintain the voltage on the node. This drives Mp2 into cutoff, holding the output at $Q = 0v$. In the opposite case (shown in the lower drawings) D is initially high when $\phi \rightarrow 0$, V_X is at $0v$ and the output Q is maintained high by conduction through Mp2. If D falls low, Mp1 turns on, causing V_X to rise to V_{DD}; this in turn drives Mp2 into cutoff, isolating Q from the circuit gives a Hi-Z state.

The operation of the active-low latch is summarized in Figure 8.99. For this circuit, a clock of ϕ=0 defines the transparent state as in drawing (a). Since both clocking pFETs Mc1 and Mc2 are active, the input D determines whether Mp1 or Mn1 is conducting, thus establishing the voltage at node X as either high or low. This voltage is applied to Mp2 and Mn2 which are connected as an

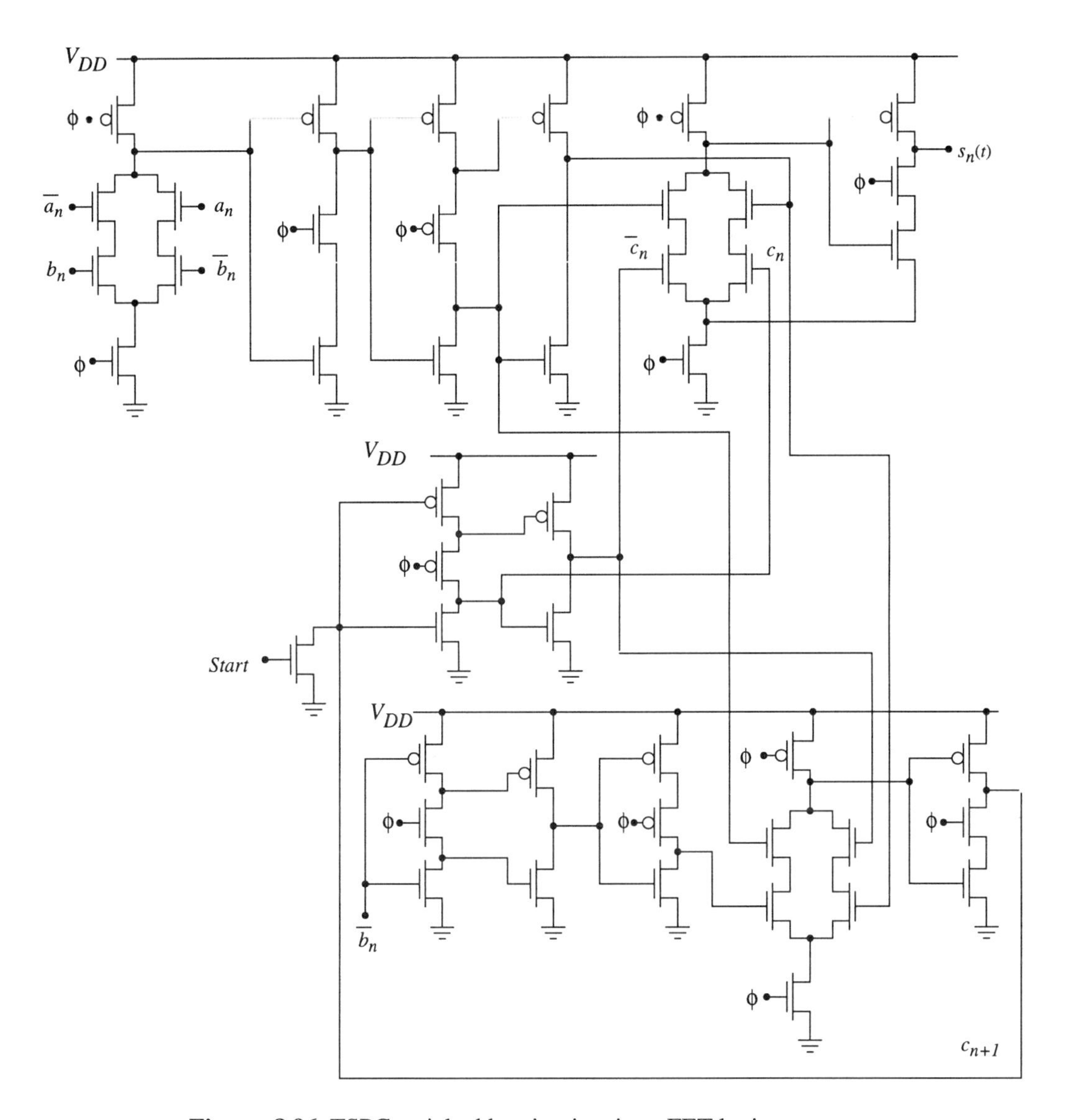

Figure 8.96 TSPC serial adder circuit using nFET logic

inverter, giving $Q = D$ during this time interval. If D is high, then X is low and the output is $Q = 1$; if $D = 0$, then X is high and the output is $Q = 0$. When ϕ rises to a 1, the circuit is opaque and latches onto the previous value. The main features are shown in Figure 8.99(b). Suppose that D was high when the clock changed. If D falls towards 0, Mn1 is biased off, but the feedback transistor Mn3 is active and holds $X = 0$. This maintains the value $Q = 1$ using charge storage at the output. Conversely, if $D = 0$ when the clock changes then node X is high which turns on Mn2. If D rises towards a 1 value, Mn1 goes active, which in turn drives Mn2 into cutoff, isolating the output. This keeps $Q = 0$ on the tri-stated output node.

The general circuit structures may be modified to provide additional logic functions. Figure 8.100 shows AND2 latches for both active-high (a) and active-low (b) clocking. In both circuits, A and B drive complementary nFET/pFET pairs. Careful examination will reveal that the elements of a static NAND2 gate are embedded in the circuitry. The additional transistors are used to insure that

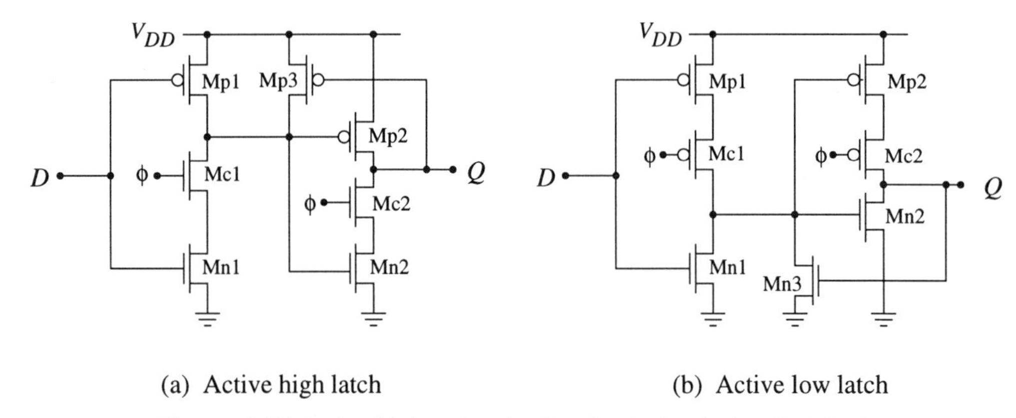

(a) Active high latch (b) Active low latch

Figure 8.97 Active high and active low latch circuits in TSPC logic

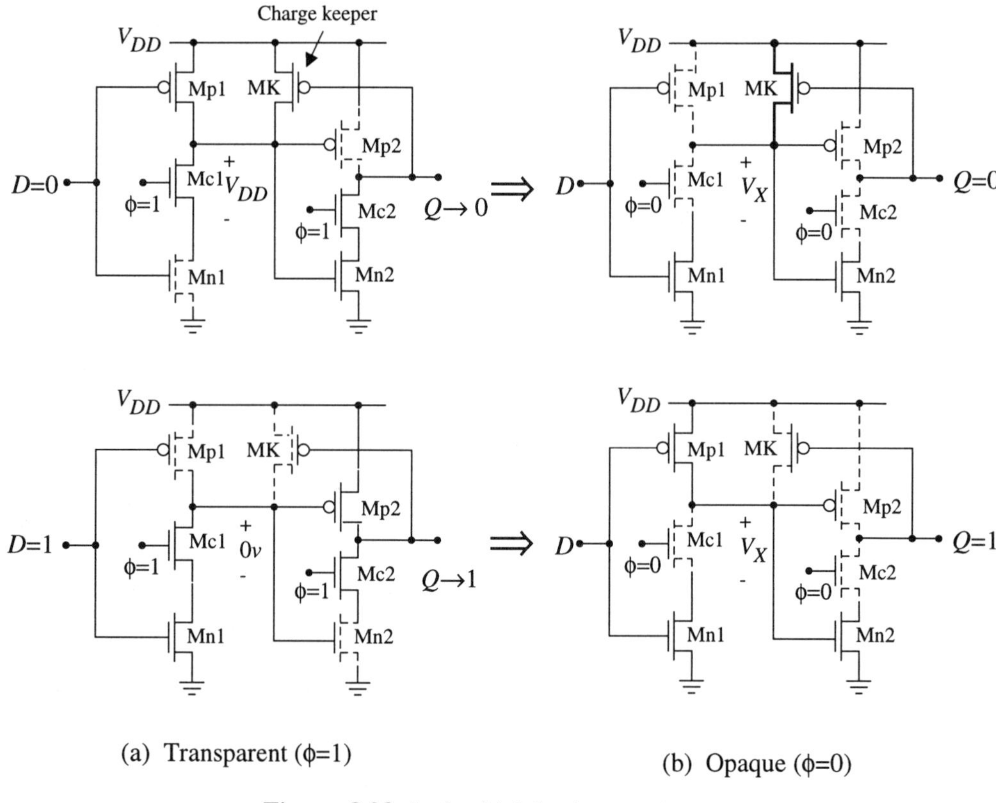

(a) Transparent (ϕ=1) (b) Opaque (ϕ=0)

Figure 8.98 Active high latch operation

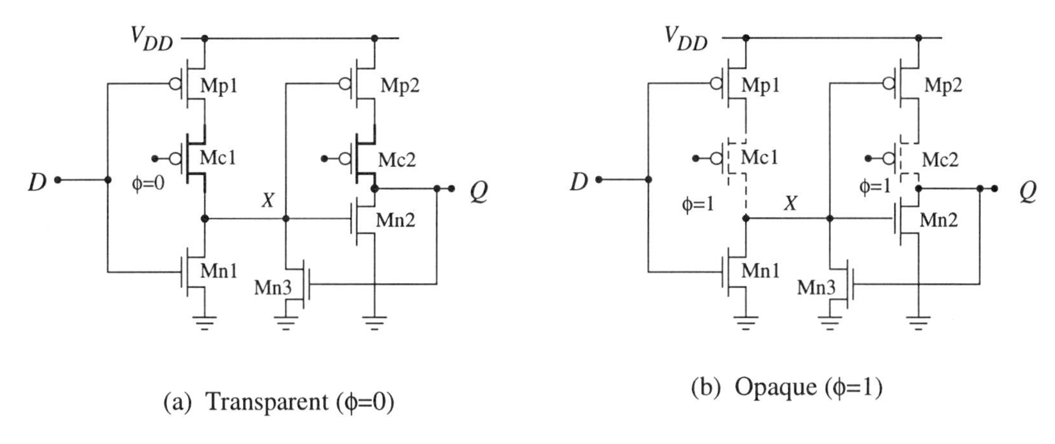

(a) Transparent (ϕ=0)

(b) Opaque (ϕ=1)

Figure 8.99 Active low latch operation

the node voltages can be maintained when the latches are driven into the opaque condition. The NAND functions can be obtained by simply adding a static inverter to the output. The OR2 latches in Figure 8.101 are created in a similar manner using static NOR2 gates as a basis. In both the AND and the OR gates, the primary difference between an active-high latch and an active-low latch is the placement of the clock ϕ: active-high circuits use clocked nFETs while active-low circuits employ clocked pFETs. Logic design symbols for the gates are shown in Figure 8.102. Again, the important characteristic is that an active-high gate is transparent with ϕ=1 and opaque with ϕ=0, while an active-low gate is exactly opposite. This allows us to create the single-phase logic cascades discussed at the beginning of this section.

As our final example of single-phase logic styles, let us examine what are termed **All-NFET Logic** (**ANL**) dynamic circuits. The main idea is to develop a set of single-phase stages that do not employ any pFET logic transistors, and use these to build race-free pipelines. Figure 8.103 shows the general form of what is termed an "N1-type" gate. This circuit is created by cascading the output of a dynamic nMOS P/E gate into a clocked latching circuit. Applying the clock ϕ to Mp1 and Mn1 defines the precharge (ϕ=0) and evaluate (ϕ=1) intervals for the logic. The output latch is syn-

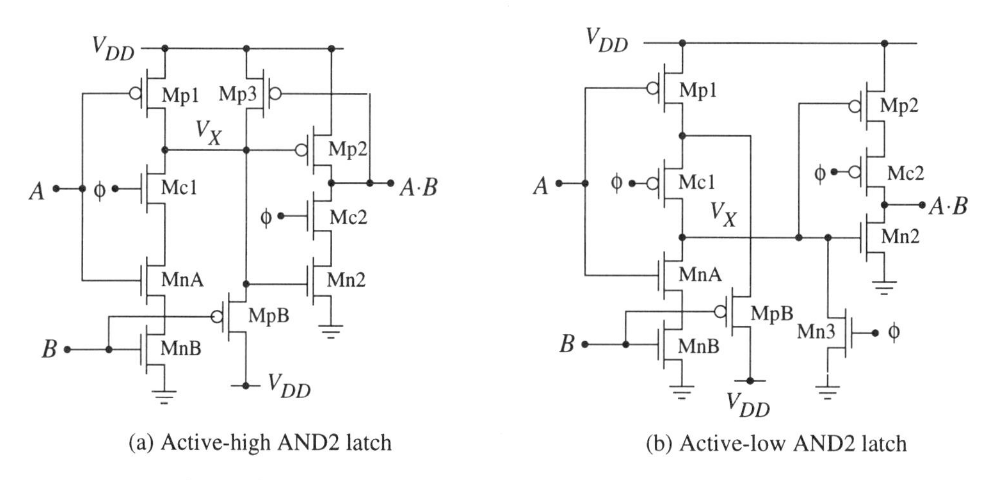

(a) Active-high AND2 latch

(b) Active-low AND2 latch

Figure 8.100 AND2-latching circuits

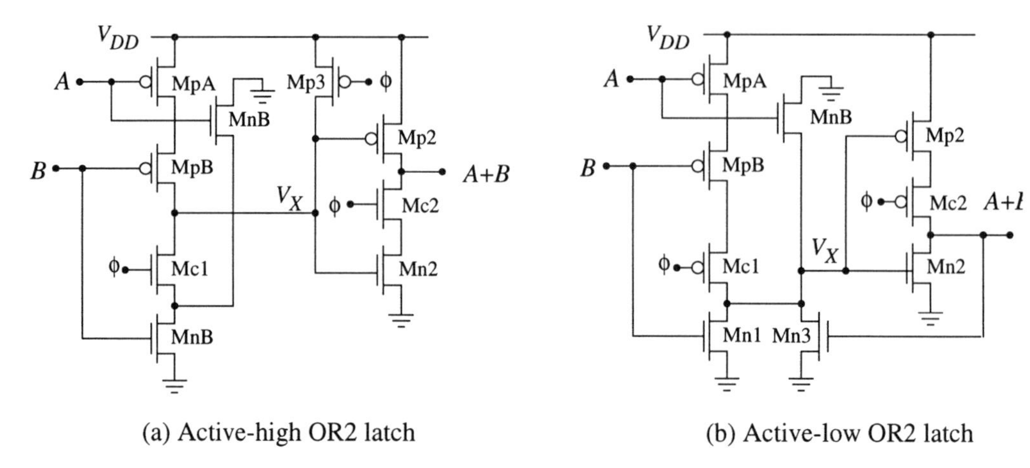

(a) Active-high OR2 latch (b) Active-low OR2 latch

Figure 8.101 OR2-latching circuits

chronized with ϕ at the gate of Mn2, which makes the circuit opaque when ϕ=0, and transparent when ϕ=1. Tracing the signal path shows that when the gate is transparent, Mp2 and Mn3 act as an inverter. This makes the gate non-inverting, i.e., f cannot provide the NOT function. In terms of canonical forms, this limits N1-type gates to AO and OA structures.

The complementary gate is termed an "N2-type" circuit and has the general structure shown in Figure 8.104. This also consists of a dynamic nMOS P/E gate, but has a modified output circuit. In this case, the output latching is dictated by the clock signal ϕ that is applied to pFET Mp3, giving it clocking characteristics that are opposite to that of an N1-type gate: the N2 circuit is transparent

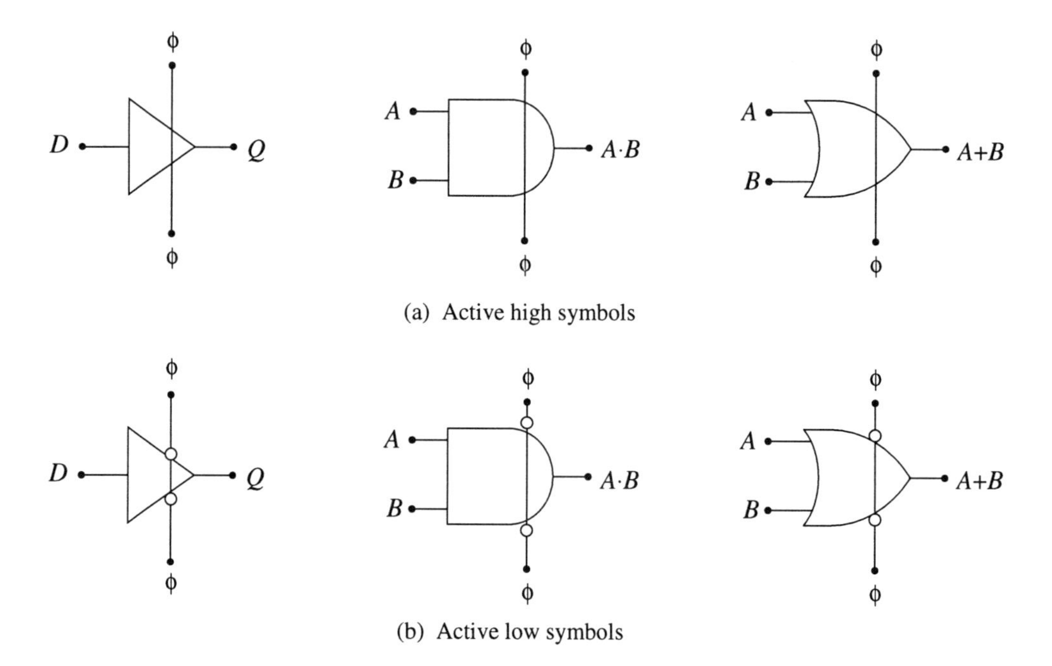

(a) Active high symbols

(b) Active low symbols

Figure 8.102 Latch symbols

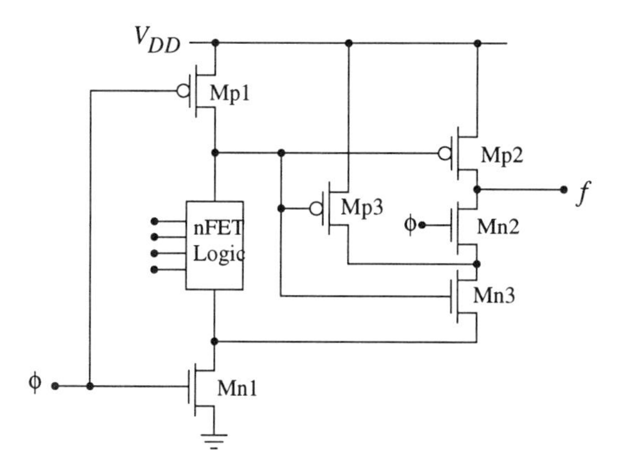

Figure 8.103 N1-type circuit structure

when ϕ=0 and opaque when ϕ=1. Also note that the gates of the output transistors Mp2 and Mn2 are driven by the node between the nFET logic block and evaluate transistor Mn1, limiting the maximum gate voltage to (V_{DD} -V_{Tn}). Logically, this arrangement implies that Mp2 and Mn2 do not act to invert the signal. Overall, the gate is therefore inverting in nature and can be used to provide, for example, AOI and OAI output functions at g. Both N1- and N2-type gates have additional MOSFETs that are added to insure that the output circuits can hold the output when they are opaque. Their function can be deduced by the usual analysis, and will not be presented here.

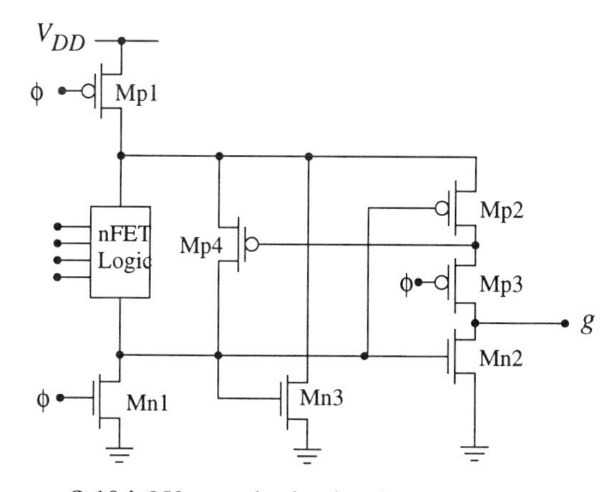

Figure 8.104 N2-type logic circuit

An ANL pipeline structure can be built using the general scheme illustrated in Figure 8.105. This alternates N1- and N2-type stages to eliminate races. The philosophy for creating a system pipeline that only uses nFET logic is that it may run faster than one based on alternating nMOS/ pMOS logic arrays, such as is the case in NORA logic. An example of an ANL circuit is the 8-bit CLA carry circuit shown in Figure 8.106. The first block is an N1-type circuit that calculates the output carry c_4 from the 0-3 bits when ϕ=1. This is cascaded into an N2-type gate that calculates the eight carry bit c_8 when ϕ=0. The entire function thus requires one clock cycle to complete.

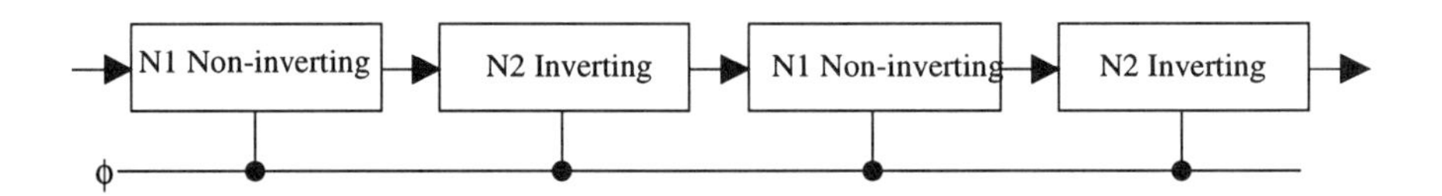

Figure 8.105 An ANL pipeline

8.8 An Overview of Dynamic Logic Families

In this chapter we have examined several of the dynamic CMOS logic families that have been developed in CMOS. The treatment was by no means exhaustive; several other design styles have appeared in the literature, but were not included in the treatment here for lack of space. When attempting to decide which of the many logic families to include in the discussion, an effort was made to include those that would best provide a basic understanding of the important circuit aspects and could be used for describing more advanced techniques. As we have seen in our analysis, every logic family has characteristics that make it unique. Regardless of the approach, however, all of the dynamic circuits are based on the same principles. The following are especially important concepts, and are worthwhile to summarize.

- Dynamic logic circuits use the clock ϕ to control the internal operation of a gate.
- The clock also synchronizes the data flow by controlling the output timing.
- Dynamic logic techniques are applied to groups of cascaded gates, not to isolated individual gates. The complexity of the cascade is closely related to the allowed clock frequency.

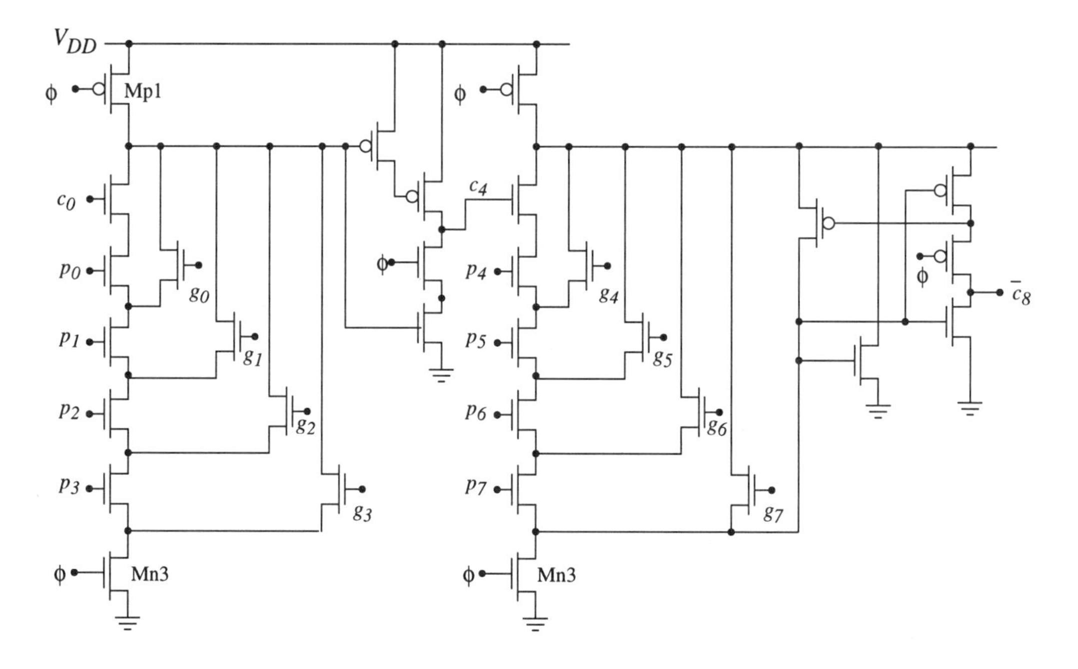

Figure 8.106 An ANL 8-bit CLA circuit

- Dynamic logic gates use isolated storage nodes that are subject to charge leakage and charge sharing problem. Additional circuity may be required to overcome these problems.
- The ultimate throughput of a dynamic cascade is limited by the slowest circuit in the chain.

While these may seem obvious, they provide a common link that unites all CMOS dynamic logic design styles.

8.9 Problems

[8-1] Construct a dynamic nMOS gate for the function

$$f = \overline{x \cdot y + x \cdot (z + w)}$$

[8-2] Consider the charge leakage described for the P/E NAND3 gate in equation (8.19). This assumes that charge sharing has reduced all capacitors in the circuit of Figure 8.8 to the same voltage as in equation (8.14) . Write the equations that describe the leakage for the case where the final value of the output voltage satisfies $V_f > (V_{DD} - V_{Tn})$ [equation (8.17)], and describe the leakage event.

[8-3] Draw the circuit diagram for an AOI full adder using a 2-stage nMOS-nMOS cascade. Suppose that a single reference nFET is characterized by an aspect ratio of $(W/L)=4$ such that a series chain of m-nFETs is made up of transistors that have and aspect ratio of $m(W/L)=4m$. Would this design be glitch free if used in this circuit? Make any reasonable assumptions that are necessary to quantify your answer.

[8-4] Construct a dynamic pMOS gate for the logic function

$$f = \overline{x \cdot y + x \cdot (z + w)}$$

[8-5] Consider a logic cascade where the first set of stages produce the functions

$$f_1 = \overline{a + b \cdot c}$$

$$f_1 = \overline{x \cdot (y + z)}$$

which are then used as inputs into the next stage that gives

$$g = \overline{f_1 \cdot f_2}$$

Design the logic chain using either alternating nMOS-pMOS logic stages.

[8-6] Consider the functions defined in Problem [8-5]. Can this be designed in an nMOS-nMOS cascaded arrangement?

[8-7] Consider the logic glitch problem portrayed in Figure 8.14. What happens to the behavior of the logic cascade if we change the clocking scheme so that ϕ is applied to the first stage, but $\overline{\phi}$ is applied to the second stage? Does this help solve the glitch problem?

[8-8] Discuss the problem of glitches in a 3-stage cascade of dynamic nMOS NAND3 gates. Assume exponential output voltages of the form

$$V_{out,j}(t) = V_{DD} e^{-t/\tau_{dis,j}}$$

for j=1, 2, 3, and assume a value of $V_{IH}=V_{Tn}$ for each gate.

[8-9] Draw the circuits for a 3-stage domino cascade that implements the function

$$f = f_1 \cdot f_2 \cdot f_3$$

where

$$f_1 = x \cdot y \cdot z$$
$$f_2 = (a + b + c) \cdot d$$
$$f_3 = u \cdot (v + w)$$

[8-10] Consider the dynamic NAND3 circuit shown in Figure P8.1 below. Use the process specifications stated in Problem [7-9] on pages 346-347 of the previous chapter to perform the following analysis.

(a) Calculate the output capacitance for the gate. Assume that the fan-out is into a single nFET with the same dimensions as that used in the gate.

(b) Estimate the precharge time for the circuit.

(c) Construct the RC-equivalent discharge path, and estimate the discharge time.

Now suppose that $x = 0$ during the evaluation portion for the clock cycle.

(d) Calculate the final voltage at the output after charge sharing takes place.

(e) The leakage current densities are estimated to be $J_{g,nFET} = 50\ fA/\mu m^2$ and $J_{g,pFET} = 275 fA/\mu m$. Estimate the hold time for the circuit.

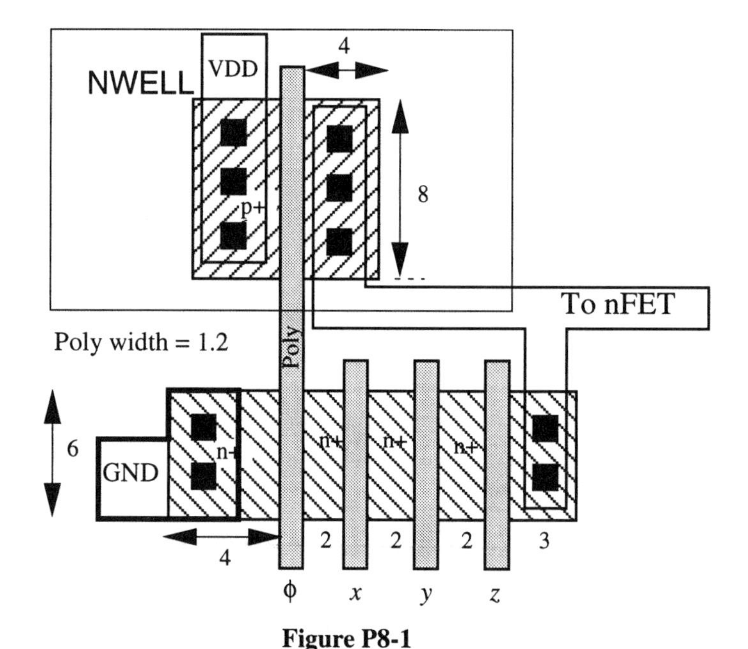

Figure P8-1

[8-11] Consider the domino AND3 circuit shown in Figure P8.2. Use the process specifications stated in Problem [7-9] on pages 346-347 of the previous chapter to perform the following analysis.

(a) Calculate the internal node capacitance C_X for the gate.

(b) Estimate the precharge time for the internal node.

(c) Construct the RC-equivalent discharge path, and estimate the discharge time.

(d) Find the value of V_{IH} for the inverter.

(e) Calculate the final voltage at the output after charge sharing takes place if $x = 0$ during the evaluation.

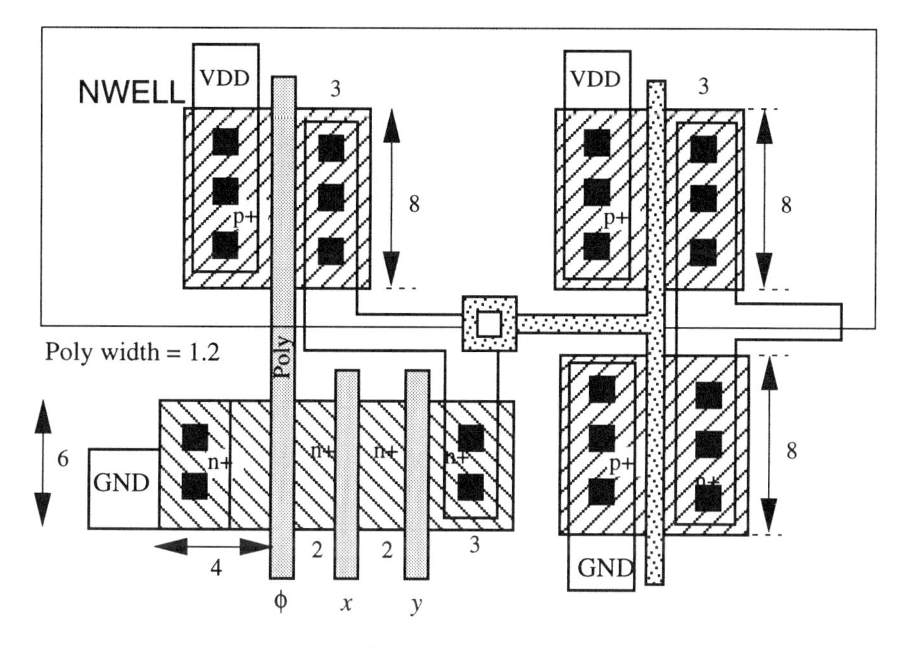

Figure P8.2

(f) The leakage current densities are estimated to be $J_{g,nFET} = 50\ fA/\mu m^2$ and $J_{g,pFET} = 275 fA/\mu m^2$. Estimate the hold time for the circuit.

[8-12] Suppose that the scaled FET chain in Figure 8.45(a) is designed with the $S_{p\text{-}a} = 3\ \mu m$. with $L' = 1.2\ \mu m$ and $L_o = 0.1\ \mu m$. The top FET (M1) has a width of $W_1 = 8\ \mu m$, and a scaling factor of $\alpha = 1.25$ is used. The junction parameters are given by $C_j = 0.4 fF/\mu m^2$ and $C_{jsw} = 0.32 fF/\mu m$ with $m = 0.5$ for both bottom and sidewalls. Assume $V_{DD} = 3.3v$ and a built-in voltage of $0.9v$ for all junctions. The oxide thickness is 150 $\AA$, $V_{Tn} = 0.7v$, and $k'_n = 130\ \mu A/V^2$. Ignore body bias in all calculations.

(a) Construct the RC equivalent circuit as shown in Figure 8.44(b) with numerical values for all components.

(b) Use your circuit in (a) to estimate the discharge time constant.

(c) Now suppose that the layout is changed to the simper scheme in Figure 8.45(b) where all FETs have the same width $W = 8\ \mu m$. The poly-poly spacing is $S_{p\text{-}p} = 4 \mu m$. Construct the RC equivalent network for this circuit and then calculate the time constant. Which design is better with these numbers?

[8-13] Solve Problem [8-12] with modified spacing values of $S_{p\text{-}a} = 3\ \mu m$ and $S_{p\text{-}p} = 5 \mu m$.

8.10 References

[1] D.W. Dobberpuhl, *et al*, "A 200 MHz 64-bit Dual-issue CMOS Processor," *Digital Technical Journal*, vol. 4, no. 4 , pp. 35-50, 1992.

[2] V. Friedman and S. Liu, "Dynamic Logic CMOS Circuits," *IEEE J. Solid-State Circuits*, vol. SC-19, no. 2, pp. 263-266, April, 1984.

[3] N.F. Goncalves and H.J. DeMan, "NORA: A Racefree Dynamic CMOS Technique for Pipelined Logic Structures," *IEEE J. Solid-State Circuits*, vol. SC-18, no. 3, pp. 261-266, June, 1983.

[4] R.X. Gu and M.I. Elmasry, "All N-Logic High-Speed True Single-Phase Dynamic CMOS

Logic," *IEEE J. Solid-State Circuits*, vol. 31, no. 2, pp. 221-299, Feb., 1996.
[5] I.S. Hwang and A.J. Fisher, "Ultrafast Compact 32-bit CMOS Adder in Multiple-Output Domino Logic," *IEEE J. Solid-State Circuits*, vol. 24, no. 2, pp. 358-369, April, 1989.
[6] Y. Ji-ren, I. Karlsson, and C. Svenson, "A true single phase clock dynamic CMOS circuit technique," *IEEE J. Solid-State Circuits*, vol. SC-22, pp. 899-901, 1987.
[7]
[8] R.H. Krambek, C.M. Lee, and H-F. Law," High-speed compact circuits with CMOS," *IEEE J. Solid-State Circuits*, vol.SC-17, no. 3, pp. 614-618, April, 1982.
[9] J.A. Pretorius, A.S. Shubat, and C.A.T. Salama, "Analysis and design optimization of domino CMOS logic with applications to standard cells," *IEEE J. Solid-State Circuits*, vol. SC-20, no. 2, pp. 523-530, April, 1985.
[10] J.A. Pretorius, A.S. Shubat, and C.A.T. Salama, "Charge distribution and noise margins in domino CMOS," *IEEE Trans. Circuits and Systems*, vol. CAS-33, no. 8, pp. 786-793, August, 1986.
[11] J. Rabaey, **Digital Integrated Circuits**, Prentice-Hall, Upper Saddle River, NJ, 1996.
[12] M. Shoji, **CMOS Digital Circuit Technology**, Prentice-Hall, Englewood cliff, NJ, 1988.
[13] M. Shoji, "FET Scaling in Domino CMOS gates," *IEEE J. Solid-State Circuits*, vol. SC-20, pp. 1067-1071, Oct., 1985.
[14] J. Yuan and C. Svensen, "High-Speed CMOS Circuit Technique," *IEEE J. Solid-State Circuits*, vol. 24, no. 1, pp. 62-70, Feb., 1989.

Chapter 9

CMOS Differential Logic Families

This chapter introduces the basic concepts of CMOS differential logic circuits. Differential logic networks are more complicated to design from the viewpoint of both the circuits and the layout. However, this type of logic is of great interest because it can provide striking improvements in the switching speed.

9.1 Dual Rail Logic

All of the circuits discussed up to this point have employed what is known as **single rail logic** where a logic state is represented by a single variable a that is either a 0 or a 1. Given a logic gate, the output is specified to be a function f, such that $f(a)$ acts as the input to the next gate(s) in the chain. Single rail logic is used in the great majority of logic circuits due to its simplicity.

Dual rail logic circuits are quite different from single rail implementations in that they use both the variable and its complement $(a, \bar{a})$ as an input pair. Using this philosophy, the output of a dual rail circuit is also a pair $(f, \bar{f})$ that drives the next gate(s) in the logic cascade. However, dual rail logic interprets the **difference** $(f - \bar{f})$ as the logic variable instead of just one or the other. Some circuits can be made to react to small differences, giving us the term **differential logic**. When viewed at the level of Boolean algebra, the use of both the variable and its complement is superfluous; the result is the same as that found using a single-rail circuit. Moreover, dual rail networks are inherently more complicated to wire since interconnect lines must be allocated for twice as many signals. Figure 9.1 illustrates the basic difference between single and dual rail logic gates.

The real advantage to using dual rail logic lies in the fact that an **electronic** dual rail circuit can be faster than an **electronic** single rail circuit. In other words, it is the circuit design, not the logic, that produces the improved performance. Since increased speed is often the most important goal of high-performance design, many dual rail logic networks have been developed for use in various system.

The speed advantage can be understood by using simple arguments. Consider the single rail variable $f(t)$ that is obtained at the output of a "regular" single rail logic gate as shown in Figure 9.2. The logic 0 and logic 1 ranges are shown in the drawing for both the input and the output variables. Consider the output $f(t)$, which we will interpret as a voltage signal. The **slew rate** is simply the rate

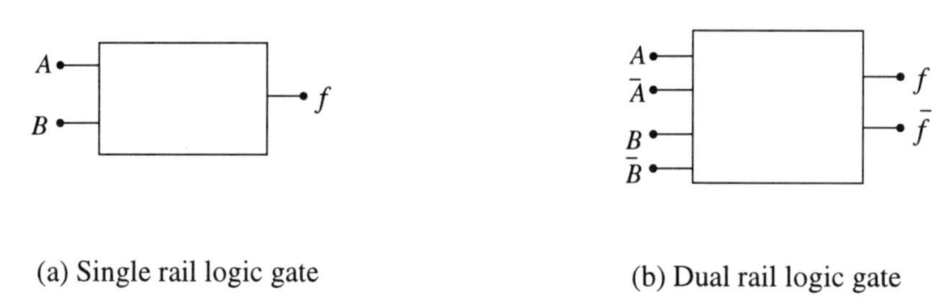

(a) Single rail logic gate (b) Dual rail logic gate

Figure 9.1 Single rail and dual rail logic gates

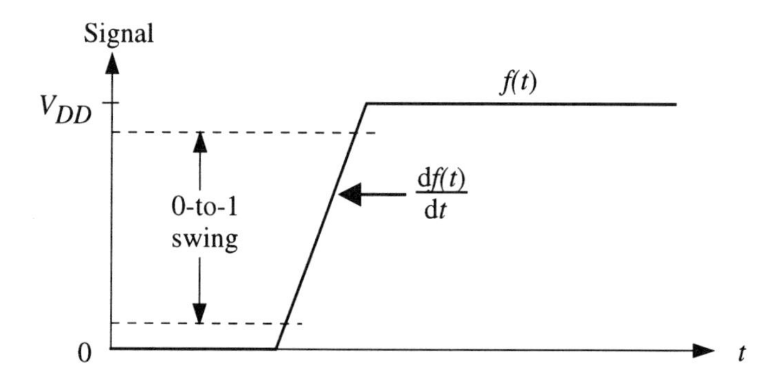

Figure 9.2 Switching waveform for a single rail network

of change of the output voltage in time, i.e., the derivative (df/dt). This determines all of the output switching times including t_{HL} and t_{LH}. A large slew rate implies a fast switching speed. Also note that the switching times depend upon the values of the initial and final voltages.

Now let us examine the output of a dual rail logic gate as shown in Figure 9.3. In a circuit of this type, both $f(t)$ and $\bar{f}(t)$ are generated as outputs of the gate. The logic variable is taken to be the **difference signal** $[f(t) - \bar{f}(t)]$. The effective slew rate of the difference signal is then

$$\frac{d}{dt}(f - \bar{f}) \approx 2\left|\frac{df}{dt}\right| \tag{9. 1}$$

where we have assumed that $(df/dt) \approx -(d\bar{f}/dt)$. This illustrates that a dual rail circuit intrinsically exhibits faster switching speed than that possible in a single-rail network.

Although dual rail logic can provide the basis for fast logic circuits, several features make this approach more difficult to implement than simpler single-rail networks. Some of the problems are

- Increased circuit complexity
- Increased interconnect required in the layout
- Timing issues become critical

These problems have been investigated using both static and dynamic circuit techniques. Our introduction to the subject will start with **cascode voltage switch logic** which is one of the best characterized CMOS differential logic families. By the natural laws of evolution, it also serves as the basis for other approaches.

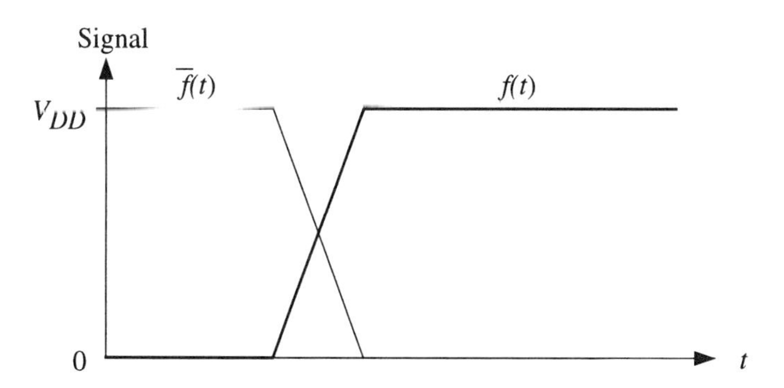

Figure 9.3 Dual rail logic signals used to create $(f\text{-}\bar{f})$

9.2 Cascode Voltage Switch Logic (CVSL)

Cascode voltage switch logic provides the basis for many dual-rail circuits.[1] The general structure of a CVSL gate is shown in Figure 9.4. It consists of two major sections. The pFETs Mp1 and Mp2 are cross-coupled to form a simple latch that provides complementary outputs f and $\bar{f}$; the latch section allows us to hold a result. The latch is driven by an nFET network that can be viewed as two complementary switching blocks; when one block acts as a closed circuit (from top to bottom), the other is open. Closing a switch block pulls the corresponding output to ground, forcing it to a logic 0 level, while the complementary output is set to a logic 1 value by the latching action.

9.2.1 The pFET Latch

Let us first examine the operation of the pFET latching circuit that consists of transistors Mp1 and Mp2. There are two stable states as shown in Figure 9.5. The source-gate voltages on the devices that control the conduction are given by

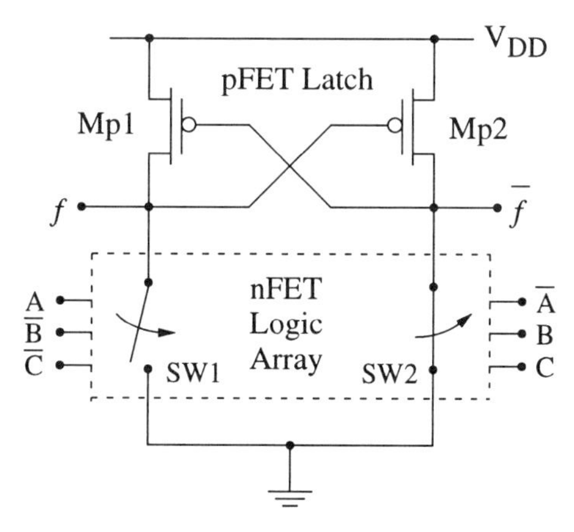

Figure 9.4 Basic structure of a CVSL logic gate

[1] In the literature, CVSL is also known as DCVS logic (for differential CVS).

$$V_{SG1} = V_{DD} - V_2$$
$$V_{SG2} = V_{DD} - V_{\bar{1}} \quad (9.2)$$

The behavior of the latch is understood by noting that $V_{\bar{1}}$ and V_2 are voltage complements in this circuit, so that one is high while the other is low.

Latching is induced by the nFET switching network, which has been split into two distinct blocks labelled SW1 and SW2 in the drawing. Suppose first that SW1 is open and SW2 is closed, as in Figure 9.5(a). SW2 pulls $V_2 = 0v$ as shown, which biases Mp1 into conduction since

$$V_{SG1} = V_{DD} \quad (9.3)$$

With Mp1 conducting, $V_{\bar{2}}$ rises to V_{DD}, which drives Mp2 into cutoff since

$$V_{SG2} = 0 \quad (9.4)$$

This represents one stable state of the latch. The opposite case shown in Figure 9.5(b) is where SW1 is closed and SW2 is open. The voltage $V_{\bar{1}}$ is pulled to $0v$, which gives $V_{SG2} = V_{DD}$ and biases Mp2 into conduction. This in turn pulls V_2 to V_{DD} which drives Mp1 into cutoff. Note that there is no direct path for current flow from V_{DD} to ground for either situation, so that only leakage currents exist.

9.2.2 CVSL Buffer/Inverter

Many of the basic feature of CVSL can be studied using the simple Buffer/Inverter circuit shown in Figure 9.6. This uses complementary inputs A and $\bar{A}$ that are associated with voltages of V_A and $V_{\bar{A}}$, respectively. Ideally, the two are related by

$$V_A + V_{\bar{A}} = V_{DD} \quad (9.5)$$

The outputs are denoted by $\bar{f}$ and f and are defined as shown. These are described using respective voltages $V_{\bar{f}}$ and V_f where

$$V_f + V_{\bar{f}} = V_{DD} \quad (9.6)$$

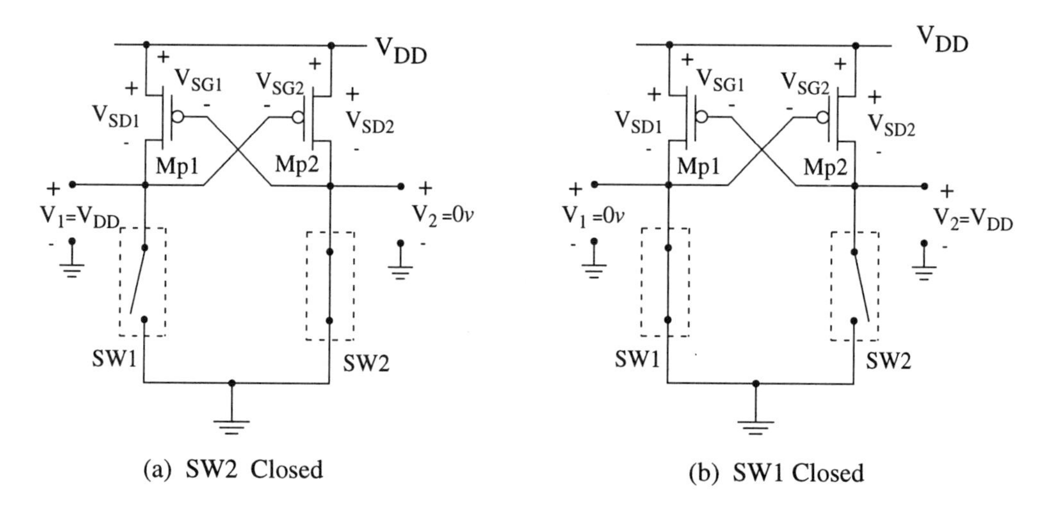

Figure 9.5 Stable states of the pFET latch

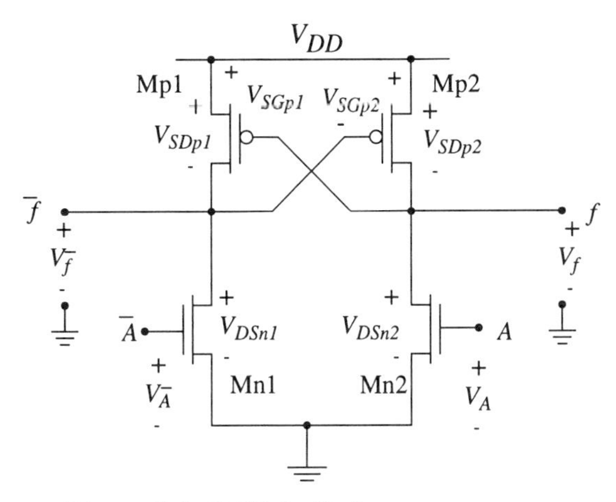

Figure 9.6 CVSL buffer/inverter

gives the ideal relationship between the two. Throughout the analysis, it is important to remember that the switching of this circuit is based on the difference signals (V_A - $V_{\overline{A}}$) and (V_f- $V_{\overline{f}}$), not on the individual voltages.

Since the logic FETs Mn1 and Mn2 can be modelled as voltage controlled switches, the operation is straightforward. Suppose that $A = 1$ and $\overline{A} = 0$. The input voltages are $V_A = V_{DD}$ and $V_{\overline{A}} = 0v$, so that Mn2 is active and Mn1 is in cutoff. Since Mn2 is conducting, $V_{DSn2} \rightarrow 0v$ and the output variable f is zero. The feedback action of the latch biases Mp2 into conduction, which gives $V_{\overline{f}} = V_{DD}$ corresponding to $\overline{f} \rightarrow 1$. This identifies

$$f = \overline{A} \tag{9.7}$$

and $\overline{f} = A$. To verify the operation, we reverse the inputs such that $V_A = 0v$ and $V_A = V_{DD}$ describing the case where $A = 0$ and $\overline{A} = 1$. Mn1 is now active giving $V_{DSn1} \rightarrow 0$ and $\overline{f} = 0$. Mn2 is in cutoff and Mp2 is active which pulls the right side up to a voltage $V_f = V_{DD}$ corresponding to $f = 1$.

Switching Transients

Let us examine the main factors that determine the switching speed of the logic gate. The presence of the feedback loop in the pFET latch makes the analysis of the non-linear network quite complicated; although simplifications can be made, the results are not very illuminating. In practice, the performance would be analyzed via a computer simulation, which is the best approach. However, it is possible to understand the overall characteristics of the switching problem using a simple overview.

Consider the inverter/buffer switching pair shown in Figure 9.7(a). The input voltages $V_x(t)$ and $V_y(t)$ are taken to be ideal voltage complements as in Figure 9.7(b). With the voltages shown, Mn1 is initially in cutoff while Mn2 is ON. The voltages are applied to reverse this situation, with the threshold voltage V_{Tn} being the critical value. Mn1 turns at time t_1 such that

$$V_x(t_1) = V_{Tn} \tag{9.8}$$

As $V_x(t)$ increases, Mn1 becomes more and more conductive as it attempts to pull down the drain node voltage V_1 to ground. However, Mn2 is active until a time t_2 when $V_y(t)$ falls to a value of V_{Tn}. Once Mn2 is OFF, the switching can proceed without hinderance.

This simplified discussion ignores the fact that the pFET latch has triggering characteristics that aid in the switching. This implies that Mn2 need not be completely off to switch the holding state of

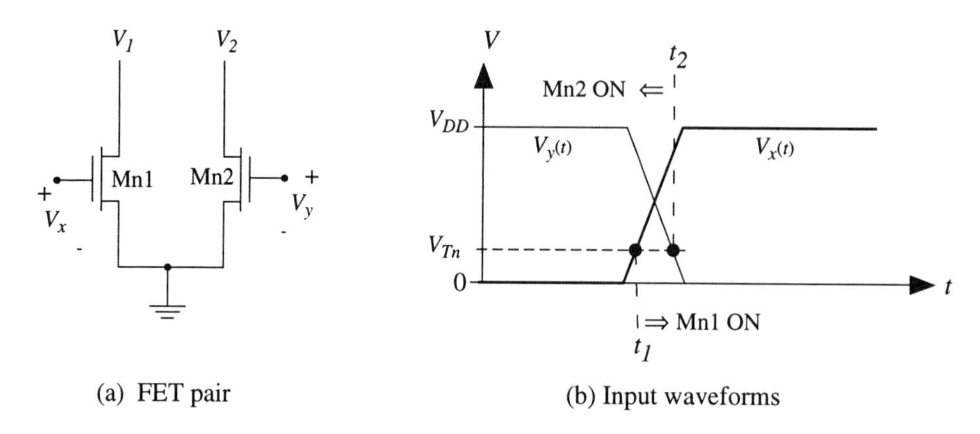

(a) FET pair (b) Input waveforms

Figure 9.7 Switching in an nFET pair

the latch. The triggering voltage itself is determined by the aspect ratios of the pFETs in the latch when driven by the nFET logic transistors. If both pFETs are chosen to have the same aspect ratio $(W/L)_p$, then we would expect the intrinsic triggering voltage for the latch (without the nFETs) to be at $(V_{DD}/2)$. This sets the critical values for V_1 and V_2 in the pull-down network.

One should always test the design of latching network using computer simulations that provide accurate device models. Timing and the shape of input signals are very important to producing valid results.

9.2.3 nFET Switching Network Design

The nFET switching networks used in CVSL can be designed using structured techniques that relate the logic function to the topology of the transistor array. We will examine two approaches in this section. Both are designed to use complementary input pairs $(A, \overline{A})$, $(B, \overline{B})$, etc., and produce the complementary outputs $(f, \overline{f})$.

AOI/OAI Logic

One approach to designing CVSL logic is to use AOI or OAI logic forms as a starting point for one side of the gate, and then use the DeMorgan relations and logic reductions to create the switching network for the opposite side. The nFET placement rules developed for static logic gates in Chapter 5 are applicable here, and provide the starting point of the technique.

To illustrate the technique, suppose that we start with the OAOI function

$$f = \overline{(A+\overline{B})\cdot C+\overline{D}} \tag{9.9}$$

that is described by the logic diagram in Figure 9.8(a). The basic rules developed in Chapter 5 may be used directly to construct the nFET logic array in Figure 9.8(b). The opposite side of the logic gate requires an nFET network that

- Uses the complements of the inputs, i.e., $\overline{A}$, B, $\overline{C}$, and D
- Provides the function

$$\overline{f} = (A+\overline{B})\cdot C+\overline{D} \tag{9.10}$$

To accomplish this task, we start with the original logic specification and apply DeMorgan reductions until the simplest form is reach. This may be done using equations, or with logic symbols. From the viewpoint of Boolean algebra, the steps are given by

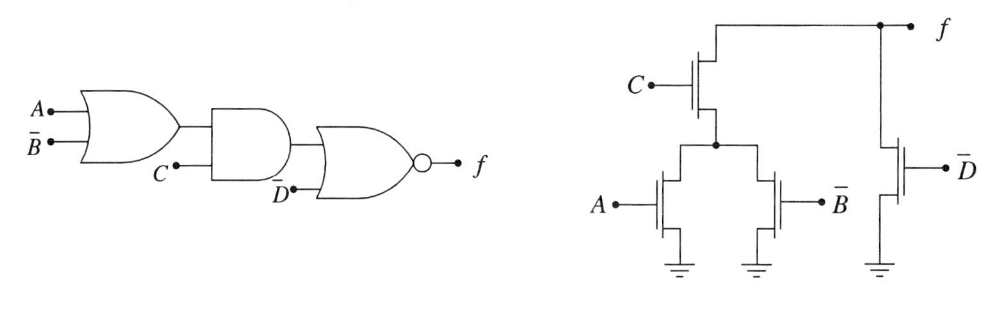

(a) OAOI logic diagram (b) nFET logic equivalent

Figure 9.8 FET switching network design

$$\begin{aligned} f &= \overline{(A+\bar{B})\cdot C+\bar{D}} \\ &= \overline{[(A+\bar{B})\cdot C]}\cdot D \\ &= \left(\overline{(A+\bar{B})}+\bar{C}\right)\cdot D \\ &= ((\bar{A}\cdot B)+\bar{C})\cdot D \end{aligned} \tag{9.11}$$

Figure 9.9(a) shows the logic diagram for f, where the complemented inputs are performed by the input bubbles. It is easily verified that this yields the same result as graphically "pushing the bubble" backwards through each gate. Taking the complement yields the AOAI expression

$$\bar{f} = \overline{((\bar{A}\cdot B)+\bar{C})\cdot D} \tag{9.12}$$

as shown in Figure 9.9(b), which is the required form. This can be used to create the nFET logic array in Figure 9.9(c). Since we have both sides of the logic array, the two may be combined to give the complete gate shown in Figure 9.10. This approach is very general, and can be used to implement any basic AOI or OAI function.

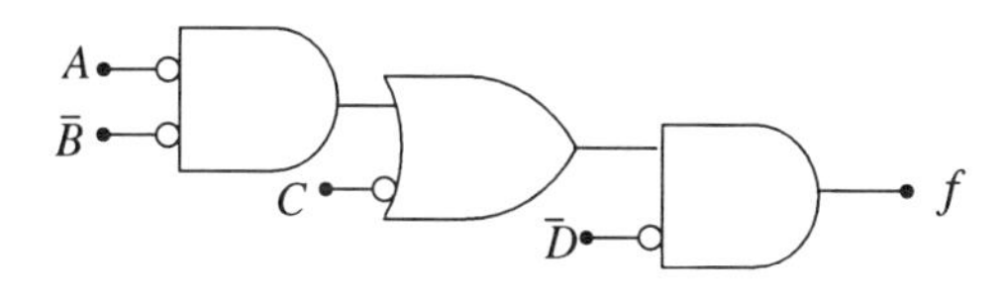

(a) After DeMorgan reductions

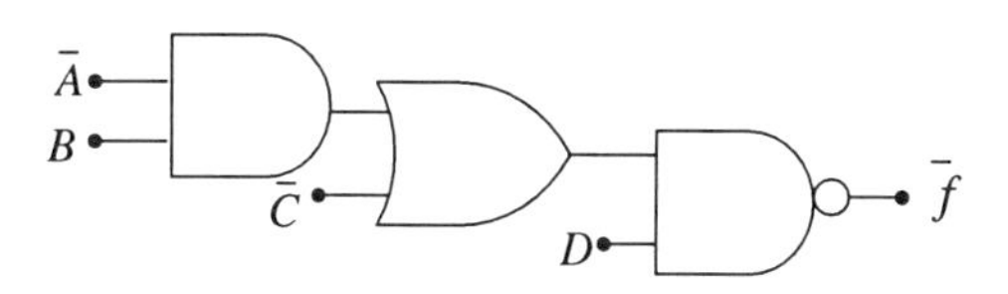

(b) Complemented input variables

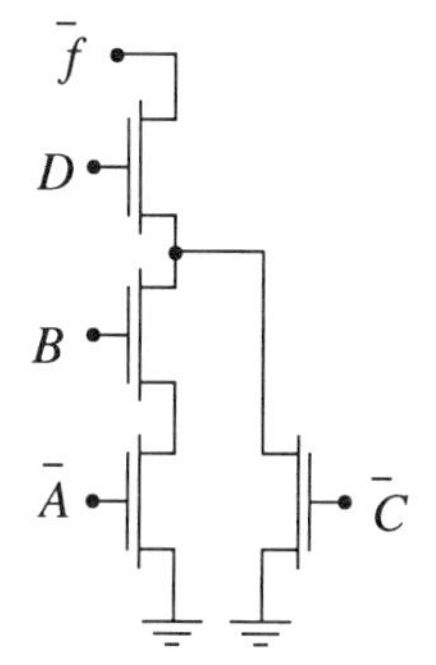

(c) nFET logic equivalent

Figure 9.9 Construction of the complementary logic array

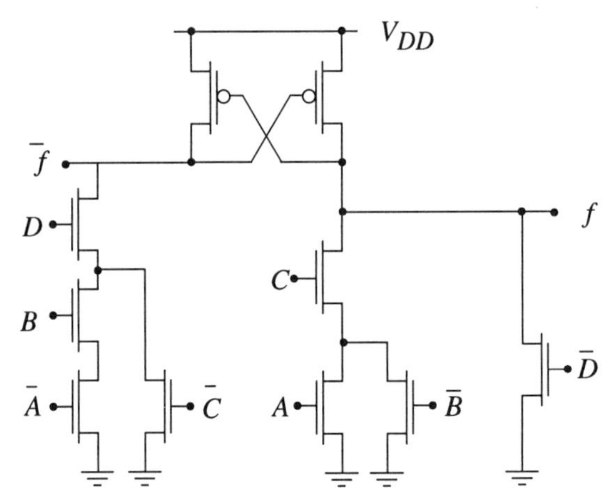

Figure 9.10 Completed logic gate

Structured Logic Trees

A more structured approach to designing CVSL logic has been developed by Pulfrey and Chu. This technique uses the function table as the basis for deriving a single logic tree that provides both f and $\bar{f}$. This is illustrated by the gate circuit shown in Figure 9.11. In the brute force AOI-OAI technique above, the logic arrays are taken to be separate circuits. Using a single logic tree that has two pull-down nodes is attractive because is allows for the possibility of sharing transistors, reducing the complexity of the network and (perhaps) saving area and gaining speed.

The technique itself can be understood by examining a function table such as that shown in Figure 9.12. A horizontal format has been used to more clearly illustrate the characteristics of the input variables. Following the input variables A, B, and C across the table shows that each has a distinct sequencing order. These are given by

$$A : 01\ 01\ 01\ 01$$

$$B : 00\ 11\ 00\ 11$$

$$C : 0000\ 1111$$

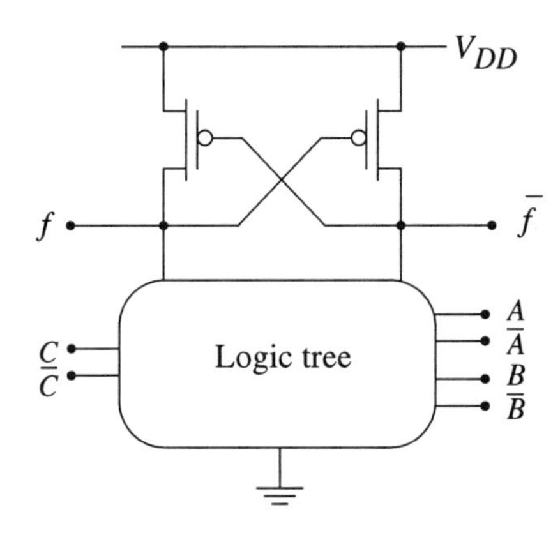

Figure 9.11 Use of a logic tree in CVSL

f	1	0	0	1	0	1	1	1
A	0	1	0	1	0	1	0	1
B	0	0	1	1	0	0	1	1
C	0	0	0	0	1	1	1	1

Figure 9.12 Truth table specification of the function

Logically, an input of "1" corresponds to the variable, while an input of "0" implies the complement of the variable. To cast this into a switching network, we use a pair of source-coupled nFETs as shown in Figure 9.13(a). One transistor is directly driven by the input variable, while the other is connected to the complement of the input. Each transistor has a distinct drain/source input a, b, but the pair produce a single output u. For the example shown,

$$u = \bar{x} \cdot a + x \cdot b \tag{9. 13}$$

by using the general rules developed in Chapter 4. In this configuration, the nFET pair is used as a 2:1 multiplexor that selects either a or b for the output u, depending upon the value of x.The simplified symbol in figure 9.13(b) will be used to represent a pair of nFETs of this type. The side with the minus (-) implies an nFET with $\bar{x}$ applied to its gate, while the plus side (+) represents the nFET with x controlling the gate.

Now then, the structure of the function table implies that the logic tree can be created by cascading nFET pairs as implied by the 0-1 sequencing of each variable. Figure 9.14 shows the general structure of a tree obtained in this manner. We have following the sequencing in the table from the top to the bottom, with the values of the output f used as inputs into the A-level FET pairs. Since there are four groupings of 01 01 01 01, we use 4 independent pairs. The next level down has B inputs, and only requires 2 FET pairs corresponding to the input sequence 00 11 00 11, while the third and final level with the C input only has a single FET pair due to the 0000 1111 sequencing. It can be shown that directly substituting FETs at this point can result in a functional logic network. However, it is possible to perform some simplifications that reduce the number of transistors before the FETs are actually inserted into the network. This is particularly important to improving the switching performance of the circuit.

Simplification rules are straightforward to find by analyzing the logic function of special cases.

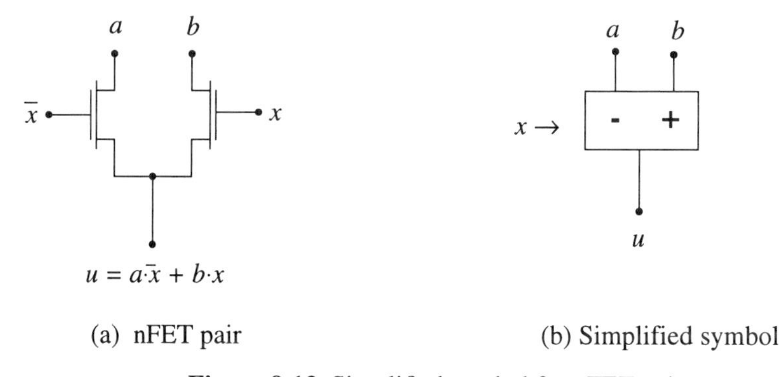

(a) nFET pair (b) Simplified symbol

Figure 9.13 Simplified symbol for nFET pair

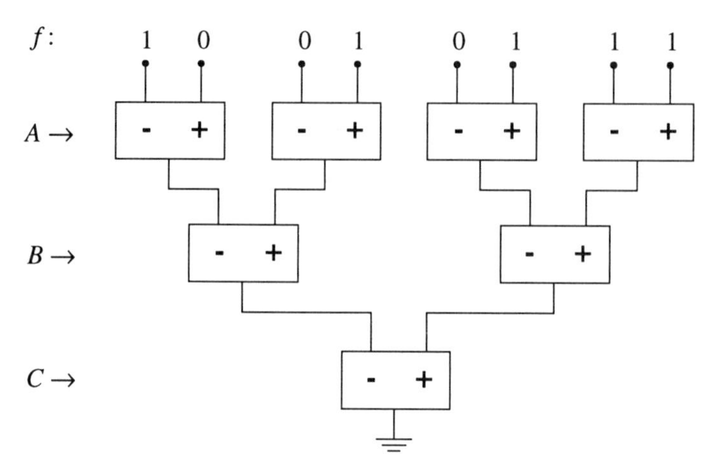

Figure 9.14 General structure of 3-input logic tree

Consider first the nFET pair shown in Figure 9.15 where the inputs (at the top) are identical. Since the output is given by

$$\begin{aligned} u &= x \cdot a + \bar{x} \cdot a \\ &= a \cdot (x + \bar{x}) \\ &= a \end{aligned} \tag{9. 14}$$

the transistors do not perform any logical operation (except a pass through) and the entire block may be "shorted" as shown; this eliminates the pair entirely from the logic tree. Another observation is that if two FET pairs (at the same input level) have the same inputs, then one can be eliminated and the outputs shorted together. This is shown in Figure 9.16. Since both nFET blocks have inputs of (α,β), then the outputs are equal: $u_1=u_2$. One block is thus superfluous, and can be discarded completely.

These rules may be applied to the present example to give the simplified tree shown in Figure 9.17. Replacing the blocks with nFET pairs yields the network in Figure 9.18. The complete logic gate is created by attaching the logic network to the pFET latching circuit. Since the latch is defined to have complementary outputs, we simply connect all of the 0's to one side of the latch, and the 1's to the other side. This results in the finished design shown in Figure 9.19.

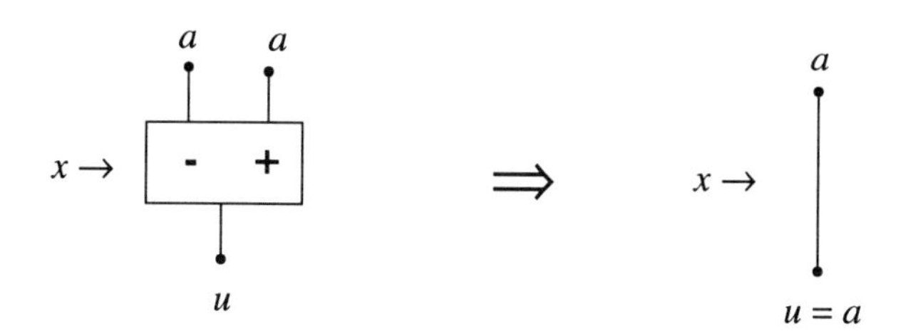

Figure 9.15 Elimination of an nFET pair

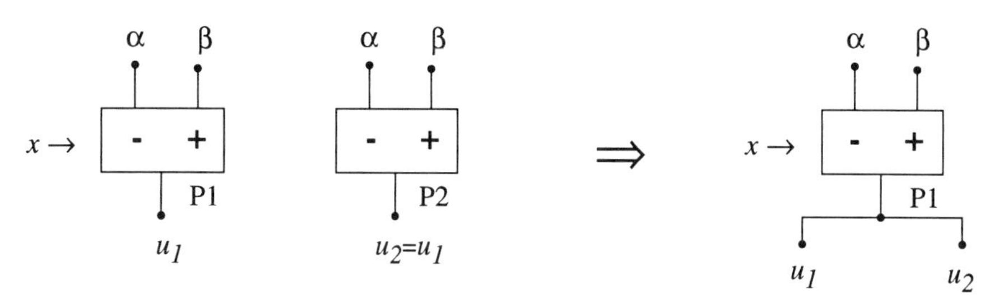

Figure 9.16 Elimination of a redundant pair

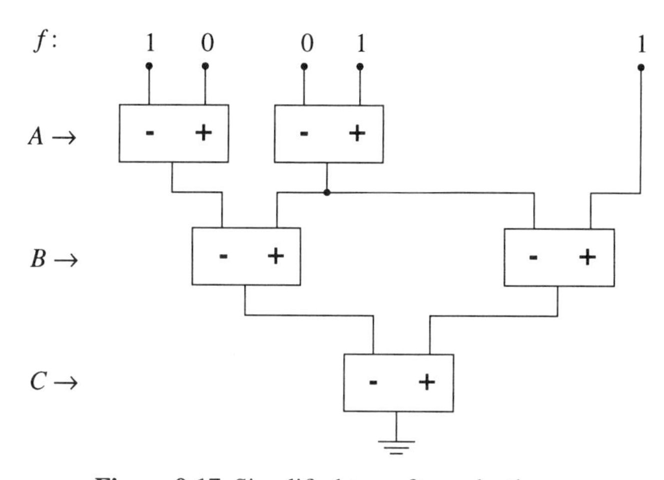

Figure 9.17 Simplified tree after reduction process

It is seen that the tree-reduction technique provides a straightforward method for designing CVSL logic arrays. It can be applied to arbitrary functions by following the same procedure: construct a generic tree, then use the outputs as defined in the function table to eliminate FET pairs where possible.

9.2.4 Switching Speeds

Although the design of the logic network is straightforward, we should note the issues involved in creating a fast switching network. Logic trees have the characteristic that the topology of the simplified tree depends directly on the details of the logic function. This tends to yield a non-symmetric tree as in the example above. Creating an RC ladder model for the possible discharge paths shows that the worst-case pull-down event occurs through the longest chain. The situation is helped by the trigger voltage of the pFET latch, but one should always examine the chain delay in the context of the triggering event itself. Once again, accurate computer simulations are critical to obtaining a complete understanding of the effects in this case.

9.2.5 Logic Chains in CVSL

CVSL is intrinsically a dual-rail logic family, so that creating logic chains requires that we route each variable and its complement as a pair. Once this is recognized, building a logic chain in CVSL follows the same approach as in any logic family. The presence of input and output pairs does seem

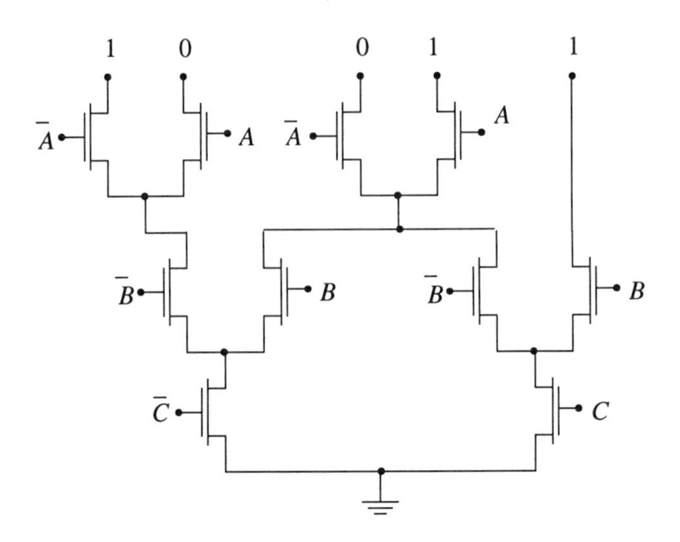

Figure 9.18 nFET wiring for the reduced tree

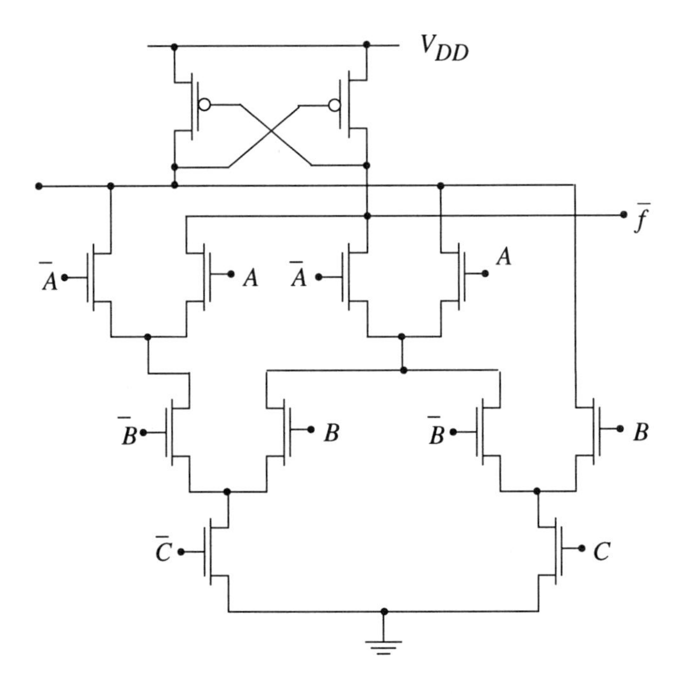

Figure 9.19 Completed CVSL logic gate

to complicate the drawings, but the wiring is straightforward. An example is shown in Figure 9.20. This shows output pairs such as $(f, \overline{f})$ and $(g, \overline{g})$ being used as complementary inputs to the following stage.

One aspect of dual rail logic is that the switching speed along the chain is enhanced by the fact that both signals are involved in carrying the logic. This is equivalent to the statement that it is the difference signal $(f - \overline{f})$ that supplies the results. Switching speed can be maintained even if there is a short time delay between the individual components in the pair (such as that due to non-symmetrical trees or unbalanced loading). The critical voltage level in this configuration is the FET threshold voltage V_{Tn} as this is the border between an ON and OFF transistor. Extending this observation to the chain implies that the switching can become truly differential in that the signals do not have to swing over the full range of possibilities, but can carry the logic information with reduced amplitude swings.

9.2.6 Dynamic CVSL

The static CVSL logic gate can be transformed into a dynamic circuit by rewiring the pFET latch to the clock-driven arrangement shown in Figure 9.21. This eliminates the feedback loop and changes Mp1 and Mp2 into precharge devices that are controlled by the clock ϕ. A value of ϕ=0 drives both pFETs into conduction, resulting in precharging of the output nodes. To avoid DC current flow during this event, an evaluation nFET Mn is added as shown. This is also controlled by ϕ so it is OFF during the precharge.

The operational modes of the CVSL dynamic gate are summarized in Figure 9.22. The precharge is shown in (a); the clock is at ϕ=0 which allows the voltages across both C_1 and C_2 to precharge to values of

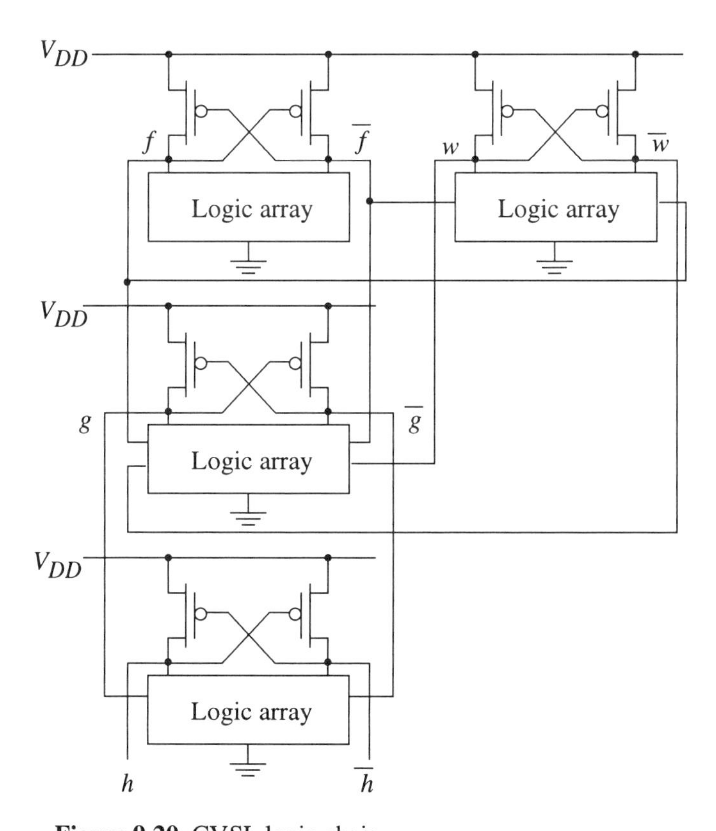

Figure 9.20 CVSL logic chain

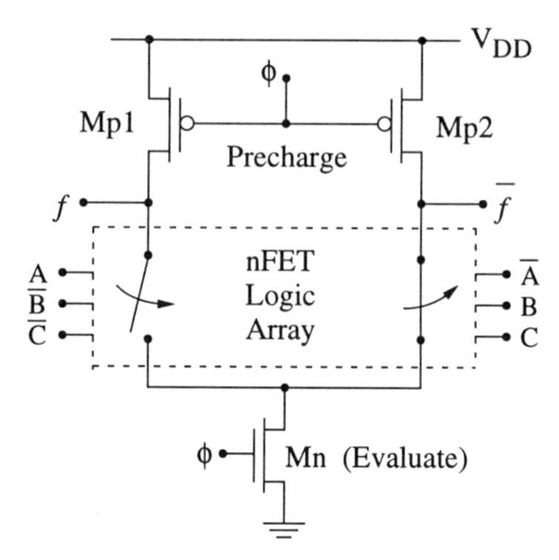

Figure 9.21 Dynamic CVSL gate

$$V_1 \rightarrow V_{DD}$$
$$V_2 \rightarrow V_{DD} \tag{9. 15}$$

using the precharge FETs. When the clock changes to the value ϕ=1, the circuit is driven into the evaluation phase. Mn is ON, and the input signals are valid. For the case shown in Figure 9.22(b), switch SW1 is open and V_1 is held high while SW2 is closed and discharges C_2 to

$$V_2 \rightarrow 0v \tag{9. 16}$$

corresponding to a logic 0 output there. The output voltages are initially complementary. However, the left output voltage V_1 is subject to the usual dynamic problems of charge sharing and charge leakage, which reduces its value in time. As with all dynamic logic circuits, this gives rise to a minimum clock frequency.

One solution to the dynamics problems is to add charge keeper pFETs to maintain the voltage at a high level when needed. A circuit that provides this action is shown in Figure 9.23. The weak charge keeper pFETs Mk1 and Mk2 are controlled by the output states f and $\bar{f}$. Since the gate signals are complements by definition, one of the FETs will be OFF while the other is ON. To work in the present circuit, both of the keeper FETs must be weak devices with small aspect ratios to allow for charge compensation without excessive current flowing onto the node.

9.3 Variations on CVSL Logic

CVSL provides the basis for several alternative differential CMOS logic design styles that have appeared in the literature. Although the circuits have evolved from the CVSL structure, each has features that make it unique. In this section we will briefly examine three of these logic families with emphasis on learning circuit design techniques. The interested reader is referred to the literature for more detailed discussions.

9.3.1 Sample-Set Differential Logic (SSDL)

SSDL is a clocked differential logic style that was developed to overcome the delay problem that originates in the discharge of a capacitor using a chain of series-connected logic FETs. Consider the

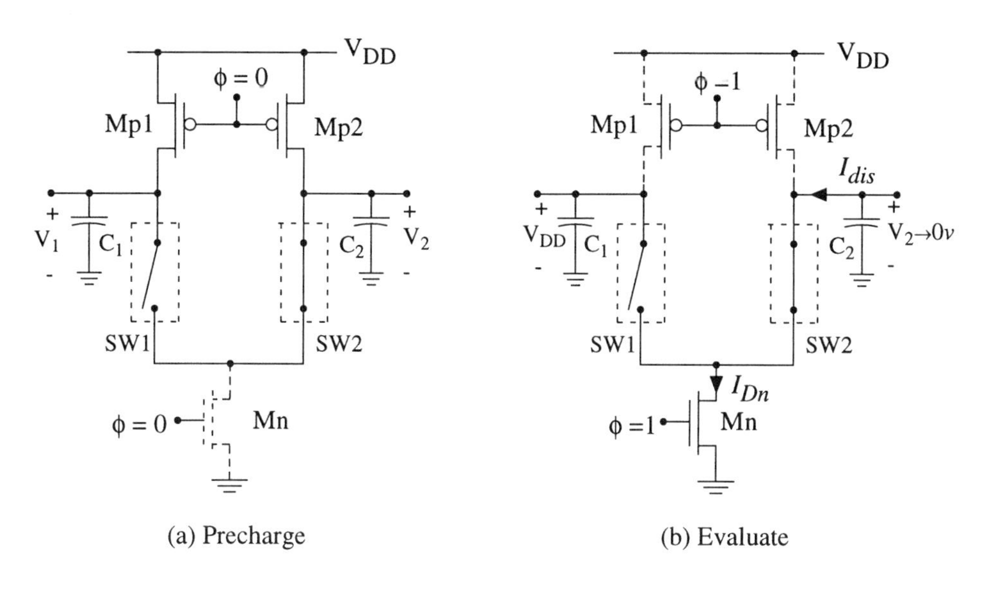

Figure 9.22 Operation of the dynamic circuit

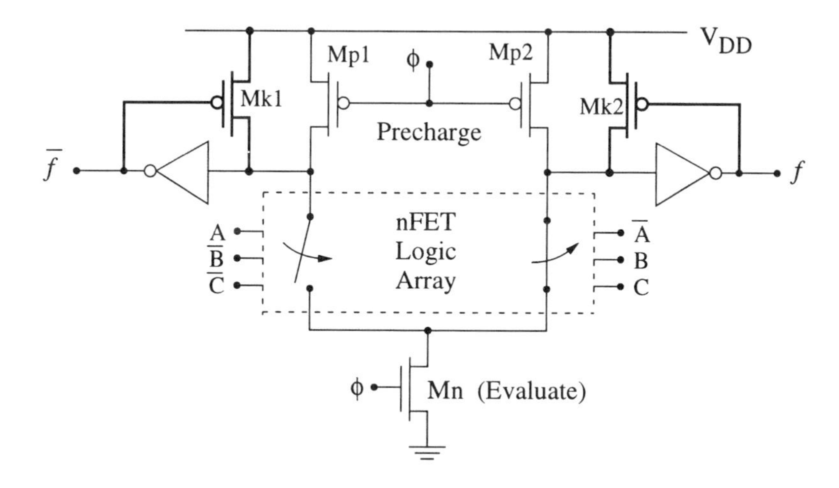

Figure 9.23 Dynamic CVSL circuit with charge keepers added

dynamic cross-coupled nFET latch made up of Mn1 and Mn2 that acts as a differential sense amplifier. Note that the nFET logic array is parallel to the latch lines (defined by the voltages V_1 and V_2).

The operation of the circuit can be understood by examining the effect of the clocking signal ϕ on the state of the network. A clock signal of ϕ=0 defines the precharge portion of the logic cycle. The state of the circuit during this time is summarized in Figure 9.25. Precharge pFETs Mp1 and Mp2 conduct to charge the output node capacitances C_1 and C_2 to high values. Note, however, that the presence of the clock-controlled nFET at the bottom of the latch (Mn3) and nFET Mn at the bottom of the logic array. With ϕ=0, Mn3 is OFF, so that the latch is floating at this time. On the other hand, Mn is controlled by $\overline{\phi}$=1, and is active during this time interval. This means that the

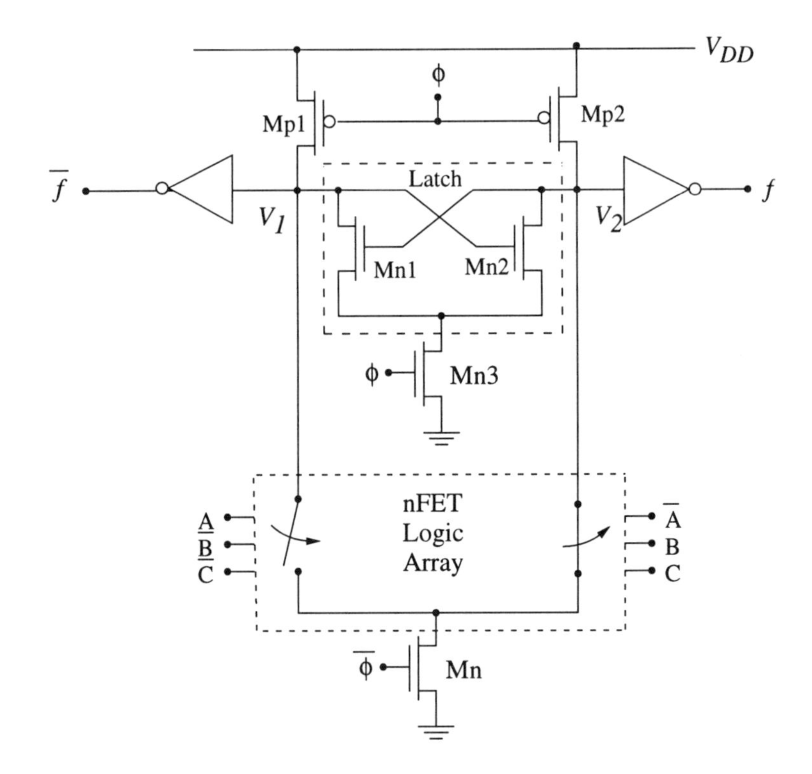

Figure 9.24 Basic SSDL circuit

state of the logic array determines the values of V_1 and V_2 that can be achieved on C_1 and C_2. For example, if the left side switch is OPEN and the right side switch is CLOSED during this time as shown in the drawing, V_1 can reach a final precharge voltage of V_{DD}, but V_2 will be below this value because of the conduction path to ground on the right side. When the clock makes a transition to ϕ=1, the circuit goes into evaluation and the latch nFET Mn3 conducts. Evaluation is helped by the cross-coupled action of Mn1 and Mn2. Owing to the feedback, one transistor conducts to quickly discharge the lower voltage node to ground. For example, if $V_1 = V_{DD} > V_2$ is at the gate of Mn2, it conducts to quickly discharge C_2 to a final voltage of $V_2 = 0v$. The fact that $V_1 > V_2$ implies that although Mn1 may initially be biased active, it will not discharge C_1 as quickly.

The most important aspect to understand about SSDL is that logic array is switched into the circuit only during the precharge time. Evaluation is achieved entirely by the nFETs Mn1 and Mn2 in the latch. The circuit thus completely avoids the problem of a discharge delay through an nFET logic chain. In other words, the complexity of the logic circuitry has no effect on the evaluation time. This makes the idea very attractive in high-density designs.

One of the obvious problems with the SSDL circuit is that establishes a DC current flow path between the power supply V_{DD} and ground during the precharge phase, giving power dissipation during that time. This is important to the design, as it reduces the voltage on one of the latch nodes as required for proper operation during the evaluation phase. A variation of the circuit that does not exhibit DC power dissipation is shown in Figure 9.26. In this approach, the dynamic latch has been replaced by a static RAM cell made up of two static inverters, and additional clocking transistors have been added to control the DC current flow paths. During precharge (ϕ = 0), both of the output nodes try to charge to equal values of $V_1 = V_{DD} = V_2$ which is not a stable state of the circuit. Since the logic array is disconnected from the circuit during this time, the RAM will be in an unstable

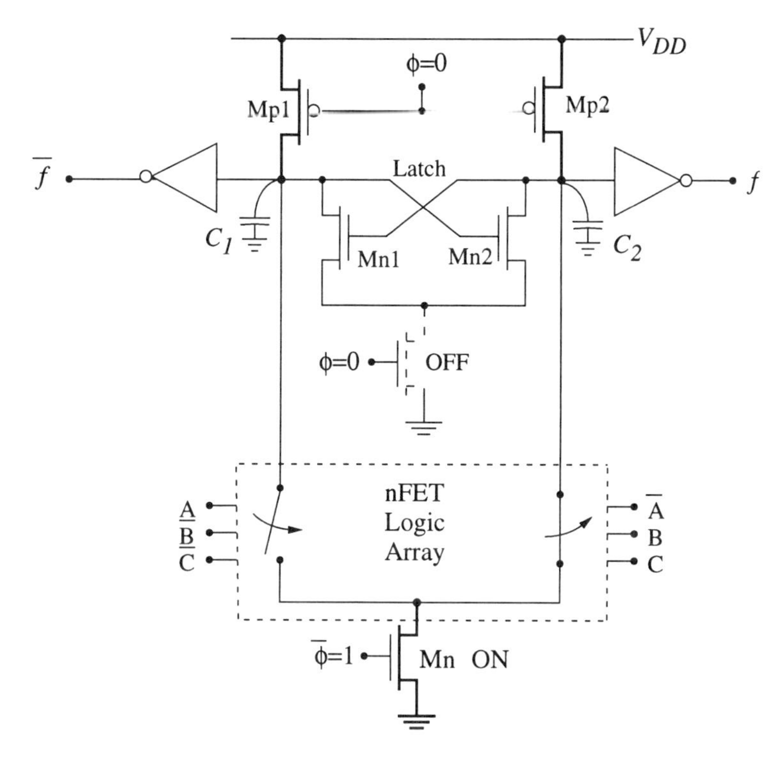

Figure 9.25 SSDL precharge circuit

state. When the clock makes a transition to φ=1, the logic tree switches into the circuit and dictates which side discharges to 0. In summary, the aim of this circuit is to preserve the speed while lowering the power dissipation.

9.3.2 ECDL

ECDL is an acronym that is derived from **Enable/disable CMOS Differential Logic**, another differential logic design style. This approach was developed to overcome the static power dissipation problem in SSDL by eliminating the DC current flow path and reducing the FET count to reduce the real estate requirements. The general ECDL circuit is shown in Figure 9.27. The latching is accomplished by a pair of cross-coupled static inverters (i.e., an SRAM storage cell). The clock controls the precharge pFET Mp that connects the power supply to the latch. In addition, two nFETs Mn1 and Mn2 are included at the outputs; these provide the enable/disable feature of the circuit.

Consider first the case where the clock has a value of φ=1. Both Mn1 and Mn2 are active, which disables the latch by setting both sides to 0*v*. From the logic viewpoint, these act to reset the state of the circuit to ground. When the clock makes a transition to φ=0, Mn1 and Mn2 are turned off, allowing the nodes to achieve other voltages. During this time pFET Mp is biased into conduction and supplies power to the latching circuit. The state of the parallel-connect logic array determines the state that the latch will settle in.

The circuit does in fact eliminate the DC current flow path between the power supply and ground, but the circuit is designed to discharge the output nodes every half-cycle (during the reset) which increases the dynamic component of power dissipation. In addition, the circuit relies on logic chains to discharge one of the output nodes, so that the RC delays may be a limiting factor. It does have the advantage of being simple to design with reduced interconnect requirements.

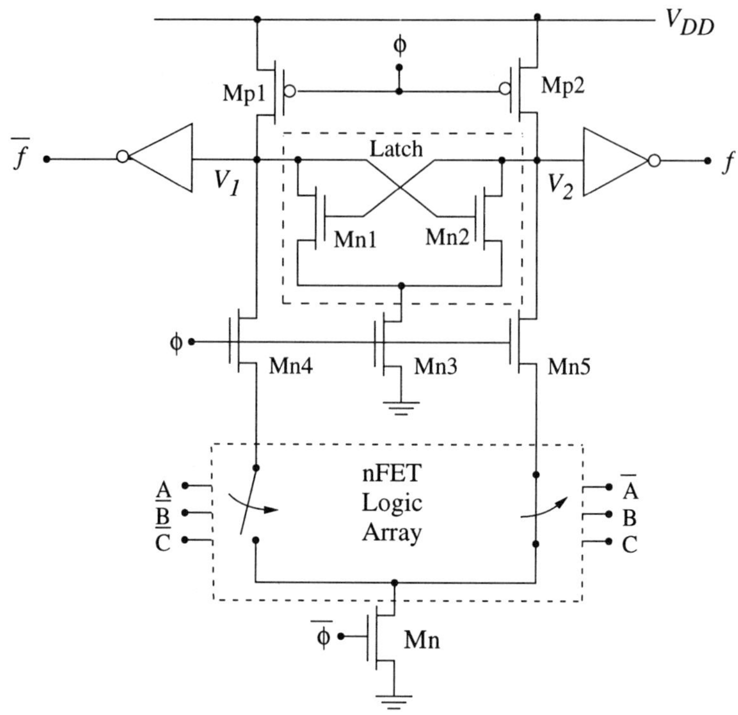

Figure 9.26 Variation on SSDL core circuitry

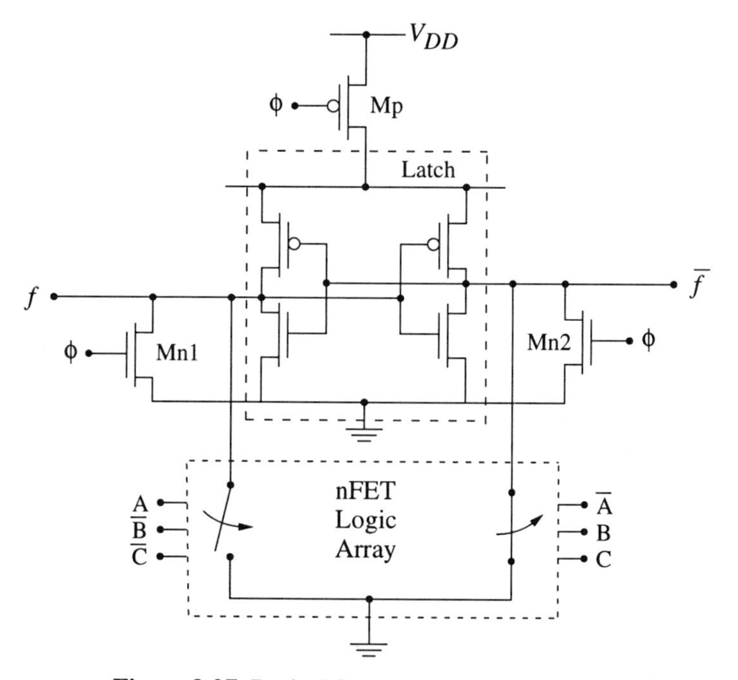

Figure 9.27 Basic ECDL circuit with nFET logic

9.3.3 DCSL

Our last example will be a very brief look at **Differential current switch logic** (DCSL) which has been proposed as a low-power approach to dual rail CMOS logic. It achieves the low power objective by limiting the voltage swings on internal nodes, but requires a relatively high FET count and the circuits are sensitive to noise and circuit imbalances. It is discussed here as an example of how the basic ideas embedded in the CVSL circuits have been expanded to address new problems such as power dissipation.

Figure 9.28 shows a basic DCSL circuit. Although it is somewhat complicated at first sight, one can easily pick out the static latch in the center. Several clocking transistors have been added to the output nodes. Mp1 and Mp2 are precharge devices, Mn is a latch enable FET, and Mn3 and Mn4, control the current between the latch and logic array. Mn1 and Mn2 are unique to this design. They are controlled by the state of the latch, and help to limit the internal voltages in an effort to reduce the power dissipation. The detailed behavior of this circuit is somewhat involved and is best understood by studying the results of a computer simulation; the interested reader should track down the original paper for more information. For our purposes, it has been mentioned solely as an example of how various design styles have evolved from the basic CVSL structure.

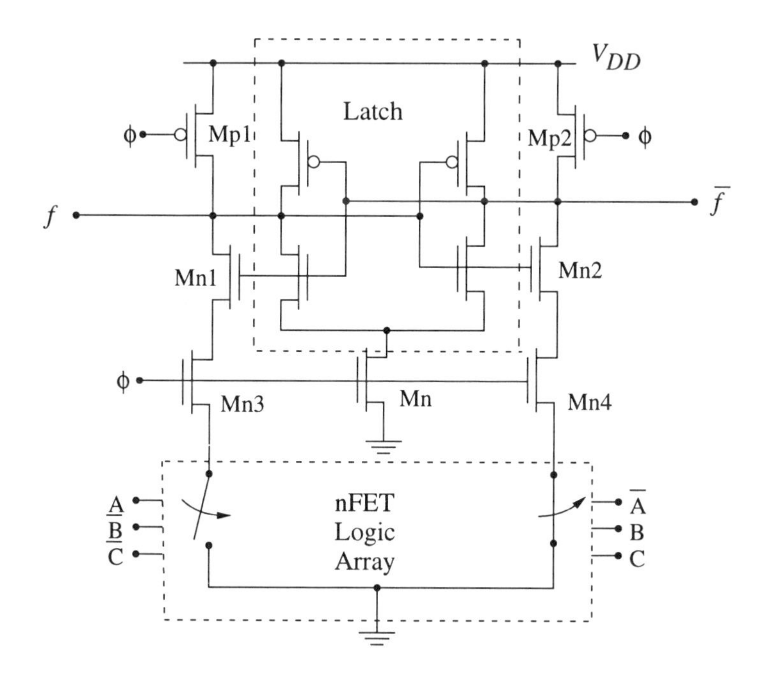

Figure 9.28 Basic DCSL circuit structure

9.4 Complementary Pass-Transistor Logic (CPL)

Complementary pass-transistor logic uses many features of CVSL to implement a very simple and compact approach to high-performance design. CPL is based on the use of nFET multiplexors to construct logic functions. In general, input variables are applied to both the gate and the drain/source connections of FETs as implied by the system drawing in Figure 9.29, with the output taken from the other side. Transistors are arranged in arrays that provide gate-level functions. One major advantage of CPL is that nFETs are used exclusively in the data path. Another interesting feature of this design style is that the logic function is changed by redefining the inputs while keeping the array topology[2] constant. These and other characteristic combine to make CPL a potentially power-

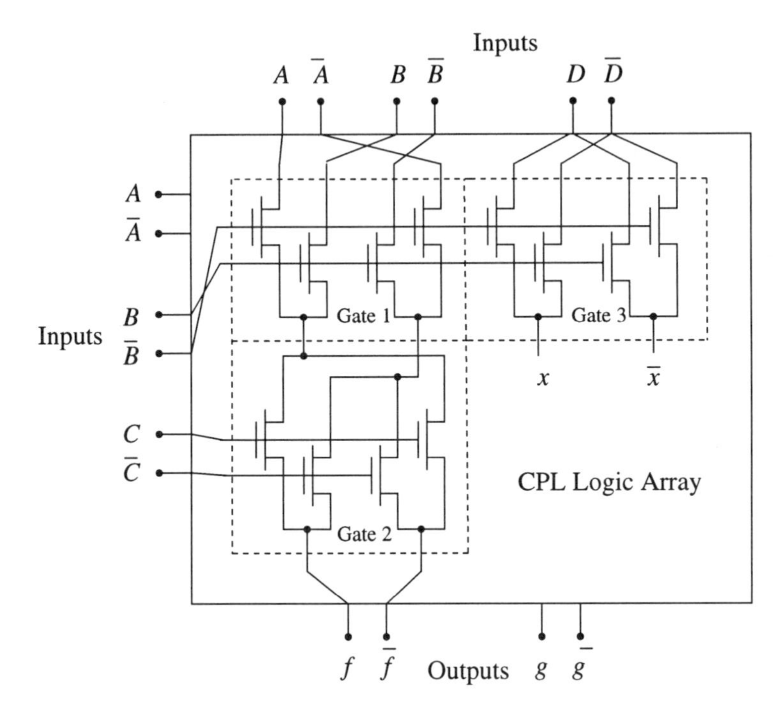

Figure 9.29 General structure of a CPL logic network

ful approach to high-density VLSI.

The basis of CPL is the use of transistors as fundamental logic elements. Consider the nFET illustrated in Figure 9.30(a). Logic formation is based on the observation that the switching of the transistor is described by the AND operation[3]

$$f = A \cdot B \tag{9. 17}$$

except that this expression is not defined for the case $B = 0$ when the transistor is in cutoff. To avoid the problem of a floating node at the output, another transistor is added to form the 2:1 MUX shown in Figure 9.30(b). The output of the MUX is described by

$$\begin{aligned} f &= A \cdot B + B \cdot \overline{B} \\ &= A \cdot B \end{aligned} \tag{9. 18}$$

where the second term evaluates to a logical 0, but is included in the electrical circuit to insure that $f = 0$ is at zero volts when B=0 as required. The simplicity of CPL logic is apparent: transistors are used as basic logic gates, with additional circuitry added to preclude the possibility of an undefined voltage.

The alert reader will have already questioned the use of single nFETs to perform logic due to the problems of

---

2 i.e., the circuit

3 Recall from Section 4.X that the output of a MOSFET switch is the logical AND of the signals applied to the input and the gate.

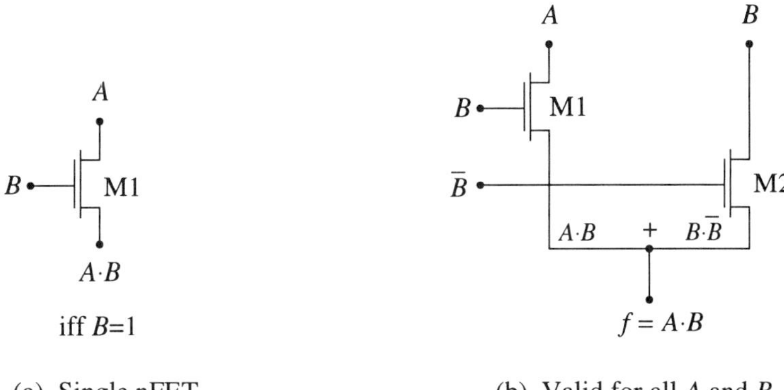

(a) Single nFET (b) Valid for all A and B

Figure 9.30 nFET implementation of the AND operation

- threshold voltage loss, and,
- slow logic 1 transfers.

These concerns are addressed by adding a static inverter at the output as shown in Figure 9.31(a). The inverter is used for two primary tasks: to restore the logic 1 voltage level to V_{DD}, and to increase the switching speed. It also increases the drive current for the next stage.

The design of the inverter can be an important consideration for this type of logic circuit. Since the input voltage V_X to the inverter is in the range $[0,V_{max}]$, where

$$\begin{aligned} V_{max} &= V_{DD} - V_{Tn} \\ &= (V_{DD} - V_{T0n}) - \gamma\left(\sqrt{2|\phi_F| + V_{max}} - \sqrt{2|\phi_F|}\right) \end{aligned} \tag{9. 19}$$

We see that the noise margins are an important DC consideration to insure that the logic 1 level can be detected. This is particularly true since the logic 1 transfer through the pass transistor is given by

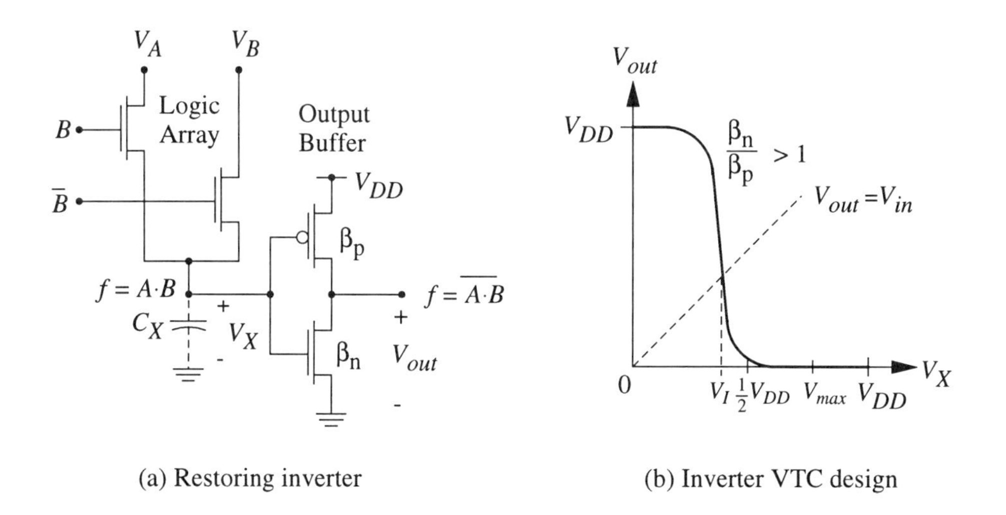

(a) Restoring inverter (b) Inverter VTC design

Figure 9.31 Output inverter design considerations

$$V_X(t) = V_{max}\left[\frac{t/2\tau_n}{1+t/2\tau_n}\right] \tag{9.20}$$

is a relatively slow transition. To design the inverter for this situation, recall the inverter threshold voltage

$$V_I = \frac{V_{DD} - |V_{Tp}| + \sqrt{\frac{\beta_n}{\beta_p}}V_{Tn}}{1 + \sqrt{\frac{\beta_n}{\beta_p}}} \tag{9.21}$$

gives the point where the voltage transfer curve intersects the unity gain line [as shown in Figure 9.31(b)]. To insure that the maximum logic 1 input voltage of $V_X = V_{max}$ is correctly interpreted, we want to shift the value of V_I to the left. One simple solution is to use identical sized transistors for the inverter with $(W/L)_n = (W/L)_p$, so that $(\beta_n / \beta_p) = (k'_n / k'_p) > 1$. This results in a value $V_X < 0.5V_{DD}$, with the actual value dependent upon the process parameters.

Once the inverter has been designed, then the important switching time for a logic 1 input can be estimated by solving for the time t_I needed for V_X to reach V_I. This is given by

$$t_I = 2\tau\left(\frac{V_I/V_{max}}{1+V_I/V_{max}}\right). \tag{9.22}$$

Alternately, the value t_I can be used as a design specification for the inverter through the condition

$$V_X(t_I) = V_I = V_{max}\left[\frac{t_I/2\tau_n}{1+t_I/2\tau_n}\right] \tag{9.23}$$

Once V_I is determined, then the inverter transistors are chosen using

$$\frac{\beta_n}{\beta_p} = \left(\frac{V_{DD} - V_I - |V_{Tp}|}{V_I - V_{Tn}}\right)^2 \tag{9.24}$$

Although this sets the proper DC characteristics, it is important to remember that the switching times t_{HL} and t_{LH} at the output of the inverter are individually dependent upon the values of β_n and β_p, respectively.

9.4.1 2-Input Arrays

Let us examine how CPL ideas can be used to construct 2-input logic gates for use in a dual-rail logic network. Figure 9.32 shows the AND/NAND array that is based on the simple circuit discussed above. The right side creates the NAND function by means of

$$\begin{aligned} f &= \overline{A} \cdot B + \overline{B} \cdot \overline{B} \\ &= \overline{A} + \overline{B} \\ &= \overline{A \cdot B} \end{aligned}, \tag{9.25}$$

where we have used redundancy to perform the first simplification, and the DeMorgan relation to arrive at the final result. Output inverters have been provided on both sides to overcome the threshold voltage loss problem and increase the drive capacity of the network.

The general structure of the FETs in the AND/NAND circuit define the structure of 2-input CPL

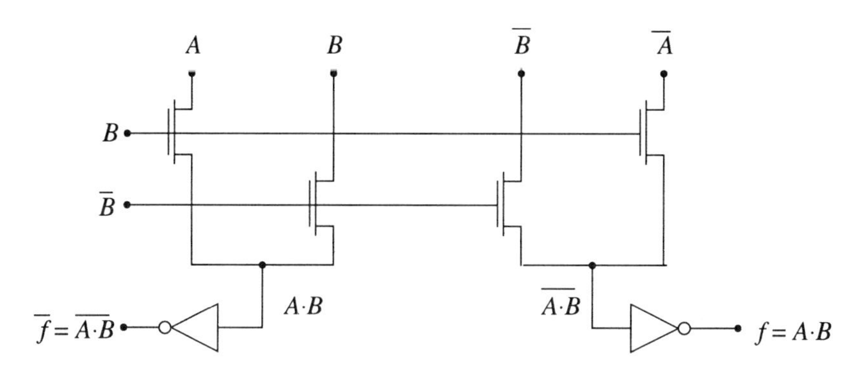

Figure 9.32 CPL AND/NAND 2-input array

arrays. A useful characteristic of CPL is that the function can be changed by rearranging the inputs to the arrays. If we reverse the gate inputs B and $\overline{B}$ in the AND/NAND network we obtain the OR/NOR circuit shown in Figure 9.33. The operation can be verified by applying the rules of FET logic formation. For the left side, the output is given by

$$\begin{aligned} A \cdot \overline{B} + B \cdot B &= A \cdot \overline{B} + B \\ &= A + B \end{aligned} \tag{9. 26}$$

while the right-hand array evaluates the NOR operation as seen in

$$\begin{aligned} \overline{A} \cdot \overline{B} + B \cdot \overline{B} &= \overline{A} \cdot \overline{B} \\ &= \overline{A + B} \end{aligned} \tag{9. 27}$$

by using the DeMorgan theorem. Once again, we have provided output inverters on both side.

Since the 2-input arrays are based on the a 2:1 multiplexor network, it is a simple matter to create the exclusive-OR and equivalence functions

$$\begin{aligned} A \oplus B &= A \cdot \overline{B} + \overline{A} \cdot B \\ \overline{A \oplus B} &= A \cdot B + \overline{A} \cdot \overline{B} \end{aligned} \tag{9. 28}$$

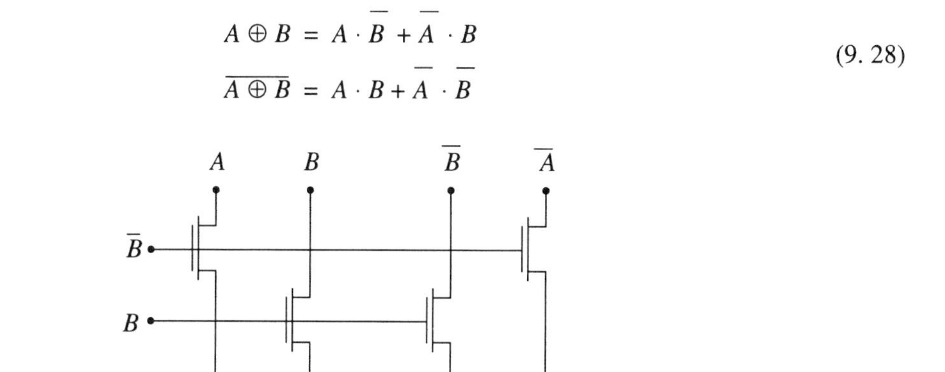

Figure 9.33 CPL OR/NOR 2-input array

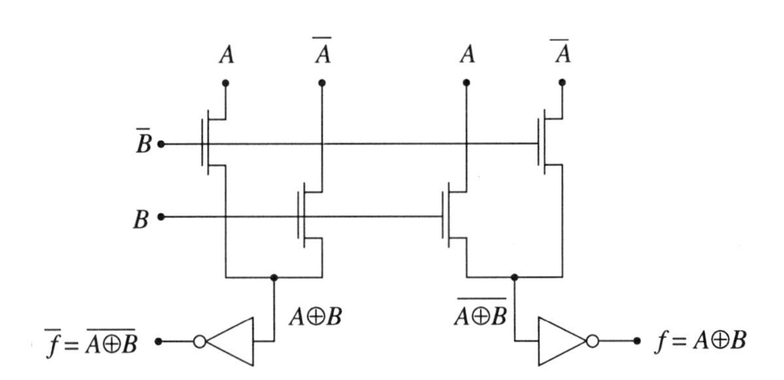

Figure 9.34 CPL XOR/XNOR 2-input array

using the networks shown in Figure 9.34.

It is possible to create other functions using the 2-input array as a basis, but it is important to remember that dual-rail logic requires that both the output f and its complement $\overline{f}$ need to be formed. Consider the array in Figure 9.35. By applying the logic rules, the left side evaluates to

$$f_1 = \overline{A} \cdot B + \overline{B} \cdot \overline{C} \tag{9. 29}$$

while the right side gives

$$f_2 = A \cdot B + \overline{B} \cdot C \tag{9. 30}$$

We see that f_2 is in fact the complement of f_1; this observation can be verified by using a simple truth table listing. However, caution must be exercised when designing logic functions as the placement of the input variables becomes critical. In a more general case, two outputs g_1 and g_2 should be checked to insure that they are in fact complements of one another. If not, we must either generate $\overline{g}_1$ and $\overline{g}_2$, using separate circuitry, or include the stages in a logic cascade where the final result at the end of the chain produces complementary outputs.

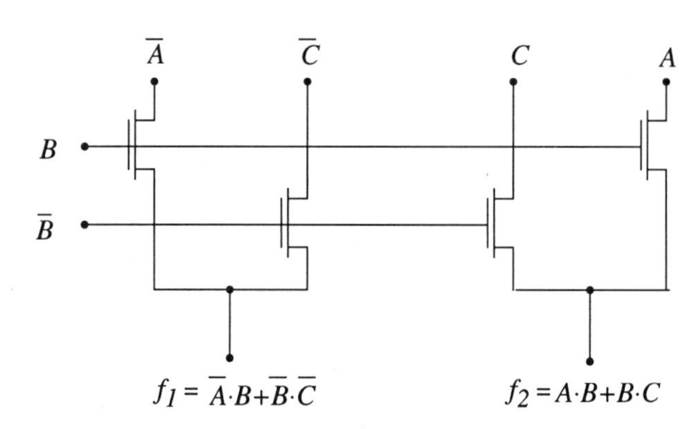

Figure 9.35 General logic arrangement in CPL

Layout

One striking feature of CPL arrays is the simplicity of the layout. One layout strategy for 2-input arrays is shown in Figure 9.36. This approach uses horizontal-oriented FETs and allows simple linear gates for the control variables X and $\overline{X}$. The metal routing is arbitrary at this point. If one is designing a library cell, then the input and output port locations should carefully selected so as to allow simple cascades and wiring. CPL has the advantage that the 2-input array layout can be used for any function pair AND/NAND, OR/NOR, or XOR/XNOR by routing the input signals to the proper nodes (a, b, c, d, and X, $\overline{X}$). In other words, the circuit topology is invariant, and the signal placement determines the actual logic function that the circuit performs. This aspect is even more intriguing when we note that we can optimize the circuit for switching time, and then maintain many aspects of the speed for every function.

9.4.2 3-Input Arrays

CPL also provides for 3-input gates using a structured approach. Figure 9.37 shows the switching network for a 3-input AND/NAND array. To understand the logic construction, consider first the left array. Applying the rules gives

$$A \cdot B \cdot C + A \cdot \overline{A} + B \cdot \overline{B} = A \cdot B \cdot C \tag{9.31}$$

directly. The right hand array evaluates to

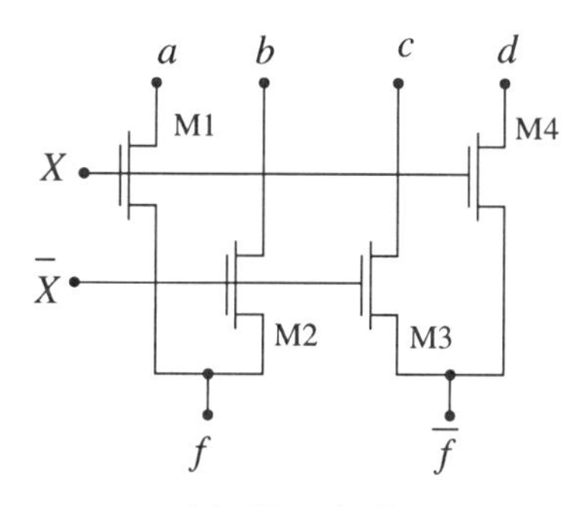

(a) Circuit diagram

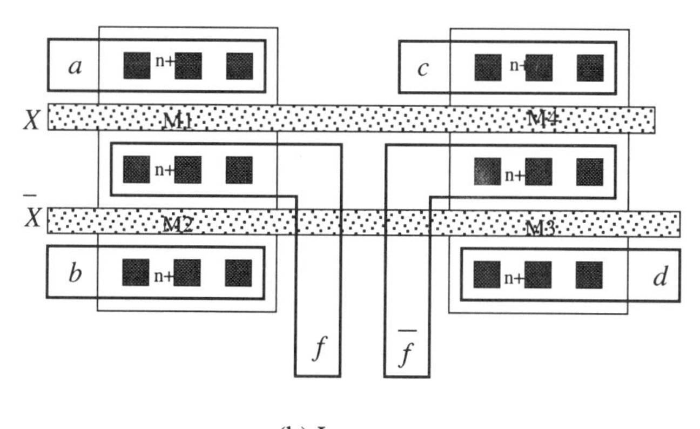

(b) Layout

Figure 9.36 2-input array layout example

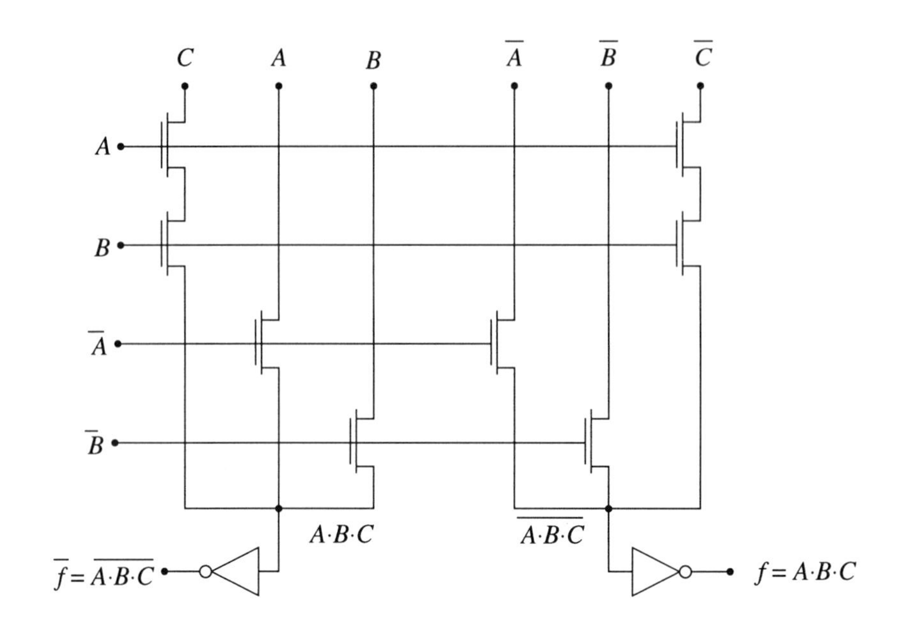

Figure 9.37 3-input AND/NAND array in CPL

$$\begin{aligned} \bar{A} \cdot \bar{A} + \bar{B} \cdot \bar{B} + A \cdot B \cdot \bar{C} &= \bar{A} + \bar{B} + A \cdot B \cdot \bar{C} \\ &= \bar{A} + \bar{B} + \bar{C} \\ &= \overline{A \cdot B \cdot C} \end{aligned} \tag{9. 32}$$

which is the NAND3 operation. We thus see that the AND/NAND pair can be built using only 8 nFETs to provide the logic. Inverters may be added at the output to restore the logic 1 voltage and speed up the circuit response.

A 3-input OR/NOR array can be obtained by simply interchanging the variables applied to the gates, which results in the circuit shown in Figure 9.38. The left side provides the OR function as verified by the reduction

$$\begin{aligned} \bar{A} \cdot \bar{B} \cdot C + A \cdot A + B \cdot B &= \bar{A} \cdot \bar{B} \cdot C + A + B \\ &= A + B + C \end{aligned} \tag{9. 33}$$

Similarly, the right hand array gives

$$\begin{aligned} A \cdot \bar{A} + B \cdot \bar{B} + \bar{A} \cdot \bar{B} \cdot \bar{C} &= \bar{A} \cdot \bar{B} \cdot \bar{C} \\ &= \overline{A + B + C} \end{aligned} \tag{9. 34}$$

and represents a NOR3 logic gate. It is worth mentioning again that the OR/NOR array uses the same transistors arrangement as the AND/NAND. The only difference between the two arrays is in the order of the input variables, just as in the case of the 2-input arrays.

One important aspect of the 3-input arrays is that the longest signal path on either side requires the transmission through 2 series-connected nFETs. This leads to the circuit shown in Figure 9.39.

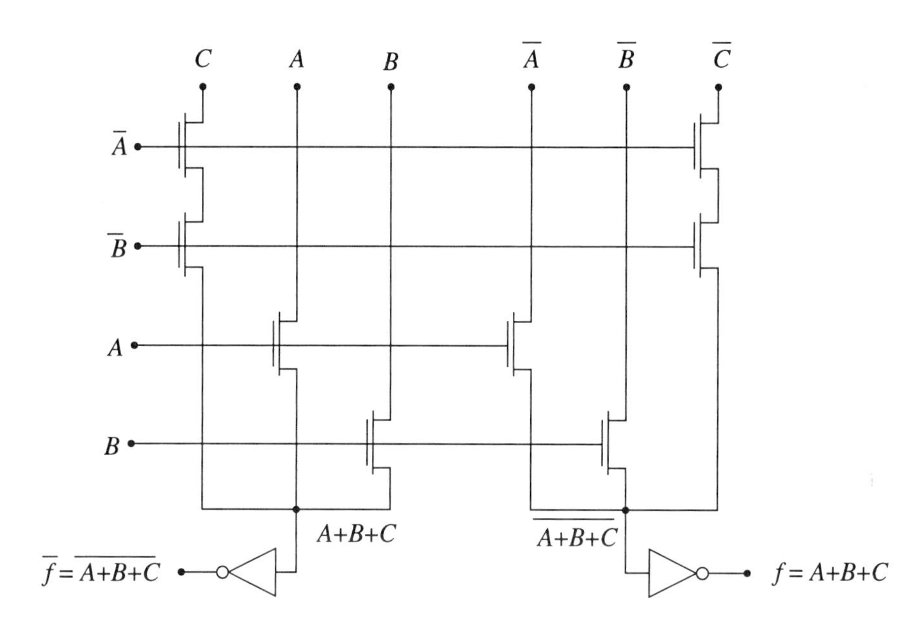

Figure 9.38 3-input CPL OR/NOR array

The delay can be estimated using the RC ladder time constant

$$\tau = (R_1 + R_2)\, C_X + R_1 C_1 \tag{9.35}$$

such that the inverter input voltage is approximated using

$$V_X(t) = V_{max}\,[1 - e^{-t/\tau}] \tag{9.36}$$

for a simple exponential logic 1 transfer. A better approximation is given by the nFET logic 1 analysis which says that

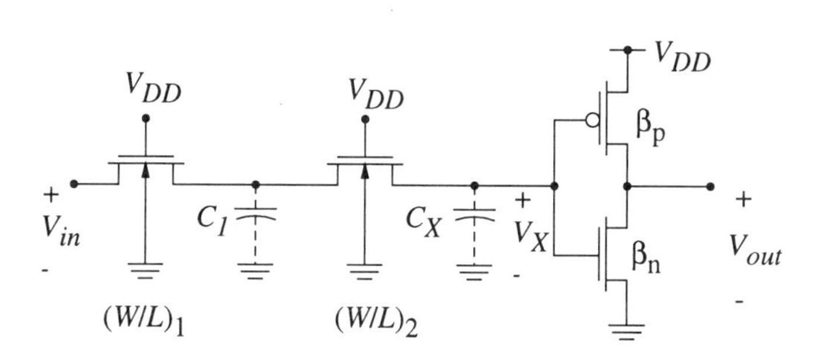

Figure 9.39 Delay circuit for 3-input array

$$V_X(t) = V_{max}\left[\frac{\left(\frac{t}{2\tau_n}\right)}{1+\left(\frac{t}{2\tau_n}\right)}\right] \tag{9.37}$$

indicating the poor logic 1 transfer characteristics. The critical design parameters are the aspect ratios and the layout parasitics, as always. The circuit should always be simulated on a computer, with particular attention paid to the parasitics.

9.4.3 CPL Full-Adder

A CPL full-adder circuit can be built entirely from 2-input arrays. The final circuits are shown in Figure 9.40. The sum circuit in (a) is easily seen to use two XOR/XNOR arrays to give

$$s_n = a_n \oplus b_n \oplus c_n \tag{9.38}$$

by a simple cascade. The carry-out circuit in (b) is a little more complicated. The first (upper) stages are AOI-type networks that provide output for the second (lower) stage. Let us examine the right side of the circuit to understand how c_{n+1} is formed before the level-restoring inverter. The right-most line gives

$$\bar{a}_n \cdot (a_n \cdot b_n + b_n \cdot c_n) = \bar{a}_n \cdot b_n \cdot c_n \tag{9.39}$$

where the first term has evaluated to 0. The remaining (left) line provides the terms

$$a_n \cdot (\bar{b}_n \cdot c_n + a_n \cdot b_n) = a_n \cdot b_n + a_n \cdot \bar{b}_n \cdot c_n \tag{9.40}$$

Combining the two by ORing yields

$$\begin{aligned} a_n \cdot b_n + a_n \cdot \bar{b}_n \cdot c_n + \bar{a}_n \cdot b_n \cdot c_n &= a_n \cdot b_n + c_n \cdot (a_n \oplus b_n) \\ &= c_{n+1} \end{aligned} \tag{9.41}$$

which is the required expression. A little algebra will verify that the output from the left side is c_{n+1} (before the inverter). This example illustrates how AOI functions can be handled in CPL.

9.5 Dual Pass-Transistor Logic (DPL)

Dual-pass transistor logic is also based on the use of nFETs as logic gates, but uses additional transistors (pFETs) to overcome some of the electrical problems found in CPL.

Recall that CPL uses nFETs as logic devices for passing both logic 0 and 1 voltages. At this point in our discussion, it is second nature to realize that the maximum voltage that can pass through an nFET is limited to

$$V_{max} = V_{DD} - V_{Tn} \tag{9.42}$$

because of the threshold voltage drop. CPL compensates for this drop by using static inverters at the outputs to restore the logic swing to a full rail value. DPL uses a different approach: create logic arrays where pFETs provide the logic 1 output values. While this may sound like an obvious solution,[4] DPL also addresses loading problems that arise in the manner in which CPL uses signals.

The main features of DPL can be understood by analyzing the AND array in Figure 9.41(a)

---

[4] After all, this entire book is about CMOS where pFETs are introduced to pass high voltages!

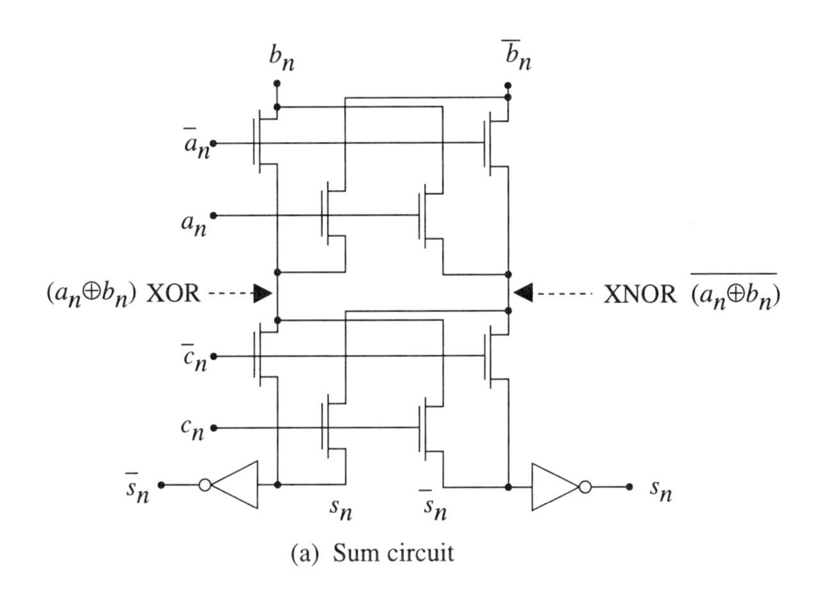

(a) Sum circuit

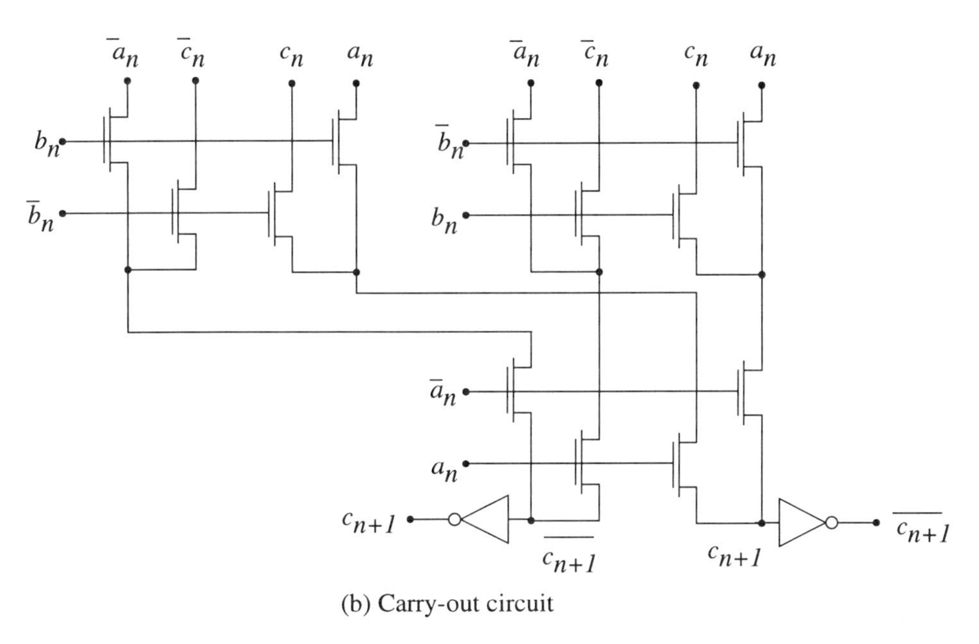

(b) Carry-out circuit

Figure 9.40 Full adder circuits designed in CPL

which consists of 4 transistors. A moment's inspection reveals that DPL uses two pFETs Mp1 and Mp2 to pass a high voltage to the output when

$$A \cdot B = 1 \tag{9. 43}$$

As with CPL, the output voltage for this case is derived from the signals A and B directly. If both are in the range $[0, V_{DD}]$, then the output will be

$$V_{out} = V_{DD} \tag{9. 44}$$

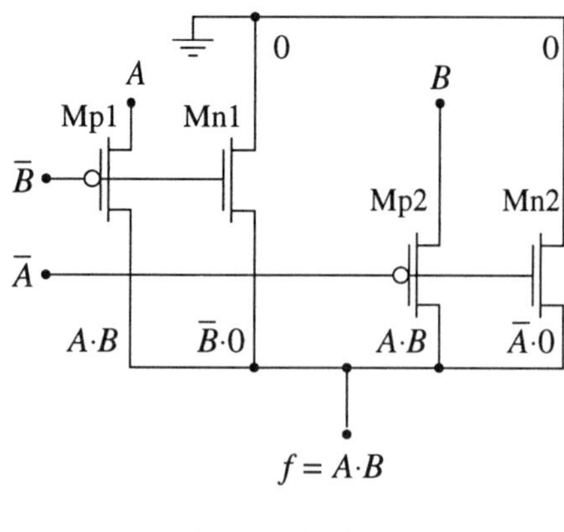

(a) AND logic array

A	B	Mp1	Mn1	Mp2	Mn2	f
0	0	off	ON	off	ON	0
0	1	ON	off	off	ON	0
1	0	off	ON	ON	off	0
1	1	ON	off	ON	off	1

(b) Operation summary

Figure 9.41 Operation of a DPL AND gate

for this case. Since there is no threshold voltage loss when we use a pFET to pass a logic 1, this is more reliable for low values of the power supply. Although this is the main point of introducing the additional transistors, the circuit operation for the case where

$$A \cdot B = 0 \tag{9.45}$$

is also worth studying. The three cases $A = 0 = B$, $A = 0$ and $B = 1$, and $A = 1$ and $B = 0$ are summarized in Figure 9.41(b). We see that if one input is 0, then the output is due to the transmission of the ground voltage through an nFET (even though a pFET is on, it cannot pass 0v level). If both inputs are 0, then the output is a result of 2 nFETs transmitting the ground voltage.

These results seem to nullify the usefulness of DPL in high-density designs. Compared to CPL, it uses twice as many transistors (and the associated increase in chip area) with more complicated interconnect wiring. At first sight, it seems that the only advantage is the ability to pass the power supply voltage V_{DD}. There is, however, another aspect that enters into the picture. Note that the inputs to the DPL gate are $(A, \overline{A})$ and $(B, \overline{B})$ and that each variable is only used once. This means that the driving gates have equal loading and can be identical. CPL circuits do not have this characteristic. Instead, the placement of the variables tends to be unbalanced in usage which makes high-speed design more complicated because of increased problems in timing and skew. This may (or may not) be more important than increased transistor count and interconnect complexity.

With this in mind, let us study other gates with this structuring. DPL is also a dual rail logic family, so that we really need to use the AND/NAND array shown in Figure 9.42. The NAND circuit on the right is the electrical complement of the AND circuit, and has the characteristics that it can pass the full range of voltages and exhibits balanced input usage. Figure 9.43 provides the DPL arrays for the OR/NOR operations; it is easily verified that these circuit possess the same primary characteristics. Finally, the XOR/XNOR pair is shown in Figure 9.44. Because of the AO structure of the basic equations, both gates exhibit a high degree of symmetry.

The techniques used in DPL are of interest to us because they illustrate the trade-offs that must be made when attacking high-speed, compact, logic design problems. DPL itself has been used to implement basic circuits such as adders, and is representative of this type of an approach to CMOS circuit design.

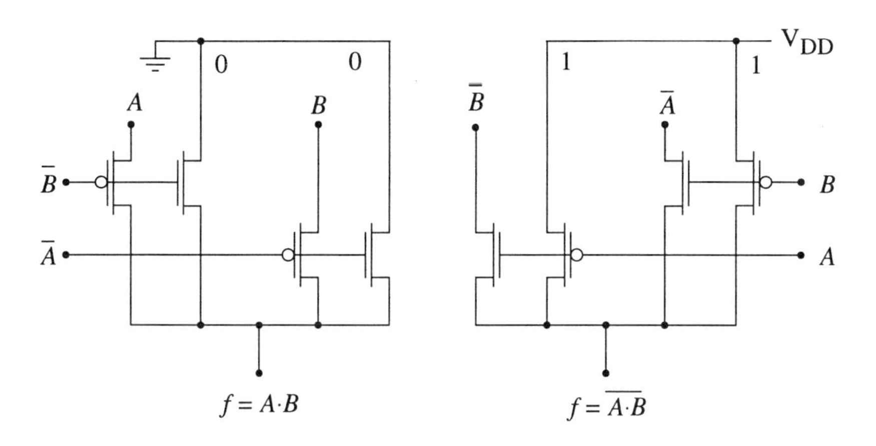

Figure 9.42 2-input DPL AND/NAND arrays

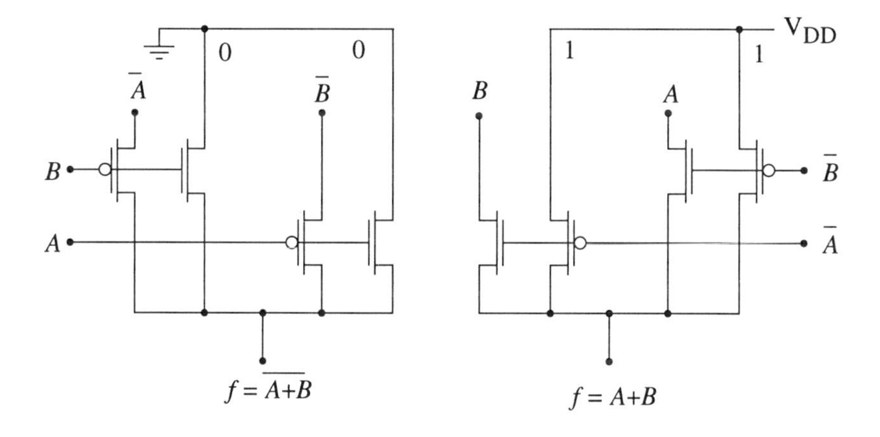

Figure 9.43 2-input OR/NOR arrays in DPL

9.6 Summary of Differential Design Styles

Thus far, this chapter has been directed towards studying various CMOS design styles for differential logic. The presentation started with CVSL and then evolved to concentrating on the two main sections of a CVSL gate, namely, nFET logic arrays and a latch/restoring circuit. In hindsight, it is straightforward to see how the various approaches are related to one another and tend to share one or more important characteristics. Perhaps the most interesting aspect of the chapter is that it illustrates how powerful CMOS really is: the basic idea of complementary transistors has provided the basis for an impressive list of various circuit techniques. [5]

To complete our trek down the "MOS-sy" road,[6] let us briefly examine one final design style that

[5] And, from the author's perspective, it made this book a viable project!

[6] Yes, this section is being written late at night...

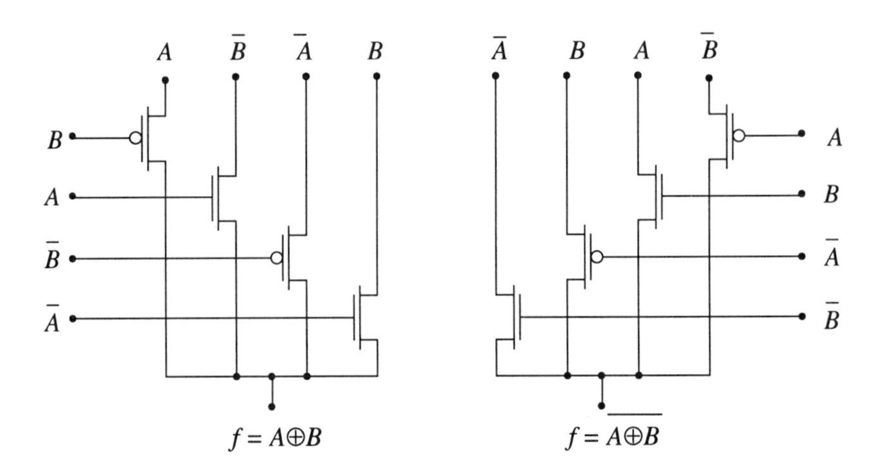

Figure 9.44 XOR and XNOR gates in DPL

illustrates the evolution and relationship of the various techniques. This approach has been named DCVSPG logic (for **Differential Cascode Voltage Switch with Pass-Gate** logic) by its developers, and is a combination of CVSL and CPL design styles.

Figure 9.45 shows the AND/NAND gate in this approach. The presence of the pFET latch and the nFET logic transistors Mn1 and Mn4 clearly shows that the circuit's roots are in CVSL. However, note that Mn1 is connected to the power supply voltage V_{DD}, so it acts as a pull-up device, not a pull down FET like Mn4. Moreover, two of the logic transistors Mn2 and Mn3 use input variables at both the gate and drain/source, indicating the inclusion of pass-gate logic ideas. The operation is straightforward to analyze. Consider the left side logic network. Denoting the power supply as a logic 1 gives us the basic equation

$$\overline{f} = \overline{A} \cdot 1 + A \cdot \overline{B} \tag{9.46}$$

This reduces to

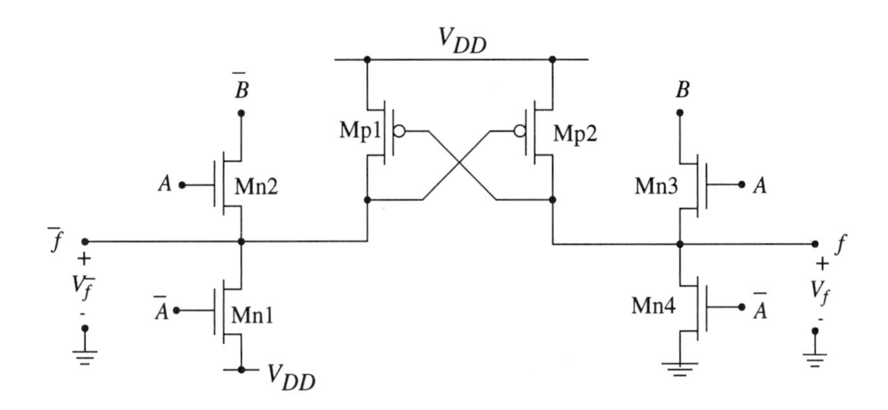

Figure 9.45 AND/NAND gate is DCVSPG logic

$$\begin{aligned} \bar{f} &= \bar{A} + \bar{B} \\ &= \overline{A \cdot B} \end{aligned} \tag{9.47}$$

so that f is the AND operation. This is verified by analyzing the logic circuit on the right side (with ground as a logic 0):

$$\begin{aligned} f &= \bar{A} \cdot 0 + A \cdot B \\ &= A \cdot B \end{aligned} \tag{9.48}$$

The electrical characteristics are unique in that the output may be driven by a connection to ground, the power supply, or directly by the logic voltage B or $\bar{B}$.

The basic structure of the AND/NAND gate provides the rules for constructing the OR/NOR in Figure 9.46. Comparing this with the AND/NAND circuit we see that the variables and their complements have been interchanged, as have the ground and power supply connections in the logic arrays. To verify the operation, we can construct the right-side logic as

$$\begin{aligned} f &= A \cdot 1 + \bar{A} \cdot B \\ &= A + B \end{aligned} \tag{9.49}$$

as advertised. Similarly, the logic array on the left side gives

$$\begin{aligned} f &= A \cdot 0 + \bar{A} \cdot \bar{B} \\ &= \overline{A + B} \end{aligned} \tag{9.50}$$

where we have used a DeMorgan reduction in writing the second line.

The DCVSPG XOR/XNOR gate is shown in Figure 9.47. A brief examination of the logic network shows that there are no connections between the logic FETs and the power supply or ground. In fact, both the right and left logic arrays are simple pass-FET groups that yield the desired functions as verified by writing the right-side logic equation

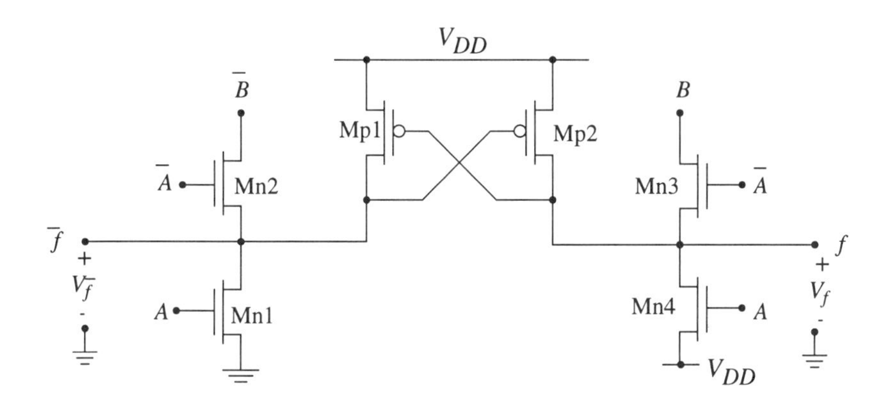

Figure 9.46 The OR/NOR gate in DCVSPG logic

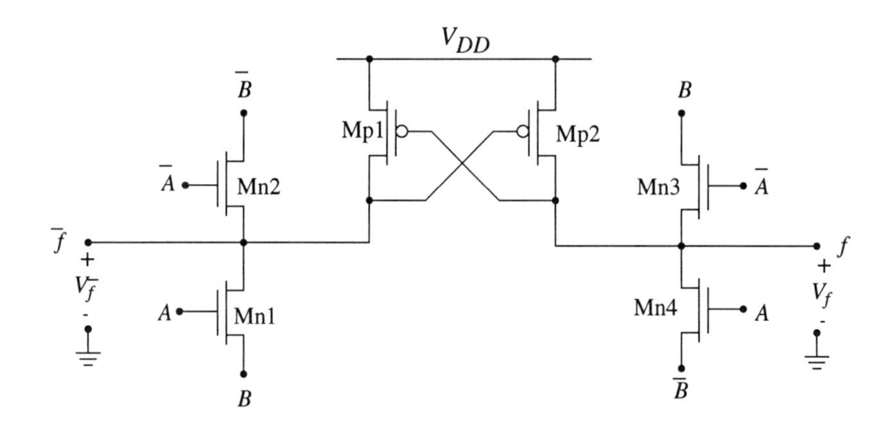

Figure 9.47 DCVSPG XOR/XNOR logic gate

$$f = \overline{A} \cdot B + A \cdot \overline{B} \\ = A \oplus B \tag{9. 51}$$

A similar equation can be written for the left-side logic by translating the FET circuits into logic expressions.

This discussion has been included to make a point: all CMOS logic design styles are based on a relatively simple set of fundamental principles. The evolution of the distinct approaches to achieving circuits that falls within the critical budget limits of speed, area, or power are all very similar, yet each has characteristics that make it unique. If one takes the time to study the subject from the invention of CMOS in the 1960's to the present, it is clear that the development of different approaches has been spurred on by developments in fabrication technology and the increased sophistication of the concept of a "system" on a chip. Since neither effort is fading, we should expect to see continued work in this field.

9.7 Single/Dual Rail Conversion Circuits

Dual rail logic circuits require complementary signals at every stage. Since all input/output formats are single rail by design, we need to examine the circuitry needed to convert between the two formats.

9.7.1 Single-to-Dual Rail Conversion

This problem centers around taking a single input variable a and providing both a and $\overline{a}$ at the output. The simplest approach is to use an inverter to generate the complementary signal using the circuitry shown in Figure 9.48. This approaches induces a small delay of t_p between the two outputs. If this is a problem, then a D-type latch can be used with slightly less skew.

9.7.2 Dual-to-Single Rail Conversion

This problem is more difficult. If we use dual rail logic circuits, then our result will be in the form of complementary outputs f and $\overline{f}$. Although we could just use f as the single rail quantity, we will lose some of the speed that we gained by going to differential logic circuits in the first place.

A more efficient approach is to create a circuit that can sense the difference $(f - \overline{f})$ at the high data rates, and translate that into a single output F with the improved slew rate. This is illustrated in Fig-

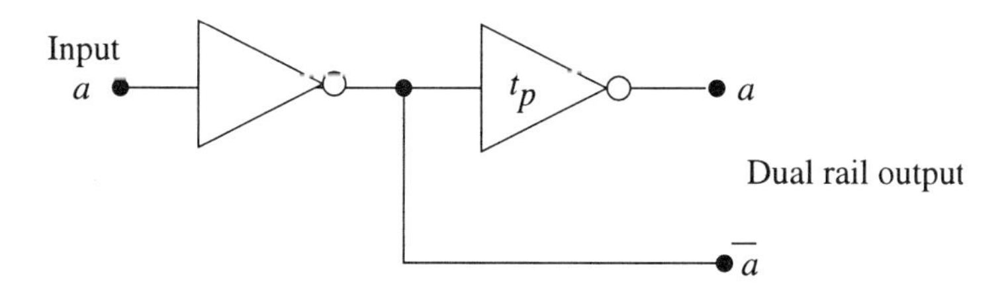

Figure 9.48 Single-to-dual rail conversion

ure 9.49. A straightforward approach to solving this problem is to use a differential amplifier circuit.

Differential Amplifier

The differential amplifier accepts two input voltages V_1 and V_2 and produces an output voltage that depends upon the difference

$$V_D = V_1 - V_2 \tag{9.52}$$

This type of circuit can be used to convert a dual-rail logic signal back into a single-rail value that can be interfaced with standard logic circuits.

A basic differential amplifier is shown in Figure 9.50. This uses a **source-coupled pair** of nFETs Mn1 and Mn2 as the input devices. The pFETs Mp1 and Mp2 are used as **active-load** devices to provide a pull-up path to the power supply voltage V_{DD}. A single output voltage V_o is shown; it is a single-rail variable corresponding to the value associated with the input voltage V_1. The output voltage is determined by the current I_{D2} as it produces a voltage drop across Mp2. To analyze the circuit, assume that the nFETs are both in saturation. The currents I_{D1} and I_{D2} are given by

$$\begin{aligned} I_{D1} &= \frac{\beta_n}{2}(V_{GS1} - V_{Tn})^2 \\ I_{D2} &= \frac{\beta_n}{2}(V_{GS2} - V_{Tn})^2 \end{aligned} \tag{9.53}$$

where we will assume that Mn1 and Mn2 have the same aspect ratio so that β_n applies to both. At the source, the currents must sum to

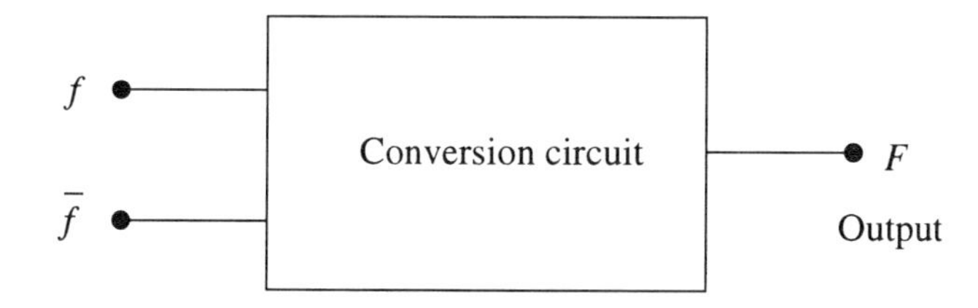

Figure 9.49 Conversion network for a single-rail output

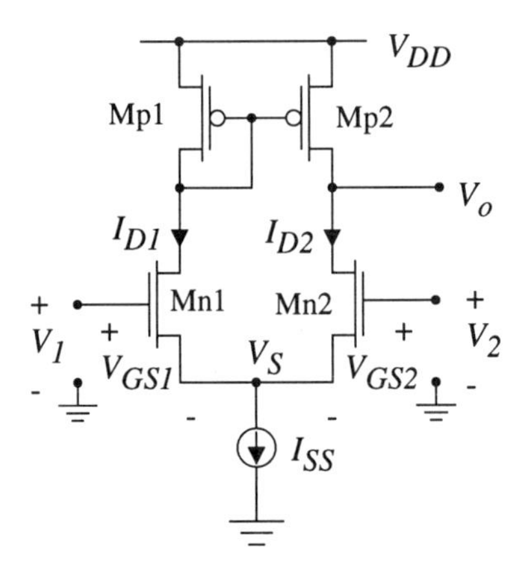

Figure 9.50 Differential amplifier circuit

$$I_{D1} + I_{D2} = I_{SS} \tag{9.54}$$

due to the current source there. The behavior of the circuit revolves around finding I_{D1} and I_{D2} as functions of the difference voltage V_D.

Let us calculate the currents by first noting that both nFETs have the same source voltage, so that

$$\begin{aligned} V_{GS1} - V_{GS2} &= V_1 - V_2 \\ &= V_D \end{aligned} \tag{9.55}$$

provides the important relationship between the input voltages V_1 and V_2 and the current equations. Next, we may rearrange the expression for I_{D1} and I_{D2} to give

$$\begin{aligned} V_{GS1} &= \sqrt{\frac{2I_{D1}}{\beta_n}} + V_{Tn} \\ V_{GS2} &= \sqrt{\frac{2I_{D2}}{\beta_n}} + V_{Tn} \end{aligned} \tag{9.56}$$

so that

$$V_D = \sqrt{\frac{2I_{D1}}{\beta_n}} - \sqrt{\frac{2I_{D2}}{\beta_n}} \tag{9.57}$$

This may be used to find the desired relations. For example, I_{D2} may be eliminated by writing

$$V_D = \sqrt{\frac{2I_{D1}}{\beta_n}} - \sqrt{\frac{2(I_{SS} - I_{D1})}{\beta_n}} \tag{9.58}$$

Squaring both sides gives

$$V_D^2 = \frac{2I_{D1}}{\beta_n} + \frac{2(I_{SS} - I_{D1})}{\beta_n} - 2\sqrt{\frac{4}{\beta_n^2} I_{D1}(I_{SS} - I_{D1})} \qquad (9.59)$$

Expanding and simplifying,

$$\frac{\beta_n V_D^2}{4} = \frac{I_{SS}}{2} - \sqrt{I_{D1}(I_{SS} - I_{D1})} \qquad (9.60)$$

which, upon squaring and rearrangement, gives

$$I_{D1}^2 - I_{SS} I_{D1} + \frac{1}{4}\left[I_{SS}^2 + \frac{\beta_n^2 V_D^4}{4} - I_{SS}\beta_n V_D^2\right] = 0 \qquad (9.61)$$

This is a quadratic equation for I_{D1} with a solution of

$$I_{D1} = \frac{I_{SS}}{2}\left[1 + \sqrt{\frac{\beta_n V_D^2}{I_{SS}} - \frac{\beta_n^2 V_D^4}{4I_{SS}^2}}\right] \qquad (9.62)$$

where we have chosen the positive root to insure that I_{D1} increases as V_1 increases. Similarly, the I_{D2} is found to be

$$I_{D2} = \frac{I_{SS}}{2}\left[1 - \sqrt{\frac{\beta_n V_D^2}{I_{SS}} - \frac{\beta_n^2 V_D^4}{4I_{SS}^2}}\right] \qquad (9.63)$$

It is seen by inspection that $I_{D1} + I_{D2} = I_{SS}$. Also note that when V_D=0 corresponding to $V_1 = V_2$,

$$I_{D1}\big|_{V_D = 0} = I_{D2}\big|_{V_D = 0} = \frac{I_{SS}}{2} \qquad (9.64)$$

This corresponds to equal inputs giving balanced current flow.

The general behavior of both currents I_{D1} and I_{D2} are shown in Figure 9.51. As V_D increases from a negative number ($V_1 < V_2$) to a positive value ($V_1 > V_2$), I_{D1} increases from 0 to the maximum value of I_{SS}, while I_{D2} has the opposite behavior. The value of the difference voltage V_D needed to obtain $I_{D2} = 0$ can be calculated by setting

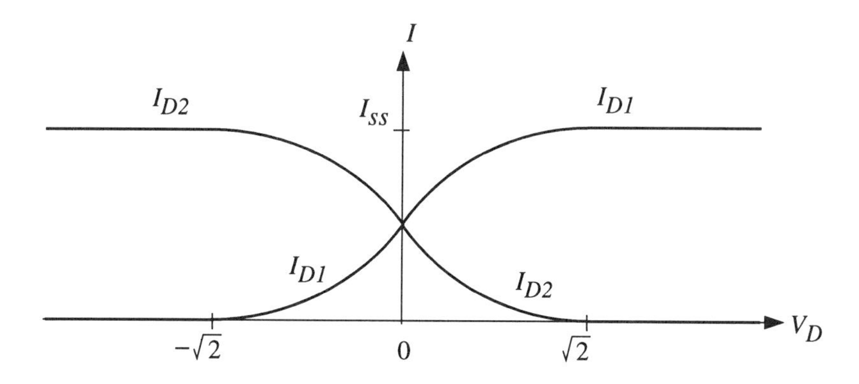

Figure 9.51 Currents in a differential amplifier

$$\frac{\beta_n V_D^2}{I_{SS}} - \frac{\beta_n^2 V_D^4}{4I_{SS}^2} = 1 \quad (9.65)$$

which is a quartic for V_D. Solving for V_D^2 and then taking the square root gives the value as

$$V_D = \sqrt{\frac{2I_{SS}}{\beta_n}} \quad (9.66)$$

The same approach can be used to show that

$$V_D = -\sqrt{\frac{2I_{SS}}{\beta_n}} \quad (9.67)$$

gives $I_{D1} = 0$, so that the total voltage change needed to divert the current from one FET to the other is

$$(\Delta V_D) = 2\sqrt{\frac{2I_{SS}}{\beta_n}} \quad (9.68)$$

This shows that the width of the input transition is set by the ratio of (I_{SS}/β_n). This provides a basis for the circuit design since

$$\beta_n = k'_n\left(\frac{W}{L}\right)_n \quad (9.69)$$

can be chosen according to the value of I_{SS}.

9.7.3 A Basic Current Source

The above analysis shows that the sensitivity of the differential amplifier depends upon the value of the current source I_{SS}. Although several types of current source circuits have been published in the literature, the simple one illustrated in Figure 9.52 illustrates the important points. This circuit uses a FET MnC to provide the current. It is biased with a gate-source voltage of V_R, where V_R is a reference voltage supplied by the voltage divider circuit made up of MpR and MnR. Assuming that MnC is biased into saturation, we can estimate

$$I_{SS} \approx \frac{\beta_{nC}}{2}(V_R - V_{Tn})^2 \quad (9.70)$$

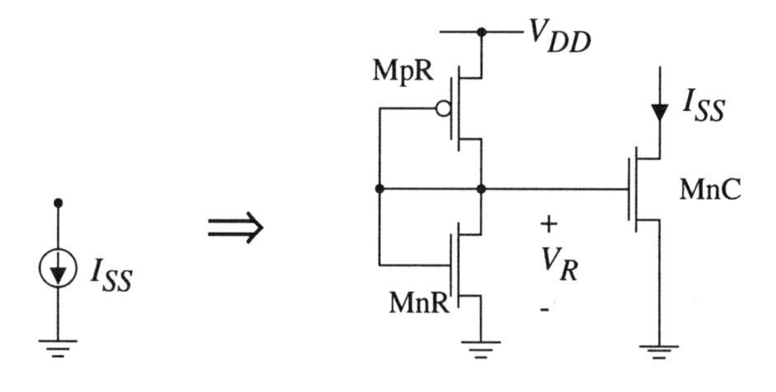

Figure 9.52 Simple current source design

Channel length modulation effects may be included by multiplying this expression by the factor $[1+\lambda(V_{DnS}-V_{sat})]$.

To determine the value of V_R, first note that MpR is defined by the terminal voltages

$$V_{SGp} = V_{DD} - V_R = V_{SDp} \tag{9.71}$$

while the nFET MnR has

$$V_{GSn} = V_R = V_{DSn}. \tag{9.72}$$

This shows that both transistors are saturated, so equating drain currents gives

$$\beta_{nR}(V_R - V_{Tn})^2 = \beta_{pR}(V_{DD} - V_R - |V_{Tp}|)^2. \tag{9.73}$$

Rearranging gives

$$V_R = \frac{V_{DD} - |V_{Tp}| + \sqrt{\frac{\beta_{nR}}{\beta_{pR}}}V_{Tn}}{1 + \sqrt{\frac{\beta_{nR}}{\beta_{pR}}}} \tag{9.74}$$

This shows that the ratio (β_{nR}/β_{pR}) can be used to set V_R, which in turn biases MnC to provide I_{SS}. The alert reader will have noticed that this identical to the formula for the inverter threshold voltage V_I; this is due to the fact that the voltage divider circuit made up of MnR and MpR is simply an inverter with the input shorted to the output!

9.8 Problems

[9-1] Design the CVSL logic gate for the function

$$f = \overline{(\bar{x} + y) \cdot (w + z)} \tag{9.75}$$

and its complement using the AOI/OAI logic network design approach.

[9-2] Design the CVSL logic gate for the function

$$g = \overline{\bar{A} \cdot B + C} \tag{9.76}$$

and its complement using the AOI/OAI logic network design approach.

[9-3] Design the CVSL logic gate for the function

$$h = \overline{(x + \bar{y}) \cdot (a \cdot \bar{b} + c)} \tag{9.77}$$

using AOI/OAI design approach.

[9-4] Create the CVSL logic tree network for the 2-input function described by the table shown in Figure P9.1.

[9-5] Design the CVSL gate by using the function in Figure P9.2 to construct the logic tree.

[9-6] Design the CVSL gate by using the information provided in the truth table of Figure P9.3 to construct the logic tree for the function G and $\overline{G}$.

f	0	1	1	0
a	0	1	0	1
b	0	0	1	1

Figure P9.1

F	1	1	0	0	0	1	0	1
A	0	1	0	1	0	1	0	1
B	0	0	1	1	0	0	1	1
C	0	0	0	0	1	1	1	1

Figure P9.2

G	0	0	1	0	0	1	1	1
A	0	1	0	1	0	1	0	1
B	0	0	1	1	0	0	1	1
C	0	0	0	0	1	1	1	1

Figure P9.3

[9-7] Consider the logic function

$$F = A \cdot \bar{B} + B \cdot \bar{C} \tag{9. 78}$$

(a) Construct the function table for this function using a horizontal format where the output F and the input variables are in one column with the top-to-bottom order of *F, A, B, C*.

(b) Construct the CVSL logic tree as discussed in the text.

[9-8] Use 2-input CPL arrays to implement the NAND4

$$\begin{aligned} f &= a \cdot b \cdot c \cdot d \\ \bar{f} &= \overline{a \cdot b \cdot c \cdot d} \end{aligned} \tag{9. 79}$$

[9-9] Use CPL gates to construct the circuit for the logic function

$$G = a \cdot b + c \tag{9. 80}$$

and its complement.

[9-10] Create the DPL circuit for the odd function

$$T = x \oplus y \oplus z \tag{9. 81}$$

and its complement using basic DPL cascades. Then compare your circuit with the CPL equivalent by looking at device count and electrical operation.

9.9 References

[1] K.M. Chu and D.L. Pulfrey, "A Comparison of CMOS Circuit Techniques: Differential Cascode Voltage Switch Logic Versus Conventional Logic," *IEEE J. Solid-State Circuits*, vol. SC-22, no. 4, pp. 528-532, August, 1987.
[2] K.M. Chu and D.L. Pulfrey, "Design Procedures for Differential Cascode Voltage Switch Logic Circuits," *IEEE J. Solid-State Circuits*, vol. SC-21, no. 4, pp. 1082-1087, December , 1986.
[3] T. A. Grotjohn and B. Hoefflinger, "Sample-Set Differential Logic (SSDL) for Complex High-Speed VLSI," *IEEE J. Solid-State Circuits*, vol. SC-21, no. 2, pp. 367-369, April, 1986.
[4] L.G. Heller, *et al.*, "*Cascode voltage switch logic: a differential CMOS logic family*, ISSCC84 Digest, pp. 16-17, February, 1984.
[5] N. Kanopoulos and N. Vasanthavada, "Testing of Differential Cascode Voltage Switch (DCVS) Circuits," *IEEE J. Solid-State Circuits*, vol. SC-25, no. 3, pp. 806-812, June, 1990.
[6] F-S. Lai and W. Hwang, "Design and Implementation of Differential Cascode Voltage Switch with Pass-Gate (DCVSPG) Logic for High-Performance Digital Systems," *IEEE J. Solid-State Circuits*, vol. 32, no. 4, pp. 563-573, April, 1997.
[7] S-H. Lu, "Implementation of Iterative Networks with CMOS Differential Logic," *IEEE J. Solid-State Circuits*, vol. 23, no. 4, pp. 1013-1017, August, 1988.
[8] D. Somasekhar and K. Roy, "Differential Current Switch Logic: A Low Power DCVS Logic Family," *IEEE J. Solid-State Circuits*, vol. 31, no. 7, pp. 981-991, July, 1996.
[9] M. Suzuki, *et al.*, "A 1.5ns 32-b CMOS ALU in Double Pass-Transistor Logic," *IEEE J. Solid-State Circuits*, vol. 28, no. 11, pp. 1145-1151,November, 1993.
[10] K. Yano, *et al.*, "A 2.8-ns CMOS 16x16-b Multiplier Using Complementary Pass-Transistor Logic," *IEEE J. Solid-State Circuits*, vol. 25, no. 2, pp.388-395, April, 1990.

Chapter 10

Issues in Chip Design

Designing a CMOS integrated circuit requires more than just understanding the logic circuits. Items such as interconnect delay on the chip, and interfacing the circuit to the outside world require special considerations. In this chapter we will study some important circuit problems that occur at the chip level and affect the internal operations. The introduction here is designed to provide a solid background for more specialized studies.

10.1 On-Chip Interconnects

It is interesting to examine the evolution of MOS technology in recent years. The typical channel length in a transistor has shrunk to a nominal value of less than 0.2 microns using the best manufacturing technology. With this type of resolution, the footprint area required for an FET has shrunk to the point where it is almost insignificant when compared with the surface area needed for contacts, vias, and interconnect routing. This leads to the conclusion that modern CMOS chip design is **interconnect-limited**. In other words, we usually don't worry about the number of FETs on a chip; in most cases, the wiring complexity is much more important to the real estate consumption.

Modern CMOS process flows provide several metal layers for use as interconnect wiring. Although 3 or 4 interconnect layers were sufficient for networks with a million or so FETs, the high-density compact systems being designed at the start of the 21st century require the use of 7-to-10 or more interconnect layers. Obviously, accurate modelling of on-chip wiring is important to the circuit designer. Parasitic-induced delays and stray coupling may require "tweaking" or re-design at the circuit level to make a chip operational. We will therefore direct our attention to this important topic from the viewpoint of electrical modelling of the interconnect structures.

10.1.1 Line Parasitics

Let us examine the basic geometry shown in Figure. 10.1. This is representative of an interconnect line that is described by a width w, and has a distance of d. The material layer itself has a height (or thickness) h. Parasitic electrical elements include resistance and capacitance; although the wire also

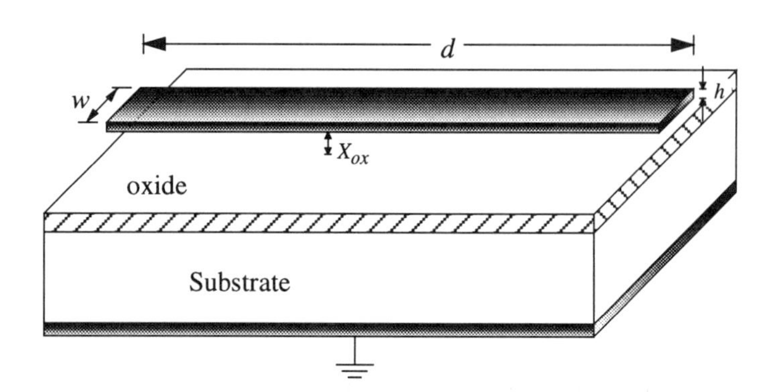

Figure 10.1 Geometrical structure of the an interconnect line

has inductance associated with it, we usually do not encounter magnetic effects at local circuit level.[1] Our program at this point is two-fold. First, we want to determine the values of the parasitic elements that are introduced by the interconnect. Once these have been calculated, we can then proceed to evaluate their effects on the performance.

Line Resistance

The resistance of the line from one end to the other is computed by using the standard equation

$$R_{line} = \frac{\rho d}{A} = \frac{\rho d}{hw} \tag{10.1}$$

where ρ is the resistivity in units of [Ω-cm] and $A = hw$ is the cross-sectional area of the line. Every material is characterized by a value of ρ. When choosing interconnect lines, metals dominate due to their small values of resistivity. Although this expression can be used directly, a more useful formulation for use in chip design is based on the use of the **sheet resistance** R_s which has units of ohms (Ω) for the layer. This is defined by

$$R_S = \frac{\rho}{h} \tag{10.2}$$

and is the end-to-end resistance of a square section of material wit $d = w$ as seen from the top. The sheet resistance is useful as it can be directly measured on a test structure in the laboratory. Once R_s is known, then the total resistance of a line that has a width w and spans a distance d is given by

$$R_{line} = R_s n \tag{10.3}$$

where

[1] This is due to the small current flow levels. One exception (among several) to this statement are the power supply and ground lines, which can exhibit inductive effects. For example, if the current changes very quickly in a power supply line then there is a temporary voltage drop of $v = L(di/dt)$ on the line, which reduces the voltage reaching the circuit.

$$n = \frac{d}{w} \tag{10. 4}$$

is the **number of squares** of dimensions ($w \times w$) encountered by the current. This can be seen by the top view of the interconnect shown in Figure 10.2. Owing to this observation, R_s is sometimes labeled as having units of "ohms per square" in processing jargon.It is worthwhile to note that the sheet resistance is quite sensitive to the height h of the interconnect. As the minimum linewidth has decreased, the thickness of the material layer has remained fairly large. In fact, most of the interconnects in a state-of-the art process are thicker than they are wide.

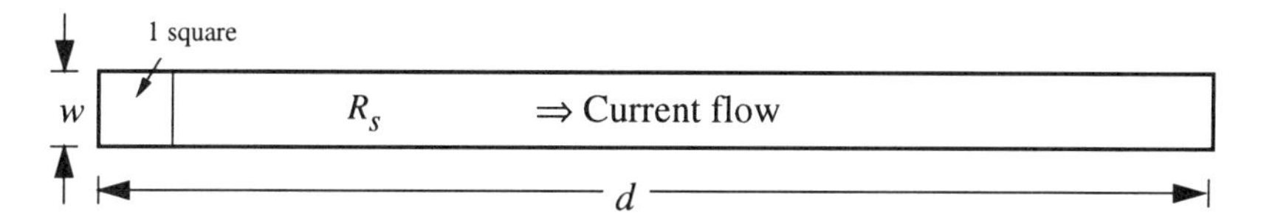

Figure 10.2 Top view of interconnect geometry

Example 10-1

Doped polysilicon has a best-case sheet resistance of about 20-to-25 Ω per square. Consider a poly line that has a width of 0.4 μm and a length of 20 μm . The number of squares of size (0.4 $\mu m \times$ 0.4 μm) is

$$n = \frac{20}{0.4} = 50 \tag{10. 5}$$

so with $R_s = 20\ \Omega$ we have a line resistance of

$$R_{line} = 20\,(50) = 1k\Omega \tag{10. 6}$$

To show the relative significance of this results, consider an nFET with $k' = 200\mu A/V^2$, an aspect ratio of 10, and a threshold voltage of 0.7 volts. With a 3.3v power supply, the linearized resistance is

$$R = \frac{1}{(200\times10^{-6})\,(5)\,(3.3-0.7)} = 192\Omega \tag{10. 7}$$

so that the line resistance is about five times larger than the FET drain-source resistance. If we used this type of interconnect at the output of a logic gate, the parasitic resistance would dominate the delay times.

Line Capacitance

The capacitance of an interconnect line can be the limiting factor in high-speed signal transmission. Consider the cross-sectional geometry shown in Figure 10.3(a). Most formulations are based on the capacitance per unit length c' with units of farads per centimeter such that the total capacitance of the line in farads is given by

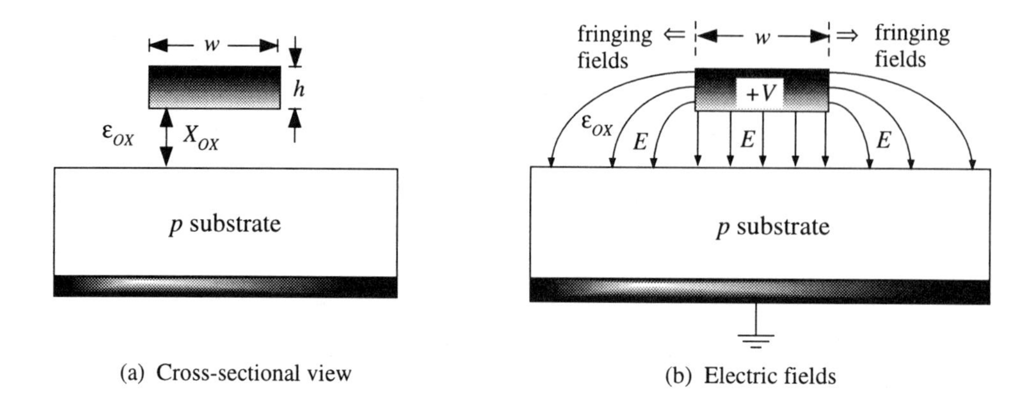

Figure 10.3 Cross-sectional view of the interconnect line

$$C_{line} = c'd. \tag{10. 8}$$

The simplest expression for c' is obtained from the parallel-plate capacitor as

$$c' = \frac{\varepsilon_{ox} w}{X_{ox}} \tag{10. 9}$$

in units of F/cm.. However, this formula ignores the existence of fringing electric fields at the edges, and is therefore too small.

A more accurate empirical expression is given by

$$c' = \varepsilon_{ox}\left[1.15\left(\frac{w}{X_{ox}}\right) + 2.8\left(\frac{h}{X_{ox}}\right)^{0.222}\right] \ F/cm. \tag{10. 10}$$

This gives a value that is larger than the simple parallel-plate formula since it accounts for the fringing fields shown in Figure 10.3(b). The first term is the standard parallel plate formula that has been increased by 15% to account for fringing fields at the bottom, while the second term accounts for fringing field lines that originate from the side of the layer with a height h.

In modern interconnects, we have reached the point where the line width w is smaller than the thickness (height) h. This implies that we cannot ignore the fringing capacitance contributions without incurring significant errors.

10.1.2 Modelling of the Interconnect Line

Now that the parasitics have been identified, let us examine the electrical characterization of the interconnect line. The first task we need to address is the construction of an electrical model for the interconnect structure itself. Consider the network illustrated in Figure 10.4 in which the output of Inverter 1 is used as the input to Inverter 2 via the interconnect line. The important relationship that must be determined is how the voltage $V_1(t)$ is related to $V_2(t)$ by the characteristics of the interconnect parasitics. There are essentially three approaches that are found in the literature.

Lumped Element Approximation

The simplest approach is to model the interconnect using a resistor of value R_{line} and a capacitor of

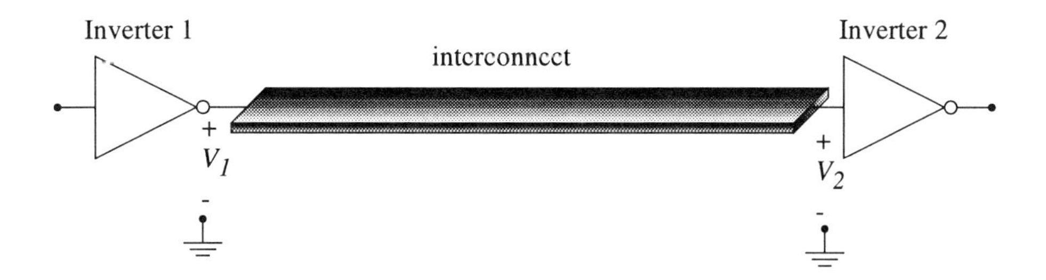

Figure 10.4 Model of interconnect application

value C_{line} as shown in Figure 10.5(a). To understand the effect of the parasitic line elements, let us analyze the circuit shown in Figure 10.5(b) which includes the important parasitics of the inverter circuits. In particular, we note that the FET capacitance C_{FET} in the circuit represents the device contributions due to Inverter 1 transistors Mp1 and Mn1, while C_{in} is the input capacitance of stage 2 due to the transistors in Inverter 2. Using this network we see that the important voltage is V_2 into the second stage.

To analyze the effects of the line parasitics, we assume that $V_2(t)$ can be approximated as an exponential and use the Elmore formula calculate the time constants. For a high-to-low transition we write

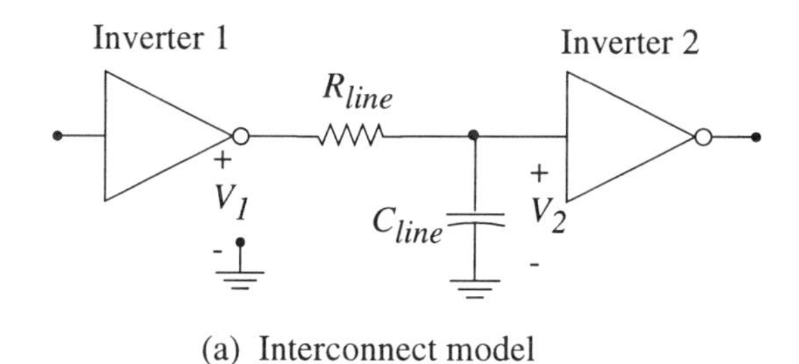

(a) Interconnect model

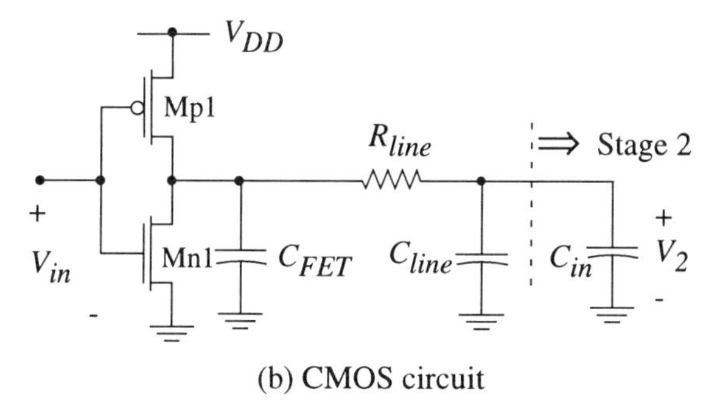

(b) CMOS circuit

Figure 10.5 Lumped element modelling of the interconnect

$$V_2(t) = V_{DD}e^{-t/\tau_n} \tag{10. 11}$$

where the discharge time constant is

$$\tau_n = (C_{in} + C_{line})(R_{n1} + R_{line}) + C_{FET}R_{n1} \tag{10. 12}$$

with R_{n1} the resistance of nFET Mn1. This gives a high-to-low time of

$$\begin{aligned} t_{HL} &\approx 2.2\tau_n \\ &= 2.2\left[(C_{in} + C_{line})(R_{n1} + R_{line}) + C_{FET}R_{n1}\right] \end{aligned} \tag{10. 13}$$

Similarly, a low-to-high transition is modelled as

$$V_2(t) = V_{DD}\left[1 - e^{-t/\tau_p}\right] \tag{10. 14}$$

where

$$\tau_p = (C_{in} + C_{line})(R_{p1} + R_{line}) + C_{FET}R_{p1} \tag{10. 15}$$

is the charging time constant with R_{p1} the resistance of pFET Mp1. This gives

$$\begin{aligned} t_{LH} &\approx 2.2\tau_p \\ &= 2.2\left[(C_{in} + C_{line})(R_{p1} + R_{line}) + C_{FET}R_{p1}\right] \end{aligned} \tag{10. 16}$$

In both cases, the interconnect parasitics increase the switching time by a factor of

$$\Delta\tau = 2.2\left[R_{line}C_{in} + R_{line}C_{line} + R_{FET}C_{line}\right] \tag{10. 17}$$

where R_{FET} is the appropriate FET resistance. The overall increase in delay can be estimated by examining each time constant in this sum. These equations may also be used to provide initial design criteria for the circuit.

RC Ladder Network

The next level of modelling for the interconnect is to replace the single RC lumped element model by a multi-stage RC ladder that has m rungs. Defining the individual element values by

$$R_m = \frac{R_{line}}{m}, \qquad C_m = \frac{C_{line}}{m} \tag{10. 18}$$

allows us to construct the desired ladder network. Figure 10.6 shows the equivalent circuits for the cases of $m = 3$, 4, and 5. Interest is then directed towards the parasitic-induced delay from the left side (A) to the right side (B) of the line. These may be estimated by using the Elmore formula for each case. With $m = 3$, the delay time constant is

$$\begin{aligned} \tau_3 &= C_3(3R_3) + C_3(2R_3) + C_3R_3 \\ &= 6C_3R_3 \end{aligned} \tag{10. 19}$$

For $m = 4$, we have

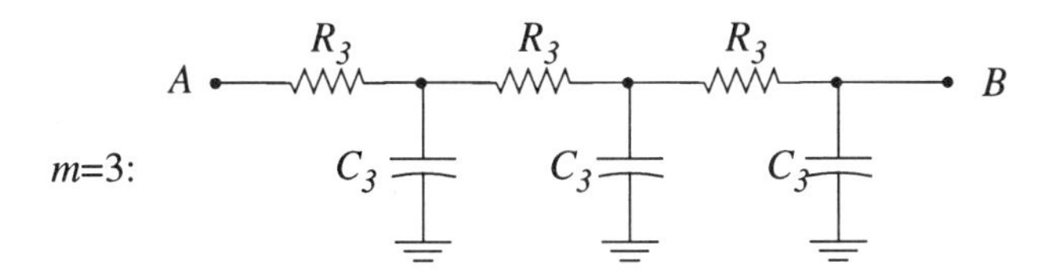

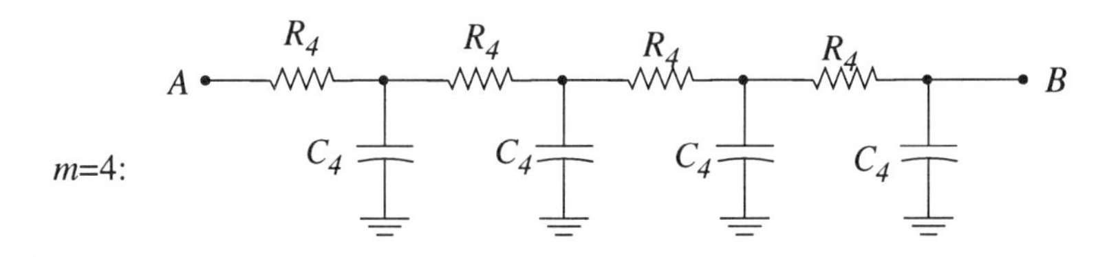

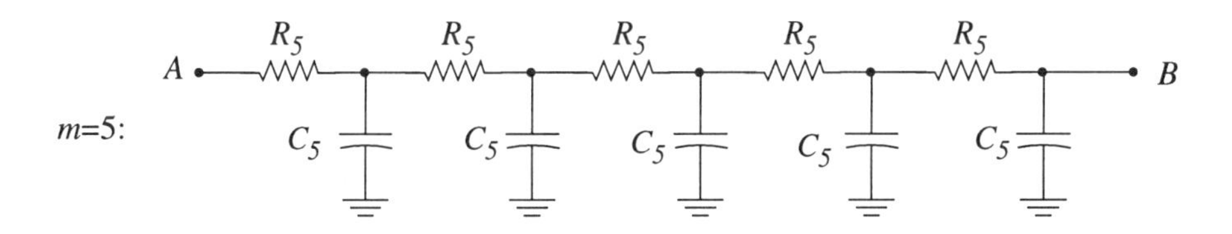

Figure 10.6 RC ladder modelling for the interconnect line

$$\begin{aligned}\tau_4 &= C_4(4R_4) + C_4(3R_4) + C_4(2R_4) + C_4R_4 \\ &= 10C_4R_4\end{aligned} \tag{10. 20}$$

while the case where $m = 5$ gives

$$\begin{aligned}\tau_5 &= C_5(5R_5) + C_5(4R_5) + C_5(3R_5) + C_5(2R_5) + C_5R_5 \\ &= 15C_5R_5\end{aligned} \tag{10. 21}$$

and so on. From the series, we see that a ladder with m-rungs has a time constant of

$$\tau_m = \frac{m(m+1)}{2}C_mR_m \tag{10. 22}$$

that represents the delay due to the line parasitics.

To understand the meaning of this results, first note that we are attempting to model an interconnect line by breaking it into an arbitrarily number of segments. The actual value of m that we choose affects how well the equivalent circuit will model the delay. Substituting for R_m and C_m into the equation gives

$$\begin{aligned}\tau_m &= \frac{m(m+1)}{2}\left(\frac{C_{line}}{m}\right)\left(\frac{R_{line}}{m}\right) \\ &= \frac{m(m+1)}{2m^2}C_{line}R_{line}\end{aligned} \tag{10. 23}$$

For large m, this predicts that

$$\tau_m \to \frac{1}{2} C_{line} R_{line} \tag{10. 24}$$

which is only one-half the value of the time constant in the simple lumped element (single RC) model. This tells us that the ladder will yield more accurate results. However, since the introduction of additional nodes into a circuit simulation program increases the CPU time, the ladder simulation will take longer to execute. For example, SPICE models an N-node circuit by using matrices of dimensions $N \times N$, so that using ladders with large values of m increases the matrix size quadratically. This results in increased computer time for the simulations.

Distributed Analysis

Let us examine the interconnect structure shown in Figure 10.7. Since the resistance and the capacitance are distributed along the line, the analysis should reflect this fact and be based on the resistance per unit length r' and the capacitance per unit length c' as shown. To understand the problems involved in signal delay, we will first analyze the structure itself to obtain the equation for voltage transmission, and then solve the equation for a specified input.

Consider the differential segment of the line with length dz that is shown in the lower half of Figure 10.7. The left side is a some point z, and is characterized by a voltage $V(z,t)$ and a current $I(z,t)$; similarly, the right side of the segment is at the point $(z+dz)$ and is described by $V(z+dz,t)$ and $I(z+dz,t)$. As with any circuit. the relationships among the voltages and currents are obtained by applying the Kirchhoff laws. In the present context, this will result in the differential equation that describes both the voltage and the current as functions of position z and time t.

To analyze the structure, we first note that the voltage across the resistor is

$$\begin{aligned} V_R &= V(z+dz, t) - V(z, t) \\ &= -I(z, t)\, r'dz \end{aligned} \tag{10. 25}$$

where we have used Ohm's Law and the fact that the resistance of the segment is $r'dz$. Since dz is a differential length, let us expand around the point $z = 0$ to obtain.

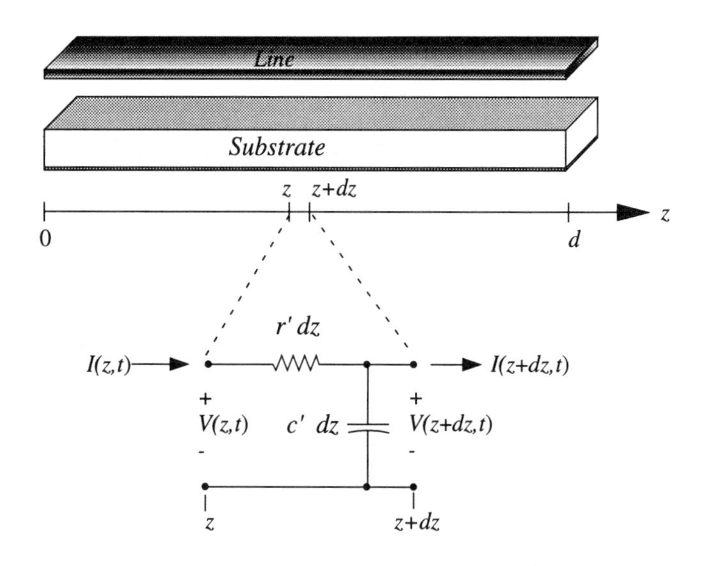

Figure 10.7 Distributed modelling of the interconnect line

$$V(z+dz,t) = V(z,t) + \frac{\partial V(z,t)}{\partial z}dz + \frac{1}{2}\frac{\partial^2 V(z,t)}{\partial z^2}(dz)^2 + \ldots \qquad (10.26)$$

Substituting into the V_R expression and ignoring terms of order $(dz)^2$ and higher gives the KVL equation

$$\frac{\partial V(z,t)}{\partial z} = -r'I(z,t) \qquad (10.27)$$

as our first relation. Next, note that the current flow into the capacitor is given by

$$I_C = c'dz\frac{\partial V(z,t)}{\partial t} \qquad (10.28)$$

since the capacitance in farads is given by $c'dz$. Applying KCL to the segment gives

$$\begin{aligned} I(z+dz,t) - I(z,t) &= -c'dz\frac{\partial V(z,t)}{\partial t} \\ &= \frac{\partial I(z,t)}{\partial z}dz \end{aligned} \qquad (10.29)$$

where we have one again performed an expansion. Cancelling dz yields

$$\frac{\partial I(z,t)}{\partial z} = -c'\frac{\partial V(z,t)}{\partial t} \qquad (10.30)$$

as the second relation.

A single differential equation for $V(z,t)$ can be obtained by combining the two expressions. To this end, we first take the spatial derivative of the KVL result [equation (10.24)] to obtain

$$\frac{\partial^2 V(z,t)}{\partial z^2} = -r'\frac{\partial I(z,t)}{\partial z}. \qquad (10.31)$$

Substituting for the current derivative using the KCL expression [equation (10.27)] then yields

$$\frac{\partial^2 V(z,t)}{\partial z^2} = r'c'\frac{\partial V(z,t)}{\partial t} \qquad (10.32)$$

which is a well-studied expression from mathematical physics known as the **diffusion equation**.

Let us examine the problem of launching a voltage pulse onto the line at the point $z = 0$ as illustrated by the circuit in Figure 10.8. The voltage source is taken to be a unit step in the form

$$V(z,0)\big|_{z=0} = V_o u(t) \qquad (10.33)$$

which will serve as both a boundary condition and the initial condition for the differential equation. A straightforward approach to solving the diffusion equation is to recognize that it is first-order in time, so that we may define the Laplace transform voltage $v(z,s)$ that satisfies the s-domain equation

$$\frac{\partial^2 v(z,s)}{\partial z^2} - sr'c'v(z,s) = -r'c'V(z,0) \qquad (10.34)$$

showing that the voltage source is the driving function for the voltage. The homogeneous solution to this equation is given by

$$v(z, s) = V^{+}e^{-\alpha z} + V^{-}e^{+\alpha z} \tag{10. 35}$$

where

$$\alpha = \sqrt{sr'c'} \tag{10. 36}$$

By physical reasoning, the voltage must remain finite as $z \to \infty$, so we set $V^{-} = 0$ to avoid the situation where the expression grows without bound. The remaining constant V^{+} can be found by applying the initial condition at z=0 by using the Laplace transform

$$\begin{aligned} V(0, s) &= \frac{V_o}{s} \\ &= V^{+} \end{aligned} \tag{10. 37}$$

so that the s-domain solution is

$$v(z, s) = \frac{V_o}{s} e^{-\sqrt{sr'c'}\, z} \tag{10. 38}$$

Inverse transforming back to time domain then results in

$$V(z, t) = V_o erfc\left(\frac{z}{2}\sqrt{\frac{r'c'}{t}}\right) u(t) \tag{10. 39}$$

where $erfc(y)$ is the **complementary error function.**

The complementary error function is based on the **error function** $erf(y)$ from statistical analysis as defined by the integral representation

$$erf(y) = \frac{2}{\sqrt{\pi}} \int_0^y e^{-\eta^2} d\eta \tag{10. 40}$$

that has limiting values of

$$\begin{aligned} erf(0) &= 0 \\ erf(\infty) &= 1 \end{aligned} \tag{10. 41}$$

as can be verified by substitution. The complementary error function is given by

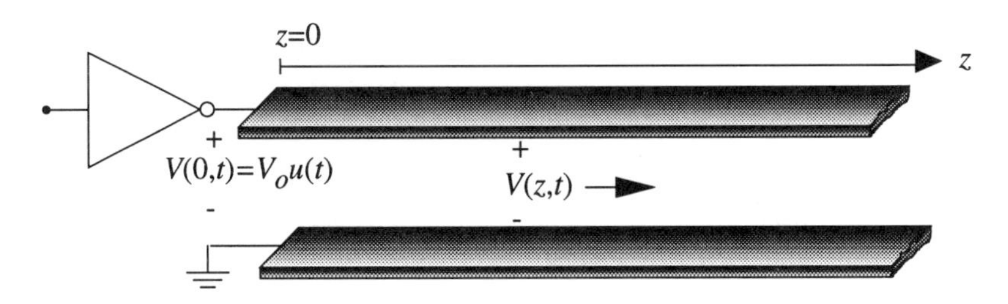

Figure 10.8 Pulsing launching on an interconnect line

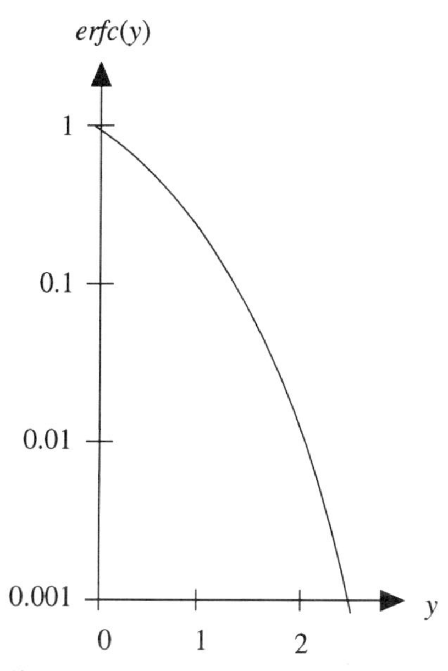

Figure 10.9 General shape of the complementary error function

$$erfc\,(y) = 1 - erf\,(y) = \frac{2}{\sqrt{\pi}}\int_{y}^{\infty} e^{-\eta^2}\,d\eta \quad (10.42)$$

and has limiting values of

$$erfc\,(0) = 1$$
$$erfc\,(\infty) = 0 \quad (10.43)$$

which are the complements of the error function. This shows that *erfc*(y) decreases monotonically with *y*. A general plot of *erfc*(y) is shown in Figure 10.9; this shows the basic shape but should not be used to read numerical values (use a table or a CAD program).

The signal delay associated with launching the voltage on a parasitic RC line can be studied by noting that the argument

$$y = \frac{z}{2}\sqrt{\frac{r'c'}{t}} \quad (10.44)$$

that determines the value of *erfc*(*y*) is itself a function of both z and t. Once we have launched the voltage at time $t = 0$, it will progress down the line as described by the *erfc*(*y*) function. To see this motion, consider the situation where we choose to follow a particular voltage V_x

$$V_x = V_o erfc\,(y_x) \quad (10.45)$$

down the line. The value of the argument is specified, and may be written as

$$y_x = K\frac{z}{\sqrt{t}} \quad (10.46)$$

where

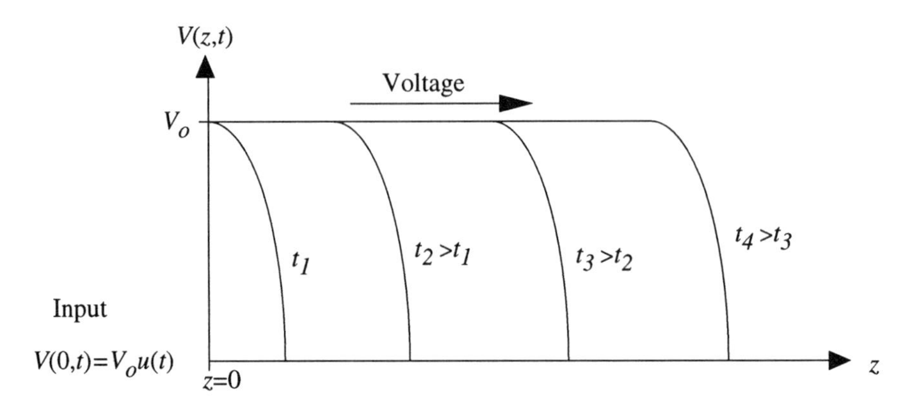

Figure 10.10 Movement of voltage down an RC line

$$K = \frac{1}{2}\sqrt{r'c'} \tag{10. 47}$$

is a constant determined by the line. As t increases, the value of z must also increase (at a different rate) in order to keep the same value for V_x. This type of behavior is portrayed in Figure 10.10. In this drawing, each curve represents the voltage if we would sample the voltage along the line at different times. The progression of the voltage down the line is governed by the above equations which make the *erfc* shape prevalent.

The general behavior of a voltage diffusing down a line can be understood by rearranging the argument to the form

$$t = az^2 \tag{10. 48}$$

where $a = (K/y)^2$ is a constant. This important result shows that the delay time increases as the **square** of the distance. In other words, doubling the length of an interconnect quadruples the time delay. The time delay t_d is plotted in Figure 10.11 and is an extremely important characteristic to remember.

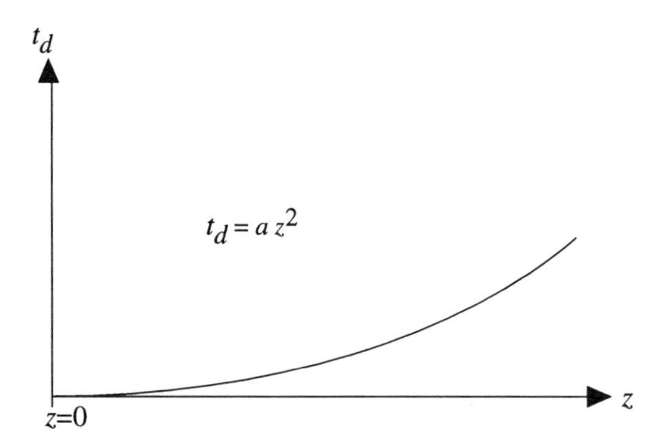

Figure 10.11 Quadratic delay on an RC line

Example 10-2

Suppose that we have a resistive interconnect line of length d_o which exhibits a signal delay of t_o. If we double the length of the line to $2d_o$, then delay increases to a value of $t_1 = 4t_o$ as justified by the calculation below:

$$a = \left(\frac{t_o}{d_o^2}\right) \tag{10. 49}$$

so that

$$\begin{aligned} t_1 &= \left(\frac{t_o}{d_o^2}\right)(2d_o)^2 \\ &= 4t_o \end{aligned} \tag{10. 50}$$

Similarly, if we extend the line to a length $4d_o$, then delay increases to a value of $t_4 = 16t_o$ and so on. This behavior is important to remember when interconnect delays are important to the system timing of the chip.

Although our result was derived from a more rigorous analysis, this result is very similar to that obtained using the m-rung RC ladder discussed above. To understand this comment, recall that the time constant for the RC ladder was found to be

$$\tau_m = \frac{m(m+1)}{2m^2} C_{line} R_{line}. \tag{10. 51}$$

The values of R_{line} and C_{line} are given by

$$C_{line} = c'd \qquad R_{line} = r'd \tag{10. 52}$$

where d is the length of the line. Substituting,

$$\begin{aligned} \tau_m &= \frac{m(m+1)}{2m^2}(c'd)(r'd) \\ &= \frac{m(m+1)}{2m^2} r'c'd^2 \end{aligned} \tag{10. 53}$$

so that

$$\tau_m = Bd^2 \tag{10. 54}$$

where B is a constant.

This analysis demonstrates that the length of an interconnect is critical, both in itself and relative to other lines. The latter case is especially important if we are working with binary word circuits where the bits must be synchronized as they traverse the chip from section to section.

10.1.3 Clock Distribution

One problem that is directly associated with interconnect delay is the distribution of clock signals over the chip. Consider the structure of a chip floorplan; the important characteristics are illustrated in the drawing of Figure 10.12. In a modern design, all major sections of the chip will be synchronized by one or more clocking signals. The simplest timing technique is to have all sections in phase with each other, as this allows data flow control at the highest system level. In the notation of the drawing, the clock driver circuit produces a primary clocking signal ϕ, which is distributed to other units of the system. The receiver positions in each section see clocks ϕ_A, ϕ_B, etc., that are delayed from the original signal because of interconnect effects. Since each distribution line is a different length, the received clock signals will all be slightly out of phase.

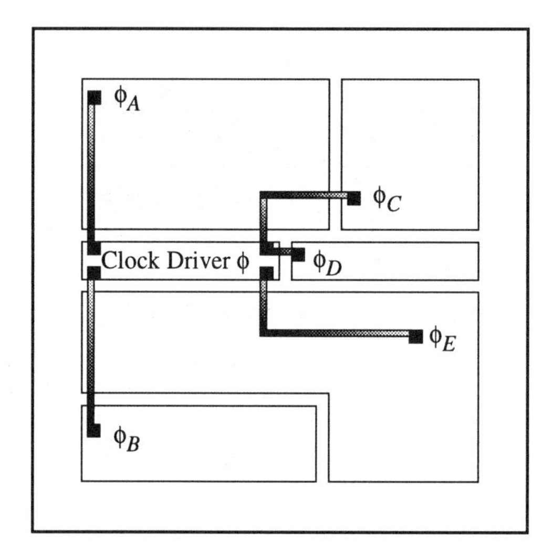

Figure 10.12 Chip floorplan used to illustrate clock distribution problems

Figure 10.13 shows the skew times induced by interconnect delay[2]. Since the broadcast line for ϕ_D is the shortest, the skew delay t_{sD} will be small. Since the delay is proportional to the square of the length, the skew time t_{sA} for ϕ_A will be much longer. The importance of this type of delay depends upon the clock period T. It is easily seen that if the skew times satisfy

$$t_s \ll T \tag{10. 55}$$

then there won't be much effect on the timing. However, this limits the clock frequency, which in turn limits the system throughput.

One particularly attractive solution to this type of problem is to use **H-tree** structuring for the clock distribution lines. An H-tree pattern is one that replicates the shape of a letter "H" in the interconnect patterning. Figure 10.14(a) illustrates the concept. If an input signal is applied to the center of the "H" and the outputs are taken at the tips as shown, then every path length is the same. This means that the delay between the input and each output is identical. Using the H as a basic geometrical shape allows us to construct an H-tree network with the main features illustrated in Figure 10.14(b). The central clock driver is placed in the center of the chip, and nested H-trees are used to

[2] This means that we are assuming that the receivers inputs all present the load capacitance to the lines, so that the delay is entirely due to the parasitic line parameters.

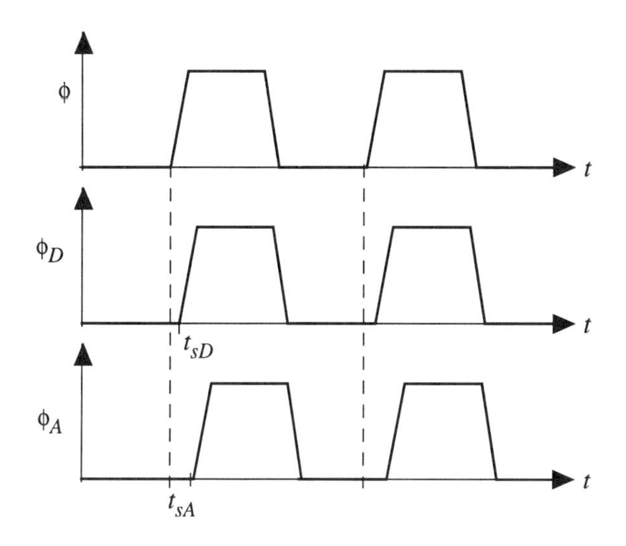

Figure 10.13 Interconnect-induced clock skew

distribute the signal to various points on the chip. If we place receivers at the tips of the smallest "H" line, then all of the received signals will be in phase.[3] A driver tree can be created so long as we place a driver at every equivalent point in the geometry. This overcomes the problem of clock skew among the various units in the system, but usually requires the use of a dedicated interconnect layer to insure that the geometry can be maintained.

Other techniques have been developed to control clock skew. In the drawing shown in Figure 10.15, the clock driver circuits are placed in the center of the chip. Conceptually, we visualize the

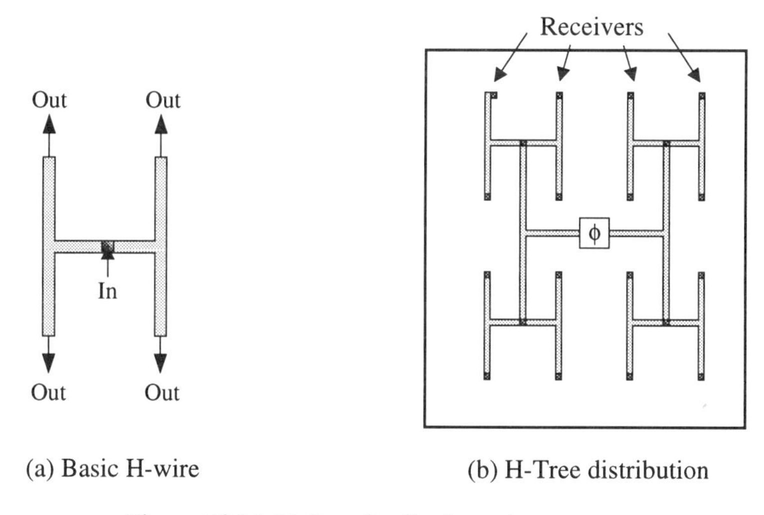

(a) Basic H-wire (b) H-Tree distribution

Figure 10.14 H-Tree distribution scheme

[3] Note that the system clocks will be delayed from the central driver circuit, but the rest of the chip will be in phase.

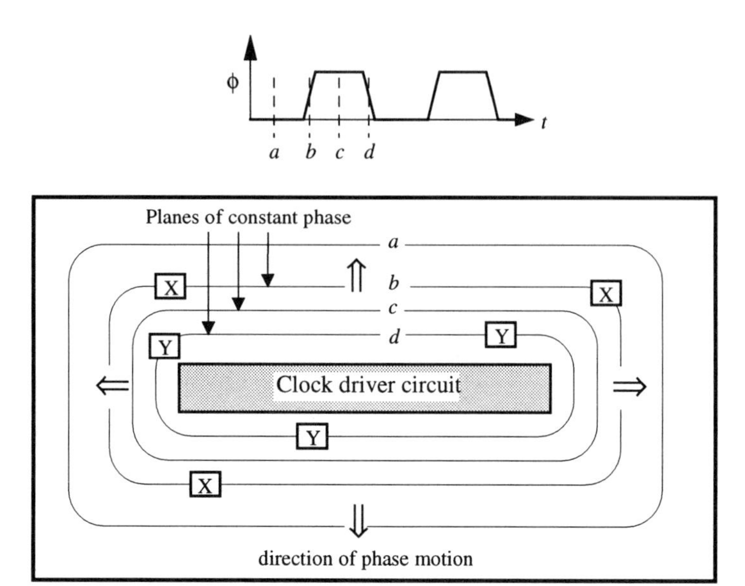

Figure 10.15 Central clock placement and distribution

driver circuits as "broadcasting" the clock signal $\phi(t)$ as away from the center towards the surrounding circuits. These are shown as contours of constant phase such that each phase line represents a specific value of the clock. For example, the phase contour labelled "a" represents the value of the clock at the time "a" shown in the timing diagram. Of course, interconnect must be added to broadcast the signal, but the idea is clear: all receiver circuits that are place on the same timing contour will be in phase. This means that the receivers "X" are all in phase with one another; similarly, the group of receivers labelled "Y" will be in phase. This approach is more difficult to implement than using a structured distribution geometry such as the H-tree. Particular care must be applied to the design of the interconnect routing and the receiver circuitry to insure that the signals are indeed in phase with one another.

10.1.4 Coupling Capacitors and Crosstalk

Another problem associated with interconnects is that of **crosstalk**. Crosstalk is the term given to the situation where energy from a signal on one line is transferred to a neighboring line by electromagnetic means. In general, both capacitive and inductive coupling exist. At the chip level, however, the currents through the signal lines are usually too small to induce magnetic coupling, so that parasitic inductance is ignored here.

Capacitive coupling, on the other hand, depends on the line-to-line spacing S as illustrated in the general situation portrayed in Figure 10.16. Since capacitive coupling between two conducting lines is inversely proportional to the distance between the two lines, a small value of S implies a large coupling capacitance C_c exists. Because of this dependence, it is not uncommon to find a minimum layout spacing design rule for critical lines that is actually larger than which could be created in the processing line. Also, the capacitive coupling increases with the length of the interaction, so it is important that the interconnects not be placed close to one another for any extended distance.

Let us use the geometry in Figure 10.17 to estimate the coupling capacitance. This cross-sectional view shows the spacing S between two identical interconnect lines. An empirical formula that provides a reasonable estimate for the coupling capacitance c_c' per unit length is given by

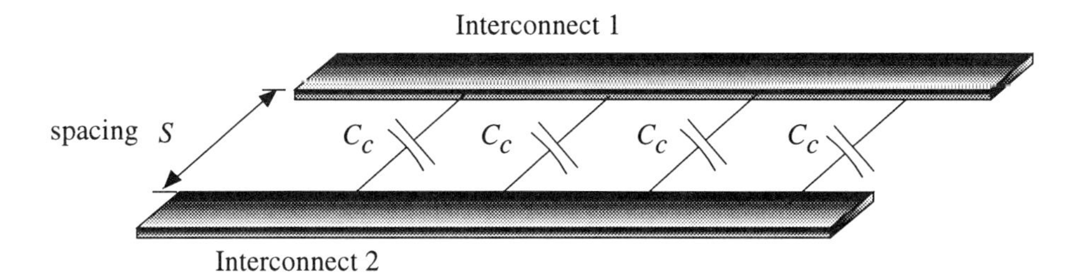

Figure 10.16 Capacitive coupling between two adjacent interconnect lines

$$c_c' = \varepsilon_{ox}\left[0.03\left(\frac{w}{X_{ox}}\right) + 0.83\left(\frac{h}{X_{ox}}\right) - 0.07\left(\frac{h}{X_{ox}}\right)^{0.222}\right]\left(\frac{X_{ox}}{S}\right)^{4/3} \tag{10. 56}$$

in units of *F/cm* which can be applied directly to the geometry. The total coupling capacitance in farads of a line that has a length d is calculated from

$$C_c = c_c'd \tag{10. 57}$$

This shows explicitly the fact that C_c increases as the separation distance S decreases.

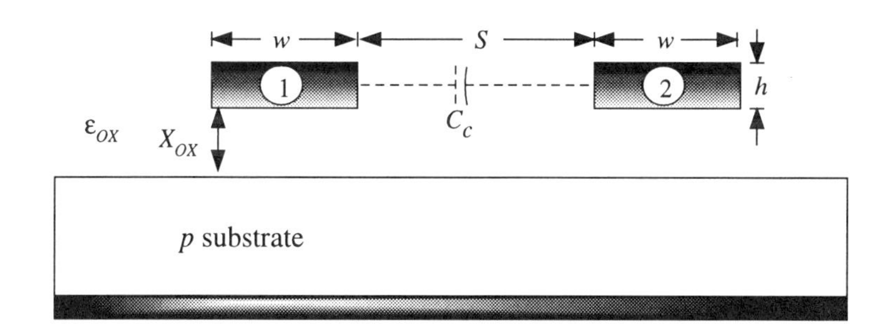

Figure 10.17 Geometry for calculating the coupling coefficient

The importance of C_c becomes evident when we examine how two circuits can interact via electric field coupling. Consider the situation shown in Figure 10.18 where two independent lines interact through a coupling field E_c. Line 1 is at a voltage $V_1(t)$ at the input to inverter B, while line 2 has a voltage V_2 (t) which is the input of inverter D. The field is supported by the difference in voltages $(V_1 - V_2)$. At the circuit level, we analyze the situation by introducing lumped-equivalent transmission line models as in Figure 10.19. The electric field interaction is included through the coupling capacitor C_c. The placement of C_c in the circuit corresponds to the simplest type of single-capacitor coupling model; a more accurate analysis might add two capacitors, one on each side of the resistors. The current through the capacitor is calculated from the relation

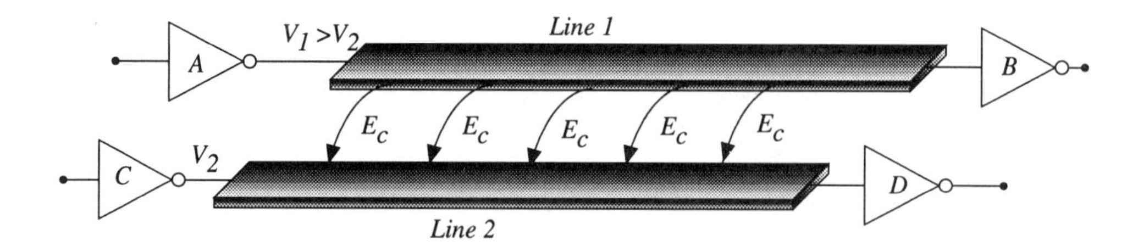

Figure 10.18 Electric field coupling between two lines

$$\begin{aligned} i_c &= C_c \frac{dV_c}{dt} \\ &= C_c \frac{d(V_1 - V_2)}{dt} \end{aligned} \tag{10. 58}$$

and is assumed to flow from line 1 to line 2 by the choice of voltages. If the difference $(V_1 - V_2)$ changes in time, then the two lines become electrically coupled and the voltages are different from the case where they are independent.

Circuit-Level Modelling

To understand the effects of this type of coupling, consider the equivalent circuit model shown in Figure 10.20. Line 1 is excited with a source voltage

$$V_s(t) = V_o u(t) \tag{10. 59}$$

where $u(t)$ is the unit step function. The input to Line 2 is at 0*v*. Crosstalk can be seen by analyzing the response at $V_2(t)$; in the absence of coupling, this voltage would remain at 0. We have taken the two lines to be identical with a line resistance of R and a line capacitance of C. Loading effects are including by terminating both lines with load capacitances C_L.

Let us analyze the network by writing the time-domain node equations as follows:

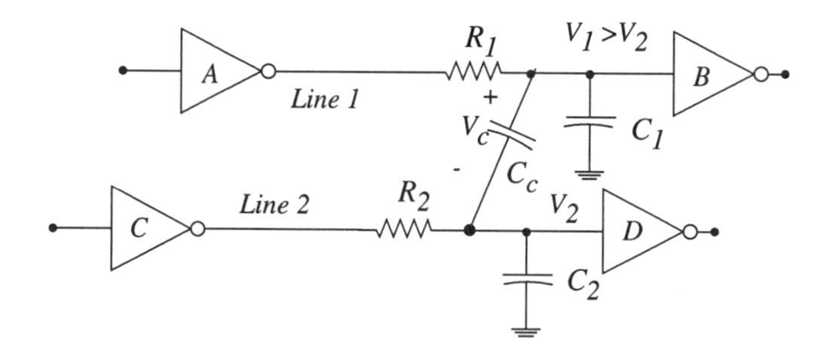

Figure 10.19 Circuit model for the electric field interaction

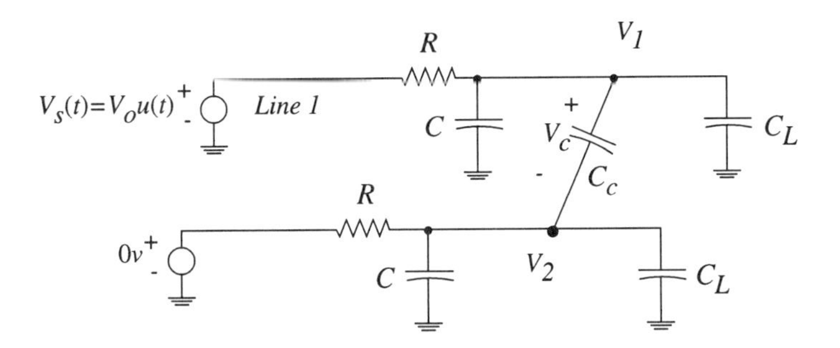

Figure 10.20 Reduced circuit for coupling calculation

$$
\begin{aligned}
\frac{V_1 - V_o u(t)}{R} + (C + C_L)\frac{dV_1}{dt} + C_c\frac{d(V_1 - V_2)}{dt} &= 0 \\
\frac{V_2}{R} + (C + C_L)\frac{dV_2}{dt} + C_c\frac{d(V_2 - V_1)}{dt} &= 0
\end{aligned}
\qquad (10.60)
$$

Defining the s-domain voltages $v_1(s)$ and $v_2(s)$ that represent the time domain functions $V_1(t)$ and $V_2(t)$, respectively, gives the s-domain equation set

$$
\begin{aligned}
v_1(s)\left[\frac{1}{R} + s(C + C_L + C_c)\right] - sC_c v_2(s) &= \frac{V_o}{sR} \\
v_1(s)\,[-sC_c] + v_2(s)\left[\frac{1}{R} + s(C + C_L + C_c)\right] &= 0
\end{aligned}
\qquad (10.61)
$$

Let us define

$$C_T = C + C_L + C_c \qquad (10.62)$$

Then we may write the equation pair in the matrix form

$$\begin{bmatrix} \frac{1}{R} + sC_T & -sC_c \\ -sC_c & \frac{1}{R} + sC_T \end{bmatrix} \begin{bmatrix} v_1(s) \\ v_2(s) \end{bmatrix} = \begin{bmatrix} \frac{V_o}{sR} \\ 0 \end{bmatrix} \qquad (10.63)$$

which clearly shows the driving function on the right side.

To solve this set, we first calculate the determinant

$$\Delta = \det \begin{bmatrix} \frac{1}{R} + sC_T & -sC_c \\ -sC_c & \frac{1}{R} + sC_T \end{bmatrix} \qquad (10.64)$$

as

$$\Delta = \left(C_T^2 - C_c^2\right)s + 2\left(\frac{C}{R}\right)s + \frac{1}{R^2} \qquad (10.65)$$

We may then calculate $v_2(s)$ using Cramer's Rule, which results in the determinant

$$\begin{aligned} v_2(s) &= \frac{1}{\Delta}\det\begin{bmatrix} \frac{1}{R} + sC_T & \frac{V_o}{sR} \\ -sC_c & 0 \end{bmatrix} \\ &= \frac{V_o C_c / R}{\Delta(s)} \end{aligned} \qquad (10.66)$$

Now then, the poles indicated by the condition $\Delta(s) = 0$ are found to be given (after a little algebra) by the simple expressions

$$\begin{aligned} s_1 &= -\frac{1}{R(C_T + C_c)} = -\frac{1}{\tau_1} \\ s_2 &= -\frac{1}{R(C_T - C_c)} = -\frac{1}{\tau_2} \end{aligned} \qquad (10.67)$$

where we have defined two time constants

$$\begin{aligned} \tau_1 &= R(C + C_L + 2C_c) \\ \tau_2 &= R(C + C_L) \end{aligned} \qquad (10.68)$$

The values clearly show that

$$\tau_1 > \tau_2 \qquad (10.69)$$

so we have two different transient events taking place.

The final step in solving the problem is to perform a partial fraction expansion of the form

$$v_2(s) = \frac{K_1}{(s - s_1)} + \frac{K_2}{(s - s_2)} \qquad (10.70)$$

which results in the s-domain expression

$$v_2(s) = \frac{V_o/2}{(s - s_1)} - \frac{V_o/2}{(s - s_2)} \qquad (10.71)$$

Inverse transforming back to time-domain yields the solution

$$V_2(t) = \frac{V_o}{2}\left[e^{-t/\tau_1} - e^{-t/\tau_2}\right]u(t) \qquad (10.72)$$

as our final result. This is the difference between a slow decay (due to τ_1) and a fast decay (due to τ_2) as plotted in Figure 10.21. It is easily seen that the effect of the coupling capacitance is to create a spurious pulse. Since this is the voltage of the undriven line, we must insure that the magnitude of the induced voltage is still within the logic 0 voltage of the next logic stage. In terms of the layout problems, the coupling capacitance C_c is directly proportional to the interaction length, so that long parallel lines need to be avoided.

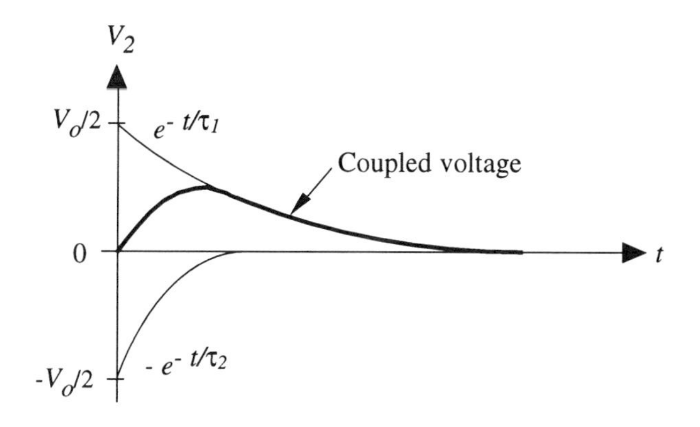

Figure 10.21 Coupled voltage behavior

Theoretical Basis of Crosstalk

Crosstalk is actually the modern manifestation of a classical problem in electrostatics. Consider two conducting bodies as show in Figure 10.22. Conductor 1 has a voltage V_1 with respect to ground. If it were completely isolated from all other conducting bodies, then we would describe its relationship to ground by the **self-capacitance** C_{11} shown. Similarly, conductor 2 is at a voltage V_2 with respect to ground, and has a self-capacitance C_{22}. When the two interact, a coupling capacitance C_c must be added to describe the situation. In classical electrostatics, the emphasis is placed upon relating the charges Q_1 and Q_2 to the voltages V_1 and V_2. This is done by introducing a 2×2 matrix $[C]$ whose elements are called the **coefficients of capacitance**. Then we write in matrix form that

$$[Q] = [C]\,[V] \tag{10.73}$$

where $[Q]$ and $[V]$ represent two component vectors (i.e., column matrices). Explicitly, the equation is written as

$$\begin{bmatrix} Q_1 \\ Q_2 \end{bmatrix} = \begin{bmatrix} C_{11} & C_{12} \\ C_{21} & C_{22} \end{bmatrix} \begin{bmatrix} V_1 \\ V_2 \end{bmatrix} \tag{10.74}$$

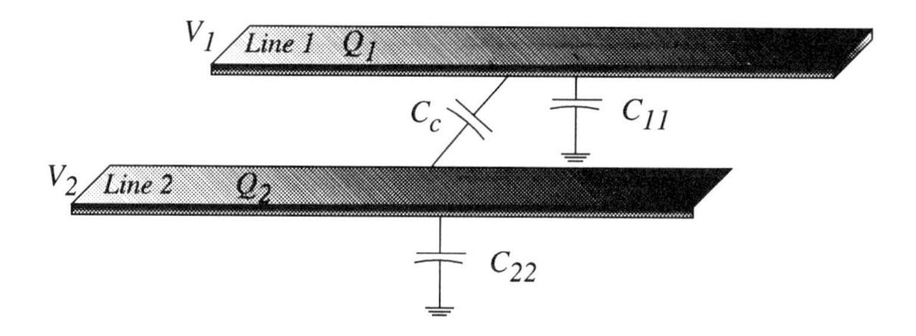

Figure 10.22 Electric coupling of two conductors

where we can show that the matrix is symmetric with

$$C_{12} = C_{21} = C_c \tag{10. 75}$$

being the coupling capacitance. In component form, the charge equations are

$$\begin{aligned} Q_1 &= C_{11}V_1 + C_cV_2 \\ Q_2 &= C_cV_1 + C_{22}V_2 \end{aligned} \tag{10. 76}$$

by simply expanding the matrices. The coupling is a result of voltages that change in time. Differentiating gives

$$\begin{aligned} \frac{dQ_1}{dt} &= C_{11}\frac{dV_1}{dt} + C_c\frac{dV_2}{dt} = i_1 \\ \frac{dQ_2}{dt} &= C_c\frac{dV_1}{dt} + C_{22}\frac{dV_2}{dt} = i_2 \end{aligned} \tag{10. 77}$$

These equations clearly show the interaction between the two bodies. For example, if V_2 changes in time, then both i_1 and i_2 flow. Moreover, it is the time rate-of-change (dV_2/dt) that is important. In high-speed switching networks, the derivative may be quite large, making the coupling problem more difficult to deal with.

10.2 Input and Output Circuits

CMOS circuits are constructed using sub-micron geometries on a silicon substrate and are designed to implement digital functions using tiny amounts of charge and relatively small voltage swings. Capacitances are measured in *fF* and timing delays are in units of *ns*. These orders of magnitude are intrinsic to the internal operation of the electronics, but are quite different from the values we work with at the board and system level. These considerations lead us to examine the special requirements placed on both input and output circuits that are needed to communicate with the outside world.

10.2.1 Input Protection Networks

Modern CMOS technology provides for the use of extremely thin gate oxides. Small values of x_{ox} are desirable because the device transconductance equation

$$k' = \frac{\mu\varepsilon_{ox}}{x_{ox}} \tag{10. 78}$$

illustrates that more sensitive MOSFETs will result. Decreasing the oxide thickness is not without costs, however. Ultra thin oxides lead to higher gate capacitance levels and increased manufacturing difficulty (and lower yields). In addition, a thin oxide MOSFET at a pad input is more susceptible to being destroyed by an electrostatic discharge (ESD) event where an excessive amount of charge is dumped onto the gate.

Consider a MOSFET whose gate is connected to an input pad a shown in Figure 10.23. Applying a gate voltage V_G is to the device gives an oxide electric field that can be estimated to first order by

$$E_{ox} \approx \frac{V_G}{x_{ox}}. \tag{10. 79}$$

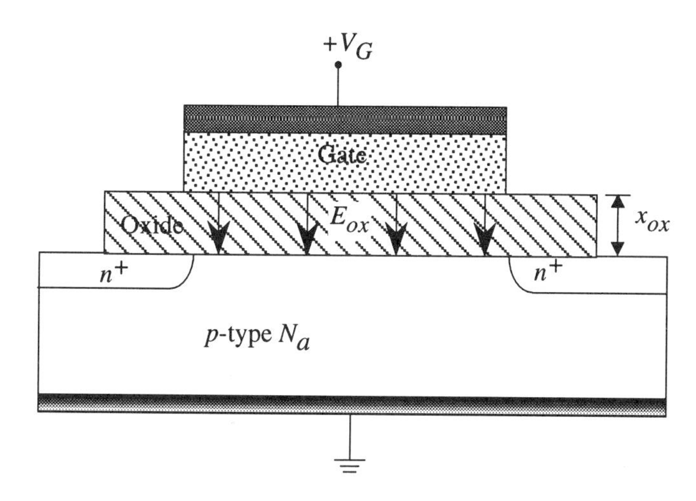

Figure 10.23 Oxide breakdown in a MOSFET

Since x_{ox} is on the order of tens of Angstroms, even moderate voltages lead to large electric fields. The problem arises from noting that every dielectric material is characterized by a maximum electric field E_{max} that it can withstand. If an applied field exceeds this value, the material undergoes breakdown where it no longer acts as an insulator. In the case of silicon dioxide, this leads to current flow, heating, and local melting of the device which destroys it. And, if the input circuits do not work, then the entire chip is bad.

Static electricity problems have always plagued the semiconductor industry. Friction can induce static charges of several thousand volts on surfaces; these can destroy the chip if they reach the input stages. Sources of ESD are numerous and include normal operations such as testing the die, chip insertion, and normal handling of the packaged devices. Since ESD cannot be eliminated, **input protection networks** are included whenever static electricity is a problem. As shown in Figure 10.24(a), this problem occurs if the input pad is directly connected to the gate of a MOSFET. Owing to this, we provide the protection circuits as shown in Figure 10.24(b). These networks are designed to provide alternate charge flow paths to keep excessive charge levels away from the transistor gates. All CMOS chips utilize some type of protection circuitry on input circuits. The actual designs themselves are usually contained as a standard cell in the circuit library. In this section, we will examine two techniques for achieving input protection.

Diode Clamps

One straightforward approach is to use reverse biased pn junctions to keep the voltage level at the MOSFET gate below the breakdown voltage V_{BD}. The basis for this technique is the fact that a reverse biased pn junction exhibits a well-defined breakdown phenomena as shown by the *I-V* plot in Figure 10.25. Let us denote the reverse voltage by $V_R = -V$, where V is the forward voltage shown in the plot. When V_R is small, the diode blocks conduction and only leakage currents flow. However, if V_R is increased to the Zener voltage V_Z, the diode undergoes breakdown and can no longer block the current flow. Because of the shape of the transition, this is called the "Zener knee" region.

The breakdown phenomena shown in the *I-V* curve is due to avalanche breakdown in the semiconductor in which the electric field gets large and accelerates electrons (and holes) such that they smash into atoms and liberate more charges; these liberated charges do the same so that the current flow can get quite large. However, if we remove the voltage and then reapply it, the diode still

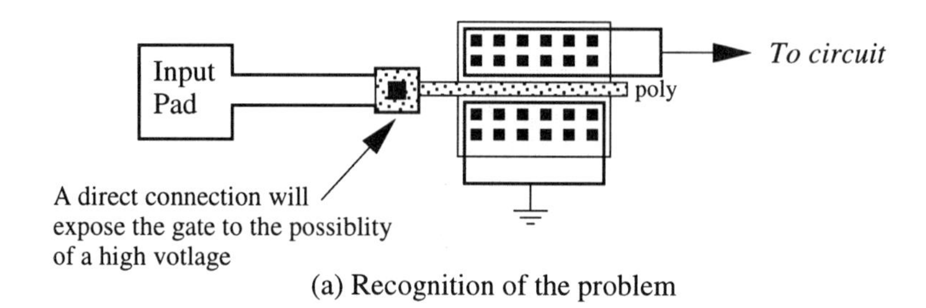

(a) Recognition of the problem

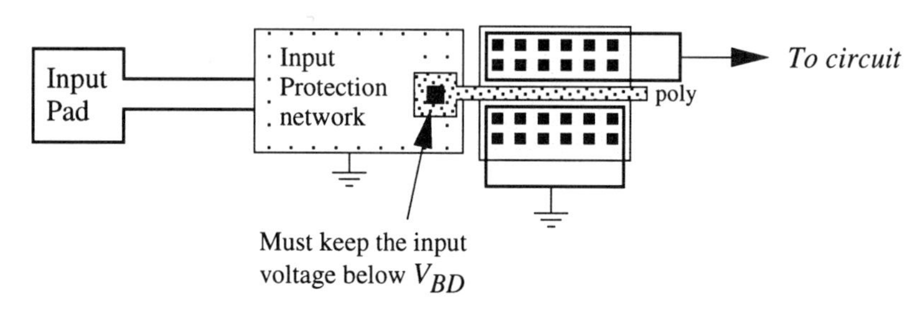

(b) Input protection network

Figure 10.24 Introduction of an input protection network

exhibits the same *I-V* characteristics. In other words, the breakdown is not destructive. From the device viewpoint, the value of V_Z is set by the doping levels in the semiconductors.

The Zener breakdown voltage can be used to solve our problem of protecting the MOSFET gate by insuring that $V_Z < V_{BD}$ in the fabrication, and then constructing a circuit like that shown in Figure 10.26. In this circuit, the gate of the MOSFET is isolated from the input pad by a resistor-diode network that is designed to operate normally if the input voltages are at nominal levels, but conducts excess charge to ground if the voltage is excessive. Protection is achieved by using diodes that have a reverse breakdown voltage $V_z < V_{BD}$ of the MOSFET.

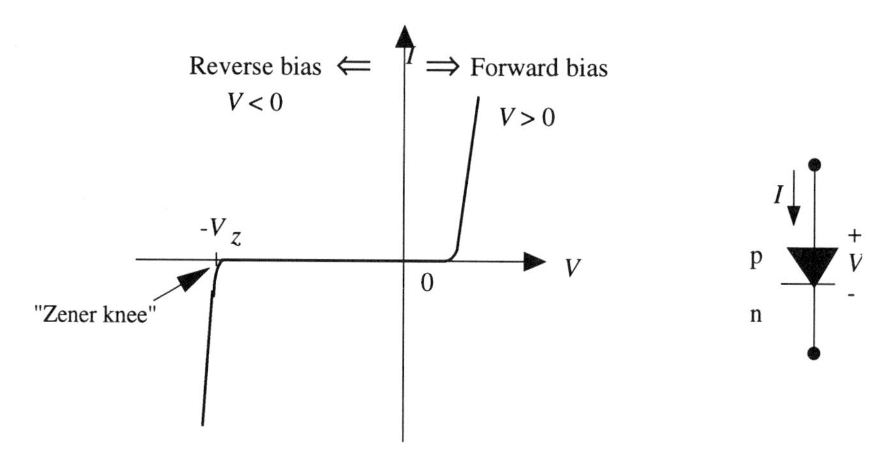

Figure 10.25 Diode *I-V* characteristics

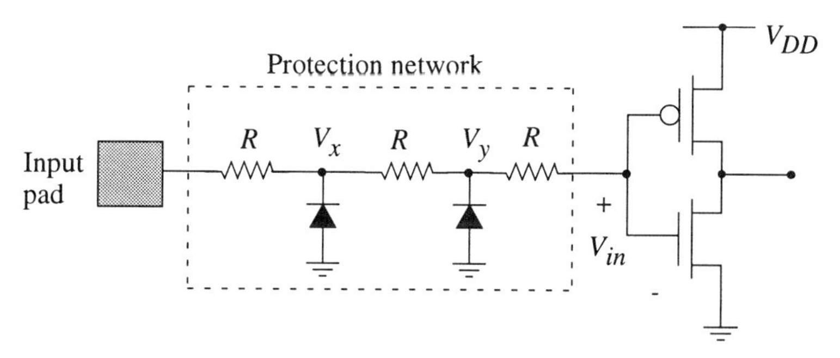

Figure 10.26 Diode-resistor input protection network

Consider the case where the input pad is subjected to a large positive voltage. The resistors in the signal path help to drop some of the voltage. If either V_x or V_y reaches the Zener voltage V_Z, then breakdown occurs across the respective diode, providing a conduction path for the excess charge to ground. If the input pad sees a large negative voltage, then the diodes are forward biased and charge is directed accordingly.

At the CMOS integrated circuit level, a resistor-diode protection network may be created using an n^+ to p-substrate junction as indicated by the cross-sectional drawing in Figure 10.27. This forms a natural junction because of the doping. Moreover, the n^+ region has a sheet resistance R_s (typically about 25 Ω) that provides a resistive drop between the input on the left side and the output on the right side. Note that this structure is really a distributed network (not discrete as in the schematic drawing). It is also worth mentioning that the pn junction introduces parasitic depletion capacitance between the signal line and ground.

Figure 10.28 shows a serpentine pattern that can be used to implement the protection circuit. The total resistance of the line is given by

$$R_{in} = R_s n \tag{10. 80}$$

where a corner section counts as about 0.65 squares. The serpentine pattern allows us to "squeeze" a long path into a rectangular region for better fitting on the chip.

A variation on this type of protection scheme is shown in Figure 10.29. This approach adds diodes between the power supply and the input line. In the discrete representation of the schematic, the new diodes are denote by D1 and D2, while D3 and D4 represent the diodes discussed above. This adds an extra degree of protection at the added expense of a more complex geometry and

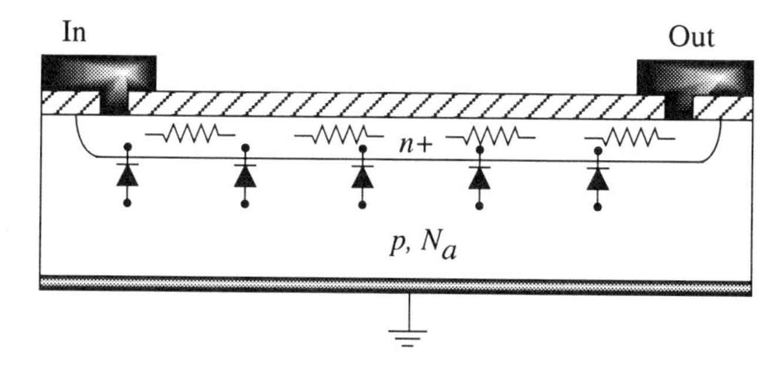

Figure 10.27 Diode-resistor implementation

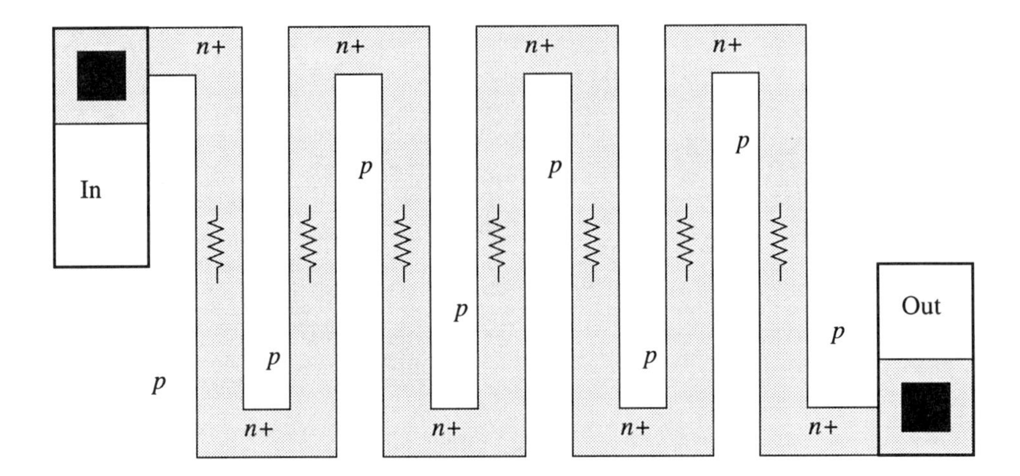

Figure 10.28 Serpentine pattern for protection network

increased parasitic capacitance.

The operation of the modified circuit is easily understood by examining the behavior of the diodes under different input conditions. If the input voltage rises above

$$V_{DD} + V_{on} \tag{10. 81}$$

where V_{on} is the on voltage of the diode, then D1 and D2 are biased into conduction and charge can be dumped to the power supply. If the input transient is very fast, then D3 and D4 may undergo breakdown to help keep the charge away from the input.

Thick Oxide MOSFET Network

Another popular input protection scheme is based on the introduction of a "thick-oxide" MOSFET that is specially designed for this purpose. Recall that the LOCOS CMOS process flow discussed in Chapter 2 grpws two thermal oxides of different thicknesses on the silicon substrate. The gate oxide has a thickness of x_{ox} and is used to create MOSFETs using the oxide capacitance per unit area

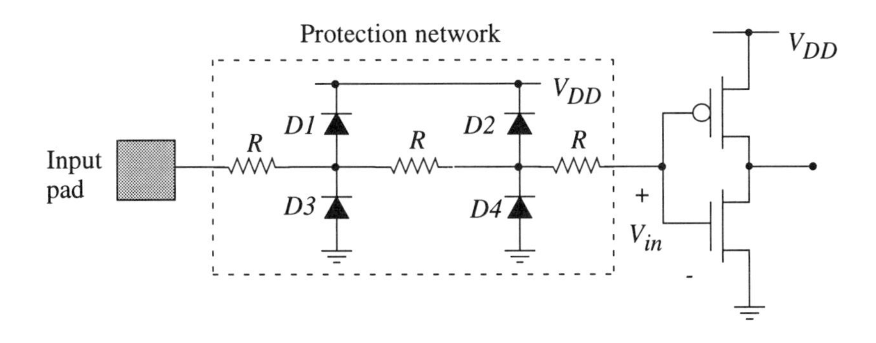

Figure 10.29 Dual-diode protection circuit

$$C_{ox} = \frac{\varepsilon_{ox}}{x_{ox}} \tag{10. 82}$$

x_{ox} is very small, typically less than about 100Å (10 *nm*) in a modern process. The other thermal oxide grown during the LOCOS process is the thick field oxide (FOX) that has a thickness $X_{FOX} >> x_{ox}$. This may be as large as 0.5-0.7 μm thick to provide the recessed isolation. The capacitance per unit area of the FOX is calculated from

$$C_{FOX} = \frac{\varepsilon_{ox}}{x_{FOX}} < C_{ox} \tag{10. 83}$$

and is designed to provide electrical isolation between devices even if an interconnect runs over it.[4]

Now consider the situation shown in Figure 10.30. A "normal" MOSFET with a gate oxide (x_{ox}) is shown on the right side. However, we have added a special field-oxide FET (FOXFET) on the left side that uses the FOX for a gate insulator. The reduced coupling through the thick oxide means that the threshold voltage $V_{T,F}$ of the FOXFET is much larger than the threshold voltage V_{Tn} of a regular FET:

$$V_{T,F} > V_{Tn} \tag{10. 84}$$

Using the field implant, we can adjust $V_{T,F}$ to a value of around 6-10 volts such that

$$V_{T,F} < V_{BD} \tag{10. 85}$$

is satisfied.

The thick-oxide FET can be used to construct the protection network shown in Figure 10.31. In this circuit, the input voltage controls V_F, which is the gate-source voltage of the FOXFET. If $V_F < V_{T,F}$ then the transistor is in cutoff and the signal can pass through. If the input voltage rises to a value $V_F \geq V_{T,F}$, then the FET is biased into the active region and conducts current I_D to ground and away from the inverter gate. In practice, this scheme is often used in addition to the resistor-diode clamp discussed above.

Input protection circuits generally require specialized techniques and circuits. As such, they are usually available as standard cells in the CAD library for use by the circuit designers. One should also investigate the available circuits and understand how they affect the signal flow.

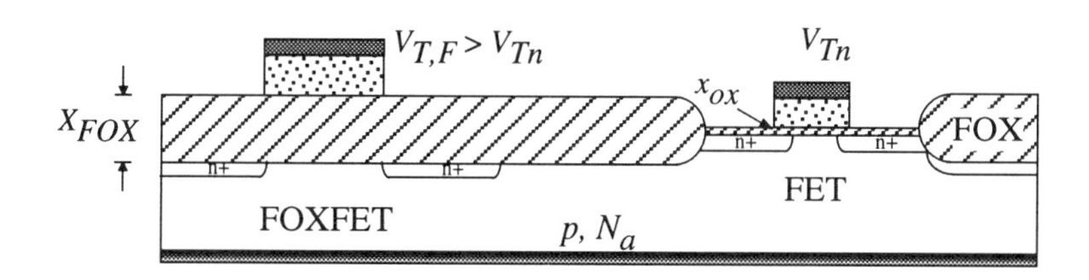

Figure 10.30 Field-Oxide FET (FOXFET)

[4] This was discussed previous in Chapter 2.

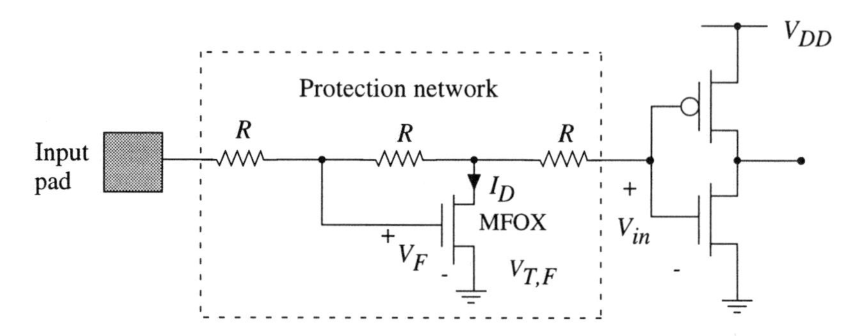

Figure 10.31 Input protection using a field-oxide FET

10.2.2 Output Circuits

Output driver circuits must be designed to provide large current flow levels to charge and discharge the large capacitance seen at an output pad. The problem is illustrated schematically in Figure 10.31. Since the output pad is connected to a pin, which is in turn connected to a printed circuit (PC) board, the load capacitance C_L seen by the CMOS driver is at the picofarad (pF) level. For example, chip testers typically exhibit loads of 70-80 pF. If the output circuitry is not designed to drive these large capacitive loads, then all of the speed of the internal logic will be lost.

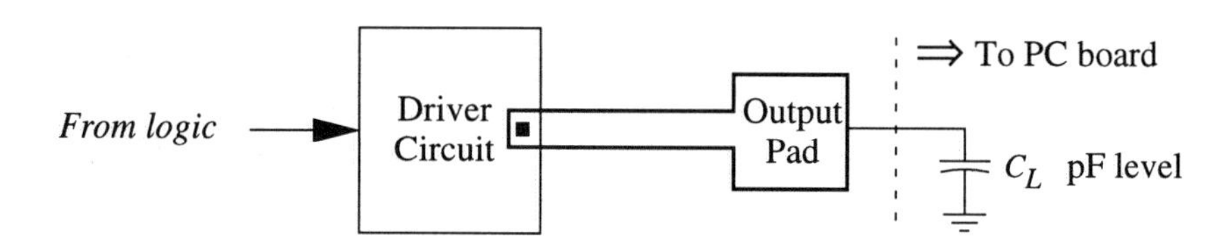

Figure 10.32 Output driver problem

This is one of the classical problems of high-speed chip design. It has been well-known that on-chip speeds can be increased by improving technology. However, the hard part is getting the signals into and out of the chip without losing the switching speed. Because of its importance to modern system design, several approaches have been developed for attacking the problem. In this section, we will examine a few techniques to give an idea of the methodology. Cell libraries usually have specialized output pad drivers for use by the chip designer, but it is still worthwhile to examine some of the circuits to understand the basic concepts.

Driver Chains

One approach to output circuits is to use a cascade of scaled logic gates as discussed in Section 3.6. The general idea is shown in Figure 10.33. The large capacitive load C_L requires us to use large aspect ratios to keep the switching times small. However, this results in a gate with a large input capacitance, so that we must use a larger-than-normal gate to drive it. This leads to the driver chain idea shown in the drawing. The number of gates in the chain and their relative sizing depends upon the desired characteristics, but all have the feature that the device sizes increase from the input to

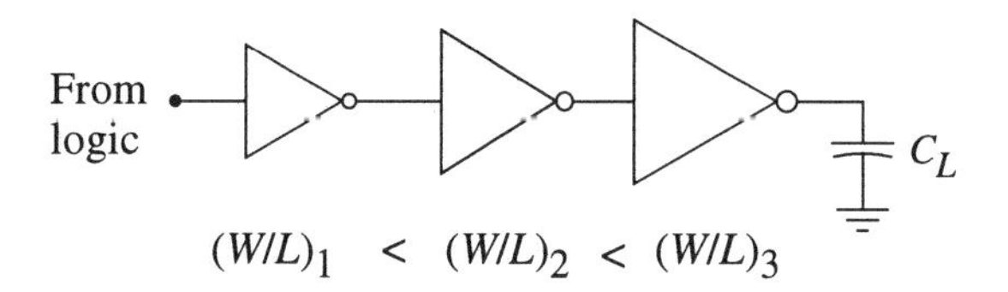

Figure 10.33 Driver chain

the output such that

$$\left(\frac{W}{L}\right)_1 < \left(\frac{W}{L}\right)_2 < \left(\frac{W}{L}\right)_3 \tag{10. 86}$$

With modern bus requirements at $100MHz$ or greater, the sizes of the output transistors in the final stage can get large; values of (W/L)=100 are not out of the question.

Tri-State Output

A tri-state output driver is very useful for applications where we need to decouple the chip from the external bus. An example of a tri-state pad driver circuit is shown in Figure 10.34. Data input D is fed through circuitry that splits the signal into two paths. The upper path produced the pFET gate voltage V_p, while the lower path controls the nFET gate voltage V_n. The inverters provide the proper phase and act as scaled driver chains for the large output FETs Mp and Mn. The Hi-Z state is activated by Z. With this signal at a value of $Z = 1$, the lowest nFET Mnz is ON and the upper and lower circuits form separate NAND2 gates. If $Z = 0$, Mnz is off while Mpx and Mpy are switched ON. This forces $V_p = V_{DD}$ and $V_n = 0v$ so that both Mp and Mn are OFF, giving the Hi-Z state.

Figure 10.35 shows another approach to creating a tri-state output circuit. D is the data input

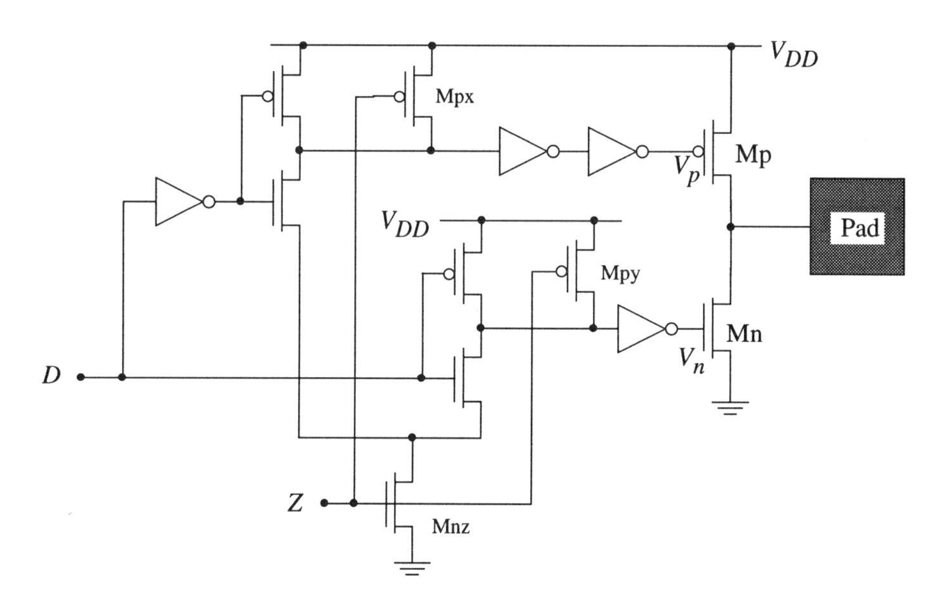

Figure 10.34 Tri-state output driver circuit

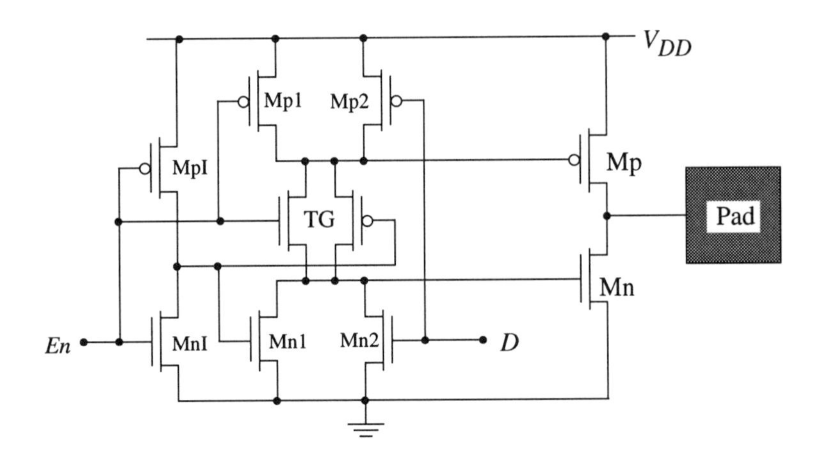

Figure 10.35 Tri-state driver with TG switching

that is fed into Mp2 and Mn2. The tri-state control is labelled as *En* (for Enable) and controls the transmission gate (TG) and FETs Mn1 and Mp1. If $En = 1$, the transmission gate is ON and the FET pair Mp2 and Mn2 form an inverter with an output to the inverter pair Mp and Mn. This gives a normal output, and the FETs are scaled to handle the large drive currents. Note that Mp1 and Mn1 are both OFF and do not affect the signal flow.

An Enable value of $En = 0$ induces the Hi-Z state by turning placing the TG into an OFF state while simultaneously turning Mp1 and Mn1 ON. With Mp1 ON, the gate of output FET Mp is at V_{DD} placing it in cutoff. Similarly, the drain of Mn1 pulls the gate of Mn to ground, and it too is in cutoff. Since both output FETs are OFF, the Hi-Z state is achieved.

BiCMOS Drivers

BiCMOS drivers use CMOS logic circuits but employ npn bipolar junction transistors (BJTs) at the output. BiCMOS techniques can also be used internally for driving high capacitance lines such as data buses and clock distribution trees. However, the standard CMOS a process flow must be modified to incorporate the BJTs into the wafer, which increases the complexity of the fabrication.

Let us briefly review the characteristics of the bipolar npn transistor. The symbol is shown in Figure 10.36(a), and we have identified the collector (C), the base (B), and the emitter (E) termi-

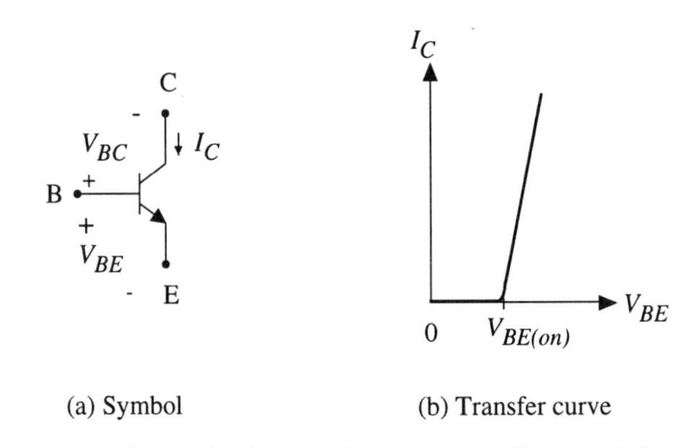

Figure 10.36 Bipolar junction transistor characteristics

nals. Current flow through the device is controlled by the base-emitter voltage V_{BE} and the base-collector voltage V_{BC}. There are 4 regions of operation depending upon the polarity of these voltages. For our purposes, we note that the combination of

$$V_{BE} > 0 \qquad V_{BC} < 0 \tag{10. 87}$$

defines **forward-active bias** where the collector current is controlled by the base-emitter voltage by the transfer equation

$$I_C = I_S e^{(V_{BE}/V_{th})} \tag{10. 88}$$

In this equation, I_S is the saturation current of the device and $V_{th} = (kT/q)$ is the thermal voltage. This is plotted in Figure 10.36(b). Note that a voltage of $V_{BE(on)}$ is required to turn on the transistor, and the base-collector junction is assumed to remain reverse biased. The exponential behavior implies that the BJT can switch very quickly; moreover, the current flow is very large in a BJT, with several hundred milliamperes easily attainable. BiCMOS recognizes this characteristic of bipolar transistors, and uses this high current flow to quickly charge or discharge large capacitive loads. Finally, an npn bipolar transistor can be driven into **cutoff** with $I_C = 0$ by dropping the base-emitter voltage to a value $V_{BE} \leq V_{BE(on)}$.[5]

A basic inverting BiCMOS driver is drawn in Figure 10.37. This technique employs MOSFETs at the inputs, but uses two BJTs Q1 and Q2 at the output for driving the output capacitance C_{out}. To understand the operation of the circuit, we first make note of the fact that Mn and Mp are logic transistors, while M1 and M2 act as pull-down devices to shut off the output BJTs. With this in mind, consider the case where we apply and input voltage of $V_{BE} = 0v$. The operation may be understood by referring to the drawing in Figure 10.38. With this input voltage, we see that Mp is conducting while Mn and M1 are OFF. The base of Q1 is thus at a voltage of V_{DD} through Mp, and this same voltage turns on M2. With M2 conducting, the base of Q2 is pulled down to ground potential, placing it in cutoff. The drawing thus shows that Q1 can charge the output capacitance via

$$I_C = C_{out}\frac{dV_{out}}{dt} = I_S e^{(V_{BE}/V_{th})} \tag{10. 89}$$

Note that the minimum base-emitter voltage on Q1 is

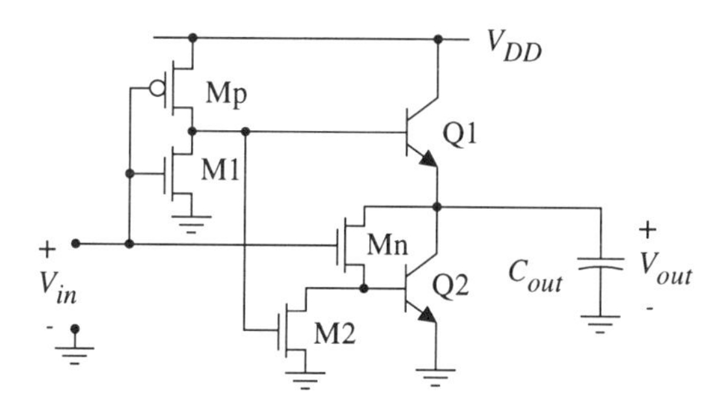

Figure 10.37 BiCMOS driver circuit

[5] This is a "relaxed" definition of cutoff in a bipolar transistor since it is assumed that the reader has some knowledge of these devices.

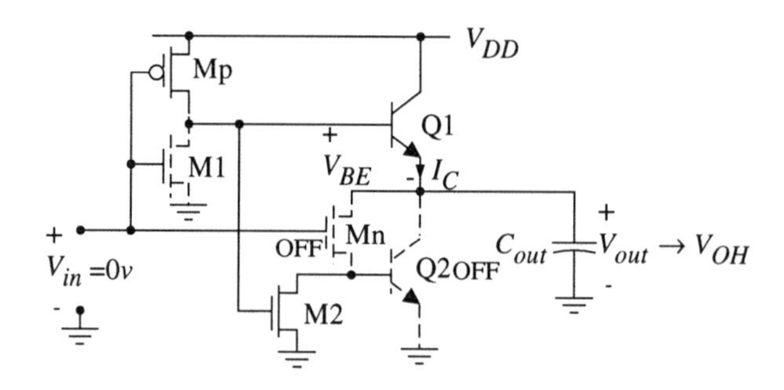

Figure 10.38 Calculation of V_{OH}

$$V_{BE} \approx V_{BE(on)} \tag{10.90}$$

Thus, the output high voltage of the circuit is about

$$V_{OH} \approx V_{DD} - V_{BE(on)} \tag{10.91}$$

Treating I_C as being approximately constant allows us to estimate the output voltage as

$$V_{out}(t) = V_{OL} + \frac{I_C}{C_{out}}t \tag{10.92}$$

so that the full swing low-to-high time is

$$t_{LH} \approx \frac{C_{out}}{I_C}(V_{OH} - V_{OL}) \tag{10.93}$$

The speed advantage is seen from this equation: I_C is very large compared to MOSFET levels, which yields faster charging.

The output low voltage and high-to-low time can be estimated in the way. Figure 10.39 shows the circuit for $V_{in} = V_{DD}$. This places M1 in the active region operation while Mp is in cutoff. M1 pulls the base voltage of Q1 to 0v so that it is OFF as indicated. Logic FET Mn is biased ON, and provides base current to Q2 from the output node which is initially at a high voltage V_{OH}. C_{out} can then discharge to ground through Q2 as described by

$$I_C = -C_{out}\frac{dV_{out}}{dt} = I_S e^{(V_{BE}/V_{th})} \tag{10.94}$$

Applying KVL to the Mn-Q2 loop gives the smallest output voltage as

$$V_{OL} \approx V_{BE(on)} \tag{10.95}$$

If we once again treat I_C as a constant for simplicity, the equation integrates to give us

$$V_{out}(t) = V_{OH} - \frac{I_C}{C_{out}}t \tag{10.96}$$

so that t_{HL} is also of the form

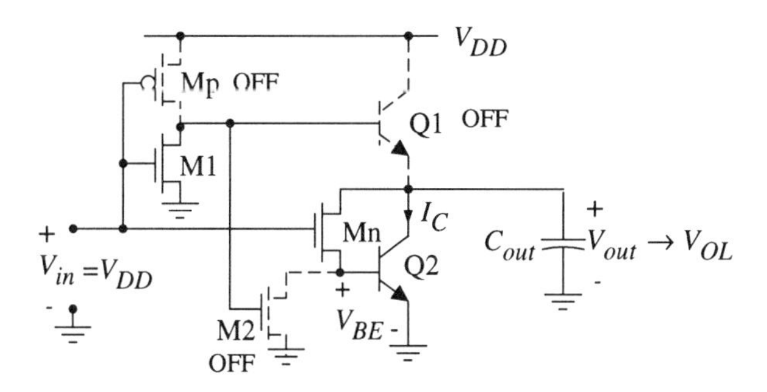

Figure 10.39 Circuit for calculating V_{OL}

$$t_{HL} \approx \frac{C_{out}}{I_C}(V_{OH} - V_{OL}) \tag{10.97}$$

since the discharge and charge circuits are symmetrical. Note that the output logic swing of this circuit is

$$V_{OH} - V_{OL} = V_{DD} - 2V_{BE(on)} \tag{10.98}$$

with the low value caused by the $V_{BE(on)}$ drops.

Several other techniques can be applied to overcome the problem of the reduced logic swing voltage at the output. Perhaps the simplest approach is to recognize that the worst problem in using a CMOS inverter is the size needed for the pFET; although the nFET is also large, the higher transconductance of n-channel devices reduces the requirements on $(W/L)_n$. This leads us to the solution illustrated in Figure 10.40 which uses only one BJT as a replacement for a pFET but keeps the nFET pull-down in the circuit to achieve

$$V_{OL} = 0v \qquad V_{OH} = V_{DD} - V_{BE(on)} \tag{10.99}$$

This results in a better logic swing. The operation of the circuit is easily seen. Mp and Mn are logic devices, while M1 is used to pull charge out of the base of Q1 and turn it off. The main difference is that we would use the equation

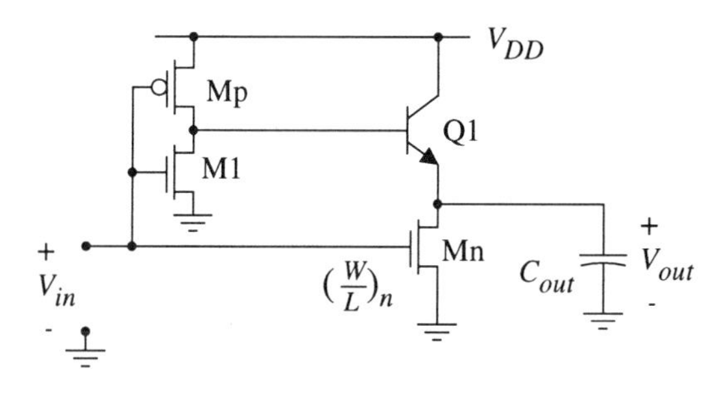

Figure 10.40 Modified circuit with improved logic swing

$$t_{HL} = s_n \tau_n \tag{10. 100}$$

as in the case of the inverter.

Interest in digital BiCMOS technology has waned in recent years because bipolar transistors cannot overcome the $V_{BE(on)}$ drops of about 0.75*v*, which are too high for low-voltage CMOS designs that operate at 1.5*v* or less. However, if bipolar transistors are available to the designer in a specific process, they can be used to provide unique solutions to speed problems.

10.3 Transmission Lines

Once the signal leaves the chip, it is guided by a **transmission line** structure that cannot be modeled as a simple wire. A good example of a transmission line is seen by a close-up examination of a printed circuit board as in Figure 10.41. Voltages and currents are transmitted on thin metal **traces** that are above a **ground plane** using a dielectric material such as epoxy or teflon. Although this looks identical to the on-chip interconnect shown in Figure 10.1, it is much larger in size and carries high current levels among the chips. This leads to the situation where the parasitic capacitance per unit length c' and the inductance per unit length l' (in units of *H/cm*) determine the energy transfer properties. This structure has been well studied in the context of microwave networks where it is known as a **microstrip** geometry.

10.3.1 Ideal Transmission Line Analysis

Let us analyze the **ideal transmission line** shown in Figure 10.41 where the resistance is small enough to be ignored. Applying KVL to the differential segment *dz* as in the analysis of the RC line gives

$$\frac{\partial V(z,t)}{\partial z} = -l' \frac{\partial I(z,t)}{\partial t}, \tag{10. 101}$$

while KCL yields

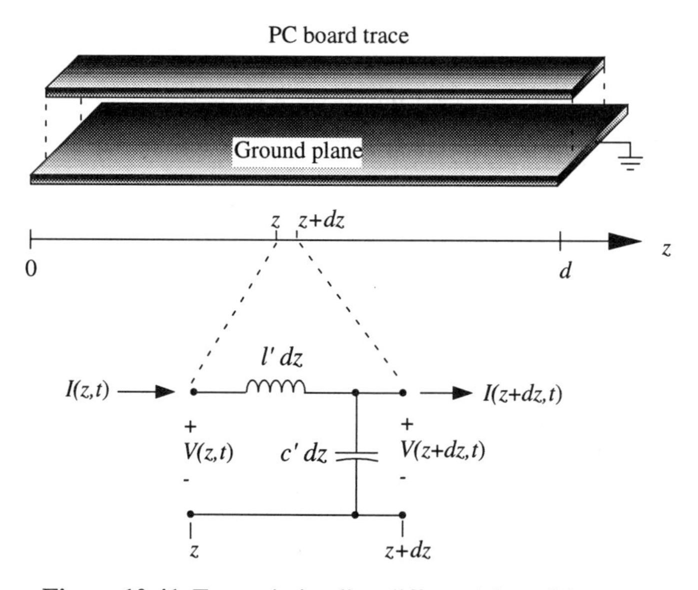

Figure 10.41 Transmission line differential model

$$\frac{\partial I(z,t)}{\partial z} = -c'\frac{\partial V(z,t)}{\partial t} \qquad (10.102)$$

as before. Differentiating the first equation with respect to z gives

$$\frac{\partial^2}{\partial z^2}V(z,t) = -l'\frac{\partial^2}{\partial z \partial t}I(z,t) \qquad (10.103)$$

Differentiating the second equation with respect to t yields

$$\frac{\partial^2}{\partial t \partial z}I(z,t) = -c'\frac{\partial^2}{\partial t^2}V(z,t) \qquad (10.104)$$

Combining these two expressions results in the **wave equation**

$$\frac{\partial^2 V(z,t)}{\partial z^2} = l'c'\frac{\partial^2 V(z,t)}{\partial t^2} \qquad (10.105)$$

for the line voltage $V(z,t)$. By changing the differentiation steps, it is easily shown that the current $I(z,t)$ obeys the same equation.

In general, a **wave** can be defined as a disturbance that moves in both space and time. It is an interesting exercise to perform a dimensional analysis on the equation. Note that the left side has units of volts per square centimeter, so that the right side must have the same units. This simple observation shows that we can define a special velocity v_p with units of *cm/sec* by

$$v_p = \frac{1}{\sqrt{l'c'}} \qquad (10.106)$$

substituting units for l' and c' (with units of *H/cm* and *F/cm*, respectively) confirms this relation. The parameter v_p is called the **phase velocity**; as shown below, it is the speed that a voltage wavefront travels along the line.[6] If we have a wave moving in a two-conductor line that is immersed in a dielectric material with a permittivity $\varepsilon = \varepsilon_r \varepsilon_o$, then the phase velocity is given by

$$v_p = \frac{c}{\sqrt{\varepsilon_r}} \qquad (10.107)$$

where $c \approx 3 \times 10^{10}$ *cm/sec* is the speed of light in free space.

The general solutions to the wave equation

$$\frac{\partial^2 V(z,t)}{\partial z^2} = \frac{1}{v_p^2}\frac{\partial^2 V(z,t)}{\partial t^2} \qquad (10.108)$$

are of the form

$$V(z,t) = f^+(z - v_p t) + f^-(z + v_p t) \qquad (10.109)$$

which respectively represent waves that travel in the $+z$ and the $-z$ directions. An interesting observation is that f^+ and f^- are *arbitrary* functions of the composite variables $(z \pm v_p t)$; any function of

6 This is only true for an electromagnetic TEM (transverse electric-magnetic) wave, but it a reasonable approximation here.

these arguments constitute solutions of the wave equation, so long as the form of the argument remains intact! For our purposes, it is useful to employ the general unit step function defined by

$$u(y) = \begin{cases} 0 & (y < 0) \\ 1 & (y \geq 0) \end{cases} \qquad (10.110)$$

to write voltage waves in the form

$$V(z, t) = V^{+} u(z - v_p t) + V^{-} u(z + v_p t) \qquad (10.111)$$

These represent voltage wavefronts that move along the line.

Once the form of the voltage waves have been established, the line current $I(z,t)$ is given by

$$I(z, t) = \frac{1}{Z_o} f^{+}(z - v_p t) - \frac{1}{Z_o} f^{-}(z + v_p t) \qquad (10.112)$$

where

$$Z_o = \sqrt{\frac{l'}{c'}} = \frac{V(z, t)}{I(z, t)} \qquad (10.113)$$

has units of Ohms, and is called the **characteristic impedance** of the line. This expression describes current waves moving in the $+z$ and $-z$ directions, and can be obtained using either the KCL or the KVL equation[7]. In practice, Z_o values in the range of about 30Ω to 100Ω are typical. If the voltage wavefronts are described by step functions as in eqn. (10.62), then the current waves are given by

$$I(z, t) = I^{+} u(z - v_p t) - I^{-} u(z + v_p t) \qquad (10.114)$$

where

$$I^{+} = \frac{V^{+}}{Z_o}, \qquad I^{-} = \frac{V^{-}}{Z_o} \qquad (10.115)$$

are the amplitudes.

The meaning of voltage and current waves can be understood using the simple circuit shown in Figure 10.42 where an ideal voltage source excites the input of a line with a step voltage

$$V_{in} = V_o u(t) \qquad (10.116)$$

This launches a wave moving in the $+z$ direction that is described by

$$V(z, t) = V_o u(z - v_p t) \qquad (10.117)$$

The current wave is given by

$$I(z, t) = \frac{V_o}{Z_o} u(z - v_p t) \qquad (10.118)$$

[7] The minus sign on the second term merely indicates that the wave is moving in the $-z$ direction.

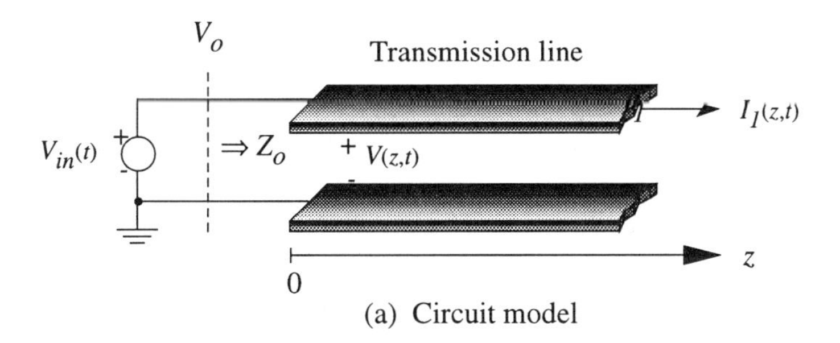

(a) Circuit model

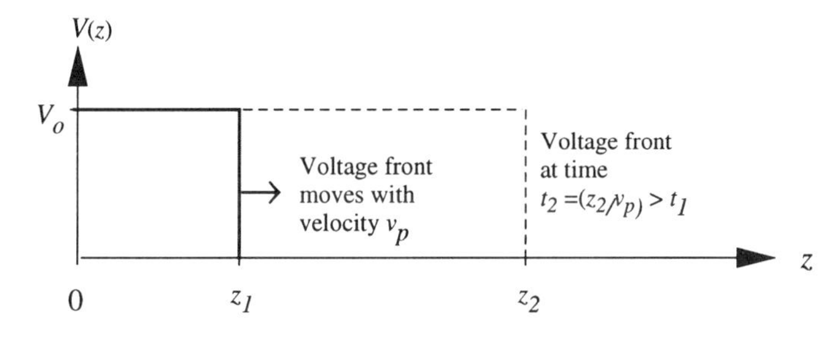

(b) Motion of voltage front along line

Figure 10.42 Voltage pulse on an infinite line

Since the step function $u(z - v_p t)$ is zero for $(z - v_p t) < 0$, these equations represent wavefronts that move along the line (increasing z) as time t increases. If an observer is situated at some distance z_1 from the input, no voltage or current will be measured until a time

$$t_1 = \frac{z_1}{v_p}. \tag{10. 119}$$

This delay is a characteristic of the transmission line.

10.3.2 Reflections and Matching

The definition of the characteristic impedance

$$Z_o = \frac{V(z, t)}{I(z, t)} \tag{10. 120}$$

should be taken very literally as being the ratio of the voltage to the current at every point along the line. This leads to several important consequences when applied to a practical system. If the wavefront should encounter a region that has a different impedance, then some of the energy will be transmitted through the discontinuity but some will be reflected back to the source.

This can be understood by referring to the circuit shown in Figure 10.43. In this situation, a positively travelling wave front with voltage V^+ and current I^+ is moving along a finite-length line that has a characteristic impedance of Z_o. The end of the line is connected to a load with an impedance of Z_L. The load voltage V_L and current I_L are related by

$$V_L = I_L Z_L \tag{10. 121}$$

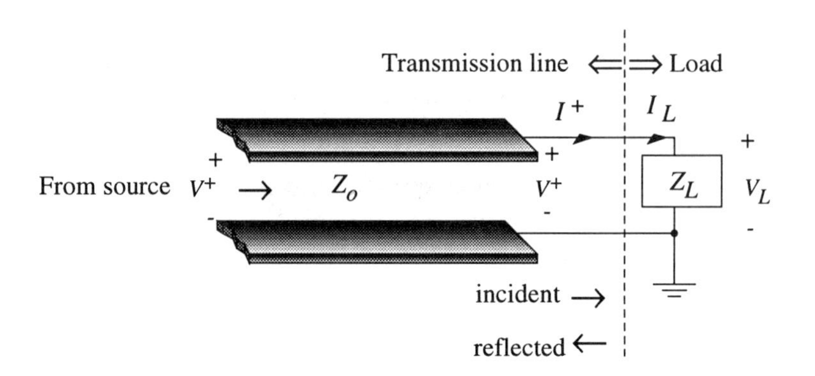

Figure 10.43 Transmission and reflection at a load

Now then, at the interface between the transmission line and the load, the voltage and the current must be continuous. Mathematically, this means that

$$\begin{aligned} V_{line} &= V_L \\ I_{line} &= I_L \end{aligned} \tag{10. 122}$$

For the incident wave, the voltage and current are related by

$$\frac{V^+}{I^+} = Z_o \tag{10. 123}$$

If we try to match these values for the voltage and current to those of the load, we see that it cannot be done except in the special case where $Z_o = Z_L$ is true, i.e., a matched load. This means that we must modify our initial guess for the line quantities V_{line} and I_{line} by noting that the wave equation has negatively travelling wave solutions V^- and I^- that we can use to construct the superposition

$$\begin{aligned} V_{line} &= V^+ + V^- \\ I_{line} &= I^+ + I^- \end{aligned} \tag{10. 124}$$

as the total values. Since these are defined on the line with impedance Z_o, we can write

$$I_{line} = \frac{V^+}{Z_o} - \frac{V^-}{Z_o} \tag{10. 125}$$

and apply the load continuity conditions

$$\begin{aligned} V_{line} &= V^+ + V^- = V_L \\ I_{line} &= \frac{V^+}{Z_o} - \frac{V^-}{Z_o} = \frac{V_L}{Z_L} \end{aligned} \tag{10. 126}$$

Physically, these equations tell us that when a wavefront V^+ hits an impedance discontinuity, a reflected wavefront V^- will be generated and move back towards the source. The amplitude of the reflected wave may be found by solving these equations for the **reflection coefficient** Γ given by

$$\Gamma = \frac{V^-}{V^+} = \frac{Z_L - Z_o}{Z_L + Z_o} \tag{10. 127}$$

This says that an incident wave voltage V^+ will generate a reflected wave amplitude of

$$V^- = \Gamma V^+ \tag{10. 128}$$

where the value of Γ (and hence, V^-) is determined by the value of the load impedance Z_L relative to Z_o. Note that if $Z_L > Z_o$ then Γ is a positive number, while $Z_L < Z_o$ gives a negative Γ. Only a matched load with $Z_L = Z_o$ eliminates reflections as indicated by Γ=0.

To understand the consequences of the reflection coefficient, consider the terminated line shown in Figure 10.44. We will assume that the input voltage is

$$V_{in} = V_o u(t) \tag{10. 129}$$

as before. This has the effect of launching a positively-travelling wave from the source to the load as described by

$$V^+ = V_o u(z - v_p t) \tag{10. 130}$$

The transit time from one end of the line to the other is given by

$$t_d = \frac{v_p}{d} \tag{10. 131}$$

so that a wavefront of amplitude V_o reaches the load at t_d. Now, let us assume that the load Z_L is a simple resistor R_L. The load reflection coefficient is then given by

$$\Gamma_L = \frac{R_L - Z_o}{R_L + Z_o} \tag{10. 132}$$

such that the reflected wave has an amplitude of

$$V^- = \Gamma_L V_o \tag{10. 133}$$

The load voltage at this time is given by KVL as

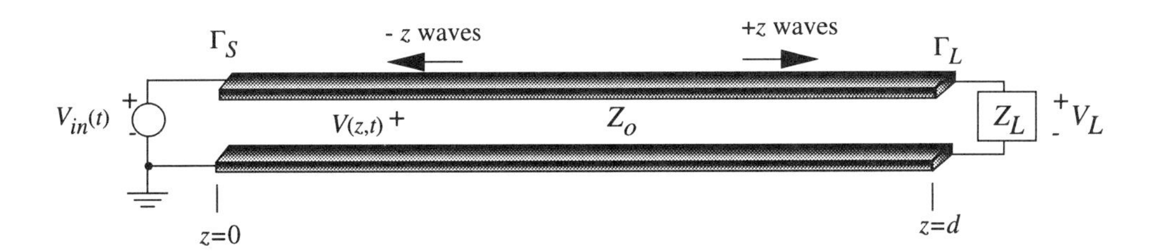

Figure 10.44 Terminated transmission line

$$V_L = V^+ + V^- \\ = V_o(1+\Gamma_L) \quad (10.\ 134)$$

where we again note that Γ_L can be positive or negative. The reflected wave reaches the source at a time $2t_d$ where it will be subject to reflection (back towards the load). From the drawing we see that the source impedance has been assume to be $Z_S = 0$, so the source reflection coefficient is

$$\Gamma_S = \frac{Z_S - Z_o}{Z_S + Z_o} = -1 \quad (10.\ 135)$$

This indicates a perfect reflection with inversion. The reflected wave at the source is really the second wave travelling to towards the load, so we will call it V^{2+}. It has a general value of

$$V^{2+} = \Gamma_S V^- \\ = \Gamma_S(\Gamma_L V_o) \quad (10.\ 136)$$

The second positive wave V^{2+} reaches the load at a time $3t_d$ and is again subject to a reflection. The second reflected wave V^{2-} has an amplitude of

$$V^{2-} = \Gamma_L V^{2+} \\ = \Gamma_L(\Gamma_S \Gamma_L V_o) \quad (10.\ 137)$$

The load voltage at this time is

$$V_L = V^+ + V^- + V^{2+} + V^{2-} \\ = V_o(1 + \Gamma_L + \Gamma_S\Gamma_L + \Gamma_L\Gamma_S\Gamma_L) \quad (10.\ 138)$$

As time increases, the back-and-forth bouncing gives rise the infinite series

$$V_L = V_o\left[1 + \Gamma_S\Gamma_L + (\Gamma_S\Gamma_L)^2 + (\Gamma_S\Gamma_L)^3 + \ldots\right] \\ + V_o\Gamma_L\left[1 + \Gamma_S\Gamma_L + (\Gamma_S\Gamma_L)^2 + (\Gamma_S\Gamma_L)^3 + \ldots\right] \quad (10.\ 139)$$

which sums to

$$V_L = V_o\left(\frac{1+\Gamma_L}{1-\Gamma_S\Gamma_L}\right) \quad (10.\ 140)$$

in the limit where $t \to \infty$. For our problem where Γ_S=-1, we can substitute for Γ_L and compute

$$V_L = V_o\left(\frac{1 + \left(\frac{R_L - Z_o}{R_L + Z_o}\right)}{1 + \left(\frac{R_L - Z_o}{R_L + Z_o}\right)}\right) \quad (10.\ 141)$$

which reduces to

$$V_L = V_o \quad (10.\ 142)$$

as it should! This result becomes obvious when we note that the transmission acts like a simple wire after the transients decay away. The main point to be remembered here is that it takes a finite amount of time for the transmission line effects to converge into the final value. This consideration becomes very important at high frequencies.

Capacitive Load

Now let us analyze the situation shown in Figure 10.45 where a transmission line with a length d is terminated with a load capacitor C_L. This problem is of interest because it models the situation where we are driving the input of a CMOS gate that is intrinsically capacitive. We have chosen a source with an internal impedance of $Z_S = Z_o$ to get $\Gamma_S = 0$ for simplicity. The source voltage is once again chosen to be

$$V_{in} = V_o u(t) \tag{10. 143}$$

However, the presence of the source impedance changes the value that actually makes it to the transmission line. At the input to the line given by $z = 0$ we have

$$Z_o = \frac{V(0, t)}{I(0, t)} \tag{10. 144}$$

so that Z_o represents the input impedance seen by the source circuitry (which is the series combination of the voltage source and $Z_S = Z_o$). The internal impedance and input impedance form a voltage divider with

$$\left(\frac{Z_o}{Z_o + Z_S}\right)V_o = \frac{V_o}{2} \tag{10. 145}$$

so that the launched pulse is given by

$$V^+(t) = \left(\frac{V_o}{2}\right)u(z - v_p t) \tag{10. 146}$$

as it propagates towards the capacitive load.

To analyze the effect of C_L, we want to compute the load reflection coefficient

$$\Gamma_L = \frac{Z_L - Z_o}{Z_L + Z_o} \tag{10. 147}$$

However, since the load is a pure capacitor, we will transform to s-domain where

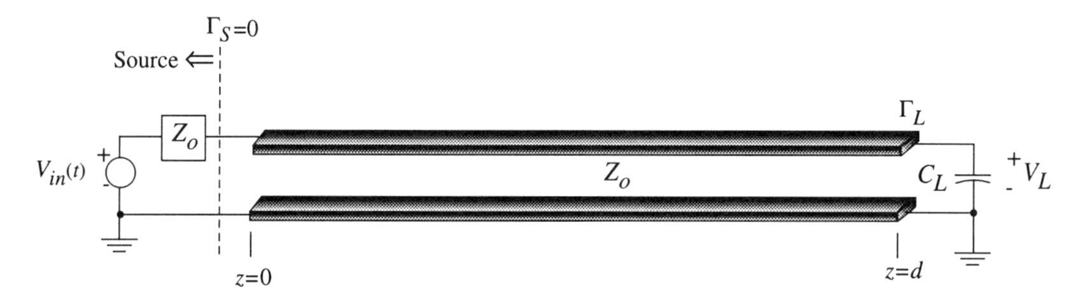

Figure 10.45 Transmission line with a capacitive load

$$Z_L = \frac{1}{sC_L} \tag{10. 148}$$

and assume that the characteristics impedance Z_o is purely real. The s-domain reflection coefficient $\rho_L(s)$ is then given by

$$\rho_L(s) = \frac{\left(\frac{1}{sC_L}\right) - Z_o}{\left(\frac{1}{sC_L}\right) + Z_o} \tag{10. 149}$$

such that

$$v^-(s) = \rho_L(s)\, v^+(s) \tag{10. 150}$$

where $v(s)$ are the s-domain voltages.

To complete the analysis, we note that the launched pulse reaches the load capacitor after a transit delay t_d. The s-domain value is

$$v^+(s) = \left(\frac{V_o}{2s}\right) e^{-st_d} \tag{10. 151}$$

so that the reflected pulse is described by

$$v^-(s) = \left[\frac{\left(\frac{1}{sC_L}\right) - Z_o}{\left(\frac{1}{sC_L}\right) + Z_o}\right] \left(\frac{V_o}{2s}\right) e^{-st_d} \tag{10. 152}$$

Rearranging and inverse transforming back to time-domain gives the reflected voltage as

$$V^-(t) = V_o\left[\frac{1}{2} - e^{-(t-t_d)/\tau}\right] \quad (t \geq t_d) \tag{10. 153}$$

where

$$\tau = Z_o C_L \tag{10. 154}$$

is the time constant. The total load voltage across the capacitor is calculated as

$$V_L(t) = V^+ + V^- \tag{10. 155}$$

which gives

$$V_L(t) = V_o\left[1 - e^{-(t-t_d)/\tau}\right] u(t - t_d) \tag{10. 156}$$

This is plotted in Figure 10.46 where we see that this is just the charging of the capacitor through the source impedance with a delay due to the transit time on the transmission line. Since this is a highly idealized analysis with a matched source assumed, we must be careful about extending the behavior to a CMOS system. However, it does illustrate the general problems involved.

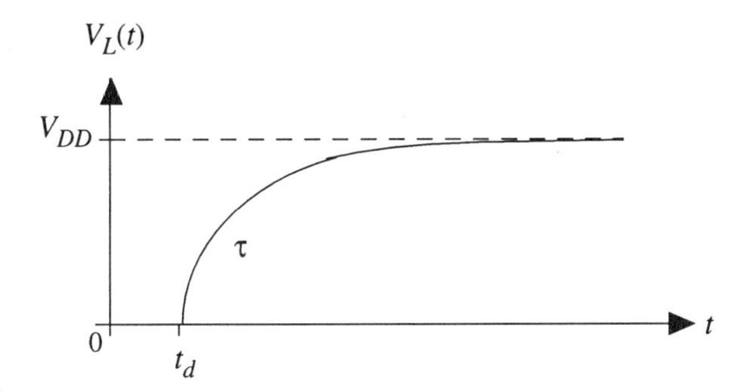

Figure 10.46 Charging delay through a transmission line

MOSFET Driver Matching

To apply our understanding of transmission line analysis to the problem of CMOS drivers, consider the situation illustrated in Figure 10.47(a) where a logic gate is connected to a transmission line. At the FET level [Figure 10.47(b)], we are concerned with trying to drive the line impedance Z_o with the transistor circuitry. Since Z_o is a real number, we might be tempted to use the linearized FET resistances

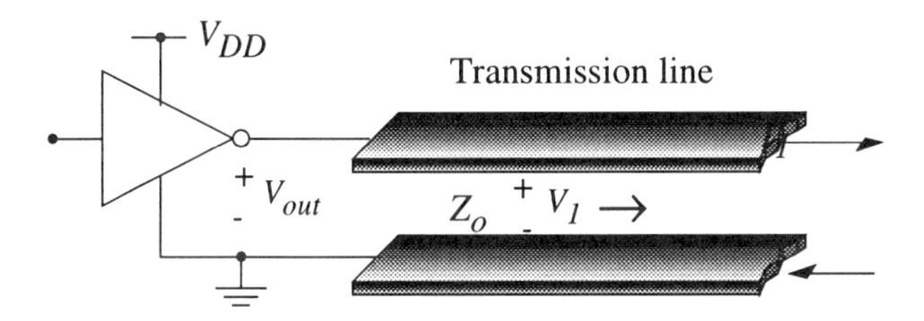

(a) Transmission line driving problem

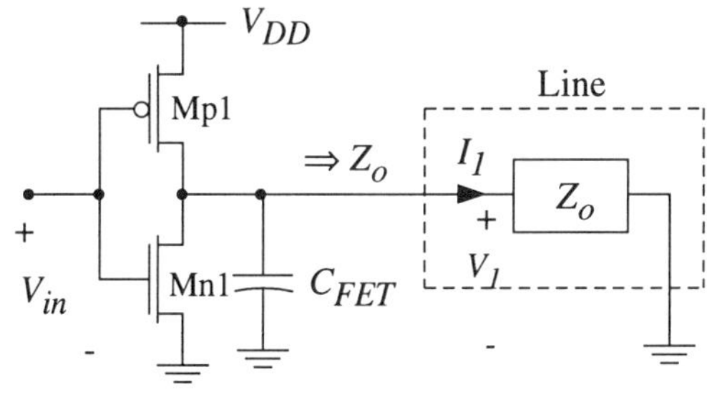

(b) Circuit model

Figure 10.47 CMOS inverter driving a transmission line

$$R_n = \frac{1}{\beta_n (V_{DD} - V_{Tn})}, \qquad R_p = \frac{1}{\beta_p (V_{DD} - |V_{Tp}|)} \tag{10. 157}$$

to choose the device aspect ratios $(W/L)_n$ and $(W/L)_p$ that would match the line impedance. The problem with this approach, of course, is that R_n and R_p are defined as linear resistances, while the MOSFETs are intrinsically nonlinear devices. Because of this problem, some designs insert a resistance R in between the driver and the line to swamp out some of the nonlinear variations. This is illustrated in Figure 10.48. The actual value of R depends upon the effect of the FETs, but since these will generally be quite large (to handle the large capacitance), $R \approx Z_o$ as shown is a reasonable first estimate.

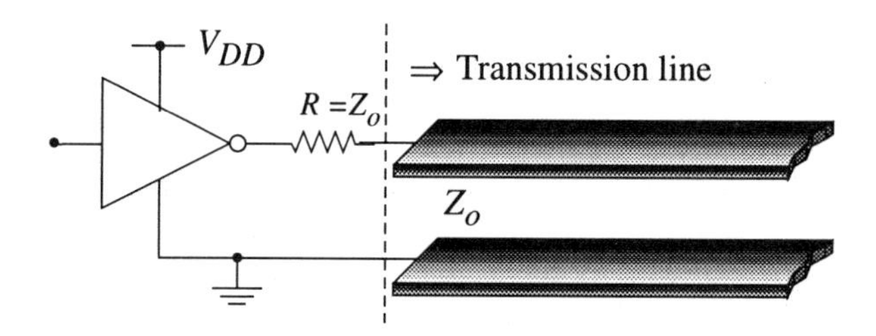

Figure 10.48 Line matching using a series resistor

Damped Driver

One problem with driving an open line is that noise problems can destroy the integrity of the data transmission. Because of this problem, terminated bus designs like those used in high-speed bipolar ECL systems have been proposed for multi-signal CMOS transceiver arrangements. The main idea is shown in Figure 10.49; resistors are added to the ends of the lines to prevent reflections; V_t is the terminator supply voltage. Each unit can act as a transmitter (X-section) or as a receiver (R-section), with arbitration performed by the system controller. The resistors provide pull-up action to V_t for a high voltage, while the line voltage is pulled down by a transmitter circuit toward ground if a low voltage is to be placed on the line. This is quite different from the situations discussed above where the chip power supply was responsible for driving the line high.

When a transmitter sends a signal to the bus, the line will react with and some reflections may occur. One way to overcome this is using the circuit shown in Figure 10.50 which is used to drive the line and provide damping.[8] To understand the operation of the circuit, suppose that initially we have $V_{in} = 0v$. With this applied, Mn1 and Mn2 are in cutoff, while Mp, Mn3 and Mn4 are ON. This gives an output voltage of $V_{out} = 0v$. When V_{in} is switch to a high voltage $V_{in} = V_{DD}$, the output from the inverter made up of Mp1 and Mn1 goes low. Note that the series combination of Mn2 and Mn3 connect the gate and drain of Mn4. The inverter chain is used to delay the turnoff of Mn3. If we design Mn1 to have a small aspect ratio, then Mn4 will continue to conduct for a short time after the input transition. This provides damping to ground that is eventually turned off when Mn3 goes into cutoff. Once Mn4 is off, the output is pulled to the terminator voltage $V_{out} = V_t$.

Problems such as these tend to be more critical as the bus speeds and number of connected units increase. It is clear that much research will be devoted towards the problems in the future.

---

[8] This design technique has been termed *GTL* by its developers for "Gunning Transistor Logic"

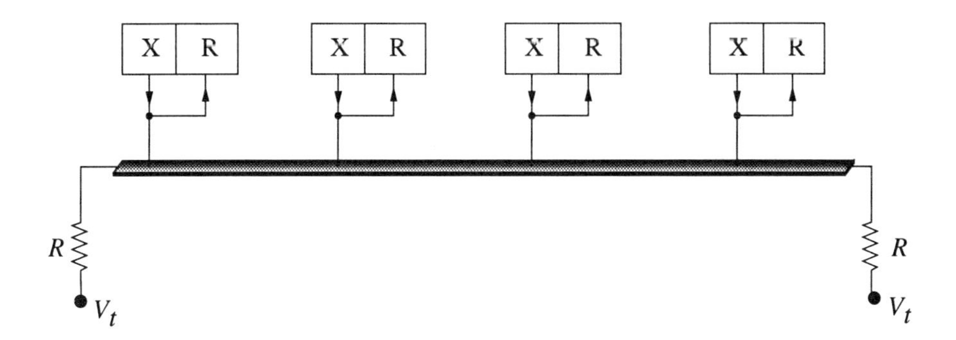

Figure 10.49 Terminated-line bus connection scheme

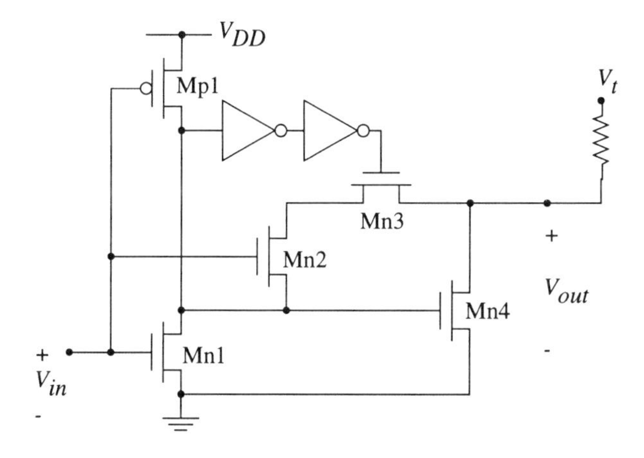

Figure 10.50 Bus driver circuit with damping

10.4 Problems

[10-1] Consider a simple doped poly interconnect line that is 0.8μm wide and has a sheet resistance of 25 Ω. A FET is made with an aspect ratio of (W/L)=4 such that the process variables are k'_n = 100 $\mu A/V^2$ and V_{Tn} = 0.70v.

(a) Find the length of the line where the line resistance is equal to one-half of the LTI nFET resistance.

(b) Suppose that we coat the poly with W which produces a silicide that has a sheet resistance of 0.035Ω. What is the length of the line in this new material that gives one-half of R_n?

[10-2] The interconnect line described in Problem [10-1] has a thickness of 0.5μm and is routed over an isolation oxide that has a thickness of X_{FOX} = 0.8μm.

(a) Find the capacitance for the line using the ideal parallel-plate formula.

(b) Find the line capacitance accounting for contributions from fringing fields.

(c) What is the percentage error in the total line capacitance if fringing fields are ignored?

[10-3] Consider a layer of doped polysilicon that has a resistivity of $\rho = 0.0015\ \Omega\text{-}cm$.

(a) Calculate the sheet resistance if the layer is 6000Å thick.

(b) The thickness of the layer is chosen to be 5500Å. The lithographic resolution of the process sets the minimum width at 0.4 microns. Find the length d (in microns) for an interconnect that has a line resistance of 1 $k\Omega$.

[10-4] An interconnect line is described by the cross-sectional view shown in Figure 10.3(a) with dimensions of $X_{ox} = 0.8\ \mu m$, $h = 0.6\ \mu m$, and $w = 0.4\ \mu m$. Silicon dioxide is used as the insulator. The line has a length of 35 μm and a sheet resistance of 0.05 Ω.

(a) Find C_{line} by including fringing fields as in equation (10.10).

(b) Find the line resistance R_{line}.

(c) Construct RC ladder equivalent networks for $m = 2$ and $m = 7$. Then find the time constant fo each.

(d) Assume that the $m = 7$ model is more accurate. find the percentage error if the $m = 2$ result is used instead.

[10-5] An interconnect line has a sheet resistance of $R_s = 0.05$ ohms and a line capacitance per unit length of $c' = 6 \times 10^{-9}\ F/cm$. The line has a width of 0.8 μm. The results of the distributed analysis given in equation (10.45) are used with V_o=5 volts and V_x=4.5 volts as our reference. We know that $erfc(0.09) \approx 0.9$.

(a) Find the equation for the interconnect-induced time delay as a function of the line length d.

(b) Plot the delay for distances up to 50 μm.

[10-6] Prove explicitly that any function $f(z\text{-}v_p t)$ is a solution to the wave equation, i.e., that

$$\frac{\partial^2 f}{\partial z^2} = \frac{1}{v_p^2}\frac{\partial^2 f}{\partial t^2}. \tag{10. 158}$$

Hint: First define the composite variable $\xi = z\text{-}v_p t$, and then calculate the z- and t- derivatives using the chain rule. For example, the start the time derivative calculation with

$$\frac{df}{dt} = \left(\frac{df}{d\xi}\right)\left(\frac{d\xi}{dt}\right), \tag{10. 159}$$

and then use this to compute the second derivative.

[10-7] Show explicitly that the function

$$f(z, t) = A^{(z - v_p t)} \tag{10. 160}$$

with A = constant satisfies the wave equation.

[10-8] Two coupled interconnect lines are described by the cross-sectional view shown in Figure 10.17 with dimensions of $X_{ox} = 0.8\ \mu m$, $h = 0.6\ \mu m$, and $w = 0.4\ \mu m$. Silicon dioxide is used as the insulator. The line has a length of 65 μm.

(a) Calculate the coupling capacitance C_c if $S = w$.

(b) Calculate the coupling capacitance C_c if the spacing is increased to $S = 3w$.

[10-9] Calculate the phase velocity v_p of a line that uses silicon dioxide as an insulator.

[10-10] Silicon nitride has a relative permittivity of about ε_r=7.8. What is the phase velocity of a signal if nitride is used as the dielectric?

[10-11] An oxy-nitride (silicon dioxide-silicon nitride combination) has a relative permittivity of ε_r=5.1.

(a) Calculate the phase velocity v_p of a line that uses this material as an insulator

(b) Find the signal delay in units of $ps/\mu m$.

[10-12] Consider the oxy-nitride line describe

[10-13] Consider the parallel RC transmission line termination shown in Figure P10.1. A shunt resistor is sometimes used in this manner to help match the load to the transmission line. Find the s-domain reflection coefficient $\rho_L(s)$ for this case.

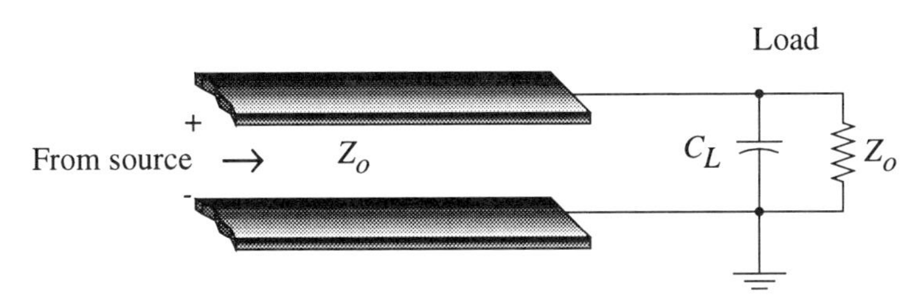

Figure P10.1

10.5 References

The books below provide detailed discussions on the topics presented in this chapter.

[1] H. B. Bakoglu, **Circuits, Interconnection and Packaging for VLSI**, Addison-Wesley, Reading, MA, 1990.

[2] A. K. Goel, **High-Speed VLSI Interconnections**, John Wiley & Sons, New York, 1994.

[3] B. Gunning, L. Yuan, T. Nguyen, and T. Wong, "A CMOS Low-Voltage-Swing Transmission-Line Transceiver," *ISSCC92 Technical Digest*, pp.58-59, 1992.

[4] E. G. Friedman (ed.), **Clock Distribution Networks in VLSI Circuits and Systems**, IEEE Press, New York, 1995.

[5] C. R. Paul, **Multiconductor Transmission Lines**, John Wiley & Sons, New York, 1994.

[6] R. K. Poon, **Computer Circuits Electrical Design**, Prentice-Hall, Englewood Cliffs, NJ, 1994.

[7] M. Shoji, **High-Speed Digital Circuits**, Addison-Wesley, Reading, MA, 1996.

[8] S. Ramo, T. Van Duzer, and J. Whinnery, **Fields and Waves in Communication Electronics**, 3rd ed., John Wiley & Sons, New York, 1994.

Index

T

U

V

W

X

Y

Z